Two

ISOLATED SHALLOW MARINE SAND BODIES: SEQUENCE STRATIGRAPHIC ANALYSIS AND SEDIMENTOLOGIC INTERPRETATION

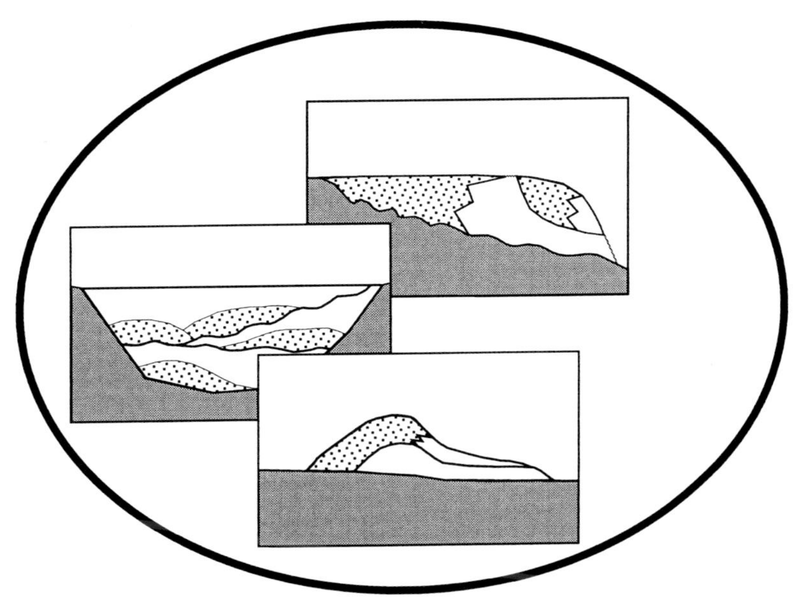

Edited by:

KATHERINE M. BERGMAN
Department of Geology, University of Regina, Regina, SK S4S 0A2, Canada

AND

JOHN W. SNEDDEN
Mobil Exploration & Producing Technical Center, PO Box 650232, Dallas, TX 75265 U.S.A.

Robert W. Dalrymple, Editor of Special Publications
SEPM Special Publication Number 64

Tulsa, Oklahoma, U.S.A. *December, 1999*

A Publication of
SEPM (Society for Sedimentary Geology)

ISBN 1-56576-057-3

1731 E. 71st Street
Tulsa, Oklahoma 74136-5108
Printed in the United States of America

Isolated Shallow Marine Sand Bodies: Sequence Stratigraphic Analysis and Sedimentologic Interpretation

Editors: Katherine M. Bergman and John W. Snedden

TABLE OF CONTENTS

Part 4: Biological and Chemical Aspects of Isolated Shallow Marine Sandbodies

Part 5: Sequence Stratigraphic Analysis of other Ancient Shallow Marine Sandbodies

Part 6: Summary and Conclusions

ISOLATED SHALLOW MARINE SAND BODIES: SEQUENCE STRATIGRAPHIC ANALYSIS AND SEDIMENTOLOGIC INTERPRETATION

FOREWORD

Our understanding of isolated shallow marine sand bodies has changed dramatically over the last fifteen years. These deposits are common in the Cretaceous Western Interior seaway and are significant reservoirs of hydrocarbons elsewhere. Understanding sand body genesis and geometry is critical to successful exploration and exploitation of these reservoirs. Sequence stratigraphy has provided the necessary tools for new correlation ideas and interpretations in ancient shelf regimes. Recent improvements in technology allow for more in-depth analysis and interpretation of modern sand bodies.

With the advent of sequence stratigraphy, new and conflicting models for isolated shallow marine deposits have appeared in the literature. Instead of resolving the controversy over their origin, the debate has shifted focus as researchers disagree on the correlation and genesis of key surfaces and sand bodies contained between them, adding fuel to the debate concerning genesis of the sand bodies. The models offer very different predictions in terms of both exploration and exploitation.

It was in this setting that Donald J. P. Swift brought proponents of the differing interpretations together to discuss facts and principles of these models and their relationship to isolated shallow marine sand bodies using the controversial Lower Campanian Shannon Sandstone as the focus for discussion. The 1995 SEPM-sponsored research conference was held in Casper, Wyoming and entitled "Tongues, Ridges and Wedges". It was convened by Donald J. P. Swift (Old Dominion University), John W. Snedden (Mobil E&P Technical Center) and A. Guy Plint (University of Western Ontario). While the conference focused upon isolated shallow marine sand bodies encased in marine mudstone that were previously interpreted as shelf-ridge complexes, analogous modern sand bodies were also discussed. Great effort was made to ensure that there was an equal exchange of ideas and that all models were discussed in an open forum through the technical program, core workshop, and field trips.

This volume is an outgrowth of that research conference and expands upon the theme of presenting differing interpretations for the same sand bodies in a non-judgmental manner. No preference for one model over another is given in this volume. The preferred interpretation is left for the reader to decide based on the facts and arguments presented in the papers. The differing interpretations are supplemented by a variety of papers presenting possible analogs for isolated shallow marine sand bodies from the modern and by additional papers providing data based on basin modeling, geochemistry and ichnology. A synthesis by John R. Suter (Conoco, Inc.) and H. Edward Clifton (Conoco, Inc.), summarizing the strengths and weaknesses of each of the differing interpretations for isolated shallow marine sand bodies, is included to provide an impartial perspective on this well known but important debate.

Katherine M. Bergman　　　　John W. Snedden

ISOLATED SHALLOW MARINE SAND BODIES: SEQUENCE STRATIGRAPHIC ANALYSIS AND SEDIMENTOLOGIC INTERPRETATION

ACKNOWLEDGMENTS

We are pleased to acknowledge the support of SEPM for the 1995 targeted research conference held in Wyoming that focused on the problems associated with interpreting isolated shallow marine sand bodies. We would like to express our gratitude to the conveners of this technical conference, Don Swift, Guy Plint and John Snedden as well as the support staff and volunteers. This volume is an outgrowth of that conference.

The reviewing of manuscripts is critical to the success of any publication. This time consuming task often goes unnoticed and unrewarded. We would like to take this opportunity to thank each of our reviewers for their contribution toward this volume.

John Anderson
John Armentrout
Bill Arnott
Katherine Bergman
Janok Bhattacharya
Roger B. Bloch
John Carey
Rick Cheel
Janis Dale
Bob Dalrymple
David Eberth
Ashton Embry
Ray Fitzsimmons
Neal Gaynor
Martin Gibling
Bruce M. Kofron
Lee Krystinik
Daniel Labelle
Dale Leckie
Randi Martinsen
Dag Nummedal
Judith Parrish
Guy Plint
Karen Porter
Bruce Power
Louise Quinn
Mogens Ramm
Jim Rine
John Snedden
Morgan Sullivan
Gary Yeo
Brian Zaitlin

The quality of the volume has been greatly improved by the professional editing skills of Catherine White. Robert T. Clarke (Mobil E&P Technical Center) provided his extensive publication expertise. We would like to express our gratitude to these individuals for the countless hours they have contributed toward the production of this volume. Mobil E&P Technical Center (Dallas, Texas) is acknowledged for allowing Dr. Snedden the time to work on this volume as well as incidental costs. Additional funding for this volume was provided by an NSERC Research grant to K. M. Bergman. We would also like to thank SEPM for their advice and encouragement in support of this project.

Finally we would like to thank the authors who have contributed manuscripts to this volume. Their efforts have ensured that the volume was produced in a timely fashion.

Katherine M. Bergman

John W. Snedden

ISOLATED SHALLOW MARINE SAND BODIES: DEPOSITS FOR ALL INTERPRETATIONS

JOHN W. SNEDDEN
Mobil Exploration & Producing Technical Center, PO Box 650232, Dallas, TX 75265 U.S.A.
AND
KATHERINE M. BERGMAN
Department of Geology, University of Regina, Regina, SK S4S 0A2, Canada

Abstract: Isolated shallow marine sand bodies detached from coeval coastal deposits and encased in marine mudstone have been a focus of research and exploration in sedimentary geology for over 30 years. Yet, their origin and occurrence continues to spark debate and motivate explorationists interested in accessing the hydrocarbons within these reservoirs. Interpretations of these complex and enigmatic sand bodies have evolved from genetic models based on physical and biological sedimentary structures to regional sequence stratigraphic models invoking allocyclic processes. This evolution of interpretation parallels the stepwise progression of sedimentary geology as a whole. The Shannon Sandstone of Wyoming is a well-documented example of this type of sand body that was originally interpreted as a shelf sand ridge on the basis of sedimentary structures and external geometry. Addition of local and regional stratigraphic studies have generated new ideas involving incised shoreface and estuarine valley-fill models. Determination of the correct model is critical, as the various interpretations imply significant differences in exploration approach and production characteristics.

INTRODUCTION

Shallow marine sand bodies have been defined as accumulations of sand deposited between the shoreline and shelf-edge (Johnson and Baldwin, 1986). In this realm, sediments are affected by short term dynamic processes such as waves, currents, and tides which are superimposed on longer term changes in relative sea level. Relative sea level changes are, in turn, a function of eustasy, tectonics, and sediment supply. Under certain conditions, sand bodies formed in a shallow marine setting are spatially and temporally isolated, detaching from more strike-continuous coastal deposits and becoming encompassed within offshore marine shale. In this setting, the origin and depositional environment of the sandstone deposits are enigmatic and challenging to sedimentologists and stratigraphers attempting to make genetic interpretations.

A certain exploration mystique has also surrounded these features, partly because of the large hydrocarbon reserves contained in these isolated subsurface reservoirs. This is particularly true of the Rocky Mountain Basin, where stratigraphic trap discoveries such as Bell Creek Field in 1967 completely reversed declining exploration interest (McGregor and Biggs, 1968). In 1975, ensuing exploration resulted in the discovery of the giant Hartzog Draw oil field (350 MMBO original-oil-in place) in the Shannon Member of the Cody Shale (Fig. 1). The primary reservoir here was interpreted as an isolated shallow marine sand body (Tillman and Martinsen, 1987).

In the mid- to late-1980s, interest in isolated shallow marine sand bodies reached an apex, with numerous publications devoted to the study of outcrop, subsurface, and modern examples (e.g., Tillman and Siemers, 1984; Tillman et al., 1985; Knight and McLean, 1986; Morton and Nummedal, 1989). Studies were focused primarily upon lithofacies characteristics and depositional interpretations. Subsequent to this, in parallel with the declining fortunes of domestic U.S. and Canadian oil companies, fewer studies of these enigmatic sandstones were published.

In June 1995, an SEPM research conference focused on isolated shallow marine sand bodies was convened in Casper, Wyoming by Donald J. P. Swift, A. Guy Plint, and John W. Snedden. Interest was renewed by recent stratigraphic studies that suggested novel interpretations for the Shannon and similar units (Plint, 1988; Walker and Bergman, 1993). Advances in sequence stratigraphic interpretation, with models of lowstand wedges and "forced regressions" (Van Wagoner et al., 1990; Posamentier et al., 1993), provided new viewpoints for understanding isolated shallow marine sand bodies. In addition, studies of modern Atlantic shelf sand ridges were being published (e.g., Rine et al., 1991; McBride and Moslow, 1991; Snedden et al., 1994).

While this volume is an outgrowth of this 1995 SEPM research conference, focus has been broadened considerably to

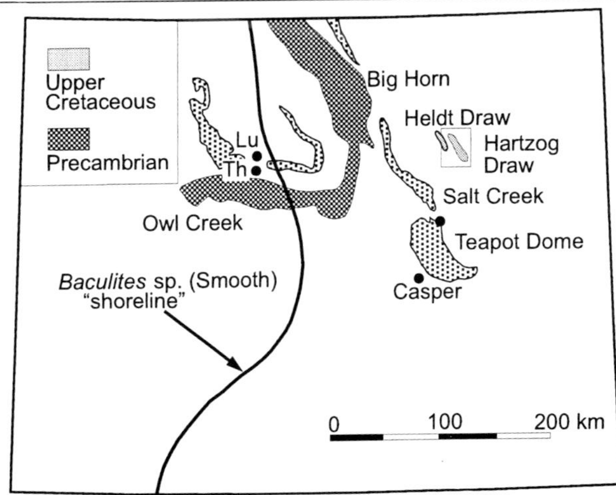

Fig. 1.—Map of Wyoming showing location of study area (rectangle) and Shannon time equivalent shoreline as proposed by Gill and Cobban (1973). Lu, Lucerne; Th, Thermopolis. The distribution of Upper Cretaceous rocks is shown only in the Lucerne and Salt Creek areas. The Owl Creek and Big Horn Mountains are Precambrian uplifts. Open box shows study area. Modified from Walker and Bergman (1993) and Bergman (1994).

Isolated Shallow Marine Sand Bodies: Sequence Stratigraphic Analysis and Sedimentologic Interpretation.
SEPM Special Publication No. 64, Copyright © 1999
SEPM (Society for Sedimentary Geology), ISBN 1-56576-057-3, p. 1-11.

incorporate studies of modern shallow marine sand bodies (Catteneo and Trincardi, this volume; McBride et al., this volume; Rodriguez et al., this volume; Snedden et al., this volume). In addition, overview papers providing new unified models for modern ridge sands (Snedden and Dalrymple, this volume) and isolated shallow marine sand bodies in general (Suter and Clifton, this volume) are included. In compiling this volume, we have sought to provide a balanced and impartial viewpoint for understanding these controversial sand bodies.

Having addressed the purpose and thrust of this volume, several other topics pertinent to the subject of isolated shallow marine sand bodies should be discussed: 1) terminology and definitions; 2) documentation of recent global occurrences, particularly since the crescendo of publications in the 1980s; 3) depositional and stratigraphic models for relevant ancient sand bodies (e.g., Shannon, Viking and Tocito Sandstones); and 4) comparison of the differing implications of the various models for hydrocarbon exploration and production.

TERMINOLOGY

Much terminological confusion exists regarding isolated shallow marine sandstones, both in North America and worldwide. Early papers referred to lenticular shallow marine bodies apparently formed at some distance seaward of coeval shoreline deposits as "marine-bars," (Exum and Harms, 1968). While quite popular among explorationists in the 1970s, this did invite unfortunate comparisons with barrier bars and/or breaker bars on beaches. Later papers used the term "offshore, shallow-marine sand bodies" (e.g., Spearing, 1976).

The term "shelf sand(stone)" came into vogue in the late 1970s and early 1980s to describe siliciclastic deposits that are interpreted to have formed seaward of the shoreline and landward of the shelf edge (Tillman et al., 1985; Knight and McLean, 1986). However, sequence stratigraphers would later employ the same term to describe any deposit formed landward of the slope, thus including shoreline sand bodies (Vail, 1987). In addition, further complexity was added with models for isolation of shoreface sands through lowstand bypassing and/or transgressive ravinement (Plint, 1988; Posamentier et al., 1993). This mechanism thus allows the formation of shallow marine sand bodies as far offshore as the upper slope (lowstand wedge prograding complexes of Mitchum et al., 1994), considerably broadening the defined zone of shelf sandstone occurrence.

In this volume, we focus on "isolated" shallow marine sand bodies, thus placing emphasis on their discrete temporal and spatial occurrence rather than any arbitrary zone or process of deposition. Sand body isolation, like hydrocarbon entrapment, can occur because of both stratigraphic and depositional factors. We concede that a myriad of depositional processes will influence shallow marine sand bodies and that varying amounts of post-depositional modification can take place. As discussed below, we believe in first establishing a stratigraphic framework for any sand body of interest before deriving a specific nomenclature or classification. Our attention is centered on "isolated" shallow marine sand bodies because they are challenging to interpret and have their own self-contained petroleum systems (trap, seal, source, and reservoir).

GLOBAL ANCIENT OCCURRENCES

Isolated shallow marine sand bodies are widely recognized in surface outcrops and subsurface penetrations (Table 1). The examples provided are interpretations made since interest in siliciclastic shelf sandstones peaked in the late-1980s. Continued recognition of discrete shallow marine sand bodies is partly due to development of high resolution sequence stratigraphic approaches that emphasize identification of subtle sand body pinchouts and stratigraphic isolation in general (Posamentier et al., 1993). Perhaps another reason is the enhanced resolution of seismic (two- and three-dimensional) and well logging tools.

One trend apparent from this list is the increase in the number of papers describing "lowstand wedge prograding complexes" (terminology of Mitchum et al., 1994) in recent years. This parallels the overall decline in the popularity of the shelf sand ridge model.

Many of the examples listed in Table 1 are found in Jurassic and Cretaceous strata; far fewer are described from Tertiary units. This may be due to the relative frequency of shallow marine versus more proximal paleoenvironments: the Jurassic and Cretaceous periods are known as worldwide "greenhouse" episodes, with relatively high sea levels and a profusion of intracratonic shallow marine deposition. By contrast, Tertiary strata are dominated by global "icehouse" deltaic progradation and accumulation of large amounts of nonmarine and marginal marine sediments on continental margins.

The economic significance of these isolated shallow marine sand bodies should not be overlooked. The Magnus, Munin, Fulmar, and Cook reservoirs are charged with large volumes of hydrocarbons and continue to provide attractive exploration objectives in the North Sea (Ravnas and Steel, 1997). The Rogn sandstone is the primary reservoir of the giant Draugen Field, the first field to produce hydrocarbons in the mid-Norway sector (Provan, 1993). The Tocito sandstone in the giant Bisti Field represents the largest reservoir of oil in the gas-prone San Juan Basin, New Mexico (Sabins, 1963; Nummedal and Riley, this volume). Large accumulations of oil at Pembina (Alberta) and Hartzog Draw (Wyoming) have demonstrated the economic significance of the Cardium and Shannon reservoirs, respectively.

EVOLUTION OF MODELS FOR ISOLATED SHALLOW MARINE SAND BODIES

The evolution of interpretations for isolated shallow marine sand bodies parallels the stepwise progression of sedimentary geology research since the late 1960s. The shift in focus from depositional processes (as inferred from relatively small-scale sedimentary structures) to regional stratigraphy is superbly illustrated by studies of the Lower Campanian Shannon Sandstone (Powder River Basin, Wyoming; Fig. 1). Similar cases can be described in the Cretaceous Viking and Tocito Formations of North America.

The Lower Campanian Shannon Sandstone is composed of a series of *en echelon* northwest-southeast trending, elongate sand bodies encased in marine mudstone. In this setting, sand bodies were originally interpreted as part of a shelf ridge complex located some 160 km seaward of their contemporaneous shoreline deposits (Spearing, 1976; Seeling, 1978; Tillman and Martinsen, 1984, 1987; Tye et al., 1986; Gaynor and Swift, 1988). The arguments supporting the shelf ridge interpretation are summarized in Swift and Parsons (this volume), and Tillman (this volume). Despite extensive description of the Shannon deposits both in outcrop and the subsurface, several questions remain unresolved about the apparent transport of coarse sediment several tens to hundreds of kilometers offshore to depositional sites on the mid-to outer-shelf and the mechanism for subsequent reworking into long, linear, coarsening-upward ridges.

Table 1.—Recent interpretations of ancient isolated shallow marine sand bodies.

Unit/Location	Age	Interpretation and Citation
Lower sandstone member of Thermopolis Shale, Montana	Cretaceous	Shelf sand ridges (Stine and Schmitt, 1987)
Mooreville Chalk, east-central Alabama	Cretaceous	Shelf sand bar (King, 1987)
Gog Group, Yoho National Park, Canada	Cambrian	Shelf sand ridge complex (Hein, 1987)
Cardium Formation, Pembina Field, Alberta	Cretaceous	Lowstand shoreface (Plint, 1988); Shelf sand ridge (Krause and Nelson, 1991);
Cook Formation, Oseberg Field, North Sea	Jurassic	Tidal sand ridge (Livbjerg and Mjos, 1989)
Munin Formation, Statfjord Nord Field, North Sea	Jurassic	Syntectonic lowstand wedge (Barnes et al., 1992)
Holder Formation, Sacramento Mountains, New Mexico	Pennsylvanian	Shelf sand ridges (Carr and Scott, 1990)
Olmos Formation, Webb County, Texas	Cretaceous	Shelf ridge and sheet sands (Snedden and Jumper, 1990)
Winnipeg Group, Manitoba, Saskatchewan, North Dakota	Ordovician	Shelf sand ridges (Kessler, 1991)
Wilcox Group, Duval and Jim Hogg Counties, Texas	Paleocene-Eocene	Lowstand wedge prograding complex (Snedden et al., 1991)
Oppdalsata Member, Agardhfjellet Formation, Spitsbergen, Svalbard	Jurassic	Shelf sand ridge (Dypvik et al., 1991)
Tocito Sandstone, San Juan Basin, New Mexico	Cretaceous	Shelf sand ridge (Nummedal et al., 1993)
Semilla Sandstone, San Juan Basin, New Mexico	Cretaceous	Shelf sand ridge (Nummedal et al., 1993)
Rogn Formation, Draugen Field, Mid-Norway	Jurassic	Lowstand wedge prograding complex (Provan, 1993)
Shannon Member of Cody Shale, Hartzog Draw Field, Wyoming	Cretaceous	Lowstand shoreface (Walker and Bergman, 1993; Bergman, 1994); Tidal-estuarine valley-fill (Sullivan et al., 1997)
Viking Formation, Alberta, Canada	Cretaceous	Lowstand shoreface (Walker and Wiseman, 1995)
Kenilworth Member, Book Cliffs, Utah	Cretaceous	Lowstand shoreface (Pattison, 1995)
Haystack Mountains Formation, southeast Wyoming	Cretaceous	Incised valleys and estuaries (Mellere and Steel, 1995)
Oseberg Sandstone, North Sea	Jurassic	Lowstand wedge prograding complexes (Muto and Steel, 1997)
Kremmling and Muddy Buttes Sandstone, Colorado	Cretaceous	Shelf ridges (Krystinik, 1995)

Recognizing these troubling issues, Gaynor and Swift (1988) suggested that the Huthnance (1982) model of ridge development could be applied to the formation of the Shannon sand body at Hartzog Draw (Fig. 1). The Huthnance mechanism, discussed in greater detail by Snedden and Dalrymple (this volume), allows for the development of ridge-like sand bodies from an initial precursor during transgressive evolution of a shoreline deposit. In this model, sediment-bearing current flows interact with bathymetric highs to form large, shoreline-oblique sand bodies. However, this model still requires the availability of shoreline sediment near or at the site of ridge development. For the Shannon, this shoreline or shoreface deposit is not immediately apparent. In addition, the shelf ridge interpretation did not explain the common co-occurrence of glauconite and siderite, somewhat incompatible mineralogies at the top of these deposits (Stonecipher, this volume).

The source of the sand in this large reservoir could not be determined from a purely sedimentological approach based on the observed physical and biological sedimentary structures and local sand body geometry. At this point, the focus of studies began to shift toward understanding the effects of allochthonous controls on sand body deposition and preservation through sequence stratigraphic analysis. At this time, the fundamentals of high-resolution sequence stratigraphy were published by Exxon Research and Production (Van

Wagoner et al., 1990), and these writings would profoundly impact the way all shallow marine deposits would subsequently be described and interpreted. High-resolution sequence stratigraphy places less emphasis on somewhat equivocal primary sedimentary structures and focuses instead on the stratigraphic framework developed through correlation of regionally extensive discontinuities formed in response to allocyclic controls (Plint et al., 1986; Posamentier et al., 1993). These techniques provide a high resolution chronostratigraphic framework. From this context, the reconstructed depositional history is constrained by the sedimentary structures, textures, facies, and facies associations.

Reflecting this emphasis on allocyclic controls, more recent studies invoking a shelf ridge interpretation suggested that these sand bodies form the basal part of the transgressive systems tract and that ridge sediment was supplied and reworked from the lowstand shoreface during ravinement (Nummedal and Swift, 1987; Nummedal and Riley, this volume). This view of sediment supply to the shelf was supported by new research on similar deposits in the Canadian portion of the Western Interior Seaway (e.g., Plint et al., 1986; Bergman and Walker, 1987, 1988; Downing and Walker, 1988; Plint and Walker, 1987).

Plint et al. (1986) proposed a regional stratigraphic framework for the Cardium Formation in the Alberta Basin based on the recognition and correlation of multiple regionally extensive erosion surfaces. These surfaces were interpreted to have formed in response to fluctuations in relative sea level (Fig. 2). Bergman and Walker (1987, 1988) suggested that the 20-m-thick Cardium conglomerates at Carrot Creek field, Alberta, were deposited in incised shorefaces that formed during pauses in an overall transgression. The bases of the sand bodies were shown to be erosional and were interpreted as a surface formed at fairweather wave base during lowstand and initial transgression. The sandstone was deposited at a time when the shoreline rapidly prograded basinward. Transgresssion ensued and the upper part of the coarsening upward sand body was removed by erosion at fairweather wave base and overlain by a fining upward succession of transgressive mudstone. On these flat, broad shelves, relatively minor variations in relative sea level (± 10 m) could temporally and spatially isolate large sand bodies (Fig. 2). The fundamental differences between the incised shoreface and the transgressive shelf ridge interpretation are: 1) the implicit suggestion in the shelf ridge interpretation that the sand bodies are unique to the transgressive systems tract; and 2) the shelf ridge interpretation requires the reworking or evolution of the incised shoreface deposits into a marine ridge.

A similar model has been suggested for shoreface incision during regression and is referred to as "forced regression" (Plint, 1988). Subsequently, Posamentier et al. (1992) examined this concept in a sequence stratigraphic framework, illustrating the model with examples from the Viking Formation (Alberta) and discussing the exploration implications. Additional conceptual models were presented that demonstrated how stratigraphic isolation could occur by proximal lowstand incision and sediment bypass transitioning to downdip deposition characterized primarily by the lowstand and transgressive systems tracts (Posamentier and Allen, 1994). Nummedal et al. (1994) then proposed the concept of a falling stage systems tract to include deposits formed during the fall of relative sea level.

The results of the Cardium work (Plint et al., 1986; Bergman and Walker, 1987, 1988), coupled with the evolving concepts of sequence stratigraphy, provided the necessary platform for a reevaluation of the shelf ridge hypothesis. The Shannon Sandstone was viewed as the classic ancient example of a shelf ridge sand body and became the focus of subsequent studies employing sequence stratigraphic techniques. However, rather than solving the problems associated with interpreting these complex and enigmatic sandstone bodies, sequence stratigraphy has added fuel to the debate as researchers argue about the significance of key surfaces and how these surfaces should be correlated to produce a chronostratigraphic framework. Coupled with this is the increased understanding of biologic structures and their significance (MacEachern et al., this volume) and the increased recognition of tidal signatures in these deposits.

Three different interpretations of the Shannon and other isolated shallow marine sand bodies now exist: a) shelf ridge (Tillman, this volume; Swift and Parsons, this volume); b) incised shoreface (Bergman and Walker, this volume; Burton and Walker, this volume) and c) incised valley-fill (Sullivan et al., 1997; MacEachern et al., this volume). Each of these interpretations has its strengths and weaknesses. The debate led to a SEPM research conference held in June 1995, in Casper, Wyoming, which was designed to bring together the different schools and to critically evaluate each interpretation. This volume is an attempt to provide an impartial documentation of these contrasting viewpoints and background on the continuing debate from which readers may draw their own conclusions.

Although the above discussion has focused on the Shannon Sandstone, parallel controversies and similar discussions surround the Tocito Sandstone, San Juan Basin, New Mexico (McCubbin, 1969; Campbell, 1971, 1979; Jennette et al., 1991; Nummedal and Riley, 1991; Nummedal and Riley, this volume). The Viking Formation of the Alberta Basin also has triggered debate (Koldijk, 1976; Beaumont, 1984; Hein et al., 1986; Walker and Wiseman, 1995; Burton and Walker, this volume; MacEachern et al., this volume). Discussion is still underway regarding the Cardium Formation of Alberta (Swagor et al., 1976; Krause and Nelson, 1984, 1991; Bergman and Walker, 1987, 1988; Arnott, 1991, 1992). The arguments for these different sand bodies will not be repeated here. However, readers are referred to key references and relevant papers contained in this volume as a guide to the ongoing debate for sand bodies other than the Shannon.

IMPORTANCE OF RECENT ANALOGS FOR UNDERSTANDING ISOLATED SHALLOW MARINE SAND BODIES

Paralleling the increasing impact of seismic and sequence stratigraphy are the rapid advances in seismic imaging of seafloor and near-surface modern sediments. High resolution seismic data can now be acquired from many difficult locations not previously studied (e.g., estuaries, rivers, high-energy coastal settings). Research on recent analogs, previously limited to description of surficial processes and sediments, has expanded into the early Holocene and Pleistocene (Cattaneo and Trincardi, this volume).

Furthermore, recent studies from a variety of depositional settings are providing evidence that post-depositional modification of shallow marine sand bodies is common. Work on the continental shelves of Texas, Mississippi-Alabama, New Jersey-Atlantic, and the Canadian-Atlantic shows that sand bodies evolve considerably following their initial genesis as shoreline-associated features (Snedden and Dalrymple, this volume; Rodriguez et al., this volume; McBride et al., this volume). The long (>15 ky) relative sea level rise since the Late Wisconsin lowstand has allowed ample time for waves and currents to rework submerged nearshore features. Evidence suggests that these shoreline-associated features become increasingly more marine with time (Dalrymple and Hoogendoorn, 1997; Snedden et al., this volume).

Such sand body evolution is likely to have occurred in ancient analogs. It is intriguing to consider the possibility that isolated shallow marine sand bodies like the Shannon Sandstone may have originated as shoreface sands or incised valley-fill tidal bars, as advocated by various workers, but with transgression evolved into more marine elements like shelf sand ridges (cf., Swift and Parsons, this volume; Nummedal and Riley, this volume).

IMPLICATIONS OF DIFFERENT INTERPRETATIONS FOR ISOLATED SHALLOW MARINE SAND BODIES

One obvious question arises after reviewing the various discussions about the genesis of isolated shallow marine sand bodies like the Shannon, Tocito, Viking, and Cardium: does it really matter? The answer depends on whether the user intends to employ stratigraphic and depositional models as a tool for local or regional sand body prediction and exploitation. It is clear that an appropriate model must be formulated before hydrocarbons from ancient reservoirs can be explored and produced (Barwis, 1989). Forward seismic modeling, often used to calibrate two- or three-dimensional seismic data, requires a valid stratigraphic framework. Reservoir simulation, used to justify economic development of reservoirs, is based on an interpreted architecture of sedimentary facies, vertical and lateral arrangement of reservoir volume, and permeability (grain size). It is worthwhile to compare various models for shallow marine sandstones from an exploration and production standpoint.

As mentioned, the three most popular models for isolated shallow marine sand bodies, like the Shannon, are: 1) shelf sandstone ridge; 2) incised valley-fill; and 3) incised shoreface; The different models provide contrasting predictions for ex-

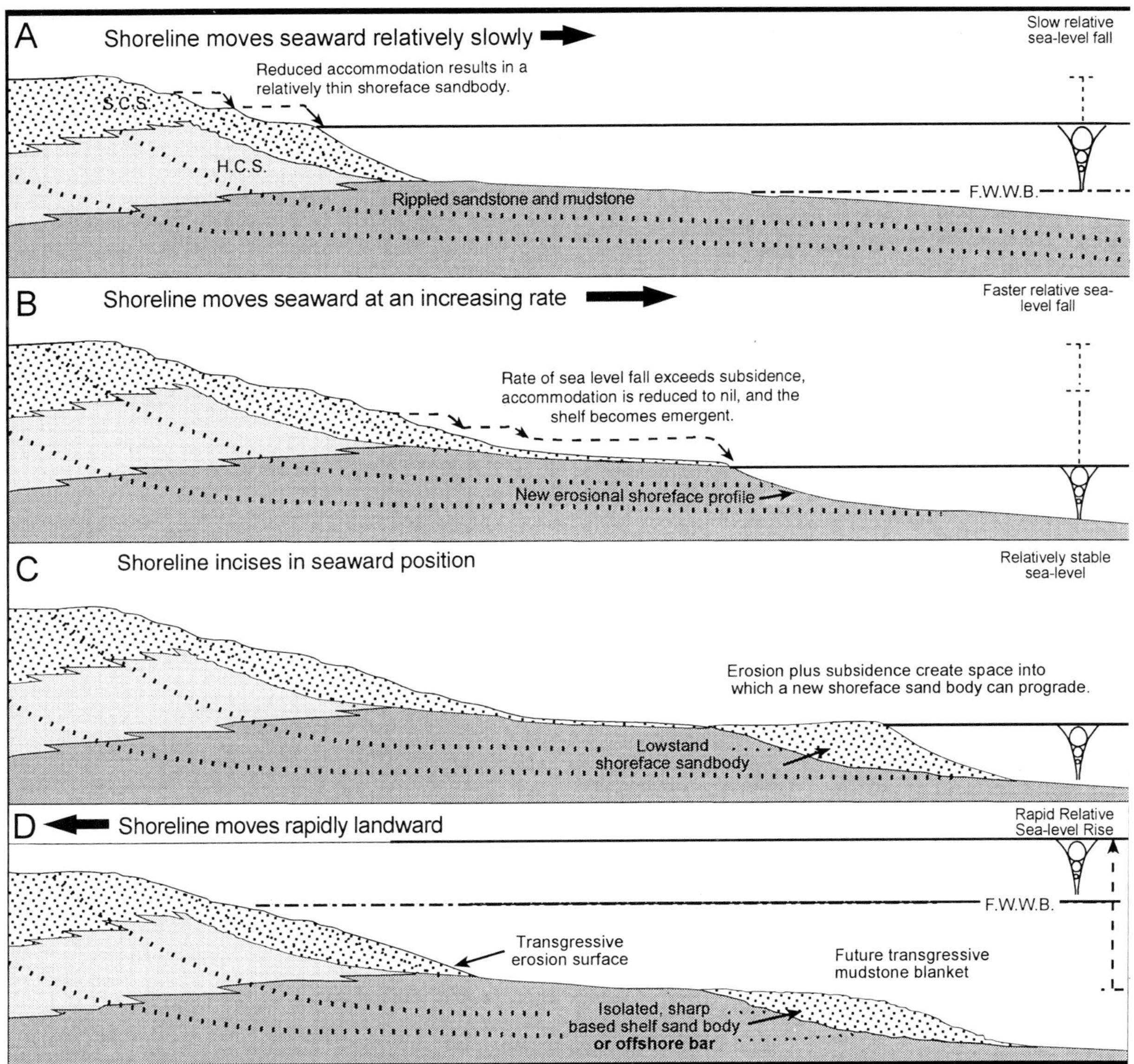

Fig. 2.—Model of formation of isolated shallow marine sand bodies via relative sea level changes. Modified from Plint et al. (1988).

ploration (Fig. 3) and production (Fig. 4). In terms of exploration, the shelf sand ridge model tends to emphasize accumulation of sandstone in the transgressive systems tract (Snedden and Dalrymple, this volume; Fig. 3A). Thus, subtle stratigraphic indications like strong onlap of reflections onto the underlying sequence boundary might indicate the presence of these types of sand bodies (Mitchum et al., 1994). Because ridges often form on or close to preexisting bathymetric or topographic highs with the proper orientation relative to the prevailing combined flow (Huthnance, 1982), detailed isopach mapping of the underlying intervals to identify stratal thinning may be useful for pre-drilling prediction.

The incised valley-fill model emphasizes the lowstand systems tract and thus stratigraphic indications of erosion such as truncation and toplap (e.g., Snedden et al., 1997; Fig. 3B). Identification of topographic lows are critical as these may control the orientation of paleodrainage systems. Underlying structural trends are more important in the formation of lowstand incised valley-fills than for shelf ridge sands developed during the transgressive flooding of the shelves (Sullivan et al., 1997).

The incised shoreface model would also emphasize subtle stratal terminations such as toplap and erosional truncation of reflections (Bergman, 1994; Fig. 3C). However, the identification of pronounced seaward-dipping, shoreline-parallel declivities in an otherwise nearly flat shelf might indicate the presence of associated lowstand shoreface sandstone (e.g., Plint, 1988). In addition, the complex arrangement of high frequency erosion and transgressive surfaces may only be resolvable at the three-dimensional seismic-scale (Bergman and Walker, this volume). The incised shoreface sand bodies would largely constitute the lowstand systems tract, though some could be incorporated within the overlying transgressive systems tract.

From a production standpoint, the various models are quite different in application. Shelf sand ridge reservoirs, like their modern analogs, tend to have a strong permeability anisotropy, with the coarsest grain sizes (and thus highest permeability) located on the "high-energy ridge margin," typically on the steep flank of these asymmetric sand bodies (Fig. 4A). Depending upon the completion strategy, this strong anisotropy can affect hydrocarbon recovery, as low permeability zones may not contribute to fluid flow until pressure has dropped below suitable levels. The association with underlying tidal inlet channels or ridge precursors such as ebb-tidal bars may cause upward coning of fluids into the ridge sand reservoir with a strong water drive. However, this effect is mitigated if mudstone drapes the contact between the ridge sand and the precursor facies or the ridge reservoir is stratigraphically isolated from underlying water-bearing units.

Incised valley-fills may exhibit a variable distribution of grain size and sand percentage, depending on the spatial development of tidal bars and tidal mud within the valley (Fig. 4B). The presence of dipping, mud-draped, internal accretion surfaces (clinoforms) may greatly impede fluid flow (Sullivan et al., 1997). Erosional truncation associated with various sequence boundaries may also affect the occurrence of sand-prone facies and thus reservoir volume (Fig. 4B).

Incised shorefaces also show strong horizontal and vertical changes in grain sizes, and thus permeability, as a function of progradation and decreasing wave energy in an offshore direction (Fig. 4C). However, the complex arrangement of bounding discontinuities may impart considerable heterogeneity if these surfaces are flow barriers or baffles between reservoir compartments. Conversely, these surfaces may prevent upward or downward fluid coning during production.

To summarize, attempting to understand the genesis of isolated shallow marine sand bodies like the Shannon, Tocito, Viking, and Cardium is not an exercise in futility. These units are important producing intervals in the Western Interior Seaway and worldwide (Table 1). When properly interpreted, these genetic models provide powerful tools for reservoir prediction and exploitation.

CONCLUSIONS

Isolated shallow marine sand bodies have been a focus of sedimentary geology research and hydrocarbon exploration for over 30 years. The presence of large sand bodies encased in marine shale and separated from coeval coastal sediments continues to spark interest and debate. Studies of these complex and enigmatic features have evolved from genetic interpretations to regional sequence stratigraphic models.

Isolated shallow marine sand bodies differ from more widespread, strike-continuous strandplain or lobate deltaic systems whose facies associations indicate proximity of coastal-plain and fluvial systems. These features are common throughout the geologic record, particularly where there is a predominance of broad, shallow marine paleoenvironments and relatively coarse sediment is available.

The succession of interpretations made for the Shannon Sandstone of Wyoming are illustrative of the increasing use of integrated studies in reconstructing the geologic history of a shallow marine sand body. Building upon earlier studies of physical sedimentary structures and sand body geometry are advances in regional sequence stratigraphic correlation, ichnology, inorganic geochemistry, and paleodynamical modeling. In addition, work on potential modern analogs has revealed that many shoreline sand bodies evolve into more marine elements over time. This allows for the possibility that units like the Shannon may have been deposited as shoreline-associated sand bodies (e.g., shoreface or estuary bar-form) and were modified during the transgression into a shelf sandstone ridge. The degree of subsequent modification is a subject of continued debate.

The genetic interpretation applied for isolated shallow marine sand bodies is critical, because the various models have significantly different implications for the exploration for and production of contained hydrocarbons. It is worthwhile to weigh the various arguments and come to your own conclusion regarding these often complex and enigmatic sandstone deposits.

REFERENCES

Arnott, R.W.C., 1991, The Carrot Creek "K" Pool, Cardium Formation, Alberta: A conglomeratic reservoir retated to a wave-reworked distributary mouth-bar complex: Bulletin of Canadian Petroleum Geology, v. 39, p. 43-53.

Arnott, R.W.C, 1992, The role of fluvial processes during deposition of the (Cardium) Carrot Creek/Cyn-Pem conglomerates: Bulletin of Canadian Petroleum Geology, v. 40, p. 356-362.

Barnes, K.R., Snedden, J.W. and McAdow, D., 1992, An integrated exploration search for additional Upper Jurassic Prospectivity in the North Viking Graben, Norwegian North Sea (abstract): American Association of Petroleum Geologists Bulletin, v. 76, p. 772.

Barwis, J.H., 1989, The explorationist and shelf sand models: Where do we go from here?, *in* Morton, R.A. and Nummedal, D., eds., Shelf Sedimentation, Shelf Sequences and Related Hydrocarbon

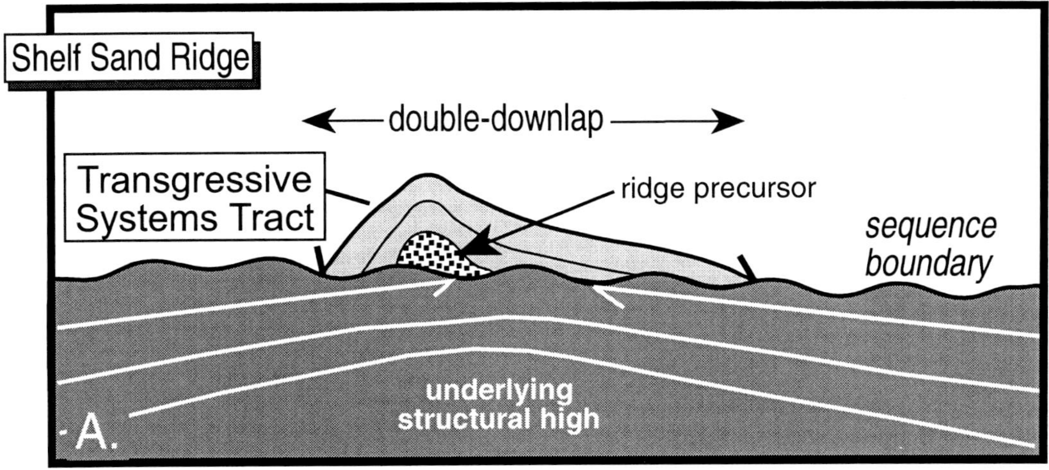

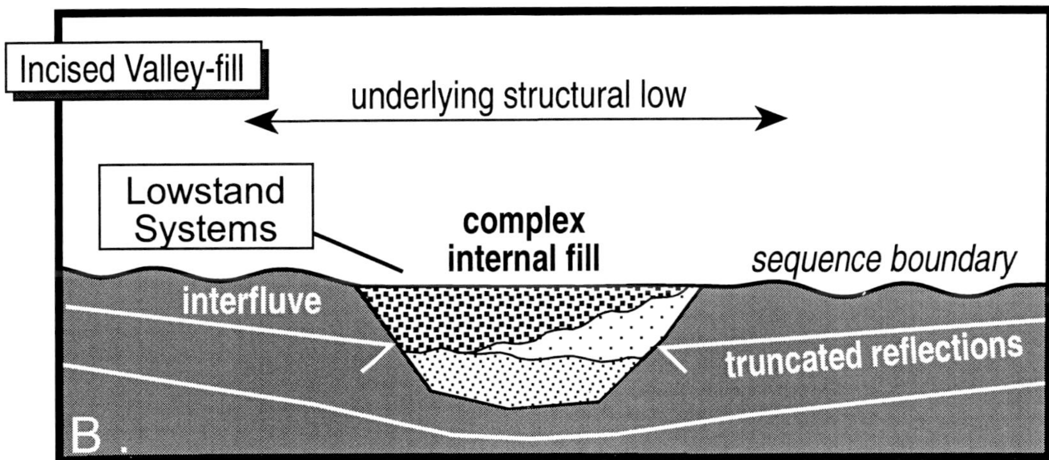

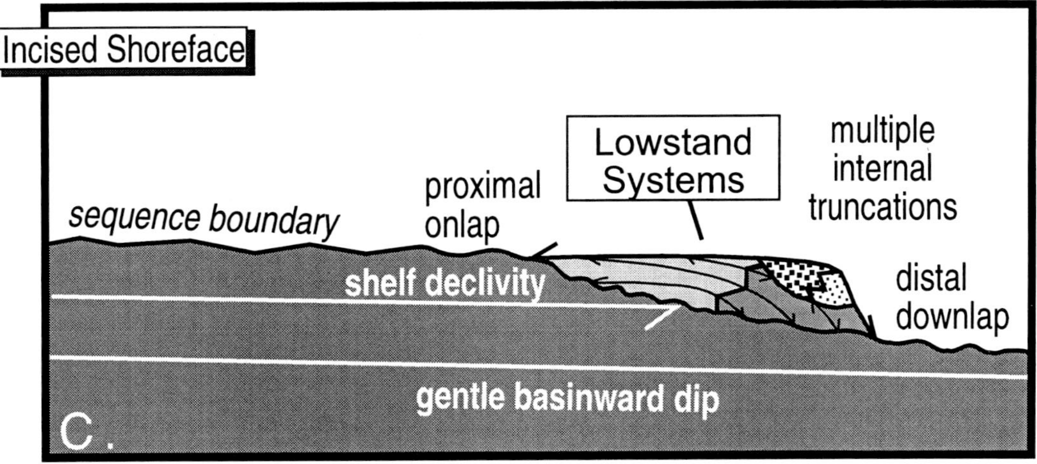

Fig. 3.—Differing stratal geometries and exploration aspects of three types of isolated shallow marine sand bodies: A) Shelf sand ridge; B) Incised valley-fill; C) Incised shoreface.

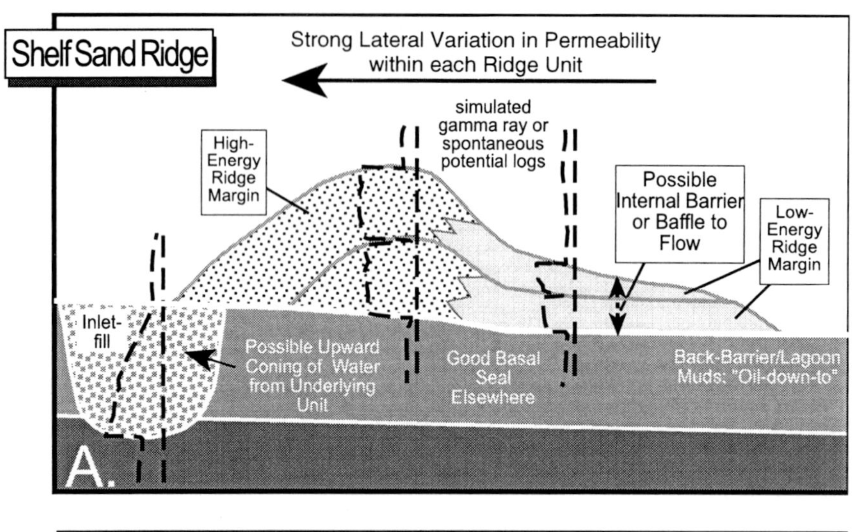

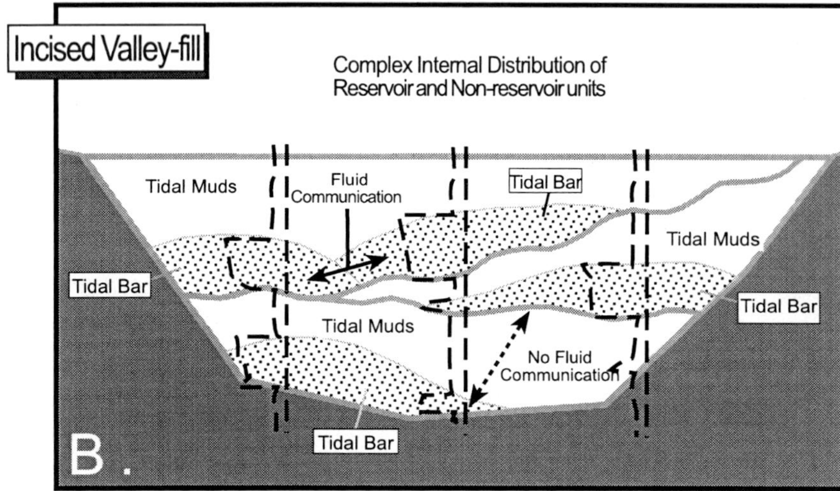

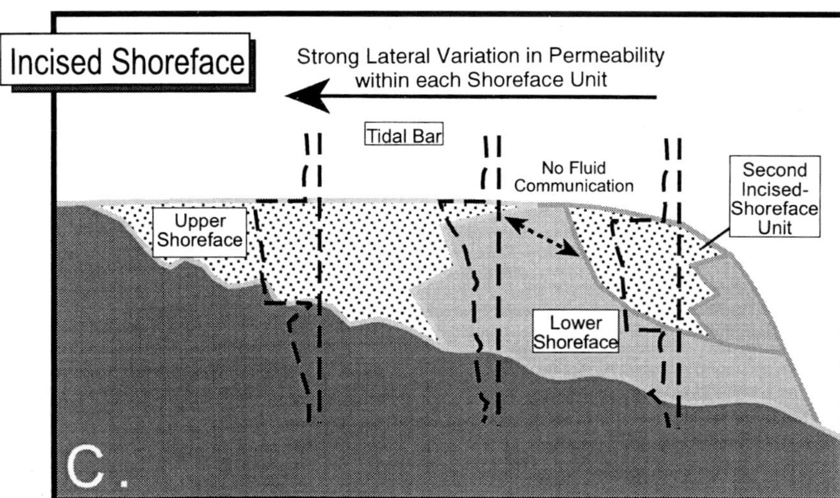

Fig. 4.—Differing log response and production aspects of three types of isolated shallow marine sand bodies: A) Shelf Sand Ridge; B) Incised Valley-Fill; C) Incised Shoreface.

Accumulation: Tulsa, Gulf Coast Section-SEPM (Society for Sedimentary Geology) Annual Research Conference Proceedings, p. 1-15.

BEAUMONT, E.A., 1984, Retrogradational shelf sedimentation: Lower Cretaceous Viking Formation, central Alberta, *in* Tillman R.W. and Seimers, C.T., eds., Siliciclastic Shelf Sediments: Tulsa, SEPM (Society for Sedimentary Geology) Special Publication 34, p. 163-177.

BERGMAN, K.M. AND WALKER, R.G., 1987, The importance of sea-level fluctuations in the formation of linear conglomerate bodies; Carrot Creek Member of Cardium Formation, Cretaceous western interior seaway, Alberta, Canada: Journal of Sedimentary Petrology, v. 57, p. 651-665.

BERGMAN, K.M. AND WALKER, R.G., 1988, Formation of Cardium erosion surface E5, and associated deposition of conglomerate; Carrot Creek Field, Cretaceous Western Interior Seaway, Alberta, *in* James, D.P. and Leckie, D.A., Sequences, Stratigraphy, Sedimentology: Surface and Subsurface: Calgary, Canadian Society Petroleum Geologists Memoir 15, p. 15-24.

BERGMAN, K.M., 1994, Shannon Sandstone in Hartzog Draw-Heldt Draw fields (Cretaceous, Wyoming, USA) reinterpreted as lowstand shoreface deposits: American Association of Petroleum Geologists Bulletin, v. B64, p. 184-201.

CAMPBELL, C.V., 1971, Depositional model—Upper Cretaceous Gallup beach shoreline, Ship Rock area, northwestern New Mexico: Journal of Sedimentary Petrology, v. 41, p. 395-404.

CAMPBELL, C.V., 1979, Model for beach shoreline in Gallup Sandstone (Upper Cretaceous) of northwestern New Mexico, in Circular 164: Socorro, New Mexico Bureau of Mines and Mineral Resources, 32 p.

CARR, D.L. AND SCOTT, A.J., 1990, Late Pennsylvanian storm-dominated shelf sand ridges, Sacramento Mountains, New Mexico: Journal of Sedimentary Petrology, v. 60, p. 592-607.

DOWNING, K.P. AND WALKER, R.G., 1988, Viking Formation, Joffre Field, Alberta: Shoreface origin of a long narrow sandbody encased in marine mudstones: American Association of Petroleum Geologists Bulletin, v. 72, p. 1212-1228.

DYPVIK, H., NAGY, J., EIKELAND, T.A., BACKER-OWE, K. AND JOHANSEN, H., 1991, Depositional conditions of the Bathonian to Hauterivian Janusfjellet Subgroup, Spitsbergen: Sedimentary Geology, v. 72, p. 55-78.

EXUM, F.A. AND HARMS, J.C., 1968, Stratigraphic traps, Western Nebraska: American Association of Petroleum Geologists Bulletin, v. 52, p. 1851-1869.

GAYNOR, G.C. AND SWIFT, D.J.P., 1988, Shannon Sandstone depositional model: Sand ridge dynamics on the Campanian Western Interior shelf: Journal of Sedimentary Petrology, v. 58, p. 868-880.

HEIN, F.J., 1987, Tidal/littoral offshore shelf deposits—Lower Cambrian Gog Group, Southern Rocky Mountains, Canada: Sedimentary Geology, v. 52, p. 155-182.

HEIN, F.J., DEAN, M.E., DELURE, A.M., GRANT, S.K., ROBB, G.A. AND LONGSTAFFE, F.J., 1986, The Viking Formation in the Caroline, Garrington, and Harmattan East fields, western south-central Alberta: Sedimentology and paleogeography: Bulletin of Canadian Petroleum Geology, v. 34, p. 91-110.

HUTHNANCE, J.M., 1982, On one mechanism forming linear sand banks: Estuarine and Coastal Marine Science, v. 14, p. 79-99.

JENNETTE, D.C., JONES, C.R., VAN WAGONER, J.C. AND LARSEN, J.E., 1991, High-resolution sequence stratigraphy of the Upper Cretaceous Tocito Sandstone: The relationship between incised valleys and hydrocarbon accumulation, *in* Van Wagoner, J.C., Jones, C.R., Taylor, D.R., Jennette, D.C., Nummedal, D. and Riley, G.W., eds., Sequence stratigraphy applications to shelf sandstone reservoirs—outcrop to subsurface examples: Tulsa, American Association of Petroleum Geologists Field Conference.

JOHNSON, H.D. AND BALDWIN, C.T., 1986, Shallow siliciclastic seas, *in* Reading, H.G., ed., Sedimentary Environments and Facies: Oxford, Blackwell Scientific Publications, p. 229-283.

KESSLER, L.G., 1991, Subsidence controlled stratigraphic sequences and the origin of shelf sand ridges, Winnipeg Group (Middle Ordovician), Manitoba, Saskatchewan, North Dakota, *in* Christopher, J.E. and Haidl, F.M., eds., Sixth Williston Basin Symposium: Regina, Saskatchewan Geological Society, v. 6, p. 1-13.

KING, D.T., 1987, Sedimentary facies, depositional environments, and sea-level history—Mooreville Chalk, Lower Campanian of East-central Alabama: Southeastern Geology, v. 27, p. 141-154.

KNIGHT, R.J. AND MCLEAN, J.R., 1986, Shelf Sands and Sandstones: Calgary, Canadian Society of Petroleum Geologists Memoir 11, p. 580 p.

KOLDIJK, W.S., 1976, Gilby Viking "B": A storm deposit, *in* Lerand, M.M., ed., The sedimentology of selected clasitic oil and gas reservoirs in Alberta: Calgary, Canadian Society of Petroleum Geologists Core Conference Proceedings, p. 62-77.

KRAUSE, F.F. AND NELSON, D.A., 1984, Storm event sedimentation: Lithofacies association in the Cardium Formation, Pembina area, west-central Alberta, Canada, *in* Stott, D.F. and Gass, D.J., eds., The Mesozoic of Middle North America: Calgary, Canadian Society of Petroleum Geologists Memoir 9, p. 485-511.

KRAUSE, F.F. AND NELSON, D.A., 1991, Evolution of an Upper Cretaceous (Turonian) shelf sandstone ridge; analysis of the Crossfield-Cardium Pool, Alberta, Canada, *in* Swift, D.J.P., Oertel, G.F., Tillman, R.W. and Thorne, J.A., eds., Shelf Sand and Sandstone Bodies; Geometry, Facies, and Sequence Stratigraphy: Oxford, International Association of Sedimentologists Special Publication 14, p. 427-456.

KRYSTINIK, L.F., 1995, Lateral facies relationships in the Kremmling and Muddy Buttes Sandstones of Northern Colorado, *in* Swift, D.J.P., Snedden, J.W. and Plint, A.G., eds., Tongues, Ridges and Wedges: Highstand versus lowstand architecture in marine basins: Tulsa, SEPM (Society for Sedimentary Geology) Research Conference Program.

LIVBJERG, F. AND MJOS, R., 1989, The Cook Formation, an offshore sand ridge in the Oseberg area, northern North Sea, *in* Collinson, J., ed., Correlation in Hydrocarbon Exploration: London, Graham & Trotman, p. 299-312.

MCBRIDE, R.A. AND MOSLOW, T.F., 1991, Origin, evolution, and distribution of shoreface sand ridges, Atlantic inner shelf, U.S.A.: Marine Geology, v. 97, p. 57-85.

MCCUBBIN, D.G., 1969, Cretaceous strike-valley sandstone reservoirs, northwestern New Mexico: American Association of Petroleum Geologists Bulletin, v. 53, 2114-2140.

MCGREGOR, A.A. AND BIGGS, C.A., 1968, Bell Creek Field, Montana: American Association of Petroleum Geologists Bulletin, v. 52, p. 1869-1887.

MELLERE, D. AND STEEL, R., 1995, Facies architecture and sequentiality of nearshore and "shelf" sandbodies; Haystack Mountains Formation, Wyoming, USA: Sedimentology, v. 42, p. 551-574.

MITCHUM, R.M., SANGREE, J.B., VAIL, P.R. AND WORNARDT, W.W., 1994, Recognizing sequences and systems tracts from well logs, seismic data, and biostratigraphy: Examples from the Late Cenozoic of the Gulf of Mexico, *in* Posamentier, H.W. and Weimer, P., eds.: Sequence Stratigraphy: Tulsa, American Association of Petroleum Geologists Memoir No. 58, p. 163-197.

MORTON, R.A. AND NUMMEDAL, D., 1989, eds., Shelf Sedimentation, Shelf Sequences and Related Hydrocarbon Accumulation: Tulsa, Gulf Coast Section-SEPM (Society for Sedimentary Geology) Annual Research Conference Proceedings, 212 p.

MUTO, T. AND STEEL, R., 1997, The Middle Jurassic Oseberg Delta, northern North Sea: A sedimentological and sequence stratigraphic interpretation: American Association of Petroleum Geologists Bulletin, v. 81, p. 1070-1086.

NUMMEDAL, D., MOLENAAR, C.M. AND RILEY, G.W., 1994, Sequence stratigraphy of the Gallup and Tocito sandstones, San Juan Basin, New Mexico: Denver, Rocky Mountain Association Geologists, Trip # 11, 41 p.

NUMMEDAL, D. AND RILEY, G.W., 1991, Origin of late Turonian and Coniacian unconformities in the San Juan Basin, *in* Van Wagoner, J.C., Jones, C.R., Taylor, D.R., Jennette, D.C., Nummedal, D. and Riley, G.W., eds., Sequence stratigraphy applications to shelf sandstone reservoirs—outcrop to subsurface examples: Tulsa, American Association of Petroleum Geologists Field Conference.

NUMMEDAL, D. AND SWIFT, D.J.P., 1987, Transgressive stratigraphy at sequence-bounding unconformities: Some principles derived from Holocene and Cretaceous examples, *in* Nummedal, D., Pilkey, O.H. and Howard, J.D., eds., Sea-level Fluctuation and Coastal Evolution: Tulsa, SEPM (Society for Sedimentary Geology) Special Publication 41, p. 241-260.

NUMMEDAL, D., WOLTER, N., FLEMING, T.F. AND BERGSOHN, I., 1993, Lowstand, shallow marine sandstones in Upper Cretaceous strata of the San Juan Basin, New Mexico, *in* Caldwell, W.G.E. and Kaufmann, E.G., eds., Evolution of the Western Interior Basin: St Johns, Geological Association of Canada Special Paper 39, p. 199-218.

PATTISON, S.J., 1995, Sequence stratigraphic significance of sharp-based lowstand shoreface deposits, Kenilworth Member, Book Cliffs, Utah: American Association of Petroleum Geologists Bulletin, v. 79, p. 444-462.

PLINT, A.G., 1988, Sharp-based shoreface sequences and "offshore bars" in the Cardium Formation of Alberta: Their relationship to relative changes in sea level, *in* Wilgus, C.K., Hastings, B.S., Kendall, C.G.St. C., Posamentier, H.W., Ross, C.A. and Van Wagoner, J.C., eds., Sea-level Changes: An Integrated Approach: SEPM (Society for Sedimentary Geology) Special Publication 42, p. 357-370.

PLINT, A.G. AND WALKER, R.G., 1987, Morphology and origin of an erosion surface cut into the Bad Heart Formation during major sea level change, Santonian of west central Alberta, Canada: Journal of Sedimentary Petrology, v. 57, p. 639-650.

PLINT, A.G., WALKER, R.G. AND BERGMAN, K.M., 1986, Cardium Formation 6. Stratigraphic framework of the Cardium in subsurface: Bulletin of Canadian Petroleum Geology, v. 34, p. 48-64.

POSAMENTIER, H.W. AND ALLEN, G.P., 1994, The role of tectonics and eustasy in the development of stratal architecture (abstract): Tulsa, 1994 American Association of Petroleum Geologists Annual Convention, p. 237.

POSAMENTIER, H.W., ALLEN, G.P., JAMES, D.P. AND TESSON, M., 1992, Forced regressions in a sequence stratigraphic framework: Concepts, examples and exploration significance: American Association of Petroleum Geologists Bulletin, v. 76, p. 1687-1709.

POSAMENTIER, H.W., SUMMERHAYES, C.P., HAQ, B.U. AND ALLEN, G.P., 1993, Sequence Stratigraphy and Facies Associations: Oxford, International Association of Sedimentologists Special Publication 18, 350 p.

PROVAN, D.M.J., 1993, Draugen oil field, Haltenbanken Province, Offshore Norway, *in* Halbouty, M.T., ed., Giant Oil & Gas Fields of the Decade 1978-1988: Tulsa, American Association of Petroleum Geologists Memoir 54, p. 371-382.

RAVNAS, R. AND STEEL, R.J., 1997, Contrasting style of Late Jurassic syn-rift turbidite sedimentation: A comparative study of the Magnus and Oseberg Areas, northern North Sea: Marine and Petroleum Geology, v. 14, p. 417-449.

RINE, J.M., TILLMAN, R.W., CULVER, S.J. AND SWIFT, D.J.P., 1991, Generation of late Holocene sand ridges on the middle continental shelf of New Jersey, USA—evidence for formation in a mid-shelf setting based on comparisons with a nearshore ridge, *in* Swift, D.J.P., Oertel, G.F., Tillman, R.W. and Thorne, J.A., eds., Shelf Sand and Sandstone Bodies: Geometry, Facies and Sequence Stratigraphy: Oxford, International Association of Sedimentologists Special Publication 14, p. 395-426.

SABINS, F.F., 1963, Anatomy of a stratigraphic trap, Bisti field, New Mexico: American Association of Petroleum Geologists Bulletin, v. 47, p. 193-228.

SEELING, A., 1978, The Shannon Sandstone, a further look at the environment of deposition at Heldt Draw Field, Wyoming: Mountain Geologist, v. 15, p. 133-144.

SNEDDEN, J.W. AND JUMPER, R.S., 1990, Shelf and shoreface reservoirs, Tom Walsh-Owen Field, Texas, *in* Barwis, J.H., McPherson, J.G. and Studlick, J.R.J., eds., Sandstone Petroleum Reservoirs: Casebooks in Earth Sciences: New York, Springer-Verlag, p. 415-436.

SNEDDEN, J.W., COOKE, J.C., JOHNSON, R.K. AND CONRAD, K.T., 1991, Sequence stratigraphy and sedimentology of a shelf-margin lowstand wedge in the deep Wilcox Flexure Trend of South Texas (abstract): American Association of Petroleum Geologists Bulletin, v. 75, p. 673.

SNEDDEN, J.W., TILLMAN, R.W., KREISA, R.D., SCHWELLER, W.J., CULVER, S.J. AND WINN, R.D., 1994, Genesis and stratigraphy of a shoreface-attached sand ridge, Peahala Ridge, New Jersey: Journal of Sedimentary Research, v. B64, p. 560-581.

SNEDDEN, J.W., JOHANSEN, O.K., VARHAUG, P., HVIDSTEN, G.E. AND BAKKEN, K., 1997, Seismic, sedimentologic, and sequence stratigraphic criteria for recognition of subsurface incised-valley fills: Mesozoic strata of the Barents Shelf, Norway, *in* Shanley, K.W. and Perkins, B.F., eds., Shallow Marine and Nonmarine Reservoirs, Sequence Stratigraphy, Reservoir Architecture, and Production Characteristics: Tulsa, Gulf Coast Section-SEPM (Society for Sedimentary Geology) Annual Research Conference Proceedings, v. 18, p. 303-317.

SPEARING, D.R., 1976, Upper Cretaceous Shannon Sandstone: An offshore shallow marine sand body, in 28th Annual Wyoming Geological Association Field Conference: Casper, Wyoming Geological Association Guidebook, p. 65-72.

STINE, A.D. AND SCHMITT, J.G., 1987, Storm-influenced shelf deposition of the lower sandstone member, Lower Cretaceous Thermopolis Shale, southwestern Montana: Contributions to Geology, University of Wyoming, v. 25, p. 35-53.

SULLIVAN, M.D., VAN WAGONER, J.C., JENNETTE, D.C., FOSTER, M.E., STUART, R.M., LOVELL, R.W. AND PEMBERTON, S.G., 1997, High resolution sequence stratigraphy and architecture of the Shannon Sandstone, Hartzog Draw Field, Wyoming, *in* Shanley, K.W. and Perkins, B.F., eds., Shallow Marine and Nonmarine Reservoirs, Sequence Stratigraphy, Reservoir Architecture, and Production Characteristics: Gulf Coast Section-SEPM (Society for Sedimentary Geology) Annual Research Conference Transactions, v. 18, p. 331-344.

SWAGOR, N.S., OLIVER, T.A. AND JOHNSON, B.A., 1976, Carrot Creek Field, Central Alberta, *in* Lerand, M.M., ed., The sedimentology of selected clastic oil and gas reservoirs in Alberta: Canadian Society of Petroleum Geologists Bulletin, p. 78-95.

TILLMAN, R.W. AND MARTINSEN, R.S., 1984, The Shannon shelf ridge sandstone complex, Salt Creek Anticline area, Powder River basin, Wyoming: *in* Tillman, R.W. and Siemers, C.T., eds., Siliciclastic Shelf Sediments: SEPM (Society for Sedimentary Geology) Special Publication 34, p. 1-34.

TILLMAN, R.W. AND SIEMERS, C.T., 1984, eds., Siliciclastic Shelf Sediments: Tulsa, SEPM (Society for Sedimentary Geology) Special Publication 34, 267 p.

TILLMAN, R.W., SWIFT, D.J.P. AND WALKER, R.G., 1985, Shelf Sands and Sandstones: Tulsa, SEPM (Society for Sedimentary Geology) Short Course Notes 13, 708 p.

TILLMAN, R.W. AND MARTINSEN, R.S., 1987, Sedimentologic model and production characteristics of Hartzog Draw field, Wyoming, a Shannon shelf-ridge sandstone, *in* Tillman, R.W. and Weber, K.J., eds., Reservoir Sedimentology: Tulsa, SEPM (Society for Sedimentary Geology) Special Publication 40, p. 15-112.

TYE, R.S., RANGANATHAN, V. AND EBANKS, W.J., JR., 1986, Facies analysis and reservoir zonation of a Cretaceous shelf sand ridge: Hartzong Draw field, Wyoming, *in* Moslow, T.F. and Rhodes, E.G., eds., Modern and ancient shelf clastics: A core workshop: SEPM (Society for Sedimentary Geology) Core Workshop 9, p. 169-216.

VAIL, P.R., 1987, Seismic stratigraphy: American Association of Petroleum Geologists Short Course Notes.

VAN WAGONER, J.C., MITCHUM, R.M., CAMPION, K.M. AND RAHMANIAN, V.D., 1990, Siliciclastic sequence stratigraphy in well logs, cores, and outcrops: Tulsa, American Association of Petroleum Geologists Methods in Exploration Series 7, 55 p.

WALKER, R.G. AND BERGMAN, K.M., 1993, Shannon sandstone in Wyoming: a shelf ridge complex reinterpreted as lowstand shoreface deposits: Journal of Sedimentary Petrology, v. 63, p. 839-851.

WALKER, R.G. AND WISEMAN, T.R., 1995, Lowstand shorefaces, transgressive incised shorefaces and forced regressions: Examples from the Viking Formation, Joarcam Area, Alberta: Journal of Sedimentary Research, B65, p. 132-141.

MODERN SHELF SAND RIDGES: FROM HISTORICAL PERSPECTIVE TO A UNIFIED HYDRODYNAMIC AND EVOLUTIONARY MODEL

JOHN W. SNEDDEN

Mobil Exploration and Production Technical Center, P.O. Box 650232, Dallas, TX, 75265, U.S.A.

AND

ROBERT W. DALRYMPLE

Department of Geological Sciences, Queen's University, Kingston, Ontario, K7L 3N6, Canada

ABSTRACT: Shelf sand ridges are common features on many modern continental shelves but have not been widely cited as analogs for ancient sand bodies. Modern shelf sand ridges are elongate, coastal- to shelf-sand bodies that are larger than subaqueous dunes, with lengths of the order of 10 km and heights that are more than 20% of the water depth. These ridges are also longer-lived than smaller bedforms, persisting for thousands to tens of thousands of years. Progress in understanding these features has gone from early studies of their morphology and surficial-sediment characteristics to recent investigations of current and wave dynamics and establishment of their internal stratigraphy through vibracoring. From these studies, it is clear that once formed, most ridges are maintained and even enlarged by present-day shelf-flow dynamics. Ridges existing in tide- and storm-dominated shelf regimes share many similarities, which suggests that both are formed by the same fundamental interaction between flow and an initial bathymetric irregularity. The Huthnance (1982) hydrodynamic model appears to offer an adequate description of that interaction. Synthesis of previous work indicates that storm- and tide-dominated ridges pass through three stages of development: 1) an initial irregularity (the ridge nucleus) forms due to coastal or shelf processes; 2) nearshore and/or shelf currents (either storm- or tide-driven) interact with the irregularity in the manner described by the Huthnance model; and 3) the ridge evolves as a result of continued current action. These processes and events are most likely to occur during transgressions, because sandy coastal deposits, the most common ridge nuclei, are reworked on the shelf as relative sea level rises. Thus, most ridges are expected to overlie the transgressive ravinement surface.

Understanding the nature of the initial irregularity and the amount of subsequent migration is the key to explaining the diversity of ridge characteristics. Based on the degree of ridge evolution, we categorize shelf ridges into three classes: Class I- juvenile or stationary ridges that retain their initial nucleus; Class II- ridges that have migrated somewhat but retain part of their nucleus; and Class III- "fully evolved" ridges that have migrated sufficiently that they contain no trace of their origin. The degree of ridge evolution greatly affects the economic potential of ancient subsurface analogs, as ridge reworking/remolding during migration results in cleaner sand and increased sand-body volume. Transgressive conditions are most favorable for ridge development (e.g., Transgressive Systems Tracts), given the presence of widened shelf areas, drowned bathymetric irregularities that act as ridge nuclei, and loose sand derived from ravinement.

INTRODUCTION

Isolated sand bodies surrounded by marine shales (e.g., the Shannon Sandstone; sand bodies in the Cardium Formation, Alberta) have traditionally been interpreted as shelf sand ridges that formed long distances seaward of the contemporaneous shoreline (e.g., Tillman and Martinsen, 1984). However, difficulties in explaining how sand reached such sand bodies, coupled with the present focus on shoreline movement in response to relative sea-level change (i.e., sequence stratigraphy), has led to reinterpretation of such sand bodies either as lowstand or transgressive shorefaces (e.g., Bergman, 1994) or valley fills (Sullivan et al., 1997). Indeed, the entire notion that shelf ridges might exist in the stratigraphic record has fallen into disfavor, as evidenced by recent papers on foreland basin sequence stratigraphy (e.g., Van Wagoner and Bertram, 1995) and by the lack of reference to them in some recent textbooks (e.g., Emery and Myers, 1996). Certainly the number of well-accepted, ancient shelf ridges is small.

By contrast, shelf sand ridges are present and even abundant on some, but not all, modern continental shelves (Fig. 1). This raises several questions. Is the modern a valid analog for the ancient? If not, why not? If it is, under what conditions is the analogy appropriate and when and where would we expect sand ridges to form? The purpose of this paper is to examine these questions and speculate on the occurrence, distribution, and internal characteristics of shelf sand ridges on the basis of modern shelf ridges. Our intent is to develop a model that integrates our understanding of ridge genesis and evolution in a way that can be applied more easily to ancient deposits. Clearly, any investigation based on a single period of time (i.e., modern, recently transgressed shelves) is limited by the specific history of the areas in question. However, our analysis is grounded on the fundamental physical processes responsible for ridge genesis, growth, and subsequent evolution. Because of this theoretical underpinning, we believe that our conclusions are generally applicable.

In detail, our goals are to: 1) summarize previous work on modern shelf sand bodies; 2) synthesize present knowledge into a genetically based model for the stratigraphic and geographic occurrence of such sand bodies; and 3) propose an evolutionary progression of sand body types. We include ridges from both tide- and storm-dominated settings in our discussion, because, as we will argue below, there is reason to believe that all ridges have some general genetic features in common and because we will obtain a more complete understanding of these ridges by considering the widest possible set of examples. We also consider ridges presently attached to the shoreline, though these are technically part of the shoreface system. However, shoreface-attached sand ridges exhibit characteristics similar to those of shelf sand ridges and are believed to evolve into them during transgressions. Clearly it is important to understand all ridge types, because these large sand bodies have the potential to be sizable reservoirs of oil and gas or significant resources for beach nourishment.

DEFINITIONS

In this paper we define sand ridges as all elongate coastal to shelf sand bodies that are larger and longer-lived than

Isolated Shallow Marine Sand Bodies: Sequence Stratigraphic Analysis and Sedimentologic Interpretation.
SEPM Special Publication No. 64, Copyright © 1999
SEPM (Society for Sedimentary Geology), ISBN 1-56576-057-3, p. 13-28.

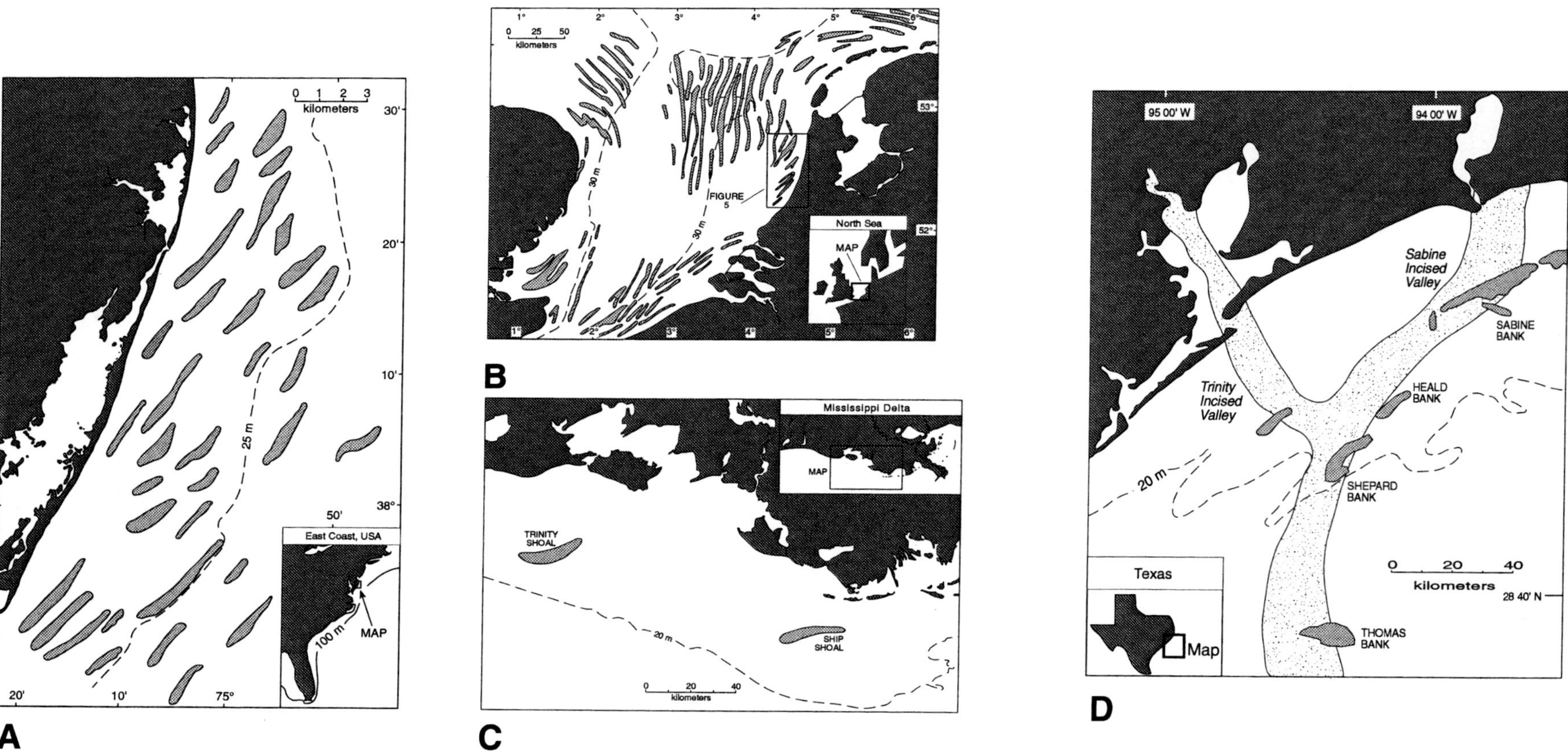

Fig. 1.—Examples of shelf sand ridges. A) Atlantic shelf, East Coast USA (Swift and Field, 1981). B) North Sea (Van Den Meene, 1994). Box shows area studied and ridge depicted in Fig. 5. C) Louisiana shelf (Penland et al., 1986). D) Gulf of Mexico (Thomas and Anderson, 1994).

subaqueous dunes (cf., Swift, 1985). Sand ridges typically have heights that are more than 20% of the water depth and lengths of the order of 10 km (Table 1). While several papers (Belderson et al., 1982; Huthnance, 1982; Amos and King, 1984) describe ridges as a type of bedform, the fact that ridges do not scale with flow depth as dunes do suggests that they are better categorized as "barforms" (Dalrymple and Rhodes, 1995). Ridges are generally oriented at an oblique angle to the strongest currents but, unlike dunes, are more nearly parallel to the flow than at right angles to it. Also, like dunes, ridges are commonly asymmetrical, although their steeper face only rarely dips at more than 1°. They typically occur in groups, with the spacing of individuals of the order of 250 times the water depth, but solitary ridges do occur.

Although many workers differentiate between storm- and tide-built ridges because of differences in the nature of the currents that generate them (e.g., Swift, 1985), we note that both ridge types are morphologically similar in most regards. The only notable difference is that tide-built ridges are commonly longer than storm-built ridges (10-60 km long versus <15 km long; Fig. 1B). Furthermore, theoretical considerations suggest that both ridge types are formed and maintained by the same fundamental interaction between the flow and the bed. As a result, we do not distinguish between the two types of ridges in this paper.

PREVIOUS WORK ON MODERN SAND RIDGES

Topographic irregularities exceeding 10 m in relief have long been recognized on continental shelves (e.g., Van Veen, 1935; Veatch and Smith, 1939), sparking numerous hypotheses regarding possible origin and present maintenance. In general, investigations into the character and genesis of these features can be grouped into three phases related to the technology used to study, sample, and monitor ridges and the physical processes operating in the surrounding area.

Morphology and Surficial Sediment Studies

Much of the early work on modern ridges centered on their morphology and characteristics of the surficial sediments. Even on early depth recordings, ridges and their adjacent swales could be differentiated (e.g., Veatch and Smith, 1939). Later bathymetric mapping on the U.S. Atlantic margin suggested that the ridge and swale topography actually occurs in several distinct groupings: radiating clusters near the mouths of estuaries; arcuate, seaward-convex ridge systems near cuspate forelands; shoreface ridge and swale systems; and broadly spaced ridges and swales on the open shelf (Fig. 1A; Uchupi, 1968).

Mapping efforts in the North Sea revealed that tidal sand ridges (also called sand banks) have a comparable morphological zonation: they can be grouped into sets of parallel to radiating ridges in funnel-mouthed estuaries, parabolic nearshore ridges, and groups of linear offshore ridges (Fig. 1B). Side-scan sonar records show that many of these banks are covered by smaller scale dunes (Stride, 1963). It was suggested that the latter form perpendicular to the dominant tidal-current flow, while the larger banks are largely oriented parallel to that flow (Off, 1963). Later studies of the flow dynamics would suggest a more flow oblique trend for the ridges, as discussed below.

Bathymetric mapping also indicated considerable changes in ridge morphology as a function of water depth, sampling period (summer fair-weather versus winter storm periods), and longer time intervals. Swift and Field (1981) noted a systematic change in ridge morphology from shoreface to offshore ridges: ridge cross-sectional area increases while the length/width declines progressively offshore. Investigators related this to changes in hydraulic regime from shoreface to offshore. In the case of storm-built shoreface ridges, variations between fair-weather and storm periods were noted when comparing successive bathymetric profiles (McHone, 1973). This suggests that crestal and seaward accretion of the ridge occurs during summer months while erosion of the steeper landward margin takes place during winter and spring storms (McKinney et al., 1976; Swift and Field, 1981). Longer term variations in ridge morphology were also observed. For example, comparison of surveys taken 50 years apart over some shoreface ridges showed considerable deepening of swales, deposition on ridge crests, and some erosion of their seaward flanks (Swift et al., 1972a).

During the late 1970s, little was known of the internal structure of storm- or tide-built ridges. Surficial sediment

Table 1.—Modern shelf sand ridge characteristics*.

Criteria	Characteristics
Orientation	flow-oblique
Symmetry	dominantly asymmetrical
Vertical Scale (relief)	5 to 40 m
Horizontal Width Scale	0.7 to 8 km
Barform side slopes	typically <1° to a maximum of 7°
Grain Size	fine to coarse sand
Lateral Trends	stoss side coarser than lee side
Superimposed Bedforms	medium to large 2D and 3D dunes, hummocky megaripples, current ripples

*From Belderson et al. (1982); Amos and King (1984); Swift (1985); Dalrymple and Hoogendoorn (1997).

sampling did, however, reveal a distinct textural zonation. In both tide- and storm-built ridges, coarse sediment was noted in swales and was thought to be derived from incision into underlying Pleistocene sediment (Swift et al., 1972b; 1977). Storm-built shoreface and nearshore ridges, presumably the most active ridges, were found to exhibit an asymmetric grain-size distribution relative to topography with the coarsest grains located on the steep up-current flank of the ridges, not on the crest (e.g., Fig. 2; Swift et al., 1977; Swift and Field, 1981). By contrast, later studies of tide-influenced shoreface ridges along the Belgian coast found a consistent coarsening of grain size on the seaward flank (De Maeyer and Wartel, 1988). Other studies of ridges off the Dutch central coast showed little correlation of grain-size and topography but a distinct improvement in sorting on the crests of the ridges (Van den Meene, 1994). Some variation in surface texture was also noted between storm and fair-weather periods (McHone, 1973). Much of the textural variation in surficial sediments over ridges is probably due to the relative strength of waves and currents around these shallow ridges (Snedden et al., this volume).

Given the available data, it is not surprising that early interpretations of the genesis of sand ridges were based heavily upon observations of their morphology. The obvious similarity of the ridges and swales to adjacent barrier islands and lagoons led to hypotheses of in-place drowning of barriers to form "offshore bars" (Shepard, 1963; Sanders and Kumar, 1975; Stubblefield et al., 1984a, b). The swales were thought to be related to erosion by stream systems (McKinney and Friedman, 1970) or combined marine and terrestrial processes (Emery, 1968).

Hydrodynamic Origin

The next advances in understanding modern sand ridges came with the measurements of current- and wave-generated flows on and around modern ridges, leading to formulation of fluid-dynamic models for ridge construction and maintenance. Early work was mainly confined to offshore areas, where equipment was less susceptible to the stresses associated with shoaling waves. Numerous papers documented strong, near-bottom, along-shelf currents in the Middle Atlantic Bight during winter storms (Beardsley and Butman, 1974; Gadd et al., 1978). Given the known orientation of the mid-shelf ridges, it was concluded that southward-directed, geostrophically balanced flows were clearly passing obliquely across the ridges (Swift, 1985). By contrast, it was readily apparent that fair-weather waves were largely ineffective over these mid-shelf ridges (McClellen, 1973).

Similarly, tide-built ridges in the North Sea were shown to display a flow-oblique orientation (Kenyon et al., 1981). It was determined that crest lines of most linear banks are oriented counter-clockwise relative to the peak tidal flow and that superimposed dunes also turn to the right of the current. The degree of asymmetry of the banks was observed to increase in proportion to the inequality of ebb and flood tidal currents (Kenyon et al., 1981).

These observations led Huthnance (1982) to formulate a fluid-dynamical model explaining how these oblique bottom currents can bring about ridge growth, given a sufficient supply of sand either carried in from distant areas by the currents or excavated from the local swales. Previous to Huthnance (1982), ridge genesis and maintenance were typically ascribed to the presence of counter-rotating, flow-parallel vortices, with ridges underlying the rising limbs and swales underlying the descending limbs (e.g., Houbolt, 1968). The observation that all ridges are oblique to the most intense

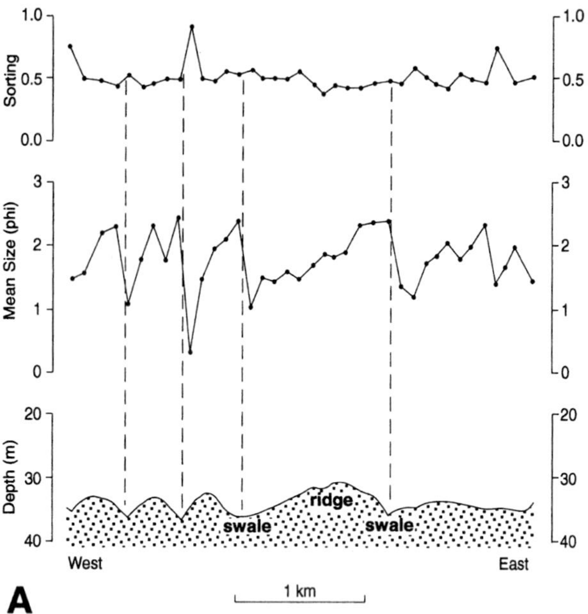

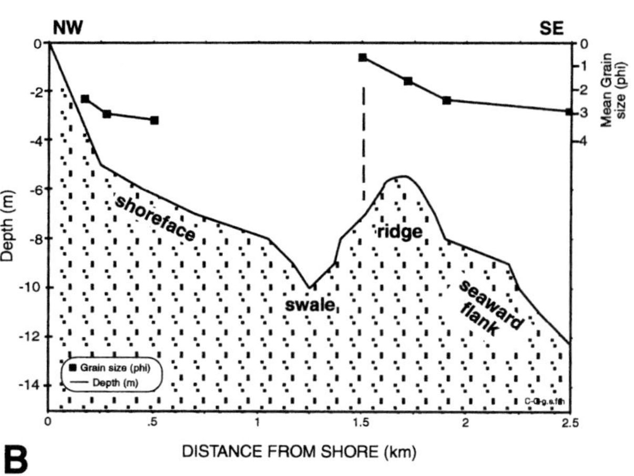

Fig. 2.—Trends in texture of surficial sediments over shoreface sand ridges. A) Sable Island, Nova Scotia. B) Peahala Ridge, New Jersey. Note offset between the bathymetric and grain-size profiles.

sediment-transporting currents (Swift et al., 1978; Kenyon et al., 1981) and that near-bed and surface current directions on each side of a ridge do not reveal the presence of such vortices (McCave, 1979; Snedden et al., 1994) has led to the gradual abandonment of the helical flow model of ridge genesis. Thus, the simple yet elegant Huthnance model is widely accepted as an explanation for ridge construction and maintenance in both storm- and tide-dominated settings (Swift, 1985; Dalrymple and Rhodes, 1995; Dalrymple and Hoogendoorn, 1997) and is discussed in greater detail in subsequent sections.

As instrumentation improved to withstand greater wave loads, studies of shallow-water ridges and adjoining shorefaces were initiated. In this setting, storm and even many fair-weather waves are capable of resuspending sandy sediments. However, sediment movement is thought to be more complicated than simple Stokes drift, which describes landward-directed bottom motion forced by shoaling waves. During storms, a strong downwelling current is measured (Neidoroda and Swift, 1985). This current flow, sustained over several days, is capable of moving sand from the shoreface onto the shelf at rates far exceeding long-term fair-weather Stokes drift. Subsequent monitoring of the shoreface-attached Peahala Ridge documented currents from the shoreface flowing obliquely seaward across the ridge, as well as intense shear stresses in the swale separating the ridge from the adjacent shoreface (Snedden et al., 1994).

This line of scientific inquiry revealed that most shoreface and shelf sand ridges are not moribund features, but are frequently modified by waves and storm- and tidal-currents, actively accreting and in some cases actively migrating across the shelf surface (e.g., Hoogendoorn and Dalrymple, 1986). The rates of ridge accretion and migration and swale excavation are known to vary widely as a function of the magnitude, frequency, and orientation of the waves and currents (Swift, 1985).

Internal Ridge Structure and Stratigraphy

The present phase of ridge research is largely based on the success of deep vibracoring and high-resolution seismic imaging of storm- and tide-built ridges. Vibracoring in tide-built ridges has been much less successful than in storm-dominated ridges because of ambient wave and current conditions as well as the apparent hardness of the ridge substrate (cf., Davis and Balson, 1992).

Vibracoring of storm-built ridges was initially undertaken in the early 1970s, but technological advances permitting preservation of delicate sedimentary structures were not available until the 1980s. These early coring efforts pointed to a complex stratigraphy. For example, coring of Beach Haven Ridge, near Atlantic City, New Jersey, revealed that Pleistocene sediments were incised by both contemporaneous streams and later by modern swales and are overlain by Holocene sediments, including valley-fill, ebb-tidal delta, and modern ridge sands (Alpine, 1974; Fig. 3).

Other attempts to core storm-built ridges (Figueiredo, 1984; Rine et al., 1991) recognized a similar complicated stratigraphy. The presence of microfauna diagnostic of the present water depths, and what was then interpreted as reworked macrofauna, led to the conclusion that the shoreface and nearshore ridges were formed in their present water depths (Rine et al., 1991). The authors thus saw no direct link between the underlying stratigraphy and the genesis of modern ridges, an oversight in view of later work.

The Atlantic Shelf Coring Project, initiated in 1984, was the first attempt to obtain relatively undisturbed, long vibracores from multiple sites on a storm-dominated shelf ranging from the shoreface to the mid-shelf (Snedden et al., this volume). The four ridges studied contained a relatively similar set of stratigraphic units, though different in age and microfaunal content (Fig. 4). These observations suggest a similar initial genesis but considerable subsequent evolution (Culver and Snedden, 1996). The evidence supports the model discussed by Figueiredo (1984) and McBride and Moslow (1991): initiation as ebb-tidal delta deposits, with evolution to shoreface-attached ridges and then to detached mid-shelf ridges. The work on the New Jersey shelf indicates that once detached, ridges continue to evolve and adjust their morphology, texture, and volume to the prevailing shelf flow regime (Snedden et al., this volume).

Rapidly migrating shoreface ridges near Sable Island, Nova Scotia, were recently described by Dalrymple and Hoogendoorn (1997). Strong, alongshore storm currents and waves force ridge migration at rates that may reach 50 m/yr and cause considerable swale erosion. Vibracores from the upper portion and lee sides of these ridges reveal the presence of thick, normally graded storm beds that dip downcurrent and are conformable with the lee side of the ridge. Unlike ridges on the New Jersey shelf (Snedden et al., this volume), little evidence of a ridge precursor was observed, but this is not surprising, given the rapid migration and turnover of sediments in this high-energy regime.

Other vibracore studies of ridges, banks, and shoals have also been undertaken. Coring of Ship Shoal on the Louisiana shelf revealed development of a 5 m-thick transgressive succession above regressive deltaic sediments in 3-10 m of

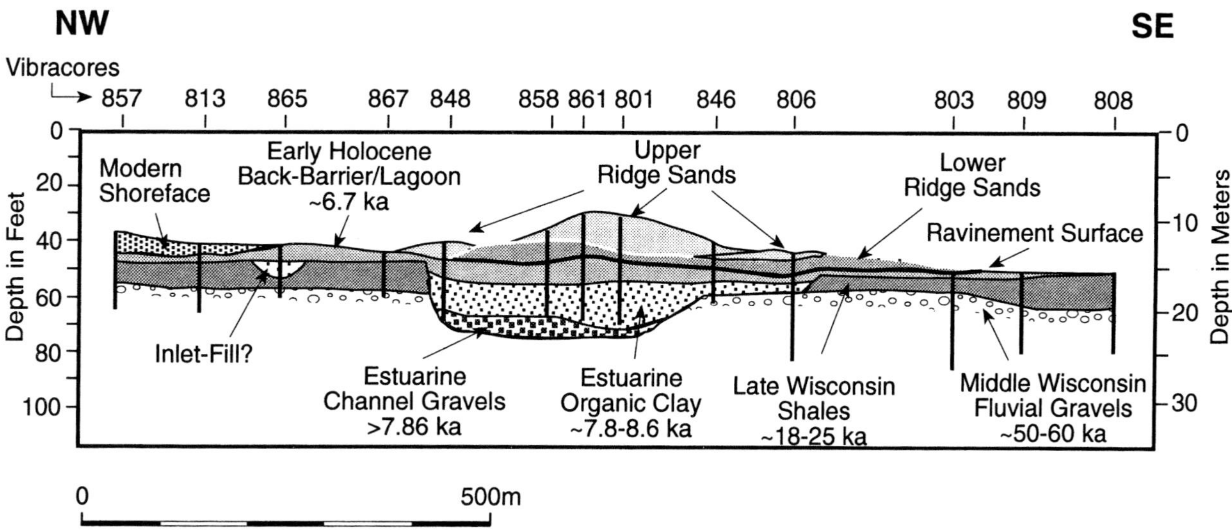

Fig. 3.—Beach Haven ridge (modified from Alpine, 1974).

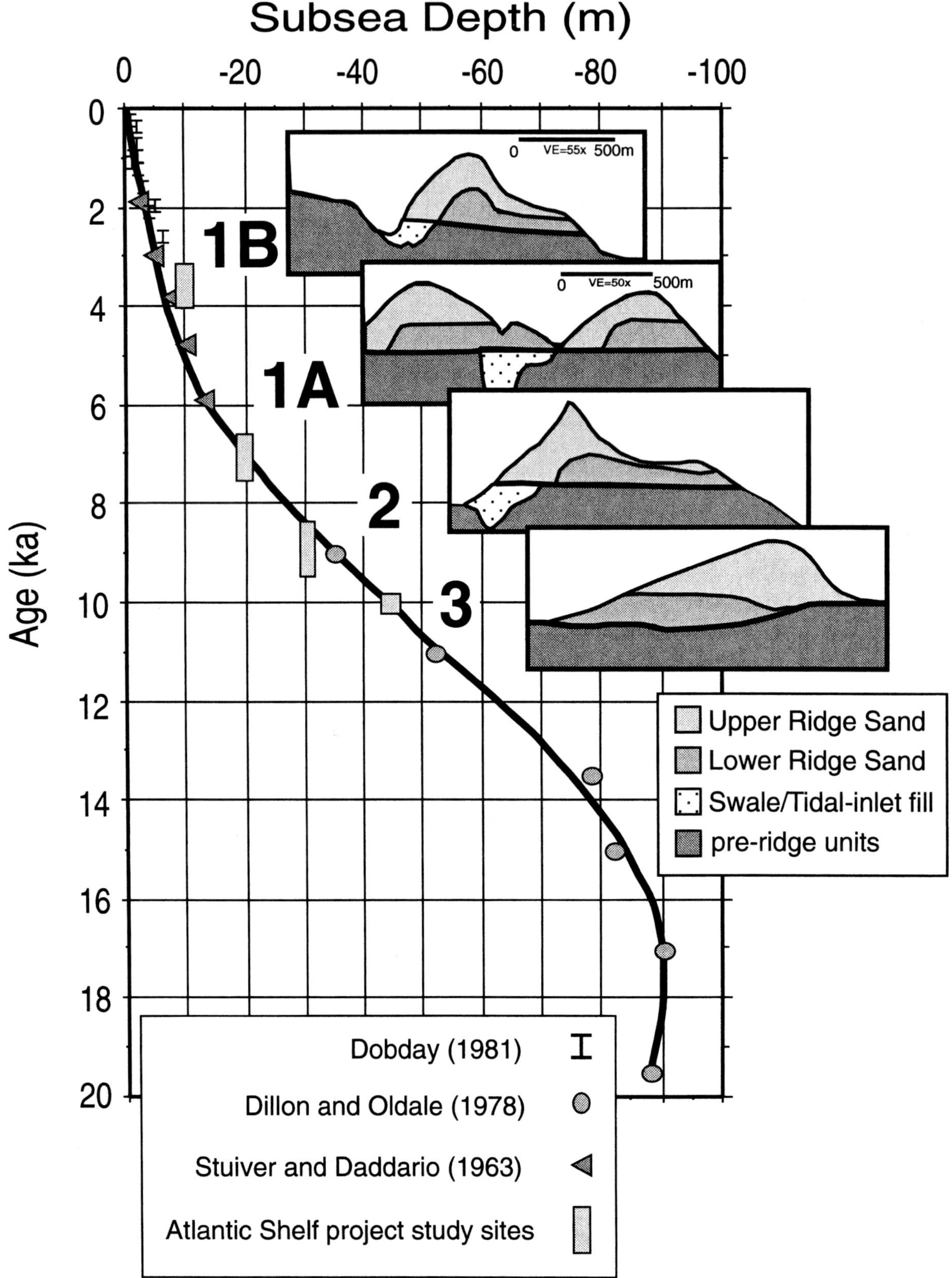

Fig. 4.—Comparison of ridge stratigraphy and relative sea-level curve for the New Jersey Atlantic shelf. Width of rectangle indicates amount of uncertainty in age of the transgression at each study site.

water (Penland et al., 1986; Fig. 1C). Coarsening-up successions of sandy shoal crest, front, and base deposits that form landward-dipping bedsets have developed in spite of limited wave and current action and high rates of mud deposition. Like the shoreface ridges observed on the Atlantic Shelf, Ship Shoal exhibits a steep northern (landward) flank, suggesting onshore sediment transport by waves or northerly directed storm currents. Importantly, Ship Shoal is an example of a shelf sand body that originated through in-place drowning of a transgressive barrier island, rather than by evolution from a smaller-scale, shoreline-associated feature as is the case with the Atlantic shelf ridges.

Further west, on the Texas shelf where sediment supply is significantly lower, Rodriguez et al. (this volume) document considerable post-transgressive reworking/remolding of sand bodies associated with the Trinity and Sabine incised fluvial valleys (Fig. 1D). Thomas, Shepard, Heald, and Sabine banks are seen to consist of modern shelf storm-deposited sands overlying slightly older coastal sands including shoreface/ebb-tidal delta and back-barrier/flood-tidal delta deposits (Rodriguez et al., this volume). Comparison of the four banks suggests that the shelf sands become progressively thicker offshore at the expense of underlying units. This can be explained by the greater duration of ridge reworking and evolution in deeper water since the passage of the transgressing shoreline. The association of the ridges with incised-valley systems is likely due to the local abundance of sand and higher compactional subsidence associated with the valleys (Thomas and Anderson, 1994).

In northeastern Gulf of Mexico, a laterally extensive sand sheet or blanket has developed in Holocene sediments (McBride et al., this volume). This sheet-like deposit includes several topographic elements: 1) nearshore oblique sand ridges; 2) shoreline-parallel, mid-shelf ridges, and 3) shelf-edge deltas. Both the nearshore and mid-shelf ridges lie above the transgressive ravinement surface and contain marine faunal assemblages, suggesting post-transgressive reworking and ridge evolution on the shelf. Low subsidence, medium grain size, and limited deltaic-mud input to this area have facilitated development of this sand-rich shelf (McBride et al., this volume). The frequent passage of both tropical storms and winter cold fronts subjects this area to intense reworking.

Shallow water, shoreface-attached ridges in the mesotidal central Dutch coast (Van Den Meene, 1994) exhibit sedimentary structures much like their storm-built counterparts on the New Jersey coast (Fig. 5; cf., Snedden et al., 1994). Cross-bedded sands dominate the high-energy landward margin, while burrowed sands are found on the seaward side. Shell lags are present in the swale that separates the ridge from the shoreface (Van den Meene, 1994). Although tidal currents are substantial (commonly >70 cm/s), these ridges are still greatly influenced by shoaling waves and storm-generated currents, like nearshore ridges of the Atlantic shelf.

Offshore in deeper water tide-dominated settings, where vibracoring has proven difficult, high-resolution seismic studies have revealed much about the internal structure and stratigraphy of both shelf and shoreface sand ridges. Early work documented the presence of internal low-angle reflections that dip subparallel to the steep (i.e., down-current) bank face, in ridges of the southern North Sea (Houbolt, 1968). However, more recent studies in this and other tide-dominated settings indicate a more complex structure, with features inherited from preceding lowstand deposits.

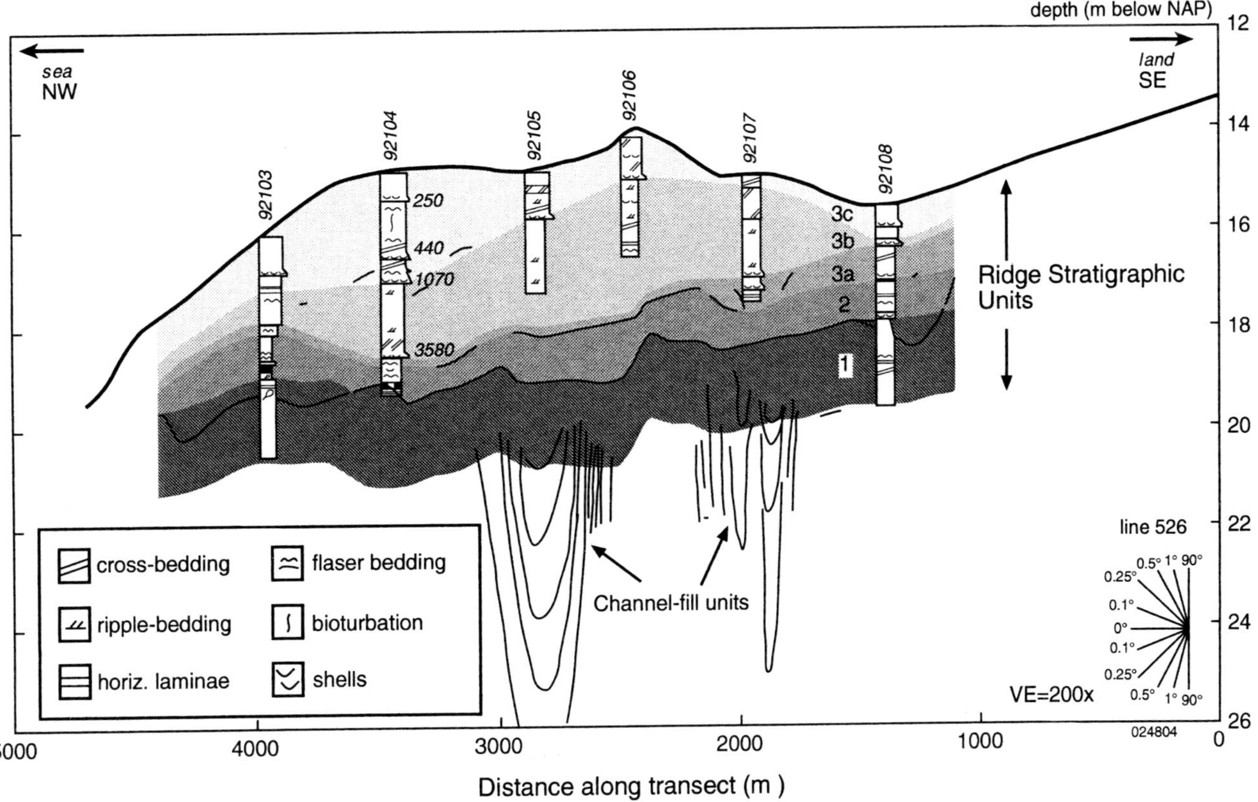

Fig. 5.—Cross-section across a shoreface-attached sand ridge from the North Sea (modified from Van den Meene, 1994).

Rabineau and Berné (1997) describe "erosional ridges" in the Gulfe de Lion, France, that consist almost entirely of erosional remnants of preexisting lowstand estuarine and coastal deposits. Ridge-like features in the southern Yellow Sea of Korea also consist of preexisting deposits, in this case lowstand deltaic sediments (Yang, 1989).

The previous work demonstrates that the internal structure of shelf sand ridges is inherently variable. All modern ridges, including those associated with high sediment flux of the Mississippi delta, overlie a transgressive ravinement surface. A few consist almost entirely of previously deposited coastal sediments and are largely erosional in origin (Gulfe de Lion, Rabineau and Berné, 1997; Yellow Sea, Yang, 1989) while others are composed mostly of modern shelf sediments (e.g., some North Sea ridges, Houbolt, 1968; Sable Island shelf, Dalrymple and Hoogendoorn, 1997). This suggests that there is an evolutionary spectrum of ridges. However, before discussing this point in detail, we must reconcile the hydrodynamic and evolutionary perspectives of ridge genesis.

UNIFIED MODEL FOR RIDGE GENESIS AND MAINTENANCE

As discussed above, previous research has generally focused on either the hydrodynamic or stratigraphic/evolutionary aspects of modern shelf ridges. No concerted effort has been made to integrate these different, but related, visions into a unified model of ridge origin and maintenance. However, such a synthesis is essential if we are to develop a model that is transferable to the widest range of settings and has the greatest predictive capability for ancient settings.

Ridge Genesis and Maintenance: The Huthnance Process

As already discussed, most workers now accept the Huthnance (1982; Hulscher et al., 1993) model of sand ridge genesis and growth. In brief, the Huthnance (1982) model is based on the assumption that one or more initial irregularities of unspecified origin exist on a sandy bed. Currents flow around and over these features, typically accelerating as they flow over the up-current side because of flow constriction, but weakening over the crest and down-current side (Fig. 6). As a result of this interaction between the flow and the irregularity, the up-current side is eroded and sand is deposited on the crest and lee side, causing upward growth and/or down-current migration. In the numerical models, maximum upward growth occurs when the irregularity is oriented oblique to the dominant flow (± 20-30° under normal conditions); initial irregularities that are parallel to flow are generally eroded and disappear, while perturbations perpendicular to flow evolve into dunes.

If sufficient sand is available, the sand ridges grow upward and laterally until reaching an equilibrium profile. At this point, the Huthnance process ceases to promote active upward growth, because the flow over the ridge is retarded rather than accelerated (cf., Fig. 6), leading to deposition on the up-current side and widening of the ridge. Wind waves may also limit ultimate vertical growth by preferentially

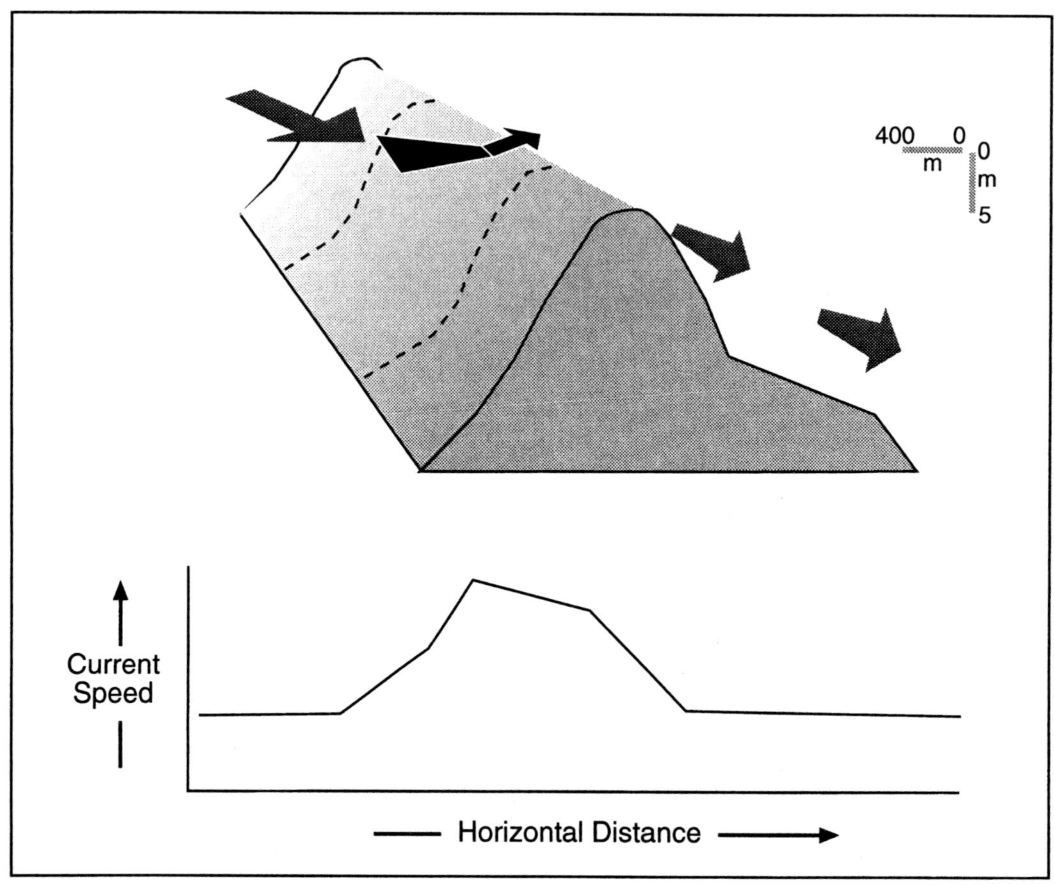

Fig. 6.—Schematic diagram illustrating variation of current flow passing across a shelf sand ridge. Modified from Swift (1985).

suspending sand from the ridge crest, but any decrease in ridge height is counteracted by the Huthnance process. In fact, continuous operation of the Huthnance process is essential to the continued existence of a ridge, because, once formed, wave action would gradually flatten the ridge unless some process counteracted this tendency. The preferred spacing of the resulting ridges depends upon several factors, among which the water depth and current speed are most important. If sand supply is limited during growth, ridges can still accrete vertically but will decrease in width, with relatively large interridge areas. Swales also deepen in this case, until they reach non-erodible substrates.

Although initially developed to explain tidal sand ridges, the Huthnance model does not require reversing flows. In fact, the formation of ridges may be more pronounced and rapid when one direction and speed of flow predominates, because the competing growth of features with different orientations and/or spacings is reduced. Although this process operates in both tide- and storm-dominated settings, only minor differences in ridge characteristics occur because of the different flow patterns. For example, in areas with consistently oriented, unidirectional flows (e.g., the south side of Sable Island, Dalrymple and Hoogendoorn, 1997), the ridges will migrate consistently and relatively rapidly in one direction and have a pronounced asymmetry, with the steeper face dipping in the migration direction. By contrast, ridges in areas with nearly equal, reversing tidal currents tend to be more symmetrical and migrate more slowly, if at all.

Thus, the Huthnance model provides a simple yet elegant hydrodynamic explanation for the genesis and growth of shelf sand ridges in diverse geographic and hydrographic settings, including detached sand ridges on both storm- and tide-dominated shelves, shoreface-attached ridges, channel-margin linear bars and other tide-dominated estuary mouth sand bodies, and even "banner banks" associated with bedrock irregularities such as islands and promontories. Stripped to its essentials, the Huthnance model requires the following conditions to form and maintain a ridge: a sufficient quantity of loose sand; currents of any origin that are at least periodically capable of moving that sand; and a preexisting irregularity that initiates the flow-bed interaction that leads to and then maintains the irregularity. Solitary ridges will form where there is only a single, initial perturbation (and limited sand supply), while fields of ridges develop where there are many nuclei and an abundance of sand.

The Search for Ridge Precursors

The Huthnance (1982) model and its refined version (Hulscher et al., 1993) provide the hydrodynamic process component of a unified model of ridge genesis, but explicitly avoid addressing the nature of the "initial irregularity" required to form a ridge. However, we must understand the types of potential ridge precursors if we are to develop the ability to predict ridge occurrences in ancient successions. Swift (1985), in discussion of the Huthnance (1982) model, indicates that an initial topographic irregularity could conceivably be as small as a few centimeters in height, given the proper conditions. We believe, however, that the initial irregularities from which ridges develop are typically much larger. Modern ridges provide clues to the types of likely precursors that may lead to the development of ridges.

Several studies have suggested that shoreline-associated features may act as the nucleus for sand ridges. Indeed, as noted in the discussion of previous work, coastal sands are commonly found within the lower portion of nearshore and open-shelf ridges. One logical precursor for ridges that are initiated in the nearshore region of storm-dominated areas is the deposits left behind as an ebb-tidal delta migrates alongshore while the shoreline transgresses (Figueiredo, 1984; McBride and Moslow, 1991). This association is logical since ebb-tidal deltas contain large amounts of sand—as much as 77% of the available clean sand in some areas (Sexton and Hayes, 1996). Migration of inlets also provides a ready mechanism for cutting swales. Ridge-like protuberances growing from ebb-tidal bars have been noted (see Snedden et al., this volume) and ebb-tidal delta deposits have been recorded within a modern shoreface-attached ridge (Snedden et al., 1994). Thus, this origin is attractive for many ridges on mesotidal coasts with abundant tidal inlets. However, such an origin is not applicable in all instances. For example, at Sable Island there are far more ridges than can possibly be explained by the ebb-tidal-delta mechanism (Dalrymple and Hoogendoorn, 1997). In the southern North Sea, ridges are present in deeper water but are not actively forming in shallower water adjacent to modern tidal inlets. This suggests that some other precursor may be responsible for the ridges (Antia, 1994). In such situations the ridge precursor remains unknown, although possibilities include bars at the seaward end of rip channels and wave-generated nearshore bars. Presumably a topographic irregularity of any origin would serve equally well as the initiator of ridge growth.

In the relatively sand-starved Gulf of Mexico, transgressive drowning of barrier islands provides the necessary sand-rich precursor for ridge growth. Such barrier sand bodies are associated either with eroding deltaic headlands (Penland et al., 1986; Fig. 1C) or the mouths of incised-valley estuaries (Thomas and Anderson, 1994; Rodriguez et al., this volume; Fig. 1D). These sand bodies are submerged by rapid rises of relative sea level, in this case largely due to the high rate of compactional subsidence. That these ridges continue to exist in the presence of hurricane-produced waves suggests that they are maintained by ridge-oblique flow.

In tide-dominated coastal areas, elongate tidal bars associated with the mouths of funnel-shaped estuaries may form the nucleus of shelf ridges (Yang and Sun, 1988; Dalrymple, 1992). These bars are analogous to the channel-margin linear bars of ebb-tidal deltas (Hayes, 1975; Dalrymple and Rhodes, 1995). They originate in the area of low shear stress between mutually evasive sediment-transport streams but are maintained by the Huthnance process. Rocky islands and coastal promontories also establish patterns of mutually evasive tidal flow, leading to the formation of "banner banks," ridges anchored at one end to a bedrock obstacle between zones of opposing sediment transport (Belderson et al., 1982).

Ridges nucleated on any of these coast-associated features may end up lying on the open shelf as a result of transgression. Provided there is still sufficient current flow on the shelf to episodically move sand, the ridge may be maintained against wave action and could continue to move and grow. Many open-shelf ridges appear to originate in this fashion, as indicated by the presence of coastal facies within the ridge. However, it is also possible for ridges to be generated "from scratch" (*de novo*) on the shelf, provided there is sufficient sand and current action. Previous work does not offer insight into possible precursors for such ridges, but general considerations suggest that random topographic irregularities on the ravinement surface (e.g., features formed by differential erosion of the substrate) or particularly large dunes could serve as the nucleus of shelf-formed ridges.

Thus, application of the general conditions imposed by the Huthnance hydrodynamic model of ridge genesis—sufficient sand, sand-transporting currents, and an initial irregularity—leads to the conclusion that the coastal zone is the most likely place for the initiation of ridges. This is so because the coastal zone commonly contains more sand and stronger currents than areas further offshore. Furthermore, the coastal zone possesses a large number of potential ridge-precursor irregularities, including ebb-tidal delta deposits, nearshore bars of various types, estuary-mouth sand bodies, and barrier islands. Although ridges can theoretically form further out on the shelf, this would appear to be less likely, given the limited range of potential precursors, the scarcity of sand, and the weaker currents relative to nearshore areas.

Evolutionary Progression of Sand Ridges

Observations of modern ridges indicate that a sand ridge can form in geologically short time periods, from hundreds to a few thousands of years, especially if the ridges originate in the energetic nearshore to inner-shelf area (Penland et al., 1986; Snedden et al., 1994). Once formed, however, the ridge will continue to exist as long as the hydrodynamic regime is favorable; that is, as long as there is a positive to neutral feedback between the flow and the ridge topography and the Huthnance process continues to operate. During this "mature" phase of the ridge existence, which may be of longer duration than the growth phase, the ridge will respond somewhat like a smaller-scale bedform and will migrate in the direction of residual sediment transport (e.g., Dalrymple and Hoogendoorn, 1997). Rates of migration depend upon the imbalance of transport in opposing directions over the ridge: rates are negligible in areas where opposing tidal or storm-generated currents are equal in magnitude and frequency. By contrast, pronounced directional consistency of storm currents in some areas (e.g., Sable Island, Dalrymple and Hoogendoorn, 1997) leads to relatively rapid, unidirectional migration of the ridges.

The degree to which a ridge has migrated has important stratigraphic implications for the internal structure of the ridge. As ridges migrate, sediment is eroded from the up-current flank (relative to the net transport direction) and deposited on the down-current side. These deposits commonly take the form of gently dipping strata that conform to the lee side of the ridge (cf., Houbolt, 1968; Dalrymple and Hoogendoorn, 1997). Stoss-side erosion will gradually remove the initial irregularity, progressively destroying evidence of how the ridge formed. The extent to which a ridge retains evidence of its precursor irregularity provides a stratigraphically useful means of classifying ridges (Table 2). This subdivision of ridges into three classes represents an arbitrary compartmentalization of what is presumably a continuum of increasing ridge reworking and migration.

Class I ridges retain all of their original nucleus, which may constitute a significant fraction—up to 100%—of the ridge's volume (Fig. 7). This could occur for one of three reasons: a) the ridge formed in an area where there was no net sediment transport and did not migrate; b) the energy regime decreased abruptly, soon after the ridge formed, abandoning the ridge in an "immature" state; and/or c) the ridge was buried by sediments before it had time to migrate. Ridges that experience considerable swale erosion during formation are more likely to fall into this class than ridges with significant crestal aggradation. Such ridges may be more common in tide-dominated settings because the reversing currents probably produce slower net migration than is experienced on storm-dominated shelves. Overall low energy levels which lead to limited migration also favor the occurrence of Class I ridges.

The spacing, orientation, grain-size distribution, internal architecture, and age of the ridge deposits largely reflect the nature of the precursor (Table 2), although there may be some newly added shelf deposits in ridges that have a constructional component. Possible modern examples of Class I ridges include ridge-like features developed by modern erosion of lowstand to early transgressive barrier sand bodies. Examples include tide-built ridges in the Gulfe de Lion (Rabineau and Berné, 1997) and Peahala Ridge, a shoreface-attached ridge on the storm-dominated New Jersey coast that retains its ebb-tidal-delta nucleus (Snedden et al., 1994). Similarly, the tide-built ridges lying seaward of the Yangtze River (East China Sea) appear to contain a large nucleus of transgressive-age, delta-mouth deposits (Yang and Sun, 1988; Yang, 1989) and thus are also assigned to Class I.

Class II ridges are partially evolved sand bodies that have migrated less than their width and thus contain recognizable evidence of their precursor irregularity (Fig. 7). However, a significant proportion of the ridge consists of sediment deposited in a shelf setting. As a result, Class II ridges are expected to have a composite and potentially complex internal stratigraphy/architecture, with sediment characteristics that reflect the depositional environments of both the precursor core and overlying shelf sediments (Table 2). Because such ridges have been reworked more than Class I ridges, their spacing and orientation are more strongly controlled by the Huthnance process: they have spacings of the order of 250 times the water depth and orientations skewed ±20-30° relative to the currents (i.e., oblique to the shoreline in storm-dominated settings).

A large number of modern ridges fall into this class. For example, Ship Shoal (Louisiana shelf; Fig. 1C) contains a nucleus of transgressive barrier deposits overlying older deltaic sediments; however, modern ridge deposits (shoal crest and shoal-front facies) dominate the ridge superstructure (Penland et al., 1986). On the Atlantic shelf, small remnants of coastal deposits are present within some ridges, the relative proportion decreasing in an offshore direction (Fig. 4). Thus, this area may illustrate the evolutionary transition from Class I ridges in nearshore areas (Peahala Ridge), through Class II ridges, to Class III ridges in mid-shelf regions (Snedden et al., 1994; this volume). In addition, many of the shoreface-attached to inner-shelf ridges along the southern North-Sea coasts of France, Belgium, and the Netherlands (Fig. 1B) contain cores that are interpreted to consist of transgressive coastal facies (Berné et al., 1994; Van den Meene, 1994; Tessier et al., 1995)

Class III ridges may be thought of as fully evolved barforms that have migrated a distance equal to or more than their own width, eroding virtually all evidence of their precursor irregularity (Fig. 7). Such ridges are fully evolved in that all of their morphological and internal characteristics reflect the shelf setting in which they exist (Table 2). Thus, their spacing and orientation are controlled by the dynamics of the Huthnance process and their architecture typically consists of gently dipping strata that were deposited on the lee flank. An upward-coarsening trend should characterize storm-built ridges, while an upward-fining trend may occur in some tide-built ridges. Examples of Class III storm-built ridges include those south of Sable Island (Dalrymple and Hoogendoorn, 1997), as well as some mid-shelf ridges on the U.S. East Coast (Snedden et al., this volume). Active tidal sand ridges in the

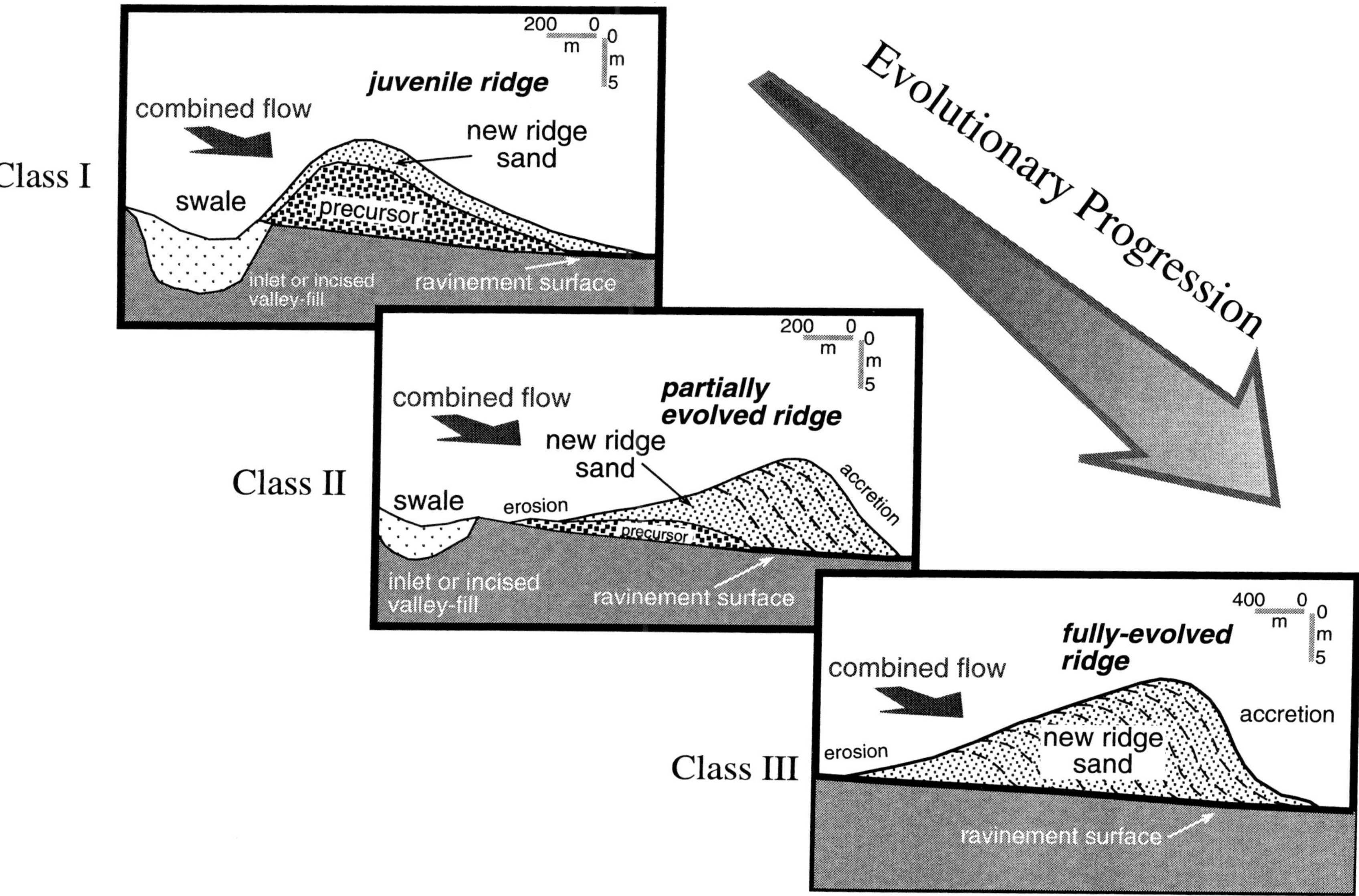

Fig. 7.—Schematic diagram of ridge classes. The precusor in the case of the Class I and II ridges is a pre-existing bathymetric feature, sometimes associated with a shoreline or inlet, which provides the nucleation point for the ridge via the Huthnance process. Subsequently, this precursor may be removed or reduced in size through current erosion and ridge migration. Accretion on the landward side of the juvenile ridge (Class I) is largely induced by fairweather wave transport from the ridge crest and is not expected to occur in ridges developed in deeper water, as with Classes II and III. New ridge sand is primarily deposited in shelf waters by combined flows associated with storm passage.

Table 2.—Classes of shelf sand ridges.

Class	I	II	III
Type	juvenile/static ridge	partially evolved ridge	fully-evolved ridge
Precursor	largely preserved	partially preserved	not preserved
Dynamics	ridge stationary or rapidly buried	ridge migrates a short distance	ridges migrates more than its original width
Signs of pre-transgressive state	strong or dominant	partial or subtle	minor or non-existant
Grain-size distribution	inherited from precursor	cross-ridge and vertical size segregation	cross-ridge and vertical size segregation
Deposit Age	largely pre-dates transgressive ravinement	partially contemporaneous with transgressive ravinement	post-dates transgressive ravinement
Multi-lateral sand bodies	solitary or in groups	commonly in groups	commonly in groups
Spacing	determined by precursor spacing	determined by Huthnance process	determined by Huthnance process
Internal Stratigraphy	same as precursor, simple to complex	compound—precursor overlain by new deposits	simple—new shelf deposits throughout
Probability of muddy burial	high	moderate	low
Holocene Examples	Gulfe de Lion[1], East China Sea[2], Peahala Ridge, New Jersey[3]	New Jersey Atlantic Shelf (Areas 1A and 2)[4], East Texas shelf[5], Ship Shoal[6]	New Jersey Atlantic Shelf (Area 3)[4], Sable Island[7], Dutch Central Coast[8]

[1]Rabineau and Berné (1997); [2]Yang (1989); [3]Snedden et al.. (1994);
[4]Snedden et al. this volume; [5]Rodriguez et al. this volume; [6]Penland et al. (1986)
[7]Dalrymple and Hoogendoorn (1997); [8]Van den Meene (1994)

North Sea that fall in this class include offshore banks described by Houbolt (1968), as well as some nearshore ridges along the central Dutch coast (Van den Meene, 1994). All of these areas have experienced high to very high energy conditions throughout the Holocene transgression. As a result, net sediment transport rates are high, thereby permitting relatively rapid migration.

MODERN SHELF SAND RIDGES AS ANALOGS FOR ANCIENT SAND BODIES

As illustrated above, shelf sand ridges are common on many modern continental shelves. One reason for this abundance relates to the global Late Pleistocene to Holocene (18 to 6 ka) sea-level rise, which produced a predominance of continental margins that are mantled by sandy sediments left behind as the shorelines moved landward. Indeed, we believe that transgressive conditions, in general, are particularly favorable for ridge genesis. As noted above in discussion of the Huthnance model, three conditions are necessary for ridge development: sufficient loose sand; currents capable of moving that sand; and a preexisting irregularity. In addition, sand must remain exposed on the seafloor long enough for the currents to mold it into a ridge.

In general, regressive areas and episodes do not provide these conditions. Regressive shelves are commonly mud-covered because sand is trapped in nearshore areas and is not transported out onto the shelf except during storms. Fairweather mud deposition prevents the accumulation of thick sandy beds and limits wave action because of energy attenuation (Snedden and Nummedal, 1991). In addition, muddy regressive shelves are typically smooth, except near river mouths, and thus lack the topographic irregularities that might act as ridge precursors. Furthermore, the high rates of deposition typical of regressive, deltaic settings lead to the burial of potential precursors before they have time to form ridges (cf., Coleman and Prior, 1982).

By contrast, transgressive shelves are commonly sand-covered and topographically irregular. Muddy sediments that might bury the irregularities are generally trapped in drowned-valley estuaries, allowing long periods of time for reworking of sand into ridges (Swift, 1985). The association of shelf ridges with abandonment and transgression of Mississippi delta lobes in an overall regressive setting (Penland et al., 1986) provides an excellent illustration of these ideas. For this reason we believe that shelf ridges will be preferentially developed in the transgressive portions of ancient successions.

The requirement that there be sufficient sand for ridge development places additional constraints on where ridges are likely to occur. At the most general level, the abundance of ridges on the U.S. East Coast as compared to the U.S. Gulf Coast probably reflects the relatively limited accommodation potential and sediment-starved nature of the East-Coast shelf. In more detail, most transgressive shelves, including both tide- and storm-dominated settings, experience net sediment transport in one direction (e.g., Swift, 1985). Sand is removed from the up-current end of these "sediment transport paths" and preferentially accumulates at the down-current end, forming large sand accumulations termed "shoal retreat massifs" offshore from transgressing capes and estuary mouths (Swift, 1975). Ridges are extensively developed in such regimes on both the U.S. East Coast and in the North Sea. A similar situation might be expected in the ancient.

On the U.S. East Coast, ridges are also preferentially developed in an along-strike portion of the mid shelf where the shoreline is believed to have sat for a considerable time during the post-glacial transgression (Stubblefield et al., 1984a; Swift et al., 1984). On the sand-starved Gulf of Mexico shelf, ridges are associated either with incised valleys (Thomas and Anderson, 1994; Fig. 1D) or with the distributary channels of transgressed delta lobes (Penland et al., 1986; Fig. 1C). These associations may occur for two reasons: a) there is a greater sand supply in and around the incised valleys, distributary channels, and paleoshorelines; and/or b) there is preferential preservation because of greater compactional subsidence associated with these features.

For the reasons outlined above, we expect ancient shelf ridges to occur within transgressive systems tracts (TST) of third-, fourth-, or fifth-order (e.g., cycles of less than 1.0 m.y. duration) depositional sequences. These sand bodies are likely to lie between the basal sequence boundary and the maximum flooding surface (Fig. 8). In addition, modern studies demonstrate that ridge sands, particularly those of the Class II and Class III variety, are located above the transgressive ravinement or flooding surface. In some cases, little sediment occurs between the basal sequence boundary and the transgressive ravinement surface; here ridge sands lie directly upon the sequence-bounding unconformity (Nummedal and Swift, 1987). In flat, wide shelves such as the Western Interior Cretaceous, it may be difficult to differentiate between these transgressive ridges and isolated, lowstand to transgressive, incised-shoreface sand bodies because both occupy a position between the sequence boundary and the maximum flooding surface. This may partly explain the considerable difficulty in determining the origin of isolated shallow marine sandstones like the Shannon of Wyoming that lie in such a stratigraphic position.

The characteristics of sediments that form the ridge depend on the class to which the ridge belongs. The nucleus within Class I and II ridges will most likely consist of coastal deposits that are slightly to significantly older than any shelf deposits within the ridge. In tide-dominated settings, large shelf dunes might comprise the nucleus. The relative proportion of sand deposited in a shelf environment (that is, sand overlying the transgressive flooding surface) will increase from negligible in Class I ridges, to 100% in Class III ridges (Fig. 7). In most cases, we expect these ridge sands to lie erosionally on the older coastal to shelf sediments, the exact stratigraphic relationship depending on the amount of erosion produced by migration of the inter-ridge swale. Gradational upward passage from shelf mud into ridge sand is probably rare, except possibly in the last stages of ridge movement.

Shelf sediments within the ridge will consist of relatively clean sands that may become slightly muddier later in the life of the ridge. Sands may consist largely of graded storm beds, but there should be some evidence of current action (that is, cross-bedding) because either tidal or storm-generated flows are necessary for ridge maintenance according to the Huthnance process. Bedding should be gently inclined, typically less than one degree, because of deposition on the side (usually the downcurrent, but sometimes the upcurrent flank) of the ridge. The degree of bioturbation may range from rare in continuously high-energy settings to pervasive in ridges formed by more episodic currents. Fauna within such sediments should largely reflect the shelf setting in which the post-transgressive ridge deposits accumulated, with older, more marginal marine forms being progressively destroyed (Culver and Snedden, 1996).

The preservation potential of modern sand ridges is, of course, an open question. In general, ridge preservation should be high in situations where distal muds of prograding siliciclastic shorelines bury them (Fig. 8). Indeed, the burial of isolated sand bodies by downlapping, muddy, highstand clinoform packages is postulated to have occurred throughout the Cretaceous (Asquith, 1970). This situation is extremely favorable for hydrocarbon entrapment, as the impermeable muds act as top seal to the sand-body reservoirs (Fig. 8). The distal deltaic muds may also provide an organic-rich source for hydrocarbons. Underlying incised-valley fluvial, inlet-fill, delta-distributary deposits, and/or shoreface deposits or other sandy sediments associated with the basal sequence boundary/ravinement surface may also act as carrier beds to enhance the petroleum potential of this system. It must be noted, however, that Class III ridges may have migrated far away from any previously associated valley, tidal inlet, or distributary channel, so the connection may have been broken (Fig. 7). In transgressive settings, the ridges located furthest seaward formed first and may have undergone more reworking, thereby increasing their sand content and volume with time. Thus, the ridges located furthest downdip may form the best overall reservoirs in the succession (Snedden et al., this volume; Fig. 8).

CONCLUSIONS

Sand ridges are abundant on modern shelves but show considerable variability in their external and internal characteristics. A synthesis of previous work suggests that a unified model of ridge origin and evolution can be developed to encompass and explain this variability, including both storm- and tide-built ridges. This model consists of three components or stages: 1) formation of an initial irregularity by coastal or shelf processes; 2) interaction of nearshore and/or shelf currents (either storm- or tide-driven) with the irregularity in the manner described by the Huthnance (1982) hydrodynamic model of ridge development; and 3) subsequent evolution of the ridge as a result of the continued action of the currents.

The wide range of potential precursor irregularities and variations in the degree of evolution account for the diversity of ridge characteristics and occurrences. Based on the degree of evolution, we categorize shelf ridges into three classes (Table 2; Fig. 7): Class I—juvenile or stationary ridges that retain their initial nucleus; Class II—ridges that have migrated somewhat but retain part of their nucleus; and Class III—"fully evolved" ridges that have migrated sufficiently that they contain no trace of their origin.

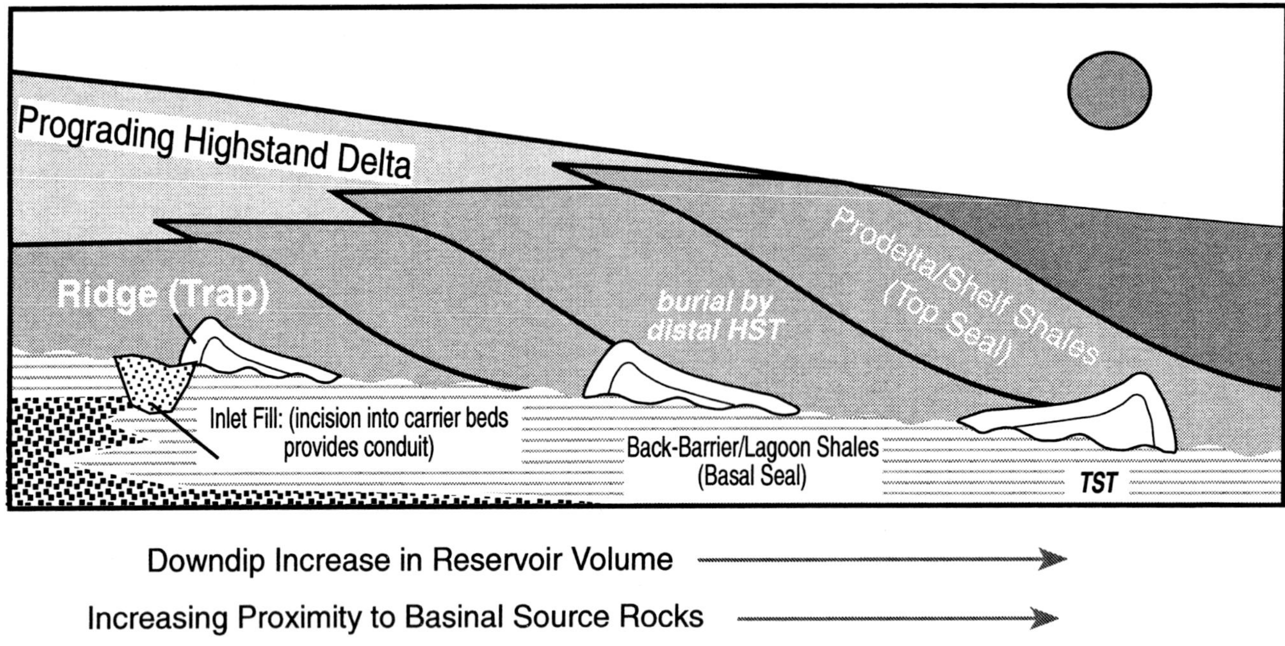

Fig. 8.—Preservation and petroleum potential of shelf sand ridges.

The universal applicability of the Huthnance process leads to the identification of four required conditions for the existence of shelf ridges: an initial irregularity that is most commonly developed in the nearshore region; a sufficient supply of sand; nearshore and/or shelf currents capable of transporting the sand; and sufficient time for the sand to be molded into a ridge. These conditions dictate that ridges will form almost exclusively in transgressive settings—that is, they will form part of the transgressive systems tract—and will lie on, or a short distance above, a ravinement surface. The requirement that sand be available also dictates that ridges will be most abundant on sediment-starved shelves, at the down-current end of sediment transport pathways—on shoal-retreat massifs—and in the vicinity of overstepped/transgressed barrier complexes. There may also be a spatial association with an underlying incised valley, tidal inlet, delta-distributary channel, or incised shoreface sands. Fully evolved ridges are likely to have an erosional base and may coarsen upward. These ridges may consist largely of storm-deposited sediment, but should contain evidence of current action, as the existence of currents is required for the Huthnance process to operate.

We believe that these conclusions are directly applicable to ancient settings, because they are based on an understanding of the fundamental physical processes operating in the nearshore and shelf environment. When applying these results to a particular ancient succession—for example, when developing specific hydrocarbon play types or trying to understand the origin of specific shelf sand bodies—it is important to carefully consider the factors controlling the distribution of sand on the shelf and the particular types of precursors that might have occurred. Most ridges originate near the coast but are reworked on the shelf as the transgression continues. When such ridges are subsequently buried by prograding, organic-rich, shelf muds—as is very likely—they have the potential to form ideal hydrocarbon reservoirs.

REFERENCES

Alpine Geophysical Associates, 1974, Geophysical investigation of Atlantic generating station site and offshore region: Unpublished report to Public Service Electric and Gas Company, Newark, New Jersey, 56 p.

Amos, C.L. and King, E.L., 1984, Bedforms of the Canadian Eastern Seaboard: A comparison with global occurrences: Marine Geology, v. 57, p. 167-208.

Antia, E.E., 1994, The ebb-tidal delta model of shoreface ridge origin and evolution: Appraisal and applicability along the southern North Sea barrier island coast: A discussion: Geo-Marine Letters, v. 14, p. 59-64.

Asquith, D.O., 1970, Depositional topography and major marine environments, Late Cretaceous, Wyoming: American Association Petroleum Geologists Bulletin, v. 54, p. 1184-1224.

Beardsley, R.C. and Butman, B., 1974, Circulation on the New England continental shelf, response to strong winter storms: Geophysical Research Letters, v. 1, p. 181-184.

Belderson, R.H., Johnson, M.A. and Kenyon, N.H., 1982, Bedforms, *in* Stride, A.H., ed., Offshore Tidal Sands, Processes and Deposits: London, Chapman and Hall, p. 27-55.

Bergman, K.M., 1994, Shannon Sandstone in Hartzog Draw-Heldt Draw fields (Cretaceous, Wyoming, USA) reinterpreted as lowstand shoreface deposits: Journal of Sedimentary Research, v. B64, p. 184-201.

Berné, S., Trentesaux, A., Stolk, A., Missiaen, T. and De Batist, M., 1994, Architecture and long term evolution of a tidal sand bank: The Middelkerke Bank (southern North Sea): Marine Geology, v. 121, p. 57-72.

Coleman, J.M. and Prior, D.B., 1982, Deltaic sand bodies: Tulsa, American Association Petroleum Geologists Short Course Notes 15, 170 p.

Culver, S.J. and Snedden, J.W., 1996, Foraminiferal implications for the formation of New Jersey shelf sand ridges: Palaios, v. 11, p. 161-175.

DALRYMPLE, R.W., 1992, Tidal depositional systems, *in* Walker, R.G. and James, N.P., eds., Facies Models Response to Sea Level Change: St. John's, Geological Association of Canada, p. 195-218.

DALRYMPLE, R.W. AND HOOGENDOORN, E.L., 1997, Erosion and deposition on migrating shoreface-attached ridges, Sable Island, Eastern Canada: Geoscience Canada, v. 24, p. 25-36.

DALRYMPLE, R.W. AND RHODES, R.N., 1995, Estuarine dunes and bars, *in* Perillo, G.M.E., ed., Geomorphology and Sedimentology of Estuaries: Developments in Sedimentology 53, p. 359-422.

DAVIS, R.A. AND BALSON, P.S., 1992, Stratigraphy of a North Sea tidal sand ridge: Journal of Sedimentary Petrology, v. 62, p. 116-121.

DE MAEYER, PH. AND WARTEL, S., 1988, Relation between superficial sediment grainsize and morphological features of the coastal ridges off the Belgian coast, *in* De Boer, P.L., Van Gelder, A. and Nio, S.D., eds., Tide-influenced Sedimentary Environments and Facies: Dordrecht, D. Reidel Publishing Company, p. 91-100.

DILLON, W.P. AND OLDALE, R.N., 1978, Late Quaternary sea-level curve: Reinterpretation based upon glaciotectonic influence: Geology, v. 6, p. 56-60.

DOBDAY, M.P., 1981, The Holocene geologic history of the Great Egg Harbor River Estuary: Unpublished M.S. Thesis, Temple University, 200 p.

EMERY, D. AND MYERS, K.J., 1996, Sequence Stratigraphy: Oxford, Blackwell Science, 297 p.

EMERY, K.O., 1968, The Atlantic Continental Shelf and Slope of the United States (Geologic Background): Washington, D.C., U.S. Geological Survey Professional Paper 529-A, p. 1-23.

FIGUEIREDO, A.G., 1984, Submarine sand ridges: Geology and development, New Jersey, USA: Unpublished Ph.D. dissertation, University of Miami, 385 p.

GADD, P.E., LAVELLE, J.W. AND SWIFT, D.J.P., 1978, Calculations of sand transport on the New York shelf using near bottom current meter observations: Journal of Sedimentary Petrology, v. 48, p. 239-252.

HAYES, M.O., 1975, Morphology of sand accumulations in estuaries, *in* Cronin, L.E., ed., Estuarine Research, v. 2: New York, Academic Press, p. 3-22.

HOOGENDOORN, E.L. AND DALRYMPLE, R.W., 1986, Morphology, lateral migration and internal structures of shoreface-connected sand ridges, Sable Island Bank, Nova Scotia, Canada: Geology, v. 14, p. 400-403.

HOUBOLT, J.J.H.C., 1968, Recent sediments in the southern Bight of the North Sea: Geologie En Mijnbouw, v. 47, p. 245-273.

HULSCHER, S.J.M.H., DE SWART, H.E. AND DE VRIEND, H.J., 1993, The generation of offshore tidal sand banks and sand waves: Continental Shelf Research, v. 13, p. 1183-1204.

HUTHNANCE, J.M., 1982, On one mechanism forming linear sand banks: Estuarine and Marine Coastal Science, v. 14, p. 79-99.

KENYON, N.H., BELDERSON, R.H., STRIDE, A.H. AND JOHNSON, M.A., 1981, Offshore tidal sand banks as indicators of net sand transport and as potential deposits, *in* Nio, S.D., Shuttenheim, R.T.C. and van Weering, T.C.E., eds., Holocene Marine Sedimentation in the North Sea Basin: Oxford, International Association Sedimentologists Special Publication 5, p. 257-268.

MCBRIDE, R.A. AND MOSLOW, T.F., 1991, Origin, evolution and distribution of shoreface sand ridges, Atlantic inner shelf, U.S.A.: Marine Geology, v. 97, p. 57-85.

MCCAVE, I.N., 1979, Tidal currents at the North Hinder Lightship, southern North Sea: Flow directions and turbulence in relation to the maintenance of sand banks: Marine Geology, v. 31, p. 101-114.

MCCLELLEN, C.E., 1973, New Jersey shelf near bottom current meter records and recent sediment activity: Journal Sedimentary Petrology, v. 43, p. 371-380.

MCHONE, J.F., JR., 1973, Morphologic time series from a submarine sand ridge on the southern Virginia coast: Unpublished M.S. Thesis, Old Dominion University, Norfolk, Virginia, 59 p.

MCKINNEY, T.F., DEALTERIS, J., CHAO, Y.Y., STAHL, L. AND RONEY, J., 1976, Offshore sedimentary processes and responses near Beach Haven-Little Egg Inlets, New Jersey: Proceedings of Fifteenth Coastal Engineering Conference, American Society of Civil Engineers, p. 1899-1918.

MCKINNEY, T.F. AND FRIEDMAN, G.M., 1970, Continental shelf sediments off Long Island, New York: Journal of Sedimentary Petrology, v. 40, p. 213-248.

NEIDORODA, A.W. AND SWIFT, D.J.P., 1985, Barrier island evolution, Middle Atlantic Shelf, USA: Part II: Evidence from the shelf floor: Marine Geology, v. 63, p. 363-396.

NUMMEDAL, D. AND SWIFT, D.J.P. , 1987, Transgressive stratigraphy at sequence bounding unconformities, some principles derived from Holocene and Cretaceous examples, *in* Nummedal, D., Pilkey, O.H. and Howard, J.D., eds., Sea-level Fluctuation and Coastal Evolution: Tulsa, SEPM (Society for Sedimentary Geology) Special Publication 41, p. 241-260.

OFF, T., 1963, Rhythmic linear sand bodies caused by tidal currents: American Association of Petroleum Geologists Bulletin, v. 47, p. 324-341.

PENLAND, S., SUTER, J.R. AND MOSLOW, T.F., 1986, Inner-shelf shoal sedimentary facies and sequences: Ship Shoal, Northern Gulf of Mexico, *in* Moslow, T.F. and Rhodes, E.G., eds., Modern and Ancient Shelf Clastics: A Core Workshop: Tulsa, SEPM (Society for Sedimentary Geology) Core Workshop 9, p. 73-122.

RABINEAU, M. AND BERNÉ, S., 1997, Modern and Pleistocene sand bodies of the Gulfe de Lion, (abs.), *in* Shanley, K. and Perkins, B.F., eds., Non-marine and Shallow Marine Environments: Houston, Proceedings of the Gulf Coast Section SEPM (Society for Sedimentary Geology) Research Conference.

RINE, J.M., TILLMAN, R.W., CULVER, S.J. AND SWIFT, D.J.P., 1991, Generation of Late Holocene ridges on the middle continental shelf of New Jersey, USA—Evidence for formation in a mid-shelf setting based upon comparison with a nearshore ridge, *in* Swift, D.J.P., Oertel, G.F., Tillman, R.W. and Thorne, J.A., eds., Shelf Sand and Sandstone, Geometry, Facies, and Sequence Stratigraphy: Oxford, International Association Sedimentologists Special Publication 14, p. 395-426.

SANDERS, J.E. AND KUMAR, N., 1975, Evidence of shoreface retreat and in-place drowning during Holocene submergence of barriers off Fire Island, New York: Geological Society of America Bulletin, v. 86, p. 65-76.

SEXTON, W.J. AND HAYES, M.O., 1996, Holocene deposits of reservoir-quality sand along the central South Carolina coastline: American Association Petroleum Geologists Bulletin, v. 80, p. 831-855.

SHEPARD, F.P., 1963, Submarine Geology (2nd ed.): New York, Harper and Row, 557 p.

SNEDDEN, J.W. AND NUMMEDAL, D., 1991, Origin and geometry of storm-deposited sand beds in modern sediments of the Texas continental shelf, *in* Swift, D.J.P., Tillman, R.W. and Oertel, G.F., eds., Shelf Sand and Sandstone, Geometry, Facies, and Sequence Stratigraphy: Oxford, International Association Sedimentologists Special Publication 14, p. 283-308.

SNEDDEN, J.W., KREISA, R.D., TILLMAN, R.W., SCHWELLER, W.J., CULVER, S.J. AND WINN, R.D., 1994, Stratigraphy and genesis of a modern shoreface-attached sand ridge, Peahala Ridge, New Jersey: Journal of Sedimentary Research, v. B64, p. 560-581.

STRIDE, A.H., 1963, Current-swept sea floors near the southern half of Great Britain: Quarterly Journal Geological Society of London, v. 119, p. 175-199.

STUBBLEFIELD, W.L., MCGRAIL, D.W. AND KERSEY, D.G., 1984a, Recognition of transgressive and post-transgressive sand ridges on the New Jersey continental shelf, *in* Tillman, R.W. and Siemers, C.T., eds., Siliciclastic Shelf Sediments: Tulsa, SEPM (Society for Sedimentary Geology) Special Publication 34, p. 1-24.

STUBBLEFIELD, W.L., MCGRAIL, D.W. AND KERSEY, D.G., 1984b, Recognition of transgressive and post-transgressive sand ridges on the New Jersey continental shelf—a reply, *in* Tillman, R.W. and Siemers, C.T., eds., Siliciclastic Shelf Sediments: Tulsa, SEPM (Society for Sedimentary Geology) Special Publication 34, p.37-41.

STUIVER, M. AND DADDARIO, J.J., 1963, Submergence of the New Jersey coast: Science, v. 142, p. 951.

SULLIVAN, M.D., VAN WAGONER, J.C., JENNETTE, D.C., FOSTER, M.E., STUART, R.M., LOVELL, R.W. AND PEMBERTON, S.G., 1997, Lowstand architecture and sequence stratigraphic control on Shannon incised valley distribution, Hartzog Draw Field, Wyoming, *in* Shanley, K.W. and Perkins, B.F., eds., Shallow Marine and Nonmarine Reservoirs, Sequence Stratigraphy, Reservoir Architecture, and Production Characteristics: Gulf Coast Section, SEPM (Society for Sedimentary Geology) Annual Research Conference Transactions, v. 18, p. 331-344.

SWIFT, D.J.P., 1975, Tidal sand ridges and shoal retreat massifs: Marine Geology, v. 18, p. 105-134.

SWIFT, D.J.P., 1985, Response of the shelf floor to flow, *in* Tillman, R.W., Swift, D.J.P. and Walker, R.G., eds., Shelf Sands and Sandstone Reservoirs: Tulsa, SEPM (Society for Sedimentary Geology)Short Course Notes 13, p. 135-241.

SWIFT, D.J.P. AND FIELD, M.E., 1981, Evolution of a classic sand ridge field: Maryland Sector, North American inner shelf: Sedimentology, v. 28, p. 461-482.

SWIFT, D.J.P., HOLLIDAY, B., AVIGNONE, N. AND SHIDELER, G., 1972a, Anatomy of a shoreface ridge system, False Cape, Virginia: Marine Geology, v. 12, p. 59-84.

SWIFT, D.J.P., KOFOED, J.W., SAULSBURY, F.P. AND SEARS, P., 1972b, Holocene evolution of shelf surface, central and southern Atlantic Shelf of North America, *in* Swift, D.J.P., Duane, D.B. and Pilkey, O.H., eds., Shelf Sediment Transport, Process and Pattern: Stroudsburg, Pennsylvania, Dowden, Hutchison and Ross, p. 447-498.

SWIFT, D.J.P., MCKINNEY, T.F. AND STAHL, L., 1984, Recognition of transgressive and post-transgressive sand ridges on the New Jersey continental shelf—discussion, *in* Tillman, R.W. and Siemers, C.T., eds., Siliciclastic Shelf Sediments: Tulsa, SEPM (Society for Sedimentary Geology) Special Publication 34, p. 25-36.

SWIFT, D.J.P., NELSEN, T., MCHONE, J., HOLLIDAY, B., PALMER, H. AND SHIDELER, G., 1977, Holocene evolution of the inner shelf of southern Virginia: Journal of Sedimentary Petrology, v. 47, p. 1454-1474.

SWIFT, D.J.P., PARKER, G., LANDFREDI, N.W., PERILLO, G. AND FIGGE, K., 1978, Shoreface-connected sand ridges on American and European shelves: A comparison: Estuarine and Coastal Marine Science, v. 7, p. 257-273.

TESSIER, B., CORBAU, C. AND AUFFRET, J.P., 1995, Internal structures as dynamics pattern indicators of shoreface bank moving in a macrotidal coastal environment (Dunkerque area, North France) (abs.): Book of Abstracts, 16th International Association Sedimentologists Regional Meeting Publication 22, p. 141.

THOMAS, M.A. AND ANDERSON, J.B., 1994, Sea-level controls on the facies architecture of the Trinity/Sabine incised-valley system, Texas continental shelf, *in* Dalrymple, R.W., Boyd, R. and Zaitlin, B.A., eds., Incised-Valley Systems: Origin and Sedimentary Sequences: SEPM (Society for Sedimentary Geology) Special Publication 51, p. 63-82.

TILLMAN, R.W. AND MARTINSEN, R.S., 1984, The Shannon shelf ridge sandstone complex, Salt Creek Anticline area, Powder River Basin, Wyoming, *in* Tillman, R.W. and Siemers, C.T., eds., Siliciclastic Shelf Sediments: SEPM (Society for Sedimentary Geology)Special Publication 34, p. 1-34.

UCHUPI, E., 1968, The Atlantic Continental Shelf and Slope of the United States (Physiography): Washington, D.C., U.S. Geological Survey Professional Paper 529-I, 30 p.

VAN DEN MEENE, J.W.H., 1994, The shoreface-connected ridges along the central Dutch coast: Universiteit Utrecht, Nederlandse Geografische Studies, v. 174, 222 p.

VAN VEEN, J., 1935, Sand waves in the North Sea: Hydrographic Review, v. 12, p. 21-28.

VAN WAGONER, J.C. AND BERTRAM, G.T., 1995, Sequence stratigraphy of foreland basin deposits: Outcrop and subsurface examples from the Cretaceous of North America: Tulsa, American Association of Petroleum Geologists Memoir 64, 490 p.

VEATCH, A.C. AND SMITH, P.A., 1939, Atlantic submarine valleys of the United States, and the Congo submarine Canyon: Boulder, Geological Society of American Special Paper 7, 101 p.

YANG, C.S., 1989, Active, moribund, and buried tidal sand ridges in the East China Sea and the southern Yellow Sea: Marine Geology, v. 88, p. 97-116.

YANG, C.S. AND SUN, J., 1988, Tidal sand ridges on the East China Sea shelf, *in* DeBoer, P.L., van Gelder, A. and Nio, S.D., eds., Tide-influenced Sedimentary Environments and Facies: Boston, D. Reidel Publishing Company, p. 23-38.

THE SHANNON SANDSTONE: A REVIEW OF THE SAND-RIDGE AND OTHER MODELS

RODERICK W. TILLMAN

Consultant, 2121 E. 51st St., Ste. 112, Tulsa, OK 74105, U.S.A.

ABSTRACT: The Shannon sandstones in the Powder River Basin of Wyoming have been interpreted by Tillman and Martinsen and others as having been deposited as sand-ridge complexes during transgression on a wide shelf in the Western Interior Seaway, which stretched from Alaska to Mexico in early Campanian time. In recent papers by other workers, the Shannon is interpreted to be a series of incised lowstand shoreface deposits or estuarine valley-fill deposits. The sand-ridge complexes are considered by different workers to have been deposited in a variety of sequence stratigraphic settings ranging from regression to sea level still-stand to transgression. The shelf sand-ridge model infers offshore deposition, possibly as far as 110-160 km from shore at middle shelf depths by south to south-southwest-flowing, shore-oblique currents intensified frequently by storms. A probable source of sediments for the ridges probably does not include shoreline sandstones to the west, but instead the Eagle Sandstone deltaic deposits of southern Montana which lie 320 km to the north-northwest.

A model inferring that Hartzog Draw Field, a major Shannon sandstone oil producer, is a remnant of several sequence boundary incisions was recently developed by the current operators of the field. They recognize a sequence boundary at the base of the Shannon and two sequence boundaries within the Shannon. They infer that deposition of the Shannon is primarily by tidal processes. The incisions at the sequence boundaries are projected to cover tens of km laterally and if these are truly incisions they are believed by this author to have formed large open bays. To accommodate the presence of marine fossils that occur below, above, and between the thick fine- to coarse-grained Shannon Sandstone accumulations, where they occur on the Salt Creek anticline, several abrupt changes in sea-level would be required prior to, during, and following Shannon deposition.

A third model has been developed by Canadian authors who have used shoreface models developed for the Cretaceous in Alberta, Canada to explain the Shannon. Their model infers that the Shannon sandstones are a series of lowstand incised shorefaces.

This paper gives a historical perspective of interpretations of the Shannon during the last 25 years. The pros and cons of the various models are considered, and some of the problems with each model are discussed.

INTRODUCTION

The Cretaceous Campanian-age Shannon Sandstone (Fig. 1) has been extensively studied in the subsurface and in outcrop. It crops out and produces oil in the Powder River Basin of eastern Wyoming and Montana. One of the earliest interpretations of the Shannon was by Parker in 1958. Using ammonites and other types of paleontological data, primarily from outcrop, the location of the easternmost preserved Shannon "shoreline" was plotted in 1973 by Gill and Cobban (Fig. 2). They indicate that prior to deposition of the Shannon the Campanian shorelines were generally moving basinward, but during deposition of the Shannon there was a reversal of the shoreline movement and transgression occurred. The Shannon lies stratigraphically immediately below the Ardmore Bentonite, which has been dated by Obradovich (1993) to be 81 m.y. old. The Shannon is equivalent in age to the lower part of the Mesa Verde Formation farther to the west (Fig. 3).

Spearing (1976) added greatly to the understanding of the Shannon in his paper discussing the shelfal origins of the Shannon outcrops. Seeling (1978) was among the first to discuss the details of a Shannon oil field, Heldt Draw Field. Following the work of Tillman and Martinsen (1984, 1987), Gaynor and Swift (1988) also interpreted many of the outcrops of the Shannon as shelf sand ridges. A summary of the Upper Cretaceous in the Western Interior was done by Merewether (1996). Among the most recent papers on the Shannon are those by Walker and Bergman (1993), Bergman (1994), Bergman and Walker (1995), and Sullivan et al. (1997). These last four papers propose origins for the Shannon that differ significantly from those of all previous authors. It is also worth mentioning that the Shannon apparently differs from any of the published interpretations of modern shelf sand ridges, which have had the benefit of core and/or shallow seismic analysis.

Numerous fields in the Powder River Basin produce from the elongate northwest-southeast trending Shannon (and Sussex) reservoirs in the Powder River Basin (Fig. 4). Following terminology originally proposed by Don Swift, these reservoirs are designated as sand ridges. The term ridge is used for features significantly larger than bars. The Shannon ridges range in size from a few km in length to over 30 km in length, and from 1 to 5 km wide. The Shannon reservoirs are interpreted to have been deposited during a transgressive expansion of the seaway during the period of *Baculites* sp. (smooth) and prior to the deposition of *Baculites* sp. (weak flank ribs), which prevailed during deposition of the Sussex Sandstone (Fig. 1).

All of the workers prior to the mid 1980s ascribed a shelfal origin to the Shannon sand ridges (Fig. 5). Hansley and Whitney (1990) studied the petrology and diagenesis of the Shannon and suggested that it was a shoreline sandstone. The Sussex sandstone, which almost immediately overlies the Shannon, has been discussed by Berg (1975) and Hobson et al. (1982), and they also attribute it's origin to marine shelf deposition similar to that of the earlier models for the Shannon.

With the advent of more general use of sequence stratigraphy, recognition of the importance of the ravinement process, and a better understanding of incised valley formation, several different origins have recently been proposed for the Shannon. The early emphasis in the 1970s on models developed from outcrops shifted to those based on subsurface data in the 1980s, and in the 1990s new models were proposed using both of these data sets.

This paper initially discusses the building blocks (facies) that form the Shannon sand ridges. A discussion of the facies is based on studies of Shannon outcrops and cores and detailed subsurface studies of several fields in the Powder River Basin. Hartzog Draw Field (Figs. 4, 5), the largest of the

Isolated Shallow Marine Sand Bodies: Sequence Stratigraphic Analysis and Sedimentologic Interpretation.
SEPM Special Publication No. 64, Copyright © 1999
SEPM (Society for Sedimentary Geology), ISBN 1-56576-057-3, p. 29-54.

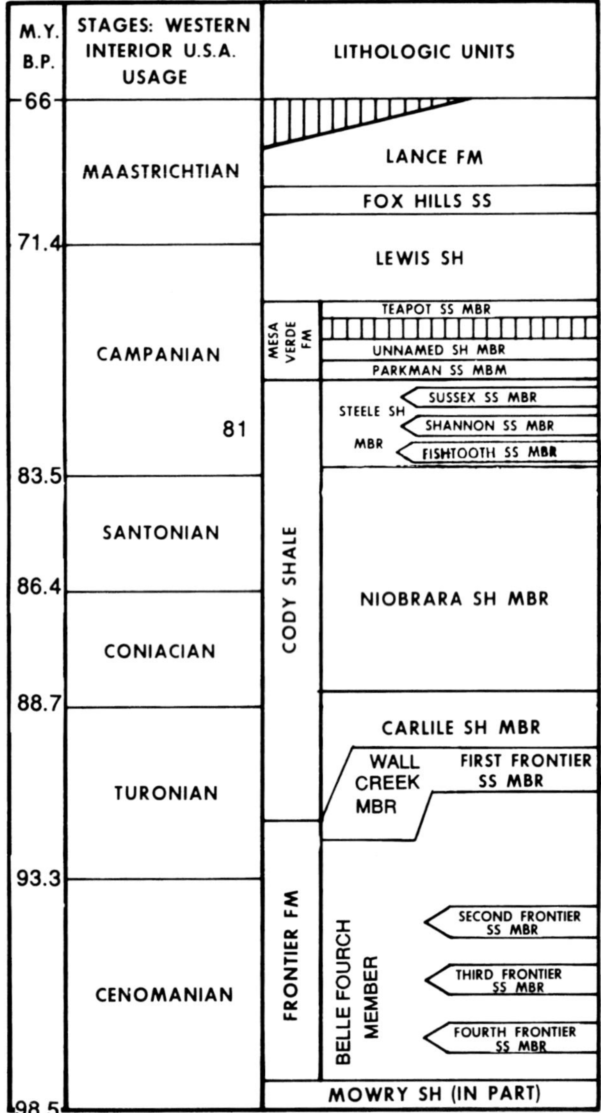

Fig. 1.—Upper Cretaceous stratigraphic section, western Powder River Basin, Wyoming. The Shannon Sandstone is a part of the Steele Shale Member of the Cody Shale Formation. Modified from Merewether (1996). Ages are from Obradovich (1993).

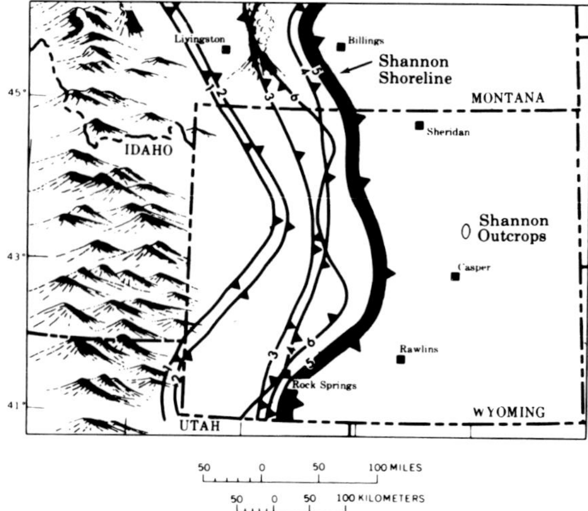

Fig. 2.—Interpreted "shoreline locations" for lower Campanian-Upper Santonian age rocks. These shorelines are in what was commonly referred to as the Telegraph Creek-Eagle Regression. Note that following and presumably during the deposition of the Shannon, the area was being transgressed (Shorelines 5-6). Barbs indicate the direction of each shoreline movement. Shoreline 1 is 83 m.y.a. (*Desmoscaphites erdamanni* ammonite zone) at the base of the Montana Group (Santonian). Shoreline 5 is *Baculites* sp. (smooth), which includes the Shannon. Shoreline 6 (transgressive) is in *Baculites* sp. (weak flank ribs) and includes the Sussex Sandstone. Modified from Gill and Cobban (1973).

Shannon oil fields, has been most thoroughly studied by the author. The distribution of facies in Hartzog Draw Field fit into a reasonable model of a sand ridge. An understanding of the distribution of the geologic facies in Shannon sand ridges is also important in understanding the varying rates of oil production from individual wells in Shannon oil fields. The initial subsurface shelf sand-ridge model was discussed by Martinsen and Tillman (1978) and Tillman and Martinsen (1984, 1987) and is summarized herein. Comments are also included concerning two recently proposed origins for the Shannon.

SHANNON FACIES

Eleven Shannon facies were defined in outcrop and in the subsurface primarily on the basis of physical and biological sedimentary structures and lithology by Tillman and Martinsen (1984, 1987) as summarized in Table 1. At least two methods have been used by geologists to designate sandstone facies. One is based, at least in part, on genesis and the other relies primarily on criteria observable in the rocks. A modification of the genetic-descriptive facies designations first formally published for Cretaceous sandstones by Porter (1976) is used here for Shannon facies. More descriptive facies names have been applied to the Shannon by Walker and Bergman (1993), Bergman (1994), and other authors. They emphasize only the sedimentary features in the names given to Shannon Sandstone facies. The evolution of Shannon facies names is summarized in Table 2.

A hierarchical scheme was used to designate the Shannon facies. The initial subdivision was based on lithology (shale, silty shale, shaley sandstone, and sandstone). The second level of subdivision was based on internal sedimentary structures (cross-stratified, rippled, lenticular bedding, and degree of bioturbation). A third level of subdivision involved inferred energy of deposition based on: a) mean grain size, b) amount of shale clasts and siderite clasts, c) amount of interlaminated shale, d) uniformity of laminasets (interbedded ripple-form beds and cross-stratified beds versus stacked cross-stratified beds), and e) the amount and degree of concentration in laminasets of glauconite. The attributes for each of the facies are summarized in Table 3.

To assure consistent classification and determine where to draw the boundaries between facies, quantitative volume percentages of each attribute were recorded for each facies observed in outcrop and core as discussed in detail previously by Tillman and Martinsen (1984, 1987). The mean percentage of each of these attributes is tabulated in Table 3.

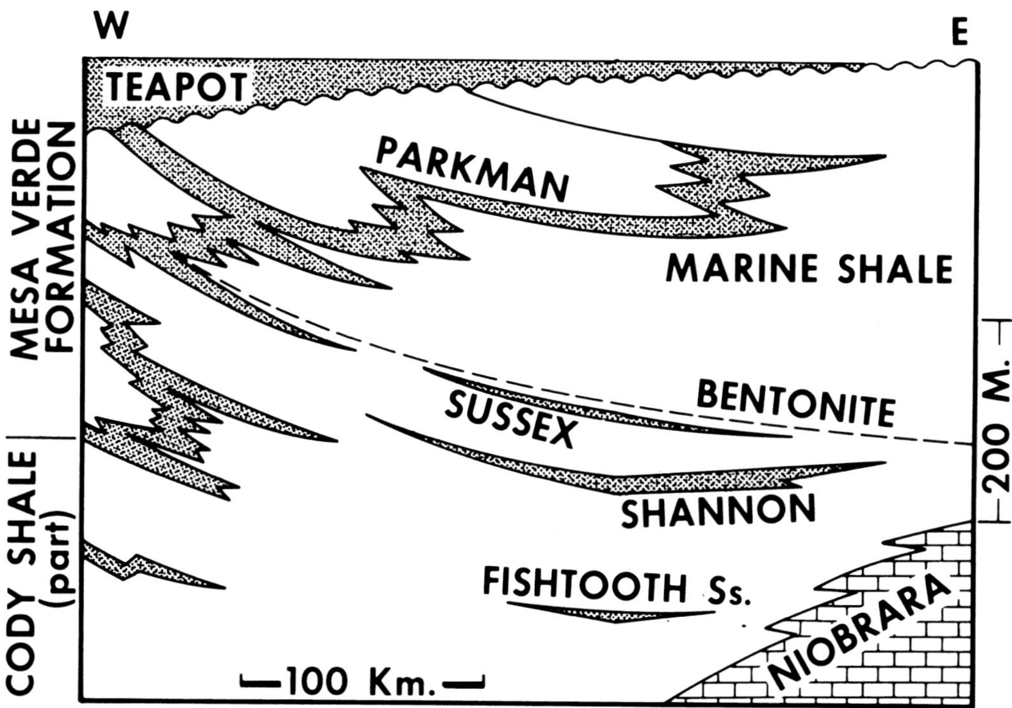

Fig. 3.—Stratigraphic reconstruction of the Upper Cretaceous in central Wyoming. Note that the Shannon is seaward of the lower part of Mesa Verde Formation and pinches out to the west and to the east. Dashed line is one of several bentonites that bracket the Shannon and are interpreted to be time synchronous. Nearly 100 km separates the Shannon from the most easterly preserved, nearly time equivalent, shoreline. After Gill and Cobban (1966)

Five sandstone facies form most of the Shannon sand ridges. The Central Ridge Sandstone is the most quartz-rich, and contains the least glauconite and the lowest percentage of shale clasts and siderite clasts of any of the cross-stratified facies (Fig. 6). This facies is most common near the crest of the sand ridges and is absent on the west side of the ridges studied.

The High-Energy Ridge-Margin Sandstone (HERM) is cross-stratified (Fig. 7), has the coarsest mean grain size, and the highest percentages and the greatest local concentrations of glauconite (Fig. 8). It also contains more shale clasts (Fig. 9) and siderite clasts (Fig. 8) than any other facies, as well as thin, lenticular beds of both shale and siderite (Fig. 10). This facies is most common on the east flanks of the ridges and is apparently absent on the west flanks.

The Low-Energy Ridge-Margin Sandstone (LERM) is recognized primarily by the successive interbedding of 0.1-0.3 m thick cross-stratified sandstone and rippled sandstone beds. This facies is almost entirely limited to the west flank of the ridges. A Planar-Laminated Sandstone forms less than 2% of the volume of the ridges.

The other major sandstone facies in the Shannon is one that occurs primarily below and between the sand-ridges. It has been termed the Inter-Ridge Facies. Two subfacies are recognized. In the outcrops this facies is almost entirely sandstone, has horizontal ripple-form bedding surfaces, and has rippled to horizontally laminated internal features (Fig. 11). In the subsurface, at Hartzog Draw Field, the overall morphology of this ripple-form facies is similar to that in outcrop, but the facies contains abundant thin lenses and drapes of clay on bedding surfaces (Fig. 12). This sub-facies is designated as Inter-Ridge (Shaley) Sandstone. Its occurrence between the sand-ridges is important as evidence for a possible open-bay/estuary origin for the Shannon. Details of these and other Shannon facies are given in Table 1 and in Tillman and Martinsen (1984, 1987).

HARTZOG DRAW FIELD

Hartzog Draw Field, which was discovered in 1975, lies west of the center of the Powder River Basin in Wyoming (Fig. 4). The field produces from the Shannon Sandstone at depths from 2750 to 3000 m. The field is approximately 35 km long and up to 5 km wide. The Shannon interval at Hartzog Draw is nearly 30 m thick, and the net sand producing interval ranges up to 20 m (Tillman and Martinsen, 1987). The reservoir is designated by the author as a sand ridge.

Development of Hartzog Draw Field was strongly dependent on a subsurface sand-ridge model that was developed by Martinsen and Tillman (1978) and Tillman and Martinsen (1987) and which was refined periodically during development of the field as core and other subsurface data were accumulated. In the 1970s it was rare to have abundant core data available from relatively closely spaced wells in a field, and it was even more uncommon that drilling models based on the results of core interpretations and calibration of nearby log suites were applied during a significant portion of the development of a field. Fortunately all these events occurred during the development of Hartzog Draw Field. To develop a predictive model ahead of the bit, characteristics and distribution of the facies were established, their distribution within the field was plotted, and recommendations to company management were made to acquire additional acreage along the 35 km trend that is

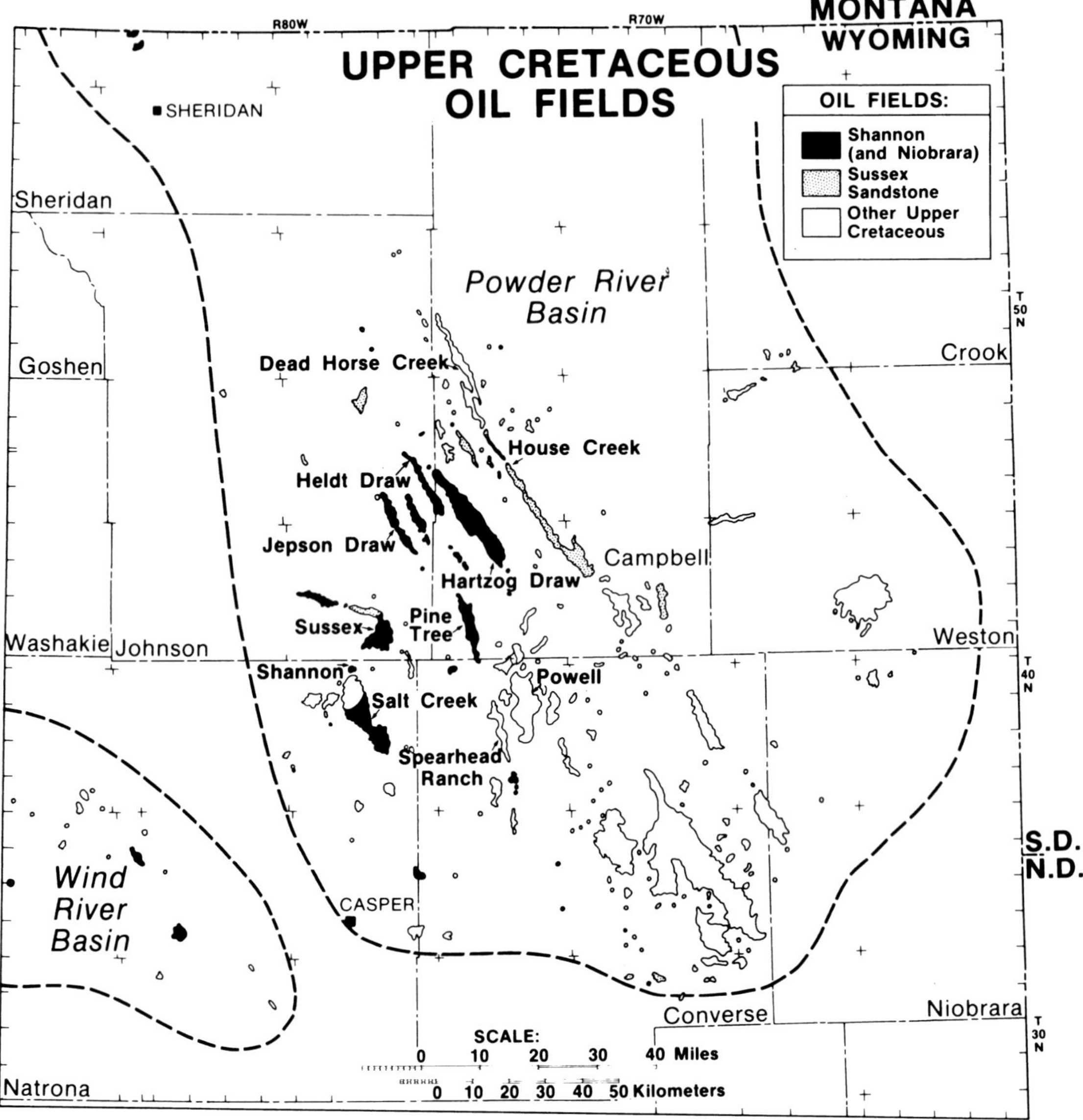

Fig. 4.—Upper Cretaceous oil fields, Powder River Basin, Wyoming. Note the parallelism and elongation of the fields. Fields that produce from the Shannon are black.

now Hartzog Draw Field. The recommendation to buy acreage more than a dozen km (along trend) from the nearest production seemed bold at the time since the longest previously discovered field (Heldt Draw) was less than 15 km long. Fortunately for the company several of the large blocks that were acquired turned out to be in the heart of the field. The production and geological characteristics of the field are summarized by Martinsen (1981) and Tillman and Martinsen (1987).

Both gross and net sand isopach maps of Hartzog Draw Field, which trends north-northwest/south-southeast, show a symmetry with the thickest interval near the center, and a relatively uniform rate of thinning northeast and southwest across the ridge (see Tillman and Martinsen, 1987). A relatively uniform rate of thinning also occurs in northwest and southeast directions, parallel with the ridge elongation (see Fig. 34 of Tillman and Martinsen, 1987). In strong contrast to the symmetry described above, a map of cumulative oil production in the field has a very strong asymmetry (Fig. 13).

As described in the facies discussion, cross-stratified sandstones (Fig. 6) containing only small amounts of glauconite and very few rip-up clasts form a large part of the central and eastern part of the sand ridge (Central Ridge Facies). The higher energy sandstone deposits are concentrated in the central area and in the eastern part of the field (Fig. 14). Commonly, wells in the most easterly part of the field have the thickest, most highly glauconitic, coarsest-grained sandstone, and the greatest abundance of siderite and shale rip-up clasts (High-Energy Ridge-Margin Sandstone). This facies is interpreted to be the result of erosion and redeposition by storms, which impinged more strongly on the eastern flank of the ridge than the west. Erosion on the northeast side of the field is also suggested by the fact that the east side of the ridge is much steeper than the west side (Fig. 15). Silty marine or open-bay shales are present on the east side of Hartzog Draw

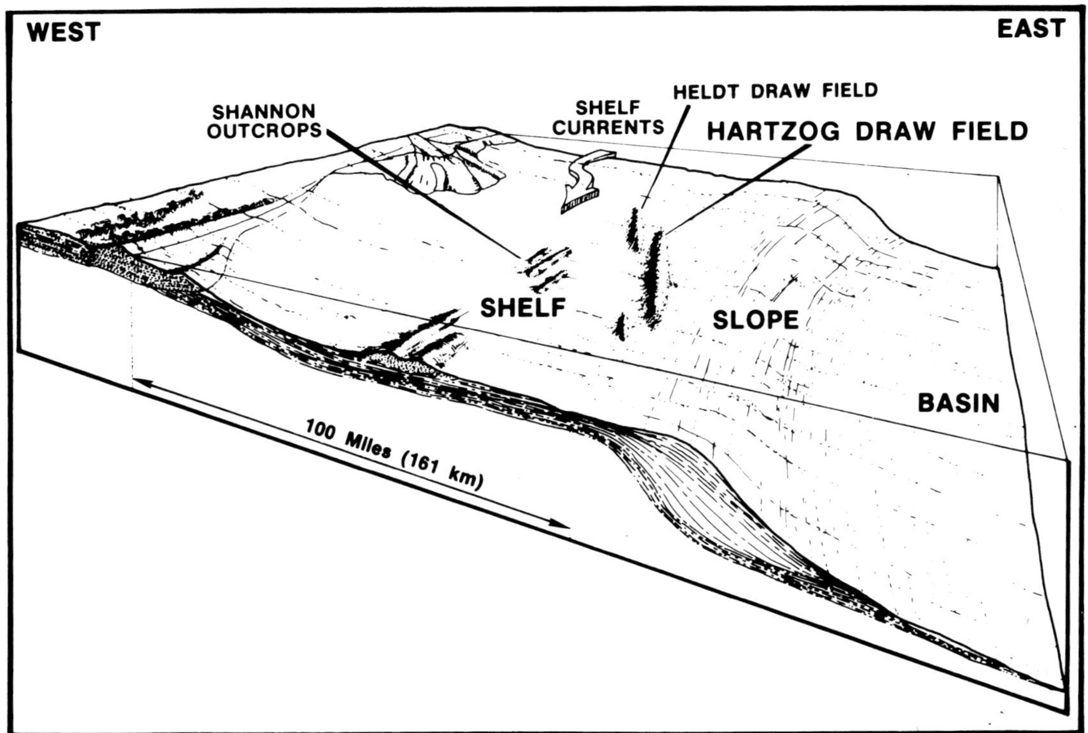

Fig. 5.—Wide shelf model for the Shannon Sandstone on the western flank of the Powder River Basin, Wyoming. North-northwest/south-southwest-trending Shannon sand ridges are superimposed on the model developed by Asquith (1970). South-flowing shelf currents may have redistributed sands supplied from the northwest from the Eagle Sandstone delta complex in southern Montana. Transportation eastward of significant amounts of sand from a distant western shoreline is unlikely.

Field; Inter-Ridge (Shaley) Sandstone occurs in a similar position on the west side of the field (Fig. 15). Both the Central Ridge Sandstone and the High-Energy Ridge-Margin Sandstone are essentially absent in the western one-third of the ridge(s) that form Hartzog Draw Field (Fig. 14). Low-Energy Ridge-Margin Sandstone forms the major production interval in the western third of the field.

Bentonites bracket the Shannon both in outcrop and at Hartzog Draw Field. The Ardmore bentonite overlies the Shannon, and several unnamed bentonites occur immediately below the Shannon. The bentonites are assumed to have been deposited horizontally and to have remained so except where differential compaction at the edges of the sand ridge caused some bending (Fig. 15). The Hartzog Draw and outcrop sand-ridge models were both constructed using horizontal bentonites below the sand-ridge as datums.

Each of the Shannon facies has relatively consistent reservoir properties as discussed by Tillman and Martinsen (1987). A summary of the reservoir properties is given in Table 4. The production history of Hartzog Draw Field during primary production is discussed by Hearn et al. (1984) and Ranganathan and Tye (1986).

SHELF-RIDGE MODEL

Tillman and Martinsen (1984,1987) interpreted the Shannon Sandstone to be a series of sand ridges resulting primarily from storm activity that formed strong bottom currents that were directed in a southerly direction along bathyl contours in an offshore marine setting (Fig. 16). The source of the sand was assumed to be the Eagle Sandstone delta 320 km to the north-northwest in southern Montana. Concentration of the sand into ridges was assumed to be the result of winnowing over low-relief topographic highs in marine waters up to several hundred feet deep.

The base of the Shannon was assumed to be conformable and to have a slightly transitional boundary. This type of apparently transitional boundary occurs in outcrops immediately south of measured section W2A (Fig. 17). Advocates of a sharp basal boundary for the Shannon point to the sharp boundaries below the Inter-Ridge Sandstones found in several measured sections, including the lower Shannon Inter-Ridge Sandstone at Measured Section W2B (Figs. 18, 19) and at Measured Section W2A (Fig. 20). However, at all of these locations the Inter-Ridge Sandstone was not deposited directly on fissile, dark gray marine shales, but instead on Bioturbated Siltstone, and the basal contact of the siltstone on marine shale is covered. The true base of the Shannon complex is at the top of fissile marine Cody Shale (Fig. 17).

Outcrop studies suggested that individual ridges may contain no significant time breaks; however, the predominantly horizontal bedding throughout the Shannon (Fig. 21) suggests that most of the individual bed surfaces were planed off periodically and that most of the beds may represent less than two-thirds of their original thickness. Storm waves may have been responsible for the planar erosion surfaces.

Precise correlation of Shannon outcrop sections is difficult because of the numerous lateral and vertical facies changes

Table 1.—Shannon facies characteristics. Facies names modified from Tillman and Martinsen (1984, 1987).

SHANNON FACIES SUMMARY

	CENTRAL-RIDGE SANDSTONE	PLANAR-LAMINATED SANDSTONE	HIGH-ENERGY RIDGE-MARGIN SANDSTONE	LOW-ENERGY RIDGE-MARGIN SANDSTONE
LITHOLOGY	Fine- to medium-grained quartzose sandstone, moderately glauconitic; rare siderite clasts and shale rip-up clasts.	Fine- to medium-grained quartzose sandstone	Predominately medium-grained sandstone, abundant shale and limonite rip-up clasts and lenses. Commonly very glauconitic.	Fine-grained sandstone with only rare shale interbeds. Fewer clasts and lenses and less glauconitic than High-Energy Ridge-Margin Facies.
SEDIMENTARY STRUCTURES	Predominantly moderate-angle trough and planar-tangential cross-bedding Trough sets commonly horizontally truncated.	Mostly sub-horizontal plane-parallel laminated sandstone. 0.5-thick laminasets. Minor shale and sandstone ripples	Mostly moderate angle troughs. Some current ripples, shale clasts rarely show preferred orientations	Sequences of several beds of troughs interbedded with sequences of several rippled beds.
BURROWING	Sparse	Sparse	Sparse	Sparse
RESERVOIR POTENTIAL	Excellent	Limited?	Good	Moderate to Good
SUBSURFACE OCCURENCES HARTZOG DRAW FIELD	Common	Very uncommon	Common	Common

	INTER-RIDGE FACIES (SHALEY)	INTER-RIDGE SANDSTONE	BIOTURBATED SANDSTONE	BIOTURBATED SILTSTONE	SILTY SHALE
LITHOLOGY	Thinly interbedded fine- to very fine-grained silty sandstone and silty shale. Slightly glauconitic.	Fine-grained sandstone. Virtual absence of silty shale. Slightly glauconitic	Silty, fine-grained sandstone. Up to 15% shale. Primarily associated with burrows. Slightly glauconitic.	Shaly, slightly sandy dark gray siltstone. Traces to moderate amounts of glauconite.	Silty dark gray shale; rare thin (1/8" thick) silty sandstone lenses.
SEDIMENTARY STRUCTURES	Predominantly horizontal ripple-form bedding surfaces marked by interbedded shales. Trace of wave ripples; current ripples predominate	Predominantly horizontal ripple-form bedding surfaces. Bedding commonly indistinct. Trace of wave ripples; current ripples predominate.	Few physical structures preserved. Mottled appearance. Some ripple-form horizontal beds up to 8" thick. Trace of distinct ripples and small troughs.	Few physical structures preserved. Scattered thin rippled sand and horizontal laminasets. Bedding commonly destroyed.	Common sub-horizontal laminae. Bedding surfaces indistinct. horizontal. Rare current ripples.
BURROWING	Moderate to locally high	Low to moderate	Mottled to distinctly burrowed. More than 75% burrowed.	More than 75% burrowed	Low to moderate
RESERVOIR POTENTIAL	Limited	Limited	Limited	None	None
SUBSURFACE OCCURENCES HARTZOG DRAW FIELD	Common	Uncommon	Common	Moderately common	Common

(Fig. 19). The presence of small faults with less than one km spacing and offset of less than 10 m also contributes to the difficulty of correlating outcrop sections.

Transport directions in outcrop and in oriented cores from Hartzog Draw Field were determined from moderate to high-angle cross beds (mostly dunes or troughs) and indicate a consistent south-southwest transport direction for currents that deposited the higher energy cross-stratified facies. A mean transport direction of 188° was determined for outcrop measurements and several oriented cores (Fig. 16). The overall range of variation in transport directions at any one location is remarkably small (usually less than 60°). Many current ripples in the Inter-Ridge Sandstone also indicate a southerly transport direction, although ripples are much less

Table 2.—Comparisons of Shannon Sandstone facies nomenclatures from 1976 to 1997. Nomenclature for this paper is in the left-hand column. For explanations of varying facies nomenclatures, refer to original papers. Note that change from Tillman and Martinsen (1984) to Tillman (1997) include changes from "bar" to "ridge" and elimination of the term "shelf" from facies names. The latter change is made to allow for the possibility that the ridges were deposited in broad estuaries.

SHANNON SANDSTONE FACIES NOMENCLATURE, 1976-1995

Tillman 1997	Bergman & Walker, 1995	Tillman & Martinsen, 1984	Ranganathan & Tye, 1986	Spearing 1976
CENTRAL-RIDGE SANDSTONE	F6, Cross-bedded Sandstone	Central Bar Facies	Facies A (and C)	Cross-bedded Sandstone Facies
PLANAR-LAMINATED SANDSTONE		Central Bar (Planar Laminated Facies)	Facies C	
HIGH-ENERGY RIDGE-MARGIN SANDSTONE	F4, Glauconitic Med. to Coarse SS	Bar Margin Facies (Type 1)	Facies B	
LOW-ENERGY RIDGE-MARGIN SANDSTONE		Bar Margin Facies (Type 2)	Facies C and A	
INTER-RIDGE FACIES	F5, Thin bedded SS	Interbar Facies	Facies D	Ripple-bedded Sandstone Facies
BIOTURBATED SANDSTONE	Coarse Bioturbated Sandy Mudstone	Bioturbated Shelf-Sandstone	Facies E	

reliable for determining paleocurrent directions (Fig. 16C). Flow directions that have other than a southerly component are rare in the Shannon and commonly only occur at the top of a bed sequence.

In the subsurface at Hartzog Draw Field the steeper eastern side of the ridge is assumed to be the result of storms preferentially attacking the eastern side and causing erosion. Much of the eroded material was almost simultaneously reincorporated into the sand ridge as it built upward. The ridge was assumed to have been a positive feature on the seafloor with some possibly contemporaneous shale deposition around the edges of the sand ridge. Any cross-ridge stratigraphic section will generally show a decrease in grain size from east to west and a coarsening from the bottom toward the top. The mean grain size changes occur at facies boundaries rather than within facies. Since only a trace of sandstones in the Shannon exhibit wave or combined-flow sedimentary features (SCS, HCS, planar lamination, symmetrical ripples) it was assumed that the ridges formed below fair-weather wave base.

A summary of the observations made during 1975-1987 and considered to be important by the author in developing the sand-ridge model hypothesis are listed below:

1. Regional southeast-northwest correlations along the west flank of the Powder River Basin fit Asquith's (1970) model, which places the Shannon landward of the shelf-slope break and a long distance away from any preserved contemporaneous shoreline.
2. A large delta system, the Eagle Sandstone, contemporaneous with the Shannon was located in southern Montana approximately 320 km to the northwest.
3. Shannon sand ridges are asymmetrical sand bodies (steeper on the northeast side) with no preserved shorelines or back barrier deposits located within tens of km west of the ridges.
4. More than 30 m of accommodation space is required to allow deposition of the accumulation typical of ridges such as those that form Hartzog Draw Field and Shannon Sandstone outcrops in the Powder River Basin.
5. The highest energy Shannon deposits were concentrated on the "seaward side" of the ridges. In a typical shoreface, the seaward side would be in deeper water and hence consist of lower energy deposits.
6. Quiet water fissile shales always overlie and underlie the Shannon ridges, which implies that the environment immediately before and following ridge formation was either relatively deep or protected.
7. Detrital components of the sand ridge sandy facies are present as numerous thin "lenses" in the dark-gray, fissile marine shales on the east side of the ridge (Diamond Shamrock 14-4 core; arrowed well on southeast corner of Hartzog Draw Field, Fig. 13), suggesting that the shales were at least in part contemporaneous with ridge deposition. See Tillman and Martinsen (1987, Fig. 19).
8. Trace fossils in the Shannon are unlike most Cretaceous shoreface deposits. Only a very few trace fossil individu-

Table 3.—Quantitative summary of Shannon Sandstone facies characteristics. Numbers are mean percentages (by volume) from all outcrop units observed in measured sections in the Salt Creek Field area. Under "Sandstone" mean grain sizes are given in parentheses. Grain sizes are in microns; 200 microns is equivalent to 0.2 mm. Values under "Reworked" include shale percentages plus other reworked material, primarily siderite clasts. The generally coarsening-upward vertical sequence observable in the table is typical of most of the Shannon, and almost always is present in the central part of the ridge that forms Hartzog Draw Field. Note that in this table the percentages of sandstone, mean grain size, and cross laminae (troughs) generally increase upward, while the percentage of ripples and total shale decrease upward.

COMPARISON OF FEATURES OF SHANNON FACIES

	Sand-stone (%)	Glauconite (%)	Cross Laminae (%)	Ripples (%)	Shale		Reworked	
					Total (%)	Clasts (%)	Total (%)	Clasts (%)
CENTRAL-RIDGE SANDSTONE	92 (200μ)	8	66	10	5	3	7	7
HIGH-ENERGY RIDGE-MARGIN SANDSTONE	75 (225μ)	16	47	11	22	15	21	17
LOW-ENERGY RIDGE-MARGIN SANDSTONE	64 (175μ)	15	25	44	30	6	14	14
INTER-RIDGE FACIES (SHALEY)	58 (150μ)	3	3	61	31	tr	4	tr

als that are typical of shorefaces are observed in the Shannon. Instead, the common Shannon trace fossils include *Terebellina*, *Paleophycus*, *Conichnus*, and *Thalassinoides*. Trace fossils common in most Cretaceous shoreface sandstones include *Ophiomorpha*, *Asterosoma*, *Rosselia*, *Rhizocorallium*, and among these only *Ophiomorpha* is sparsely present.

9. Conclusions generated from the foraminiferal data collected primarily from shales immediately above, below, and between the vertically stacked Shannon outcropping sand ridges were: 1) the forams present are mid- to outer-shelf species, and 2) the forams typical of Cretaceous shoreface deposits are absent.
10. Medium- to coarse-grained Shannon sandstone deposits, which form the bulk of the reservoirs in the subsurface, are almost all cross-stratified current deposits rather than wave or combined flow deposits.
11. No vertical and lateral facies changes like those in the Shannon (and Sussex) have been described for any other environment in the literature. The typical vertical sequence, from bottom to top, is marine shale overlain by bioturbated sandstone or siltstone that is overlain by horizontally-bedded, current-modified, wave-rippleform bedded sandstone, which in turn are overlain by up to 25 m of horizontally bedded and internally cross-stratified sandstone capped by bioturbated siltstone or marine shales. Lateral sequences commonly included shale on one side of the sand ridge, and on the other side extensive ripple-bedded sandstone, which apparently extends all the way to the next ridge.
12. The thinly bedded, rippled, and in some places shaley, Inter-Ridge Facies (Fig. 11; platform facies of Spearing, 1976) is continuous between some ridges.
13. Vertical sequences and facies of the Shannon differ significantly from classical Galveston Island and Gallup Sandstone shoreface deposits.
14. Glauconite accumulation (locally 30-50%) in some of the cross-stratified Shannon High-Energy Ridge-Margin Sandstone is significantly greater than in most shoreface deposits.
15. Current flow directions in the Shannon are almost invariably south-southwest. At any one location most of the flow directions vary no more than 30 degrees on either side of the mean flow direction, and mean flow directions also vary little from location to location.

A number of problems exist with the shelf origin for the Shannon. Some of these problems are listed below:

1. There is no nearby source for the Shannon; transport distances, based on the location of preserved deltas and shorefaces, are hundreds of km away.
2. We have no evidence of whether shorelines that were present closer to the sand ridges were removed by ravinement, nor do we know what their orientations might have been.

Fig. 6.—High-angle planar, cross-laminated sandstone in Central-Ridge Sandstone. Dark bands are glauconite. These units are essentially shale-free except for an occasional rip-up clast. All core photos in this paper are 10 cm wide and from Hartzog Draw Field. (CSOG AE-1, 9158').

Fig. 8.—Composite, rounded siderite clast consisting of numerous smaller, rounded siderite clasts. Dark colored intervals are laminae with very high concentrations of glauconite. A planar cross-laminated sandstone in High-Energy Ridge-Margin Sandstone. (Southland B.C. No. 2, 9164').

Fig. 7.—Planar-tabular dune indicating southerly flow of strong currents. Dunes such as this occur only rarely in the Shannon and at this location there is no systematic variation in laminaset thickness or angle of dip of laminae. Here it is interpreted to be a High-Energy Ridge-Margin Sandstone near the top of the lower Shannon at Measured Section W2B.

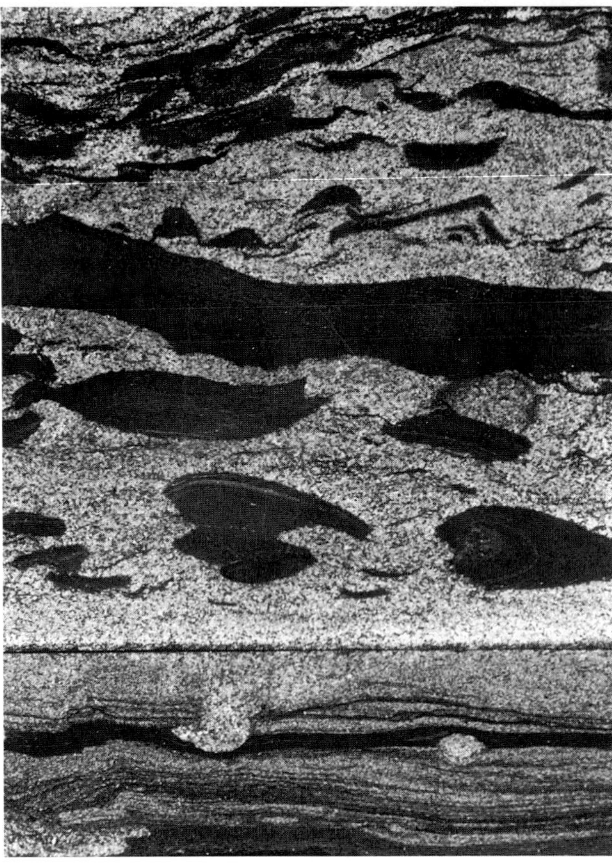

Fig. 9.—Angular shale rip-up clasts and thin-bedded, locally lenticular" shale beds typical of some cross-laminated High-Energy Ridge-Margin Sandstones. Angularity of clasts suggests very short distance of transport. (Southland B.C. No. 2, 9182.8').

3. There is no obvious reason for deposition and preservation of such a thick convex-up pile of sand on a relatively flat shelf.
4. It is unclear why the sand ridges didn't build laterally once some topographic relief was established.
5. Local relief on the sea bottom, suggested as the cause of winnowing, is less than 3 m over tens of km in a north-south direction under both the Salt Creek area and Hartzog Draw Field.
6. Patterns of shale accumulation within the sand ridge and adjacent to it are complex. The reasons for the quiet periods allowing for shale and siderite beds to accumulate in the High-Energy Ridge-Margin Sandstone and locally in the Central-Ridge Sandstone are unclear.
7. The significance of the two types of shales deposited lateral to the sand ridge and described by Tillman and Martinsen (1987) cannot be easily explained. And the degree of contemporaneity of these shales with the sand ridge cannot be conclusively demonstrated.
8. The cause of the variation of the Inter-Ridge Sandstone from shaley in Hartzog Draw and to essentially nonshaley in the outcrop is unknown.
9. Why the few modern sand ridges that have been studied using cores and shallow seismic show so few affinities to the Shannon, other than gross shape, is not understood.

SHOREFACE MODEL

Several elongate sand-ridge oil fields are parallel to Hartzog Draw Field (Fig. 22). A cross-section constructed by Tillman and Martinsen (1987) illustrates the lithologic correlation among Shannon producing fields (Fig. 23). These correlations emphasize the rise and fall of the sandy facies of the Shannon over an east-west distance of about 24 km.

Bergman (1994) has proposed a shoreface origin for these same Shannon fields. Tillman's correlations, when contrasted with correlations by Bergman (Fig. 22) using mostly the same cores and wells, illustrates the differences

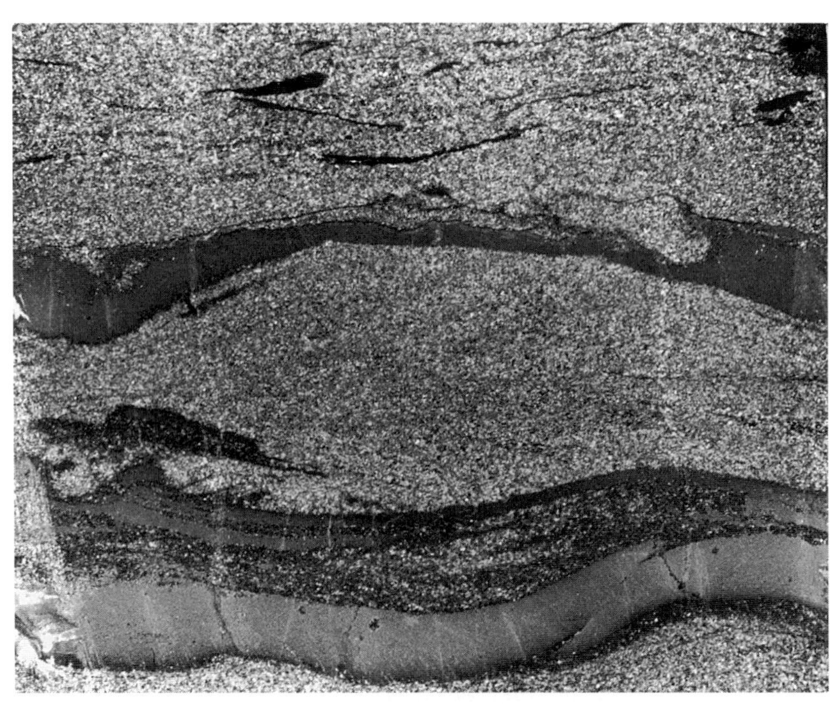

Fig. 10.—Beds of siderite and shale in High-Energy Ridge-Margin Sandstone. Where siderite and shale beds such as these are observed in outcrop they are lenticular and drape over smooth to rough underlying surfaces. Note fractures in siderite bed, which are partially filled with glauconitic sandstone from the immediately overlying bed. This relationship suggests at least moderate induration of the siderite bed prior to deposition of the overlying sandstone. (Southland B. C. No. 2, 9179.2').

Fig. 11.—Typical horizontally bedded, ripple-form bedded Inter-Ridge Sandstone. Note that little, if any, shale is interbedded with rippled sandstone. (87-101′ on Measured Section W2A, Fig. 20).

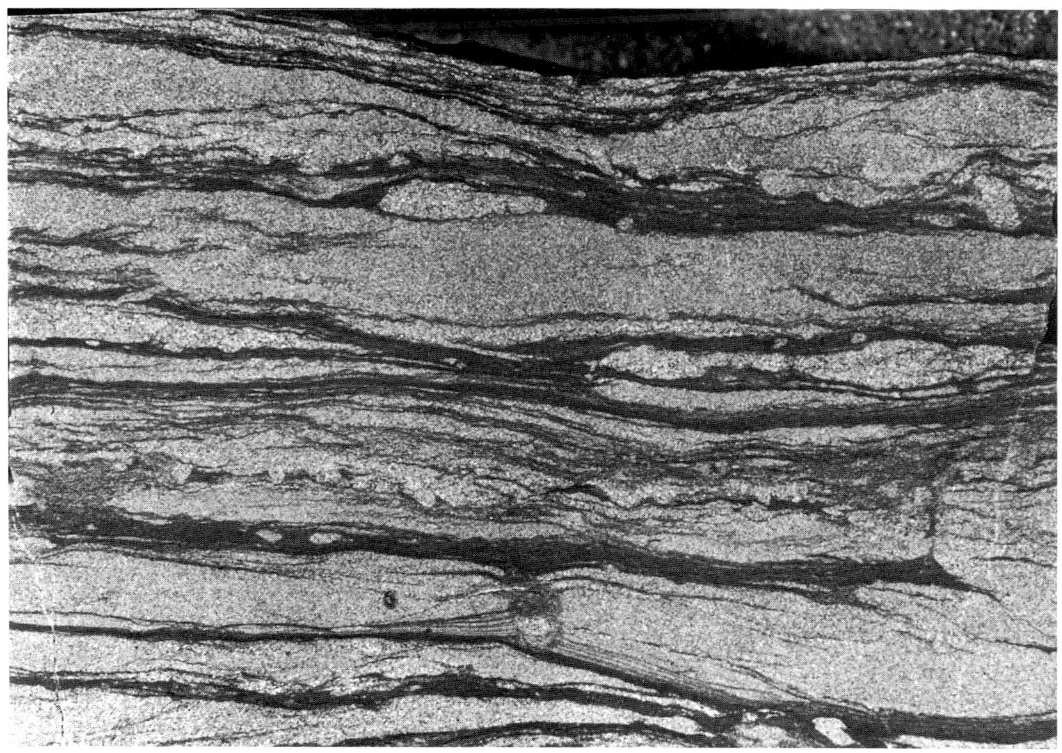

Fig. 12.—Thin, fine-grained, rippled sandstone interbedded with silty shale. A typical subsurface Inter-Ridge (Shaley) Facies. In outcrops in the Salt Creek Field area this facies is replaced by the Inter-Ridge Sandstone, which contains essentially no shale. Width of core is 4 in. (CSOG AB-1A, 9298').

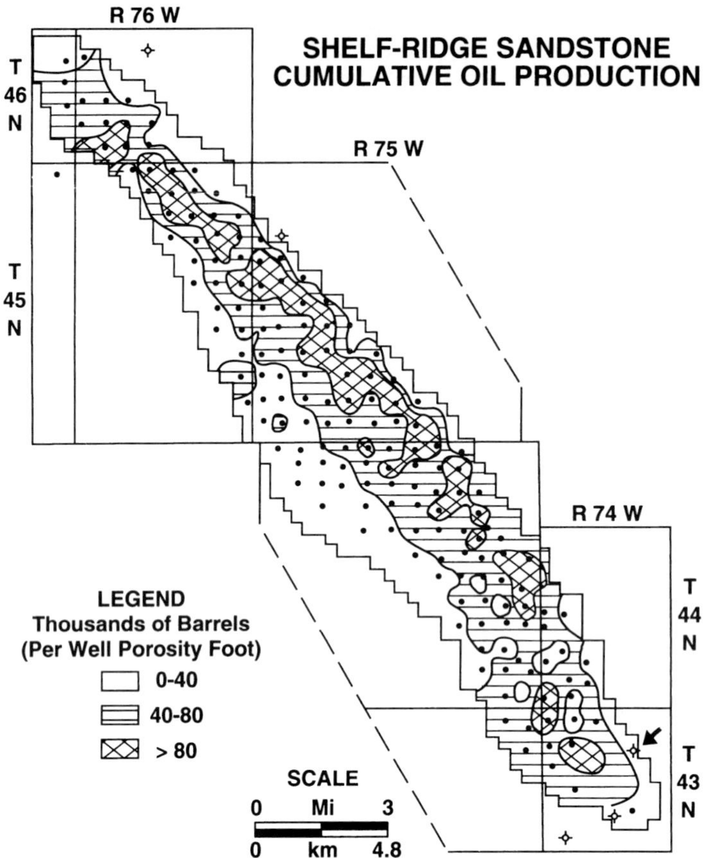

Fig. 13.—Cumulative primary oil production map (1984) for Hartzog Draw Field. Note strong asymmetry of production with highest values restricted to the northeast margin and crestal areas of the ridge. Production, on a per-well basis, along the east margin is more than twice as much as along the west margin of field. Modified from Hearn et al. (1984).

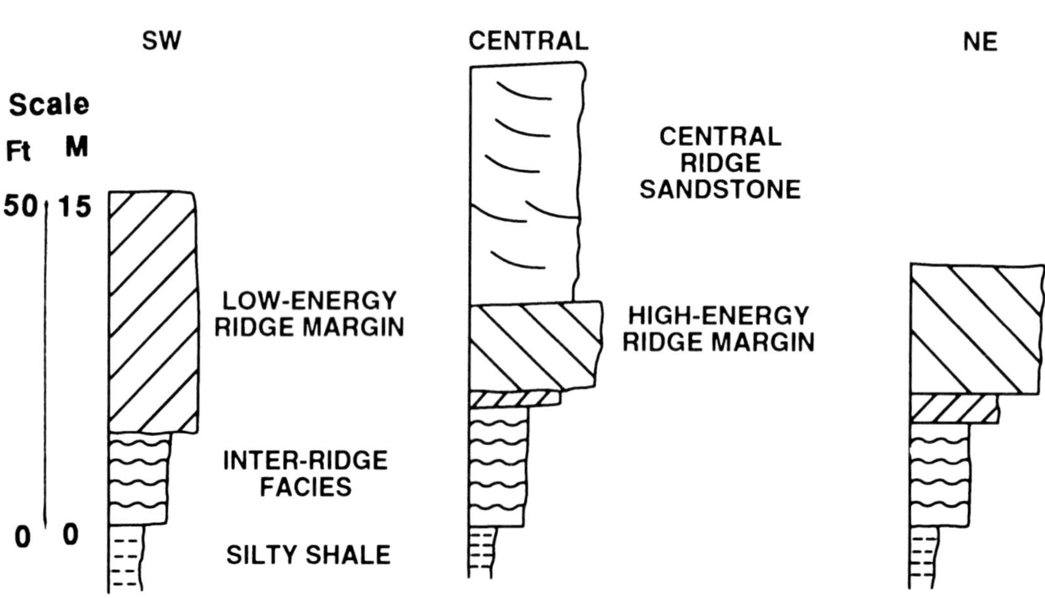

Fig. 14.—Typical lateral and vertical cross-ridge (perpendicular to elongation) facies associations for the three significantly different portions of Shannon ridges. The lateral variation of cross-bedded facies is apparent. Stratigraphic profiles include outcrop and oilfield examples; southwest-outcrop Measured Section W2A, Central — CSOG AK-1 core, Hartzog Draw Field, northeast — CSOG AS-1 core, Hartzog Draw Field.

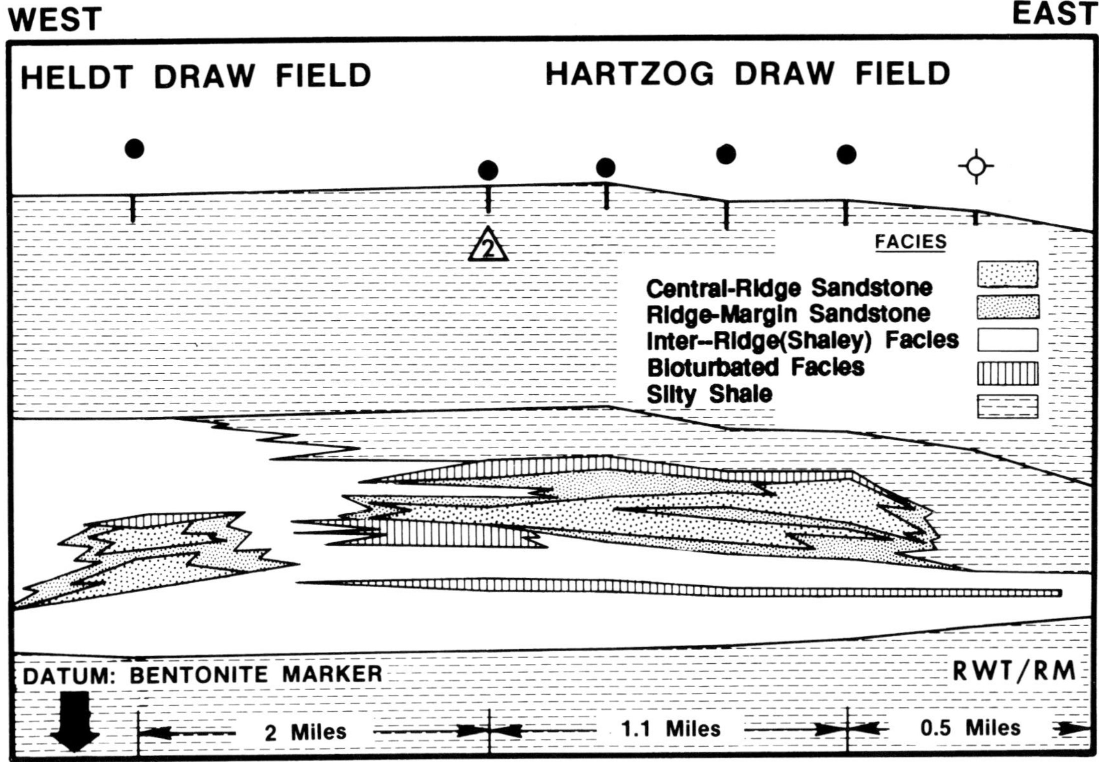

Fig. 15.—Sand-ridge facies distribution, Hartzog Draw and Heldt Draw Fields. Note that Hartzog Draw Field has a steep east flank with adjacent silty shales and a more gentle western flank. Note also that the Inter-Ridge (Shaley) Facies is inferred to be continuous between the two fields.

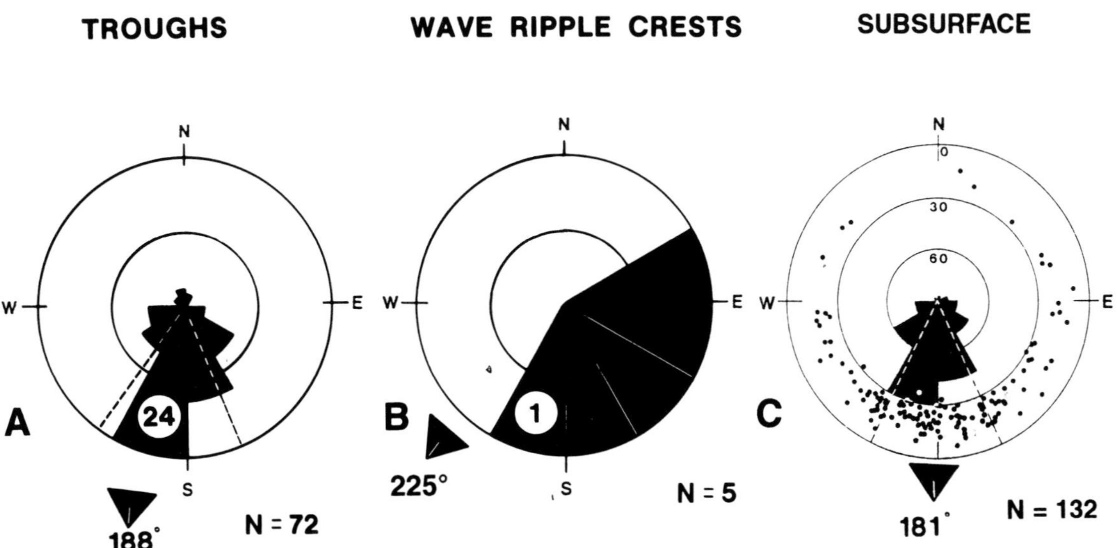

Fig. 16.—Directional current-flow data from three Measured Sections (A and B) from Shannon outcrops and from a single oriented core from Hartzog Draw Field (C). Mean flow directions are indicated by arrows and the number of samples is N. Numbers in white circles indicate the number of data points represented by the longest radius part of the rose diagram. The dashed lines encompass one standard deviation.

Table 4.—Shannon sand-ridge facies reservoir properties in Hartzog Draw Field. $\phi\bar{x}$ is mean porosity, $K\bar{x}$ is mean permeability and ϕmin, ϕmax and Kmin and Kmax are the minimum and maximum values, of porosity and permeability of the facies observed.

RESERVOIR PROPERTIES

FACIES	$\phi\bar{x}$ (%)	$K\bar{x}$ (md)	ϕ min (%)	ϕ max (%)	K min (md)	K max (md)
Central-Ridge	15.3	14.6	11.9	17.4	2.8	40.0
High-Energy Ridge-Margin	12.9	18.8	8.0	14.7	0.8	31.3
Low-Energy Ridge-Margin	11.0	5.0	5.5	12.5	0.2	13.4
Inter-Ridge (Shaley)	5.2	1.2	1.8	7.7	0.01	2.2
Burrowed Shelf-Siltstone	6.7	1.9	5.4	7.4	0.02	5.4

Fig. 17.—Apparently conformable lower contact (arrow) of Shannon sand ridge south of Measured Section W2A. Compare with sharp contact in Figure 18. A man is standing on one of the bentonites that form excellent markers below the Shannon in the area of Salt Creek Field.

Fig. 18.—Sharp lower contact of Shannon (arrow). Contrast this contact with apparent conformable basal sequence in Fig. 17. Below contact is Bioturbated Siltstone, which typically occurs 1) below Shannon Sandstone, 2) between stacked Shannon sand-ridge sandstones, and 3) lateral to some of the ridges. Interval between arrows is typical horizontally bedded, highly rippled Inter-Ridge Sandstone and above upper arrow is High-Energy Ridge-Margin Sandstone. (Measured Section W2B).

between the sand ridge model and the lowstand-shoreface model. The Bergman cross-section was designated to be a sequence stratigraphic "template." The locations of the cross-sections are shown in Figure 22.

According to K. M. Bergman (pers. commun., 1996) the key cores supporting the shoreface model are located primarily on the east side of Hartzog Draw Field.

There are a number of factors that are unexplained by a lowstand shoreface origin for the Shannon.

1. The Shannon sandstone facies are predominantly current deposits. Lowstand shorefaces should have a significant proportion of wave or combined-flow deposits; less than 5% of the Shannon sandstones are interpreted by Tillman to be wave influenced or combined flow deposits.
2. There is no evidence of soils or other exposure surfaces at a position near the base of, or below, the Shannon that suggests regressive surfaces of erosion (RSE).
3. Typical Cretaceous shoreface trace fossils such as *Ophiomorpha* are rare. *Asterosoma*, *Rosselia*, and *Rhizocorallium* are almost completely absent. Instead, a suite of trace fossils not usually found in shorefaces is observed including *Terebellina*, *Conichnus*, *Paleophycus*, and *Thalassinoides*.
4. Vertical stacking of nearly 30 m of relatively coarse-grained sediment typical of some of the larger Shannon sand ridges cannot be accomplished without significant accommodation space. Accommodation space allowing stacking of nearly 25 m of cross-statified sandstone at a shoreline would require very special conditions such as faulting or sea-level rise in which there was no lateral movement of the shoreline during an extended period of time.
5. Sandstones nearly contemporaneous with the Shannon and acknowledged by all to be shorefaces occur in the Big Horn Basin (Bergman and Walker, 1995). The sedimentary structures and overall appearance of these sandstones is distinctly different from those observed in the Shannon.

INCISED VALLEY MODEL

Sullivan et al. (1997) recently published an interpretation of the Shannon Sandstone in which they attribute the Shannon at Hartzog Draw Field to be remnants of three sequences. Erosion at the sequence boundaries is inferred to have created the topographic relief at the tops of the ridges. In addition, they conclude that tidal processes are primarily responsible for deposition of the Shannon and have subdivided the Shannon into two major facies, "Distal Tidal Bars" and "Proximal Tidal Bars" (Fig. 24). They also subdivide their bar facies into numbered subfacies (Table 5). Their grouping of rock types into subfacies is very similar to the facies of Tillman and Martinsen (1984, 1987) and those described in this paper. Facies, which apparently are equivalent in Tillman's and Sullivan's facies classifications, and the attributes for the "Tidal Bar Facies" are summarized in Table 5. The attributes of the sand-ridge model facies are summarized in Table 1. The sand ridge and incised valley-fill models differ in many respects, but also have some features in common.

A stratigraphic interpretation of the three sequences that are inferred by Sullivan to incise at the base of and within the Shannon at Hartzog Draw Field are shown in Figure 25. The inferred incision (Sequence Boundary) at the base of the Shannon is difficult to observe at most outcrop locations. In the subsurface it is not easily recognizable and commonly

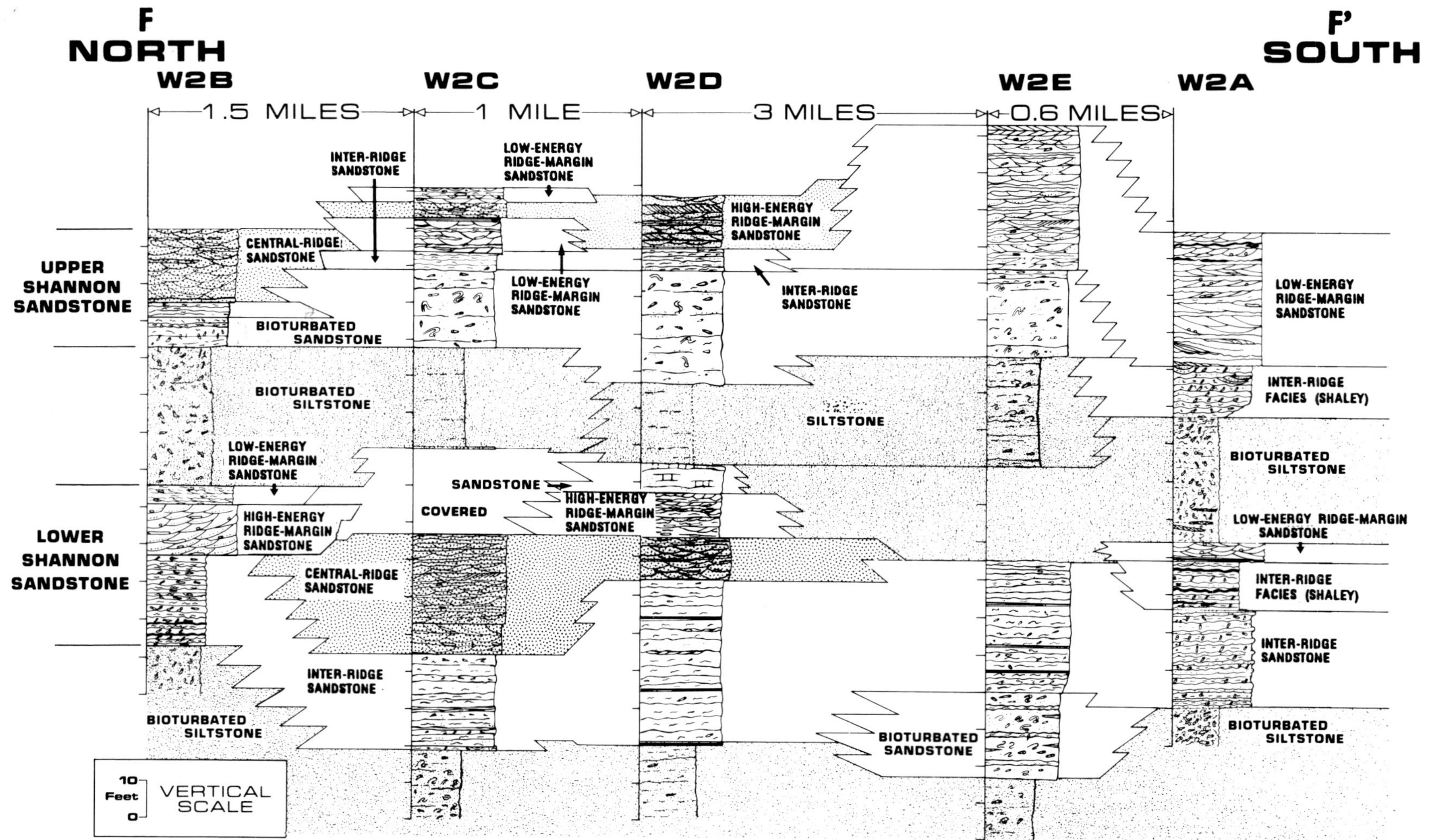

Fig. 19.—North-south stratigraphic outcrop cross-section F-F′. This outcrop section is located along the east side of Salt Creek Field (Fig. 4). Picking a marker datum for construction of this figure was particularly difficult since many of the sections are separated by small faults (not shown). If this section were to be constructed again the top or base of the siltstone separating the upper and lower Shannon sandstones might be better candidates for a datum. Note that the Central Ridge Sandstone is absent in the two more southerly sections. Subsurface mapping indicates that the cross-section is oblique to the trend of two ridges (Lower Shannon) and may cross portions of two ridges; see Figure 38, Tillman and Martinsen (1984). Modified from Tillman and Martinsen (1984).

appears to have a relief of less than 2 m in the areas surrounding Hartzog Draw Field and the Salt Creek Anticline. However, R. S. Martinsen (pers. commun., 1997) has observed significant incision at, or near, the base of the Shannon (top of the Cody Shale) north of the Salt Creek outcrop area.

Two of the sequences interpreted by Sullivan dissect the sand ridge. Significant incisions are inferred within the Shannon at the base of the "Crimson Red" and the "Canary Yellow" sequence boundaries. The Crimson Red sequence boundary produces most of the relief on the Shannon sandstones, while the Canary Yellow sequence boundary is incised only a little at Hartzog Draw. However, at Pumpkin Buttes Field the Canary Yellow incision is very deep and the whole of the Shannon is inferred to be within the Canary Yellow sequence (Fig. 25). Erosion associated with the Crimson Red sequence is inferred to have removed the entire interval around all the edges of Hartzog Draw Field.

A summary of reasons for locating the sequence boundaries at Hartzog Draw Field seems to be:

1. accumulation of clasts, in cores, immediately above a significant surface,
2. occurrence, in an organized fashion, of "Proximal Tidal Facies" sharply overlying "Distal Bar Facies,"
3. intersection of numerous longitudinally downlapping surfaces against a single surface, and
4. the need for an explanation for the progressively greater removal or thinning of reservoir facies to the north in the Crimson Red sequence.

The tidal bar interpretation by Sullivan et al. (1997) was based primarily on cores from Hartzog Draw Field. They defined the bars as having an upward coarsening series of facies. I believe that much "upward coarsening" is due to the decrease upward in the amount of shale present as clasts, drapes, and beds.

All of the rippled Inter-Ridge Sandstone and some of the Bioturbated Siltstone and Sandstone of Tillman (this paper) would probably be included in the distal tidal bar facies. At some locations the tops of the Distal Bar Facies would be intervals of interbedded rippled and cross-stratified beds (Low-Energy Ridge-Margin Sandstone) that do contain fine- to medium-grained sandstone. The relationships of the series of units that are in the Distal Tidal Bar can be seen by comparing Figures 24 and 26.

The Proximal Tidal Bar facies apparently includes all the Central-Ridge Sandstone and most of the High-Energy Ridge-Margin Sandstone of Tillman. These are all cross-stratified facies with only very limited amounts of clay and burrowing.

Sullivan et al.'s detailed well-log cross-section parallel to the ridge elongation displays strongly downlapping features that they attribute to the accretion of tidal bars (Fig. 27). They infer that the Proximal Tidal Bars grade downward and generally southward through a series of facies into Distal Tidal Bars.

This model stems in part from Pemberton's (1992) trace fossil work on the Shannon cores from Hartzog Draw Field. He concluded that the Shannon trace fossils were not typical of those found in offshore marine environments. However, the shelf trace fossil suite for the Cretaceous is still insufficiently defined.

If the incisions as described in this model are as broad as designated (up to tens of kilometers), they probably formed open bays at times. These bays would have accumulated significant amounts of mudstone and very fine-grained sandstone (Inter-Ridge Sandstone) lateral to the thick accumulations of coarser sandstone that forms the oil fields. The Inter-Ridge Sandstone accumulation is observed in cores in wells between the fields and is shown in Figure 15 between Hartzog Draw and Heldt Draw Fields.

A contrasting interpretation of Hartzog Draw Field as a shelf sand-ridge that is similarly oriented, but shorter than Sullivan's is shown in a cross section by Tye et al. (1986) (Fig. 28). The overall external geometry of the sand-ridge cross-section is similar to the incised model (Fig. 25), but the internal stratigraphy is very different. In both models the upper sandstone is separated from the lower sandstone by a shaley, nonproductive interval.

CONCLUSIONS

The Shannon Sandstone has few well-studied analogs, and the processes of deposition of the Shannon are still being debated. Outcrop and subsurface studies all indicated a marine to brackish origin for the elongate sand bodies so common in the Powder River Basin of Wyoming.

Studies of the Shannon prior to the mid-1980s all interpreted the Shannon as a series of marine offshore (or nearshore) sand bodies (ridges). However, with the advent of sequence stratigraphy and more knowledge about the importance of erosional incisions, other models were proposed. The incised lowstand shoreface model was proposed by Walker and Bergman (1993), Bergman (1994), Bergman and Walker (1995) and an incised-valley model was proposed by Sullivan et al. (1997). Proponents of all three models met in 1995 in the Powder River Basin to jointly examine the Shannon outcrops and discuss their origin. At that conference the participants failed to reach a consensus on the origin of the sand ridges. Participants at the conference were, however able to better appreciate the points of views inherent in each of the models.

This author was convinced that tides as well as storms were probably important in deposition of the Shannon sand ridges.

There seems to be agreement that evidence from interpretation of foraminifera below, between the lower and upper Shannon, and immediately above the Shannon indicates an offshore marine origin for these shales. However, considering what is known about sea-level movements in the Cretaceous it is possible that there were significant and rapid shoreline movements seaward just before deposition of both the lower and upper Shannon sand-ridges. All three of the models summarized herein incorporate a flooding surface at the top of the Shannon.

This author favors an open-bay (possibly estuarine) depositional model for the Shannon in which both tides and storms were active. It still seems probable that the source of the Shannon was the Eagle Sandstone delta in southern Montana. Whether the predominantly southerly flowing currents in the Shannon were primarily storm or tidally driven or if they resulted from combined tidal-storm flow is unclear. The near absence of other than southerly oriented flow features in the Shannon casts some doubt on a purely tidal origin for the Shannon. Detailed outcrop work also fails to show a significant number of the features listed by Sullivan et al. (1997) as occurring in cores of the Shannon as evidence of tidal origin. The presence of broad areas of horizontally bedded, ripple-formed, very fine-grained sandstone (Inter-Ridge [Shaley] Facies) between many of the thicker accumulations (ridges) is much easier to explain in an open-bay or estuarine setting.

MEASURED SECTION W2A
SHANNON SANDSTONE, SALT CREEK SANDSTONE, CASTLE ROCKS,
SW NW SEC 8 T39N R78W, NATRONA CO.,WYOMING

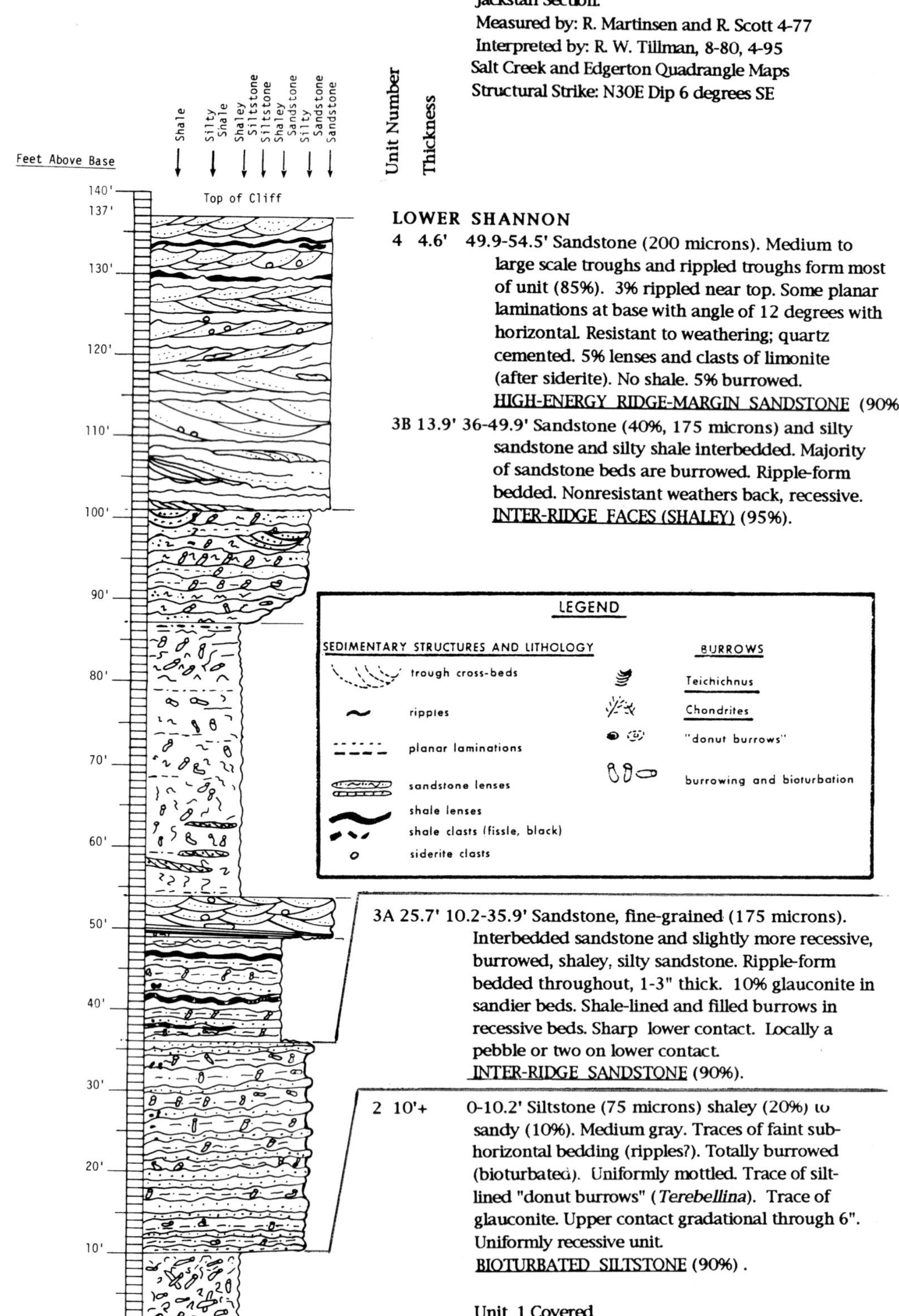

Fig. 20.— Measured Section W2A, which is typical of the Shannon in the Salt Creek Field area. Note that the stratigraphic section interval is the same on both pages, but descriptions apply to different parts of the section.

MEASURED SECTION W2A
SHANNON SANDSTONE, SALT CREEK SANDSTONE, CASTLE ROCKS,
SW NW SEC 8 T39N R78W, NATRONA CO.,WYOMING

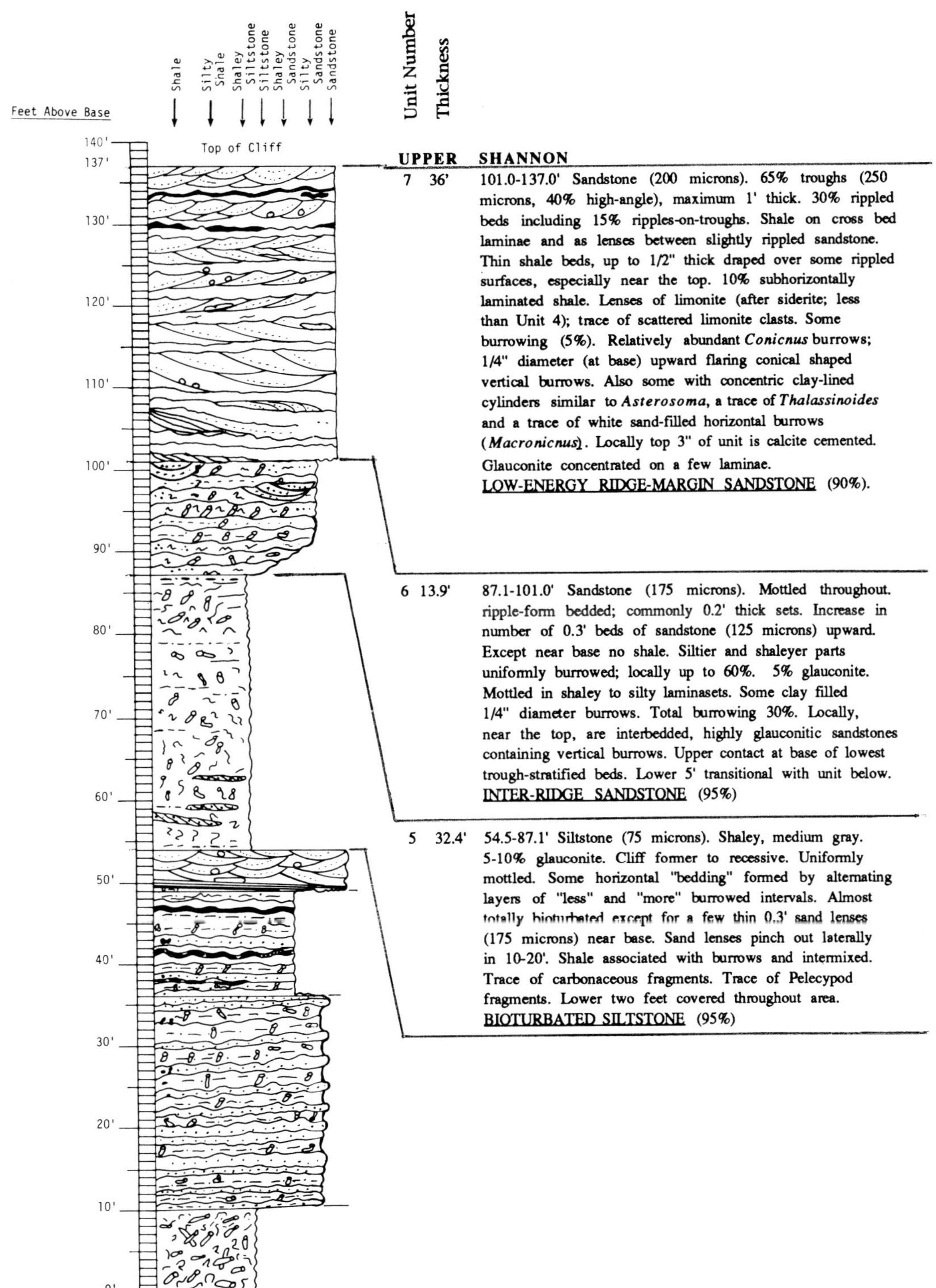

Fig. 20.—(continued).

Fig. 21.—Two stacked Shannon ridges. Arrows bracket Bioturbated Siltstone, which vertically separates the two ridges. The bases and tops of the ridges are horizontal on a local scale, as are most of the beds that form the ridge. Foreground is interbedded rippled and cross-laminated Low-Energy Ridge-Margin Sandstone. View is looking southwest from top of Measured Section W2A.

ACKNOWLEDGEMENTS

Work on the Shannon in the 1970s and early 1980s was conducted in association with Randi Martinsen and other colleagues at Cities Service. Since that time many geologists have contributed to understanding the vagaries of the Shannon during field trips, field courses, and field conferences. Randi Martinsen, Al Merewether, Bob Dalrymple, and Louise Quinn are thanked for their reviews of an earlier version of this paper. Thanks also to M. D. Sullivan and Exxon for permission to include several figures.

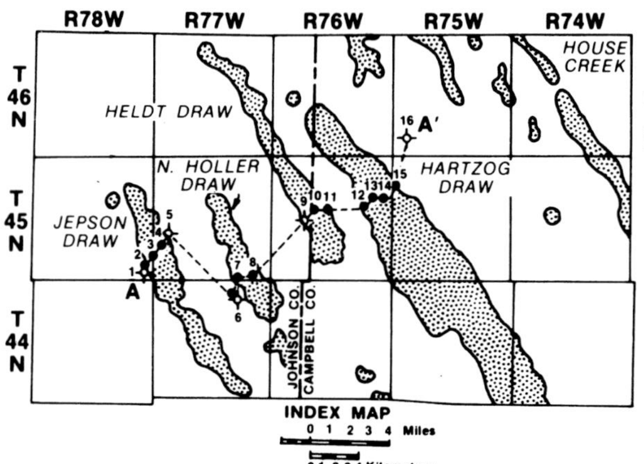

Fig. 22.—Map of Shannon oil fields showing the location of cross-section A-A'. Note northwest-southeast elongation of Shannon oil fields (Fig. 4A).

REFERENCES

Asquith, D.O., 1970, Depositional topography and major marine environments, Late Cretaceous, Wyoming: American Association Petroleum Geologists Bulletin, v. 54, p. 1184-1224.

Berg, R.R., 1975, Depositional environments of the Upper Cretaceous Sussex Sandstone, House Creek Field, Wyoming: American Association of Petroleum Geologists Bulletin, v. 59, p. 2099-2110.

Bergman, K.M., 1994, Shannon Sandstone in Hartzog Draw-Heldt Draw fields (Cretaceous, Wyoming, U.S.A.) reinterpreted as lowstand Shoreface Deposits: Journal of Sedimentary Research, v. B64, p. 184-201.

Bergman, K.M. and Walker, R.G., 1995, High resolution sequence stratigraphic analysis of the Shannon Sandstone in Wyoming using a template for regional correlation: Journal of Sedimentary Research, v. B65, p. 255-264.

Gaynor, G.C. and Swift, D.J.P., 1988, Shannon Sandstone depositional model: Sand ridge dynamics on the Campanian Western Interior Shelf: Journal of Sedimentary Petrology, v. 58, p. 868-880.

Gill, J.R. and Cobban, W.A.,1966, The Red Bird Section of the Upper Cretaceous Pierre Shale in Wyoming: Washington, D.C., United States Geological Survey Professional Paper 393A, 73 p.

Gill, J.R. and Cobban, W.A., 1973, Stratigraphy and Geologic History of the Montana Group and Equivalent Rocks, Montana, Wyoming and North Dakota: Washington D.C., United States Geological Survey Professional Paper 776, 37 p.

Hansley, P.L. and Whitney, C.G., 1990, Petrology, Diagenesis and Sedimentology of the Upper Cretaceous Shannon Sandstone Member of the Steele Shale in the Powder River Basin, Wyoming: Washington, D.C., United States Geological Survey Bulletin 1917C, 33 p.

Hearn, C.L., Ebanks, W.J., Tye, R.J. and Ranganathan, V., 1984, Geological factors influencing reservoir performance of Hartzog Draw field, Wyoming: Journal of Petroleum Technology (August 1984), p. 1335-1344.

Table 5.—Summary of facies, subfacies, lithology, sedimentary structures, and reservoir characteristics of Sullivan et al. (1997) and "Facies equivalents" of Tillman. Modified from Sullivan et al. (1997, Table 1). Facies equivalences are based solely on descriptions in Sullivan et al. (1997) Table 1 and there may be a few rocks in the Shannon that are not 100% "facies equivalent."

Sullivan et. al (1997) Descriptions						Tillman Facies
Depositional Setting	**Subfacies No.**	**Lithofacies**	**Sand/Shale Ratio**	**Porosity (%)**	**Perm. (md)**	
Proximal Tidal Bar	8	Fine to coarse-grained, cross-bedded sandstone w/ rare burrows and common mud rip-up clasts	95-100	13.6	15.9	**High-Energy Ridge-Margin Sandstone**
	7	Fine- to medium- grained, cross-bedded sandstone w/ thin mud drapes and rare burrows	80-90	12.9	13.6	**Central-Ridge Sandstone**
Distal Tidal Bar	6	Very fine- to fine-grained, subhorizontal to low-angle, planar laminated sandstone w/ thin mud drapes and common burrows	80-90	11.0	8.7	**Planar Laminated Sandstone**
	5	Medium to thick-bedded, planar-laminated, very fine- to fine-grained sandstone and interlaminated thin-bedded mudstone w/ common burrows.	70-80	8.5	6.5	**?Low-Energy Ridge-Margin Sandstone**
	4	Thin-bedded, planar-laminated, very fine- to fine-grained sandstone and interlaminated thin-bedded mudstone w/ abundant burrows.	50-60	7.0	3.2	**Inter-Ridge (Shaley) Sandstone**
	3	Intensely burrowed, planar-laminated mudstones with minor interstratified, very fine-grained sandstones.	10-20	7.0	2.4	**Bioturbated Siltstone**
Offshore	2	Planar-laminated mudstones w/ rare interstratified silty sandstones and rare to common burrows.	5-10	5.3	1.7	**Silty Shale**
	I	Planar-laminated mudstones w/ rare burrows	0	3.5	0.5	**Shale**

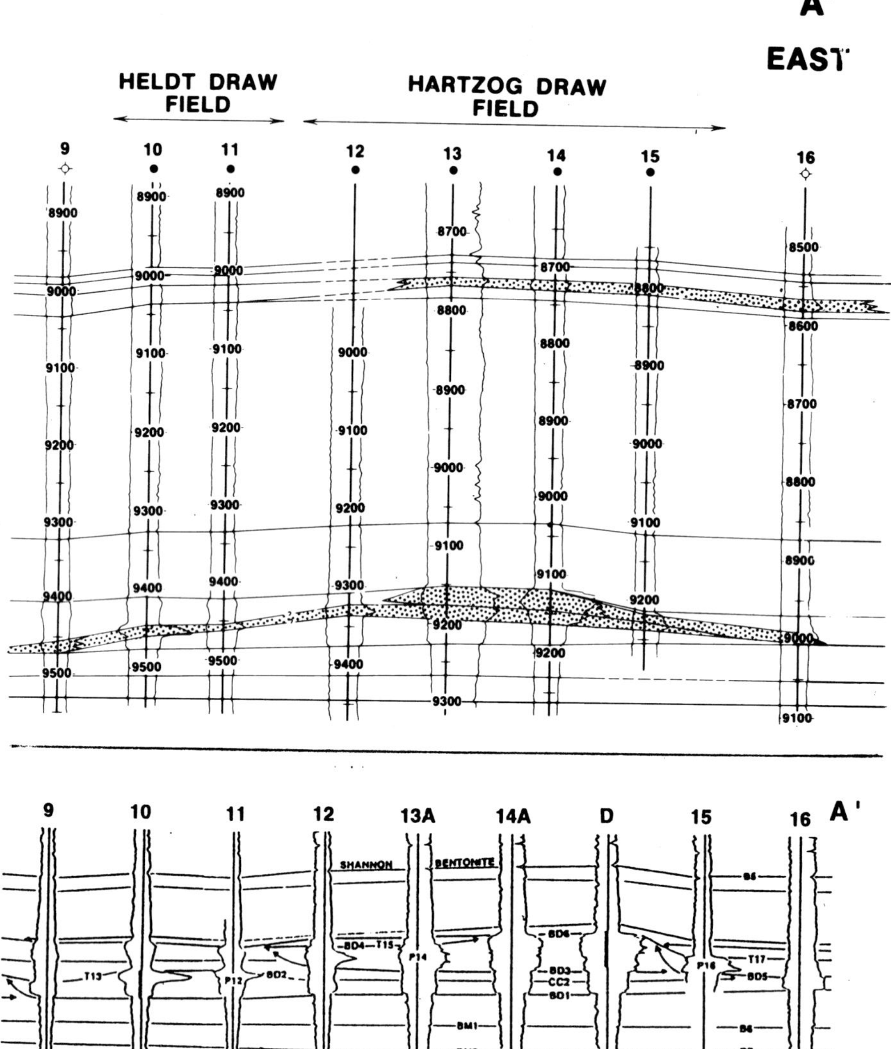

Fig. 23.—Regional west-to-east cross-sections that include four oil fields that produce from the Shannon Sandstone. Upper cross-section, from Tillman and Martinsen (1987), emphasizes correlation of cross-stratified sandstone. Contrast the horizontality of the base-markers 1 and 2 and the Ardmore Bentonite with the stratigraphic rise and fall of the sandstones in the producing fields. The lower cross-section is from Bergman and Walker (1995). Their correlations and interpretations suggest that the sands are deposited in a series of incised lowstand shorefaces. Numbered wells are the same in both cross-sections. Wells A, B, and C were not included in the upper cross-section and wells 13A and 14A lie one well location north of 13 and 14.

HOBSON, J.P., FOWLER, M.L. AND BEAUMONT, E.A., 1982, Depositional and statistical exploration models, Upper Cretaceous offshore sandstone complex, Sussex Member, House Creek Field, Wyoming: American Association of Petroleum Geologists Bulletin, v. 66, p. 689-705.

MARTINSEN, R.S. AND TILLMAN, R.W., 1978, Hartzog Draw, new giant oil field (abstract): American Association of Petroleum Geologist Bulletin, v. 62, p. 540.

MARTINSEN, R.S., 1981, Hartzog Draw, Powder River Basin Oil and Gas Fields, Volume I: Casper, Wyoming, Wyoming Geological Association, p.187, and map.

MEREWETHER, E.A., 1996, Stratigraphy and Tectonic Implications of Upper Cretaceous Rocks in the Powder River Basin, Northeastern Wyoming and Southeastern Montana: Washington, D.C., United States Geological Survey Bulletin 1917-T, 92 p.

OBRADOVICH, J.D., 1993, A Cretaceous time-scale, *in* Caldwell, W.G.E. and Kauffman, E.G., (eds.), Evolution of the Western Interior Basin: St. John's, Geological Association of Canada Special Paper 39, p. 379-396.

PARKER, J.M., 1958, Stratigraphy of the Shannon Member of the Eagle Formation and its relationship to other units in the Montana Group of the Powder River Basin, Wyoming and Montana: Wyoming Geological Association, 13th Annual Field Conference, p. 90-102.

PEMBERTON, S.G., 1992, Applications of Ichnology to Petroleum Exploration, A Core Workshop: Society For Sedimentary Geology (SEPM) Core Workshop No. 17, Calgary, Alberta, Canada, 429 p.

PORTER, K.W., 1976, Marine shelf model, Hygiene Member of the Pierre Shale, Upper Cretaceous Denver Basin, Colorado, *in* Stud-

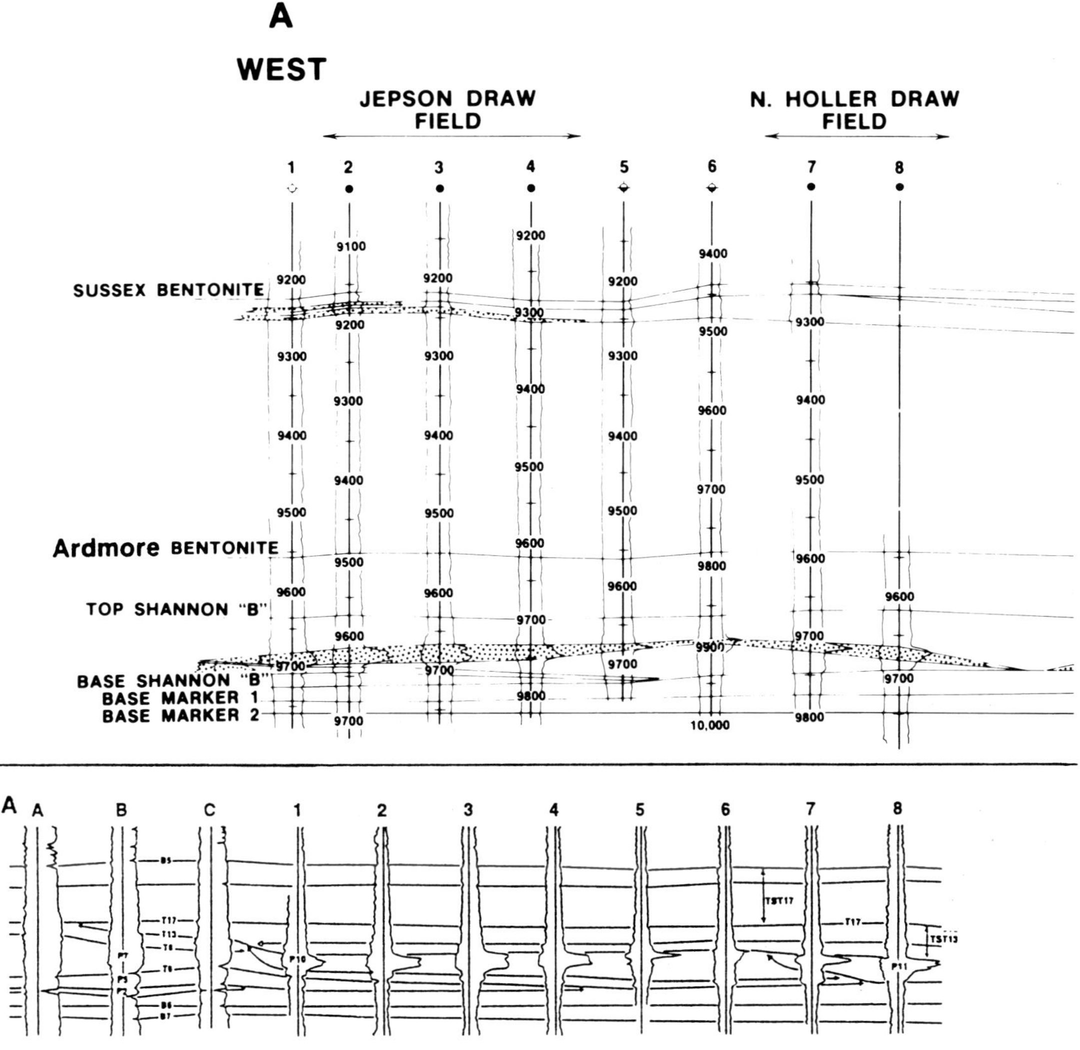

Fig. 23.—(continued).

ies of Colorado Field Geology, Professional Contributions of Colorado School of Mines, Epis, R. and Weimer, R. (eds.) Golden, Colo., p. 251-263.

Ranganathan, V. and Tye, R.W., 1986, Petrography, diagenesis and facies control on porosity in Shannon Sandstone, Hartzog Draw field, Wyoming: American Association of Petroleum Geologists Bulletin, v. 70, p. 56-69.

Seeling, A., 1978, The Shannon Sandstone, a further look at the environment of deposition at Heldt Draw Field, Wyoming: The Mountain Geologist, v. 15, p. 133-144.

Spearing, D.R., 1976, Upper Cretaceous Shannon Sandstone: An offshore, shallow marine sand body: Wyoming Geological Association, 28th Annual Field Conference Casper, Wyoming, September 1976, p. 65-72.

Sullivan, M.D., Van Wagoner, J.C., Jennette, D.C., Foster, M.E., Stuart, R.M., Lovell, R.W. and Pemberton, S.G., 1997, High Resolution Sequence Stratigraphy and Architecture of the Shannon Sandstone, Hartzog Draw Field, Wyoming: Implications for Reservoir Management: Gulf Coast Section SEPM Foundation 18th Annual Research Conference, Shallow Marine and Nonmarine Reservoirs, Houston, Texas, December 7-10, 1997, p. 331-344.

Tillman, R.W. and Martinsen, R.S., 1987, Sedimentologic Model and Production Characteristics of Hartzog Draw Field, Wyoming. Shannon Shelf-Ridge Sandstone, *in* Tillman, R.W. and Weber, K. (eds.), Reservoir Sedimentology: Tulsa, Oklahoma, Society of Economic Paleontologists and Mineralogists (SEPM) Special Publication 40, p. 15-114.

Tillman, R.W. and Martinsen, R.S., 1984, The Shannon Shelf-ridge Sandstone Complex, Salt Creek Anticline Area, Powder River Basin, Wyoming: *in* Tillman, R.W. and Siemers, C.T., (eds.), Siliciclastic Shelf Sedimentation: Tulsa, Oklahoma, Society of Economic Paleontologists and Mineralogists (SEPM) Special Publication 34, p. 85-142.

Tye, R.S., Ranganathan, V. and Ebanks, W.J., Jr., 1986, Facies and reservoir zonation of a Cretaceous shelf sand ridge: Hartzog Draw Field, Wyoming, *in* Moslow, T.F. and Rhodes, E.G. (eds), Modern and Ancient Shelf Clastics: Society of Economic Paleontologists and Mineralogists Core Workshop 9, Atlanta Georgia, June 1986, p.169-216.

Walker, R.G. and Bergman, K.M., 1993, Shannon Sandstone in Wyoming: A shelf-ridge complex reinterpreted as lowstand shoreface deposits, Journal of Sedimentary Petrology, v.63, p. 839-851.

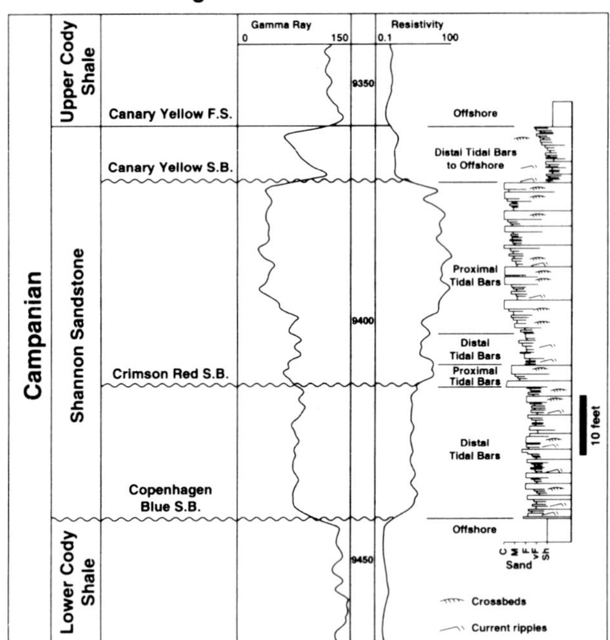

Fig. 24.—Shannon Sandstone well log from Hartzog Draw Field displaying incised-valley terminology. Inferred sequence boundaries (S. B.) and flooding surfaces (F. S.) Distal Tidal Bars are shown. Facies, in terminology of Tillman (this paper), would be designated as Inter-Ridge (Shaley) Facies (Copenhagen Blue) and Low-Energy Ridge-Margin Sandstone (near base of Crimson Red). Proximal Tidal Bars include High-Energy Ridge-Margin Sandstone (base of Crimson Red) and Central-Ridge Sandstone (top of Crimson Red). Distal Tidal Bars of the Canary Yellow Sequence include, from base to top facies as designated by Tillman as Bioturbated Siltstone, LERM, and HERM. Figure is from Sullivan et al., (1997).

Fig. 26.—Sand-ridge facies model interpretations of Hartzog Draw HDU 5308 (SESESE Sec. 30 T45N R75W). Refer to Figure 24, also HDU 5308, for comparison of sand-ridge and incised-valley facies nomenclatures. (Modified from Tye et al., 1986).

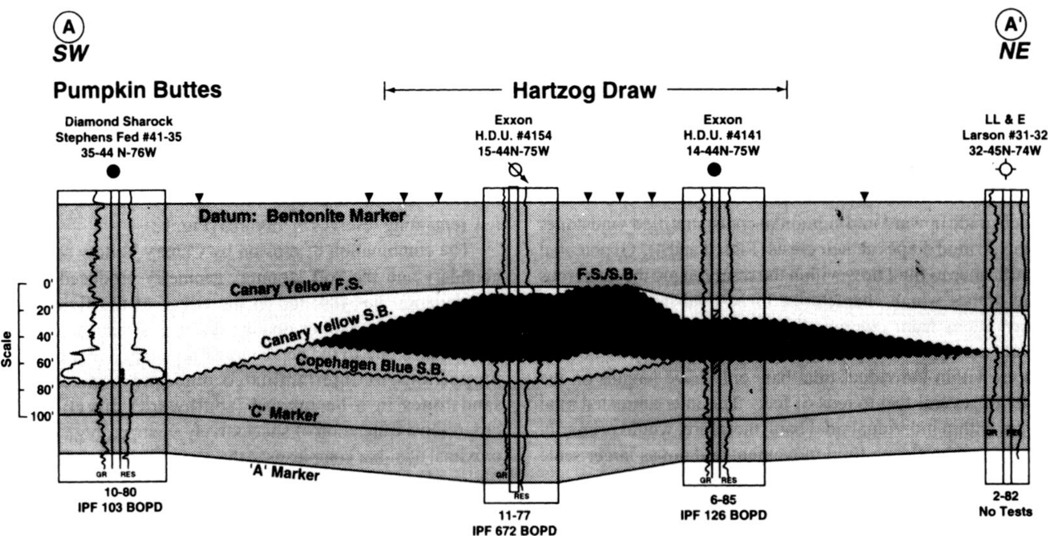

Fig. 25.—Stratigraphic section perpendicular to elongation of the field and emphasizing the three sequences, interpreted by Sullivan et al. (1997) to be controlling Shannon deposition in the Hartzog Draw and Pumpkin Buttes Oil Fields. This model interprets the sandstones in the fields to be remnants of more widespread deposits. Compare with Figure 27. (Sullivan et al., 1997).

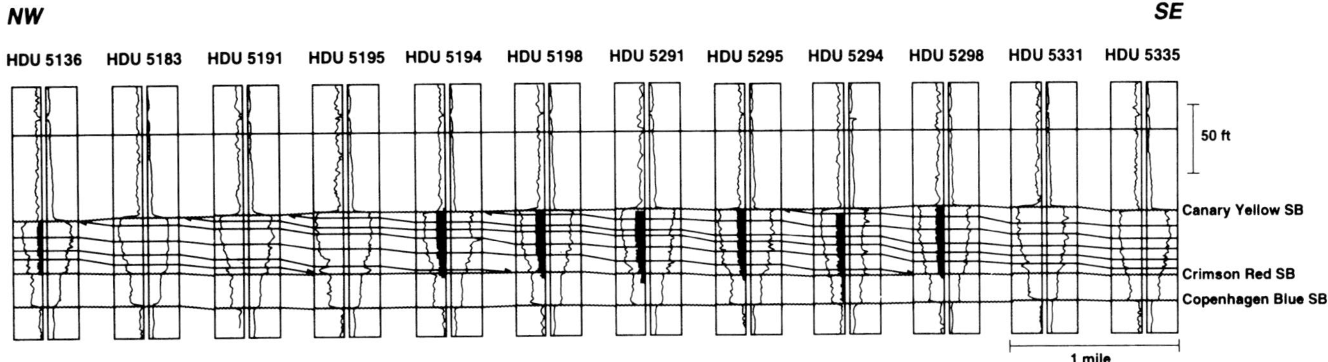

Fig. 27.—Stratigraphic well-log cross-section through Hartzog Draw Field illustrating "down lapping," shingled geometry parallel to the inferred ridge elongation direction. Sequence boundaries and bar-crests are inferred, and surfaces downlap on the Crimson Red sequence boundary (Sullivan et al., 1997). Note that there is compensation between the "Crimson Red" and "Copenhagen Blue sequences"; where one thins the other thickens and vice-versa.

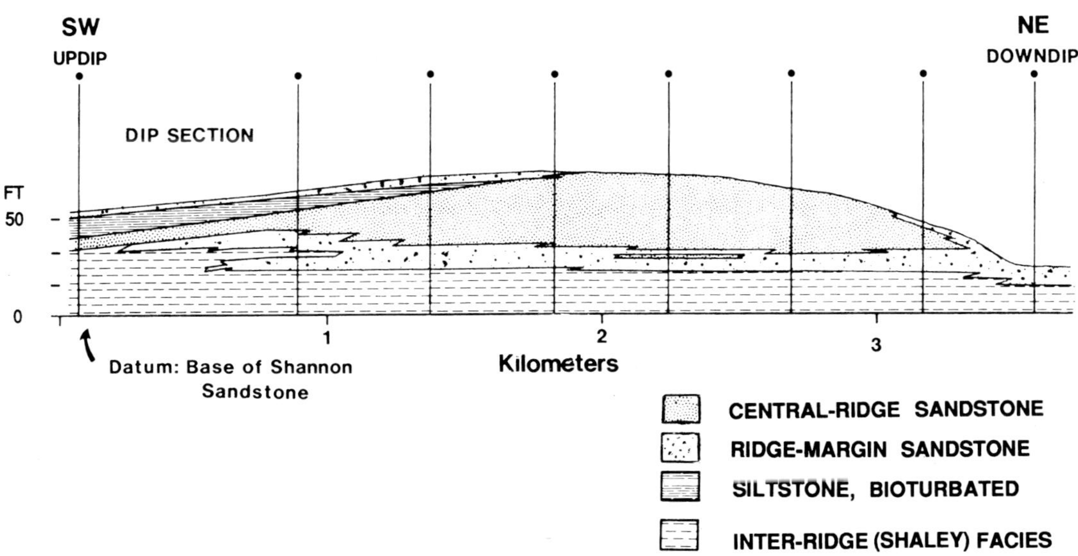

Fig. 28.—Sand ridge model stratigraphic section perpendicular to the elongation of Hartzog Draw Field. Much of Inter-Ridge (Shaley) Facies of this figure is included in Crimson Blue sequence of Figure 25. The Ridge-Margin facies here includes both the High- and Low-Energy Ridge-Margin Sandstones and probably is restricted to Sullivan's Crimson Red sequence. The Low-Energy Ridge-Margin Sandstone probably forms much of the upper part of the "Copenhagen Blue sequence" and the southwest part of the Crimson Red sequence. The Central Ridge Sandstone forms the majority of the "Proximal Tidal Bars" of Sullivan et. al. (1997). Modified from Tye et al. (1986).

SHANNON SANDSTONE OF THE POWDER RIVER BASIN: ORTHODOXY AND REVISIONISM IN STRATIGRAPHIC THOUGHT

DONALD J.P. SWIFT AND BRIAN S. PARSONS
Department of Ocean, Earth, and Atmospheric Sciences, Old Dominion University, Norfolk, VA, 23529, U.S.A.

ABSTRACT: The Shannon Sandstone, consisting of cross-stratified, heterolithic sandstone and thin-bedded sandstone-shale alternations, is a sedimentary complex of ambiguous origin. It lies seemingly "encased in marine shale," far from its time-equivalent shoreline. It has been interpreted as a shelf sand ridge complex, as a detached lowstand wedge, and as a low-stand, deltaic valley fill. The Shannon Sandstone has experienced one of the higher rates of "paradigm shift" seen within the field of stratigraphy, as successive generations have tested it for successively popular depositional models.

Interpretation of the Shannon is best undertaken in the context of a three-fold family of Cretaceous marine sandstones. Class I Sandstones cap the progradational tongues of the Cretaceous cliff-forming sandstones. They are cross-stratified with shore-normal paleocurrent orientations, high in angular deviation, have clear tidal signatures, and are of transgressive, estuarine, valley-fill origin. Class II Sandstones exhibit high-angle to hummocky cross-stratification, are amalgamated to discretely bedded, and in the latter case exhibit shore-normal, sole markings with low angular deviations. Class III Sandstones exhibit high-angle cross-stratification, low angular deviation, along-shore (southerly) paleocurrent orientations, and are enriched in intrabasinal components (phosphatic, sideritic, and glauconitic nodules). They are of transgressive, open-marine origin, and in at least some cases were deposited as sand ridges. While the paleontological evidence appears to be contradictory, the highly glauconitic phases of the Shannon provide an insurmountable problem for the lowstand shoreface, and lowstand deltaic models. Both the high glauconite content and the along-shore paleocurrent pattern point to a shelfal origin for the Class III Sandstones as a group, and for the Shannon in particular.

A generic depositional model for the Class III Shannon Sandstone can be salvaged from elements common to the diverse explanations that have been offered, and by abandoning contradictory elements. In this synthesis, the Shannon Sandstone is considered to exhibit lithologies, textures, and structures indicative of transgressive deposition in a shallow open marine setting; both estuarine and outer shelf settings are rejected. The pattern of distribution is an inherited one; it is the consequence of *in situ* reworking of sediment delivered to the site of deposition by other means. However, the result is not a simple, band-like "stranded lowstand shoreline." In the subsurface the Shannon lenses coalesce into a broad sheet of backstepping sand bodies.

INTRODUCTION

The Shannon problem

The Shannon Sandstone Member of the Cody Shale in the Powder River Basin has experienced one of the higher rates of "paradigm shift" seen within the field of stratigraphy, as successive generations have tested it for successively popular depositional models. The pattern may in part be one that students of the humanities describe as "anxiety of influence." The young scholar, reviewing the vast body of knowledge that has been inherited, thinks that there is little left to do, and sees no other way to make a mark than to attack the insights of predecessors (Bloom, 1997). In defense of our younger colleagues, however, there is more to the Shannon controversy than intergenerational conflict; the Shannon Sandstone is, in stratigraphic and sedimentological terms, an ambiguous unit.

The basic observations are simply recounted. Elongate lenses of Shannon Sandstone occur above the Lower Campanian unconformity in the Powder River Basin of Wyoming and Montana. In outcrops and well logs, a thin-bedded sandstone and shale alternation overlies, with basal unconformity, the Cody Shale. It coarsens upward into a glauconitic, heterolithic, cross-stratified sandstone. At several localities in the Powder River Basin, this sequence of lithologies is repeated several times (Fig. 1). The most detailed sedimentologic description is that of Tillman and Martinsen (1984, also Tillman and Martinsen, 1985; 1987). As noted by Asquith (1970), the Shannon Sandstone is also capped by Cody Shale, so it is "encased in shale" and further, it "is far from the time-equivalent shoreline" (Asquith, 1970). The Shannon Sandstone is well-exposed in the rimrock of the breached, doubly plunging anticline at Salt Creek, Wyoming. Likewise, its basin margin equivalent, the Gebo Member of the Eagle Formation, is well-exposed in the Eagle escarpment of the Southern Bighorn Basin (R. Fitzsimmons and S. Johnson, pers. commun., 1998). The original foreland basin, however, has been broken into successor Laramide basins, and the Big Horn Range now intervenes between the Shannon and Gebo outcrop belts (Figs. 2, 3). How do the marginward and basinward elements of the Shannon interval connect?

As a consequence of the inability to relate the Shannon to its time-equivalent shoreline, the unit has become a sort of glass slipper, to be tried on by successive stratigraphic models for proper fit. The Shannon has been of geological interest since the pioneer horseback geologist, Eldridge (1889), dodged Indians to find and include it in the Montana Group. The Salt Creek Field, with its seeps, was producing from the Shannon by 1895. When Hartzog Draw came in as a giant field in 1978 (Martinsen and Tillman, 1978), interest in the depositional environment of the Shannon accelerated abruptly, and opinions about its origin diverged widely. In the recent field conference, "Tongues, Ridges and Wedges" (Swift et al., 1995) any four participants were liable to offer five strongly held opinions concerning the Shannon's depositional setting. Suggestions have varied from the original offshore shelf setting to a regressive lowstand shoreface model to an estuarine model (See Table 1). In this paper, we review the history of Shannon studies, examine the sedimentological and stratigraphic issues of the Shannon debate, and analyze patterns of Shannon architecture.

Twenty years of Shannon studies

The stratigraphic significance of the Shannon Sandstone (Fig.2), as distinguished from its economic importance, was

Isolated Shallow Marine Sand Bodies: Sequence Stratigraphic Analysis and Sedimentologic Interpretation.
SEPM Special Publication No. 64, Copyright © 1999
SEPM (Society for Sedimentary Geology), ISBN 1-56576-057-3, p. 55-84.

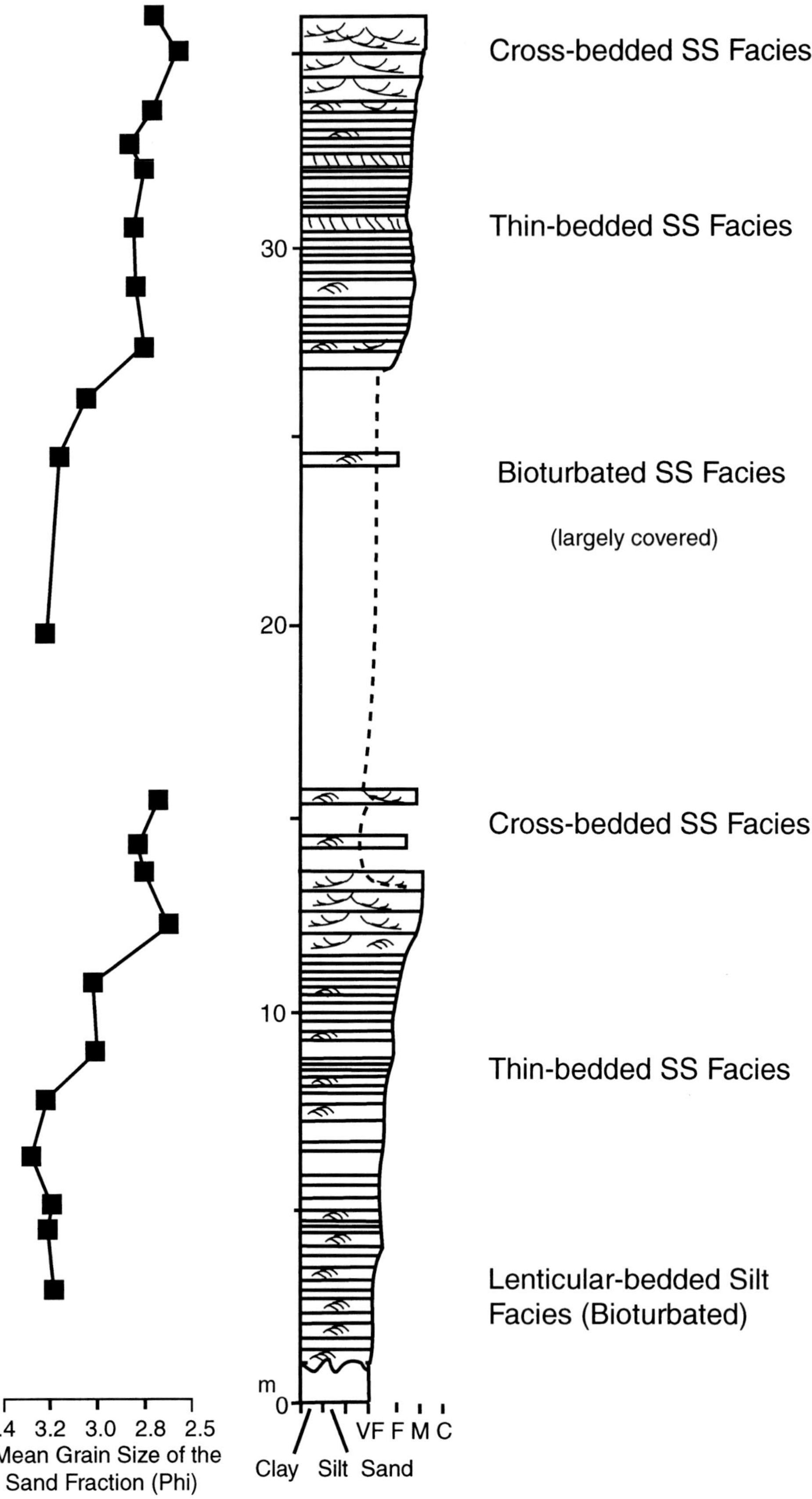

Fig. 1.—Measured section M82-1 on the eastern flank of the Salt Creek Anticline, Ship Rock locality. From Gaynor and Swift (1988).

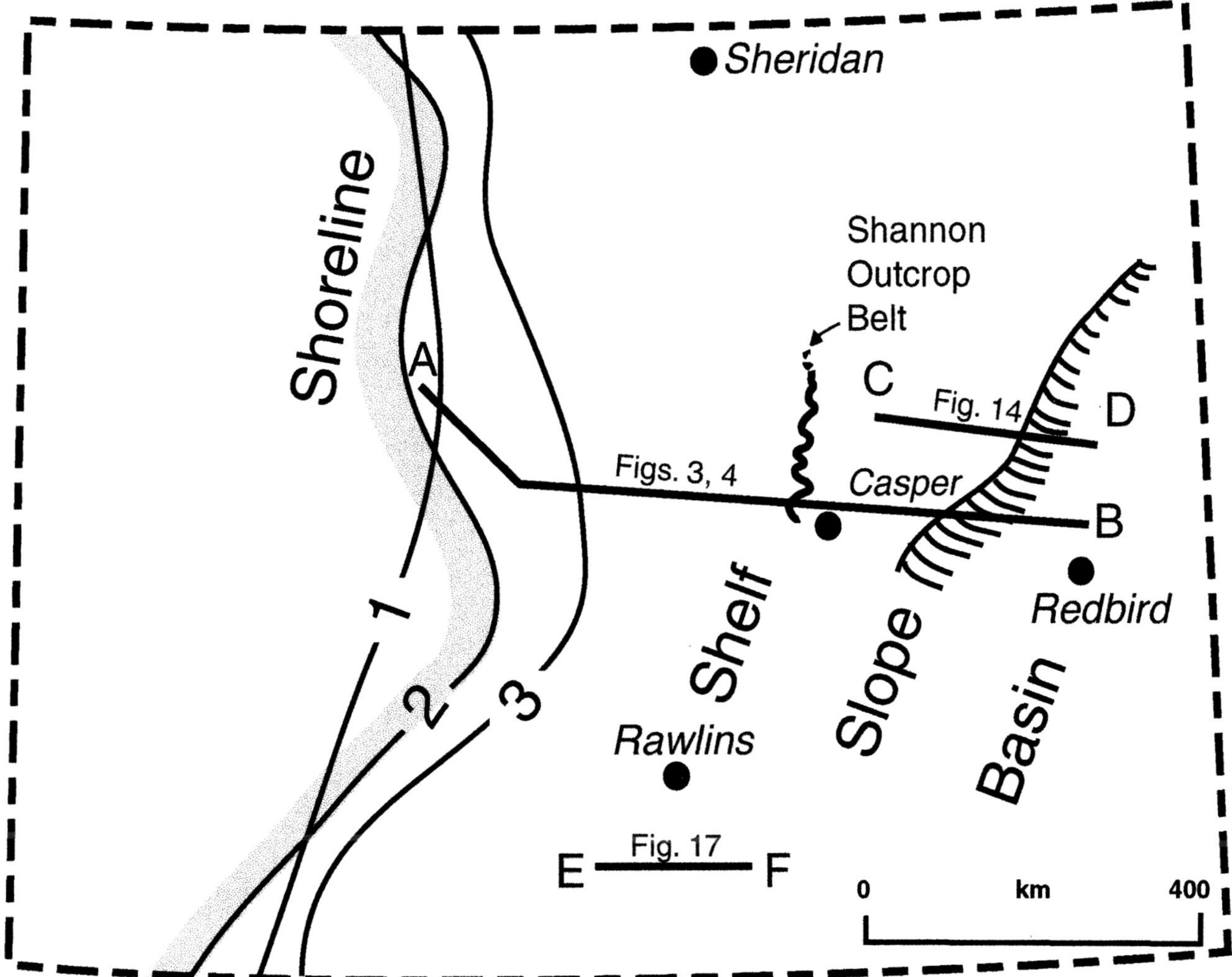

Fig. 2.—Location map, showing Early Campanian shorelines, as modified by Spearing (1976), from Gill and Cobban (1973). Shoreline 1 is the pre-Shannon position, shoreline 2 is the shoreline during Shannon time, and shoreline 3 is a post-Shannon shoreline. Map outline is the state of Wyoming. Cross-section AB is presented in Figs. 3 and 4, CD is presented in Fig. 14, and EF is presented in Fig. 17. Heavy dashed line is the Wyoming state line.

established by the biostratigraphic studies of Gill and Cobban (1966, 1973). In their key diagram, these workers show a westward-prograding and thinning "clastic wedge" (Mesaverde Formation, Cody and Pierre Shale) overlying the Niobrara Limestone. In turn, the wedge is unconformably overlain by Tertiary deposits. More recent studies (Jordan, 1982) suggest that this section was rotated down from east to west during deposition, as successive Sevier thrust plates, advancing from the west, loaded and depressed the lithosphere and shed their debris into the resulting foredeep. In the subsequent relaxation phase, the section was truncated by episodic uplift and erosion as isostatic equilibrium returned and then was buried by renewed sedimentation during the Laramide orogenic cycle.

In the cross-section (Fig. 3), Gill and Cobban (1966) drew attention to the complex interfingering of the Mesaverde sandstone with the Cody and Pierre shales. They showed that a second group of sandstones (Sussex, Shannon, and Fishtooth sandstones) occur as lenses within the shale and do not connect with the prograding tongues of the Mesaverde. In a later publication, Gill and Cobban (1973) reported that the Shannon extends west of the Big Horn Mountains at their new Nowater Creek locality, and their 1973 diagram allows the easternmost tip of the Shannon to rest with "formational contact" on the tip of a Mesaverde tongue. But, in the context of this paper, the contact is a transgressive unconformity (ravinement), and the interpretation of Shannon-Class sandstones as fundamentally lenticular is still valid.

As biostratigraphers, Gill and Cobban were preoccupied with working out the spatial and temporal framework of the Upper Cretaceous. Their efforts were paralleled by those of Asquith (1970), who examined the lithostratigraphic architecture of the Lower Campanian section in the subsurface. In a section drawn through closely spaced well logs, Asquith outlined a complex shelf-slope-basin structure in the Campanian Powder River Basin composed of repeated clinoform assemblages. The epicontinental slope that he reconstructed has slopes of 2°-3°, as steep as average values for

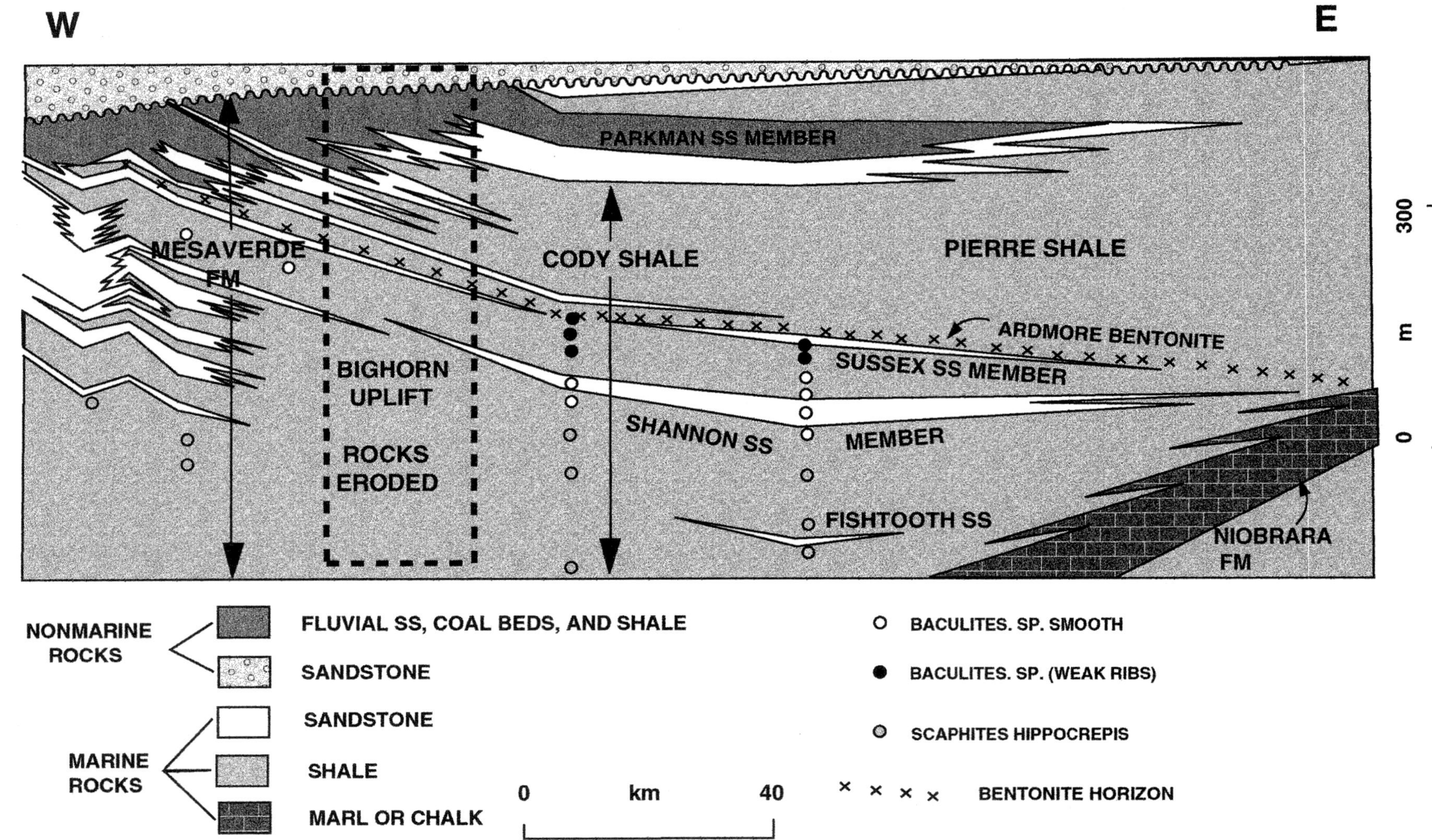

Fig. 3.—Cross-section of the Upper Cretaceous interval in the Powder River Basin, Wyoming. Modified from Gill and Cobban (1966). Only biostratigraphy relevant to the Lower Campanian is reproduced here. See Fig. 2 for location.

Table 1.—Shannon models.

Authors	Forcing	Sense of shoreline	Architecture
Asquith (1974)	Storm Currents	Regressive	Shelf Sand Ridge Complex
Tillman & Martinsen (1984, 1987)	Storm Currents	Regressive	Shelf Sand Ridge Complex
Swift and Rice (1984); Gaynor & Swift (1988)	Storm Currents	Transgressive	Shelf Sand Ridge Complex
Elliot (1987, 1988); Gaynor & Elliot (1988)	Storm and Tidal Currents	Regressive	Shelf Sand Ridge Complex
Walker & Bergman (1993); Bergman (1994)	Storm Currents	Regressive, Lowstand	Prograding Shoreface; Lowstand Wedge
Van Wagoner et al. (1990); Sullivan et al. (1995; 1996)	Tidal Currents	Lowstand	Estuarine Sand Body; Incised Valley Fill
This paper	Storm Currents	Early Transgressive	Shelf Sand Ridge Complex

the modern Gulf of Mexico slope (Shepard, 1973). Asquith reported a "wide areal distribution of shallow marine topography" with an "almost continuous presence" in at least a part of Wyoming during post-Turonian, Late Cretaceous time. His meticulously detailed well-log correlations provided a spatial framework for the complex biostratigraphic sequence of the Pierre Shale, described by Gill and Cobban (1966) from the mid-basinal Red Bird site. Asquith (1970) was able to show that the Fishtooth and Shannon intervals of Gill and Cobban (1973) terminated to the east against an epicontinental slope, as the Gammon Ferruginous Member, and an Unnamed Member of the Pierre Shale, respectively (Fig. 4).

More recently, R. Fitzsimmons and S. Johnson (pers. commun., 1998) undertook a similar reshaping of the sequence stratigraphic architecture around the basic biostratigraphic framework of Gill and Cobban (1973), this time for the Shannon-equivalent section of the basin margin (Fig 4). These workers: 1) bring Montana nomenclature down to the southern Bighorn Basin, so that Gill and Cobban's Lower Mesaverde Formation (Fig. 5) becomes the Eagle Formation as indicated in Figure 4; 2) resolved the Eagle Formation of the Southern Bighorn Basin into a lower Virgelle Member with an offlapping pattern of high-frequency sequences and an upper Gebo Member with a backstepping pattern of high-frequency sequences; 3) correlate the sequence-bounding unconformity between them with the unconformity at the base of the Shannon, and 4) find the lower sequence-bounding unconformity below the Telegraph Creek Shale and correlate it to the unconformity at the base of the Fishtooth Sandstone.

Interpretation of the Shannon Sandstone began as Asquith reflected on the results of his 1970 study and proposed a "wide shelf model" for the Lower Campanian of the Powder River Basin (Asquith, 1974) in which the Shannon sand bodies appeared as offshore shelf sand "bars." This concept had been initiated by Spearing (1976), who had looked more closely at the sedimentology and paleoflow indicators of the Shannon, and who had noted that similar shelf sand ridge deposits were reported in the modern oceanographic literature.

In several papers that more or less coincided with the development of the Hartzog Draw Field, the wide shelf model underwent further development (Tillman and Martinsen, 1984, 1987; Shurr, 1984; Swift and Rice, 1984; Gaynor and Swift, 1988). Opinion diverged on the portion of the depositional cycle in which Shannon-style sedimentation had occurred. Tillman and Martinsen (1987) saw the Shannon beds as the highstand culmination of a depositional cycle. We will style their variant of the wide shelf model a "regressive storm-built shelf ridge model," to distinguish it from the increasingly complex models that followed. However, on modern shelves, sand ridges are found in transgressive settings and tend to rest on a basal unconformity (ravinement; Swift et al., 1978a) hence Swift and Rice (1984) and Gaynor and Swift (1988) proposed a wide-shelf variant that could be titled "transgressive, storm-built shelf ridge model". Trevor Elliot and Neil Gaynor detected "tidally-produced sigmoidal cross-bedding" in the Shannon, and framed a "regressive, tidally-influenced shelf ridge model" (Elliot, 1987, 1988; Gaynor and Elliot, 1988).

As interest in sequence stratigraphic concepts intensified, several attempts were made to insert the offshore, shale-encased Shannon into sequence stratigraphic models. Walker and Bergman (1993), Bergman (1994), and Bergman and Walker (1995) retain wave-driven sedimentation and reinterpret the Shannon as a "storm-dominated, regressive lowstand wedge." Meanwhile, the sequence stratigraphic group of Exxon Production Company had become interested in the Shannon problem. Early thoughts by this group have been expressed by Van Wagoner et al. (1990), who, after examining well-log data, reinterpreted the Shannon as a "tide-built, estuarine lowstand valley fill, a view expressed more fully by Sullivan et al. (1995, 1996). These models are presented in Table 1. Choice among the models depends on issues such as the nature of the hydraulic forcing, resolution of water depth at the time of deposition, and the significance of the high glauconite content of the upper Shannon Sandstone. We consider these issues in turn below.

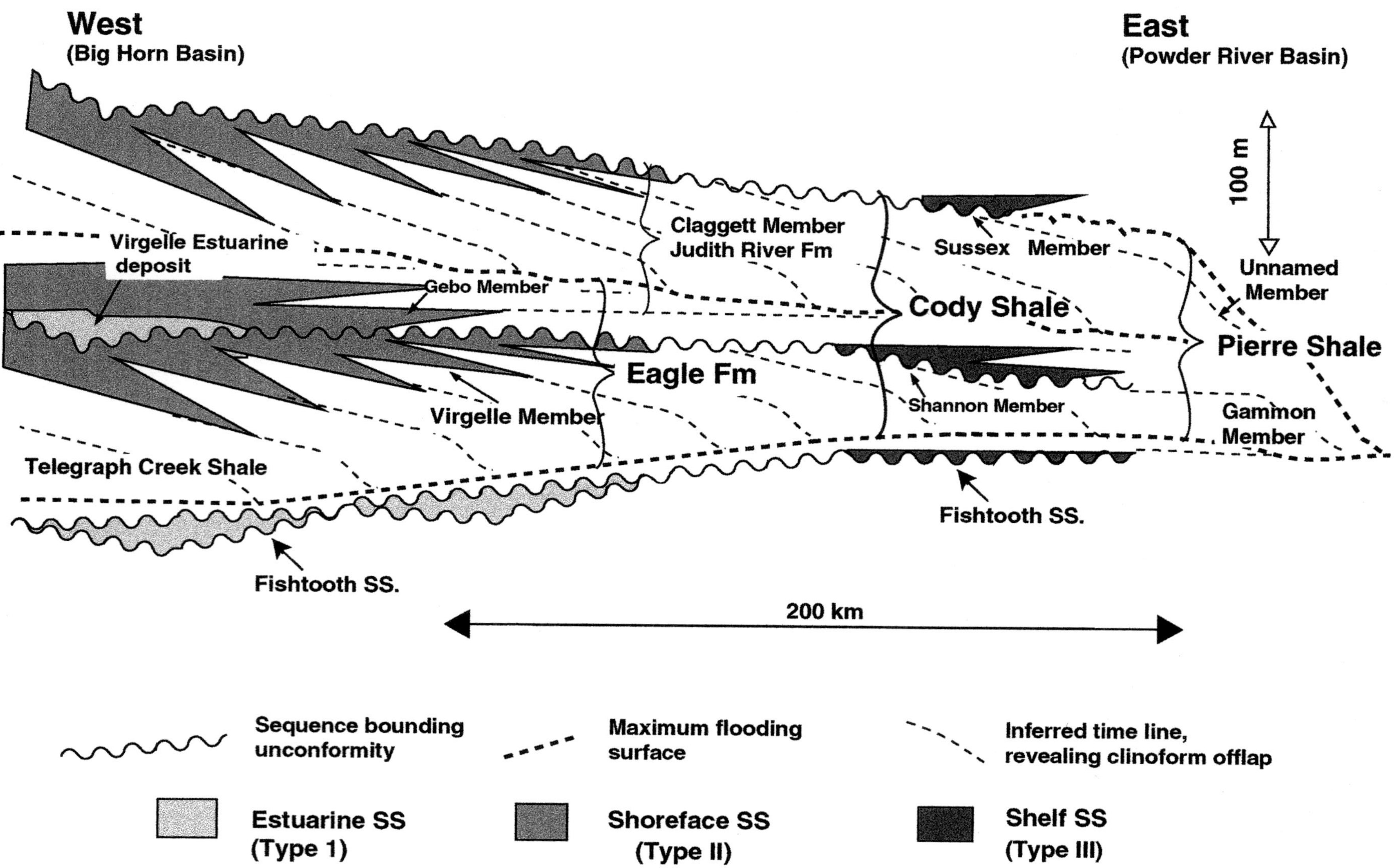

Fig. 4.—Sketch of recent modifications to Lower Campanian stratigraphy, Wyoming. R. Fitzsimmons and S. Johnson have resolved a Virgelle and a Claggett depositional sequence on the western margin of the basin, while Asquith (1970) has resolved the Lower Campanian epicontinental shelf edge and slope. See Fig. 2 for location. Sketch is not to scale.

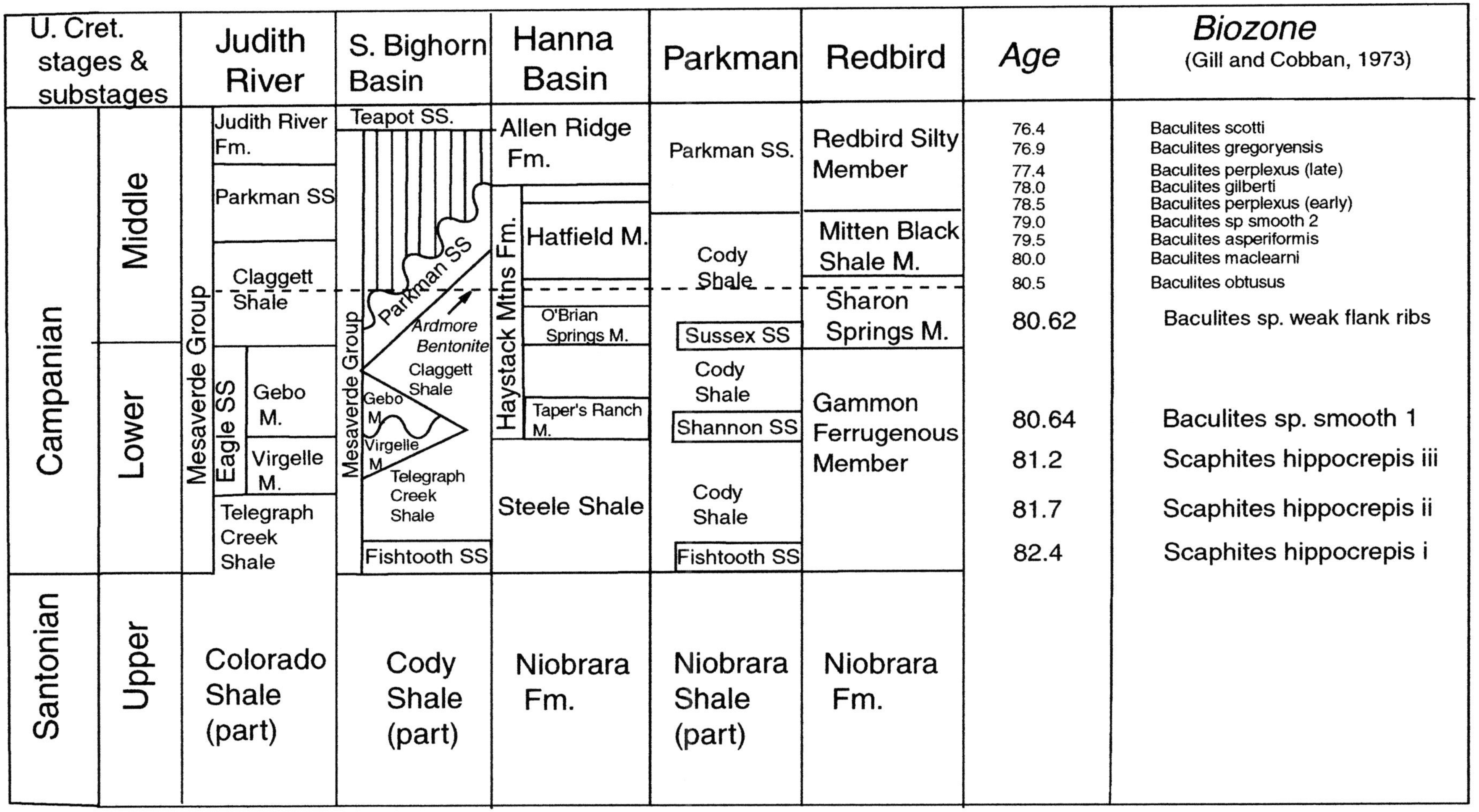

U. Cret. stages & substages		Judith River	S. Bighorn Basin	Hanna Basin	Parkman	Redbird	Age	Biozone (Gill and Cobban, 1973)
Campanian	Middle	Judith River Fm.	Teapot SS.	Allen Ridge Fm.	Parkman SS.	Redbird Silty Member	76.4	Baculites scotti
							76.9	Baculites gregoryensis
							77.4	Baculites perplexus (late)
		Parkman SS	Parkman SS				78.0	Baculites gilberti
				Haystack Mtns Fm.: Hatfield M.			78.5	Baculites perplexus (early)
					Cody Shale	Mitten Black Shale M.	79.0	Baculites sp smooth 2
							79.5	Baculites asperiformis
							80.0	Baculites maclearni
		Mesaverde Group: Claggett Shale	Mesaverde Group				80.5	Baculites obtusus
			Ardmore Bentonite	O'Brian Springs M.	Sussex SS	Sharon Springs M.	80.62	Baculites sp. weak flank ribs
	Lower	Eagle SS: Gebo M.	Claggett Shale; Gebo M.		Cody Shale			
				Taper's Ranch M.	Shannon SS	Gammon Ferrugenous Member	80.64	Baculites sp. smooth 1
		Virgelle M.	Virgelle M.				81.2	Scaphites hippocrepis iii
		Telegraph Creek Shale	Telegraph Creek Shale	Steele Shale	Cody Shale		81.7	Scaphites hippocrepis ii
			Fishtooth SS		Fishtooth SS		82.4	Scaphites hippocrepis i
Santonian	Upper	Colorado Shale (part)	Cody Shale (part)	Niobrara Fm.	Niobrara Shale (part)	Niobrara Fm.		

Fig. 5.—Correlation chart of Lower Campanian interval, Montana and Wyoming, based on Gill and Cobban's (1973) ammonite biostratigraphy. Absolute age dates from Obradovich (1993).

SHANNON ISSUES

What was the hydraulic regime?

The problem of hydraulic forcing in the Shannon environment is part of a more general problem of fluid circulation and sediment transport in the Cretaceous Western Interior Seaway (Fig. 6), a problem that has been addressed by numerical modeling techniques (Parrish et al., 1984; Ericksen and Slingerland, 1990). Perusal of these papers suggests that circulation in the Seaway was generally storm dominated. Ericksen and Slingerland (1990) report that typical winter storms generated 0.3 m s^{-1} currents that flowed southward throughout most of the event. Their investigations suggest that co-oscillating tides propagated into the seaway as progressive Kelvin waves from the Gulf of Mexico, so that tidal ranges in the seaway were mesotidal to macrotidal along the southeastern margin but microtidal everywhere else. However, local topographic restrictions, in the form of tidal inlets and estuary mouths, might have created strong tidal currents in the region of microtidal range, particularly during transgressions, when such landforms are common.

Sedimentological criteria for assessing the hydraulic regime

Decisions concerning the nature of hydraulic forcing during deposition of the Shannon Sandstone have been made on the basis of such primary structures as sigmoidal cross-stratification (originally described by Mutti et al., 1985) mud drapes (double and otherwise), mud clasts, tidal bundling, and herringbone cross-stratification (see Dalrymple, 1992, and Dalrymple et al., 1992, for reviews of these features). Shannon workers, whose papers are summarized above, tend to fall into two camps; the tidalists and the mixed-regime advocates. Tidalists argue that these features are pervasively present and indicate that the Shannon Sandstone is dominantly the product of tidal forcing. The mixed-regime advocates tend to separate these criteria for tidal sedimentation into hard and soft criteria. An "astronomical signature" is a hard criterion (Visser, 1980). For instance, in tidal bundling, 28 beds are deposited from spring tide to neap tide; or, if sand stops moving during some portion of the neap tide period, there is at least a repeating count. Double clay drapes occur when there is a "mixed tide" so that the second tidal cycle of the day is much weaker than the first and transports less sand. These are also "hard" evidence; they show that the transporting agent is an asymmetrical, semidiurnal tidal current cycle.

Soft criteria, including cross-strata sets with "sigmoidal" profiles and clay drapes are less diagnostic. Sub-tidal sand bodies on storm-dominated shelves are less accessible and have been studied less than the tide-maintained sand bodies of intertidal zones, but reports are available, and these areas exhibit many features that supposedly indicate a tidal origin. On the northern North Carolina Shelf (Albemarle Massif), storm-built sand ridges are mantled with storm-maintained sand waves (Swift et al., 1978b). This shelf surface (Fig. 7) is significant for a number of reasons; for instance, it is the top of a transgressed and modified valley fill; but for the present discussion, we note that the sand waves on the Albemarle Massif tend to have sigmoidal profiles, occasionally, clay drapes, although the fine sediment input is relatively low. The coast is microtidal. The sigmoidal profiles may be tentatively attributed to the fact that storm currents, like tidal currents, intensify, wane, cease, and reverse. Storm-driven

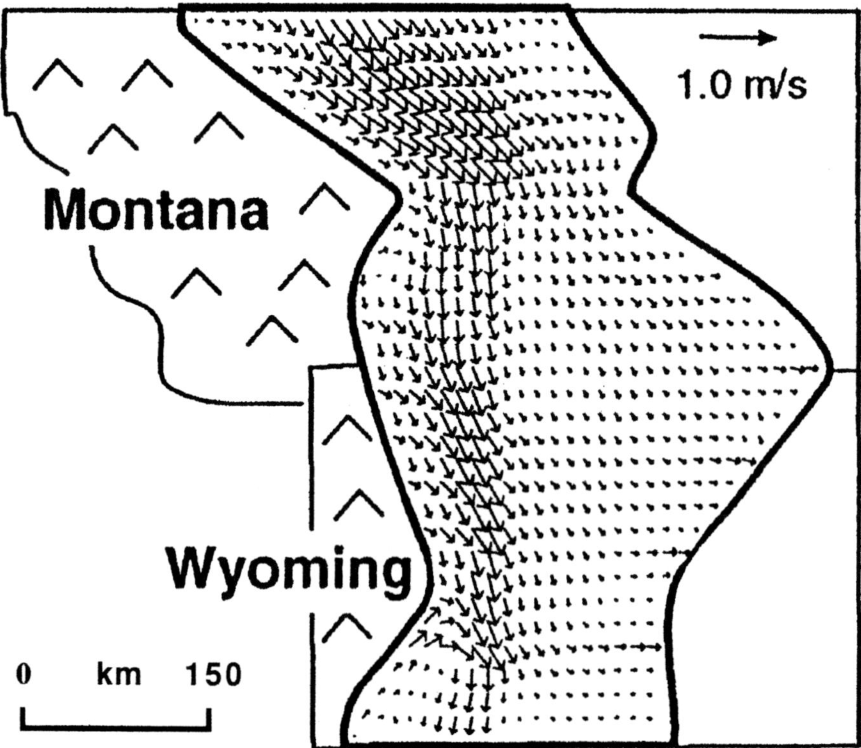

Fig. 6.—Bottom currents during the second day of a simulated storm current event on the early Campanian Shelf of Wyoming. From Y. Zhang (personal communication, 1998).

flows on the Virginia Shelf are dominantly along-shelf to the south. However, flows trending along-shelf to the north are an important though subordinate type.

It has been suggested that clay drapes are not only tidal criteria; more generally, they indicate reversing (estuarine) rather than rotary (shelf) tides (Sullivan et al., 1995, 1996). According to this interpretation, clay drapes cannot form on the open shelf because the tidal current speed never falls below the threshold speed of erosion. This interpretation is based on a misapprehension of fluid dynamics. Waning currents can deposit mud at any speed; as the current decreases less fluid power is available to generate sediment-suspending eddies, so sediment grains fall out of suspension faster than they are being entrained. Any reduction in fluid power must result in deposition, whether or not current speed falls below threshold. Clay drapes on cross-strata sets, in fact, require only two conditions: 1) sediment of sizes appropriate for both bed load and suspended load must be abundantly available, and 2) a fluctuating current must be present whose mean value exceeds the threshold for bedload transport. For instance, clay drapes and heterolithic stratification occur over vast portions of the Amazon Shelf (Adams et

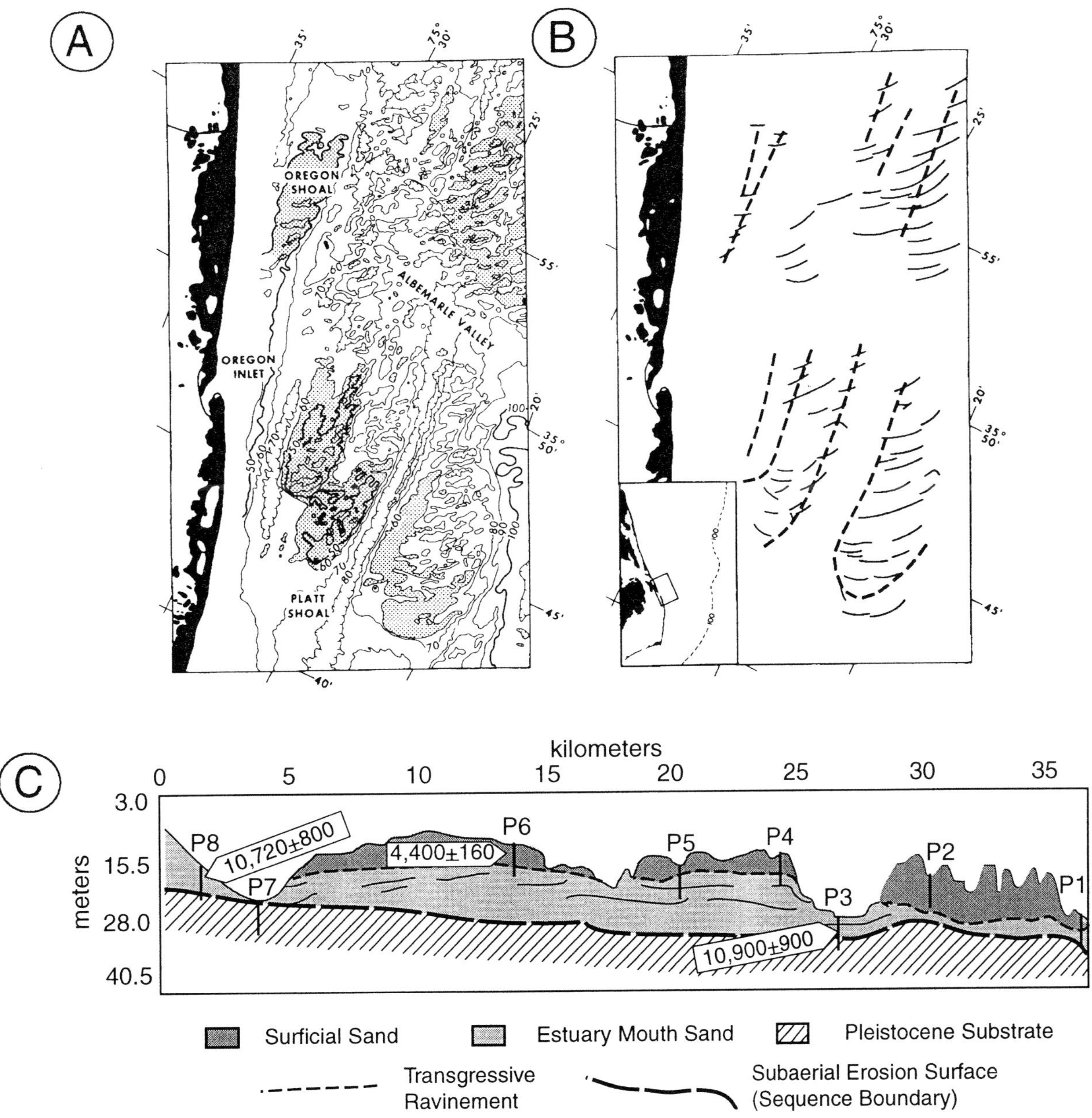

Fig. 7.—Bathymetric map of the inner North Carolina Shelf, adjacent to the Outer Banks barrier system. Sand ridges with superimposed sand waves have been molded into the shoal retreat massifs flanking an incised and filled shelf valley. Transgressive surface (ravinement) lies beneath the massif, and outcrops lie in the lows. From Swift et al. (1977). A) Topographic map; contours in feet. B) Trend of sand ridges and sand waves. C) Cross-section A - B with radiocarbon dates.

al., 1986; Kuehl et al., 1982). While tidal currents are an important component of the hydraulic regime on the Amazon Shelf, the North Brazilian Coastal Current, and wave orbital motions are the dominant forcing agent (Nittrouer et al., 1986). Clay drapes on cross-strata sets, heterolithic stratification, and clay chip conglomerates also occur in cores through levee sands on the Mississippi Fan in several thousand meters of water (Bouma and Coleman, 1985). They are not hard criteria for tidal sedimentation.

Hard criteria do exist for tidal currents in the depositional environment of the Shannon. The example of tidal bundling at the Castle Rock outcrop, cited by T. Elliot (pers. commun., 1998; Elliot, 1987; Gaynor and Elliot, 1988) is a convincing one, and there are some acceptable cases of double clay drapes. However, these features are scarce, in comparison with their frequency in, for instance, the Tocito Sandstone of New Mexico (Nummedal et al., 1989), or the Lower Greensand (Nayaran, 1971). Therefore, while tidal forcing may have contributed to the hydraulic regime, it does not appear to have played a dominant role.

How deep was the Shannon? Paleontological evidence

This issue matters a great deal for the various interpretations. If the Shannon is a lowstand wedge deposit (Walker and Bergman, 1993), it should be capped by surf or near surf deposits. If it is a tidal estuarine valley fill (Sullivan et. al., 1995, 1996), then these sandstone bodies may have been "drying shoals" that were exposed at low tide (Dalrymple, 1992). But if the Shannon was deposited as a transgressive shelf, sand-ridge complex, then the surviving beds must have been deposited in water at least as deep as the foot of the shoreface (perhaps 15 m), since transgressive shorefaces do not preserve, but are destroyed by erosional shoreface retreat (Swift et al., 1991a).

Opinions on the depth of the Shannon vary as much as opinions on the depositional agent. Tillman and Martinsen (1984, 1987), in support of their shelf sand ridge model, note that shelf trace fossils such as *Teichichnus, Thalassinoides,* and *Chondrites* are present, while "common Cretaceous shoreline fossils such as *Ophiomorpha, Asterosoma,* and *Rhizocorallium* were not observed." They report as microfossils, from shales within the Shannon interval, *Haplophragmoides* and *Bathysiphon* (modern analogs are found in transitional and inner shelf waters) and also a few *Lenticulina* and a single *Epistomina*. These latter forms are found on modern middle shelf and deeper waters. They conclude that "...combining interpretations related to stratigraphic sequence, lithology, sedimentary structures, micropaleontology (foraminifera), and trace fossils, it is suggested that the Shannon Sandstone cropping out in the Salt Creek area was deposited at a water depth greater than 60 ft (~20 m) and less than 300 ft (~100 m)."

Walker and Bergman (1993) present their lowstand wedge model with considerable stress on trace fossils. They recognize two main trace fossil assemblages; an assemblage characterized by *Helminthopsis, Schaubcylindrichnus, Thalassanoides, Chondrites,* and *Terebellina* in finer-grained settings, and an assemblage characterized by *R. Socialis, Ophiomorpha,* and *Macaronichnus* found in cross-bedded sandstones. They see *Macaronichnus* as particularly significant, since it has been described by MacEachern and Pemberton (1991) as requiring the "intense swash infiltration into the beachface within the foreshore [which] carries dissolved oxygen and nutrients well below the water sediment interface, permitting epigranular microbes and other bacteria to flourish." Walker and Bergman (1993) conclude that the cross-bedded sandstones of the Shannon were deposited in a high energy foreshore or upper shoreface environment.

Sullivan et al. (1995, 1996) also invoke trace fossils in general, and Pemberton et al. (1991) in particular, in support of their estuarine valley fill model. They recognize a proximal tidal bar ichnofacies that includes *Planolites, Teichichnus, and Macaronichnus,* and less common vertical dwelling structures (*Arenicolites and Skolithos*). A distal tidal bar ichnofacies lacks *Macaronichnus,* but includes *Thalassinoides, Paleophycus, Terebellina,* and *Chondrites.* Sullivan et al. (1995, 1996), citing Pemberton et al. (1991), see the trace fossil assemblages of the Shannon as restricted assemblages that "are typical of a brackish (tidal) setting, and are not compatible with fully marine conditions that characterize offshore to shoreface settings."

In view of the conflicting paleontological interpretations, it is possible to wonder if the ichnological literature constitutes a "monographic high." This term (of anonymous origin) refers to a situation in which the literature for a particular geological problem has, by happenstance, been restricted to a subset of the full range of existing examples, leading to an unwitting bias in interpretation. For instance, *Macaronichnus* is supposedly characteristic of shoreface settings, but elsewhere in the Cretaceous Western Interior Basin, where the ichnological work has been done, sandstones are, in volumetric terms, mostly prograding shoreface sandstones. Little in the way of transgressive offshore marine sandstone has been preserved. Except for the few thin, transgressive glauconitic sandstones, the shelf deposits are shale. Perhaps the occurrence of *Macaronichnus* in the Shannon does not mean that the Shannon was deposited in a shoreface setting, but rather that the responsible organism was viable on a mobile sand bottom at any shelf depth and it is primarily the shoreface expressions of *Macaronichnus* that have been preserved. *Macaronichnus* is, in any case, suspect as a surf zone indicator, as it is horizontal and contained within bedding planes in the Shannon Sandstone, in contrast with the vertical *Skolithos*-like burrows usually associated with turbulent environments.

How deep was the Shannon? Paleocurrent patterns and flow depth

It is also possible to approach the issue of paleodepth from the point of view of fluid dynamics. In a preceding section, numerical simulations of the hydraulic regime of the Western Interior Seaway were reviewed, and the results were compared with observations of primary structures in the Late Cretaceous deposits. In fact, fluid dynamical regimes on modern shelves and the consequent sediment dispersal regimes vary markedly as a function of depth and can be distinguished on the basis of paleocurrent indicators. We note several broad classes of sandstones within the Lower Campanian interval and can distinguish among them several contrasting paleocurrent regimes.

The prograding tongues of Mesaverde and equivalent formations produce classic shoreface sandstones in abundance. These well-sorted, fine-grained sandstones display groove casts, gutter casts, and parting lineations, oriented orthogonally with respect to wave ripples and with respect to the probable shoreline orientation (Virgelle Sandstone of Rice, 1980; R. Fitzsimmons and S. Johnson, pers. commun., 1998). Cross-bedded sandstone at the top of such progradational sandstones have been generally considered to be surf zone deposits. More recently, however, it has been shown that while surf zone deposits are present, there are in addition, thick extensive, cross-bedded to massive deposits at the tops of cliff-forming Cretaceous strata that formed in back-barrier (estuary mouth and inlet fill) environments

(Wright, 1986; Shanely et al., 1992; Nummedal and Molenaar, 1995). These capping units formed during periods of transgression, in topography incised into the underlying, progradational sandstones. Backbarrier sandstones do exhibit "hard" tidal criteria such as double clay drapes and stratal bundling, as well as other characteristics of estuarine deposits, such as heterolithic stratification and ostreid shell debris. We will refer to the transgressive, capping sandstone as Class I Sandstones, and the progradational, shoreface sandstone beneath them as Class II Sandstones. Both classes of sandstone yield paleocurrent orientations aligned at high-angles to the trend of the regional paleoshoreline (Fig. 8A).

A third class of sandstones (Class III), to which the Shannon Sandstone belongs, is seen only in the lower tongues of the Mesaverde-Eagle interval and equivalent sandstones; or more typically, to the east of the Mesaverde, as members in the Pierre or Cody Shales. These sandstones are thin, lenticular, relatively coarse-grained, and glauconitic. They include the Shannon, Sussex, and Fishtooth Sandstones of the Salt Creek Anticline (Gill and Cobban, 1973), the Tapers Ranch and O'Brien Springs Members of the Haystack Mountains Formation in Southeastern Wyoming (Tillman and Martinsen, 1985, 1987; D. Mellere and R. Steel, 1995a, 1995b, pers. commun., 1998); the glauconitic sandstone lenses of the Virgelle Member of the Eagle Formation in the southern Bighorn Basin (Swift and Zhang, 1995); the Eagle Sandstone lenses at Billings, Montana (Shelton, 1965); and the Groat Sandstone in southeastern Montana (Shurr, 1984). They are characterized by planar to trough cross-bedding. Paleocurrents trend along-shelf and slightly offshore, to the south or southeast (Fig. 8B).

Paleocurrent patterns and hydraulic regimes

These patterns can be matched by the sediment types and hydraulic regimes seen on modern continental shelves. Modern Class I sands include shoals associated with wide-mouthed estuaries (Ludwick, 1974), and the ebb-delta—flood-delta sand bodies of microtidal inlets (Boothroyd et al., 1985). They consist of medium to fine sand and produce arrays of megaripples and sand waves that are oriented generally shore-normal, albeit with relatively large angular deviations.

Modern shoreface sands (Class II Sands) are fine-grained and travel in suspension (Swift et al., 1971). They are deposited in the friction-dominated and transitional regimes of the shoreface and inner shelf (Fig. 9), where, in the equation of motion, a term for wind stress on the sea surface is balanced by the frictional drag of the bottom (Swift et al., 1985). Fine shoreface sands respond quickly to fluid accelerations, and are entrained by peak instantaneous bottom velocities during storm periods (Duke, 1990). Storm waves are refracted by bottom drag as they approach the coast, and their crests rotate into parallelism with the shoreline. Wave orbital motion therefore sweeps the shoreface with high-frequency alternations of onshore- and offshore-directed strokes. Each offshore stroke is reinforced by the downwelling, offshore component of the coastal flow, a low-frequency motion induced by wind drag on the sea surface and by the resulting coastal set-up. Modern shorefaces respond to such downwelling flows by developing rippled surfaces with offshore asymmetry (Héquette and Hill, 1995). The shore-normal groove casts, gutter casts, and parting lineations of Mesozoic shoreface sandstones, such as those described from the Virgelle sandstones, by their nature cannot be observed on modern sea floors, but they are believed to result from similar high-frequency, wave-induced motions (Duke, 1990).

Modern shelf sands (Class III Sands) occur primarily in transgressive settings, as the product of long-term erosional shoreface retreat (Swift et al., 1991a). They are of variable grain size, but are frequently medium- or coarse-grained, and seaward of 50 m are commonly glauconitic (Odin and Fullagar, 1988). On the Middle Atlantic Continental Shelf of North America, abundant cross-strata sets are oriented alongshore and slightly offshore to the south and southeast (Swift et al., 1986a; Fig. 20). The coarser sand of the shelf floor travels as bed load rather than suspended load. It consequently cannot respond to instantaneous, peak wave-orbital motions, but instead sustains a time-averaged response to the wind-driven alongshore component of flow and builds megaripples and sand waves oriented by the alongshore flow (Duke, 1990).

Shelf sands (Class III Sands) are also distinctive in the consistency of their paleocurrent orientation. Shelf currents, in the deeper zone of geostrophic flow are less affected by bottom friction and respond to a balance between the pressure term and the geostrophic term in the equation of motion (Fig. 9). Storm flows on the Middle Atlantic Shelf of North America are known to be "highly coherent and slab-like" and show little spatial variation during successive peak events (Boicourt and Hacker, 1976). The Early Campanian Shelf of Wyoming was, in dynamic terms, an analog of the modern

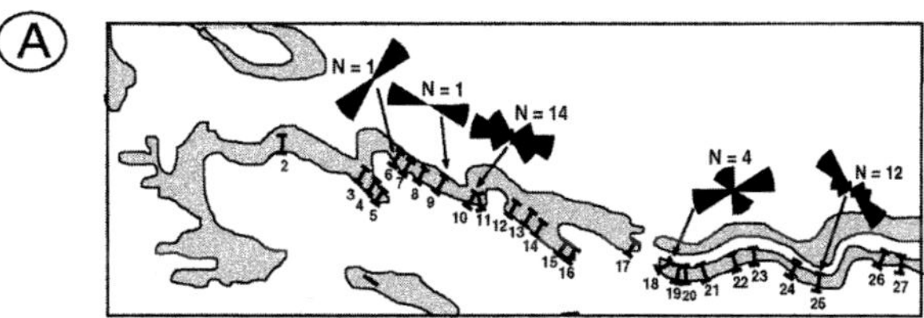

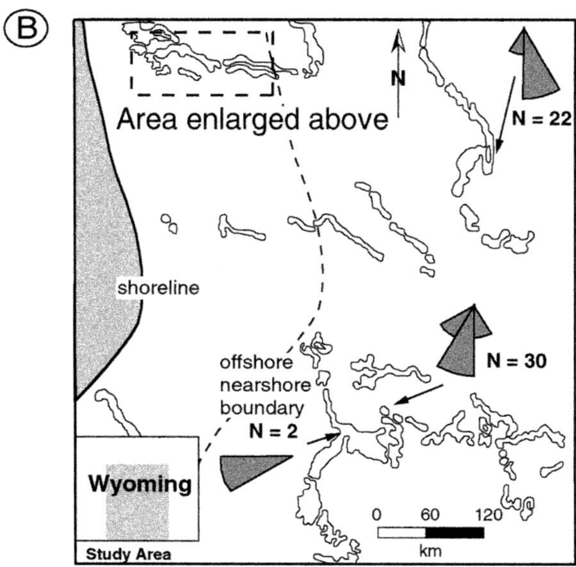

Fig. 8.—Lower Campanian paleocurrent orientations. A) Orientation of parting lineations in a shoreface and estuarine sandstones of Class II (Virgelle Member, Eagle Formation). B) Orientation of cross strata sets in glauconitic shelf sandstones of class III (Shannon and Tapers Ranch Members, Cody Shale) From Y. Zhang (personal communication, 1998).

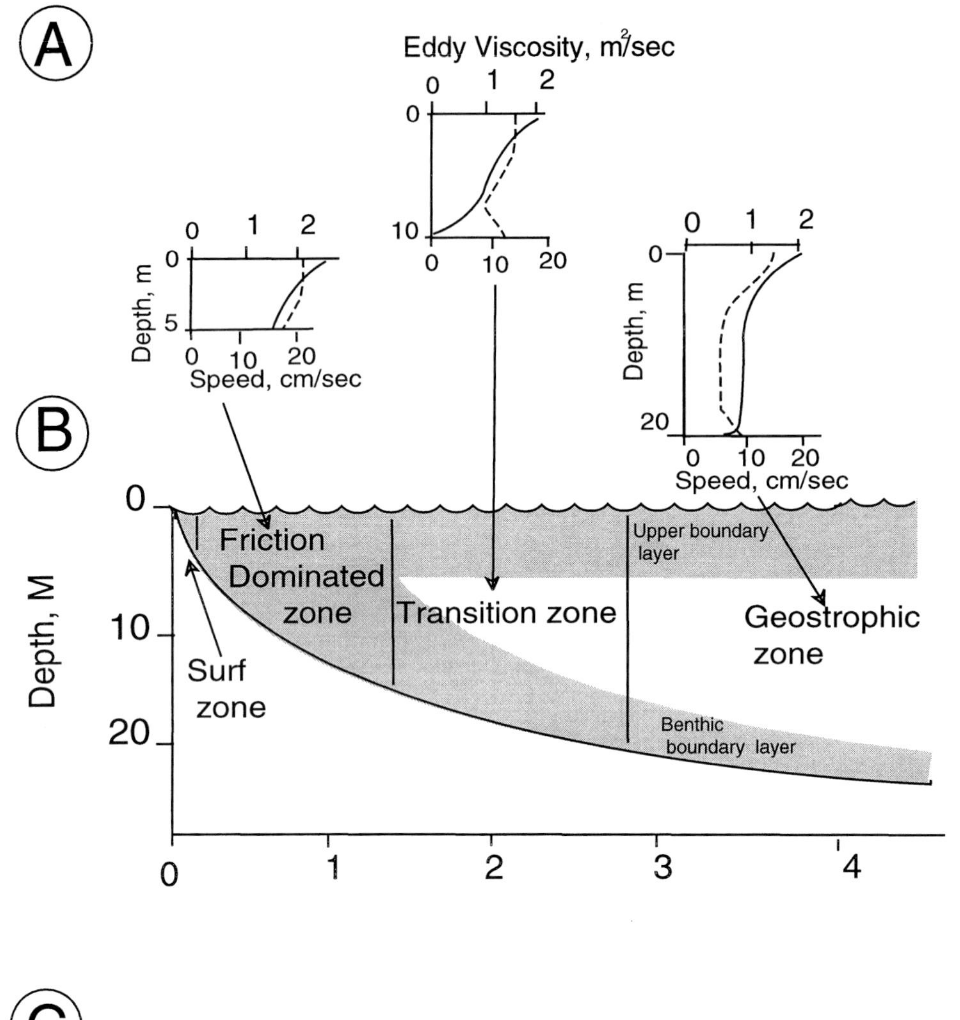

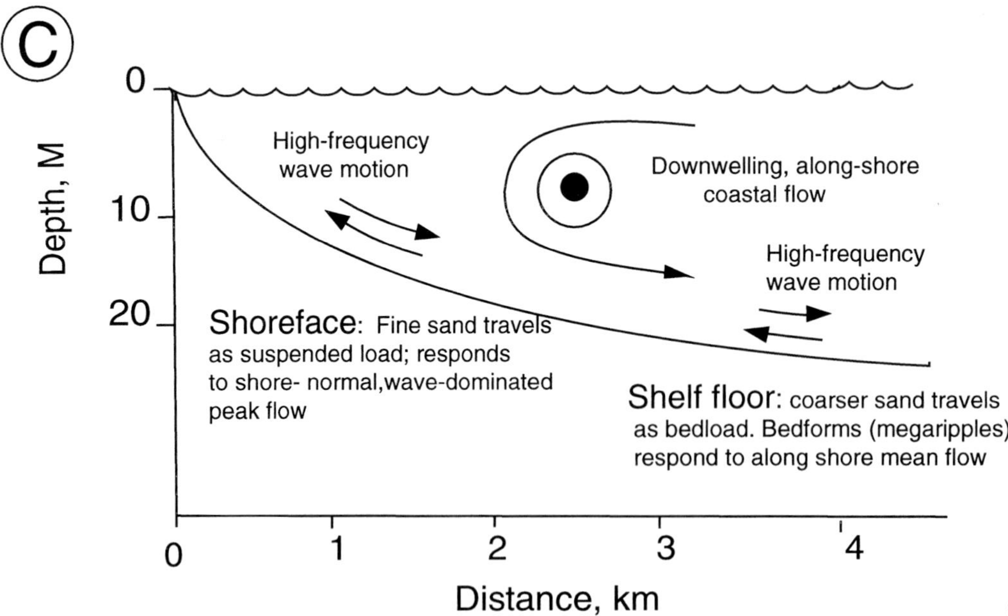

Fig. 9.—Dynamical zones of the coastal ocean during a moderate wind event. A) Speed and eddy Viscosity profiles (solid and dashed lines, respectively). B) Dynamical zones. In the friction dominated and transitional zone, eddy viscosity is relatively high. In the equation of motion, wind on the sea surface is balanced by the frictional drag of the bottom stress. Geostrophic turning of the flow is damped out. In the geostrophic zone, the friction term in the equation of motion is relatively small. In the equation of motion, an offshore oriented pressure term is balanced against an onshore-directed geostrophic term, and the current flows along-shelf. C) Fluid-sediment interactions. Modified from Swift et al. (1985).

Atlantic Shelf, in that it was a mid-latitude, north-south trending, east-facing shelf. Numerical simulations show that it responds to the passage of extra-tropical cyclones by generating strong, southerly, along-shelf flows in much the same away that the Atlantic Shelf does (compare Vincent et al., 1981, for modern Atlantic Shelf transport orientations with Figure 6, for Campanian flow simulations).

Shannon paleocurrent patterns

Paleocurrent measurements from the Shannon Sandstone strongly resemble the highly coherent offshore pattern of along-shelf flow in that the paleocurrent indicators are oriented alongshore and southerly, and their angular deviation (σ) is very low. Spearing (1976) reports a mean current trend of roughly 128° from 171 ripple crests and 192° from 254 trough cross-beds. Tillman and Martinsen (1984, 1987) calculated a mean of 188° from 72 trough cross-bed measurements and 135° from five ripple crest measurements. Gaynor and Swift (1988) report a mean of 191° from 132 measurements of linguoid ripple marks, trough, and planar cross-beds. Walker and Bergman (1993) summarized paleocurrent orientations from the capping sandstone of each of their measured sections. They calculated a mean of 165° from 59 trough cross-beds at the top of their first section, and a mean of 180° from 80 trough cross-beds from the top of the second section (unit 10). Y. Zhang (pers. commun., 1998) reports a flow average of 170° (σ = 011°) from 22 measurements of trough cross-beds and 222° (σ = 013°) from 41 measurements of ripple crests. In these studies, 830 measurements of paleocurrent indicators yield cross-strata values ranging between 165° and 191° (Gaynor and Swift, 1988). Ripple values range between a minimum of 128° (Spearing, 1976) and a maximum of 222° (Y. Zhang, pers. commun., 1998), Angular deviations are very low (11°-13°). Paleocurrent measurements from the Shannon Sandstone are comparable with cross-strata measured from other Class III Sandstones such as the Groat Sandstone of southern Montana (164°, σ = 009°; Y. Zhang, pers. commun., 1998), and the Tapers Ranch Sandstone in the Hanna basin of Wyoming (191°, σ = 017°; Y. Zhang, pers. commun., 1998).

The values indicate a southerly sediment transport direction during deposition of the Shannon and related sandstones that is parallel to the shoreline of Gill and Cobban (1973). The pattern of coherent, along-shore flow is characteristic of offshore, geostrophically balanced flows. It differs markedly from that of the nearshore outcrops of the Bighorn Basin, whose paleocurrent indicators form a high angle with the shoreline (Fig. 8). It also differs from the "...changing flow directions typical of tidal settings" (Sullivan et al., 1995, 1996). The difference in flow regime between the Shannon-Class Sandstones and their basin margin equivalents is ultimately a difference in depth, in that the outer shelf water column is sufficiently deep to sustain a geostrophically balanced response to the pressure field generated by wind stress.

How deep is the Shannon? Stratal geometry and flow depth

There is a further fluid dynamic test for depth. Storm beds decrease in thickness down the shoreface, because as water depth increases, wave surge on the seafloor weakens. The standard deviation of bed thickness also decreases, even when normalized against mean thickness (Niedoroda et al., 1989). This happens because the range of storm intensities affecting the seafloor decreases as water depth increases; with increasing depth, only the more intense storms generate waves with periods sufficiently long for orbital motion to reach the seafloor. Depth thus acts as a filter of wave motion, so that beds on the deeper shelf are thinner than those of the shoreface, but are also more uniform in thickness (Thorne et al., 1991). When the mean (*m*) of strata thickness is divided by standard deviation (*s*), the beds of the thin bedded, heterolithic facies of the Class III Shannon Sandstone are significantly more uniform in thickness than are the equivalent heterolithic shoreface beds of the Class II Virgelle. In Figure 10, we have plotted (*m/s*) against a calibration curve based on the wave climate of the New Jersey Shelf (Niedoroda et al., 1989). In this plot, the Shannon beds are deeper water beds.

How deep is the Shannon? The glauconite problem

The Shannon Sandstone is enriched in authigenic minerals, some of which serve to constrain water depth and depositional setting. Most Shannon outcrops contain at least 1-2% glauconite and the glauconite content increases to 5-30% in the bar margin facies of Tillman and Martinsen (1984, 1987). This facies contains up to 6% of clasts and nodules of sideritic shale, and limonite pseudomorphs after sideritic shale. Phosphatic nodules, some clearly of vertebrate bone origin, are also relatively abundant.

The glauconite content of the Shannon is astonishing and constitutes a major constraint for any actualistic depositional model. Glauconite is thought to be formed by the diagenetic alteration of clay mineral particles packaged as fecal pellets or as internal molds of forminiferal tests in fine-grained sediment (Odin and Fullagar, 1988), but it reaches high concentrations only after it has been winnowed into sand deposits. Glauconite is widespread on modern transgressive shelves from 50°S to 65°N, at water depths between 50 m and 300 m, where the deposits may be glauconite up to 90% by weight (Odin and Matter, 1981). However, glauconite in excess of 2% has been reported from no shoreface, beach, lagoon, or estuary anywhere on the planet (Odin and Fullagar, 1988). Recent, systematic studies of diagenetic,

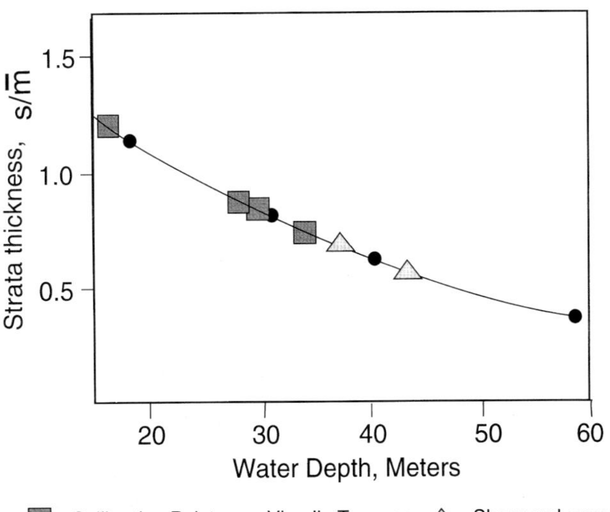

Fig. 10.—Mean bed thickness versus standard deviation for Virgelle and Shannon heterolithic stratal successions compared with calibrated depth curve, based on Niedoroda et al. (1989).

iron-bearing clays (summarized in Odin, 1988) show that some iron rich green clays do occur in nearshore settings (verdine facies; Odin et al., 1988). However, these deposits are minerologically distinct from glauconite. They only occur in the tropics (between 22° 30'S and 16° N), they occur only on muddy inner shelf surfaces where the sedimentation rate is low to negligible (Odin and Sen Gupta, 1988), and they occur in concentrations of less than 2%. Glauconite has been reported from the modern Mahakam Delta of Indonesia (Allen et al., 1979), but it occurs on a transgressed lobe (destructive phase of delta building) and in concentrations less than 2%. Glauconite is absent from modern progradational, supply-dominated shelves (shelves near major deltas; Swift et al., 1991a; Stonecipher, this volume). Granular glauconite requires very low or no detrital influx so that grains are exposed to an open marine water column for sufficiently long time (10^5 to 10^6 years for well-crystallized glauconite; Odin, 1981; Odin and Fullagar, 1988; Murav'ev, 1988). Glauconite, which is indicative of a shelf setting, coexists in the Shannon with siderite nodules. It is incorrect to assume, as Sullivan et al. (1995, 1996) do, that the siderite nodules limit the Shannon to a "brackish" setting, since Gautier (1983, 1986) reports siderite as a characteristic early diagenetic mineral in many Cretaceous marine shales of the Western Interior. Nor is it reasonable to assume that the glauconite is secondary, as do Walker and Bergman (1993), following Hansley and Whitney (1990). In modern settings, secondary glauconite (eroded from underlying source beds deposited in an earlier cycle of deposition) rarely attains concentrations of more than 2%, and in any case, there are no glauconitic source beds subcropping beneath the Shannon.

The Shannon as a glauconite concentration mill

We suggest that the glauconite in the Shannon Sandstone is entirely an authigenic mineral, and that in the Shannon, offshore sand bodies constituted "glauconite concentration mills." The concept is best understood by reference to that stratal architecture of the Shannon. In shallow marine deposits, the thickness of a storm bed is a power function of the minimum resuspension depth (a', depth to which the annual-scale storm erodes), and the return period (T) of the maximum wave height associated with the storm (Thorne et al., 1991). This relationship leads to the definition of a reworking ratio, ($r = a' / \mathring{a}$), where ($\mathring{a}$) is the accumulation per event. The parameter (r) is a dimensionless number that describes the number of times a grain is resuspended before it undergoes final burial.

Figure 11 presents a stratal dynamics model for the Shannon. In this model, the Shannon dispersal system is a subset of a larger shelf system. It begins in the source environment, upcurrent of the Shannon, as an eroding surface ($r =$). Quartz grains and glauconitized foraminifera tests and fecal pellets are eroded from this surface, or are bypassed across it from further upstream. On the surface, interaction begins between the sea bed and boundary layer flow, so that a sand ridge begins to grow as a large scale, flow-oblique bedform (Parker et al., 1982). The growing ridge migrates slowly down-current, by up-current flank erosion and down-current flank deposition. In a profile across the ridge, the reworking ratio, (r), becomes finite somewhere on the up-current side (that is, deposition begins), decreases over the crest, and takes on fractional values ($r < 1$) on the down-current side. The value of r has a major effect on the time value of the diastemic surfaces that bound the cross-strata sets of the cross-bedded facies, and the suspension-deposited beds of the thin-bedded facies. As the sand ridge is crossed from the up-current side to the down-current side, a few diastems of relatively large time values divide to become many diastems of small time value.

In this model, the most up-current facies is the most condensed ($r > 1$). It is deposited as bed load in a proximal depositional environment by migrating megaripples. In megaripple fields, each megaripple migrates up the back of the megaripple immediately downstream of it. Turbulent reattachment of the boundary layer downstream of the separation bubble results in a zone of erosion that migrates ahead of the parent megaripple, and which shears off up to one-third of the previously deposited cross-strata set. As a result, the bedding surfaces of the cross-bedded facies are diastems that contain most of the time represented by the deposit (Fig. 11B). Its beds are multiple event beds, consisting of successive bundles of cross-strata deposited during long return period events (currents associated with severe storms), which are separated by reactivation surfaces. Most of the sediment received by this depositional environment is bypassed, but the larger, heavier, fecal pellets, foram tests, and quartz grains are retained for permanent burial. Because of the very high reworking ratios in this depositional environment ($r > 1$), grains are buried, exhumed, and reburied repeatedly over thousands of years. The glauconitization process continues in this environment; the phyllosilicates in the foram test and fecal pellets are glauconitized, or further glauconitized, by contact with sea water, and are 'cured' (their crystallinity is improved) by cyclic oxidation and reduction (Odin and Fullagar, 1988). At least some of the siderite formation in the Shannon appears also to have been formed during this early stage, developing within several meters of the water-sediment interface during the reducing portion of the cycle (methanic phase; Gautier, 1983)

Successive down-current facies are less condensed, and on the down-current side of the ridge, sediments are sufficiently fine to travel as suspended load and are deposited as graded beds (thin-bedded and bioturbated facies; Fig. 11B). The bedding surfaces of these facies are diastems of very brief duration; most of the time of the deposit is represented by sediment. The beds are mainly single event beds formed by events of relatively short return period; reworking ratios are less than 1. These beds are slightly glauconitic, only because of their position immediately downstream of the glauconite mill of the cross-stratified, upstream facies.

SHANNON ARCHITECTURE

Facies architecture as a criterion

A final Shannon issue is a complex one and needs to be dealt with in sections. This is the nature of the stratigraphic architecture of the Shannon. Surely, deposits as genetically diverse as Transgressive shelf sand ridges, lowstand, prograding shoreface wedges, and incised valley fills are sufficiently different in their geometry and facies assemblages that these criteria are diagnostic. We will approach the problem in two steps. We will examine the model put forth by Sullivan et al. (1995, 1996), who propose a lowstand valley fill origin for the Shannon Sandstone, and the model of Walker and Bergman (1993); Bergman (1994); and Bergman and Walker (1995), who propose a lowstand wedge model. We will then consider two transgressive models for the Shannon; a collapsed sequence model and a shelf plume model.

Lowstand Models

Valley fill model.—Sullivan et al. (1995, 1996) have presented a model for the formation of the Shannon at Hartzog Draw as an incised valley fill. The fill is seen as a succession of deltaic sand bodies separated by high-frequency sequence boundaries. In this model, shoreline orientations, described by Gill and Cobban (1973) as trending north-south along the western basin margin during highstands, are thought to rotate into east-west orientations during lowstands, normal to the axis of the southward-plunging basin floor. The resulting subaerial drainage net is seen as controlled by a northwest-southeast fault system. Thus, at lowstand, reversing tides in the valleys would yield paleocurrent directions similar to those actually measured.

A troubling aspect of this "lowstand" model is the lack of lowstand deposits. In classical Vailian theory (e.g., Posamentier and Vail, 1988), lowstand valley fills are primarily fluvial deposits. They backstep up the erosional valley axis, even as the lowstand delta progrades seaward. Lowstand fluvial deposits extend as high in the section as the turnaround conformity, the conformity that correlates with the transgressive surface (Allen and Posamentier, 1993); subsequent backstepping fluvial beds belong to the transgressive systems tract. However, the Shannon can be traced in outcrop and in the subsurface for over 100 km to the north, but exhibits no identifiable fluvial deposits.

Estuarine deposits, which should succeed the lowstand delta during the ensuing transgression, are also missing. As the sea floods the river valley (Fig. 12), wave attack widens its margins, and a muddy estuary fill is deposited over a wave cut "bay ravinement" (Foyle and Oertel, 1997). As flooding continues, the top of the muddy estuary fill is scoured by tidal currents, and this "tidal ravinement" (Allen and Posamentier, 1993) is buried beneath the back-stepping estuary-mouth sand body. In the transgressive systems tract, central estuary muds are overlain by bayhead delta deposits, which in turn overlie the back-stepping fluvial deposits, while the original lowstand delta sandbody is transformed into an estuary mouth shoal and retreats marginward over all of these depositional systems. The result is a complex and characteristic stratigraphy (see Dalrymple, 1992; Foyle and Oertel, 1997); see also Figure 12. The depositional patterns sometimes described as a "sandwich," for which the fluvial and bayhead delta sands are the lower slice of bread, the estuary mouth sand body is the upper slice of bread, and the muddy estuary system, which is bounded below and above by the bay and tidal ravinements, respectively, is the fill. A lowstand tidal delta must thus give way to a classic transgressive estuarine sequence at some distance upstream, but as the Shannon is

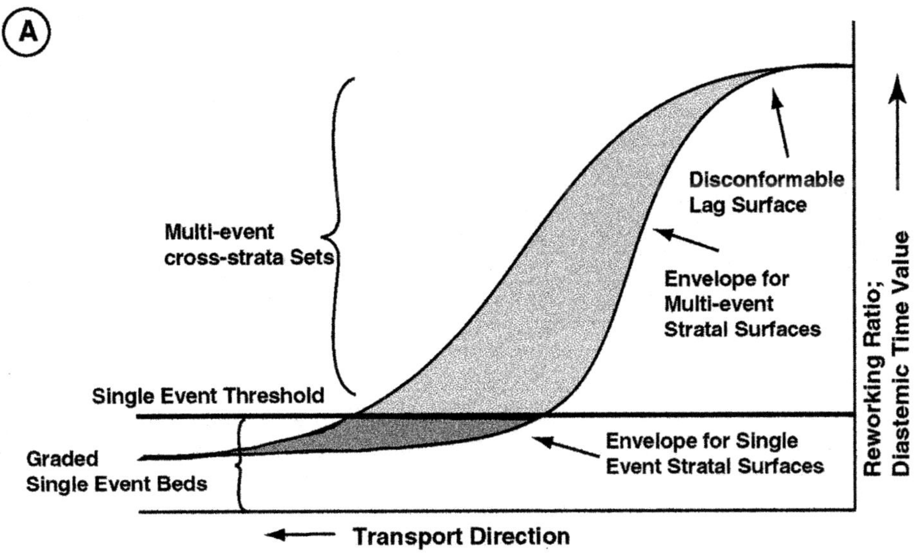

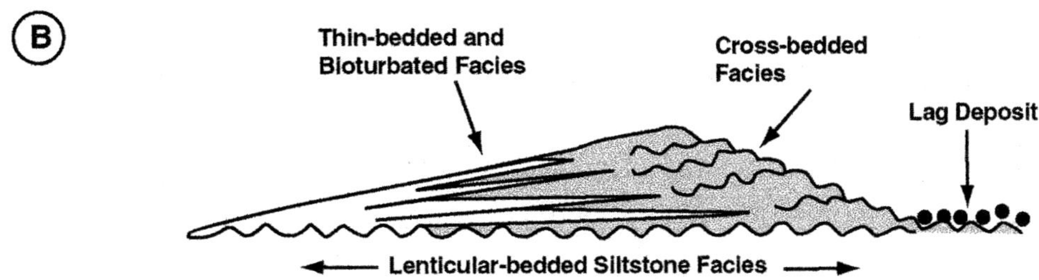

Fig. 11.—Stratal dynamics model for the Shannon. A) The reworking ratio, r, decreases from an infinitely high value in the eroding source environment upcurrent from the ridge to fractional values on the prograding downcurrent side of the ridge. B) A few diastems of relatively large time value subdivide to become many diastems of small time value on the downcurrent flank of the ridge. Modified from Gaynor and Swift (1988).

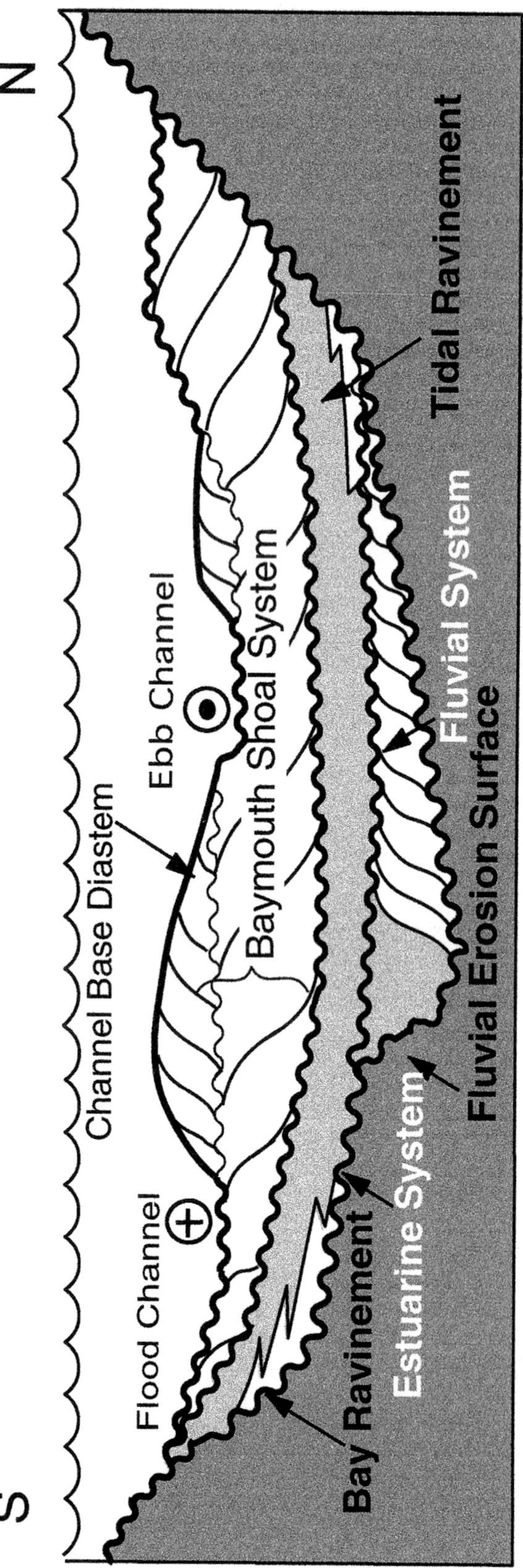

Fig. 12.—Model for an incised valley fill. Sketch of depositional systems in the Chesapeake Bay Mouth and their bounding surfaces, based on seismic work by Foyle and Oertel (1997). The "fluvial system" includes the bayhead delta. All of these systems lie above the turnaround surface and belong to the transgressive systems tract.

traced in outcrop and in the subsurface for 100 km to the north, no such pattern appears (Spearing, 1976).

Undeniable valley fills are present in the Shannon Sandstone interval, in the basin margin deposits of the Bighorn Basin and serve as calibration for identifying valley fill facies in the mud-prone downdip Shannon. The Fishtooth deposits fill a valley in the Cody Shale, while the valley-fill capping the Virgelle Member fills a valley scoured into the shoreface sands of the Virgelle (R. Fitzsimmons and S. Johnson, pers. commun., 1998). These consist mainly of well-sorted, tidal cross-strata sets (R. Fitzsimmons and S. Johnson, pers. commun., 1998). Low-energy estuary-margin deposits are abundant, including tidal point bar, tidal flat, and marsh deposits, features totally lacking in the Shannon. Nowhere in these transgressive valley fills is there more than a trace of glauconite. Tidal sand ridges in estuary mouths are fed from the beaches and surf zones of the adjacent open coasts, and surf zones do not create glauconite pellets; they destroy them (Odin and Fullagar, 1988).

Incised shoreface model.—Walker and Bergman (1993), Bergman (1994), and Bergman and Walker (1995) have presented a contrasting model for the origin of the Shannon (Fig. 13) They argue that the Shannon consists of stacked, lowstand, progradational deposits that rest in "incised shorefaces." Bergman (1994) and Bergman and Walker (1995) apply the incised shoreface model to Shannon outcrops at Salt Creek Anticline, and subsurface deposits in Hartzog Draw and Heldt Draw Fields. These surfaces are "asymmetrical, one-sided scours." They are described as regressive ravinements (In the terminology of Bergman [1994], and Bergman and Walker [1995], the surfaces are "regressive surfaces of erosion"). The scours are filled with regressive shoreface sandstones, and each is capped by a tabular transgressive deposit. The model has evolved from earlier studies of Upper Cretaceous marine sandstones in the Alberta Basin of Canada. (Bergman and Walker, 1986; Walker and Eyles, 1989).

The progradational fills within the one-sided scours are similar to the lowstand wedges described by Posamentier et al. (1992) from the Quaternary shelf edge deposits of the Rhône Continental Margin (Fig. 13), although the dynamics attributed to the Rhône margin model differ in some respects from those of the Bergman-Walker model. The Rhône margin model also shows a lowstand wedge nesting in a curved, asymmetrical surface, but according to Posamentier et al. (1992; Fig. 14), the surfaces are interpreted as correlative conformities that formed as a "drowning surface" inherited from an earlier progradation, rather than as a scour. Scour occurs only beneath the initial, sharp-based, landward portion of the lowstand shoreface succession, where the water was not yet deep enough during deposition to accommodate the full shoreface profile.

The prograding shoreface sands of the Incised Shoreface Model, can, like the deposits of the Valley Fill Model, be constrained against corresponding examples from the marginward deposits of the Bighorn Basin. Significant discrepancies exist. The tongues of the Virgelle also formed as shoreface progradations (Class II Sandstones). Cross-stratified upper shoreface facies in the Virgelle are never heterolithic, as are the cross-stratified Shannon facies. Locally, glauconitic units occur in the Virgelle, but they are transgressive, glauconitic, shelfal deposits (Class III Sandstones), which disconformably overlie both the progradational shoreface tongues and their cross-stratified estuarine caps (R. Fitzsimmons and S. Johnson, pers. commun., 1998). Furthermore, as noted above, the interbedded sandstone and shale facies of the

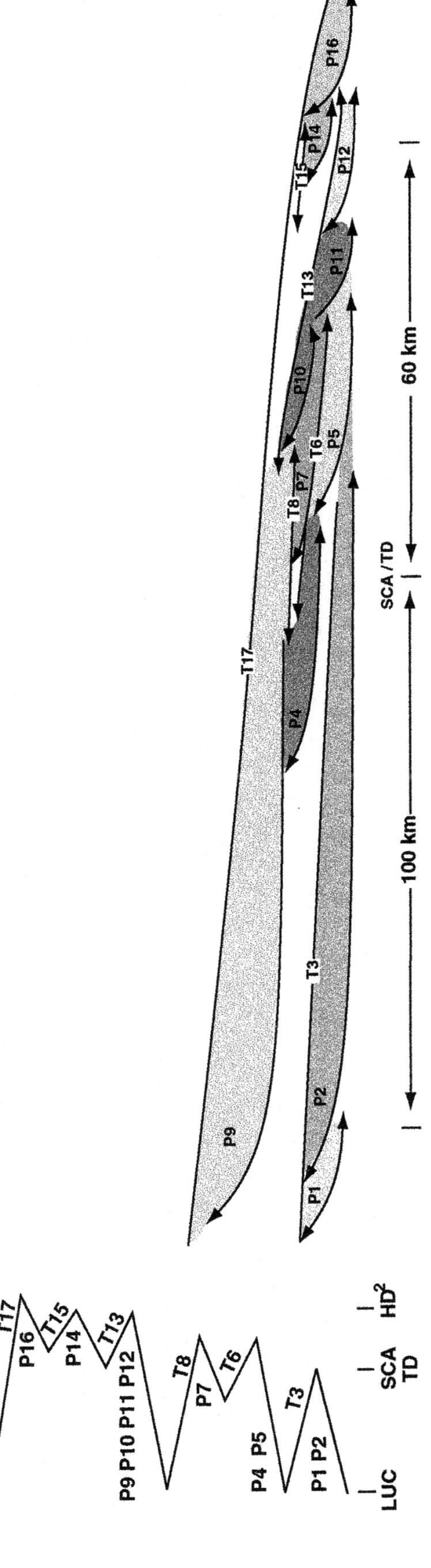

Fig. 13.—Schematic diagram (Bergman and Walker, 1995) of the relationships of the prograding (P) sandbodies in a cross-section through Lucerne (LUC), Salt Creek Anticline (SCA), Teapot Dome (TD), and Heldt and Hartzog Draws (HD2). The cross-section is not to scale. At the left, the extent of the various progradational (P) and transgressive (T) events is shown.

Shannon has notably deeper thickness characteristics than do the Virgelle tongues. Calibrations based on the New Jersey Shelf wave climate suggest 40 m water depths for the Shannon versus 25-35 m for the Virgelle (Fig. 10).

Lowstand wedges as shelf-edge deposits.—Perhaps the strongest objection to both lowstand models-lowstand tidal delta; Sullivan et al. (1995, 1996) incised shoreface; Walker and Bergman (1993), Bergman (1994), and Bergman and Walker (1995)—is that these workers are looking in the wrong place. The Vailian model for sequence stratigraphy is at present undergoing actualistic hypothesis testing, in the sense that Quaternary sections of continental margins are being used as natural laboratories in which its characteristic geometries can be identified in settings where the causative dynamics can be directly observed (J. Carey, pers. commun., 1998). Recent seismic studies on deltaic continental margins have yielded deposits very similar to Vailian lowstand wedges, but typically these are capping deposits on a well-defined shelf-slope break. Posamentier et al. (1992) give an example of a shelf-edge lowstand wedge. Other examples are described by Suter and Berryhill (1985), Milliman et al. (1990), and Tesson et al. (1990).

The Campanian interval of Wyoming, in fact, has a well-defined shelf-slope break. At the onset of the Campanian, the rate of subsidence in the Wyoming-Colorado portion of the Western Interior Basin increased by an order of magnitude (Cross and Pilger, 1978). Sediment, shed by the rising anticlines and duplexes of the Sevier front to the west, entered the basin but could not keep pace with subsidence. During periods of reduced relative sea-level rise, a shelf terrace with a clinoform internal structure would prograde seaward, building the seafloor up to an equilibrium level where wave motion and currents could mobilize the sediment, bypassing it over the topset beds, onto the clinoform front. During the next subsidence pulse, the newly formed tongue would founder, sinking faster than the sediment supply could aggrade its top; then, at the end of the pulse, a new tongue would prograde over the first. It is possible to reconstruct this history, because a detailed correlation of well logs was undertaken by Asquith (1970) in his classic paper (see Fig. 14 for a simplified version of the original Asquith cross-section). The cross-section reveals a subsurface Lower Campanian shelf edge built up by these repeated progradations.

It is clear from the cross-section that for a period of about 3 million years (Fig. 4), successive members of the Pierre Shale were able to prograde no further than the center of the Powder River Basin, where they stacked up to form a compound intracontinental shelf edge with at least 200 m of relief before being overstepped by the Mitten Black Shale Member. Both the Shannon Sandstone, and the look-alike Sussex Sandstone above it, appear in the correlation as "kicks" in the upper part of the Gammon Ferruginous Shale Member and in the upper part of an unnamed member, respectively.

Since the sandy Shannon and Sussex zones can be traced to this Early Campanian shelf edge, the Shannon and Sussex do not terminate in their outcrop zone, but continue some 80 km to the east. Here, in the basin center, not in the Salt Creek or the Hartzog Draw Fields, should lowstand architecture of the Eagle sea-level cycle be sought. In fact, the thin Sussex and Shannon zones are the sandiest horizons at the Lower Campanian Shelf edge, in what is basically a muddy sequence. Asquith believes that the front of the bentonite-rich Unnamed Member collapsed, creating a slide scar, and resulting in a fan-like deposit at the foot of the slope. However, there are no sand build-ups that might be construed as

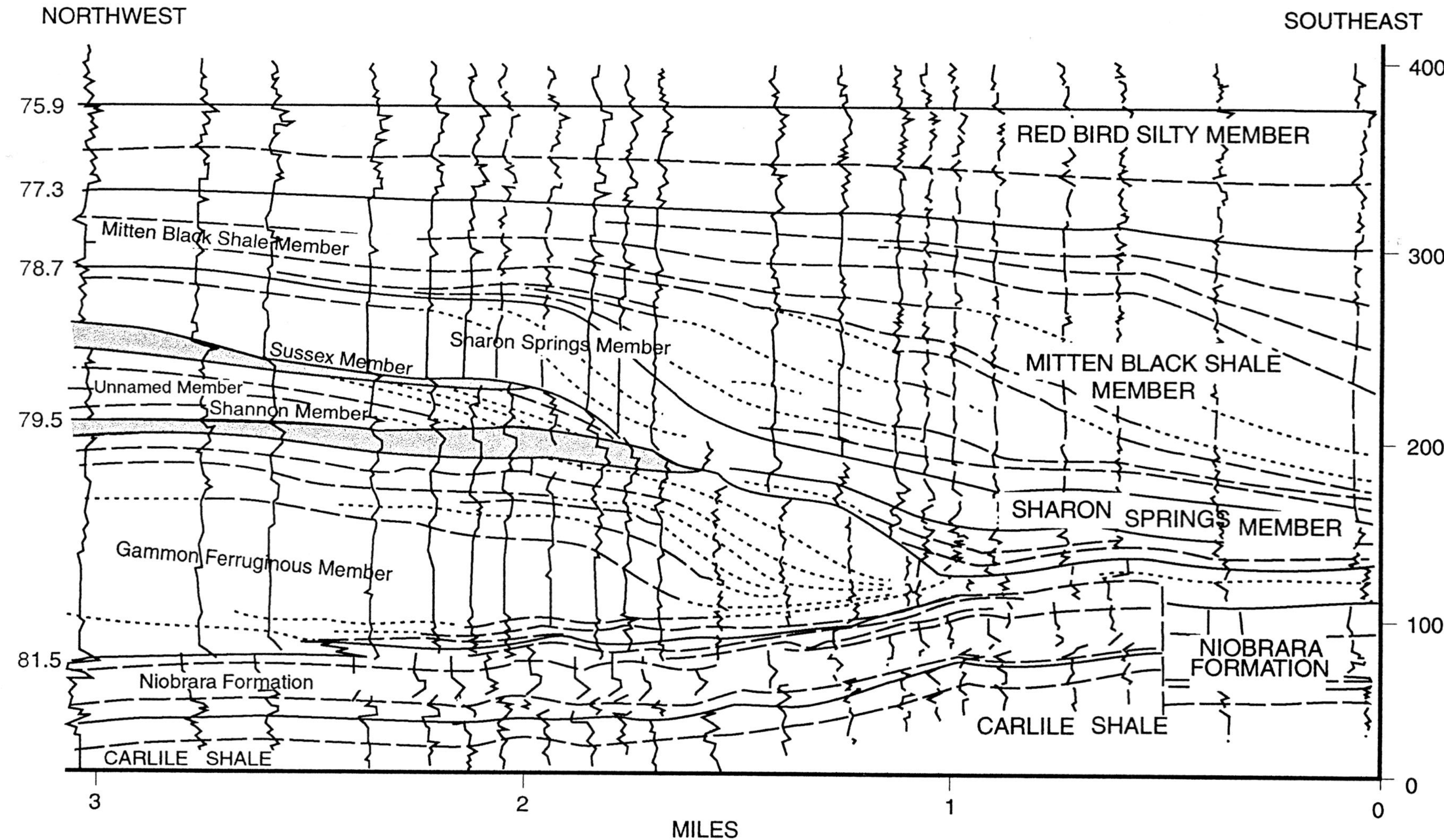

Fig. 14.—Northwest-southeast wireline log cross-section of the Lower Pierre Shale and the Niobrara Formation in the Powder River Basin, modified from Asquith (1970). Cross-section reveals an epicontinental shelf edge and slope formed during Lower Campanian time. Shannon and Sussex Members are interpreted here as transgressive systems tracts of the Sharon Springs and unnamed cycles. See Fig. 2 for location.

lowstand deltas, and no complex lowstand architecture. A "detached lowstand shoreline" (Posamentier et al., 1992) is not present, and it seems unlikely that the shoreline, with sand-supplying river mouths, ever moved this far east.

TRANSGRESSIVE MODELS

General

Transgressive depositional architecture is distinctive. We suggest that the Shannon interval is a transgressive interval. Therefore, in this section, we review transgressive sedimentation and the resulting geometries, so that we may examine the Shannon Sandstone for these characteristics.

Transgressive sedimentation

A ravinement is "an unconformity cut by a transgressive sea," a term introduced by L. D. Stamp in 1921, to describe the erosional surfaces that bounded the successive Tertiary progradations of the Paris basin. A coastal engineer, Bruun (1962), has inferred the nature of the dynamics by which the surface is cut; his concept, later encapsulated as "Bruun's Rule" (Schwartz, 1967) presupposes a coastal profile in the shape of an exponential curve, concave up, with the steeper limb being the shoreface and the gentler limb the inner shelf floor. The profile is seen as an equilibrium response to the wave and current regime of the shoreface. Bruun's Rule states that during a rise in sea level, the equilibrium profile translates landward and upward by means of shoreface erosion and that in the absence of other losses, there is an equal volume transfer of sediment from the shoreface to the ravinement surface on the inner shelf floor (Swift, 1968).

Transgressive sedimentation produces a distinctly different sedimentary deposit than does regressive sedimentation. In regressive sedimentation, rivers deliver primarily fine sediment. Sediment passes from an eroding source terrain through a graded coastal plain, in which coarse sand and any gravel is left behind on a upper braid plain; a meander plain captures the medium sand on point bars, and only fine and very fine sand, silt, and clay fractions are delivered to deltas (Swift and Thorne, 1991). The deltas, however, can be considered as "open valves," bypassing their load to the sea. Flood sands pile up on delta-mouth bars and are redistributed along coast in the surf zone, so that the shoreface progrades. Sand is swept from the upper shoreface onto the inner shelf by storms. Flood plumes rain very fine sand, silt, and clay onto the inner shelf. Sediment input is high and reworking ratios (see previous discussion) are low; a grain is subject to few, if any, resuspensions before final burial.

During transgressions, however, the "valves" are shut (Swift and Thorne, 1991). Deltas founder and become estuaries that trap river sediment as "valley fill." The source-sink relationship, between river mouths and the intervening coastal reaches, reverses. The shoreface, no longer nourished and protected by sand from the deltas, erodes. Much of the sediment thus produced moves offshore to nourish the leading edge of the transgressive sheet on the shelf in accordance with Bruun's Rule. The coarser fraction (sand) is retained in the along-shore drift. It is transported to the next estuary, where it is deposited as an estuary mouth shoal. Transgressive sediment is thus autochronous in origin; the shelf is cannibalizing is own substrate through the ravinement process. Sediment input is low and reworking ratios are high. A grain is subject to many resuspensions before final burial. Fines are bypassed offshore, and a coarse residue collects that may be made yet coarser if the ravinement process breaches an underlying shelf valley. Authigenic constituents such as shell hash and glauconite become important (Swift et al., 1991a).

Transgressive architecture: Collapsed sequences and along-shelf plume deposits

Our understanding of transgressive geometries has been greatly clarified by Vailian stratigraphic thought (Vail et al., 1977); we know transgressive sedimentation results in onlapping reflector terminations on seismic records and backstepping patterns of parasequence stacking. Beyond these baseline concepts, evidence indicates that during periods of high-frequency sea level oscillations, "collapsed sequences" may form during the transgressive phases (Swift et al., 1991b). That is, if the depth of penetration of the retreating shoreface is sufficient, the underlying, progradational, strand plain-shoreface deposit will be largely destroyed, and will be redeposited as an onlapping, backstepping sheet over a ravinement surface.

Marine transgressive deposits seaward of the shoreline are also distinctive. Low sedimentation rates and high reworking ratios lead to relatively coarse deposits that are resuspended primarily by peak, along-shelf storm flow. As a result, grain-size gradients are oriented alongshore (advective pattern), and shelf deposits take the form of "shelf floor plumes" that extend down-shelf from a rivermouth source (Columbia River; Nittrouer et al., 1979) or down-shelf from an eroding shelf area (Bothner et al., 1981). This depositional feature differs from a prodelta shelf floor plume (Swift et al., 1991; Scheihing and Gaynor, 1991) in that it is much larger in scale and results from different dynamics. Alongshelf plumes, with their cross-shelf facies boundaries, are to be distinguished from the wave-dominated, along-shelf facies boundaries characteristic of the diffusive pattern of finer-grained regressive shelves (Narayit Shelf; Curray, 1969).

Transgressive architecture: Sand ridges

The dynamics described above influence the depositional architecture. Sediments coarser than 175 µm travel as bed load, and bed load transport is largely accomplished by bedform migration. Thus, transgressive deposits, whether of storm or tidal origin, tend to contain a cross-stratified facies, while regressive deposits, whether tempestites or tidalites, consist largely of finer beds with suspensive stratal architecture (Bouma sequences). The aggrading surface of the transgressive deposit responds to flow at a series of spatial scales, resulting in such flow-normal bedforms as ripples (1 cm-1 m spacings), megaripples (1 m-10 m spacings), sand waves (100 m spacings), and sand ridges (spacings of 2-4 km; e.g., Fig. 7). Sand ridges are large-scale, flow-oblique bed forms characterized by coarse up-current facies and finer down-current facies (Fig. 15). On microtidal to mesotidal shelves, the smaller bedforms (megaripples and sand waves) occur mainly as the up-current facies of sand ridges.

The Atlantic Shelf ridges of Figure 7 are storm-built, but the fluid dynamics are thought to be similar to those of the tide-built ridges of the North Sea (Parker et al., 1982; Huthnance, 1982; Hulscher, 1996). The Atlantic Shelf sand ridges have been molded into the transgressive sand sheet formed by erosional shoreface retreat (Swift et al., 1986b). In many areas, they began as large-scale, slowly migrating bedforms on the shoreface, and are lowered onto the seafloor as the shoreface retreats out from under them. Elsewhere, they have formed more or less spontaneously on the

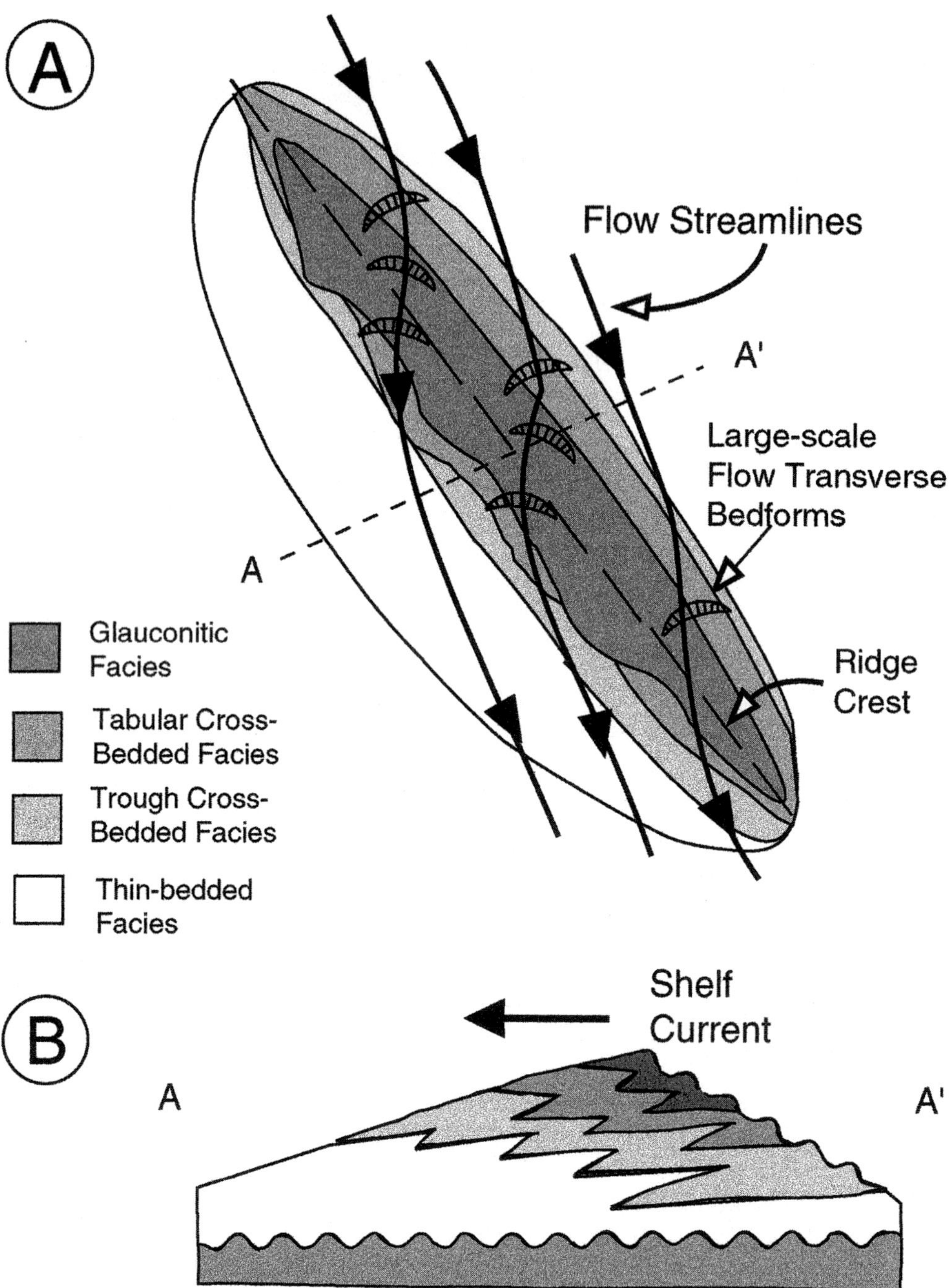

Fig. 15.—Model for a Shannon Sand Ridge. A) Plan view of lithofacies distribution on a ridge and flow stream lines. Veering of flow over ridge crest corresponds to paleocurrent observations from oriented Hartzog Draw cores (Gaynor and Swift, 1988; see also Hearn, et al., 1983). B) Flow parallel cross-section through sand ridge deposits, showing lithofacies distribution. Modified from Gaynor and Swift (1988).

shelf floor in response to southerly winter storm flows. Rine et al. (1991), report ridges on the central New Jersey shelf that either were created when their area attained mid-shelf depths or at least were remade when that depth was attained, since shallow water foraminifera and other indicators of a shallow water setting are lacking. Atlantic Shelf sand ridges have steeper down-current than up-current flanks and form northward-opening angles, with respect to the shoreline, that open into the direction of prevailing flow (Swift et al., 1986b).

Transgressive sand ridges are not confined to the open shelf; estuary mouth shoals are also molded into sand ridges, each with its complement of proximal cross-bedded facies and distal, horizontally bedded facies (Ludwick, 1974; Dalrymple, 1992). Sand ridges also occur within tidal deltas, landforms that are gradational with estuaries. Macrotidal deltas are so heavily flushed that they cannot grow a deltaic sediment prism; they have open water between their mouth shoals and the "bayhead deltas" at their apices; they are thus morphologically indistinguishable from estuaries, but differ systematically in that they are primarily river-sourced, rather than ocean-sourced as are estuaries (Wright and Coleman, 1973). To the first approximation, estuaries are transgressive landforms while deltas are regressive landforms; but open-water tidal deltas are at best products of stillstand, so the generalization, that shelfal estuarine, and tidal delta sand ridges are transgressive to stillstand morphologies is a reasonable one. It is also reasonable that Sullivan et al. (1995, 1996), and Gaynor and Swift (1988) call on sand ridges as an important element in their otherwise very different (tidal delta versus shelf) models.

Sand ridges in these diverse settings exhibit a characteristic architecture. They are aligned obliquely to flow, not parallel to it (Huthnance, 1982; Hulscher, 1996). They therefore have up-current versus down-current flanks and migrate laterally. As a consequence, they contain large-scale, low-angle inclined stratification that on seismic records resembles clinoform or shingled structure (e.g., Swift and Field, 1981). In Figure 15 we present a model for a shelf sand ridge that is adapted to the mud-prone Campanian setting and based on observations of the Shannon outcrops and the subsurface of the western margin of the Powder River Basin (Tillman and Martinsen, 1984, 1987; Gaynor and Swift, 1988). Unlike Atlantic Shelf ridges, Shannon ridges have steeper up-current than down-current flanks because the down-current flanks are mud-rich and therefore have very gentle slopes. They often make angles with the shoreline that open down-current with respect to the shelf flow. In fluid dynamical theory, this is as permissible as the other pattern (Huthnance, 1982) and also may be favored by the muddy setting.

Collapsed sequence model: Application to the Shannon

The preceding synopsis of transgressive sedimentation allows us to present two models for the Shannon-equivalent sandstones of the Cretaceous Western Interior. A Collapsed Sequence Model is probably most applicable for the inshore Shannon equivalents in the *Baculites* sp. (smooth) biozones at Billings, Montana (Shelton, 1965); in the southern Bighorn Basin (Swift and Parsons, 1995; R. Fitzsimmons and S. Johnson, pers. commun., 1998); and in the Hanna Basin region (Tillman and Martinsen, 1985; Mellere and Steel, 1995a, 1995b). We choose the Middle Atlantic Bight of the North American Atlantic Shelf as our definition sketch of the Collapsed Sequence Model. This surface bears several prominent scarps. Recent seismic studies, summarized in Milliman et al. (1990), show that the scarps and the terraces behind them were formed by brief regressions during the Holocene transgression. In each case, when the transgression resumed, relative sea level was higher and the readvancing shoreface profile sheared through the upper half of the downlapping regressive beds, slowed and rose as it approached the thick base of the regressive wedge, then slid landward across the upper terrace leaving only a truncated remant of the original regressive deposit (Fig. 16).

We are clearly seeing something similar in the Hanna Basin of Wyoming, where a succession of Mesaverde-equivalent sandstones (Haystack Mountains Formation) intertongue with Steele Shale, a Cody Shale equivalent. The careful sedimentology of Tillman and Martinsen (1985) has revealed that the basal unit (Tapers Ranch Sandstone) is, in the terminology of this paper, a Class III Sandstone. Its facies differentiation is very similar to that of the Shannon Sandstone at Salt Creek Anticline (Tillman and Martinsen, 1985; Gaynor and Swift, 1988) of which it is a time-equivalent deposit. Other workers (Mellere and Steel, 1995a, 1995b) have undertaken studies of the higher tongues. These workers also deserve praise, in this case for the exceptional quality of their measured sections (see Mellere and Steel, 1995a; Fig. 5). They have shown that higher tongues are at their bases, normal progradational sandstones, but they terminate in sandstones that are cross-stratified and glauconitic, horizontally stratified and heterolithic, or bioturbated. These overlie and cross-cut the progradational tongues and are separated from the tongues by an erosional surface (Fig. 17). Mellere and Steel note that these capping sandstones are "initially progradational," but as they are traced up-section, they become retrogradational. They further suggest a "tidally dominated delta front to estuarine depositional setting." In their synopsis in *Geology*, Mellere and Steel (1995b) describe the cross-stratified sandstone successions as lowstand wedges, presumably on the basis of the progradational initial style, but they comment that "the depositional environments encountered in these two types (the progradation tongues, and the lowstand deposits) are surprisingly different, and the lowstand wedges, which are frequently tidally influenced, show a greater affinity with the overlying transgressive systems tract than with the preceding [regressive] shoreface deposits."

We agree that the cross-stratified sandstone assemblage shows a greater affinity with the overlying transgressive tract than with the preceding falling stage systems tract. In fact, it shows an affinity with all of the shelf sandstones, from Montana to Colorado, described in this paper as Class III Sandstones. We note that the model of Mellere and Steel (1995b), in which subaerial deposits shift seaward behind the prograding, rising lowstand beach is not born out by their own data, in which the cross-stratified sandstone units are entirely marine. We suggest that the same data are more economically explained if the sharp base beneath each cross-stratified lithosome is assumed to be a transgressive ravinement rather than a regressive one, and if the lithosome itself is viewed as a transgressive, rather than a lowstand, systems tract. We note that the progradations within each cross-stratified lithosome apparently occurred in spurts, resulting in lenticular buildups of the core facies (trough cross-bedded sandstone facies), very unlike the sustained progradations of the underlying falling-stage systems tract. These lenses are easier to explain as sand ridges, with their characteristic seaward progradation pattern, than as lowstand shoreface progradations. Mellere and Steel (1995a; Figs. 6, 7) tacitly admit sand ridge structure when they refer to "sandbars in a tidal-dominated delta front," and "sandbars in outer estuary"; see discussion of sand ridge architecture on preceding pages.

In their conceptual model, Mellere and Steel (1995a; Fig. 10), suggest that the cross-stratified sandstone lithosomes are underlain by shelf valleys. Such complex estuarine-shelf transitions do exist. Figure 7 reveals a shelf sand ridge topography being incised into an overstepped estuary mouth shoal. However, the cross-section of Mellere and Steel (1995a; Fig. 4) provides little support. The high glauconite content in the cross-bedded facies is also an impediment to the estuarine interpretation. High (greater than 2%) glauconite concentrations are lacking in any modern estuary. In view of this, and in view of the consistent truncational relationships of the cross-stratified sandstone with the underlying shoreface sandstones (Fig. 17), it remains the most economical choice to attribute the entire cross-stratified lithosome to the transgressive systems tract and to view these Class III Sandstones of Central Wyoming as transgressive sandstones.

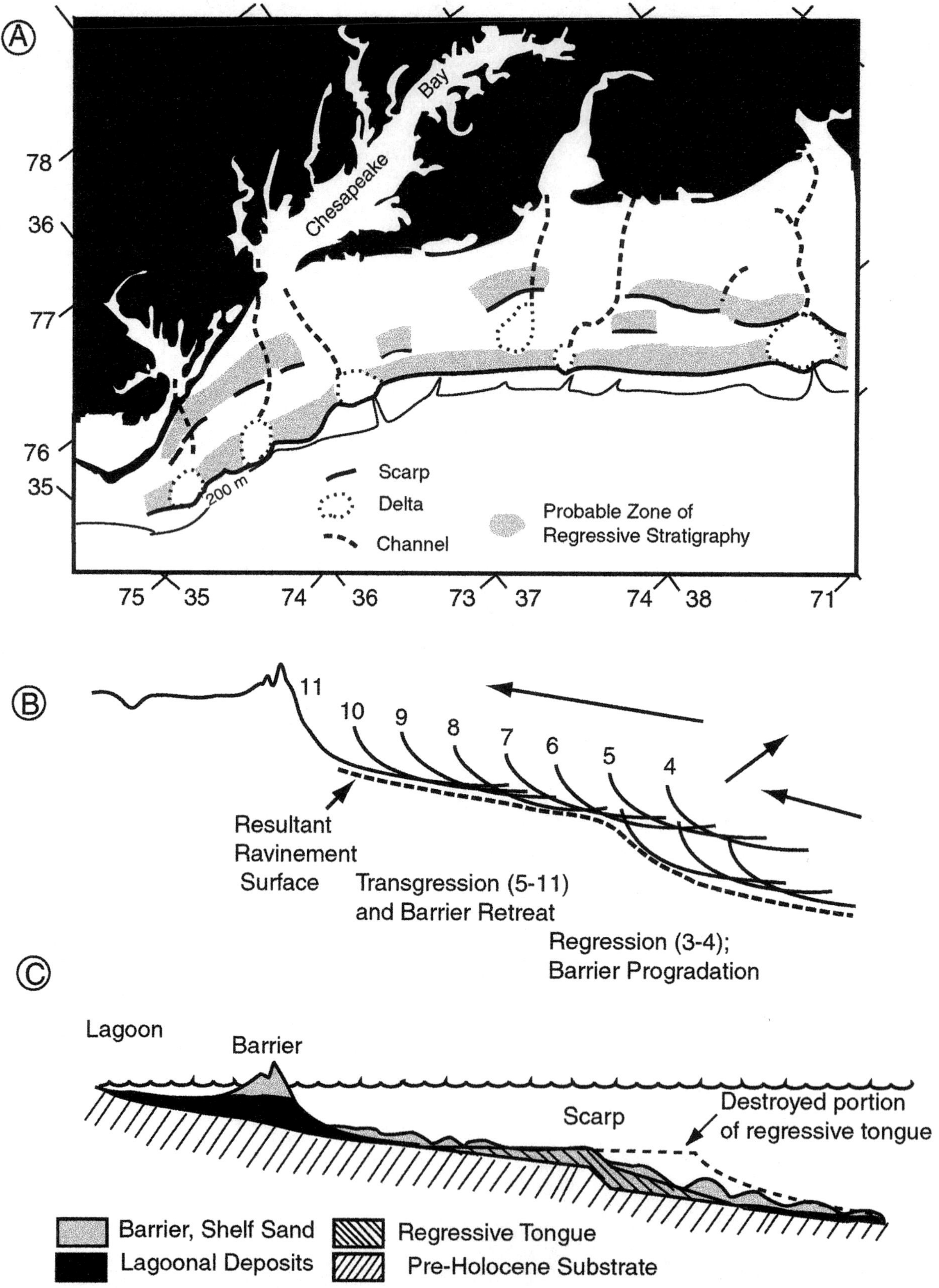

Fig. 16.—Model for a Shannon ridge field, detached shoreline version. A) Map of middle Atlantic Bight. B) Kinematics of shoreface translation, C) Resulting statigraphy. Sand is locally sourced. Here on the Atlantic shelf, transgressive oscillations left ridge fields on top of "collapsed parasequences." From Swift et al. (1991b).

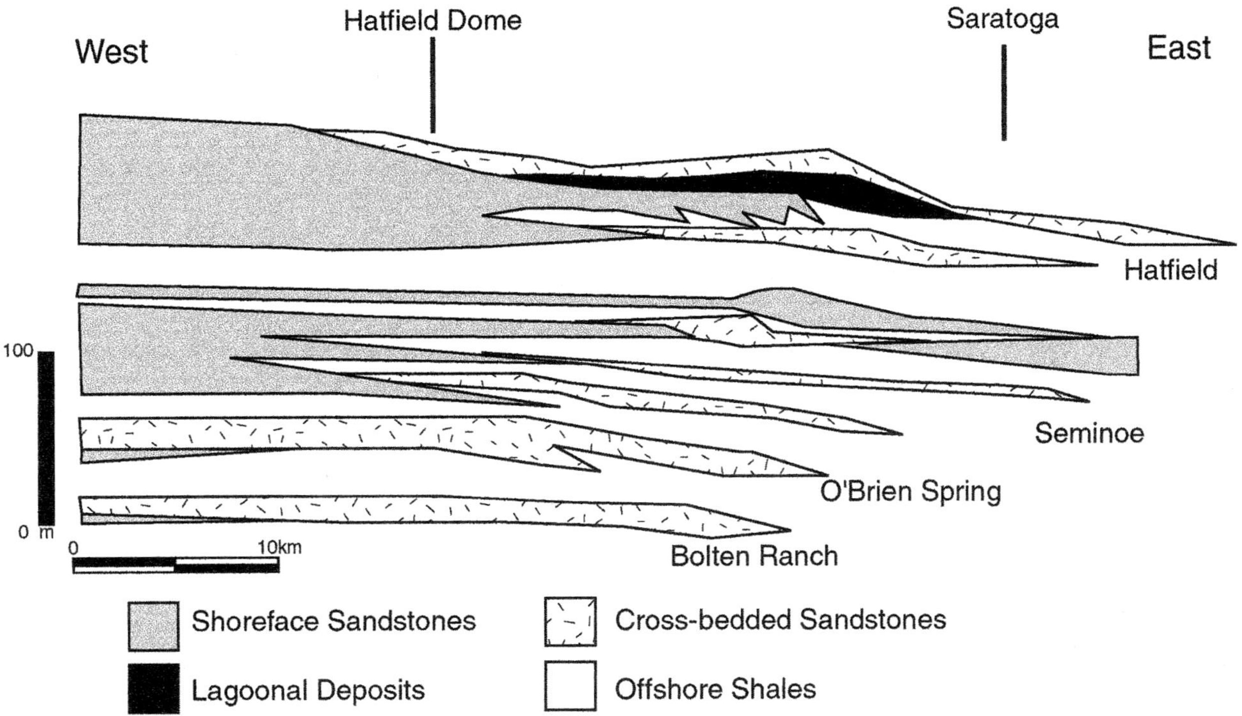

Fig. 17.—Cross-section of the lower tongues of the Haystack Mountain Formation. From Mellere and Steel (1995a). Compare with Fig. 16. See Fig. 2 for location.

Along-shelf plume model: Application to the Shannon

The collapsed sequence model described above is inappropriate for the classic Shannon Sandstone at Salt Creek Anticline and its subsurface equivalents, since truncated and transgressed shoreface sandstones are not evident. A possible transgressive model for the offshore Shannon is the wide shelf model, in which the Shannon is seen as a marine deposit that accumulates on the offshore shelf without the benefit of a lowstand shoreline.

Several Shannon workers have rejected the wide shelf model on the grounds that there are no satisfying explanations of how the sand was transported across the shelf (Walker and Bergman, 1993; Sullivan et al., 1995, 1996). In transgressive settings, however, sand does not find its way across the shelf; rather, the shelf water column advances over it and erodes sand from the substrate. It is noteworthy that the coarsest siliciclastic particles in the Shannon Sandstone are very fine to fine in grain size, a fraction which typically occurs as a coarse tail in grain size distributions of the underlying Cody Shale. Sand particles of this size are capable of traveling for long distances in short-term suspension in response to storm currents. For instance, fine sands are found many tens to hundreds of kilometers from their shoreline source on the Texas Shelf (Snedden and Nummedal, 1991), and Rice (1984) described an example of far-traveled shelf sands from the Cretaceous Western Interior Basin.

A further objection to an offshore marine setting for deposition of the Shannon Sandstone has been that it is found far from the time-equivalent shoreline (Walker and Bergman, 1993). In fact, Shannon build-ups extend into the Gebo shoreline in Figure 18B. Certainly in the Hanna Basin, the Shannon-equivalent Tapers Ranch Member of the Haystack Mountains Formation "connects" the basinal Shannon with its marginal sequence (Fig. 17). Oil-producing Shannon or Shannon equivalent sandbodies in the Wind River Basin (Lawyer et al., 1981; Cardinal et al., 1989; De Bruin and Boyd, 1990; De Bruin and Hostetler, 1991) are similarly oriented. Production from these sand bodies begins in the subsurface in the Wind River Basin due south of the termination of the Gebo outcrops in the Bighorn Basin (Cardinal et al., 1989; De Bruin and Hostetler, 1991).

Nevertheless, much of the Shannon appears to have been deposited offshore, and the transport issue must be considered. We have presented a detailed explanation of how such offshore transport might have occurred in the transgressive setting of the Lower Campanian of Wyoming (Parrish et al., 1984; Gaynor and Swift, 1988) and rephrase it here as an "along-shelf plume deposit" variant of the wide shelf model (Fig. 18). As applied to the Lower Campanian shelf of Wyoming, the along-shelf plume model requires for its source area the Central Montana Delta described by Gill and Cobban (1973) undergoing transgression at the beginning of Early Campanian time. In Figure 18A, southerly trending currents enter the narrow prodelta shelf. They are no longer receiving sediment from the flooding delta whose distributaries have become sediment-trapping estuaries. Shelf currents cannot easily cross over the shelf edge, as a consequence of the need to conserve vorticity (Casanady, 1982). As the currents approach the Montana Narrows, they must accelerate in response to the decrease in the cross-section of flow and must consequently erode the shelf floor. The wide-shelf area downstream of the Montana Narrows must conversely be a region of flow expansion, deceleration, and sediment deposition. A similar transgressive, strike-fed, shelf deposit (shelf-floor plume) is presently forming on the present-day New England Shelf south of the erosional zone of Georges Bank (Bothner et al., 1981). Figure 18B indicates that the Shannon sand ridge complex is found in the area where numerical simulations suggest similar flow expansion and deposition.

In this model, the Shannon sand ridges are seen as the product of a two-stage concentration of sand. In the first stage, erosion of the prodelta sediments of the Montana shelf produce a southward-extending, sheet-like, "fallout shadow" that overlies a marine erosion surface (Fig. 18B). In the second stage, interaction between the aggrading sheet and the repeating storm flows leads to the growth of sand ridges and the further concentration of sand (Fig. 15). The Pierre Shale beneath the Shannon Sandstone typically contains as much as 10% fine to very fine sand (Gill and Cobban, 1966). Erosion of the Pierre mud to a depth of 60 cm over 100,000 km^2 of the Montana Shelf, plus bypassing of the finer fraction and redeposition of a coarser concentrate, would produce a sandy silt or silty sand sheet tens of thousands of km^2 in extent to the south. Further winnowing and concentration of the sandy silt over hundreds of km^2 during the growth of Shannon sand ridges would lead to the cross-bedded sand facies seen in the Shannon. In this model, the Shannon is a discontinuous sheet of transgressive (Class III) sand locally molded into shelf-oblique sand ridge complexes (Fig. 18).

Note that in Figure 1, as in Figure 4 and Figure 18C, there are several Shannon horizons. High-frequency (fourth-order) sequences are well-developed in the expanded basin margin section that is equivalent to the Shannon (Claggett

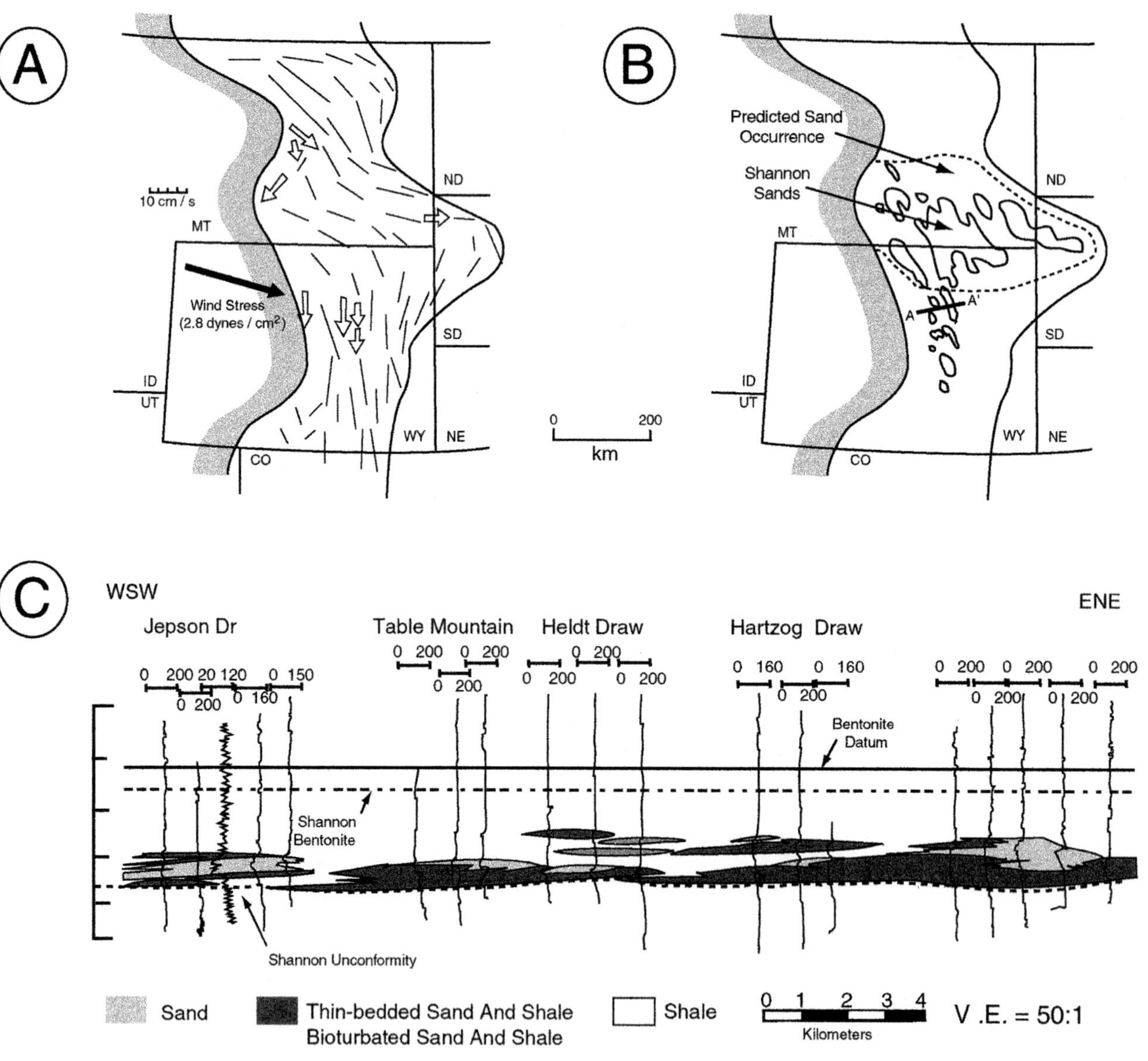

Fig. 18.—Model for a Shannon ridge field, along-shelf plume version. Sand is far-traveled. A) Numerical simulation of flow on the early Campanian Shelf in response to a northwest wind. Thin arrows are computed flow vectors; thick arrows are measured paleocurrent orientations (Parrish et al.,1984). B) Isopach map of the Shannon Sandstone (contour: greater than 2 m thickness) showing the relationship between the depositional zone predicted by the numerical simulation, and the extent of the Shannon sand-ridge complex. C) Cross-section through Shannon Fields of the western Powder River Basin showing the lenticular nature of the Shannon sand bodies above the basal unconformity. Compare the cross-section of the Hartzog Draw body with Fig. 16b. See panel B (Line A-B) for location. Modified from Gaynor and Swift (1988).

sequence; R. Fitzsimmons and S. Johnson, pers. commun., 1998) where they appear as the backstepping, progradational shoreface tongues of the Gebo Member. Asquith was not able to resolve such fine structure in the more condensed shelf-edge section of Figure 14, but the Shannon sandstones do appear to capture this high-frequency repetition. Thus, we suggest that in the Shannon Sandstone at Salt Creek Anticline we see two early oscillations of the Shannon-Gebo transgression (Fig. 4). During the subsequent interval of rapid sea-level rise, the shoreline would have migrated rapidly to the region of the Bighorn Basin. As the sea-level rise slowed again, successive Gebo tongues would have been deposited.

The argument being made here is a default argument based on the principle of parsimony, which says that when faced with multiple explanations, embrace the simplest. There is nothing inherently improbable about a "displaced shoreline" setting for the Shannon Sandstone; a sand-rich lowstand turnaround shoreline, separated from highstand equivalents by a zone of rapid transgression and regression, in which shoreline deposits are scanty or missing; see Posamentier and Vail (1988). Our objection is that there is simply no evidence for it.

DISCUSSION AND CONCLUSIONS

We have attempted in this paper to evaluate the several models for the origin of the Shannon Formation. The evidence is ambiguous; if it were not, our colleagues, persons whose competence we respect, would not hold differing views. Our review of the diverse elements of the Shannon problem has led us to several conclusions. First, while interpretations come and go, a carefully presented observation is a thing of beauty that all subsequent interpretations must explain. We refer specifically to Gill and Cobban's (1973) cross-section (Fig. 3), Asquith's (1970) cross-section (Fig. 14), the sedimentological analyses of Tillman and Martinsen (1984, 1985, 1987), and the cross-section of Mellere and Steel (1995a; Fig. 5).

More specifically, we conclude that the most important single constraint encountered in evaluating depositional models of the Shannon is the glauconite problem. Glauconite is found in open marine settings on modern continental shelves under conditions of transgressive reworking.

A further conclusion is that the Shannon Sandstone can be explained within the context of a classification of marine and marine marginal sandstones in the Cretaceous Western Interior sandstones as Class I Sandstones (fluvio-estuarine and inlet sandstones), Class II Sandstones (progradational shoreface sandstones), and Class III Sandstones (shelf sandstones). Class II Sandstones are regressive, while Class I and III Sandstones are transgressive. Only Class III sandstones are glauconitic, however. This classification has emerged in the course of this synthesis, and we have presented aspects of it in various previous sections of this paper. Here, we summarize the concept in Figure 19 and Table 2.

Then there is the issue of degrees of certainty. In a pastime such as paleoenvironmental interpretation, whose limitation, as well as its fascination, lies in its ultimately subjective nature, it is important to recognize that one is dealing with relative truths. Some explanations are merely more probable than others. The broad disagreement among our peers places our discussion firmly in this uncertain realm. The aspect of our interpretation of which we are the most confident is that the three classes of sandstone described above are coherent and meaningful divisions.

We are as confident, or almost so, that the classical Shannon Sandstone of the Salt Creek Anticline is a transgressive, open marine deposit. This means that at this point in the discussion we must abandon the Valley Fill Model as a consequence of the lack of recognizable fluvial, estuarine, or deltaic deposits (see above). We may then pause to consider a possible compromise between the remaining models. The compromise requires that some of our company of Shannon students (Mellere and Steel, 1995 a, b) would have to accede that the glauconitic lags and cross-stratified sand bodies of the distal Haystack mountain tongues are open marine deposits rather than estuarine deposits. Others (Walker and Bergman; 1993, Bergman, 1994; and Bergman and Walker, 1995) would have to agree to explain their surfaces of shoreface incision as transgressive ravinements, not regressive ones (TSE's, not RSE's), and the deposits resting on these surfaces are trangressive marine deposits. Yet others (Gaynor and Swift, 1988; Tillman and Martinsen, 1985), would have to abandon the minimalist notion that what you see is what you get; that the Mesaverde and Eagle shorelines went no further basinward than shoreline sandstones can be found. The result is a generic transgressive model in which the Shannon Sandstone consists of a series of lenses whose textures, structures, and architecture reflect an open marine sediment environment, but whose pattern of distribution is the consequence of *in situ* reworking of sediment delivered to the site of deposition by other means. However, the pattern is not a simple, band-like "stranded lowstand shoreline" (Posamentier et al., 1992). In

Table 2.—Classification of marine and marine marginal sandstones.

	Class I	Class II	Class III
Depositional Setting	Estuarine	Shoreface	Shelf
Architecture	Valley Fill	Shoreface clinoforms	Sheet or Clinoform
Paleocurrent Pattern	Shore-normal, Large angular deviation	Shore-normal, Small angular deviation	Shore-parallel, Small angular deviation
Sea-Level Cycle	Transgressive	Regressive	Transgressive
Other Criteria	Ostreid faunas	Hummocky Cross stratification	Glauconite

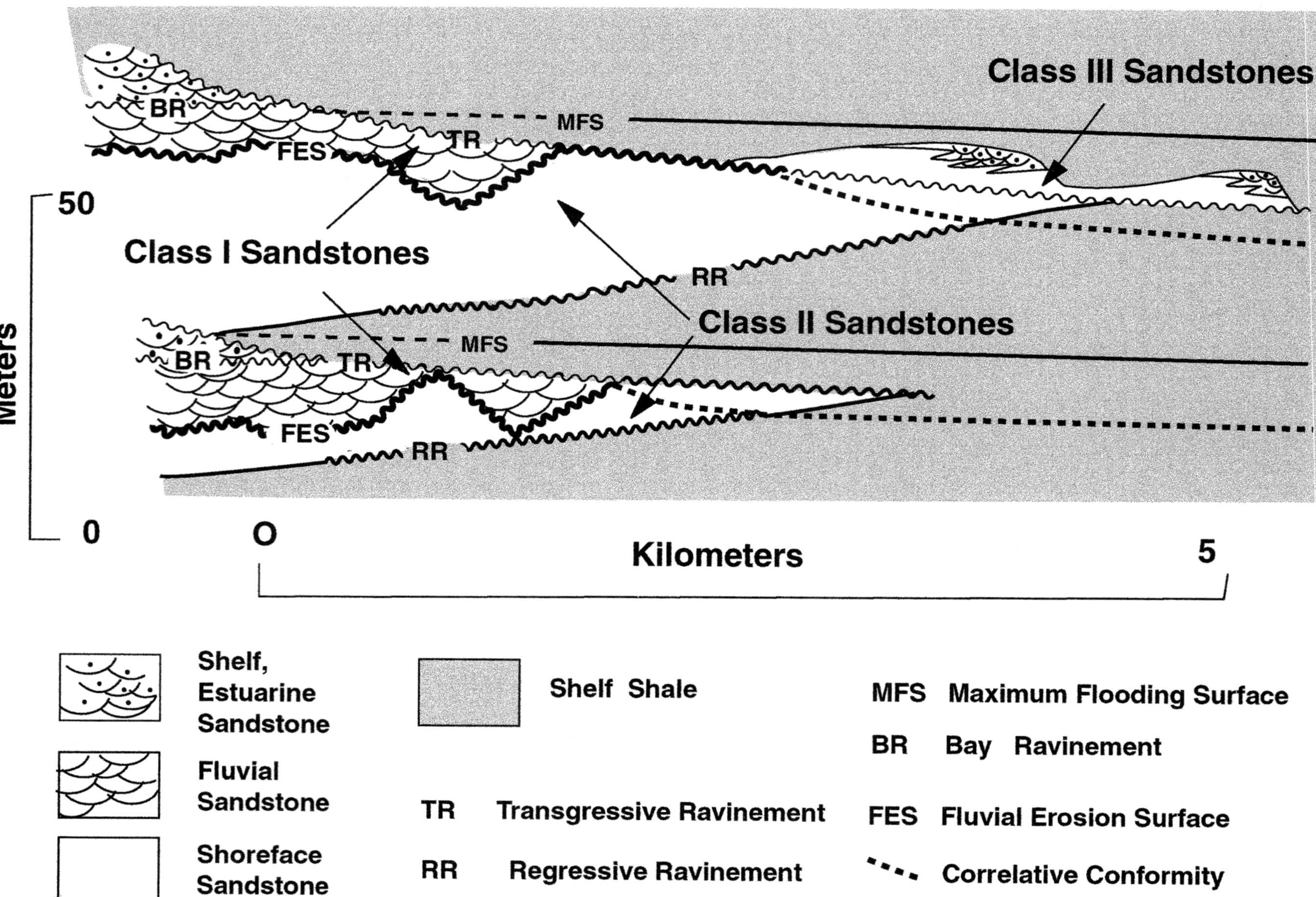

Fig. 19.—Classes of sandstone in the Cretaceous Western Interior Basin; fluvial and estuarine sandstones (Class I), shoreface sandstones (Class II), and shelf sandstones (Class III). See text for discussion. Modified from Nummedal and Molenaar (1995).

the subsurface the Shannon lenses coalesce into a broad sheet of backstepping sand bodies (Fig. 18B).

The ferment that this unusual stratigraphic unit has generated has served to sharpen the thinking of us all. We look forward to the opinions of Generation X, and the resulting opportunity to welcome our current critics into the ranks of the Old Guard.

ACKNOWLEDGMENTS

We wish to thank Cathy Sue McConaugha, who edited the manuscript. We also would like to thank L. Gifford Kessler II of Marathon Oil Company Exploration Services for not only providing BSP the opportunity to revisit the Shannon numerous times, but also for helping to fine-tune interpretations through thought-provoking conversations. Neil Gaynor and James Rine served as reviewers and provided helpful comments. We wish to acknowledge financial support for portions of the work presented here from ARCO Oil and Gas Company, Chevron Oilfield Research Company, Amoco Production Company, Mobil Research and Development Corporation, and the National Science Foundation (award EAR 9118000).

REFERENCES

ADAMS, C.E., JR., WELLS, J.T. AND COLEMAN, J.M., 1986, Transverse bedforms on the Amazon shelf: Continental Shelf Research, v. 6, p. 175-189.

ALLEN, G.P., LAURIER, D. AND THOUVENIN, J., 1979, Etude sedimentologique du delta de Mahakam: Paris, TOTAL, Compagnie Francaise des Petroles, Notes et Memoires 15, 156 p.

ALLEN, G.P. AND POSAMENTIER, H.W., 1993, Sequence stratigraphy and facies model; for an incised valley fill: The Gironde Estuary, France: Journal of Sedimentary Petrology, v. 63, p. 378-391.

ASQUITH, D.O., 1970, Depositional topography and major marine environments, Late Cretaceous, Wyoming: American Association of Petroleum Geologists Bulletin, v. 54, p. 185-202.

ASQUITH, D.O., 1974, Sedimentary models, cycles, and deltas, Upper Cretaceous, Wyoming: American Association of Petroleum Geologists Bulletin, v. 58, p. 2274-2283.

BERGMAN, K.M., 1986, The importance of sea-level fluctuations in the formation of linear conglomerate bodies; Carrot Creek Member of the Cardium Formation, Cretaceous Western Interior Seaway, Alberta, Canada: Journal of Sedimentary Petrology, v. 57, p. 651-665.

BERGMAN, K.M., 1994, Shannon Sandstone in Hartzog Draw-Heldt Draw Fields (Cretaceous, Wyoming, USA) reinterpreted as lowstand shoreface deposits: Journal of Sedimentary Research, v. 64, p. 184-201.

BERGMAN, K.M. AND WALKER, R.G., 1995, High resolution sequence stratigraphic analysis of the Shannon Sandstone in Wyoming, using a template for regional correlation: Journal of Sedimentary Research, v. B65, p. 255-264.

BLOOM, H., 1997, The Anxiety of Influence: A Theory of Poetry, (2nd edition): New York, Oxford University Press, 157 p.

BOICOURT, W.C. AND HACKER, P.C., 1976, Circulation on the Atlantic continental shelf of the United States, Cape May to Cape Hatteras: Memoires Societe Royale des Sciences de Liege, v. 6, p. 187-200.

BOOTHROYD, J.C., FRIEDRICH, N.E. AND MCGINN, S.R., 1985, Geology of microtidal coastal lagoons: The Rhode Island example: Marine Geology, v. 63, p. 35-76.

BOTHNER, M.H., SPIKER, E.C., JOHNSON, P.P., RENDIGS, R.R. AND ARUSCAVAGE, P.J., 1981, Geochemical evidence for modern sediment accumulation on the continental shelf off New England: Journal of Sedimentary Petrology, v. 51, p. 281-292.

BOUMA, A.H. AND COLEMAN, J.M., 1985, Mississippi Fan: Leg 96 program and principle results, *in* Bouma, A.H., Normark, W.R., and Barnes, N.E. (eds.), Submarine Fans and Related Turbidite Systems: New York, Springer-Verlag, p. 247-252.

BRUUN, P., 1962, Sea-level rise as a cause of shore erosion: Journal of Waterways Harbors Division, American Society Civil Engineers, v. 88, p. 117-130.

CARDINAL, D.F., MILLER, T., STEWART, W. AND TROTTER, J.F., 1989, Wyoming oil and gas symposium, Bighorn and Wind River Basins: Casper, Wyoming Geological Association, 558 p.

CROSS, T.A. AND PILGER, R.H., 1978, Tectonic controls of Late Cretaceous sedimentation, Western Interior, USA: Nature, v. 274, p. 653-657.

CSANADY, G.T., 1982, Circulation in the Coastal Ocean: Dordrecht, Holland, D. Reidel Publishing Co., 280 p.

CURRAY, J.R., 1969, Lectures 2, 3, and 6, *in* Stanley, D.J. (ed.), The New Concepts of Continental Margin Sedimentation: American Geological Institute Short Course Lecture Notes, 236 p.

DALRYMPLE, R.W., 1992, Tidal depositional systems, *in* Walker, R.G. and James, N.T. (eds.), Facies Models: St. Johns, Newfoundland, Geological Association of Canada, p. 177-218.

DALRYMPLE, R.W., ZAITLIN, B.A. AND BOYD, R., 1992, Estuarine facies models: Journal of Sedimentary Petrology, v. 62, p. 1130-1146.

DE BRUIN, R.H. AND BOYD, C.S., 1990, Oil and gas fields map of the Powder River Basin, Wyoming: Wyoming Geological Survey MS 31.

DE BRUIN, R.H. AND HOSTETLER, S.D., 1991, Oil and gas fields map of the Wind River Basin, Wyoming: Wyoming Geological Survey MS 37.

DUKE, W.L., 1990, Geostrophic circulation or shallow marine turbidity currents? The dilemma of paleoflow patterns in storm-influenced prograding shoreface systems: Journal of Sedimentary Petrology, v. 60, p. 870-883.

ELDRIDGE, G.H., 1889, Some suggestions upon the method of grouping the formations of the middle Cretaceous and the employment of an additional term in the nomenclature: American Journal of Science, v. 38, p. 313-321.

ELLIOT, T., 1987, Tidally-produced sigmoidal cross-bedding in Cretaceous shelf sandstone bodies, Western Interior Basin, USA (abstract): Tulsa, Society for Sedimentary Geology (SEPM) Annual Midyear Meeting, Aug. 20-23, 1987, Austin, Tex.

ELLIOT, T., 1988, The genesis of shelf sandstone bodies, Cretaceous Western Interior Basin, USA, and their relationship to seismic stratigraphic sequence boundaries (abstract). Shelf Sedimentation: Events and Rhythms: Society for Sedimentary Geology (SEPM) Research Conference, June 26-July 1, 1988, University of California Santa Cruz, v. 1, p. 18.

ERICSEN, M.C. AND SLINGERLAND, R., 1990, Numerical simulation of tidal and wind driven circulation in the Cretaceous Interior Seaway: Geological Society of America Bulletin, v. 102, p. 1499-1516.

FOYLE, A.M. AND OERTEL, G.F., 1997, Transgressive systems tract development and incised-valley fills within a Quaternary estuary-shelf system: Virginia inner shelf, USA: Marine Geology, v. 137, p. 227-249.

GAUTIER, D.L., 1983, Diagenesis, *in* Rice, D.D. and Gautier, D.L. (eds.), Patterns of Sedimentation, Diagenesis and Hydrocarbon Accumulation in Cretaceous Rocks of the Rocky Mountains: Tulsa, Society for Sediementary Geology (SEPM) Short Course No. 11, p. 4-1 to 4-9.

GAUTIER, D.L., 1986, Cretaceous shales from the Western Interior of North America: Geology, v. 14, p. 225-238

GAYNOR, G.C. AND ELLIOT, T., 1988, The Shannon Sandstone revisited: Evidence for tide and storm processes in sand ridge construction, Western Interior Basin, USA (abstract): Shelf Sedimentation:

Events and Rhythms; Society for Sedimentary Geology (SEPM) Research Conference, June 26-July 1, 1988, University of California, Santa Cruz, v. 1, p. 21.

Gaynor, G.C. and Swift, D.J.P., 1988, Shannon Sandstone depositional model: Sand ridge formation on the Campanian Western Interior Shelf: Journal of Sedimentary Petrology, v. 58, p. 868-880.

Gill, J.R. and Cobban, W.A., 1966, The Red Bird Section of the Upper Cretaceous Pierre Shale, in Wyoming, with a Section on a New Echinoid from the Cretaceous Pierre Shale of Eastern Wyoming by P.M. Kier: Washington, D.C., U.S. Geological Survey Professional Paper 393A, 73 p.

Gill, J.R. and Cobban, W.A., 1973, Stratigraphy and Geological History of the Montana Group and Equivalent Rocks, Montana, Wyoming and North and South Dakota: Washington, D.C., U.S. Geological Survey Professional Paper 776, 37 p.

Hansley, P.L. and Whitney, C.G., 1990, Petrology, diagenesis and sedimentology of oil reservoirs in Upper Cretaceous Shannon Sandstone beds, Powder River Basin, Wyoming: U. S. Geological Survey Bulletin, 1917C, 33 p.

Héquette, A. and Hill, P.R., 1995, Response of the seafloor to storm-generated combined flows on a sandy Arctic shoreface, Canadian Beaufort Sea: Journal of Sedimentary Research, v. A65, p. 461-471.

Hulscher, 1996, Tidal-induced, large-scale, regular bed form patterns in a three-dimensional shallow water model: Journal of Geophysical Research, v. 101, p. 20727-20744.

Huthnance, J.M., 1982, On one mechanism forming linear sand banks: Estuarine, Coastal and Shelf Science, v. 14, p. 79-99.

Jordan, T.E., 1982, Thrust loads and foreland basin evolution, Cretaceous Western United States: American Association of Petroleum Geologists Bulletin, v. 65, p. 2506-2520.

Kuehl, S.A., Nittrouer, C.A. and Demaster, D.J., 1982, Modern sediment accumulation and strata formation on the Amazon: Continental Shelf Marine Geology, v. 49, p. 279-300.

Lawyer, G., Newcomer, J. and Eger, C., 1981, Powder River Basin Oil and Gas Fields: Casper, Wyoming Geological Association, v. 1 and v. 2, 472 p.

Ludwick, J.C., 1974, Tidal currents and zig-zag sand shoals in a wide estuary mouth: Geological Society of America Bulletin, v. 85, p. 717-726.

MacEachern, J.A. and Pemberton, S.G., 1992, Ichnological aspects of Cretaceous shoreface successions and shoreface variability in the Western Interior Seaway of North America, *in* Pemberton, S.G., (ed.), Applications of Ichnology to Petroleum Exploration—A Core Workshop: Tulsa, Society for Sedimentary Geology (SEPM) Core Workshop 17, p. 54-84.

Martinsen, R.S. and Tillman, R.W., 1978, Hartzog Draw, new giant oil field (abstract): American Association of Petroleum Geologists Bulletin, v. 62, p. 540.

Mellere, D. and Steel, R., 1995a, Facies architecture and sequentiality of nearshore and 'shelf' sandbodies; Haystack Mountains Formation, Wyoming, USA: Sedimentology, v. 42, p. 551-574.

Mellere, D. and Steel, R., 1995b, Variability of lowstand wedges and their distinction from forced regressive wedges in the Mesaverde Group, Southeastern Wyoming: Geology, v. 23, p. 803-806.

Milliman, J.D., Zhou, J. and Ewing, J.I., 1990, Late Quaternary sedimentation on the Outer and Middle New Jersey Continental Shelf: Result of two local deglaciations?: Journal of Geology, v. 98, p. 966-976.

Murav'ev, V.I., 1988, Characteristics of glauconite accumulation in the Paleogene deposits of Syria; Lithology and Mineral Resources, v. 23, p. 583-591.

Mutti, E., Rosell, J., Allen, G.P., Fonnesu, F. and Sgavetti, M., 1985, The Eocene Baronia tide dominated delta-shelf system in the Ager Basin, *in* Mila, M.D. and Rosell, J. (eds.): Excursion Guidebook: Sixth Regional European Meeting, International Association of Sedimentologists, Lleida, Spain, p. 570-600.

Nayaran, J., 1971, Sedimentary structures in the lower Greensland: Sedimentary Geology, v. 6, p. 73-109.

Niedoroda, A.W., Swift, D.J.P. and Thorne, J.A., 1989, Modeling shelf storm beds: Controls of bed thickness and bedding sequence, *in* Morton, R.A. and Nummedal, D.: (eds.): Shelf Sedimentation, Shelf Sequences and Related Hydrocarbon Accumulation. Gulf Coast Section, Society for Sedimentary Geology (SEPM), Seventh Annual Research Conference Proceedings, p. 15-39.

Nittrouer, C.A., Sternberg, R.W., Carpenter, R. and Bennett, J.J., 1979, The use of ^{210}Pb geochronology as a sedimentological tool: Application to the Washington Continental Shelf: Marine Geology, v. 21, p. 297-316.

Nittrouer, C.A., Curtin, T.B. and Demaster, D.J., 1986, Concentration and flux of suspended sediment on the Amazon Continental Shelf: Continental Shelf Research, v. 6, p. 154-174.

Nummedal, D. and Molenaar, C.M., 1995, Sequence stratigraphy of ramp-setting strand plain successions: The Gallup Sandstone, Mew Mexico, *in* Van Wagoner, J.C. and Bertram, G.T. (eds.): Sequence Stratigraphy of Foreland Basin Deposits: Tulsa, American Association of Petroleum Geologists Memoir 64, p. 227-310.

Nummedal, D., Swift, D.J.P. and Kofron, B.M., 1989, Sequence stratigraphic interpretation of Coniacian strata in the San Juan Basin, New Mexico, *in* Morton, R.A. and Nummedal, D. (eds.): Shelf Sedimentation, Shelf Sequences and Related Hydrocarbon Accumulation. Gulf Coast Section, Society for Sedimentary Geology (SEPM), Seventh Annual Research Conference Proceedings, p. 175-202.

Obradovich, J.D., 1993, A Cretaceous time scale, *in* Caldwell, W.G.E. and Kauffman, E.G. (eds.), Evolution of the Western Interior Basin: St. John's, Geological Association of Canada Special Paper 39, p. 379-396.

Odin, G.S., (ed.), 1988, Green Marine Clays: New York, Elsevier, 445 p.

Odin, G.S. and Fullagar, P.D., 1988, Geological significance of the glaucony facies, *in* Odin, G.S. (ed.), Green Marine Clays: New York, Elsevier, p. 295-332.

Odin, G.S. and Matter, A., 1981, De glauconarium origine: Sedimentology, v. 28, p. 611-641.

Odin, G.S and Sen Gupta, B.K., 1988, Geological Significance of the Verdine Facies, *in* Odin, G.S. (ed.), Green Marine Clays: New York, Elsevier, p. 205-222.

Odin, G.S., Debenay, J.P. and Masse, J.P., 1988, The verdine facies deposits identified in 1988, *in* Odin, G.S. (ed.), Green Marine Clays: New York, Elsevier, p. 131-158.

Parker, G., Lanfredi, N.W. and Swift, D.J.P., 1982, Substrate response to flow in a southern hemisphere ridge field: Argentina Inner Shelf: Sedimentary Geology, v. 33, p. 195-216.

Parrish, J.T., Gaynor, G.C. and Swift, D.J.P., 1984, Circulation in the Cretaceous Western Interior Seaway of North America, a review, *in* Stott, D.F. and Glass, D.J. (eds.), Calgary, Canadian Society of Petroleum Geologists Memoir 9, p. 221-231.

Pemberton, S.G., Reinson, G.E. and MacEachern, J.A., 1991, Comparative ichnology analysis of late Albian estuarine valley-fill and shelf-shoreface deposits, Crystal-Viking Field, Alberta, *in*: Pemberton, S.G. (ed.), Applications of Ichnology to Petroleum Exploration, Society for Sedimentary Geology (SEPM) Core Workshop No. 17, p. 239-317.

Posamentier, H.W., Allen, G.P., James, D.P. and Tesson, M., 1992, Forced regressions in a sequence stratigraphic framework: Concepts, examples, and exploration significance: American Association of Petroleum Geologists Bulletin, v. 76, p. 1687-1709.

Posamentier, H.W. and Vail, P.R., 1988, Eustatic controls on clastic deposition II - Sequence and systems tract models, *in* Wilgus, C.K., Hastings, B.S., St. Kendall, C.G.C., Posamentier, H.W.,

Ross, C.A. and Van Wagoner, J.C. (eds.), Sea-level Changes: An Integrated Approach: Tulsa, Society for Sedimentary Geology (SEPM) Special Publication 42, p. 124 - 154.

Rice, D.D., 1984, Widespread, shallow marine storm-generated sandstone units in the upper Cretaceous Mosby Sandstone, central Montana, *in* Tillman, R.W. and Siemers, C.T. (eds.), Siliciclastic Shelf Sediments: Tulsa, Society for Sedimentary Geology (SEPM) Special Publication 34, p. 143-161.

Scheihing, M.H., Gaynor, A., Gerard, C., 1991, The shelf sand-plume model: A critique: Sedimentology, v. 38, p. 433-444.

Schwartz, M.L., 1967, The Bruun theory of sea-level rise as a cause of shore erosion: Journal of Geology, v. 75, p. 76-92.

Shanely, K.W., McCabe, P.J. and Hettinger, R.D., 1992, Tidal influences in Cretaceous fluvial strata from Utah, USA: A key to sequence stratigraphic interpretation: Sedimentology, v. 39, p. 905-930.

Shelton, J.W., 1965, Trend and genesis of lowermost sandstone unit of Eagle Sandstone at Billings, Montana; American Association of Petroleum Geologists Bulletin, v. 49, p. 1385-1397.

Shepard, F.J., 1973, Submarine Geology: New York, Harper ands Row Publishers, 517 p.

Shurr, G.W., 1984, Geometry of shelf sandstone bodies in the Shannon Sandstone of southeastern Montana, *in* Tillman, R.W. and Siemers, C.T. (eds.), Siliciclastic Shelf Sediments. Tulsa: Society for Sedimentary Geology (SEPM) Special Publication 34, p. 63-84.

Snedden, J.W. and Nummedal, D., 1991, Origin and geometry of Storm-deposited storm beds in modern sediments of the Texas Continental Shelf, *in* Swift, D.J.P., Oertel, G.F., Tillman, R.W. and Thorne, J.A. (eds.), Shelf Sand and Sandstone Bodies, Geometry, Facies and Sequence Stratigraphy: Oxford, International Association of Sedimentologists Special Publication 14, p. 283-308.

Spearing, D.R., 1976, Upper Cretaceous Shannon Sandstone, an offshore, shallow-marine sandbody: Casper, Wyoming Geological Association Guidebook, 28th Annual Field Conference, p. 65-72.

Sullivan, M.D., Van Wagoner, J.C., Foster, M.E., Stuart, R.M., Jennette, D.C., Lovell, R.W. and Pemberton, S.G., 1995, Lowstand architecture and sequence stratigraphic control on Shannon incised valley distribution, Hartzog Draw Field, Wyoming, *in:* Fitzsimmons, R.J., Parsons, B.S. and Swift, D.J.P. (eds.) Tongues, Ridges and Wedges: Highstand Versus Lowstand Architecture in Shallow Marine Basins: Tulsa, Society for Sedimentary Geology (SEPM) Research Conference Field Guide, Appendix A.

Sullivan, M.D., Van Wagoner, J.C., Jennette, D.C., Lovell, R.W., Foster, M.E. and Stuart, R.M., 1996, Sequence stratigraphic and tectonic controls on Shannon incised valley distribution, Hartzog Draw, Wyoming (abstract): Tulsa, American Association of Petroleum Geologists Annual Meeting Program, p. A136.

Suter, J.H. and Berryhill, H.L., 1985, Late Quaternary shelf margin deltas, Northwest Gulf of Mexico: American Association of Petroleum Geologists Bulletin, v. 69, p. 77-91.

Swift, D.J.P., 1968, Coastal and transgressive stratigraphy: Journal of Geology, v. 76, p. 444-456.

Swift, D.J.P., Sanford, R., Dill, C.E., Jr. and Avignone, N., 1971, Textural differentiation of the shoreface during erosional retreat of an unconsolidated coast, Cape Henry to Cape hatteras, Western North Atlantic Shelf; Sedimentology, v. 16, p. 221-250.

Swift, D.J.P., Nelsen, T.A., McHone, J., Holliday, B., Palmer, H. and Shideler, G., 1977, Holocene evolution of the inner shelf of Southern Virginia: Journal of Sedimentary Petrology, v. 47, p. 1454-1474.

Swift, D.J.P., Parker, G., Lanfredi, N.W., Perillo, G. and Figge, K., 1978a, Shoreface-connected and ridges on American and European Shelves: A comparison: Estuarine and Coastal Marine Research, v. 17, p. 257-273.

Swift, D.J.P, Sears, P.C., Bohlke, B. and Hunt, R., 1978b, Evolution of a shoal retreat massif, North Carolina Shelf: Evidence from aerial geology: Marine Geology, v. 27, p. 19-42.

Swift, D.J.P., Niedoroda, A.W., Vincent, C.E. and Hopkins, T.S., 1985, Barrier island evolution, Middle Atlantic Shelf, USA, Part 1: Shoreface dynamics: Marine Geology, v. 63, p. 331-361.

Swift, D.J.P., Han, G. and Vincent, C.E., 1986a, Fluid process and sea floor response on an modern storm-dominated shelf: Middle Atlantic Shelf of North America. Part 1, The storm current regime, *in* Knight, R.J. and McLean, J.R., (eds.), Shelf Sands and Sandstone Reservoirs, Calgary: Canadian Society of Petroleum Geologists Memoir 11, p. 99-119.

Swift, D.J.P., Thorne, J.A. and Oertel, G.F., 1986b, Fluid process and sea floor response on a storm dominated shelf: Middle Atlantic Shelf of North America. Part II: Response of the shelf floor, *in* Knight, R.J. and McLean, J.R. (eds.), Shelf Sands and Sandstone Reservoirs: Canadian Society of Petroleum Geologists Memoir 11, p. 191-211.

Swift, D.J.P., Phillips, S. and Thorne, J.A., 1991a, Sedimentation on continental margins IV. Lithofacies and depositional systems, *in* Swift, D.J.P., Tillman, R.W., Oertel, G.F. and Thorne, J.A. (eds.), Shelf Sand and Sandstone Bodies: Geometry, Facies and Sequence Stratigraphy: Oxford, International Association of Sedimentologists Special Publication 14, p. 89-152.

Swift, D.J.P., Phillips, S. and Thorne, J.A., 1991b, Sedimentation on continental margins V. Parasequences, *in* Swift, D.J.P., Tillman, R.W., Oertel, G.F. and Thorne, J.A. (eds.), Shelf Sand and Sandstone Bodies: Geometry, Facies and Sequence Stratigraphy: Oxford, International Association of Sedimentologists Special Publication 14, p. 153-188.

Swift, D.J.P., Snedden, J.W. and Plint, A.G., (conveners), 1995, Tongues, Ridges and Wedges: Highstand Versus Lowstand Architecture in Marine Basins: Tulsa, Society for Sedimentary Geology (SEPM) Research Conference, Powder and Bighorn Basins, Wyoming, June 24-29, 1995.

Swift, D.J.P. and Field, M.E., 1981, Evolution of a classic ridge field, Maryland Sector, North American Inner Shelf: Sedimentology, v. 28, p. 461-482.

Swift, D.J.P. and Parsons, B.S., 1995, West Cottonwood Creek: Bighorn Basin, Wyoming, *in*: Fitzsimmons, R.J., Parsons B.S. and Swift, D.J.P. (eds.), Tongues, Ridges and Wedges: Highstand Versus Lowstand Architecture in Shallow Marine Basins: Tulsa, Society for Sedimentary Geology (SEPM) Conference Field Guide, p. 46-47.

Swift, D.J.P. and Rice, D.D., 1984, Sand bodies on muddy shelves: A model for sedimentation in the Western Interior Cretaceous Seaway, *in*: Tillman, R.W. and Seimers, C.T. (eds.), Siliciclastic Shelf Sediments: Tulsa, Society for Sedimentary Geology (SEPM) Special Publication 34, p. 25-37.

Swift, D.J.P. and Thorne, J.A., 1991, Sedimentation on continental margins, I. A general model for shelf sedimentation, *in* Swift, D.J.P., Tillman, R.W., Oertel, G.F. and Thorne, J.A. (eds.), Shelf Sand and Sandstone Bodies: Geometry, Facies and Sequence Stratigraphy: Oxford, International Association of Sedimentologists Special Publication 14, p. 3-31.

Tesson, M., Gensous, B., Allen, G.P. and Ravenne, C., 1990, Late Quaternary deltaic lowstand wedges on the Rhone Continental Shelf, France: Marine Geology, v. 91, p. 325-332.

Thorne, J.A., Grace, E., Swift, D.J.P. and Niedoroda, A.W., 1991, Sedimentation on continental margins, III. The depositional fabric-an analytical approach to stratification and facies identification, *in* Swift, D.J.P., Tillman, R.W., Oertel, G.F. and Thorne, J.A. eds.), Shelf Sand and Sandstone Bodies: Geometry, Facies and Sequence Stratigraphy: Oxford, International Association of Sedimentologists Special Publication 14, p. 59-88.

Tillman, R.W. and Martinsen, R.S., 1984, The Shannon shelf-ridge sandstone complex, Salt Creek Anticline area, Powder River Basin, Wyoming, *in* Tillman, R.W. and Siemers, C.T. (eds.), Siliciclastic Shelf Sediments: Tulsa, Society for Sedimentary Geology (SEPM) Special Publication 34, p. 85-142.

Tillman, R.W. and Martinsen, R.S., 1985, Upper Cretaceous Shannon and Haystack Mountains Formation Field Trip, Wyoming: Tulsa, Society for Sedimentary Geology (SEPM) 1985 Mid-Year Meeting Field Trip Guidebook Rocky Mountain Section Field Trip, 97 p.

Tillman, R.W. and Martinsen, R.S., 1987, Sedimentologic characteristics and production model of Hartzog Draw Field, Wyoming, a Shannon shelf-ridge sandstone, *in* Tillman, R.W. and Weber, K.J. (eds.), Reservoir Sedimentology: Tulsa, Society for Sedimentary Geology (SEPM) Special Publication 40, p. 15-112.

Vail, P.R., Mitchum, R.M., Jr., Todd, R.G., Widmier, J.M., Thompson, S., III, Sangree, J.R., Bubb, J.N. and Hatlelid, W.G., 1977, Sequence stratigraphy and global changes of sea level, *in* Payton, C.E. (ed.), Seismic Stratigraphy—Applications to Hydrocarbon Exploration: Tulsa, American Association of Petroleum Geologists Memoir 26, p. 49-205.

Van Wagoner, J.C., Mitchum, R.M., Jr., Campion, K.M. and Rahmanian, V.D., 1990, Siliciclastic Sequence Stratigraphy in Well Logs, Cores and Outcrops: Techniques for High Resolution Correlation of Time and Facies: Tulsa, American Association of Petroleum Geologists Methods in Exploration 7, 55 p.

Vincent, C.E., Swift, D.J.P. and Hillard, B., 1981, Sediment transport in the New York Bight, North American Atlantic Shelf: Marine Geology, v. 42, p. 369-398.

Visser, M.J., 1980, Neap-spring cycles reflected in Holocene subtidal large-scale bedform deposits: A preliminary note: Geology, v. 8, p. 545-546.

Walker, R.G. and Bergman, K.M., 1993, Shannon Sandstone in Wyoming: A shelf-ridge complex reinterpreted as lowstand shoreface deposits: Journal of Sedimentary Petrology, v. 63, p. 839-951.

Walker, R.G. and Eyles, C.H., 1989, Topography and significance of a basinwide sequence-bounding erosion surface in the Cretaceous Cardium Formation, Alberta, Canada: Journal of Sedimentary Petrology, v. 61, p. 473-496.

Wright, R., 1986, Cycle stratigraphy as a paleogeographic tool: Point Lookout Sandstone, Southeastern San Juan Basin, New Mexico: Geological Society of America Bulletin, v. 96, p. 661-673.

Wright, L.D. and Coleman, J.M., 1973, Variations in morphology of major river deltas as functions of ocean wave and river discharge regimes: American Association of Petroleum Geologists Bulletin, v. 57, p. 370-398.

CAMPANIAN SHANNON SANDSTONE: AN EXAMPLE OF A FALLING STAGE SYSTEMS TRACT DEPOSIT

KATHERINE M. BERGMAN
Department of Geology, University of Regina, Regina, SK S4S 0A2, Canada,
AND
ROGER G. WALKER
Department of Geology, McMaster University, Hamilton, ON L8S 4M1, Canada

ABSTRACT: Recent interpretations of the lower Campanian Shannon Sandstone in the Powder River Basin, Wyoming have suggested that it was deposited in incised shorefaces of the falling stage and lowstand systems tracts rather than in offshore stacked shelf ridge complexes. A complete high resolution stratigraphic analysis of the Shannon requires that outcrops in the Bighorn and Powder River Basin be correlated with subsurface fields basinward. In the Hartzog Draw area where there is good core control, a consistent pattern of facies and erosion surfaces can be defined. From this pattern a template was constructed that could be used to interpret depositional patterns in uncored areas. The template consists of a regressive surface of erosion (RSE) that forms the base of an incised shoreface sandbody and a transgressive surface of erosion (TSE) that truncates the top of the sandbody and the RSE. Applying the template to the cross section suggests that six progradational successions and six transgressions can be defined within the Shannon. Positions of the RSEs and TSEs allow estimates of the magnitude of relative sea level fluctuations that controlled Shannon deposition (a few meters to 60 m). The correlations are consistent with the interpretation of the Shannon as incised shoreface sandbodies of falling stage and lowstand systems tracts, separated by mudstone deposits of the following transgressive systems tract.

INTRODUCTION

The purpose of this paper is to present a summary of the regional correlation (Bergman and Walker, 1995) of the Shannon Sandstone, Powder River Basin, Wyoming, (Fig. 1) from outcrops in the Bighorn Basin (Lucerne) and Salt Creek anticline to subsurface deposits at Hartzog Draw field. The Shannon Sandstone was re-interpreted (Walker and Bergman, 1993; Bergman, 1994) as incised shoreface deposits that formed during periods of stillstand in an overall regression (falling stage systems tract). It follows that a plausible correlation can be constructed only if certain principles for correlation are established. These principles are embodied in the the stratigraphic relationships described for Hartzog Draw (Bergman, 1994) and serve as a template for regional correlation (Bergman and Walker, 1995). The details of the arguments used in arriving at the correlations and interpretations are discussed in Walker and Bergman (1993), Bergman (1994), and Bergman and Walker (1995) and will not be presented here. The history of relative sea-level changes and an estimate of their magnitude follows from the sequence stratigraphy.

STUDY AREA AND DATABASE

The study area is contained in T38 to 46 and R74 to 80 (Fig. 2). There are about 1400 wells, of which 90 are cored in the study area. All of the 90 cores were described and photographed. Correlations are based on gamma-ray and resistivity log signatures, and are supplemented with available core. Stratigraphic control exists at the Salt Creek outcrops and core and well log control at Teapot Dome (Figs. 1 and 2). Core and well log control is available at Hartzog Draw, but there is relatively limited core and well log control in the 30 km separating Teapot Dome and Hartzog Draw (Fig. 2). One regional cross-section linking Lucerne to Hartzog Draw (Figs. 1 and 2) is presented. On the cross sections, P signifies progradation, T signifies transgression, and R signifies regression. Numbers are in inferred chronologic order. The facies and correlations have been described in detail for both the outcrop (Walker and Bergman, 1993) and the subsurface (Bergman, 1994) and will not be discussed here.

SHANNON SANDSTONE OVERVIEW

The Shannon Sandstone is contained in the Steele Member of the Cody Shale of the Montana Group and is stratigraphically equivalent to part of the Lower Mesaverde Group in the Bighorn Basin (Lucerne; Fig. 1). The Shannon is assigned to the *Baculites* sp. (smooth) ammonite zone, which has an estimated age of about 81 Ma (Gill and Cobban, 1973). The Shannon appears to be related to the last stages of the Telegraph-Eagle Creek regression, which began about 83.5 million years ago (Tillman and Martinsen, 1984). Gill and Cobban

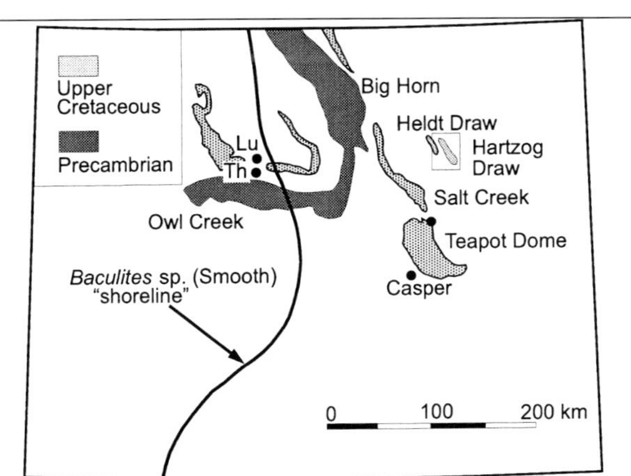

Fig. 1.—Map of Wyoming showing location of study area (rectangle) and Shannon time equivalent shoreline as proposed by Gill and Cobban (1973). Lu, Lucerne; Th, Thermopolis. The distribution of Upper Cretaceous rocks is shown only in the Lucerne and Salt Creek areas. The Owl Creek and Big Horn Mountains are Precambrian uplifts. Open oval shows Salt Creek area, and black areas show the major Shannon reservoirs at Teapot Dome, Heldt Draw, and Hartzog Draw. Modified from Walker and Bergman (1993) and Bergman (1994). Both the eastern and western Wyoming State borders trend north-south.

Isolated Shallow Marine Sand Bodies: Sequence Stratigraphic Analysis and Sedimentologic Interpretation.
SEPM Special Publication No. 64, Copyright © 1999
SEPM (Society for Sedimentary Geology), ISBN 1-56576-057-3, p. 85-93.

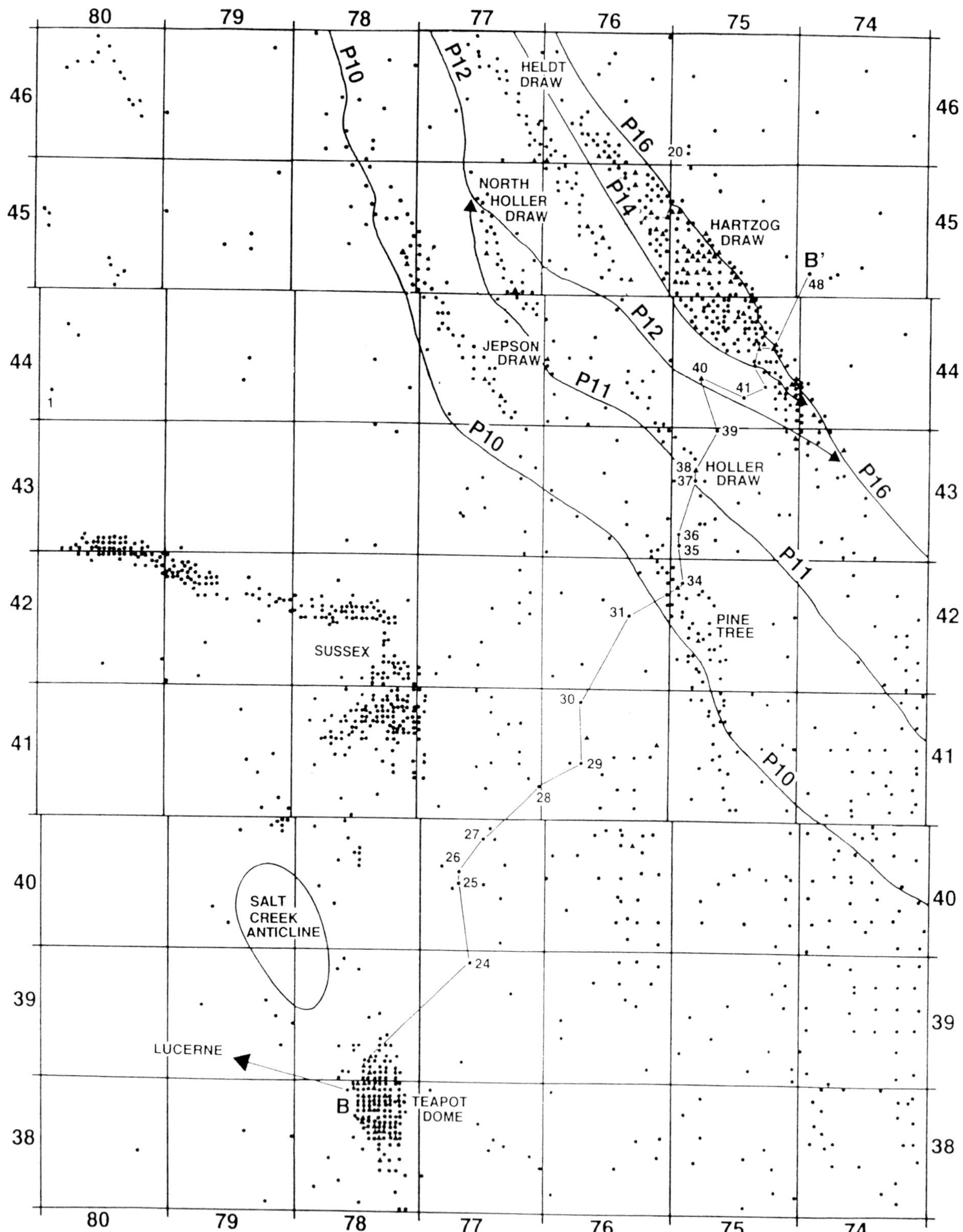

Fig. 2.—Detailed map of the study area (rectangle in Fig. 1). The location of the Shannon reservoirs and positions of the cross-section are shown. The bold lines indicate the positions where the RSEs associated with progradational events are truncated by a TSE. These lines correspond to the westward margins of the oil fields. The progradational events are labeled as explained in the text. Arrows at the end of the RSE lines indicate where one progradational event has been truncated by a later progradational event. Closed circles represent the well log control used in the study; black triangles are the cored wells examined.

(1973) used paleontologic evidence to suggest that the shoreline was moving eastward at this time.

The Shannon is a 50 m thick unit composed of isolated, coarsening upward sandbodies encased in marine mudstone and oriented parallel to subparallel to regional tectonic trends. In this setting the Shannon was previously interpreted as a shelf ridge complex (Parker, 1958; Spearing, 1976; Seeling, 1979; Tillman and Martinsen, 1984, 1987; Gaynor and Swift, 1988; Rine et al., 1991). This interpretation poses problems in terms of sediment transport, reworking, and stacking (Walker and Bergman, 1993; Bergman, 1994; Bergman and Walker, 1995). Fluctuations of relative sea level were not discussed by these earlier authors and their shelf ridge interpretation implies that all ridges were formed contemporaneously during one cycle of relative sea-level fluctuation. The shelf ridge interpretation does not allow correlation between 1) shelf ridge sandbodies and 2) contemporaneous shoreline deposits and the shelf ridges. In an attempt to solve some of these problems a regional high-resolution stratigraphic framework was constructed (Bergman and Walker, 1995). The local stratigraphic relationships and the contained facies, facies associations, and trace fossil assemblage led to the reinterpretation of the Shannon sandstone as falling stage and lowstand shoreface deposits separated by transgressive systems tract mudstone deposits (Walker and Bergman, 1993; Bergman, 1994).

CROSS-SECTION CORRELATION

A template was constructed (Bergman and Walker, 1995) based on detailed facies analysis, lateral and vertical facies relationships, and the correlation of erosional discontinuities in Hartzog Draw (Fig. 2). Three wells from Hartzog Draw are shown (Fig. 3) to illustrate the criteria and assumptions used in making the template. The four basic elements of the template are (1) a coarsening upward and sandier upward succession, (2) an underlying asymmetrical erosion surface, (3) a transgressive surface of erosion that truncates the sandbody, and (4) extension of the erosion surfaces beyond the immediate position of the incision. The application of the template is shown below the cross section (Fig. 4). In the following analysis we show how the template was used and then systematically develop the chronology of the depositional and erosional events. The template can be successfully applied in Wells 32-33 and 37-39 (Fig. 4). Note that the shoreface implied in Wells 37-39 (Fig. 4) is truncated seaward by the shoreface in Wells 40-45 (the Heldt Draw sandbody). Details of the along-strike correlations are discussed below.

The southwestern end of the cross section is more complex. In Teapot Dome and Salt Creek there are two main sandier-upward successions, from RSE4 to T3 and from RSE14 to T17. The two sandier-upward successions in the Salt Creek/Teapot Dome area have been correlated with two sandier-upward shoreface successions at Lucerne (Walker and Bergman, 1993). There is almost no control between Lucerne and Salt Creek. The rationale of this correlation is discussed in Walker and Bergman (1993).

The lower prograding succession is labeled P2. Smaller sandier-upward cycles (1 and 2 in Well 22) can be identified in some wells and outcrops in the Teapot Dome-Salt Creek area (Walker and Bergman, 1993). The top of the lower succession (T3 in Well 22) appears to drop stratigraphically basinward, and the entire sandier-upward succession is cut out between Wells 28 and 32 (Fig. 4). Transgressive systems tract deposits (TST3) are up to 14 m thick above T3, with three onlapping markers present southwest of Well 24 (Fig. 4).

Progradation P4 is interpreted from outcrop and core, where bioturbated, muddy sandstone is overlain by interbedded, hummocky cross-stratified sandstone and bioturbated mudstone (events 11 and 12 in the interpretation of Walker and Bergman, 1993). The P5 sandbody is defined by the template in Wells 28-31, where it is characterized by funnel-shaped gamma ray and resistivity logs (Fig. 4). The basal incision below P5 appears to cut out TST3 and sandbody P2 with about 12 m of incision between Wells 27-28 and 31 (Fig. 4). P5 was truncated by T6. Sandbody P7 can be defined using the template in Well 27. Here, the funnel-shaped well-log pattern is abruptly terminated by T8 (Fig. 4). Surface T8 can be traced southwestward to Well 24. It is eroded farther to the southwest by the RSE below P9. Thus, the interval between T3 and P9 at Salt Creek and Teapot Dome represents the basin expression of the coaly shale deposits that separate the two shoreface successions at Lucerne (Fig. 4).

P9 is the upper prograding shoreface at Lucerne and is the upper sandstone of succession 2 in the Salt Creek area. It appears to grade seaward into mudstone between Wells 26 and 27 (Fig. 4). Relationships between Wells 27 and 32 are partially constrained by the overlying transgressive surfaces. T17 is the final transgressive surface of the Shannon Sandstone. It appears to truncate T13 near Well 31. Below T13, events 12, 11, and 10 are characterized by prograding sandbodies.

Along-Strike Correlation (Figure 2)

The cross section shows 11 separate prograding sandbodies. Seven of these (P1, P2, P4, P9, P12, P14, and P16) are based on good outcrop and/or core control, and four (P5, P7, P10, P11) have been defined using the template. Five of the incisions (P10, P11, P12, P14, and P16) mark the landward edges of prograding sandbodies. A series of short, dip-oriented cross- sections were drawn to check the existence of the incisions and help define their trend. Further definition of the positions (shown in Fig. 2) was achieved by using all adjacent landward and seaward pairs of well logs along the entire trend of the incisions.

P10 defines the erosional edge of Jepson Draw and Pine Tree, and P11 marks the erosional edge of North Holler Draw and Holler Draw. Note that P12 cuts out P11 along strike at North Holler Draw. P12 marks the erosional edge of Heldt Draw and P14 marks the erosional edge of the main incision at Hartzog Draw. Both P12 and P14 are cut out along strike by P16 at the southern end of Hartzog Draw.

Some incisions do not appear on the map. Those associated with the Lucerne and Salt Creek sandbodies (P1 and P2 for Succession 1, and P4 and P9 for Succession 2) lie to the west of the study area shown (Fig. 2), where outcrop control is too limited to define the position of the incision.

PROGRADATION, TRANSGRESSION, AND CHRONOLOGY OF SEA-LEVEL FLUCTUATION

The numbered P and T chronology presented here is consistent with previously published chronologies in the Lucerne-Salt Creek-Teapot Dome area (Walker and Bergman, 1993), and the Hartzog Draw-Heldt Draw area (Bergman, 1994). The terminology used in these studies is given in square brackets in the following discussion. Abbreviations are RSL for relative sea level, RSE for regressive surface of erosion, TSE for transgressive surface of erosion, BD for bounding discontinuity, CC for correlative conformity, and HCS for hummocky cross-stratification.

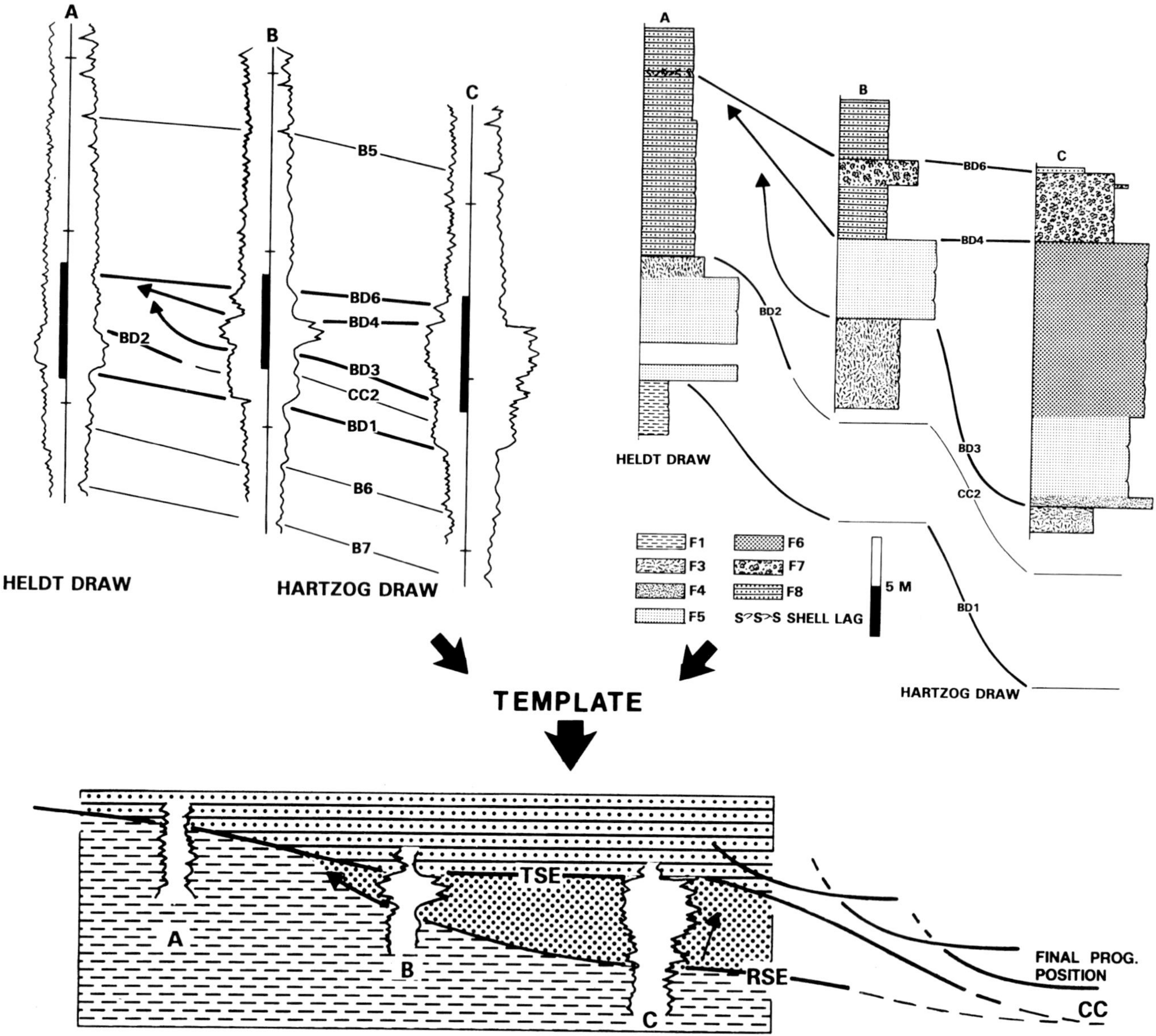

Fig. 3.—Template used for regional correlation in areas of limited core control between Hartzog Draw, Teapot Dome/Salt Creek, and the Bighorn Basin. Applications are shown in Figure 4. Facies were described by Bergman (1994). F1, Black Mudstone; F3, Bioturbated Sandstone; F4, Glauconitic Medium to Coarse Sandstone; F5, Thin-Bedded Sandstone; F6, Cross-Bedded Sandstone; F7, Coarse Bioturbated Sandy Mudstone; F8, Laminated Mudstone. BD, bounding discontinuity; CC, correlative conformity; B, correlation lines; RSE, regressive surface of erosion; TSE, transgressive surface of erosion. In Well A the position of BD6 in the core is marked by a shell lag.

The prograding sandbodies are shown diagrammatically in Figure 5 and are believed to be the result of a relative sea level fall (forced regression; Posamentier et al., 1992). The interpreted preserved lateral extent of the P and T events is shown at the left (Fig. 5), and the entire Shannon can be interpreted as a prograding stack of incised shoreface deposits separated by transgressions. Alternatively, the figure can be interpreted as three prograding stacks made up of (1) P1, 2; (2) P4, 5, and 7; and (3) P9, 10, 11, 12, 14, and 16. Although T8 appears to be eroded near Teapot Dome (Fig. 5), it presumably once underlaid P9 in the Lucerne area, making it the most extensive transgressive surface in the Shannon.

Chronological Development

The proposed chronology is consistent with application of the template to sandbodies between Hartzog Draw and Teapot Dome and presents a correlation of erosional discontinuities between Lucerne and Hartzog Draw.

Progradation 1: At Lucerne, the Cody Shale was abruptly overlain by HCS sandstone (P1). Walker and Bergman (1993) interpreted this abrupt contact as a forced regression. This contact represents the base of the Shannon Sandstone.

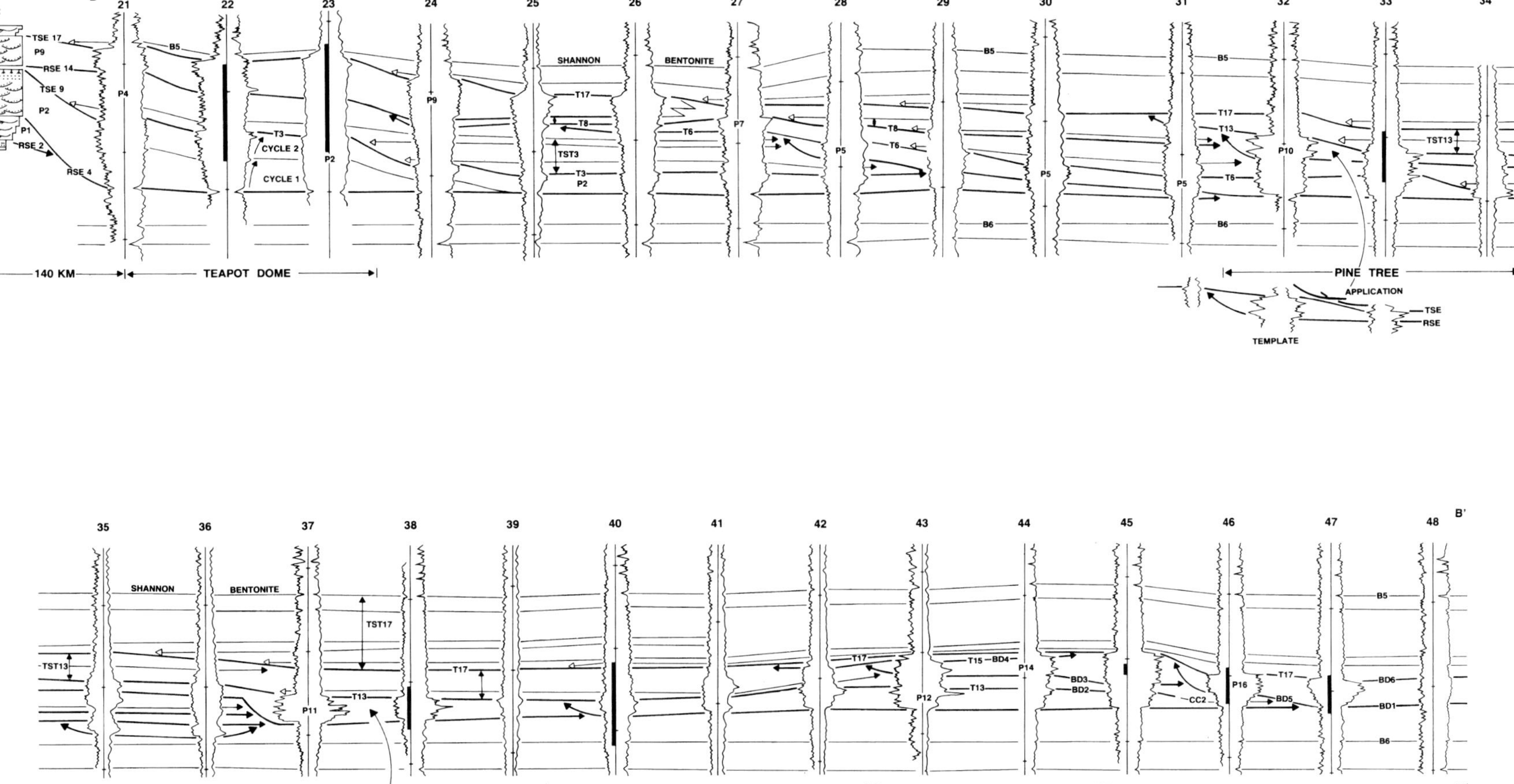

Fig. 4.—Cross-section located in Figure 2. Well logs are gamma ray and resistivity. Wells are numbered sequentially. For land locations, see Appendix 1. Ticks on the central axis are 100 ft (because the original logs are in feet; 1ft ~ 0.3 m), and there is no implied horizontal scale. The black bar between the well logs represents the cored interval. Heavy black lines represent erosional discontinuities and are labeled as referenced in the text. The thinner lines are correlation lines. Areas where the template has been used in the correlation are shown on the cross-section. Open arrows represent onlapping markers and closed arrows represent truncated horizons. The section is hung on the lower bentonite marker B6.

Progradation 2: An RSL fall resulted in P2 giving rise to the event beds of [Cycle 1] preserved at Salt Creek and Teapot Dome. [Cycle 1] was terminated by a minor RSL rise, forming a flooding surface (Wells 22, 23, and 24; Fig. 4). This surface was locally sideritized. Progradation 2 then resumed with shoreface sandstones passing seaward into HCS event beds. The abrupt appearance of the HCS beds on the event beds suggests that this progradation may have be driven by a forced regression. This minor RSL fall was followed by a larger RSL fall, causing the shoreface to prograde rapidly from Lucerne to Salt Creek. Progradation 2 was terminated by a major transgression T3 [TSE 9]. T3 was overlain by a minimum of 12.2 m of TST mudstone.

Progradations 4 and 5: An RSL fall resulted in renewed progradation, with bioturbated, gray sandstone downlapping onto the flooding surface. Another RSL fall resulted in shoreface incision (Well 28) and renewed progradation (P5). The root traces at Lucerne (Walker and Bergman, 1993) probably formed at this time. P4 and P5 were terminated by erosive transgression (T6), during an RSL rise.

Progradation 7: After T6, another RSL fall resulted in shoreface incision (Well 26) and progradation 7. P7 was terminated by an RSL rise (T8). This major flooding surface shifted the shoreline position back to Lucerne.

Progradations 9, 10, 11, and 12: During a series of RSL falls, shorefaces were incised at Wells 32 (P10), 37 (P11), and 40 (P12, Heldt Draw). Each of these shoreface incisions is associated with a progradational event. Progradation 9 [RSE14] represents the upper sandstone at Lucerne, Salt Creek, and Teapot Dome, and progradation was believed to have been driven by forced regression (Walker and Bergman, 1993). Progradational events 10, 11, and 12 formed during an overall RSL fall punctuated by periods of stillstand. Shoreface incision occurred during the stillstands, and progradation of shoreface sandstones formed the sandbodies that are now reservoirs in the areas of Jepson Draw (P10), North Holler Draw (P11), and Heldt Draw (P12), respectively (Fig. 2). North of North Holler Draw P11 was removed by subsequent erosion associated with the P12 incision (Fig. 2). P12, P11, P10, and possibly P9 are truncated by erosion (T13) during an RSL rise. The westerly extent of T13 is not known because it was truncated by T17 between Wells 30 and 31 on the cross section (Fig. 4). T13 was overlain by a well-developed transgressive system tract (minimum 11.7 m thick) that contains a number of onlapping markers.

Progradation 14: After transgression [BD2], an RSL fall [BD3] resulted in the progradation (P14) of the main Hartzog Draw Sandstone. The top of Hartzog Draw was terminated by erosional transgression [BD4] during an RSL rise (T15).

Progradation 16: The northeastern margin of Hartzog Draw is formed by an RSL fall, resulting in shoreface incision [BD5] and progradation (P16). The incision associated with this progradational event truncates both P14 and P12 at the southern end of Hartzog Draw (Fig. 2). P16 was truncated by erosional transgression (T17) during an RSL rise. T17 [TSE 17] can be traced across the basin from Hartzog Draw to Teapot Dome/Salt Creek (P9) and onto Lucerne. This surface marks the end of Shannon deposition and is characterized by a well developed transgressive systems tract (minimum 32 m thick) containing a number of onlapping markers.

MAGNITUDE OF SEA-LEVEL FLUCTUATIONS DURING SHANNON TIME

Estimates can be made of the magnitudes of relative sea-level fluctuations from the cross section (Fig. 4). Our estimates were made using compacted sediment thicknesses; decompaction would be difficult and imprecise because of the absence of lithological control between Teapot Dome and Hartzog Draw. The estimates of relative sea-level fluctuation are therefore minima and are measured as fluctuations in the position of fair-weather wave base (FWWB). For a detailed discussion of the arguments used to make the estimates see Bergman and Walker (1995).

Figure 6 was constructed from these measurements. The dashed lines shown for the transgressions indicate that their maximum extent is unknown because of subsequent truncation of the surfaces, or because the surface continues westward beyond the study area (T17). There are two anomalous segments in the diagram where the drops of sea level related to P9 and P12 both plot as rising segments. The T8 to P9 rise is due to the truncation of T8 between Wells 25 and 26 (Fig. 4). Sea level undoubtedly rose higher than shown in Figure 6, and the

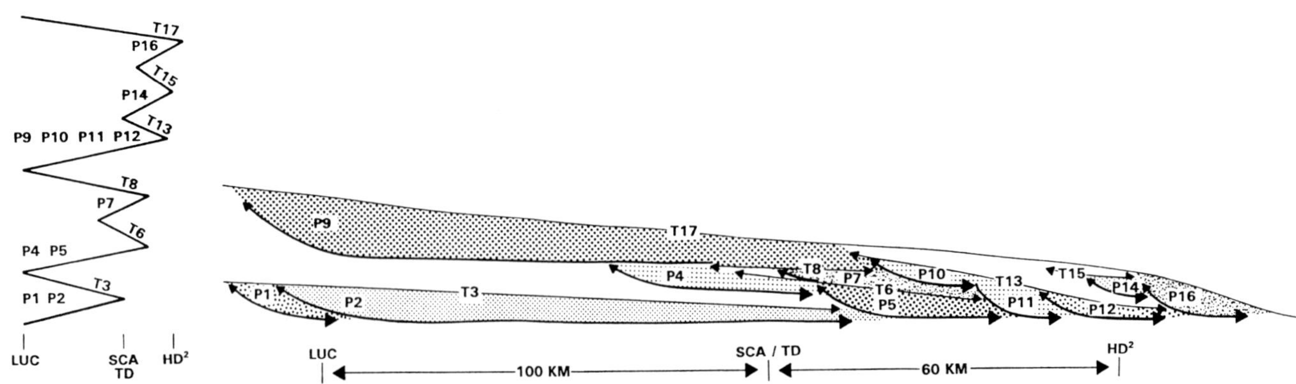

Fig. 5—Schematic diagram of the relationships of the prograding (P) sandbodies in the Lucerne (LUC)-Salt Creek Anticline (SCA)-Teapot Dome (TD)-Heldt Draw-Hartzog Draw (HD2). The cross-section is not to scale. At the left the extent of the various progradational (P) and transgressive (T) events is shown.

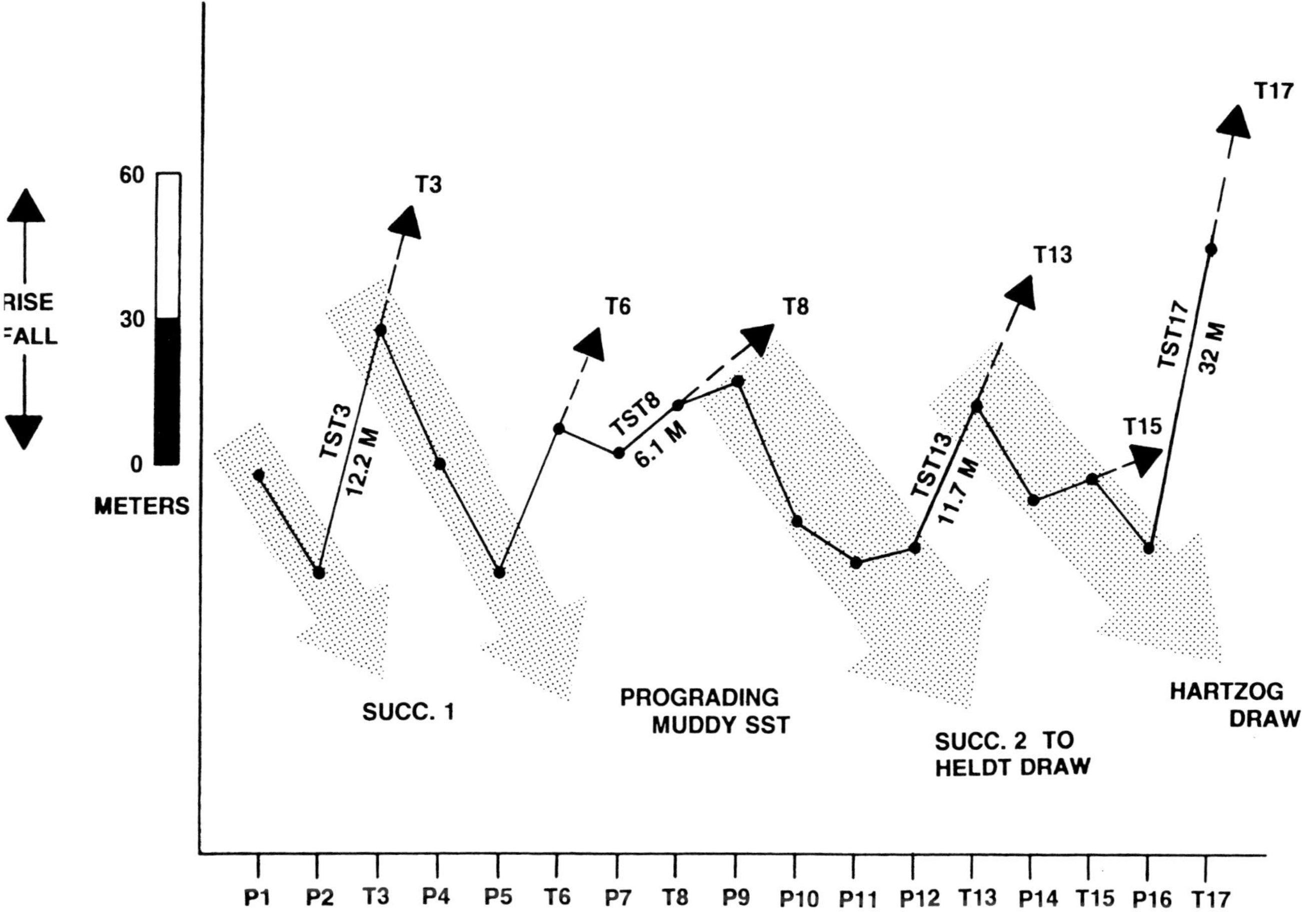

Fig. 6.—Estimated magnitudes of sea-level fluctuation between progradational (P) and transgressive (T) events. Dashed lines indicate that transgressions continued for an unknown period but that the T-surfaces are truncated on the cross-sections. The numbered TSTs indicate the preservation of Transgressive Systems Tract mudstone. The thickness of the TST is also given.

progradation of P9 represents a drop in FWWB. The P11 to P12 rise of about 2.5 m may result from our assumption that the datum (the B6 bentonite, Fig. 4) is horizontal. Bergman (1994) has argued that a transgressive surface (T13, Wells 42-40, Fig. 4) cannot dip landward. If T13 is rotated to the horizontal, implying that it was cut during a stillstand of sea level, the basinward gradient of B6 would be about 0.0007 (Bergman, 1994). Wells 37 (P11) and 43 (P12) are about 10 km apart, implying that with a gradient of 0.0007 on B6, the datum would be 7 m higher in Well 37 than in Well 43. The 2.5 m rise of FWWB between Wells 37 and 43 is actually a 4.5 m fall.

Rate of Sea-Level Fluctuation

The ammonite zones and absolute age dates presented by Gill and Cobban (1973) suggest that the Shannon was deposited in less than 1 million years. During this time there were six progradational episodes (P1, 2; P4, 5; P7; P9, 10, 11, 12; P14; P16) and six transgressive episodes. Therefore each progradational-transgressive pair took an average of less than 167,000 years to develop. During the greatest transgressions (T3 and T17), relative sea level rose a minimum of about 50 m and 60 m, respectively (Fig. 6). It is necessary to check that fluctuations of such magnitude, within short periods of time, are reasonable.

Present-day rates of eustatic rise are considered very low compared with those at the beginning of the Holocene transgression, and are of the order of 1 mm/yr (Gornitz and Lebedeff, 1987). If today's low rates are applied to the Shannon, the 50 m and 60 m rises could have taken place within 50,000 and 60,000 years, respectively. It follows that there is sufficient time during one progradational-transgressive pair (167,000 years) for transgressions of the calculated magnitude (60 m) to have taken place. There is much more time available if it is also assumed that the basin is subsiding for tectonic reasons.

The Shannon was deposited about 80 million years ago (Gill and Cobban, 1973), which coincides with a fall of eustatic sea level greater than 50 m on the short-term eustatic curve of Haq et al. (1988). This suggests that on a large scale, the Shannon Sandstone represents a third-order cycle. The six progradational-transgressive pairs within the Shannon indicate a fourth-order cyclicity superimposed on the third-order cycles. This is consistent with other fourth-order, progradational-transgressive cyclic deposits documented in the Cretaceous Western Interior Seaway. For example, in Alberta there are: (1) eight cycles with 100,000 to 150,000 year cyclicities in the Basal Belly River Formation (Campanian; Power and Walker, 1996); (2) seven cycles with average durations of about 215,000 years in the Dunvegan Formation (middle Cenomanian; Bhattacharya and Walker, 1991); (3) at least six cycles with durations of 167,000 to 333,000 years in the Cardium Formation (Turonian-Coniacian; Bergman and Walker, 1988); and (4) fifteen cycles with average durations of about 100,000 years in the Muskiki-Marshybank Formations (middle Coniacian to early Santonian; Plint, 1991). The interpretation of fluctuations of relative sea level on a scale of a few hundred thousand years is very difficult in the Cretaceous, a time generally believed to be free of continental glaciation. A recent discussion of these problems is presented by Plint (1991).

CONCLUSIONS

1. Excellent core and well-log control in Hartzog Draw permits the documentation of the facies, facies relationships, facies geometries, and erosional discontinuities. An interpretation of incised falling-stage and lowstand shorefaces has been suggested (Bergman, 1994). The erosional discontinuities and well-log characteristics can be simplified into a template that can be used for well-log interpretation in areas without core control.

2. The template can be used to make correlations and interpretations between areas with good control, specifically between the Salt Creek-Teapot Dome area and Hartzog Draw. However, these correlations can only be regarded as the simplest that are compatible with existing data.

3. Eleven progradational events can be suggested, separated by six transgressions. The progradational events can be combined into six groups; P1-P2 represents the progradation of the lower sandstone in the Lucerne-Salt Creek area; P4-P5 represents the progradation of muddy sandstones between Teapot Dome and Jepson Draw; P7 represents a minor, prograding sandbody between Teapot Dome and Pine Tree; P9-P10-P11-P12 represents a series of falling-stage, incised shorefaces, forming the upper sandstones at Lucerne and Salt Creek and continuing basinward as far as Heldt Draw; P14 represents the main sandbody at Hartzog Draw; and P16 represents a sandbody along the northeastern margin of Hartzog Draw.

4. Topographic changes in the position of RSEs below shorefaces, and in position of TSEs, allow estimates to be made of the magnitude of changes in relative sea level. These range from a few meters to more than 60 m for the progradational-transgressive cycles.

5. If the duration of the Shannon is a little less than 1 million years, the average duration of each of the six progradational-transgressive cycles would be about 167,000 years.

6. Recent interpretations in Salt Creek and Hartzog Draw (Walker and Bergman, 1993; Bergman, 1994) and the regional correlations (Bergman and Walker, 1995) suggest that the Shannon sandbodies were deposited in incised falling stage and lowstand shorefaces rather than shelf-ridge complexes (Tillman and Martinsen, 1984, 1987; Gaynor and Swift, 1988).

7. This interpretation indicates that the Shannon sandbodies formed under conditions of fluctuating relative sea level and that the individual sandbodies did not form contemporaneously. This interpretation allows the correlation between the isolated sandbodies at Salt Creek, Teapot Dome, and Hartzog Draw, and suggests ways in which the isolated sandbodies might be correlated with the highstand shoreline deposits at Lucerne. From the stacking patterns the depositional history and hence the history of relative sea level can be evaluated.

ACKNOWLEDGMENTS

This work was supported by the Natural Sciences and Engineering Research Council of Canada, through a Post-Doctoral Fellowship to KMB and research grants to KMB and RGW. We thank the Exxon Research and Production Corporation for allowing KMB access to the Hartzog Draw Cores, and the USGS Federal Core Storage Facility in Denver, where the remaining cores are stored. We would like to thank John Snedden and Bob Dalrymple who acted as external reviewers for this manuscript.

REFERENCES

Bergman, K.M., 1994, Shannon Sandstone in Hartzog Draw-Heldt Draw Fields (Cretaceous, Wyoming, USA) reinterpreted as lowstand shoreface deposits: Journal of Sedimentary Research, v. B64, p. 184-201.

Bergman, K.M. and Walker, R.G., 1988, Formation of Cardium erosion surface E5, and associated deposition of conglomerate: Carrot Creek field, Cretaceous Western Interior Seaway, Alberta, *in* James, D.P. and Leckie, D.A., eds., Sequences, Stratigraphy, Sedimentology: Surface and Subsurface: Calgary, Canadian Society of Petroleum Geologists Memoir 15, p. 15-24.

Bergman, K.M. and Walker, R.G., 1995, High-resolution sequence stratigraphic analysis of the Shannon Sandstone in Wyoming, using a template for regional correlation, Journal of Sedimentary Research, v. B65, p. 255-264.

Bhattacharya, J. and Walker, R.G., 1991, Allostratigraphic subdivision of the Upper Cretaceous Dunvegan, Shaftesbury, and Kaskapau Formations in the northwestern Alberta subsurface: Bulletin of Canadian Petroleum Geology, v. 39, p. 145-164.

Gaynor, G.C. and Swift, D.J.P., 1988, Shannon sandstone depositional model: Sand ridge dynamics on the Campanian Western Interior shelf: Journal of Sedimentary Petrology, v. 58, p. 868-880.

Gill, J.R. and Cobban, W.A., 1973, Stratigraphy and geologic history of the Montana Group and equivalent rocks, Montana, Wyoming and North Dakota: Denver, United States Geological Survey Professional Paper 776, 37 p.

Gornitz, V. and Lebedeff, S., 1987, Global sea-level changes during the past century, *in* Nummedal., D., Pilkey, O.H. and Howard, J.D., eds., Sea-Level Fluctuation and Coastal Evolution: Tulsa, Society of Economic Paleontologists and Mineralogists Special Publication 41, p. 3-16.

Haq, B.U., Hardenbol, J. and Vail, P.R., 1988, Mesozoic and Cenozoic chronostratigraphy and Eustatic cycles, *in* Wilgus, C.K., Hastings, B.S., Kendall, C.G. St. C., Posamentier, H.W., Ross, C.A. and Van Wagoner, J.C., eds., Sea-Level Changes: An Integrated Approach: Tulsa, Society of Economic Paleontologists and Mineralogists Special Publication 42, p. 71-108.

Parker, J.M, 1958, Stratigraphy of the Shannon Member of the Eagle Formation and its relation to other units in the Montana Group in the Powder River Basin, Wyoming and Montana, *in* 13th Annual Guidebook: Laramie, Wyoming Geological Association, p. 90-102.

Plint, A.G., 1991, High frequency relative sea-level oscillations in Upper Cretaceous shelf clastics of the Alberta foreland basin: Possible evidence for a glacio-eustatic control?, *in* Macdonald,

D.I.M., ed., Sedimentation, Tectonics and Eustasy: International Association of Sedimentologists Special Publication 12, p. 409-428.

Posamentier, H.W., Allen, G.P., James, D.P. and Tesson, M., 1992, Forced regressions in a sequence stratigraphic framework: Concepts, examples and exploration significance: American Association of Petroleum Geologists Bulletin, v. 76, p. 1687-1709.

Power, B.A. and Walker, R.G., 1996, Allostratigraphy of the Upper Cretaceous Lea Park-Belly River transition in central Alberta, Canada: Bulletin of Canadian Petroleum Geology, v. 44, p. 14-38

Rine, J.M., Tillman, R.W., Culver, S.J. and Swift, D.J.P., 1991, Generation of late Holocene sand ridges on the middle continental shelf of New Jersey, USA-evidence for formation in a mid-shelf setting based on comparisons with a nearshore ridge, *in* Swift, D.J.P., Oertel, G.F., Tillman, R.W. and Thorne, J.A., eds., Shelf Sand and Sandstone Bodies: Geometry, Facies and Sequence Stratigraphy: International Association of Sedimentologists Special Publication 14, p. 395-426.

Seeling, A., 1979, The Shannon Sandstone, a further look at the environment of deposition at Heldt Draw Field, Wyoming: The Mountain Geologist, v. 15, p. 133-144.

Spearing, D.R., 1976, Upper Cretaceous Shannon Sandstone: an offshore shallow marine sandbody: Wyoming Geological Association, 28th Annual Field Conference Guidebook, p. 65-72.

Tillman, R.W. and Martinsen, R.S., 1984, The Shannon shelf-ridge sandstone complex, Salt Creek anticline area, Powder River Basin, Wyoming, *in* Tillman, R.W. and Siemers, C.T., eds., Siliciclastic Shelf Sediments: Tulsa, Society of Economic Paleontologists and Mineralogists Special Publication 34, p. 85-142.

Tillman, R.W. and Martinsen, R.S., 1987, Sedimentologic model and production characteristics of Hartzog Draw field, Wyoming, a Shannon shelf-ridge sandstone, *in* Tillman, R.W. and Weber, K.J., eds., Peservoir Sedimentology: Tulsa, Society of Economic Paleontologists and Mineralogists Special Publication 40, p. 15-112.

Walker, R.G. and Bergman, K.M., 1993, Shannon Sandstone in Wyoming: A shelf ridge complex reinterpreted as lowstand shoreface deposits: Journal of Sedimentary Petrology, v. 63, p. 839-851.

Appendix 1—List of numbered well locations used in the cross sections.

Well Number	Well Identification
A	Getty # 20-11 Lone NE SW 20-44N-75W
B	Exxon HDU 4085 SW NW 8-44N-75W
C	CSOG HDU 5308 SE SE 30-45N-75W
21	Fenix and Scisson 64-AX-4 SE SW NE 4-38N-78W
22	Fenix and Scisson 31-S-3 NE NW 3-38N-78W
23	Fenix and Scisson 67-SX-27 SW SE 27-39N-78W
24	Oil Development # 1-4 Moore Federal 4-39N-77W
25	Clinton # 1 Yarbrough Federal SE SE 17-40N-77W
26	Moncrief # 17-1 Ford Draw SE NE 17-40N-77W
27	AMOCO # 9 Sherwood SW SW 3-40N-77W
28	Moncrief # 7 Crawford SE NE 25-41N-77W
29	Woods Petroleum # 8 Taylor Unit SE NE 20-41N-76W
30	Woods Petroleum # 14 Taylor Unit NE SE 5-41N-76W
31	Woods Petroleum # 1 Window Island NW SW 14-42N-76W
32	Sohio Petroleum # 12-9 Brown NE SE 12-42N-76W
33	Woods Petroleum # 7-22 Pine Tree Unit NW SE NW 7-42N-75W

Well Number	Well Identification
34	Woods Petroleum # 7-14 Pine Tree Unit NW NE 7-42N-75W
35	Woods Petroleum # 31-1 Ruby Ranch Federal SE NW 31-43N-75W
36	Davis Oil # 1 Davis Ruby Federal SW SE 30-43N-75W
37	Diamond Shamrock # 12-17 Jacobb Federal SW NW 17-43N-75W
38	Getty Oil # 1C Davis SW SW 8-43N-75W
39	Miami Oil Production Inc. # 1 Camblin NW NW 4-43N-75W
40	Getty Oil # 20-11 Lone NE SW 20-44N-75W
41	Woods Petroleum # 27-1 Tony NE SW 27-44N-75W
42	Woods Petroleum Camblin # 26-1 NE NW 26-44N-75W
43	Penzoil Federal No. 1 - 22 NE NE 22-44N-75W
44	Gulf Oil Schlautmann # 1-15 SE SE 15-44N-75W
45	CSOG HDU # 4145 SW NW 14-44N-75W
46	CSOG HDU # 4142 C NE 14-44N-75W
47	CSOG HDU # 41414 NE NE 14-44N-75W
48	Davis Oil Katie # 1 SW SE 30-45N-74W

HOLOCENE STRATIGRAPHIC ARCHITECTURE OF A SAND-RICH SHELF AND THE ORIGIN OF LINEAR SHOALS: NORTHEASTERN GULF OF MEXICO

RANDOLPH A. MCBRIDE
Department of Geography and Earth Science, George Mason University, 4400 University Drive, Fairfax, VA 22030 U.S.A.
LAURIE C. ANDERSON
Department of Geology and Geophysics, Louisiana State University, Baton Rouge, LA 70803 U.S.A.
ANDREI TUDORAN
Exxon Exploration Company, P.O. Box 4778, Houston, TX 77210 U.S.A.
AND
HARRY H. ROBERTS
Coastal Studies Institute and the Department of Oceanography and Coastal Sciences, Louisiana State University, Baton Rouge, LA 70803 U.S.A.

Abstract: Late Pleistocene and Holocene geology of the northeastern Gulf of Mexico shelf offshore Alabama and northwest Florida was investigated using 47 vibracores, foraminiferal and macrofaunal assemblages, and bathymetric data. The morphologic and stratigraphic signatures of the last rise of eustatic sea level were examined along this passive continental margin characterized by low subsidence. Major shelf features include shore-oblique sand ridges, mid-shelf linear shoals, and shelf-edge deltas. Surficial shelf sediments consist of >90% sand, <2.7% mud, and <2% granules and fine in a westerly direction from medium to fine sand. The sharp boundary that separates these two surficial sand types (Apalachicola and Mobile subprovinces) is identified for the first time in this study.

Six facies and two erosional surfaces characterize the shelf stratigraphy. Facies 1 is a Pleistocene soil horizon. This facies is truncated by a major erosional unconformity (Type 1 sequence boundary [SB]) that was created by subaerial exposure during the last sea-level lowstand and during the bay ravinement process (flooding surface [FS]) of the ensuing transgression (FS/SB). Fine-grained estuarine deposits (Facies 2, 3, or 4 [lower transgressive systems tract]) overlie the unconformity. Facies 3 or 4 are truncated by a shoreface ravinement diastem (flooding surface) and are overlain by a marine shell-bed (Facies 5; lower shoreface). Facies 5 grades into Facies 6, a quartz sand with open marine foraminifera that represents a shelf sand sheet. Facies 5 and 6 comprise the upper transgressive systems tract, which is up to 5.5 m thick.

The mid-shelf is characterized by two long (30-120 km), narrow (<6 km), shore-parallel to subparallel sand shoals that average 4 m thick. North Perdido Shoal is located 15-25 km offshore at the 20-25-m isobath, whereas South Perdido Shoal lies 20-70 km offshore at approximately the 35-m isobath. Both shoals trend southwest-northeast. The linear shoals are not *in situ* or degraded barriers (Stubblefield et al., 1984a, b), offshore shelf-ridge (bar) complexes (Tillman and Martinsen, 1984, 1987; Gaynor and Swift, 1988), or lowstand/transgressive incised shoreface deposits (Bergman, 1994; Walker and Wiseman, 1995) because the sediments that comprise the shoals lie above the shoreface ravinement diastem, and open marine species dominate the foraminiferal and molluscan assemblages. Although shelf morphology is similar to modern barrier island geomorphology, shelf morphostratigraphy is related to transgressive and post-transgressive processes. Shoal form and orientation are dictated by underlying transgressive topography (escarpments) that was cut into the Pleistocene substrate during the post-glacial transgression. During transgression, erosional shoreface retreat produced a trailing sand sheet that draped the transgressive topography. Consequently, 1) the linear nature of the shoals is derived from their formation along the shoreface (i.e., depositional strike) at lower stands of sea level during an overall transgression; 2) sediment transport from the present shoreline across the shelf appears to have little influence on shoal development; 3) the interplay between relative sea-level changes and sediment supply caused translation of the shoreface profile, thus dictating the position of the linear shoals; and 4) post-transgressive reworking and subaqueous landward migration in response to storm processes are integral parts of shoal evolution.

INTRODUCTION

Detailed investigations of modern and ancient continental shelves, especially transgressive systems tracts, have increased significantly over the past decade (e.g., Knight and McLean, 1986; Moslow and Rhodes, 1986; Nummedal and Swift, 1987; Demarest and Kraft, 1987; Wilgus et al., 1988; Swift et al., 1991a; Bergman, 1994; Dalrymple et al., 1994; Siringan and Anderson, 1994; Walker and Wiseman, 1995; Tortora, 1996; Anderson et al., 1996). Although continental-margin geology of the northern Gulf of Mexico has been studied extensively (e.g., Shepard et al., 1960; Rezak and Henry, 1972; Salvador, 1991), investigations of late Pleistocene/Holocene transgressive systems tracts have focused exclusively on the northwestern and north-central Gulf (e.g., Nelson and Bray, 1970; Penland et al., 1989; Bartek et al., 1991; Anderson et al., 1992, 1996; Thomas and Anderson, 1994; Brooks et al., 1995). In comparison, late Quaternary geology of the northeastern Gulf is relatively unexplored, especially the shelf region east of Mobile Bay, Alabama (Figs. 1 and 2).

The goal of this paper is to examine the morphologic and stratigraphic signatures of the last rise of eustatic sea level along the shelf of a passive continental margin. Specific objectives are to: 1) identify primary shelf morphologic features; 2) delineate sedimentary facies, regional erosional surfaces, depositional environments, and stratigraphic architecture of shelf deposits; 3) determine shelf sedimentary processes; 4) establish the sequence stratigraphic framework; and 5) produce a detailed geologic model synthesizing the depositional history of the shelf in response to the last eustatic rise of sea level.

Study Area

The study area encompasses approximately 10,000 km^2 of the continental shelf and inland waters along the western flank of the DeSoto Canyon between Mobile Bay, Alabama, and Pensacola Bay, Florida (Fig. 1). The continental shelf widens to the west from only 25 km along the eastern boundary of the study area (86°50') to about 110 km south of Mobile

Isolated Shallow Marine Sand Bodies: Sequence Stratigraphic Analysis and Sedimentologic Interpretation.
SEPM Special Publication No. 64, Copyright © 1999
SEPM (Society for Sedimentary Geology), ISBN 1-56576-057-3, p. 95-126.

Bay. The study area is bound on the north by the Alabama/Florida mainland shoreline and in the south by the shelf break at ~100 m water depth. This area lies on the eastern margin of the Gulf Coast depocenter (Marsh, 1966) and is blanketed by the Mississippi/Alabama/Florida (MAFLA) sand sheet (Doyle and Sparks, 1980; McBride and Byrnes, 1995). Major morphologic features in the study area include shore-oblique sand ridges, middle- to outer-shelf sand shoals, and large lobes along the continental-shelf margin (McBride and Byrnes, 1995). The Alabama/Florida shoreline is segmented by the Mobile Bay Entrance, Perdido Pass, and Pensacola Pass, which connect their estuarine systems to the Gulf of Mexico (Fig. 1).

The northeastern Gulf of Mexico is characterized by microtidal conditions, with an average tidal range of 0.4 m (National Oceanic and Atmospheric Administration, 1985). Maximum shelf tidal currents are about 15 cm/s and occur where the shelf is widest (Schroeder et al., 1994). Wind-driven currents are strongest in the winter and early spring and occur in response to 20-30 cold fronts per year that affect the northern Gulf of Mexico (Huh et al., 1984; Roberts et al., 1987). These currents can generate velocities up to 40-50 cm/s. Mean inner-shelf flow is westward and consistent with wind-driven flow (Dinnel, 1988). Hurricanes tend to make landfall in the study area about once every 7 years, generate current speeds more than 100 cm/s (Murray, 1970), and rework the seafloor out to the shelf break at ~100 m depth. In addition, the Loop Current and its associated eddies affect the outer continental shelf along the southeastern boundary of the study area. Loop Current water also has been shown to penetrate further northward within several kilometers of the Florida shoreline (Huh et al., 1981; Sturges and Evans, 1983).

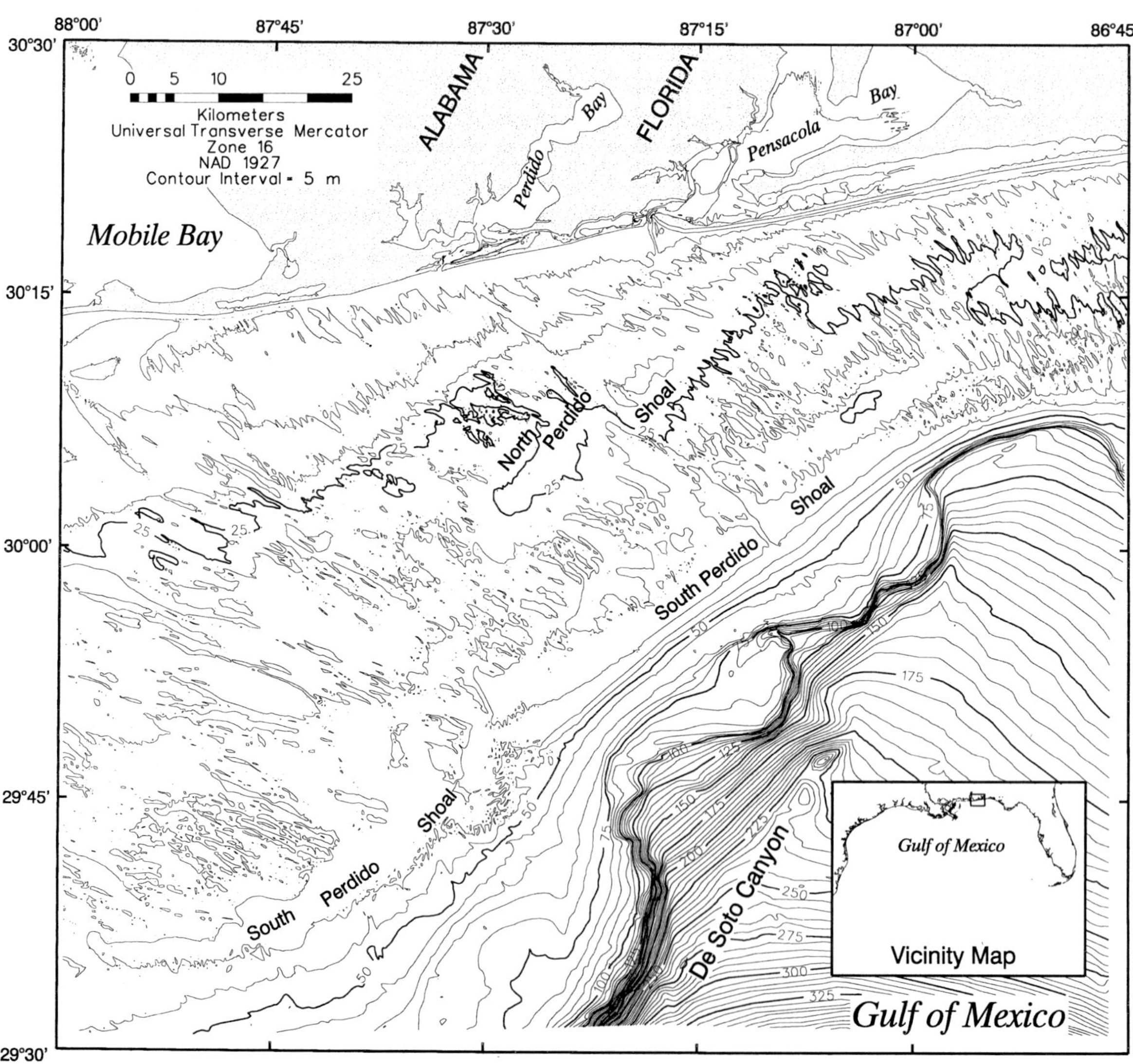

Fig. 1.—Study area in the northeastern Gulf of Mexico showing detailed bathymetry at 5-m contour intervals. Thicker contours are at 25-m intervals (modified from McBride and Byrnes, 1995). North and South Perdido Shoals trend northeast-southwest on the mid-shelf area.

PREVIOUS WORK

Seafloor Morphology

The only known studies in or adjacent to the study area that have examined seafloor morphology were conducted by Hyne and Goodell (1967) and Ballard and Uchupi (1970). Hyne and Goodell (1967) investigated the inner shelf seaward of Choctawhatchee Bay, Florida, and concluded that long, shore-parallel shoals were former barrier-island features formed at lower stands of sea level. Ballard and Uchupi (1970) conducted a regional study that covered the entire northern Gulf of Mexico shelf and upper slope. They recognized four former shoreline trends at the 20 m, 32 m, 60 m, and 160 m isobaths.

Surficial Sediments

Previous studies along the northern Gulf of Mexico document four primary sediment provinces on the continental shelf: 1) Rio Grande, 2) Western Gulf, 3) Mississippi River, and 4) Eastern Gulf (Goldstein, 1942; Gould and Stewart, 1955; Van Andel, 1960; Fairbank, 1962; Ludwick, 1964; Upshaw et al., 1966; Hyne and Goodell, 1967; Kent et al., 1976; Doyle and Sparks, 1980; Mazzullo and Bates, 1985; Arthur et al., 1986; Donoghue and Allard, 1987; Frey and Dorjes, 1988; Schroeder et al., 1988a and b; Mazzullo and Peterson, 1989; Shultz et al., 1990; Parker et al., 1992; Kennicutt et al., 1995). The shelf eastward from the abandoned St. Bernard delta complex (Mississippi River Delta) to the Apalachicola Delta is classified as the Eastern Gulf Province. Also known as the MAFLA (Mississippi-Alabama-Florida) sand sheet, this province can be further subdivided into the Mobile and Apalachicola subprovinces (Mazzullo and Peterson, 1989) (Fig. 2; subprovinces V and VI).

Sediments along the barrier beaches of Alabama and the western Florida Panhandle are dominated by fine- to medium-grained quartz sand that fines westward in the direction of net longshore sediment transport (Hsu, 1960; Gorsline, 1966; Kwon, 1969; van Wyk, 1973; Williams, 1974; Balsillie, 1975; Kent et al., 1976; Walton, 1976; Stone et al., 1992). Estuaries in the study area are repositories for fine-grained sediment delivered by fluvial systems that flow into Mobile, Perdido, and Pensacola Bays (Horvath, 1968; Parker, 1968; Ryan, 1969; Folger, 1972; George, 1988; Isphording et al., 1989), and most sand is restricted to bayhead deltas and/or concentrated around bay perimeters. Little sediment is transported directly to the Gulf of Mexico, except suspended clay and fine silt associated with the Mobile Bay plume (Abston et al., 1987; Stumpf, 1991) and during hurricane events (Isphording and Isphording, 1991).

Geologic Framework

Although previous studies lacked core data of late Quaternary deposits in the study area, numerous studies of adjacent shelf environments have been conducted using high-resolution seismic data, vibracores, piston cores, and/or long borings. Using seismic stratigraphy, Locker and Doyle (1992) investigated the Neogene to Recent shelf/slope geology offshore of northwest Florida. They delineated basic seismic packages but no details were provided for the thin Holocene deposits. For the Alabama-Mississippi coastal zone and shelf, Otvos (1982), McBride et al. (1991), Mars et al. (1992), Davies and Hummell (1994), and Kindinger et al. (1994) delineated sedimentary facies and the geologic history of Mississippi Sound and Mo-

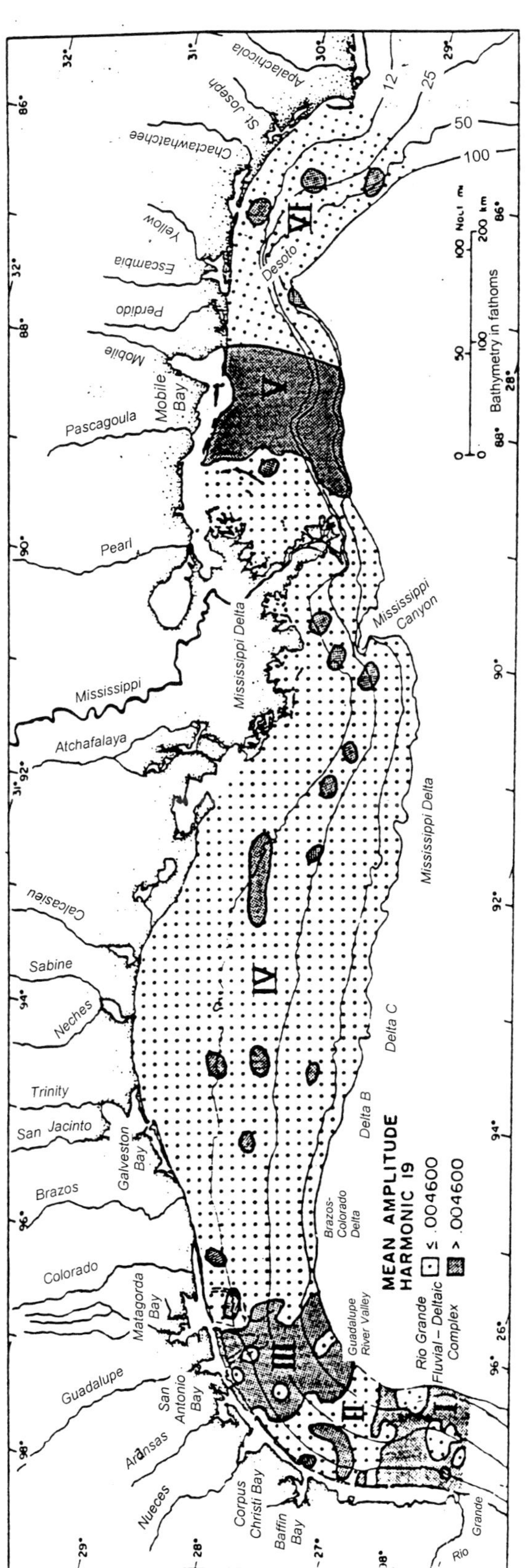

Fig. 2.—Surficial sediment subprovinces of the northern Gulf of Mexico shelf (I = Rio Grande; II = Guadalupe; III = Brazos-Colorado; IV = Mississippi; V = Mobile; and VI = Apalachicola) modified from Mazzullo and Peterson (1989).

bile Bay. Further west, studies of the Texas and Louisiana continental shelves have focused on sea-level changes, incised valley fills, and hard mineral resources (Coleman and Roberts, 1988a and b; Penland et al., 1989; Anderson et al., 1992; Donoghue, 1992; Brooks et al., 1995). Recent work on shelf-edge deltas has documented fluvial drainage patterns and depocenters during lowstands of sea level (Winker, 1982; Suter and Berryhill, 1985; Kindinger, 1988, 1989; Sydow and Roberts, 1994; Morton and Suter, 1996).

Macrofauna

Immediately to the east of the study area, Jervey (1974) compiled detailed information on sediment, living macrofauna, and death assemblages for the shoreface and inner shelf off Destin, Florida. To the west of the study area, Parker (1956, 1960) described life and death assemblages of the eastern Mississippi Delta region. A number of recent articles include information on the surface and/or stratigraphic distribution of bioclastic accumulations to the west of the study area off Alabama (Parker, 1990; Parker et al., 1992; Davies and Hummell, 1994). In addition, living communities or bioclastic remains associated with hardbottom areas on the Alabama shelf are described by Schroeder et al. (1988a and b, 1989, 1995), Gittings et al. (1990, 1992), and Parker et al. (1992).

Anderson and McBride (1996) outlined the taxonomic composition, taphonomy, and genesis of subsurface Holocene shell beds in the study area. Gangopadhyay et al. (1996) described molluscan and foraminiferal assemblages of the Pensacola and Perdido Bay estuarine systems. Anderson et al. (1997) provided evidence (based on foraminiferal, macrofaunal, and ^{14}C data sets) of a thoroughly mixed Holocene marine section and of a less seasonal middle-to-late Holocene climate in the northeastern Gulf.

Microfauna

Pioneering work on the distribution of benthic foraminifera on the continental shelf of the northeastern Gulf of Mexico was conducted by Parker (1954) and Bandy (1954). In both of these regional studies, sediment samples were obtained from several transects and the relative abundances of species in death assemblages (thanatocoenose) were examined, resulting in the recognition of depth-related foraminiferal assemblages. An extensive study of foraminiferal generic distribution patterns in the northeastern Gulf (both shallow and deep) was conducted by Walton (1964), leading to the delineation of broad faunal belts characterized by the abundance of particular genera. Bock's (1976) investigation covered much of the same area, but his focus was on the distribution of living assemblages. The comprehensive distribution of benthic foraminifera in the Gulf of Mexico, including our study area, was summarized by Poag (1981), with emphasis on abundance variations of genera (as in Walton, 1964). Another important summary of Recent benthic foraminifera distribution, which includes the Gulf of Mexico, is provided by Murray (1991). The local, brackish-water foraminifera of Mobile Bay, Alabama; Choctawhatchee Bay, Florida; and St. Andrews Bay, Florida, and the effects of salinity changes on the fauna were investigated by Lamb (1972), Pastula (1967), and Mechler and Grady (1984), respectively.

Glacio-Eustatic Changes

Although much work has been published on the last eustatic fall and rise (e.g., Emiliani, 1958; Imbrie et al., 1984; Shackleton, 1987; Pirazzoli, 1991), a consensus is lacking on a true eustatic sea-level curve for the northern Gulf of Mexico (see Pirazzoli, 1991, p. 116). Fisk (1944) recognized important links between glacio-eustatic change and continental-shelf sedimentary history. From the 1950s to 1970s, subsequent researchers investigated sea-level changes based on radiocarbon-dated material (wood, shells, peat) collected from the Gulf of Mexico and the southern U.S. Atlantic shelves (e.g., Godwin et al., 1958; McFarlan, 1961; Shepard, 1963; Curray, 1960, 1965; Milliman and Emery, 1968; Emery and Milliman, 1970; Nelson and Bray, 1970). However, one problem associated with these sea-level curves is the difficulty relating dated organic materials to an associated paleo-sea-level position (Pirazzoli, 1991). Reliable elevation versus age data (former sea levels) are obtained from reefal or encrusting marine organisms in growth position collected on stable carbonate platforms, such as coral reefs (Hopley, 1986) and coralline algae (Adey, 1986). Much recent work has concentrated on radiocarbon-dated coral with a narrow depth range from Barbados in the Caribbean, as well as Papua New Guinea and Tahiti in the South Pacific (Fairbanks, 1989, 1990; Bard et al., 1990, 1996; Chappell and Polach, 1991). Coral-based curves should provide an objective "control" for the eustatic signal. However, the Barbados and Papua New Guinea sites are located over active subduction zones, and the Tahiti data, although tectonically stable, indicate that past sea levels were deeper for comparable Gulf of Mexico dates. Contemporaneous locations around the world can show significant differences in the height of sea level because of geoidal (undisturbed sea surface) depressions and elevations (Carter, 1988; p. 245). Consequently, the best eustatic-sea-level information for the Gulf of Mexico over the past 18,000 years B.P. is probably derived from a combination of local (Curray, 1965) and worldwide studies (Fairbanks, 1989; Bard et al., 1996).

METHODS

Bathymetry

National Ocean Service (NOS) hydrographic survey data (H-sheet soundings) were compiled and modeled to provide bathymetric coverage for the shelf (McBride and Byrnes, 1995). Using the Pensacola, Florida, tide gauge and benchmark, the data were vertically adjusted to address datum shifts between mean low water (MLW) and the National Geodetic Vertical Datum (NGVD 1929) as well as local relative sea level rise of 2.4 mm/yr. Additionally, the North American Datum 1927 (NAD27) was used because horizontal datum shifts were not needed. The original projection was converted from Polyconic to Universal Transverse Mercator (zone 16). Surface modeling software was employed to produce a digital elevation model of the seafloor using a Triangulated Irregular Network (TIN) technique. Using the newly created modeled surface, continuous bathymetric profiles and a bathymetric map were produced (Fig. 1).

Vibracore Data

Between 1990 and 1993, 47 vibracores were collected to determine the stratigraphic signature of late Quaternary deposits (Fig. 3). Cores were collected in water depths ranging from 3 m to 45 m. A pneumatic vibrator attached to a submersible 7-meter-high steel tripod was lowered to the seafloor, and 6-meter-long aluminum core tubes (7.5-cm diameter) were vibrated completely into subsurface deposits or until refusal. Refusal normally resulted when the bottom of the core barrel penetrated shell beds or pre-Holocene deposits. Vibracores were logged, photographed, and sampled for grain size, ^{14}C dating, macrofossils, and microfossils. Grain-size samples were collected at 25-cm intervals and, in accordance with Folk (1980), grain size analyses were performed using a sonic sifter with size statistics (moment measures) computed.

Paleontology

Macrofauna (primarily mollusks) and benthic foraminifera were examined from selected cores to determine paleoenvironmental conditions. Macrofauna (>1.4 mm) from 10 well-developed shell beds in nine vibracores were identified, counted, and described taphonomically (Anderson and McBride, 1996; McBride et al., 1996; McBride, 1997, Appendix E). Benthic foraminifera were investigated quantitatively in 51 samples (~2 cm thick) from 12 vibracores. All representa-

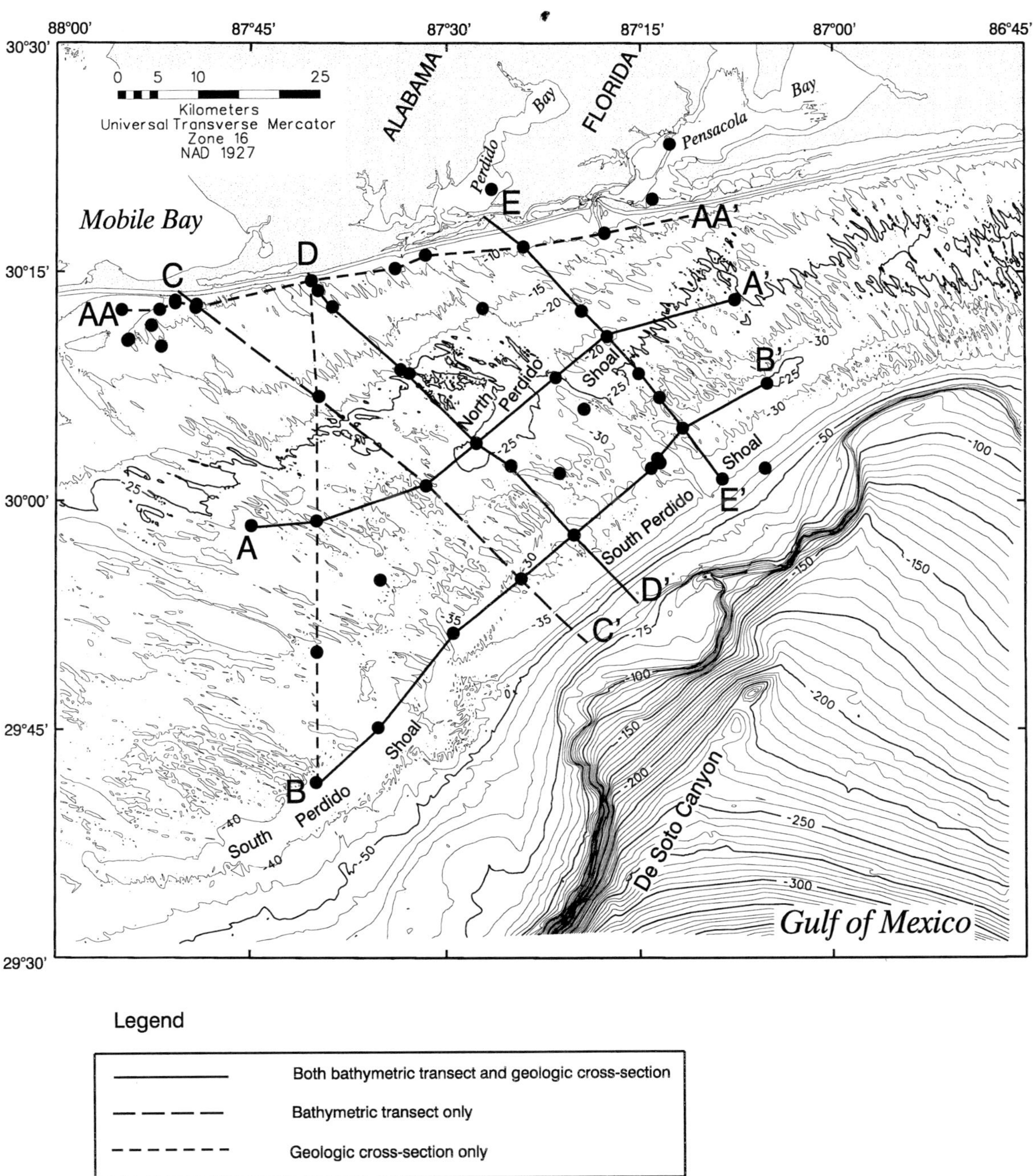

Fig. 3.—Vibracore locations (dots), bathymetric transects, and geologic cross-sections in the study area. See Figure 5 as well as Figures 11 through 16 for core identifications.

tive lithofacies were sampled. Some of the samples were floated in bromoform, but most were simply washed to recover foraminiferal residues. In both cases, a 63-micron sieve was used. The simple wash eliminates the possibility of a bias against agglutinated foraminifera of which none were found. For each sample, 300 specimens were picked, and foraminifer species were identified. The relative abundances of species representing >5% of the total number of specimens were used as data in a cluster analysis using Ward's method for non-standardized data. The proportion of the clay-silt fraction also was determined for all samples.

In performing cluster analysis, certain species belonging to the same genus or higher taxonomic entity were grouped, such as *Rosalina, Cibicidoides, Elphidium*, large benthic foraminifera (*Peneroplis proteus, Parasorites orbitolitoides*, and *Cyclorbiculina compressa*), and miliolids (*Quinqueloculina, Triloculina, Sigmoilina, Pyrgo*). Organizing certain foraminifer species into groups is justified because the taxa have common environmental preferences (Poag, 1981; Murray, 1991) as well as similar species distributions in the investigated samples. For example, the *Rosalina* group consists of *R. concinna* and *R. subaraucana* (and extremely rare specimens of *R. bahamaensis*). Whenever *R. concinna* is present in large numbers, *R. subaraucana* is also present at a constant ratio of 1:7-9. The *Elphidium* group consists of the species *Elphidium gunteri, E. poeyanum, E. fimbrilatum, E. discoidale*, and *E. galvestonense*. The *Cibicidoides* group includes the species *Cibicidoides io* and *C. mollis*.

RESULTS AND DISCUSSION

Seafloor Morphology

Detailed continental-margin morphology in the study area is characterized by a complex pattern of first-, second-, and higher-order features. First-order features are regional in extent (tens to hundreds of kilometers) and are the primary expression of the seafloor, whereas higher-order features tend to be smaller and are superimposed on first-order features (Figs. 1, 4).

Based on bathymetry, the continental shelf and upper slope in the study area can be divided into three geomorphic zones (McBride and Byrnes, 1995). Zone 1 (0-20 m) is dominated by shore-oblique shelf and shoreface sand ridges (Figs. 1, 4). These ridges are generally over 0.3 km long, have relief up to 5 m, side slopes that average less than 1°, and are 1- to 4-km wide with wavelengths of 0.5-8 km.

Zone 2 encompasses most of the middle continental shelf (20-50 m). The seafloor in this area is dominated by two long, linear shoals that are >5 m in relief and oriented parallel to the shelf break (Figs. 1, 4). Both shoals trend southwest-northeast. The most landward shoal, North Perdido Shoal, is located 15-25 km offshore in water that deepens to the southwest from 20 m to 25 m. The shoal is about 30 km long, 2-5 km wide, and narrows to the northeast (Figs. 1, 3, 5; A-A' from 25 km to 40 km and from 45 km to 55 km). In contrast, the outer shoal (South Perdido Shoal) lies 20-70 km offshore; water depths over the shoal deepen to the southwest from 25 m to 40 m (Figs. 1, 5; B-B' from 25 km to 55 km and from 60 km to 75 km). South Perdido Shoal is approximately 120 km long, 3-6 km wide, and narrows to the southwest. It is the longest, shelfbreak-parallel shoal along the northern Gulf of Mexico. The North and South Perdido Shoals tend to be asymmetrical in profile (the landward flank is the steepest) with a bathymetric low landward of, and parallel to, each linear shoal (Figs. 1, 5; C-C' at 54 km, D-D' at 29 km and 46 km, and E-E' at 22 and 36 km). As shown on Figures 1 and 4, a well-developed, narrow, shore-perpendicular bathymetric low (oriented northwest/southeast) intersects both linear shoals and appears to continue updip toward Perdido Bay (Fig. 5; A-A' at 43 km; B-B' at 58 km). In addition, the southwestern end of each shoal is truncated by a much broader bathymetric low (Figs. 1, 4, 5; A-A' from 0 to 23 km; B-B' from 15 to 23 km).

Zone 3 extends from 50 m to 150 m water depths and includes the outer continental shelf/shelf break area. The most diagnostic bathymetric features are 10-25 km wide, shelf-edge lobes that occur along strike (Figs. 1, 4). The largest shelf-edge lobe occurs along the southwest corner of the study area in water depths between 65 m and 200 m. Further east, the shelf edge lobes are much smaller and their seaward limit is in shallower water, averaging about 125 m. These smaller lobes are located seaward of the narrower, shore-perpendicular bathymetric low that bisects North and South Perdido Shoals (Figs. 1, 4; 30°00' north, 87°12' west; Fig. 5 at 43 km on A-A' and at 58 km on B-B').

Interpretation.—
The long, linear, shelf-break parallel shoals and associated bathymetric lows appear to occur on interfluves between fluvial drainage systems that tend to be shore-normal. This shelf morphology is similar to the present-day barrier/estuarine system between Mobile and Pensacola Bays. Mobile, Perdido, and Pensacola Bays represent shore-normal lows, whereas Santa Rosa Sound and Big Lagoon exemplify shore-parallel lows landward of the sand-rich barrier islands (Santa Rosa Island and Perdido Key). Furthermore, the western portion of South Perdido Shoal and the broad bathymetric low are similar in morphology to Mobile Bay and the barrier spit known as Morgan Peninsula. Also, the spatial relationship between shore-normal lows and shelf-edge lobes suggests the presence of fluvial-deltaic systems during sea-level lowstands.

Overall, shelf morphology suggests the development of shelf-edge deltas fed by fluvial systems that cut across the exposed shelf during the last sea-level lowstand. Though the South and North Perdido Shoals were initially thought to represent two barrier/estuarine systems because they closely mimic the morphology of the modern coastal system, this is not the case as discussed in the shelf evolution model below. The shoals imply periods of significant slowing in the rate of sea-level rise during the last post-glacial transgression. However, the shoals also appear to be reworked because they are asymmetrical in profile with the steepest flank facing landward. Typical barrier beaches are characterized by the steepest flank facing seaward.

Geologic Framework

The geology of transgressive deposits in the study area was delineated based on benthic foraminiferal assemblages, macrofauna, sedimentary facies, and erosional surfaces. Based on these data, six regional cross-sections were constructed to determine sand-body geometry, sequence stratigraphy, and depositional history of the shelf.

Benthic foraminifera.—
Q-mode cluster analysis of 51 benthic foraminiferal samples reveals four primary foraminiferal clusters (ES, M, R, AS) as shown in Figure 6. These clusters are critical for interpreting paleoenvironments of sedimentary deposits in

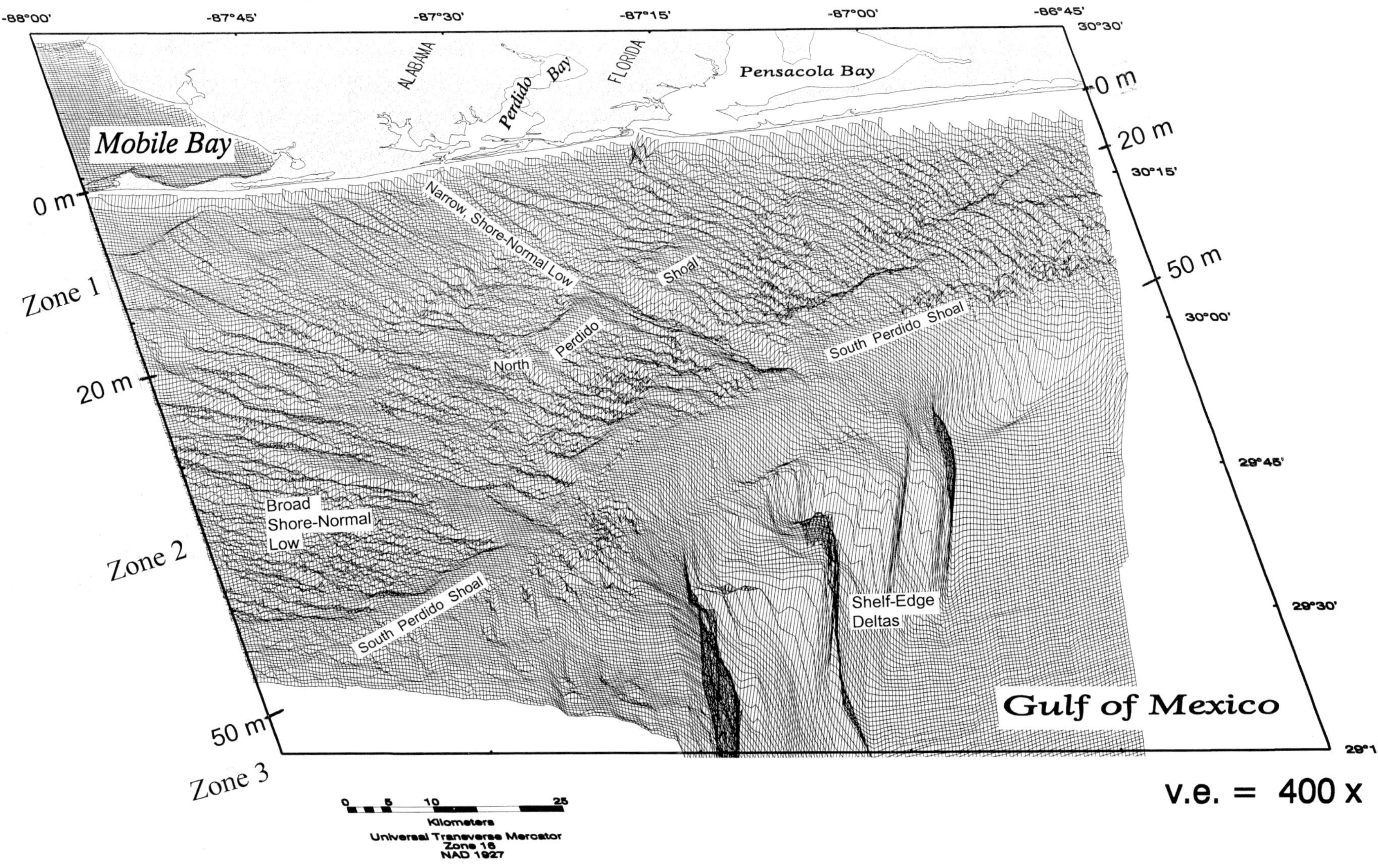

Fig. 4.—Oblique view of a digital elevation model for the Alabama and northwest Florida shelf in the northeastern Gulf of Mexico (vertical exaggeration = 400 x).

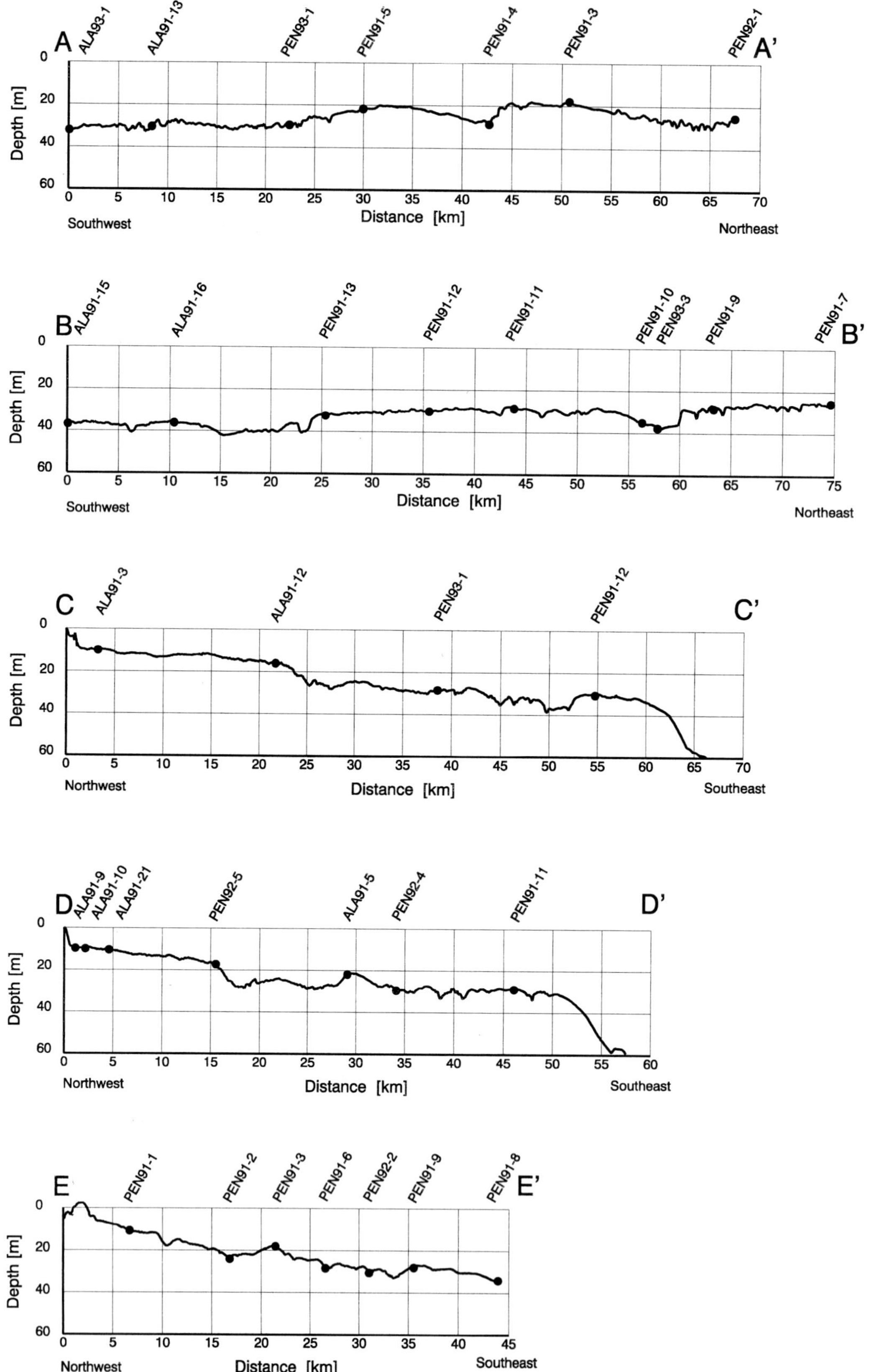

Fig. 5.—Strike-oriented bathymetric profiles A-A' and B-B' and dip-oriented profiles C-C', D-D', and E-E' with vibracore locations (black dots). Vertical exaggeration = 200 x. See Figure 3 for locations.

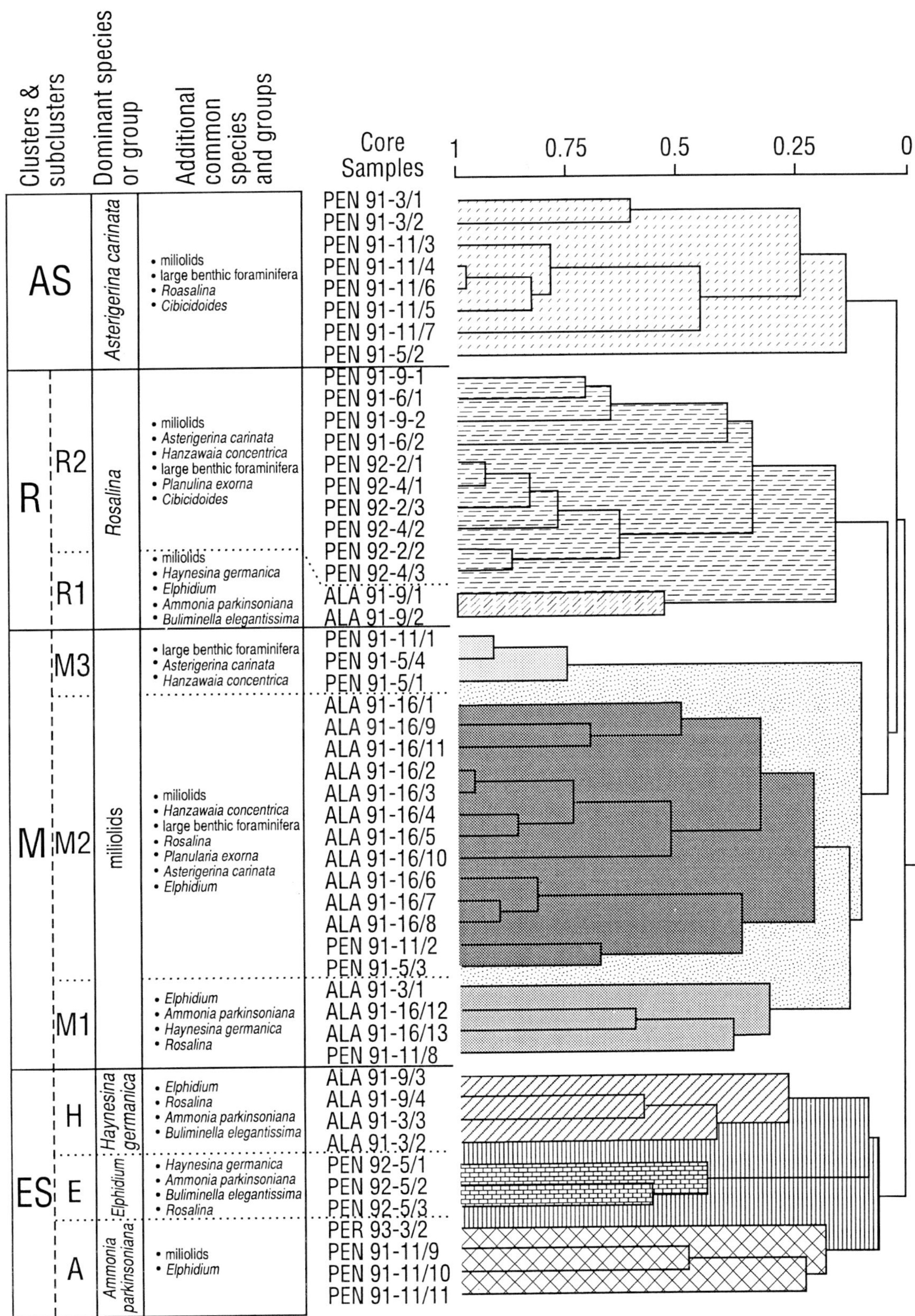

Fig. 6.—Q-mode cluster analysis of benthic foraminiferal samples from the Alabama/northwest Florida shelf. Foraminiferal assemblages/clusters are identified based on dominant species or groups, as well as additional common species and groups. See Figure 7 for interpretation of environmental settings; see Appendix 1 (following references) for additional information regarding foraminiferal sampling.

the study area because similar modern assemblages are present in Recent sediments of the Gulf of Mexico (Poag, 1981; Murray, 1991). Within primary clusters, subclusters (A, E, H, M1, M2, M3, R1, R2) occur and represent finer-scale environmental differences.

Of the 51 foraminiferal samples, 11 samples form the ES cluster (Fig. 6). These samples represent sediments deposited in brackish waters (see Pastula, 1967; Lamb, 1972; Poag, 1981; Mechler and Grady, 1984; and Murray, 1991). The ES cluster consists of three subclusters (A, E, and H) dominated respectively by the following species or genera: 1) *Ammonia parkinsoniana* (41-71%), 2) *Elphidium* (25-34%), and 3) *Haynesina germanica* (29-57%). A *Haynesina germanica*-dominated assemblage has not been reported before from the Gulf of Mexico, but it is interpreted to represent lower-than-normal marine salinity because *Elphidium* (8-17%) and *Ammonia parkinsoniana* (7-14%) also are abundant, and because *Haynesina germanica* is considered characteristic of brackish-water environments, from marshes and lagoons to the inner shelf (Murray, 1991). However, foraminiferal subcluster H is characterized by higher salinities than A and E because the *Rosalina* group (12-21%) and the species *Buliminella elegantissima* (9-13%) are present. This progressive increase in salinity from A to E to H is illustrated schematically in Figure 7.

Another large cluster (M) includes samples with abundant miliolid foraminifera (Fig. 6). Miliolids are abundant in waters with a wide range of salinities (Murray, 1991). The large M cluster is separated into three subclusters (M1, M2, and M3). In the M1 subcluster, *Elphidium*, *Ammonia parkinsoniana*, and *Haynesina germanica* are common, which indicates deposition in brackish waters (Fig. 7). In contrast, *Hanzawaia concentrica*, *Planulina exorna*, and *Asterigerina carinata*, as well as large benthic foraminifera, are indicative of normal marine waters for M2 and M3 (Figs. 6, 7).

The remaining two primary clusters are dominated by *Rosalina* (R) and *Asterigerina carinata* (AS) as shown in Figure 6. These taxa are indicative of normal-marine waters of the inner and middle continental shelf of the Gulf of Mexico (Poag, 1981; Murray, 1991). However, two samples of the *Rosalina* cluster form a distinct subcluster (R1) because miliolids (12-22%), *Haynesina germanica* (12-14%), *Elphidium* (7-10%), and *Ammonia parkinsoniana* (8-10%) are common. The presence of these taxa indicates a slightly lower normal marine salinity (shoreface) for R1 as compared to R2, M2, M3, and AS (Fig. 7).

Surficial sediments, sedimentary facies, and depositional interpretations.—

Figure 8 shows the spatial distribution of surficial sediments (facies) in the study area, which provides a sedimentologic framework for the subsurface geology. According to Walther's Law (Middleton, 1973), only those facies or environments can be superimposed which can be observed beside each other at the present time. As such, finer-grained facies dominate the central portions of estuaries, whereas sand dominates a narrow zone around their perimeters. As estuaries were transgressed in response to the post-glacial rise in sea level, the upper estuarine section was plained off through erosional shoreface retreat, thus preserving the central, stratigraphically lower, finer-grained facies. As the transgression continued, clean medium- to fine-sand of the shelf sand sheet capped the sequence.

Six primary facies characterize Holocene and late Pleistocene deposits found in the study area. These facies are identified based on primary sedimentary structures, sediment texture, microfauna, macrofauna, stratigraphic context, and position along the modern estuary-to-outer-shelf profile. The primary facies were derived from the entire vibracore data set. In this paper, facies is used as a descriptive term to characterize sedimentary deposits (see Moore, 1949; Selley 1970, 1988). Facies are described from the base to the top of section, followed by an interpretation of depositional environment.

Facies 1.—Facies 1 is a yellowish-burnt-orange and grey, massive to highly bioturbated, clayey quartz sand that is dense and mottled (Fig. 9a and b). Oxidized sediments dominate, which indicates continuous to fluctuating subaerial exposure. Large burrows and possible root traces are present; however, macro- and microfossils are absent. The lower boundary of Facies 1 is unknown but the unit is at least 0.15-m thick. The unit is truncated by an erosional surface, and the bulk density of Facies 1 is much higher than that of overlying units. The erosional unconformity is directly overlain by either Facies 2 or 3.

Facies 1 is a well-developed soil horizon produced during prolonged subaerial exposure (erosional unconformity) and it represents the top of the Pleistocene Prairie Formation. The soil horizon masks the original depositional environment, which makes paleoenvironmental interpretation difficult because micro- and macrofauna are absent and physical sedimentary structures are lacking or obliterated. The unit probably represents some combination of sandy alluvial/fluvial/strandplain deposits (Fisk, 1944; Otvos, 1985; Snedden et al., 1994). The soil horizon and erosional unconformity probably were produced by subaerial weathering during the Wisconsinan fall in sea level (especially during the oxygen isotope [$\delta^{18}O$] stage 2 lowstand about 18,000 yrs B.P.) but before the Holocene transgression.

Facies 2.—Facies 2 is a thin, matrix-supported shell bed that is typically 0.12-0.15 m thick (see Figs. 2 and 8 [lower shell bed] in Anderson and McBride, 1996). The bivalves *Chione cancellata*, *Parvilucina multilineata*, *Crassostrea virginica*, *Nuculana acuta*, and *Anomia simplex* are common, and the shell bed has a muddy-quartz-sand matrix. Most bioclasts show excellent preservation. For example, *Chione cancellata*

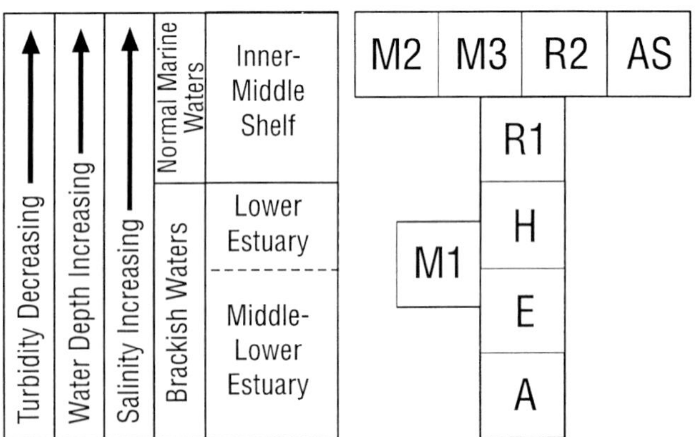

Fig. 7.—Relationship among foraminiferal subclusters, turbidity, water depth, salinity, and depositional environments in Holocene deposits of the Alabama/northwest Florida shelf. The foraminiferal data document the environmental effects of the post-glacial marine transgression on the coastal system. See Figure 6 for abbreviations.

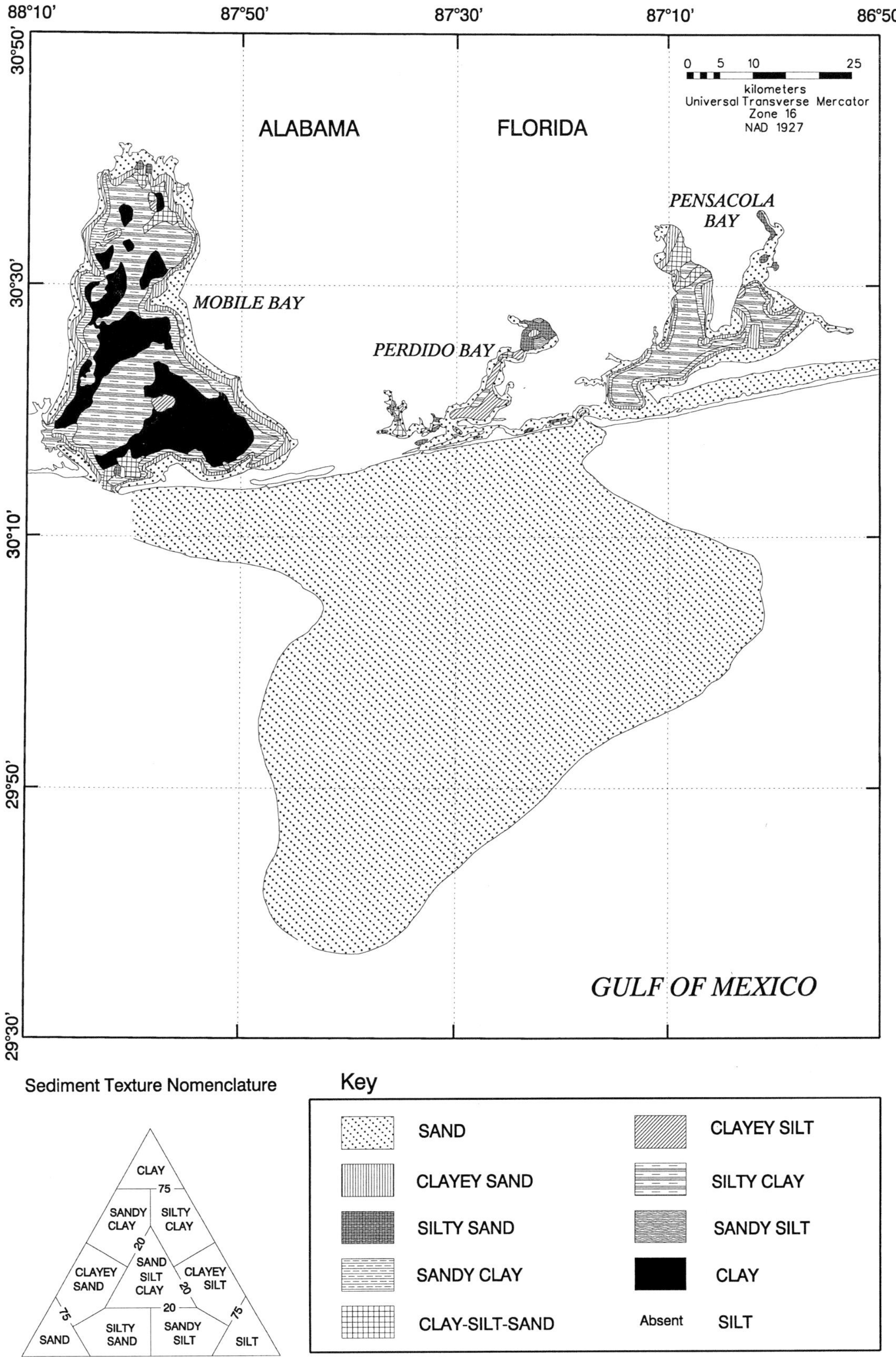

Fig. 8.—Spatial distribution of surficial sediments in the Alabama/northwest Florida area including Mobile, Perdido, and Pensacola Bays, as well as the adjacent shelf (from McBride and Byrnes, 1995).

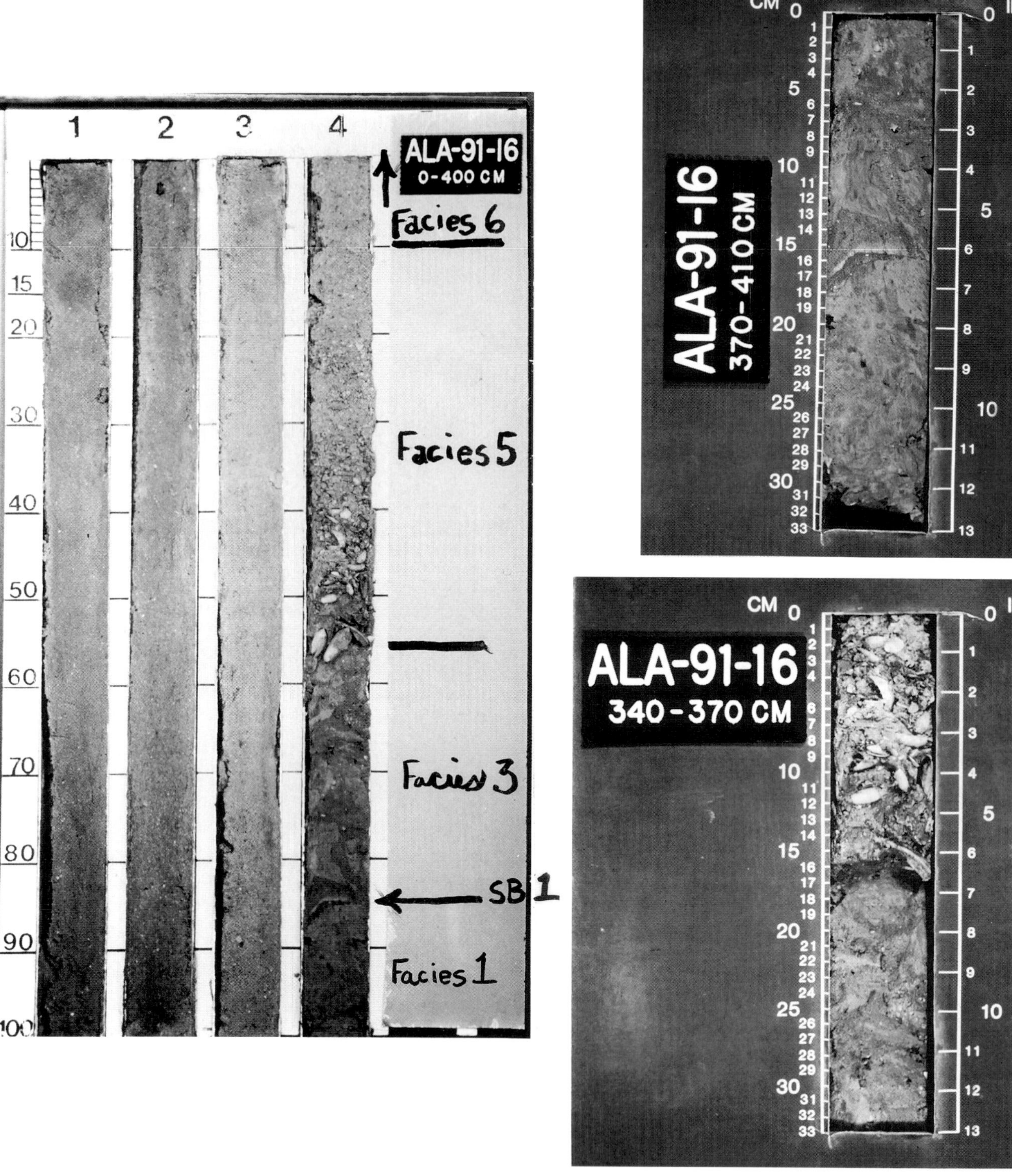

Fig. 9.—Core photographs of ALA-91-16. A) Whole core photograph showing Facies 1, 3, 5, and 6. Core top is upper left and core bottom is bottom right. Scale in centimeters. B) Close-up core photograph (interval from 370 cm to 410 cm) that shows erosional unconformity (FS/SB) truncating Facies 1 at 17 cm with overlying Facies 3 depicting rip-up clasts derived from Facies 1 at 1 and 11 cm. C) Close-up core photograph (interval from 340 cm to 370 cm) that shows shoreface ravinement diastem (second flooding surface) at 16 cm, which creates a sharp contact between Facies 3 and Facies 5. The shell bioclasts from Facies 5 have been trimmed back to allow a more detailed view of the erosional contact (compare to Figs. 9A, 10A, and 10B). For additional core photographs and data regarding shell beds in the study area, see Anderson and McBride (1996).

shells have retained their delicate ornamentation (see Fig. 6D of Anderson and McBride, 1996). Facies 2 is characterized by the foraminiferal subcluster A (*Ammonia parkinsoniana*, Fig. 6). *Elphidium* (5-45%) and miliolids (16%) represent other common foraminifer groups. Facies 2 is not always present in the study area.

Both mollusk and foraminiferal assemblages of the thin shell bed are dominated by brackish-water species (Fig. 7), which suggests that the unit formed as a lag at the bottom of a marine-influenced estuary or at the base of the bay shoreface. Although this shell lag is concentrated by wave processes on an erosional surface, the shells underwent limited transport and reworking as indicated by their generally excellent preservation.

Facies 3.—Facies 3 is a greyish-brown to tan, bioturbated to horizontally laminated, muddy-fine-quartz sand with scattered bioclasts. Well-developed, yellowish burnt-orange and grey rip-up clasts (3 x 6 cm) may occur within the lower portion of Facies 3 (Fig. 9a and b). Facies 3 is characterized by foraminiferal subcluster M1 (Fig. 6). Other common benthic foraminifera are *Ammonia parkinsoniana* (10-42%) and the genus *Elphidium* (8-16%). The foraminiferal assemblage is similar to Facies 2, but the abundance of *Ammonia parkinsoniana* decreases upward. This facies ranges from 0.25 m to 1.80 m thick and is truncated by another distinct erosional surface. Facies 3 is not always encountered in the study area.

Facies 3 was deposited in shallow, brackish waters in the middle to lower estuary (e.g., shallow subtidal area around the perimeter of lower estuary basin), based on stratigraphic context and the M1 foraminiferal subcluster. The steady decline of *Ammonia parkinsoniana* from Facies 2 up through the top of Facies 3 probably represents the increasing influence of normal-marine waters as the estuarine system undergoes transgression. Rip-up clasts incorporated within Facies 3 were derived from Facies 1. These rip-up clasts likely formed as the mainland shoreline eroded and Pleistocene deposits (Facies 1) slumped into the bay and were reworked by waves during the passage of strong cold fronts, tropical storms, and hurricanes, similar to modern mainland shoreline response in this area.

Facies 4.—Facies 4 is dominated by a dark grey, horizontally laminated silty clay with subtle bioturbation and some distinct burrows. This unit is also characterized by: 1) thick, graded, shell-rich zones with a clayey silt matrix (see Fig. 4 in Anderson and McBride, 1996), or 2) thin (1-5 cm), shelly fine- to medium-sand layers that appear to have normal and reverse grading interlaminated with clay. Facies 4 is characterized by the E and H foraminiferal subclusters (Fig. 6). The molluscan assemblage is dominated by small bivalves including *Mulinia lateralis*, *Nuculana concentrica*, and *Anadara transversa*. Facies 4 ranges between 1- and 4-m thick. When present, Facies 4 is truncated by an erosional surface.

Based on the predominance of clay and estuarine foraminiferal and molluscan assemblages, this unit represents a brackish water environment that ranges from a quiescent, partially enclosed, deeper-water estuary (central basin) to a marine-influenced open bay (Fig. 7, E and H). The former is dominated by suspension deposition except during high-energy events (tropical cyclones or strong cold fronts) that are responsible for the thin shelly sand layers, whereas the latter is located further downdip (seaward) and is influenced by both marine and estuarine processes.

Facies 5.—Facies 5 is a well-developed, clast- or matrix-supported shell bed (Figs. 9a and b, 10; see also Figs. 5, 7, 8 [upper shell bed], 9, 10, and 12 in Anderson and McBride, 1996). The mollusks *Chione intapurpurea*, *Macrocallista maculata*, *M. nimbosa*, *Ervilia nitens*, and *Anomia simplex*, as well as large soritid foraminifera and cupularid bryozoans, are common (Anderson and McBride, 1996). Although a full spectrum of preservation states is present, most bioclasts are pristine or only slightly altered. The shell bed is up to 0.88-m thick, with a fine- to medium-quartz-sand matrix and some quartz granules and pebbles found toward the base of the unit. In addition, the shell bed tends to be graded, with large (up to 6 cm) bioclasts crudely stratified (concave up, stacked, and random fabrics) at the base. Facies 5 fines upward into horizontally laminated to massive, shelly (<0.25 cm), fine quartz sand. The sand component of Facies 5 tends to be moderately well-sorted, strongly coarse-skewed, and strongly leptokurtic. Facies 5 is characterized by the foraminiferal subclusters M2, M3, and AS (Fig. 6). Other common benthic foraminifera are the hyaline species *Hanzawaia concentrica* and *Asterigerina carinata*, but the large, porcelaneous species *Cyclorbiculina compressa*, *Parasorites orbitolitoides*, and *Peneroplis proteus* also are abundant.

The bioclasts of Facies 5 are dominated by a primarily shallow-marine molluscan assemblage and, to a lesser extent, a more poorly preserved estuarine component (Anderson and McBride, 1996). The foraminiferal assemblage is dominated by normal marine species (Figs. 6, 7; M2, M3, and AS). The shell bed most likely formed at the base of the shoreface due to high-energy physical processes in response to strong cold fronts and tropical cyclones and/or large-scale bedform migration on a dynamic shelf that concentrated, vertically mixed, and amalgamated bioclasts. Shells of Facies 5 were probably concentrated initially as a transgressive lag that eroded both estuarine (Facies 2-4) and shoreface deposits, and subsequently were buried by sand deposition on the shelf (Facies 6; see below). Facies 5 was later reactivated and amalgamated by high-energy events that incorporated younger shells (see Anderson et al., 1997; McBride, 1997).

Facies 6.—Facies 6 is characterized by creamy, light tan, massive to planar-laminated, fine-to-coarse quartz sand with widely scattered shell fragments (Fig. 9). Facies 6 tends to fine and becomes better sorted upward. The size distribution of sand is leptokurtic throughout. Facies 6 is characterized by foraminiferal assemblages R1, R2, M2, M3, and AS (Figs. 6, 7). Additional benthic foraminifera include hyaline species *Hanzawaia concentrica* and *Asterigerina carinata*, but porcelaneous species *Cyclorbiculina*, *Parasorites orbitolitoides*, *Peneroplis proteus*, *Articulina mexicana*, and *Quinqueloculina agglutinans* also are common.

The *Rosalina* (R1, R2), *Asterigerina carinata* (AS), and miliolid (M2, M3) foraminiferal clusters and subclusters are interpreted as representing normal marine waters of the Gulf of Mexico shelf (Poag, 1981; Murray, 1991). In the case of the M2 and M3 foraminiferal subclusters, this interpretation is supported by the presence of *Hanzawaia concentrica*, *Asterigerina carinata*, and *Planulina exorna* as additional common species. The large porcelaneous foraminifers (especially *Cyclorbiculina* and *Parasorities*) do not currently live in the region but did inhabit coastal seagrass habitats or carbonate rubble in the middle to late Holocene under more favorable environmental conditions (Anderson et al., 1997). Consequently, the thick deposit of sand comprising Facies 6 is interpreted as a shelf sand sheet (MAFLA) deposited in normal marine waters (Fig. 7; R1, R2, M2, M3, and AS). The sand is sourced from the eroding shoreface (Swift et al., 1991b, p. 91), transported offshore, and deposited (autochthonous sedimentation; Swift and Thorne, 1991), but it is extensively reworked during high-energy events.

Shelf stratigraphy.—

Six regional cross-sections were constructed for the study area based on correlation among 34 vibracores (Figs. 11-16). Sedimentary facies and major erosional surfaces representing regional unconformities or diastems were used to determine primary stratigraphic correlation among cores. Time lines do not cross an unconformity but will cross a diastem, resulting in a diachronous surface (Nummedal and Swift, 1987).

Strike Sections.—Three strike sections were constructed along the shoreface, the middle shelf at North Perdido Shoal, and the outer shelf at South Perdido Shoal (Sections AA-AA', A-A', and B-B'; Figs. 3, 11, 12, and 13). Two regional erosional surfaces were identified and used to correlate among vibracores (Table 1). The first erosional surface is located between Facies 1 and Facies 2 or 3. Based on adjacent facies, the surface is interpreted as an erosional unconformity that formed during the last eustatic fall (Stage 2 sequence boundary)

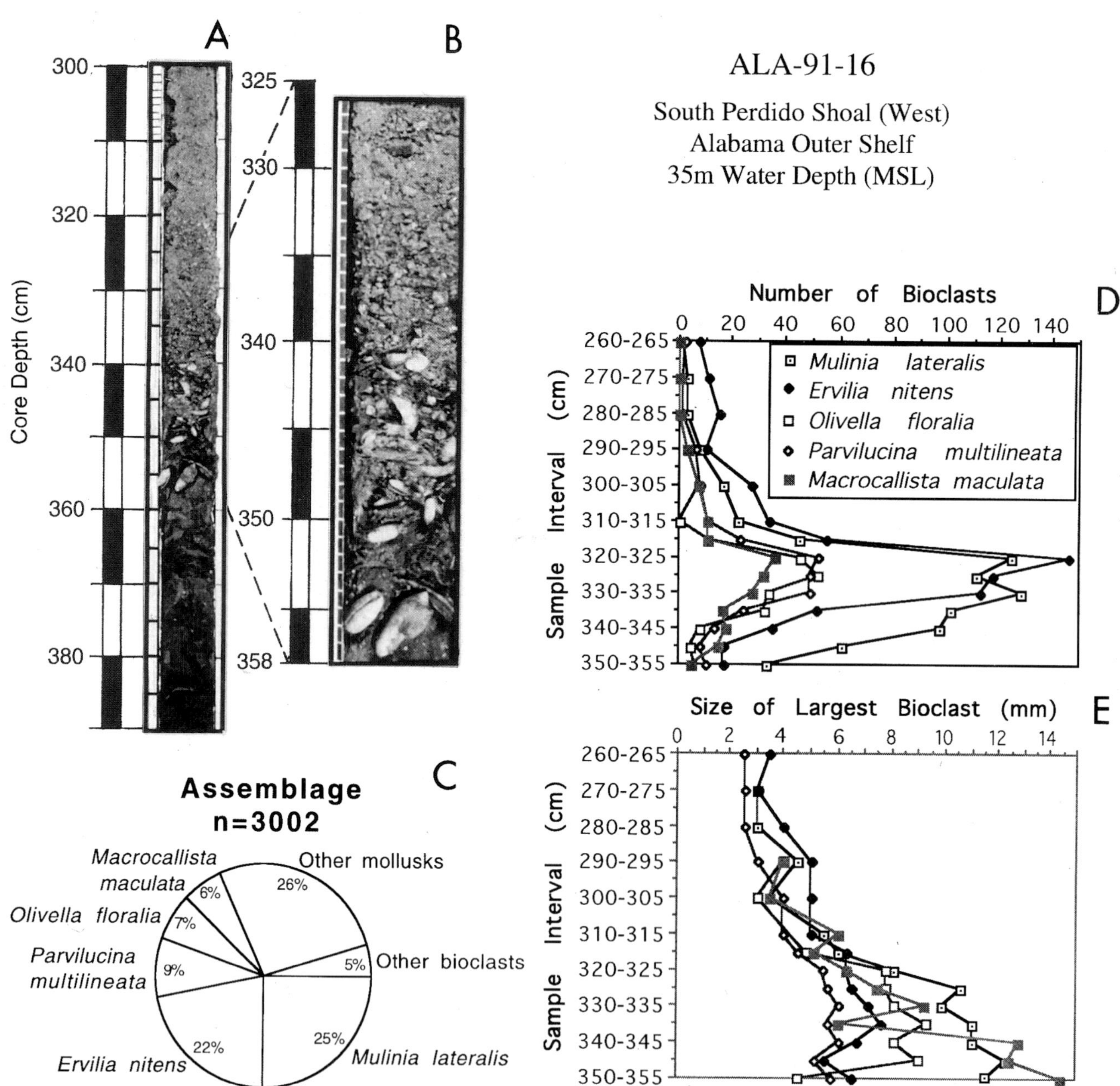

Fig. 10.—A) Partial core photograph of ALA-91-16 showing the marine shell bed (Facies 5). B) Close-up core photograph of lower portion of Facies 5 showing excellent normal grading of bioclasts. C) Pie diagram illustrating proportions of dominant bioclasts. D) Number of bioclasts per sample with depth for common species. E) Size of largest bioclast per sample with depth for common species (from Anderson and McBride, 1996).

Table 1.—Geologic framework of the transgressive systems tract offshore Alabama and northwest Florida in the northeastern Gulf of Mexico in response to the last post-glacial rise in eustatic sea level. See Figures 6 and 7 for abbreviations found in the Paleontology column.

Facies or Surface	Occurrence	Sedimentology	Paleontology	Age (kyrs. B.P.)	Primary Environment	Sub Environment	Sequence Stratigraphy	Thickness (m)
Surface 3					Modern seafloor		Future Maximum Flooding Surface	
Facies 6	PEN-91-1,2,3,4,5, 6,7,8,9,11,12,13; PEN-92-1,2, 4,5,7; PEN-93-1,5; ALA-91-3,9,12,13, 14,15,16; PER-93-1	Tan, massive to horizontally- laminated, fine to coarse quartz sand with scattered shells. Typically fines upward.	Marine foraminifera *Rosalina* (R1 & R2), *Asterigerina carinata* (AS), & miliolids (M2 & M3). Also large forams *Cyclorbiculina compressa*, *Parasorites orbitolitoides*, & *Peneroplis proteus*. Widely scattered marine mollusks.	0 - 6.01	Marine	Shelf - Surficial Sand Sheet	Upper Transgressive Systems Tract	up to 5.5
Facies 5	PEN-91-2,3,4,5,6, 8,9,11,12,13; PEN-92-4; PEN-93-5	Distinct shell bed with many pristine shells (predominately clast-supported). Some quartz granules & pebbles present.	Predominantly marine mollusks (*Macrocallista maculata*, *Ervilia nitens* most common), cupularid bryozoans, & coralline red algae. Marine foraminifera (see above).	1.15 - 10.2		Lower Shoreface		<1.0
Erosional Surface 2					Shoreface Ravinement Diastem		Flooding Surface	<0.1
Facies 4	PEN-92-5; ALA-91-3,9; PER-93-2	Dark grey, laminated to bioturbated clay. Few thin (1 to 5 cm) shelly sand layers to thick graded shell-rich zones.	*Elphidium (E)-Haynesina germanica (H)* foram assemblage. *Mulinia-Nuculana-Anadara* molluscan assemblage.		Estuarine	Central Basin of Estuary or Open Bay	Lower Transgressive Systems Tract	1-4
Facies 3	PEN-91-11; ALA-91-3,16;	Tan, bioturbated, silty to fine quartz sand; rip-up clasts.	Miliolids (M1) foraminiferal assemblage			Bay Beach (Perimeter of Estuary)		0.2 - 2
Facies 2	PEN-91-11; PER-93-3	Thin, matrix-supported shell bed.	Estuarine bivalves (*Chione cancellata*) & *Ammonia* (A) foraminiferal assemblage	6.07 - 10.2		Lower Bay Shoreface or Estuary/Bay Bottom		<0.2
Erosional Surface 1					Bay Ravinement Diastem & Erosional Unconformity		Flooding Surface & Type 1 Sequence Boundary (FS/SB)	<0.1
Facies 1	PEN-91-10; ALA-91-3,16; PER-93-3	Yellowish burnt orange & grey, massive to highly bioturbated, oxidized clayey quartz sand.	Devoid of macro- and microfauna	Pleistocene (>10.0)	Continental/ Coastal	Strandplain is masked by Soil Horizon	Highstand and Falling Stage Systems Tracts	?

and was subsequently modified by a bay ravinement diastem produced during the last post-glacial rise of sea level (Fig. 9B). The second erosional surface is situated between Facies 3 or 4 and Facies 5 (Fig. 9C). The predominant shallow-marine assemblage of Facies 5 and the stratigraphic context suggest that this surface is a shoreface ravinement diastem.

Cross-section AA-AA' is strike-oriented along the base of the shoreface in water depths that average 10 m (Fig. 11). Vibracore ALA-91-3 is virtually identical to the vertical succession found in ALA-91-16 (see Fig. 9), and thus ALA-91-3 provides key control in shallow water. The base of core ALA-91-3 penetrates the top of the Pleistocene soil horizon (Facies 1) and is overlain by a vertical succession that contains an erosional unconformity, estuarine deposits (Facies 3), a shoreface ravinement diastem, a shell bed (Facies 5), and is capped by a marine sand sheet (Facies 6). A lateral facies change occurs within the estuarine deposits from a bioturbated sandy silt (Facies 3) to a laminated to slightly bioturbated clay (Facies 4). Core ALA-91-9 penetrates estuarine facies, but remaining cores stop within or at the base of the marine sand (Facies 6).

Cross-section A-A' is strike-oriented along the crest of North Perdido Shoal (Figs. 3, 12). North Perdido Shoal is much shorter than South Perdido Shoal but it is similar in thickness. Although cores had penetration depths up to 5.4 m (Pen-93-1), no estuarine or pre-Holocene deposits were encountered. Consequently, the cores are composed entirely of marine sediments and are correlated based on basal shell beds (Facies 5) that most likely sit on top of the shoreface ravinement surface (Table 1; see also Anderson and McBride, 1996). Pristine shells from PEN-91-5 (Facies 5) yielded radiocarbon dates ranging in age from 1,720 to 1,150 yrs B.P., whereas dates from PEN-91-3 range from 3,190 to 1,460 yrs B.P. (McBride, 1997). Dates this young, at these depths, and buried by up to 4 m of sand, reveal extensive stratigraphic mixing or rapid deposition rates.

The overlying clean quartz sand (Facies 6) is dominated by an open marine foraminiferal assemblage. However, AMS dates (3,640 ± 60 yr B.P. at -22 m mean sea level [MSL] and 2320 ± 60 yr B.P. at -25 m MSL in PEN-91-5) obtained from large foraminifers are stratigraphically disordered (Anderson et al., 1997). Moreover, AMS dates of large foraminifers

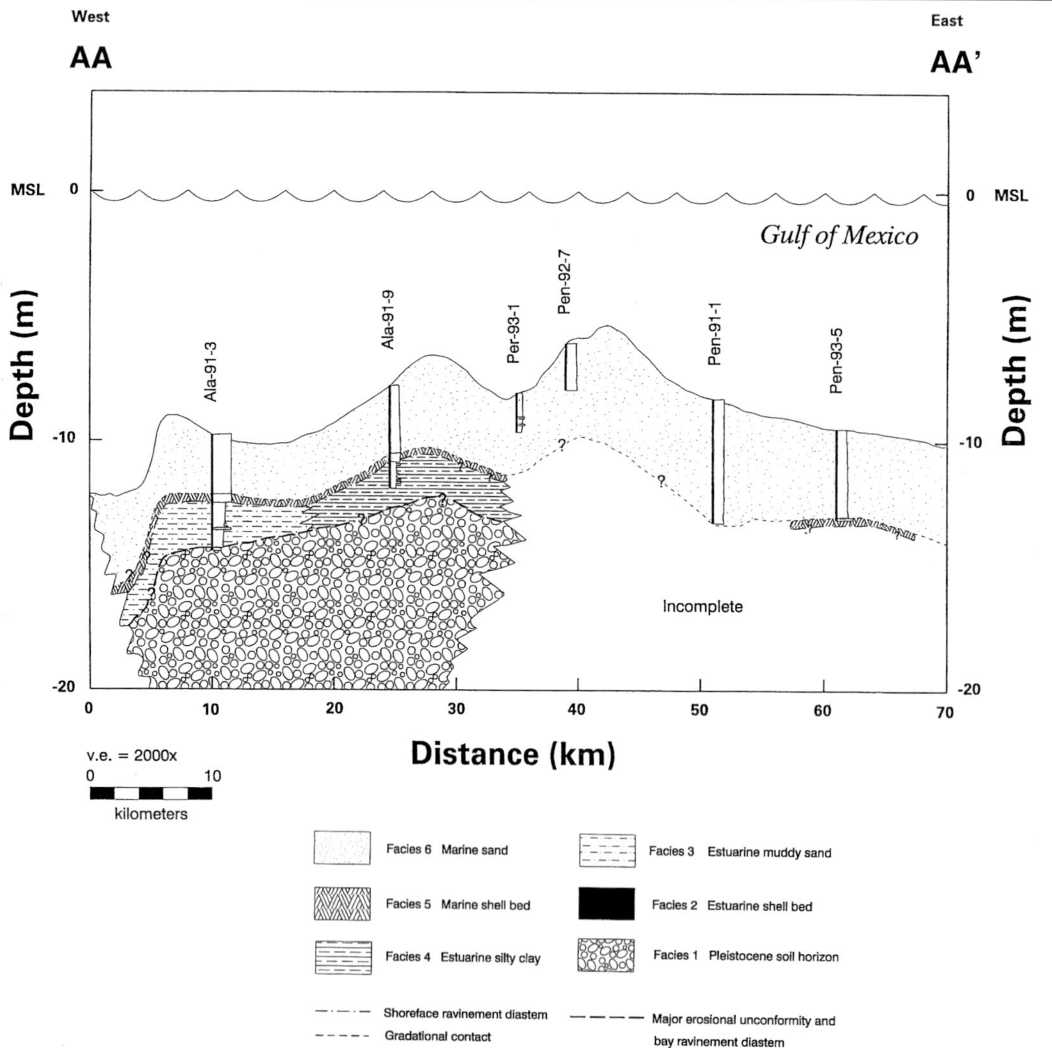

Fig. 11.—Strike-oriented geologic cross-section AA-AA' along the base of the shoreface (10-m water depth). Vertical exaggeration is 2000 x. See Figure 3 for location.

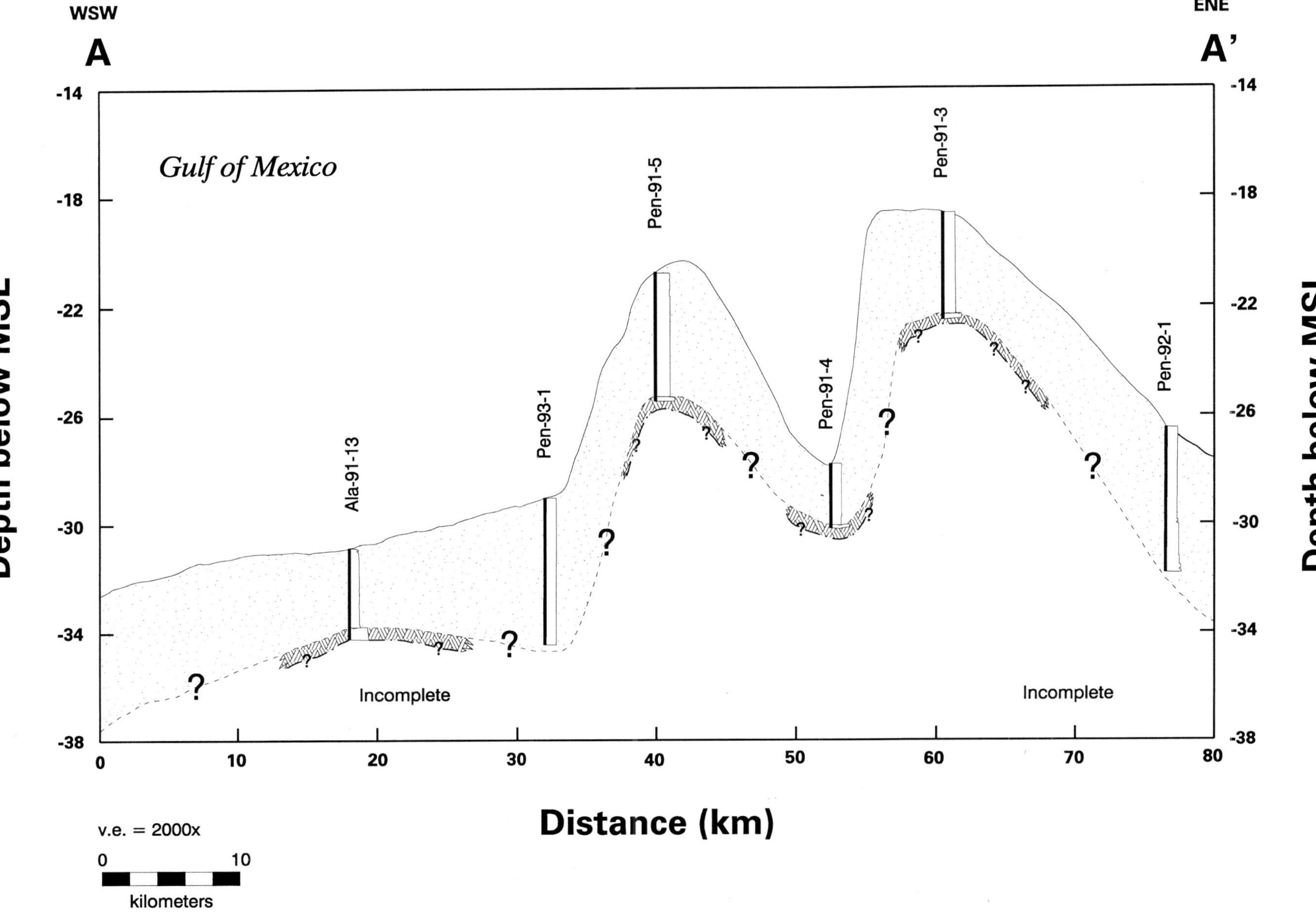

Fig. 12.—Strike-oriented geologic cross-section A-A′ along the crest of North Perdido Shoal. Vertical exaggeration is 2000 x. See Figure 3 for location and Figure 11 for legend.

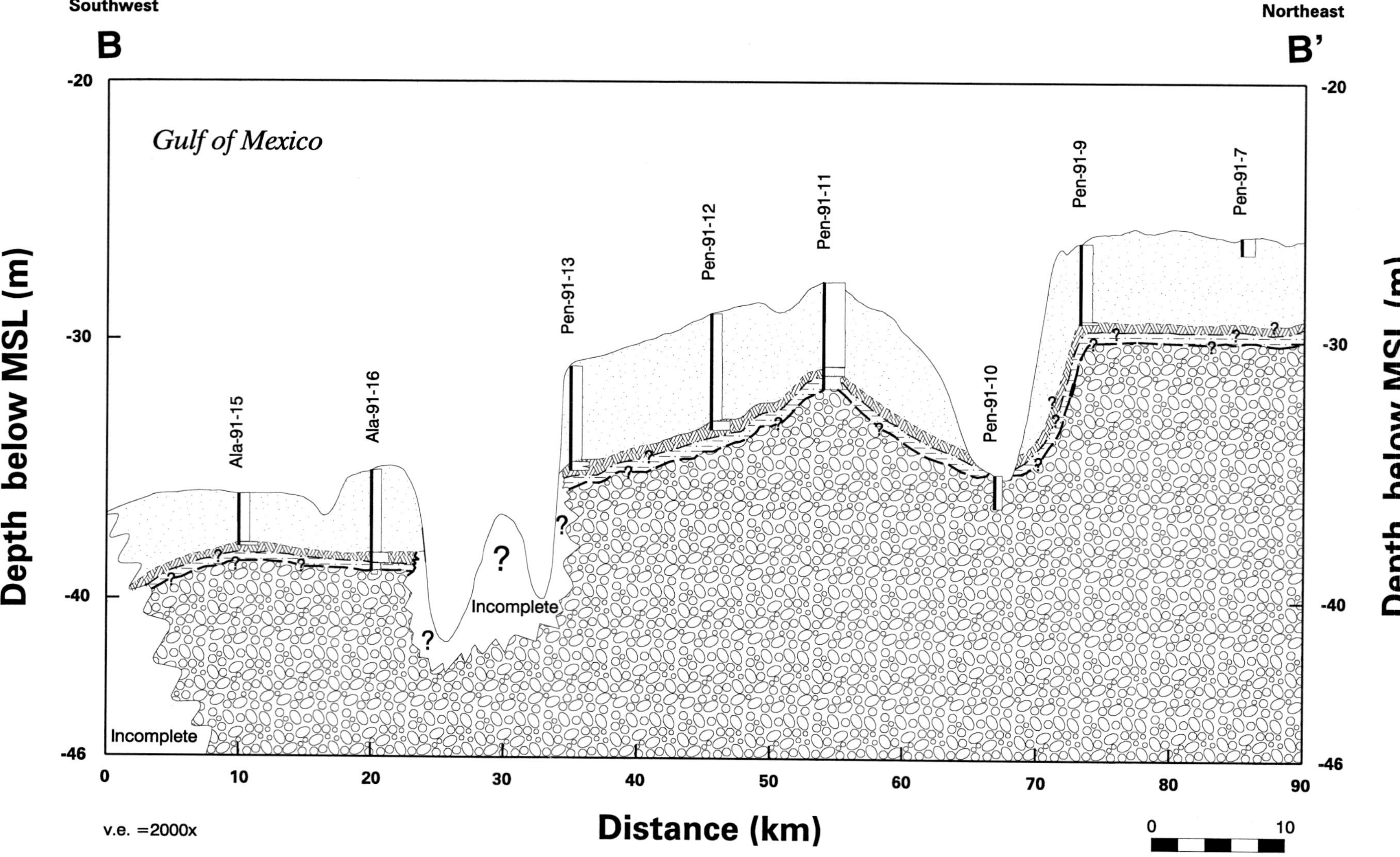

Fig. 13.—Strike-oriented geologic cross-section B-B' along the crest of South Perdido Shoal. Vertical exaggeration is 2000 x. See Figure 3 for location and Figure 11 for legend.

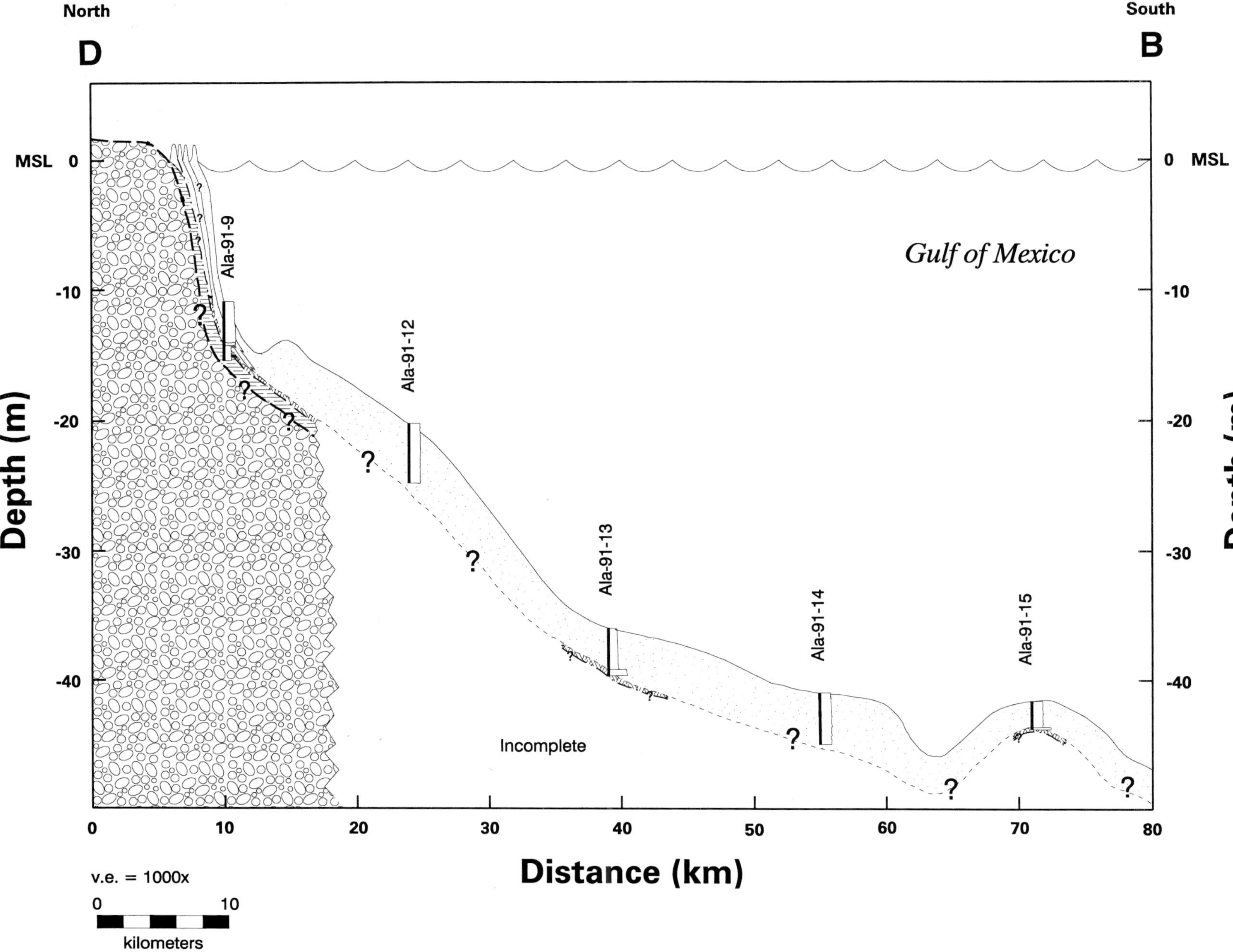

Fig. 14.—Dip-oriented geologic cross-section D-B that extends south from the Gulf Shores, Alabama area to the outer continental shelf. Vertical exaggeration is 1000 x. See Figure 3 for location and Figure 11 for legend.

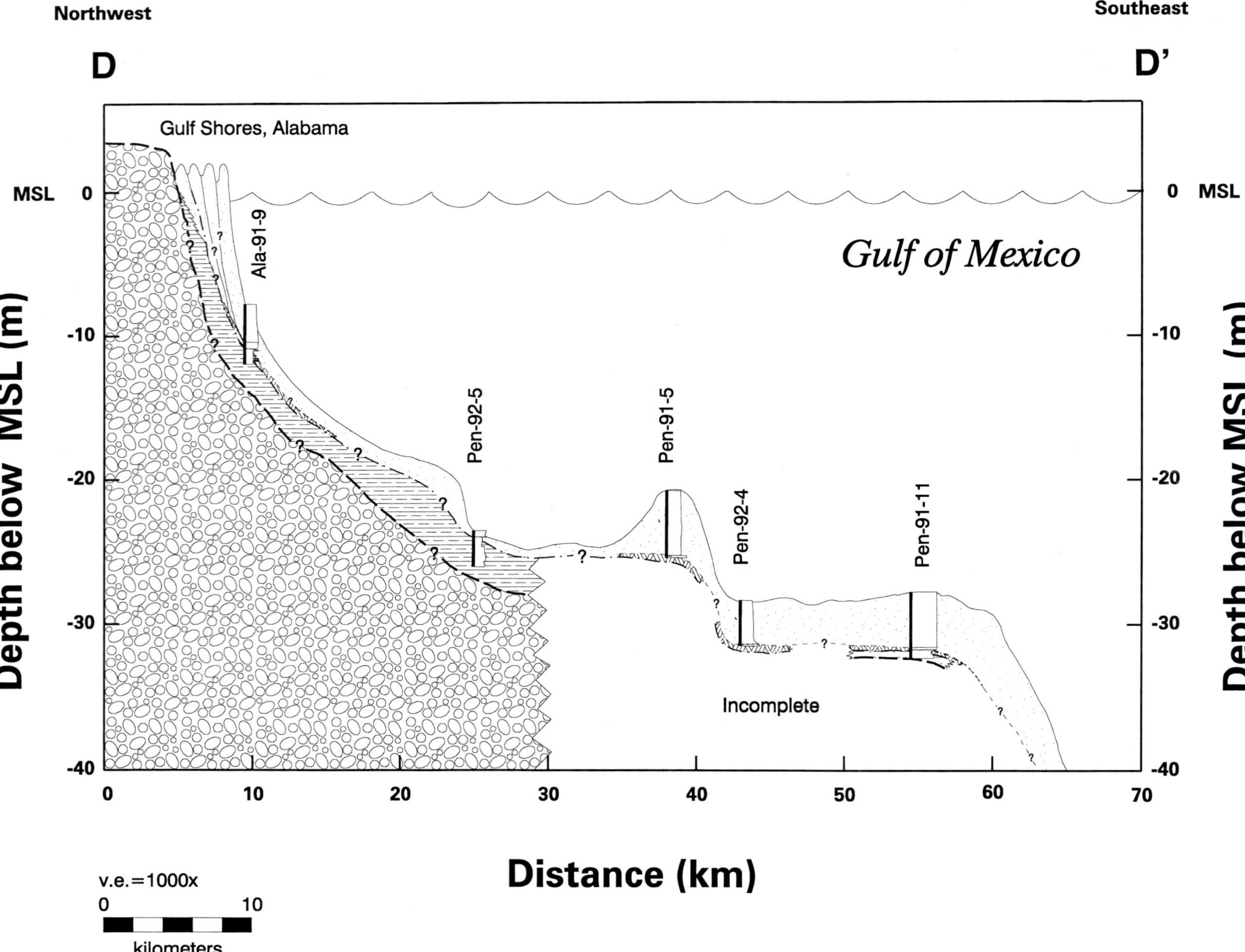

Fig. 15.—Dip-oriented geologic cross section D-D' that extends southeast from the Gulf Shores, Alabama area to the outer continental shelf. Vertical exaggeration is 1000x. See Figure 3 for location and Figure 11 for legend.

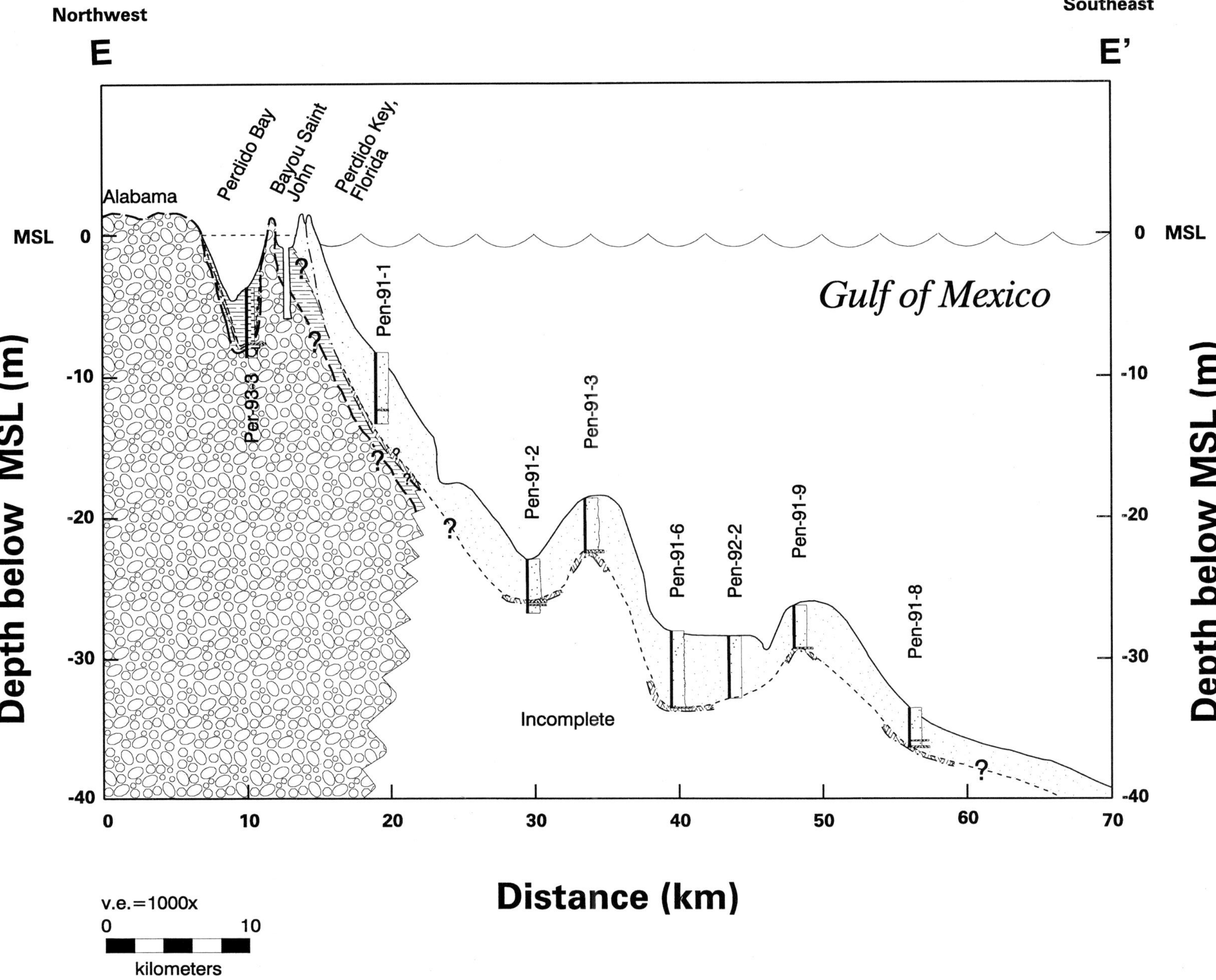

Fig. 16.—Dip-oriented geologic cross section E-E' that extends southeast from Perdido Bay, Alabama/Florida, to the outer continental shelf. Vertical exaggeration is 1000x. See Figure 3 for location and Figure 11 for legend.

are older than the underlying shell dates in Facies 5 of the same core, which indicates complete mixing of Facies 5 and 6. North Perdido Shoal, therefore, is interpreted as a reworked marine sand shoal with no *in situ* barrier shoreline deposits.

Cross-section B-B' is a 90-km strike-oriented cross-section along the crest of South Perdido Shoal (Figs. 3, 13). The soil horizon (Facies 1) and overlying erosional unconformity occur at the base of core ALA-91-16 and at the top of PEN-91-10. This erosional unconformity is overlain by estuarine deposits (Facies 2 and 3) in cores ALA-91-16 and PEN-91-11. Three almost identical AMS dates of 10,200 ± 60, 10,070 ± 60, and 10,040 ± 60 yr B.P. were obtained from pristine *Chione cancellata* shells at the base of PEN-91-11 (McBride, 1997), indicating minimal time-averaging and reworking (Table 1).

The estuarine deposits (Facies 2 and 3) are truncated by a second erosional surface that lies at the base of the distinctive shell bed (Facies 5). This shell bed was found in most cores and was used to correlate the shoreface ravinement diastem (Table 1). Radiocarbon dates of shells (Facies 5) within and among cores yielded dates ranging from 10,190 ± 150 to 1,150 ± 70 yr B.P. (McBride, 1997). Although the shell bed dips to the west and ranges from -29 m to -39 m below MSL, this wide range of dates was unexpected and suggests that the shell bed underwent, and possibly continues to undergo today, extensive mixing and is time-averaged over a period of at least 9,000 years.

The shell bed grades up into a thick (up to 4.1 m) clean quartz sand (Facies 6) that is interpreted as a large, reworked marine sand shoal (Fig. 9) because of: 1) its stratigraphic position above the shoreface ravinement diastem, 2) the abundance of an open marine foraminiferal assemblage, and 3) because it contains both stratigraphically ordered (4340 ± 60 yr B.P. at -31 m MSL; 6010 ± 60 yr B.P. at -35 m MSL in PEN-91-13) and disordered (5660 ± 50 yr B.P. at -28 m MSL and 3110 ± 50 yr B.P. at -31 m MSL in PEN-91-11) AMS dates obtained from large foraminifer (Anderson et al., 1997). No *in situ* barrier shoreline deposits were encountered. The marine sand comprising South Perdido Shoal tends to thin to the west and is part of the extensive MAFLA sand sheet.

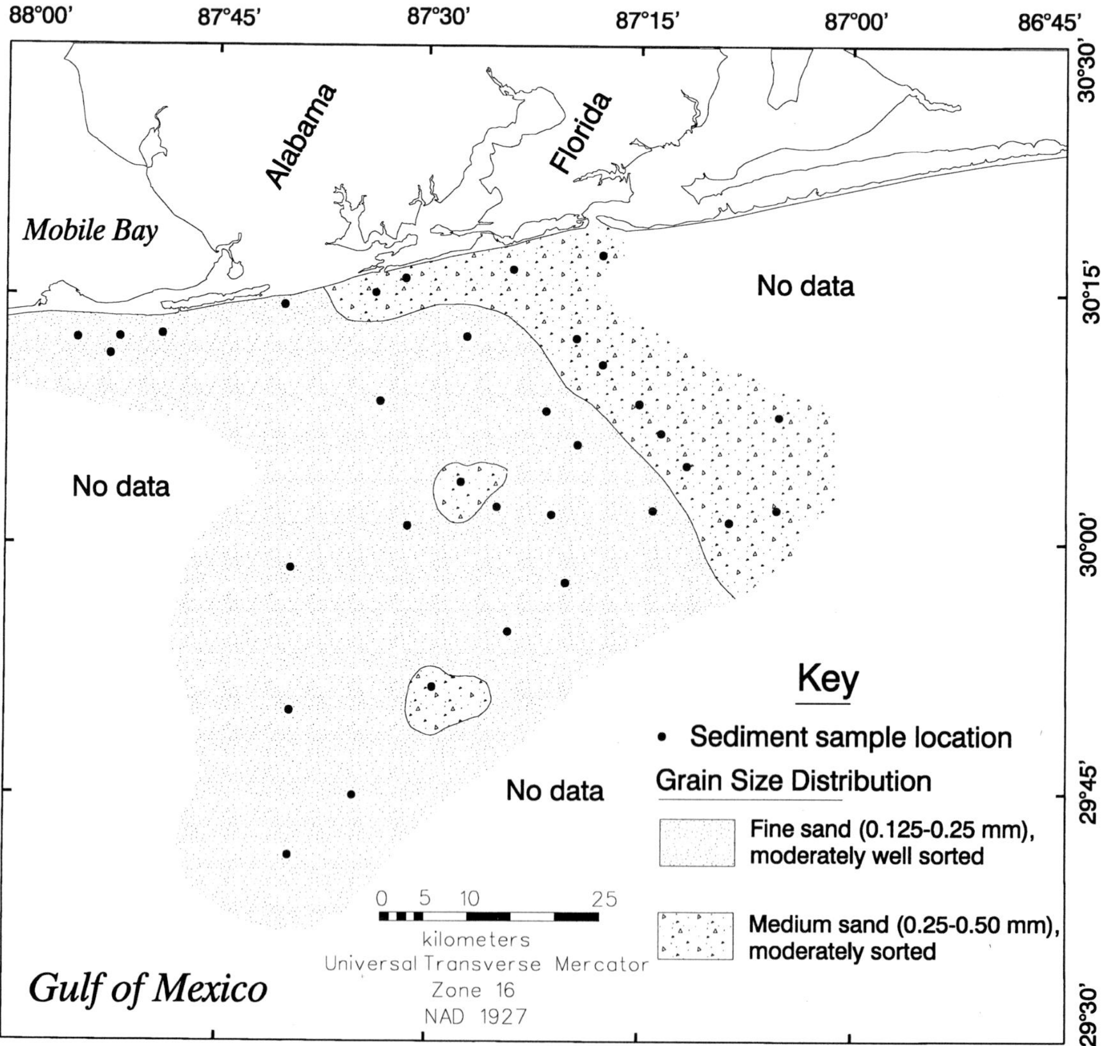

Fig. 17.—Grain size and sorting information for shelf sediments showing distinct boundary between the Apalachicola and Mobile subprovinces (modified from McBride and Byrnes, 1995).

Dip Sections.—Cross-sections D-B, D-D', and E-E' extend from the Alabama/Florida coastal region seaward to the outer continental shelf (Figs. 3, 14, 15, and 16). North Perdido Shoal is delineated by vibracores PEN-91-3 and PEN-91-5, whereas South Perdido Shoal is depicted by vibracores ALA-91-15, PEN-91-11, and PEN-91-9. The most striking characteristic of the cross sections is the thickness and amount of sand comprising the MAFLA sand sheet (Facies 6). The sand sheet maintains its thickness both in an offshore and alongshore direction except for the inner shelf area along section D-D' (vibracore Pen-92-5 at 25 km). Additionally, the sand sheet appears to drape the underlying topography.

In the study area, sediments of the MAFLA sand sheet (Facies 6) fine both vertically upward and to the west (horizontally) from a moderately sorted, medium quartz sand to a moderately well-sorted, fine quartz sand (Fig. 17). It is not clear whether this horizontal sorting trend is related to physical oceanographic processes that transport sediment to the west, similar to the littoral drift direction, or whether it is related to the natural zonation of sediment suites derived from different fluvial sources (e.g., Apalachicola and Mobile subprovinces). In addition, an isopach map of the surficial sand sheet shows two shore-normal lobes that are up to 5 m thick (Fig. 18). The two thick lobes of sand, which do not mimic surface morphology (i.e., North and South Perdido Shoals), are separated by a shore-normal thin zone (<2 m thick) that corresponds to a bathymetric low (interpreted as an antecedent fluvial valley), as shown in Figures 1 and 4. The calculated volume of this large, regional sand sheet is 4.9 billion m^3; an excellent reservoir-quality sand.

Sequence Stratigraphy

In a sequence stratigraphic framework (Posamentier et al., 1988; Posamentier and Vail, 1988; Walker, 1992), Facies 1 represents a Pleistocene soil horizon that was subaerially exposed during the last sea-level lowstand (δ^{18}O Stage 2), approximately 18,000 yr B.P. (Table 1). The magnitude of this sea-level drop was about 120 m (Curray, 1965; Fairbanks, 1989; Bard et al., 1996) causing fluvial incision as the shoreline migrated seaward to the edge of the continental shelf. Consequently, most of the shelf surface was subaerially exposed producing a Type 1 sequence boundary (SB) on top of the former highstand systems tract (HST). Shelf-margin delta deposition (see Figs. 1, 4) occurred downdip contemporaneously with sequence boundary formation (see Bart, 1997).

The sequence boundary was further reworked by a marine flooding surface, a bay ravinement diastem that was a response to the last deglaciation. Therefore, the erosional unconformity observed between Facies 1 and Facies 2 or 3 stems from a combination of subaerial exposure during sea-level lowstand and the bay-ravinement process during the

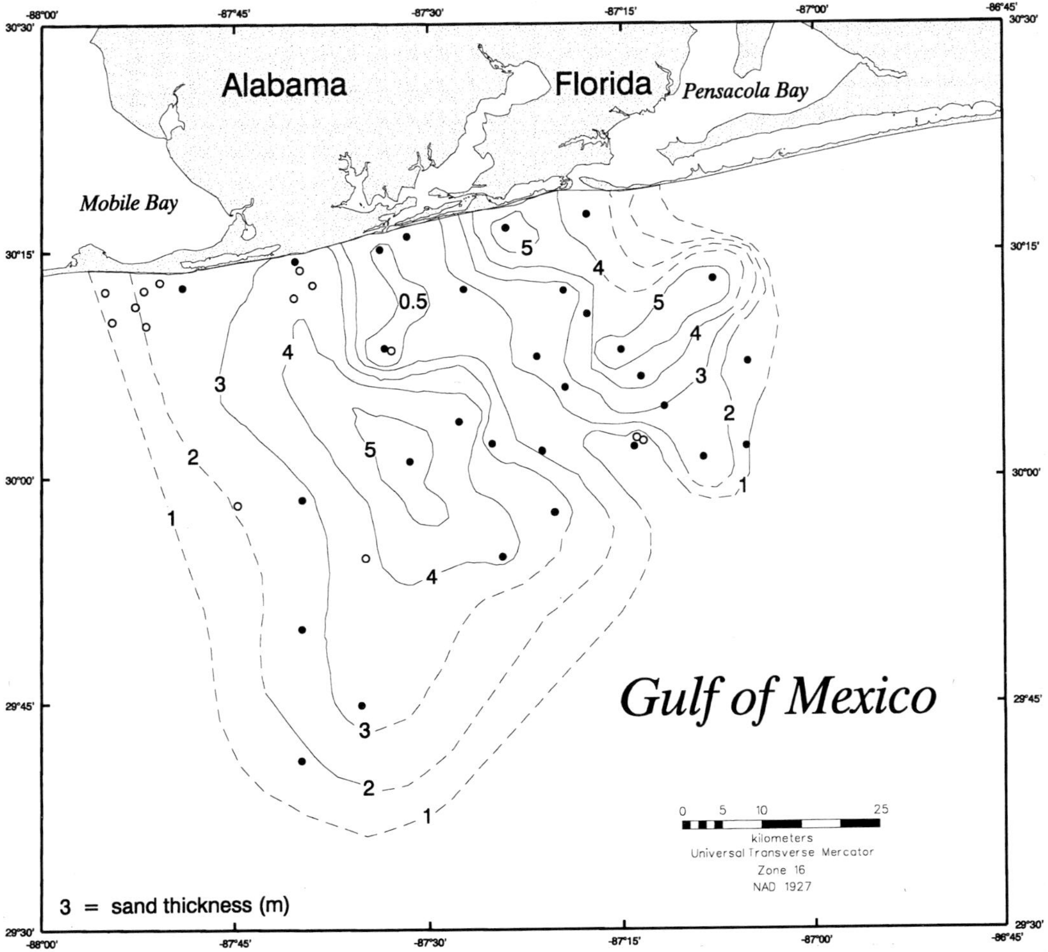

Fig. 18.—Isopach of the upper marine section that contains the shell bed (Facies 5) and the MAFLA surficial sand sheet (Facies 6).

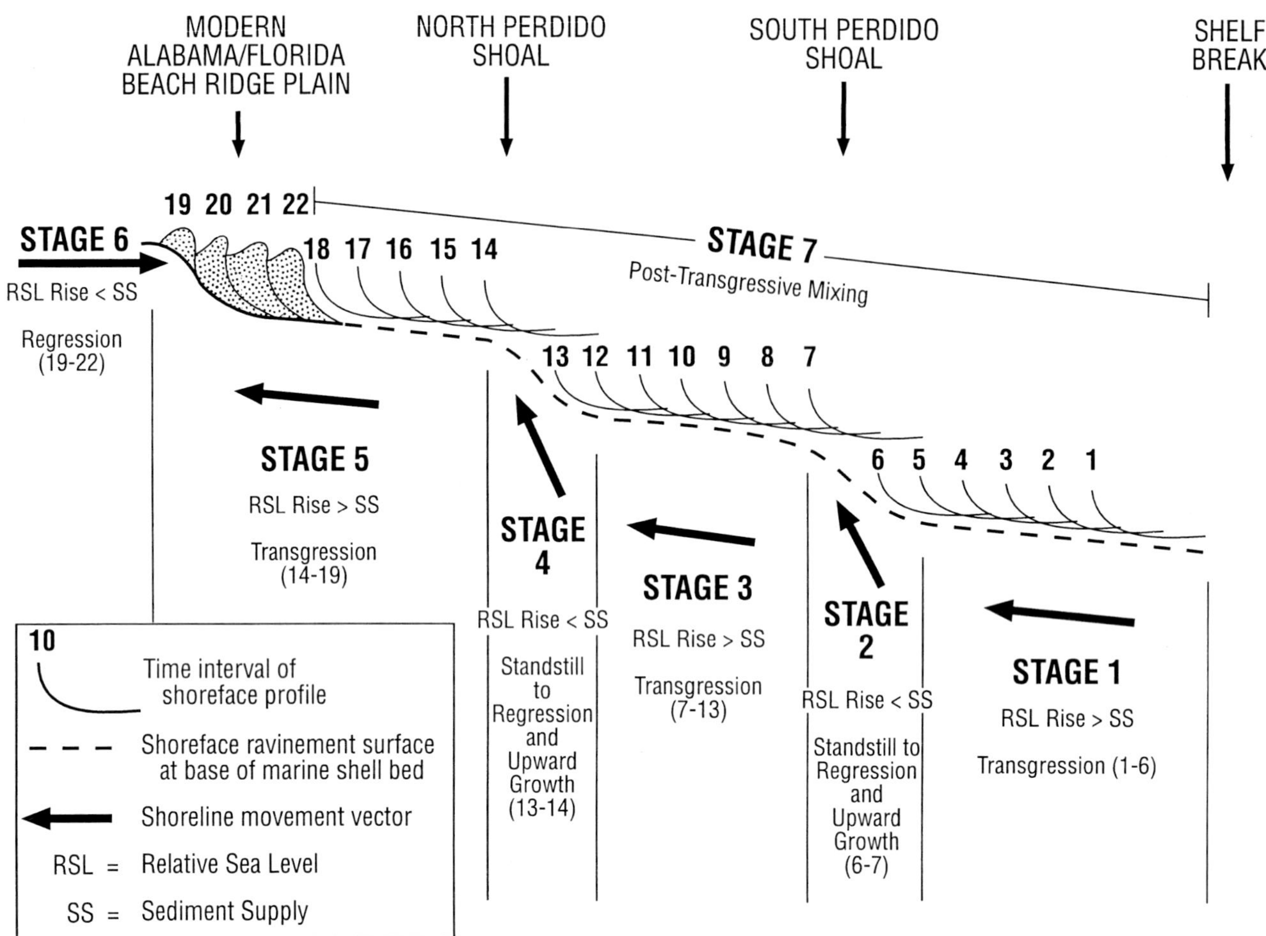

Fig. 19.—Seven-stage shelf evolution model for the Alabama and northwest Florida shelf showing the relationship between translation of the shoreface profile in response to variations in sediment supply and eustatic changes (Stages 1 through 7 are generic terms and do not refer to oxygen isotope stages). Sea level slowdowns generate escarpments. Shoreface profile locations are sequentially numbered from oldest (#1) to youngest (#22). Vertical exaggeration is approximately 1000x. See Figure 20 for resulting shelf stratigraphy. Parts of this diagram were influenced by Swift et al. (1984).

ensuing transgression (FS/SB; Fig. 9A, B). Facies 2, 3, and 4 represent the lower transgressive systems tract (TST), which consists of various estuarine environments (Table 1). As transgression continued, the outer shoreline translated landward and upward through erosional shoreface retreat and produced the second flooding surface, the shoreface ravinement diastem, which is noted at the bottom of the marine shell bed (Facies 5; Fig. 9). Facies 5 and 6 represent the upper transgressive systems tract, which is dominated by open marine environments. Over the past 3,000 years or more, the Gulf of Mexico has experienced a sea-level highstand. Consequently, the modern seafloor represents a third flooding surface, the maximum flooding surface (MFS), and is the upper boundary of the transgressive systems tract (Table 1).

As the Holocene transgression slowed to a virtual standstill, sediment supply dominated the coastal system, causing the net direction of shoreline movement to reverse and prograde seaward. As such, a majority of the present Alabama/Florida shoreline between the Mobile Bay Entrance and Pensacola Pass is characterized by extensive beach-ridge plains (see Stapor, 1975) that represent the beginning of the next highstand systems tract. The east to west littoral drift system provides the sediment source for the development of these beach-ridge plains. Therefore, the highstand systems tract will continue to prograde seaward across the maximum flooding surface, thus forming a downlap surface.

Shelf Evolution Model

A seven-stage shelf evolution model is proposed to explain the morphostratigraphy of the study area (Figs. 19, 20). The model addresses the origins of the distinct marine shell bed and overlying thick sand sheet, as well as the two prominent linear shoals. In this model, the outer shoreline is characterized by a back-stepping barrier beach that most likely consisted of a combination of Holocene barrier islands and eroding Pleistocene headlands. Although barrier-beach deposits are not preserved in the study area, resulting shelf morphology is related to changes in the migration rate of the barrier shoreline and translation direction of the associated shoreface profile in response to variations in sediment supply and relative sea-level changes (Fig. 19). In the discussion below, Stages 1 through 7 are purely generic terms referring to stages of development; they do not correspond to oxygen isotope ($\delta^{18}O$) stages.

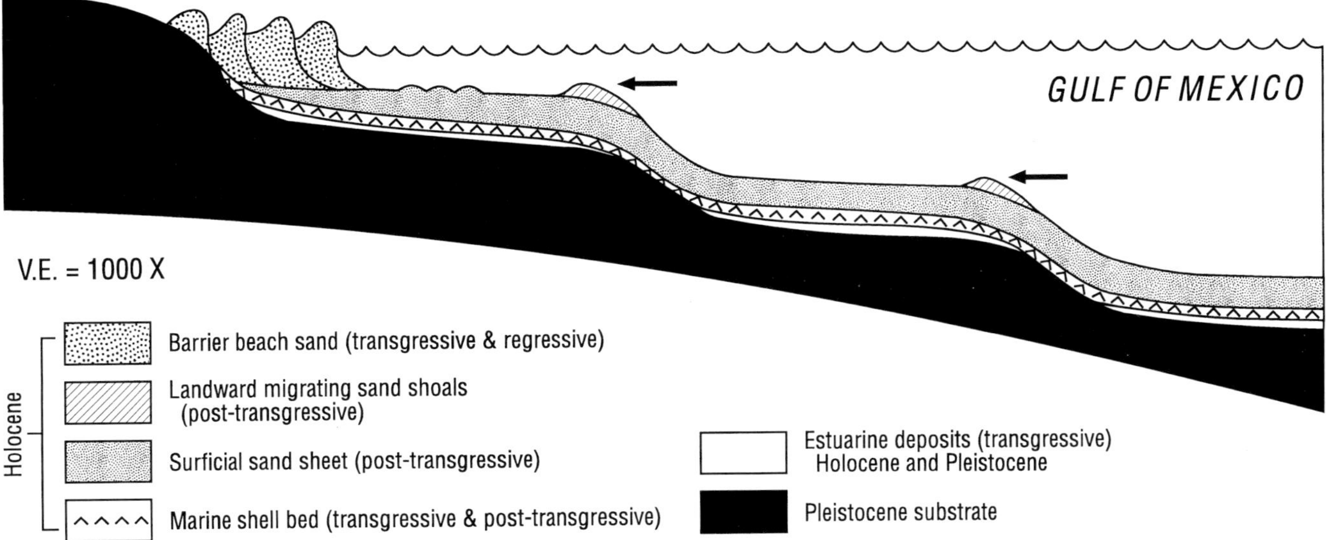

Fig. 20.—Generalized stratigraphy for the Alabama/northwest Florida shelf as explained by seven-stage model in Figure 19. Vertical exaggeration is approximately 1000 x.

According to Stubblefield et al. (1984a), shelf features and deposits can form prior to, during, or subsequent to the passage of the transgressive shoreline. Therefore, timing of the formation of shelf deposits is referred to as either pre-transgression, during transgression, or post-transgression.

During Stage 1, the barrier shoreline experienced rapid transgression (Fig. 19; shoreface profiles 1 through 6) in response to continuous rates of relative sea-level rise. The smooth and steep shelf surface seaward of South Perdido Shoal (40-75-m water depths) is geomorphic evidence of a continuous rise. During Stage 2, an escarpment below South Perdido Shoal formed as the rate of sea-level rise slowed, enabling sediment supply to dominate the coastal system. Sediment was probably supplied from updrift eroding Pleistocene headlands located east of the study area (for example, Grayton Beach, Florida) and transported westward by the net littoral drift system. In Stage 2, excess sediment caused the barrier shoreline to aggrade and possibly prograde (Fig. 19; shoreface profiles 6 and 7). Radiocarbon dates from the estuarine shell bed in PEN-91-11 indicate that the South Perdido escarpment formed just prior to 10,200 yr B.P.

Stage 3 is dominated by relative sea-level rise, which caused the barrier shoreline to experience transgression through erosional shoreface retreat as the shoreface profile translated landward and slightly upward (Fig. 19; shoreface profiles 7-13). The escarpment below North Perdido Shoal formed during Stage 4 (Figs. 19, 20). During this stage, the resultant direction of shoreface profile translation was upward and slightly landward (Fig. 19; shoreface profiles 13 and 14) due to a reduced rate of relative sea-level rise that enabled normal sediment supply to dominate or an increase in sediment supply delivered to the coast (through longshore sediment transport from the east), which kept pace with a constant rate of sea-level rise.

Transgression resumed during Stage 5 as relative sea-level rise dominated sediment supply (Fig. 19; shoreface profiles 14-19), possibly in response to a slightly higher rate of sea-level rise. Transgression continued to shoreface profile 19 on Figure 19, which represents the most landward position of the Holocene transgression (maximum flooding surface [MFS]). Shoreface profile 19 also marks the end of the transgressive systems tract and the beginning of the highstand systems tract. Stage 6 began when the direction of net shoreline movement reversed in response to an excess in sediment supply. Stage 6 was dominated by regression without upward profile translation as a result of sea-level stability over the past 3,000 years or more. However, the similarities between shoreface profile 19 and the shoreface scarps underneath North and South Perdido Shoals indicate a common origin (Fig. 19). The shoreface scarps underneath the linear shoals are degraded somewhat because they have been transgressed, whereas the shoreface profile at time interval 19 has not.

Stage 7 involves extensive post-transgressive mixing of the entire marine section (Facies 5 and 6) by shelf hydrodynamic processes in response to the passage of strong cold fronts and tropical cyclones. Post-transgressive mixing of marine sediments is a time-transgressive process and begins immediately after an area has been transgressed. Thus, Stage 7 not only commenced after Stage 6 was complete; it occured during all six stages and probably continues today (Fig. 19).

This model has important ramifications for the genesis of shore-parallel linear shoals in shallow-marine settings. The morphostratigraphy of the shoals is a result of both transgressive and post-transgressive processes. The shoals appear to be underlain by escarpments that were produced during transgression as the rate of sea-level rise slowed (Figs. 19, 20). Escarpments are preserved shoreface profiles cut by wave processes, and they represent major shoreline positions that formed at lower stands of sea level (Dillion and Oldale, 1978). The escarpments do not represent pre-transgressive topography but were cut contemporaneously with transgression into underlying Pleistocene deposits. Consequently, the escarpments exert subsurface control over geomorphic expression (i.e., bathymetry) of the seafloor in the form of linear features (Fig. 20). On the other hand, the marine shell bed and overlying sand sheet are transgressive deposits reworked by post-transgressive processes and drape the transgressive topography (e.g., escarpments). Furthermore, the shoals along the top of each escarpment typically have asymmetrical profiles with steeper landward flanks (see Fig. 5). This morphology indicates that the sand sheet was further modified and reworked landward by large-scale bedform migration that continues probably today, thus enhancing the expression of North and South Perdido shoals when combined with the influence of the underlying escarpments. This post-transgressive, modification process (i.e., large-scale bedform migration) produces the asymmetrical shape dominated by steeper landward-facing flanks (Figs. 5, 20).

Shore-Parallel Linear Shelf Shoals

Over the past 25 years, numerous depositional models have been presented to explain the genesis of long, linear shelf shoals that trend parallel or subparallel to an adjacent shoreline. These depositional models involve both modern and ancient examples and include the following: 1) in-place drowning of barrier islands or overstepped barriers (Sanders and Kumar,1975a, b; Rampino and Sanders, 1980, 1982); 2) erosional shoreface retreat and post-transgressive processes (Swift et al., 1973, 1984; Swift, 1975; Swift and Moslow, 1982); 3) degraded barriers (Stubblefield et al., 1984a, b); 4) offshore sand-ridge (bar) complex (Tillman and Martinsen, 1984, 1987; Gaynor and Swift, 1988); 5) lowstand/transgressive incised shoreface deposits (Bergman and Walker, 1987; Downing and Walker, 1988; Bergman, 1994; Walker and Wiseman, 1995); and 6) transgressive submergence (Penland et al., 1988, 1989; Pope et al., 1991). Although fiercely debated, most models can be grouped based on the timing of formation and stratigraphic position of the linear sand bodies: 1) pre-transgression (Sanders and Kumar,1975a, b; Rampino and Sanders, 1980, 1982); 2) pre- and syn-transgression (Stubblefield et al., 1984a, b; Bergman and Walker, 1987; Downing and Walker, 1988; Bergman, 1994; Walker and Wiseman, 1995); 3) syn- and post-transgression (Swift et al., 1973; Swift, 1975; Swift and Moslow, 1982; Swift et al., 1984; Rine et al., 1991); and 4) pre-, syn-, and post-transgression (Penland et al., 1988, 1989; Pope et al., 1991).

The model presented in this paper incorporates both syn-transgressive and post-transgressive processes. The upper marine section was originally deposited on top of the ravinement surface as transgressive shelf deposits that have undergone post-transgressive reworking and mixing. Thus, the linear shoals now consist solely of post-transgressive deposits and never contained *in situ* or degraded barrier-beach deposits, but were influenced by the underlying transgressive topography (escarpments). As such, the model is most similar to the concepts presented by Swift et al. (1984) and documents the extensive post-transgressive mixing of the entire surficial sand sheet down to the shoreface ravinement diastem.

SUMMARY AND CONCLUSIONS

Five primary features dominate shelf morphology between Mobile Bay, Alabama, and Pensacola Bay, Florida. Shore-normal lows and shelf-edge lobes represent lowstand fluvial systems consisting of incised river valleys and shelf-edge deltas (1st order). Shelf-break parallel linear shoals and associated lows represent former shoreline positions (2nd order) but not shoreline deposits. The linear shoals occur on interfluves between the antecedent fluvial drainage system. Shore-oblique sand ridges are found superimposed on lower order features throughout the study area, but are most prevalent between 0 m and 20 m water depths. The sand ridges form in response to shelf storm flow moving in a westerly direction.

Through quantitative analysis of the micropaleontological samples, clusters were found that correspond well to previously known foraminiferal assemblages of the Recent Gulf of Mexico shelf, and subclusters that represent finer-scale paleoenvironmental differences were identified. The *Ammonia* (A), *Elphidium* (E), part of miliolid (M1), and *Hayesina germanica* (H) subclusters represent brackish water environments with salinity increasing from the former to the latter. The *Hayesina germanica* (H) foraminiferal assemblage is reported for the first time in the Gulf of Mexico. The *Rosalina* (R1, R2), *Asterigerina carinata* (AS), and parts of the miliolid cluster (M2, M3) represent normal marine environments, with part of the *Rosalina* facies (R1) corresponding to a slightly lower normal marine salinity (shoreface).

The late Quaternary shelf stratigraphy is characterized by six facies and two erosional surfaces. These preserved facies and regional surfaces comprise the transgressive systems tract and reflect the stratigraphic signature of the last major eustatic rise. Facies 1 is a Pleistocene soil horizon characterized by an oxidized, massive to highly bioturbated, clayey quartz sand. This facies is truncated by a major erosional unconformity and is a result of subaerial exposure during the last sea-level fall and the bay ravinement process (first flooding surface) during the subsequent post-glacial rise in sea level. The unconformity represents a Type 1 sequence boundary (SB1) because the entire shelf in the study area was exposed 18,000 yr B.P. The erosional unconformity is overlain by estuarine deposits represented by Facies 2, 3, or 4, which comprise the lower transgressive systems tract. Facies 2 is a thin, matrix-supported shell bed, Facies 3 is a tan, silty to fine-grained quartz sand with rip-up clasts, and Facies 4 is a dark grey clay that can include thick, graded, shell-rich zones or very thin shelly quartz sand layers. Facies 3 or 4 is truncated by a second erosional surface that represents a shoreface ravinement diastem, signifying the second flooding surface. The diastem is overlain by a graded shell-bed (Facies 5) dominated by indigenous, well-preserved marine mollusks that were concentrated at the base of the shoreface. As shell content decreases upward, Facies 5 grades into Facies 6, which is a massive to horizontally laminated, fine- to coarse-quartz sand with open marine foraminifera. Facies 6 typically fines upward or shows no grain-size trend and represents a shelf sand sheet. Together, Facies 5 and 6 are up to 5.5-m-thick and characterize the upper transgressive systems tract.

Macrofaunal remains typically are concentrated into basal shell beds (Facies 2 and 5) that rest on bay and shoreface ravinement diastems. Taxonomic composition and environmental preferences of species permitted the bay and shoreface diastems to be distinguished and correlated. Generally, marine shell beds are clast-supported with a clean quartz sand matrix. Maximum bioclast size is about 5 cm, and some shell beds are normally graded. Additionally, some marine shell beds also contain a relict estuarine component dominated by poorly preserved *Chione cancellata*. In contrast, estuarine shell beds are supported by a muddy quartz sand or clay matrix, and maximum bioclast size is about 3.5 cm.

The entire marine section, up to 5.5 m thick, is completely reworked and mixed by post-transgressive processes down to the shoreface ravinement diastem. The depth of mixing demonstrates that the sedimentary regime on this shelf, and possibly others around the world, may be much more dynamic than previously thought (see Culver and Snedden, 1996). Sediment is mobilized during high-energy conditions, carried offshore and alongshore by shelf storm currents, and/or transported landward via large-scale bedform migration. These sedimentary processes are most active during the passage of strong cold fronts and tropical cyclones and are the primary transgressive and post-transgressive processes affecting the shelf surface.

The entire surficial sand sheet (Facies 6) consists of post-transgressive deposits, including linear shoals, because: 1) foraminiferal and molluscan assemblages are dominated by open, shallow-marine species, and 2) deposits lie above the shoreface ravinement diastem (stratigraphic context). Based on these data, the origin of the South and North Perdido shoals is related to transgressive and post-transgressive processes. The linear form and orientation of the shoals are dictated by underlying transgressive topography (i.e., escarpments) that was cut into the underlying Pleistocene substrate during slowdowns in the rate (~<15 mm/yr) of eustatic rise and enabled longshore sediment supply from the east to dominate the coastal system in the study area. This excess sediment caused the barrier shoreline to aggrade and possibly prograde. As the shoreline continued its net landward migration, erosional shoreface retreat produced a trailing surficial sand sheet (syn-transgressive) that draped the transgressive topography with up to 5.5 m of fine- to medium-quartz sand. However, the transgressive sand sheet was subsequently reworked completely by post-transgressive processes. Although shelf morphology appears to be similar to modern barrier island geomorphology, the resulting stratigraphy contains no *in situ* or degraded barrier island deposits. Hence, the morpho-stratigraphy of the Alabama and northwest Florida shelf is influenced by a combination of pre-transgressive (Pleistocene lowstand fluvial/deltaic systems) and transgressive features (escarpments) and deposits combined with extensive post-transgressive reworking by storms.

ACKNOWLEDGMENTS

This study was supported by U.S. Minerals Management Service (MMS) grant no. 1435-01-96-CT-30812 and the U.S. Department of Interior's Mineral Institute Program, which is administered by the Bureau of Mines through the generic Mineral Technology Center for Marine Minerals under grant no. G1115128-2201. In particular, we thank Barry Drucker of MMS and the Marine Minerals Technology Center (MMTC)—Continental Shelf Division at the University of Mississippi for its administrative support over the years, especially Robert Woolsey (Director), Katherine Walton, Dorothy O'Niell, Robin Buchanan, and Walter O'Niell. In addition, Doug Lockhart (MMTC), Monty Simmons (captain of the *R/V Kit Jones*), and Robert Shelton (1st Mate of the *R/V Kit Jones*) tirelessly assisted with field data collection. Additional field support was provided by Greg Stone, Cuong Nguyen, Paul Conner, and Robert Seal. Cuong Nguyen, Thuy Bui, Susan Anderson, Matthew Taylor, Traci Cash, Julie Doucet, John Ellis, Daniela Tudoran, and Ted Maul helped in core and sample processing. The authors thank Mark Byrnes, Dag Nummedal, Don Swift, John Snedden, and Alan Niedoroda for many stimulating discussions about shelf sedimentation. Cartographic assistance was provided by Kui Xu, Feng Li, Mary Lee Eggert, and Celia Harrod. Earlier drafts of this manuscript were critically reviewed by Arnold Bouma, Mark Byrnes, James Coleman, Oscar Huh, and Scott Ishman. John Anderson and Brian Zaitlin are thanked for their constructive comments in their role as official reviewers. Editorial assistance was provided by Catherine White, Claudia C. Holland, and Margaret Muller.

REFERENCES

Abston, J.R., Dinnel, S.P., Schroeder, W.W., Shultz, A.W. and Wiseman, W.J., Jr., 1987, Coastal sediment plume morphology and its relationship to environmental forcing: Main Pass, Mobile Bay, Alabama, *in* Kraus, N., ed., Coastal Sediments '87: New York, American Society of Civil Engineers, p. 1989-2005.

Adey, W.H., 1986, Coralline algae as indicators of sea-levels, *in* van de Plassche, O., ed., Sea-Level Research: Norwich, Geo Books, p. 229-280.

Anderson, L.C. and McBride, R.A., 1996, Taphonomic and paleoenvironmental evidence of Holocene shell-bed genesis and history on the northeastern Gulf of Mexico shelf: Palaios, v. 11, p. 532-549.

Anderson, L.C., Sen Gupta, B.K., McBride, R.A. and Byrnes, M.R., 1997, Reduced seasonality of Holocene climate and pervasive mixing of Holocene marine section: Northeastern Gulf of Mexico shelf: Geology, v. 25, p. 127-130.

Anderson, J.B., Thomas, M.A., Siringan, F.P. and Smyth, W.C., 1992, Quaternary evolution of the East Texas coast and continental shelf, *in* Fletcher, C.H., Jr. and Wehmiller, J.F., eds., Quaternary Coasts of the United States: Marine and Lacustrine Systems: Tulsa, Society of Sedimentary Geology (SEPM) Special Publication 48, p. 253-263.

Anderson, J.B., Abdulah, K.C., Sarzalejo, S., Siringan, F. and Thomas, M.A., 1996, Late Quaternary sedimentation and high-resolution sequence stratigraphy of the east Texas shelf, *in* De Batist, M. and Jacobs, P., eds., Geology of Siliciclastic Shelf Seas: London, The Geological Society Special Publication 117, p. 95-124.

Arthur, J.D., Applegate, J., Melkote, S. and Scott, T.M., 1986, Heavy mineral reconnaissance off the coast of the Apalachicola River Delta, Northwest Florida: Tallahassee, Florida Geological Survey Report of Investigation 95, 61 p.

Ballard, R.D. and Uchupi, E., 1970, Morphology and Quaternary history of the continental shelf of the Gulf Coast of the United States: Bulletin of Marine Science, v. 20, p. 547-559.

Balsillie, J.H., 1975, Analysis and interpretation of LEO and profile data along the western Panhandle coast of Florida: Fort Belvoir, Coastal Engineering Research Center (CERC) Technical Memorandum 49, p. 24.

Bandy, O.L., 1954, Ecology of Foraminifera in Northeastern Gulf of Mexico: Washington D.C., U.S. Geological Survey Professional Paper 274-G, p. 179-204.

Bard, E., Hamelin, B., Fairbanks, R.G. and Zindler, A., 1990, U/Th ages obtained by mass spectrometry in corals from Barbados: Calibration of ^{14}C for the last 30,000 years: Nature, v. 345, p. 405-410.

Bard, E., Hamelin, B., Arnold, M., Montaggioni, L., Cabioch, G., Faure, G. and Rougerie, F., 1996, Deglacial sea-level record from Tahiti corals and the timing of global meltwater discharge: Nature, v. 382, p. 241-244.

Bart, P.J., 1997, Seismic-stratigraphic analysis of shelf-margin delta/slope fan and basin floor fans on high-latitude and middle-latitude margins (Ross Sea, Weddell Sea, and Alabama/west Florida): Paleoclimatic and eustatic implications: Unpublished Ph.D. dissertation, Rice University, Houston, TX, 184 p.

Bartek, L.R. Anderson, J.B. and Abdulah, K.C., 1991, A case study of the preservation potential of the coastal lithosomes of small deltaic systems: Examples from the Holocene of the Texas continental shelf, *in* Coastal Depositional Systems in the Gulf of Mexico—Quaternary Framework and Environmental Issues: Austin, Gulf Coast Section-Society of Economic Paleontologists and Mineralogists (GCSSEPM) Foundation, Earth Enterprises, Inc., p. 15-25.

BERGMAN, K.M., 1994, Shannon Sandstone in Hartzog Draw–Heldt Draw fields reinterpreted as detached lowstand shoreface deposits: Journal of Sedimentary Research, v. B64, p. 184-201.

BERGMAN, K.M. AND WALKER, R.G., 1987, The importance of sea level fluctuations in the formation of linear conglomerate bodies: Carrot Creek Member, Cretaceous Western Interior Seaway, Alberta, Canada: Journal of Sedimentary Petrology, v. 57, p. 651-665.

BOCK, W.D., 1976, Distribution and significance of foraminifera in the MAFLA area: First International Symposium on Benthonic Foraminifera of Continental Margins, Part A, Ecology and Biology, Maritime Sediments Special Publication 1, p. 221-237.

BROOKS, G.R., KINDINGER, J.L., PENLAND, P.S., WILLIAMS, S.J. AND MCBRIDE, R.A., 1995, East Louisiana continental shelf sediments: A product of delta reworking: Journal of Coastal Research, v. 11, p. 1026-1036.

CARTER, R.W.G., 1988. Coastal Environments — An Introduction to the Physical, Ecological, and Cultural Systems of Coastlines: New York, Academic Press, 617 p.

CHAPPELL, J. AND POLACH, H., 1991, Post-glacial sea-level rise from a coral record at Huon Peninsula, Papua New Guinea: Nature, v. 349, p. 147-149.

COLEMAN, J.M. AND ROBERTS, H.H., 1988a, Sedimentary development of the Louisiana continental shelf related to sea level cycles. Part 1—sedimentary sequences: Geo-Marine Letters, v. 8, p. 63-108.

COLEMAN, J.M. AND ROBERTS, H.H., 1988b, Sedimentary development of the Louisiana continental shelf related to sea level cycles. Part 2—seismic response: Geo-Marine Letters, v. 10, p. 109-119.

CULVER, S.J. AND SNEDDEN, J.W., 1996, Foraminiferal implications for the formation of New Jersey shelf sand ridges: Palaios, v. 11, p. 161-175.

CURRAY, J.R., 1960, Sediments and history of Holocene transgression, continental shelf, northwest Gulf of Mexico, *in* Shepard, F.P., Phleger, F.B. and Van Andel, T.H., eds., Recent Sediments, Northwest Gulf of Mexico: Tulsa, American Petroleum Institute Project '51, American Association of Petroleum Geologists, p. 221-266.

CURRAY, J.R., 1965, Late Quaternary history, continental shelves of the United States, *in* Whuglit, H.E. and Frey, D.C., eds., The Quaternary of the United States: Princeton, Princeton University Press, p. 723-735.

DALRYMPLE, R.W., ZAITLIN, B.A. AND BOYD, R., eds., 1994, Incised Valley Systems: Origin and Sedimentary Sequences: Tulsa, Society of Sedimentary Geology (SEPM) Special Publication 51, 391 p.

DAVIES, D.J. AND HUMMELL, R.L., 1994, Lithofacies evolution from transgressive to highstand systems tracts, Holocene of the Alabama Coastal Zone: Gulf Coast Association of Geological Societies Transactions, v. 44, p. 145-153.

DEMAREST, J.M., JR. AND KRAFT, J.C., 1987, Stratigraphic record of Quaternary sea levels: Implications for more ancient strata, *in* Nummedal, D., Pilkey, O.H. and Howard, J.D., eds., Sea-Level Fluctuation and Coastal Evolution: Tulsa, Society of Economic Paleontologists and Mineralogists Special Publication 41, p. 223-239.

DILLION, W.P. AND OLDALE, R.N., 1978, Late Quaternary sea-level curve: Reinterpretation based on glaciotectonic influence: Geology, v. 6, p. 56-60.

DINNEL, S.P., 1988, Circulation and sediment dispersal on the Louisiana-Mississippi-Alabama continental shelf: Unpublished Ph.D. Dissertation, Louisiana State University, Baton Rouge, 173 p.

DONOGHUE, J.F. AND ALLARD, M.M., 1987, Heavy mineral reconnaissance on the northwest Florida shelf: Tallahassee, Final Report to the U.S. Minerals Management Service, Florida State University, Department of Geology, 126 p.

DONOGHUE, J.F., 1992, Late Quaternary coastal and inner shelf stratigraphy, Apalachicola Delta region, Florida, *in* Donoghue, J.F., ed., Quaternary Coastal Evolution: Sedimentary Geology, v. 80, p. 1-12.

DOWNING, K.P. AND WALKER, R.G., 1988, Viking Formation, Joffre Field, Alberta: Shoreface origin of long, narrow sand body encased in marine mudstones: American Association of Petroleum Geologists Bulletin, v. 72, p. 1212-1228.

DOYLE, L.J. AND SPARKS, T.N., 1980, Sediments on the Mississippi, Alabama, and Florida (MAFLA) continental shelf: Journal of Sedimentary Petrology, v. 50, p. 905-915.

EMERY, K.O. AND MILLIMAN, J.D., 1970, Quaternary sediments of the Atlantic continental shelf of the United States: Quaternaria, v. 12, p. 3-18.

EMILIANI, C., 1958, Paleotemperature analysis of core 280 and Pleistocene correlations: Journal of Geology, v. 66, p. 264-275.

FAIRBANK, N.C., 1962, Heavy minerals from the Eastern Gulf of Mexico: Deep Sea Research, v. 9, p. 307-338.

FAIRBANKS, R.G., 1989, A 17,000-year glacio-eustatic sea level record: Influence of glacial melting rates on the Younger Dryas event and deep-ocean circulation: Nature, v. 342, p. 637-642.

FAIRBANKS, R.G., 1990, The age and origin of the "Younger Dryas Climatic Event" in Greenland ice cores: Paleoceanography, v. 5, p. 937-948.

FISK, H.N., 1944, Geological investigation of the alluvial valley of the lower Mississippi River: Vicksburg, U.S. Army Engineers, Mississippi River Commission, 78 p.

FOLGER, D.W., 1972, Characteristics of Estuarine Sediments of the United States: Washington D.C., U.S. Geological Survey Professional Paper 742, 94 p.

FOLK, R.C., 1980, Petrology of Sedimentary Rocks: Austin, Hemphill Publishing Company, 182 p.

FREY, R.W. AND DORJES, J., 1988, Carbonate skeletal remains in beach-to-offshore sediments, Pensacola, Florida: Senckenbergiana Maritima, v. 20, p. 31-57.

GANGOPADHYAY, T. ANDERSON, L.C., JONES, M.H. AND MCBRIDE, R.A., 1996, Mollusca and benthic foraminifera of the Pensacola Bay and Perdido Bay estuarine systems, Florida and Alabama: Gulf Coast Association of Geological Societies Transactions, v. 46, p. 133-147.

GAYNOR, G.C. AND SWIFT, D.J.P., 1988, Shannon Sandstone depositional model: Sand ridge dynamics on the Campanian Western Interior shelf: Journal of Sedimentary Petrology, v. 58, p. 868-880.

GEORGE, S.M., 1988, Sedimentology and mineralogy of the Pensacola Bay system: Unpublished M.S. Thesis, University of Southern Mississippi, Hattiesburg, 93 p.

GITTINGS, S.R., SCHROEDER, W.W., SAGER, W.W., LASWELL, J.S. AND BRIGHT, T.J., 1990, Geological and biological surveys on topographic features of the Louisiana-Mississippi-Alabama continental shelf: New Orleans, Proceedings of the Gulf of Mexico Environmental Studies Conference, p. 29-35.

GITTINGS, S.R., BRIGHT, T.J., SCHROEDER, W.W., SAGER, W.W., LASWELL, J.S. AND REZAK, R., 1992, Invertebrate assemblages and ecological controls on topographic features in the northeast Gulf of Mexico: Bulletin of Marine Science, v. 50, p. 435-455.

GODWIN, H., SUGGATE, R.P. AND WILLIS, E.H., 1958, Radiocarbon dating of the eustatic rise in ocean-level: Nature, v. 181, p. 1518-1519.

GOLDSTEIN, A., JR., 1942, Sedimentary petrologic provinces of the northern Gulf of Mexico: Journal of Sedimentary Petrology, v. 12, p. 77-84.

GORSLINE, D., 1966, Dynamic characteristics of Gulf Coast beaches: Marine Geology, v. 4, p. 187-206.

GOULD, H.R. AND STEWART, H.R., 1955, Continental terrace sediments in the northeastern Gulf of Mexico, *in* Finding Ancient Shorelines: Society of Economic Paleontologists and Mineralogists Special Publication 3, p. 1-20.

Hopley, D., 1986, Corals and reefs as indicators of paleo-sea levels, with special reference to the Great Barrier Reef, *in* van de Plassche, O., ed., Sea-Level Research: Norwich, Geo Books, p. 195-228.

Horvath, G.J., 1968, The sedimentology of the Pensacola Bay System, northwestern Florida: Unpublished M.S. Thesis, Florida State University, Tallahassee, 89 p.

Hsu, K.J., 1960, Texture and mineralogy of the Recent sands of the Gulf Coast: Journal of Sedimentary Petrology, v. 30, p. 380-403.

Huh, O.K., Wiseman, W.J., Jr. and Rouse, L.J., 1981, Intrusion of Loop Current waters onto the west Florida continental shelf: Journal of Geophysical Research, v. 86, p. 4186-4192.

Huh, O.K., Rouse, L.J. and Walker, N.D., 1984, Cold-air outbreaks over the northwest Florida continental shelf: Heat flux processes and hydrographic changes: Journal of Geophysical Research, v. 89, p. 717-726.

Hyne, N.J. and Goodell, H.G., 1967, Origin of the sediments and submarine geomorphology of the inner continental shelf off Choctawhatchee Bay, Florida: Marine Geology, v. 5, p. 299-313.

Imbrie, J., Hays, J.D., Martinson, D.G., McIntyre, A., Mia, A.C., Morely, J.J., Pisias, N.G., Prell, W.L. and Shackleton, N.J., 1984, The orbital theory of Pleistocene climate: Support for a revised chronology of the marine $\delta^{18}O$ record, *in* Berger, A.J., Imbrie, J., Hays, J., Kukla, G. and Saltzman, B., eds., Milankovitch and Climate, Part 1: Dordrecht, Reidel, p. 269-305.

Isphording, W.C., Imsand, D.F. and Flowers, G.C., 1989, Physical characteristics and aging of Gulf coast estuaries: Gulf Coast Association of Geological Societies Transactions, v. 39, p. 387-401.

Isphording, W.C. and Isphording, G.W., 1991, Identification of ancient storm events in buried Gulf coast sediments: Gulf Coast Association of Geological Societies Transactions, v. 41, p. 339-347.

Jervey, M.T., 1974, Transportation and dispersal of biogenic material in the nearshore marine environment: Unpublished Ph.D. Dissertation, Louisiana State University, Baton Rouge, 306 p.

Kennicutt, M.C., Schroeder, W.W. and Brooks, J.M., 1995, Temporal and spatial variations in sediment characteristics on the Mississippi-Alabama continental shelf: Continental Shelf Research, v. 15, p. 1-18.

Kent, H.C., van Wyk, S.J. and Williams, S.J., 1976, Modern coastal sedimentary environments, Alabama and northwest Florida: Golden, Geological Exploration Associates Ltd., 96 p.

Kindinger, J.L., 1988, Seismic stratigraphy of the Mississippi-Alabama shelf and upper continental slope: Marine Geology, v. 83, p. 79-94.

Kindinger, J.L., 1989, Depositional history of the Lagniappe delta, northern Gulf of Mexico: Geo-Marine Letters, v. 9, p. 59-66.

Kindinger, J.L., Balson, P.S. and Flocks, J.G., 1994, Stratigraphy of the Mississippi-Alabama shelf and the Mobile River incised valley system, *in* Dalrymple, R.W., Boyd, R. and Zaitlin, B.A., eds., Incised-Valley Systems: Origin and Sedimentary Sequences: Tulsa, Society for Sedimentary Geology (SEPM) Special Publication 51, p. 83-95.

Knight, R.J. and McLean, J.R., eds., 1986, Shelf Sands and Sandstones: Canadian Society of Petroleum Geologists, Memoir 11, 347 p.

Kwon, H.J., 1969, Barrier islands of the northern Gulf of Mexico coast: Sediment source and development: Baton Rouge, Coastal Studies Institute, Louisiana State University, TR 75, 51 p.

Lamb, G.M., 1972, Distribution of Holocene foraminifera in Mobile Bay and the effect of salinity changes: Tuscaloosa, Geological Survey of Alabama Circular 82, p. 1-12.

Locker, S.D. and Doyle, L.J., 1992, Neogene to Recent stratigraphy and depositional regimes of the northwest Florida inner continental shelf: Marine Geology, v. 104, p. 123-138.

Ludwick, J.C., 1964, Sediments in northeastern Gulf of Mexico, *in* Miller, R.L., ed., Papers in Marine Geology: New York, Macmillan Co., p. 204-238.

Mars, J.C., Shultz, A.W. and Schroeder, W.W., 1992, Stratigraphy and Holocene evolution of Mobile Bay in southwestern Alabama: Gulf Coast Association of Geological Societies Transactions, v. 42, p. 529-542.

Marsh, O.T., 1966, Geology of Escambia and Santa Rosa Counties, Western Florida Panhandle: Tallahassee, Florida Geological Survey Bulletin 46, 140 p.

Mazzullo, J. and Bates, C., 1985, Sources of Pleistocene and Holocene sand for the northeast Gulf of Mexico shelf and Mississippi Fan: Gulf Coast Associations of Geological Societies Transactions, v. 35, p. 457-466.

Mazzullo, J. and Peterson, M., 1989, Sources and dispersal of late Quaternary silt on the northern Gulf of Mexico continental shelf: Marine Geology, v. 86, p. 15-26.

McBride, R.A., 1997, Seafloor morphology, geologic framework, and sedimentary processes of a sand-rich shelf offshore Alabama and northwest Florida: Northeastern Gulf of Mexico: Unpublished Ph.D. Dissertation, Louisiana State University, Baton Rouge, 509 p.

McBride, R.A. and Byrnes, M.R., 1995, Surficial sediments and morphology of the southwestern Alabama/western Florida Panhandle coast and shelf: Gulf Coast Association of Geological Societies Transactions, v. 45, p. 392-404.

McBride, R.A., Byrnes, M.R. anderson, L.C. and Sen Gupta, B.K., 1996, Holocene and Late Pleistocene sedimentary facies of a sand-rich continental shelf: A standard section for the northeastern Gulf of Mexico: Gulf Coast Association of Geological Societies Transactions, v. 46, p. 287-299.

McBride, R.A., Byrnes, M.R., Penland, P.S., Pope, D.L. and Kindinger, J.L., 1991, Geomorphic history, geologic framework, and hard mineral resources of the Petit Bois Pass area, Mississippi-Alabama, *in* Coastal Depositional Systems in the Gulf of Mexico—Quaternary Framework and Environmental Issues: Austin, Gulf Coast Section-Society of Economic Paleontologists and Mineralogists (GCSSEPM), Earth Enterprises, Inc., p. 116-127.

McFarlan, E., Jr., 1961, Radiocarbon dating of late Quaternary deposits, South Louisiana: Geological Society of America Bulletin, v. 72, p. 129-158.

Mechler, L.S. and Grady, J.R., 1984, Recent foraminifera of St. Andrew Bay, Florida: Gulf Coast Association of Geological Societies Transactions, v. 34, p. 385-394.

Middleton, G.V., 1973, Johannes Walther's law of the correlation of facies: Geological Society of America Bulletin, v. 84, p. 979-988.

Milliman, J.D. and Emery, K.O., 1968, Sea levels during the past 35,000 years: Science, v. 162, p. 1121-1123.

Moore, R.C., 1949, Meaning of facies, *in* Sedimentary Facies in Geologic History: Boulder, Geological Society of America Memoirs 39, p. 1-39.

Morton, R.A. and Suter, J.R., 1996, Sequence stratigraphy and composition of Late Quaternary shelf-margin deltas, northern Gulf of Mexico: American Association of Petroleum Geologists, v. 80, p. 505-530.

Moslow, T.F. and Rhodes, E.G., eds., 1986, Modern and Ancient Shelf Clastics—a Core Workshop: Society of Economic Paleontologists and Mineralogists, Core Workshop 9, Tulsa, OK, 459 p.

Murray, J.W., 1991, Ecology and Palaeoecology of Benthic Foraminifera: Harlow, Longman Scientific and Technical, 397 p.

Murray, S.P., 1970, Bottom currents near the coast during Hurricane Camille: Journal of Geophysical Research, v. 75, p. 4579-4582.

National Oceanic and Atmospheric Administration (NOAA), 1985, Gulf of Mexico Coastal and Ocean Zones Strategic Assessment: Data Atlas: Rockville, U.S. Department of Commerce, 187 p.

NELSON, H.F. AND BRAY, E.E., 1970, Stratigraphy and history of the Holocene sediments in the Sabine-High Island area, Gulf of Mexico, *in* Morgan, J.P., ed., Deltaic Sedimentation Modern and Ancient: Society of Economic Paleontologists and Mineralogists Special Publication 15, p. 48-77.

NUMMEDAL, D. AND SWIFT, D.J.P., 1987, Transgressive stratigraphy at sequence-bounding unconformities: Some principles derived from Holocene and Cretaceous examples, in Nummedal, D., Pilkey, O.H. and Howard, J.D., eds., Sea-Level Fluctuation and Coastal Evolution: Tulsa, Society of Economic Paleontologists and Mineralogists Special Publication 41, p. 241-260.

OTVOS, E.G., 1982, Coastal geology of Mississippi, Alabama, and adjacent Louisiana areas—Field Guidebook: New Orleans, The New Orleans Geological Society, 66 p.

OTVOS, E.G., 1985, Coastal Evolution: Louisiana to Northwest Florida—Field Guidebook: New Orleans, The New Orleans Geological Society, 91 p.

PARKER, F.L., 1954, Distribution of the foraminifera in the northeastern Gulf of Mexico: Harvard Museum of Comparative Zoology Bulletin, v. 111, p. 453-588.

PARKER, N.M., 1968, A sedimentologic study of Perdido Bay and adjacent offshore environments: Unpublished M.S. Thesis, Florida State University, Tallahassee, 57 p.

PARKER, R.H., 1956, Macro-invertebrate assemblages as indicators of sedimentary environments in the East Mississippi Delta Region: American Association of Petroleum Geologists Bulletin, v. 40, p. 295-376.

PARKER, R.H., 1960, Ecology and distributional patterns of marine macro-invertebrates, northern Gulf of Mexico, *in* Shepard, F.P., Phleger, F.B. and van Andel, T.H., eds., Recent Sediments, Northwest Gulf of Mexico: Tulsa, American Petroleum Institute 51, American Association of Petroleum Geologists, p. 302-337.

PARKER, S.J., 1990, Assessment of nonhydrocarbon mineral resources in the exclusive economic zone in offshore Alabama: Tuscaloosa, Geological Survey of Alabama Circular 147, 72 p.

PARKER, S.J., SHULTZ, A.W. AND SCHROEDER, W.W., 1992, Sediment characteristics and seafloor topography of a palimpsest shelf, Mississippi-Alabama continental shelf, *in* Fletcher, C.H., Jr. and Wehmiller, J.F., eds., Quaternary Coasts of the United States: Marine and Lacustrine Systems: Tulsa, Society of Sedimentary Geology (SEPM) Special Publication 48, p. 243-251.

PASTULA, E.J., JR., 1967, The ecology and distribution of recent foraminifera in Choctawhatchee Bay, Florida: Unpublished M.S. Thesis, Florida State University, Tallahassee, 102 p.

PENLAND, P.S., BOYD, R. AND SUTER, J.R., 1988, Transgressive depositional systems of the Mississippi River delta plain: A model for barrier shoreline and shelf development: Journal of Sedimentary Petrology, v. 58, p. 932-949.

PENLAND, P.S., SUTER, J.R., MCBRIDE, R.A., WILLIAMS, S.J., KINDINGER, J.L. AND BOYD, R., 1989, Holocene sand shoals offshore of the Mississippi River delta plain: Gulf Coast Association of Geological Societies Transactions, v. 39, p. 471-480.

PIRAZZOLI, P.A., 1991, World Atlas of Holocene Sea-Level Changes: New York, Elsevier Science Publishers, 300 p.

POAG, C.W., 1981, Ecologic Atlas of Benthic Foraminifera of the Gulf of Mexico: Woods Hole, Marine Science International, 175 p.

POPE, D.L., PENLAND, P.S., SUTER, J.R. AND MCBRIDE, R.A., 1991, Holocene geologic framework of the Trinity Shoal region, Louisiana continental shelf, *in* Coastal Depositional Systems in the Gulf of Mexico—Quaternary Framework and Environmental Issues: Austin, Gulf Coast Section-Society of Economic Paleontologists and Mineralogists (GCSSEPM), Earth Enterprises, Inc., p. 191-201.

POSAMENTIER, H.W. AND VAIL, P.R., 1988, Eustatic controls on clastic deposition II: Sequence and systems tract models, *in* Wilgus, C.K, et al., eds., Sea-Level Changes: An Integrated Approach: Society of Economic Paleontologists and Mineralogists (SEPM) Special Publication 42, p. 125-154.

POSAMENTIER, H.W., JERVEY, M.T. AND VAIL, P.R., 1988, Eustatic controls on clastic deposition I: Conceptual framework, *in* Wilgus, C.K, et al., eds., Sea-Level Changes: An Integrated Approach: Society of Economic Paleontologists and Mineralogists (SEPM) Special Publication 42, p. 109-124.

RAMPINO, M.R. AND SANDERS, J.E., 1980, Holocene transgression in south-central Long Island, New York: Journal of Sedimentary Petrology, v. 50, p. 1063-1980.

RAMPINO, M.R. AND SANDERS, J.E., 1982, Holocene transgression in south-central Long Island, New York—Reply: Journal of Sedimentary Petrology, v. 52, p. 1020-1025.

REZAK, R. AND HENRY, R.J., eds., 1972, Contributions on the Geological and Geophysical Oceanography of the Gulf of Mexico: College Station, Texas A&M University, Oceanographic Studies, v. 3, 303 p.

RINE, J.M., TILLMAN, R.W., CULVER, S.J. AND SWIFT, D.J.P., 1991, Generation of late Holocene sand ridges on the middle continental shelf of New Jersey, USA—evidence for formation in a mid-shelf setting based on comparison with a nearshore ridge, *in* Swift, D.J.P., Oertel, G.F., Tillman, R.W. and Thorne, J.A., eds., Shelf Sand and Sandstone Bodies: Geometry, Facies, and Sequence Stratigraphy: Oxford, International Association of Sedimentologists Special Publication 14, Blackwell Scientific Publications, p. 395-423.

ROBERTS, H.H., HUH, O.K., HSU, S.A., ROUSE, L.J. AND RICKMAN, D., 1987, Impact of cold-front passages on geomorphic evolution and sediment dynamics of the complex Louisiana coast, *in* Kraus, N.C., ed., Coastal Sediments '87: New York, American Society of Civil Engineering, p. 1950-1963.

RYAN, J.J., 1969, A sedimentologic study of Mobile Bay, Alabama: Tallahassee, Florida State University, Department of Geology Sedimentological Research Laboratory, Contribution 30, 110 p.

SANDERS, J.E. AND KUMAR, N., 1975a, Evidence of shoreface retreat and in-place "drowning" during Holocene submergence of barriers, shelf off Fire Island, New York: Geological Society of America Bulletin, v. 86, p. 65-76.

SANDERS, J.E. AND KUMAR, N., 1975b, Holocene shoestring sand on inner continental shelf off Long Island, New York: American Association of Petroleum Geologists Bulletin, v. 59, p. 997-1009.

SALVADOR, A., ed., 1991, The Gulf of Mexico Basin—The Geology of North America (DNAG), Volume J: Boulder, Geological Society of America, 568 p.

SCHROEDER, W.W., DARDEAU, M.R., DINDO, J.J., FLEISCHER, P., HECK, K.L., JR. AND SHULTZ, A.W., 1988a, Geological and biological aspects of hardbottom environments on the L'MAFLA shelf, northern Gulf of Mexico: Oceans '88, p. 17-22.

SCHROEDER, W.W., SHULTZ, A.W. AND DINDO, J.J., 1988b, Inner-shelf hardbottom areas, northeastern Gulf of Mexico: Gulf Coast Association of Geological Societies Transactions, v. 38, p. 535-541.

SCHROEDER, W.W., DINNEL, S.P., KELLY, F.J. AND WISEMAN, W.J. JR., 1994, Overview of the physical oceanography of the Louisiana-Mississippi-Alabama continental shelf, *in* Clarke, A.J., ed., Northeastern Gulf of Mexico Physical Oceanography Workshop: New Orleans, Proceedings, Outer Continental Shelf Study (MMS 94), U.S. Department of the Interior, Minerals Management Service and Florida State University, 248 p.

SCHROEDER, W.W., GITTINGS, S.R., DARDEAU, M.R., FLEISCHER, P., SAGER, W.W., SHULTZ, A.W. AND REZAK, R., 1989, Topographic features of the L'MAFLA continental shelf, northern Gulf of Mexico: Oceans '89, p. 54-58.

SCHROEDER, W.W., SHULTZ, A.W. AND PILKEY, O.H., 1995, Late Quaternary oyster shells and sea-level history, inner shelf, northeast Gulf of Mexico: Journal of Coastal Research, v. 11, p. 664-674.

SELLEY, R.C., 1970, Studies of sequence in sediments using a simple mathematical device: Quarterly Journal of the Geological Society of London, v. 125, p. 557-581.

SELLEY, R.C., 1988, Applied Sedimentology: New York, Academic Press, 446 p.

SHACKLETON, N.J., 1987, Oxygen isotopes, ice volume and sea level: Quaternary Science Reviews, v. 6, p. 183-190.

SHEPARD, F.P., 1963, Thirty-five thousand years of sea level, *in* Essays in Marine Geology in Honor of K.O. Emery: Los Angeles, University of South California Press, p. 1-10.

SHEPARD, F.P., PHLEGER, F.B. AND VAN ANDEL, T.H., eds., 1960, Recent Sediments, Northwest Gulf of Mexico: Tulsa, American Petroleum Institute Project 51, American Association of Petroleum Geologists, 394 p.

SHULTZ, A.W., SCHROEDER, W.W. AND ABSTON, J.R., 1990, Along-shore and offshore variations in Alabama inner-shelf sediments, *in* Tanner, W.F., ed., Coastal Sediments and Processes: Tallahassee, Proceedings of 9th Symposium on Coastal Sedimentology, Florida State University, p. 141-152.

SIRINGAN, F.P. AND ANDERSON, J.B., 1994, Modern shoreface and inner-shelf storm deposits off the east Texas coast, Gulf of Mexico: Journal of Sedimentary Research, v. B64, n. 2, p. 99-110.

SNEDDEN, J.W., TILLMAN, R.W., KREISA, R.D., SCHWELLER, W.J., CULVER, S.J. AND WINN, R.D., JR., 1994, Stratigraphy and genesis of a modern shoreface-attached sand ridge, Peahala Ridge, New Jersey: Journal of Sedimentary Research, v. B64, p. 560-581.

STAPOR, F.W., JR., 1975, Holocene beach ridge plain development, northwest Florida: Zeit für Geomorphie, p. 116-144.

STONE, G.W., STAPOR, F.W., MAY, J.P. AND MORGAN, J.P., 1992, Multiple sediment sources and a cellular, non-integrated, longshore drift system: Northwest Florida and southeast Alabama coast, USA: Marine Geology, v. 105, p. 141-154.

STUBBLEFIELD, W.L., MCGRAIL, D.W. AND KERSEY, D.G., 1984a, Recognition of transgressive and post-transgressive sand ridges on the New Jersey continental shelf, *in* Tillman, R.W. and Siemers, C.T., eds., Siliciclastic Shelf Sediments: Tulsa, Society of Economic Paleontologists and Mineralogists (SEPM) Special Publication 34, p. 1-23.

STUBBLEFIELD, W.L., MCGRAIL, D.W. AND KERSEY, D.G., 1984b, Recognition of transgressive and post-transgressive sand ridges on the New Jersey continental shelf: Reply, *in* Tillman, R.W. and Siemers, C.T., eds., Siliciclastic Shelf Sediments: Tulsa, Society of Economic Paleontologists and Mineralogists (SEPM) Special Publication 34, p. 37-41.

STUMPF, R.P., 1991, Observation of suspended sediments in Mobile Bay, Alabama from satellite, *in* Kraus, N.C., Gingerich, K.J. and Kriebel, D.L., eds., Coastal Sediments '91: New York, American Society of Civil Engineers, v. 1, p. 789-802.

STURGES, W. AND EVANS, J.C., 1983, On the variability of the Loop Current in the Gulf of Mexico: Journal of Marine Research, v. 41, p. 639-653.

SUTER, J.R. AND BERRYHILL, H.L., 1985, Late Quaternary shelf margin deltas, northwest Gulf of Mexico: American Association of Petroleum Geologists Bulletin, v. 69, p. 77-91.

Swift, D.J.P., 1975, Barrier island genesis: Evidence from the central Atlantic shelf, eastern U.S.A.: Sedimentary Geology, v. 14, p. 1-43.

SWIFT, D.J.P. AND T.F. MOSLOW, 1982, Holocene transgression in south-central Long Island, New York —Discussion: Journal of Sedimentary Petrology, v. 52, p. 1014-1019.

SWIFT, D.J.P. AND THORNE, J.A., 1991, Sedimentation on continental margins, I: A general model for shelf sedimentation, *in* Swift, D.J.P., Oertel, G.F., Tillman, R.W. and Thorne, J.A., eds., 1991, Shelf Sand and Sandstone Bodies—Geometry, Facies, Sequence Stratigraphy: Oxford, International Association of Sedimentologists Special Publication 14, Blackwell Scientific Publications, p. 3-31.

SWIFT, D.J.P., DUANE, D.B. AND MCKINNEY, T.F., 1973, Ridge and swale topography of the Middle Atlantic Bight, North America: Secular response to the Holocene hydraulic regime: Marine Geology, v. 15, p. 227-247.

SWIFT, D.J.P., MCKINNEY, T.F. AND STAHL, L., 1984, Recognition of transgressive and post-transgressive sand ridges on the New Jersey continental shelf: Discussion, *in* Tillman, R.W. and Siemers, C.T., eds., Siliclastic Shelf Sediments: Tulsa, Society of Economic Paleontologists and Mineralogists (SEPM) Special Publication 34, p. 25-36.

SWIFT, D.J.P., OERTEL, G.F., TILLMAN, R.W. AND THORNE, J.A., eds., 1991a, Shelf Sand and Sandstone Bodies—Geometry, Facies, Sequence Stratigraphy: Oxford, International Association of Sedimentologists Special Publication 14, Blackwell Scientific Publications, 532 p.

SWIFT, D.J.P., PHILLIPS, S. AND THORNE, J.A., 1991b, Sedimentation on continental margins, IV: Lithofacies and depositional systems, *in* Swift, D.J.P., Oertel, G.F., Tillman, R.W. and Thorne, J.A., eds., 1991, Shelf Sand and Sandstone Bodies—Geometry, Facies, Sequence Stratigraphy: Oxford, International Association of Sedimentologists Special Publication 14, Blackwell Scientific Publications, p. 89-152.

SYDOW, J.C. AND ROBERTS, H.H., 1994, Stratigraphic framework of a late Pleistocene shelf edge delta, northeast Gulf of Mexico: American Association of Petroleum Geologists Bulletin, v. 78, p. 1276-1312.

THOMAS, M.A. AND ANDERSON, J.B., 1994, Sea-level controls on the facies architecture of the Trinity/Sabine incised-valley system, Texas continental shelf, *in* Dalrymple, R.W., Boyd, R. and Zaitlin, B.A., eds., Incised-Valley Systems—Origin and Sedimentary Sequences: Tulsa, Society for Sedimentary Geology (SEPM) Special Publication 51, p. 63-82.

TILLMAN, R.W. AND MARTINSEN, R.S., 1984, The Shannon shelf-ridge sandstone complex, Salt Creek anticline area, Powder River Basin Wyoming, *in* Tillman, R.W. and Siemers, C.T., eds., Siliciclastic Shelf Sediments: Tulsa, Society of Economic Paleontologists and Mineralogists (SEPM) Special Publication 34, p. 85-142.

TILLMAN, R.W. AND MARTINSEN, R.S., 1987, Sedimentologic model and production characteristics of Hartzog Draw field, Wyoming, a Shannon shelf-ridge sandstone, *in* Tillman R.W. and Weber, K.J., eds., Reservoir Sedimentology: Tulsa, Society of Economic Paleontologists and Mineralogists (SEPM) Special Publication 40, p. 15-112.

TORTORA, P., 1996, Depositional and erosional coastal processes during the last postglacial sea-level rise—an example from the central Tyrrhenian continental shelf (Italy): Journal of Sedimentary Research, v. 66, p. 391-405.

UPSHAW, C.F., CREATH, W.B. AND BROOKS, F.L., 1966, Sediments and Microfauna off the Coasts of Mississippi and Adjacent States: Jackson, Mississippi Geological Survey Bulletin 106, 127 p.

VAN ANDEL, T.H., 1960, Sources and dispersion of Holocene sediments, northern Gulf of Mexico, *in* Shepard, F.P., Phleger, F.B. and Van Andel, T.H., eds., Recent Sediments, northwestern Gulf of Mexico: Tulsa, American Petroleum Institute Project 51, American Association of Petroleum Geologists, 394 p.

VAN WYK, S.J., 1973, Selected aspects of Holocene sedimentation, Pensacola area, Florida: Unpublished M.S. Thesis, Colorado School of Mines, Golden, 211 p.

WALKER, R.G., 1992, Facies, facies models, and modern stratigraphic concepts, *in* Walker, R.G. and James, N.P., eds., Facies Models—Response to Sea Level Change: St. Johns, Geological Association of Canada, p. 1-14.

WALKER, R.G. AND WISEMAN, T.R., 1995, Lowstand shorefaces, transgressive incised shorefaces, and forced regressions: Examples from the Viking formation, Joarcam area, Alberta: Journal of Sedimentary Research, v. B65, p. 132-141.

WALTON, T.L., 1976, Littoral drift estimates along the coastline of Florida: Gainesville, Florida Sea Grant, University of Florida.

WALTON, W.R., 1964, Recent foraminiferal ecology and Palaeoecology, *in* Imbrie, J. and Newell, N., eds., Approaches to Palaeoecology: New York, John Wiley, p. 151-237.

WILGUS, C.K., HASTINGS, B.S., KENDALL, C.G., POSAMENTIER, H.W., ROSS, C.A., VAN WAGONER, J.C., 1988, Sea-Level Changes: An Integrated Approach: Tulsa, Society of Economic Paleontologists and Mineralogists (SEPM) Special Publication 42, 407 p.

WILLIAMS, S.J., 1974, Aspects of barrier island shallow shelf sedimentation: West of Pensacola, Florida: Unpublished M.S. Thesis, Colorado School of Mines, Golden, 161 p.

WINKER, C.D., 1982, Cenozoic shelf margins, northwest Gulf of Mexico: Gulf Coast Association of Geological Societies Transactions, v. 32, p. 427-448.

Appendix 1.—Additional information regarding benthic foraminifera in terms of sample depth in core and water depth of core top (see Figure 6).

	Core & Sample #	Sample Depth in Core (cm)	Water Depth (m)		Core & Sample #	Sample Depth in Core (cm)	Water Depth (m)
1.	Pen-91-3/1	75	19	27.	Pen-92-4/3	300	29
2.	Pen-91-3/2	350	19	28.	Pen-92-5/1	25	23
3.	Pen-91-5/1	10	21	29.	Pen-92-5/2	75	23
4.	Pen-91-5/2	125	21	30.	Pen-92-5/3	225	23
5.	Pen-91-5/3	349	21	31.	Per-93-3/2	395	4
6.	Pen-91-5/4	453	21	32.	Ala-91-3/1	100	10
7.	Pen-91-6/1	75	28	33.	Ala-91-3/2	290	10
8.	Pen-91-6/2	525	28	34.	Ala-91-3/3	400	10
9.	Pen-91-9/1	50	26	35.	Ala-91-9/1	25	8
10.	Pen-91-9/2	300	26	36.	Ala-91-9/2	275	8
11.	Pen-91-11/1	5	28	37.	Ala-91-9/3	325	8
12.	Pen-91-11/2	50	28	38.	Ala-91-9/4	375	8
13.	Pen-91-11/3	155	28	39.	Ala-91-16/1	5	35
14.	Pen-91-11/4	248	28	40.	Ala-91-16/2	50	35
15.	Pen-91-11/5	312	28	41.	Ala-91-16/3	95	35
16.	Pen-91-11/6	332	28	42.	Ala-91-16/4	150	35
17.	Pen-91-11/7	350	28	43.	Ala-91-16/5	195	35
18.	Pen-91-11/8	366	28	44.	Ala-91-16/6	230	35
19.	Pen-91-11/9	385	28	45.	Ala-91-16/7	265	35
20.	Pen-91-11/10	392	28	46.	Ala-91-16/8	290	35
21.	Pen-91-11/11	404	28	47.	Ala-91-16/9	320	35
22.	Pen-92-2/1	25	28	48.	Ala-91-16/10	335	35
23.	Pen-92-2/2	225	28	49.	Ala-91-16/11	350	35
24.	Pen-92-2/3	425	28	50.	Ala-91-16/12	358	35
25.	Pen-92-4/1	25	29	51.	Ala-91-16/13	372	35
26.	Pen-92-4/2	175	29				

THE LATE QUATERNARY TRANSGRESSIVE RECORD IN THE ADRIATIC EPICONTINENTAL SEA: BASIN WIDENING AND FACIES PARTITIONING

A. CATTANEO AND F. TRINCARDI

Istituto per la Geologia Marina via Gobetti 101, 40129 Bologna, Italy

ABSTRACT: In the Central Adriatic basin the late Quaternary Depositional Sequence comprises four systems tracts that vary greatly in thickness and internal geometry along the margin. Each of the systems tracts was deposited during distinctive portions of the relative sea-level curve that was dominated by the eustatic signal. This paper summarizes the physical-stratigraphic framework of late Quaternary TST deposits in the Central Adriatic area, based on the interpretation of a dense network of seismic profiles, sediment cores and accelerator mass spectrometer (AMS) ^{14}C dates. The Adriatic epicontinental basin includes two contrasting shelf domains: the wide and low-gradient northern shelf and the narrower and steeper central shelf surrounding a slope basin 250 m deep. A refined stratigraphic scheme of complementary shelf and deeper water records helps to recognize the influence of past environmental change on depositional systems, internal geometry, and lateral facies changes within the late Quaternary TST.

During the late Quaternary sea-level rise, between 16,000 and 5,500 years BP, several factors influenced the deposition of the TST in the Adriatic basin. Short-term climatic instability caused repeated changes in sediment flux, while the transgressive drowning and widening of this semienclosed basin permanently changed the oceanographic regime, which affected the mechanisms of shelf sediment dispersal. The late Quaternary TST consists of three distinctive units that are easily recognized through high-resolution seismic profiles, sediment cores, and multiproxy stratigraphic data. The middle TST unit is more complex compared to the underlying and overlying TST units, which were deposited during an interval of extreme climatic instability (between approximately 14,800 and 11,300 calibrated years BP) and includes proximal progradational and distal mounded deposits on the shelf. The middle TST unit records a peak in sediment flux and mean grain size that suggest a significant change in supply regime.

This three-fold subdivision can be extended to the late Quaternary TST in the rest of the Adriatic basin in areas where the record is essentially composed of undersupplied and physically detached barrier-lagoon-estuary systems. These transgressive systems, which either drowned in place or were partially reworked, have no offshore muddy correlative on the northern shelf. This evidence indicates a marked longshore facies partitioning between a northern area with barrier-lagoon deposits and starved shelf and a southern area with thick and complex mud-dominated shelf units. This kind of facies partitioning was established during the deposition of the middle TST and affected the upper TST and the HST as well.

Variable sediment flux and oceanographic regime affect the stratigraphic record, even during an interval of rapid accommodation-dominated cyclicity like the late Quaternary. The recognition of such factors is possible through the study of mud-dominated shelf sequences on modern margins, but it is important to consider the relevance of such factors also when interpreting ancient successions, although straightforward comparisons among modern and ancient continental-margin deposits are hampered by the different investigation techniques required and the diverse scales of observation.

INTRODUCTION

Transgressions and regressions are commonly interpreted as resulting from the combination of eustatic change and constant subsidence, when a constant sediment flux is assumed. Cyclic fluctuations in sediment supply, deriving either from tectonic or climatic causes can, however, substantially alter the stacking pattern within depositional sequences (Schlager, 1993). Furthermore, the geometry of marine deposits can be greatly influenced by changes in oceanographic circulation and sediment dispersal. In particular, this kind of environmental factor tends to alter the amount of accommodation space that sediment can effectively exploit on a continental margin (Thorne and Swift, 1991). The focusing effect of currents (storm-driven, thermohaline, or tidal in origin) against coast outbulges or morphologic barriers on localized shelf sectors may either reduce or enhance accumulation of sediment in areas. In general, both sediment supply and oceanographic regime can affect the internal geometry of depositional sequences on very short time scales (decades to millennia), even during cycles that are primarily driven by sea-level oscillations (Thorne and Swift, 1991).

On modern continental margins, examples of transgressive deposits emplaced during the late Quaternary sea level rise include incised-valley fills (Anderson and Thomas, 1991; Anderson et al., 1995), starved marine mud drapes (Trincardi and Field, 1991), drowned barrier islands (Penland et al., 1988; Snedden et al., 1994), and reworked dune fields (Ikehara and Kinoshita, 1994; Correggiari et al., 1996a). In settings characterized by high sediment supply, lateral advection by coastal or shelf currents may trap sediment on the shelf and result in the deposition of thick, essentially muddy stratigraphic sections (Trincardi et al., 1996b; Cattaneo et al., 1997).

The Adriatic basin, like other modern mid-latitude epicontinental seas, is an example of a mud-dominated shelf where exceptional physical (reflection seismic, lithology, and magnetic susceptibility), biostratigraphic and chronologic resolution permit dissecting the late Quaternary depositional sequence into its composing systems tracts and parasequences. In particular, the late Quaternary transgressive sediments were deposited when accommodation space was rapidly being created, at rates up to 30 mm/year from 16,000 to 5,500 years BP (Fairbanks, 1989; Bard et al., 1996). The study of these deposits, with time resolutions up to century-scale (Asioli, 1996; Trincardi et al., 1996b; Langone et al., 1996), allows us to disentangle the effects of short-term supply fluctuations from those linked to changes in oceanographic regime and sediment dispersal on the shelf. Physical correlation to a 250 m-deep slope basin in the center of the Adriatic Sea provides a complementary (continuous muddy and marine) transgressive section for chronological reference.

The study of this kind of thick and muddy transgressive record is important because it allows a detailed reconstruction of environmental variability in the recent geologic past and because it may serve as an analogue to ancient mud-dominated shelf deposits. The analysis of ancient mud-

Isolated Shallow Marine Sand Bodies: Sequence Stratigraphic Analysis and Sedimentologic Interpretation.
SEPM Special Publication No. 64, Copyright © 1999
SEPM (Society for Sedimentary Geology), ISBN 1-56576-057-3, p. 127-146.

dominated shelf deposits is difficult where the quality of the exposures is low and key stratigraphic surfaces have poor sedimentologic expressions and very low gradients. For this reason, and because of their relevance for oil exploration, emphasis in the study of ancient and buried shelf deposit is often placed on their more proximal and sandy expressions.

Setting

The Adriatic Sea is an elongated, semienclosed basin that encompasses three distinct morphologic domains (Fig. 1). The northern Adriatic is a shallow and low gradient (~0.02°) continental shelf; the central Adriatic has a narrower and steeper shelf (~0.2°) surrounding the Mid Adriatic Deep (MAD), a small slope basin that reaches 250 m deep. The southern Adriatic is about 1200 m deep. The MAD is connected to the southern Adriatic and to the rest of the Mediterranean through the Pelagosa Sill (PS), where the maximum depth is between 160-170 m (Fig. 1). Since the last glacial maximum (LGM, ca 18,000 years BP), the Adriatic Sea has undergone two major changes in physical setting: 1) the basin widened up to seven times its LGM extent (Trincardi et al., 1994; Correggiari et al., 1996a, b), and 2) the sill connecting to the southern Adriatic—which was only a few km wide and few tens of meters deep—drowned, allowing enhanced water exchange with the rest of the Mediterranean (Jorissen et al., 1993; Asioli et al., 1996; Asioli, 1996; Trincardi et al., 1996a, b). Both basin widening and sill drowning contributed to a change in the oceanographic regime of the Adriatic that led to the onset of the modern circulation pattern. Local tectonic activity in the study area either determined the growth of a morphologic relief during the Quaternary (for example the Tremiti High; Cattaneo et al., 1997) that complicates the shelf morphology, provided a local source of clastics during times of lowered sea level, or acted as an obstacle for bottom-hugging marine currents during intervals of rising sea level.

Modern Oceanographic Regime of Adriatic Sea

Wind-push on the water's surface and the hydrostatic effect of atmospheric pressure are important factors affecting the oceanographic regime in the modern Adriatic basin. Typical current speeds are less than 10 cm/s, but may reach 1 m/s during winter along the western coast forced by the northeast-blowing Bora wind (Franco et al., 1982; Correggiari et al., 1996a). Temperature-salinity averaged profiles in the central Adriatic Sea define three water masses (Paschini et al., 1993): 1) a surficial temperature-mixed layer (0-30 m) that is affected by strong air-sea interaction with the upper 10 m of less saline and cooler waters of coastal origin, mainly Po river runoff; 2) a Levantine Intermediate-Water (LIW) layer (30-130 m) with a maximum salinity around 80-90 m deep; and 3)

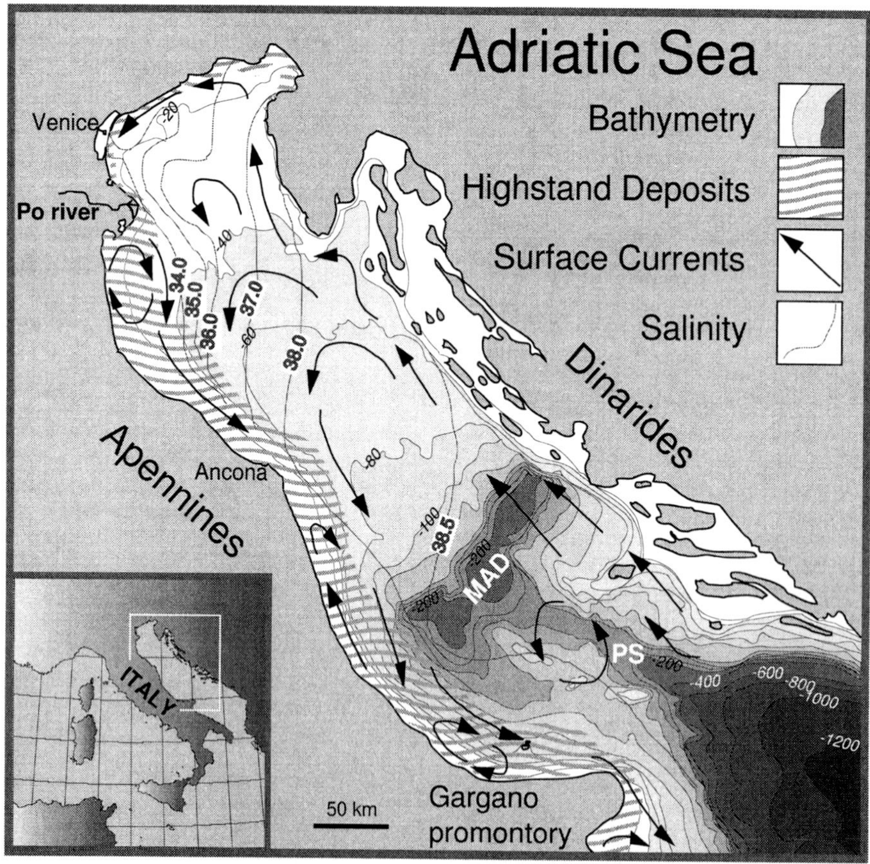

Fig. 1.—Simplified bathymetry of the Adriatic basin showing the distribution of modern surface currents (arrows) and salinity (contour lines in per mil). Note the limited seaward extent of the HST of the late Quaternary depositional sequence (wavy pattern). MAD is the mid Adriatic Deep where water depth is as much as 250 m; PS indicates the shallow Pelagosa Sill separating the Central and the Southern Adriatic basins.

a deeper-water region (130 m-bottom) with very dense waters that probably formed in the northern Adriatic and sunk to the south (Paschini et al., 1993). The surface circulation pattern of the Adriatic Sea consists of a large-scale baroclinic geostrophic structure with cyclonic gyres of thermohaline origin that appear to be seasonally modulated (Orlic et al., 1992; Malanotte-Rizzoli and Bergamasco, 1983; Artegiani et al., 1997; Fig. 1). The same cyclonic gyre affects the LIW layer and the bottom layer with a southwestward current flow along the western side of the basin (Orlic et al., 1992). Recent stable-isotope studies of Adriatic waters substantially confirmed this subdivision, tracing also the typical $\delta^{18}O$ signature of the LIW (Stenni et al., 1995).

The microtidal regime of the Adriatic is influenced by the morphology of the basin. Local tidal ranges are close to 1 m in the northern shelf (Bondesan et al., 1995). It is likely that tidal periodicities (12 and 24 hours) are amplified by the two main resonant periodicities of 11 and 22 hours in the Adriatic (Bondesan et al., 1995). Any change in water depth and basin size brought about by the late Quaternary sea-level rise may have affected the resonant period and consequently the tide amplitude. Therefore, amplification effects similar to those affecting the northern Adriatic today may have occurred in localized areas of the basin during the late Quaternary transgression.

Sediment Supply

Modern river supply (mean suspended load) to the Adriatic Sea is negligible on the eastern side, very small in the north (0.3 x 10^7 t/year), maximum on the western side (3.9 x 10^7 t/year), and is fed by the Po and other rivers draining the outer side of the Apennines (Trincardi et al., 1994). Short rivers draining the Apennines have high sediment yield and deliver large volumes of mud to the Adriatic basin (Milliman and Syvitski, 1992). High lateral variability in siliciclastic supply to the margin is expected in response to local differences in relief, substrate composition, and river runoff (Milliman and Syvitski, 1992). From direct measurements of bed and suspended load of Alpine and Apennine rivers Bartolini et al. (1996) evaluated the net sediment supply rate for present, late Holocene (last ca. 6,000 years BP), and Quaternary (above the base of *Hyalinea baltica* zone) times in the northern Apennine foredeep. These older estimates derive from reconstructions of the volume of late Holocene and Quaternary sedimentary bodies. This brought surprisingly similar values of 0.047 km^3/year, 0.037-0.046 km^3/year, and 0.045-0.047 km^3/year, respectively. These averaged data, however, do not help in identifying peaks in sediment supply that may have occurred on short stratigraphic intervals related to rapid climatic fluctuations (Langone et al., 1996; Trincardi et al., 1996a, b; Cattaneo et al., 1997).

METHODS

Seismic lines were collected by the Istituto di Geologia Marina in 1993 and 1995 using a 3.5-kHz pinger and a 500-joules Uniboom sound source. Seismic data were recorded with sampling intervals of 1/8 to 1/2 of a second and typical ship speed of 4-5 knots; the spacing between lines varied and was smaller (1 km or less) where the stratigraphic complexity was highest (Fig. 2). More than 60 piston and gravity cores composed our data base in the study area; cores were positioned on seismic lines using a shore-based positioning system or a differential GPS with a typical position uncertainty of less than 30 m. Sixteen cores gave essential information on lithology and sedimentary facies and provided the basis for multiproxy stratigraphic analyses, including foraminifera ecozonation, pollen spectra, magnetic susceptibility, stable isotopes, and alkenones (Asioli, 1996; Asioli et al., 1996; Lowe et al., 1996a, b; Trincardi et al., 1996a, b; Ariztegui et al., 1996). Tephrochronology provided an independent control to the biostratigraphic reconstructions (Calanchi et al., 1996). Chronological control comes from 45 AMS ^{14}C dates on planktic foraminifera from piston cores (Trincardi et al., 1996b; Asioli, 1996). Carbon-14 reservoir correction of 570 years was applied to all samples (Langone et al., 1996; Trincardi et al., 1996b) before they were calibrated following Stuiver and Reimer (1993).

We carried out 67 grain-size analyses to characterize possible vertical trends within the coarser-grained beds (silt to fine sand) of the transgressive record. Sampling was based on the visual recognition of the main lithologic changes. Samples were pretreated with H_2O_2 to eliminate organic material, sieved at intervals of .5 ϕ for the fraction coarser than 63 µm, and weighed. The fraction finer than 63 µm was analyzed with a 5000D Micrometric Sedigraph. Grain size analyses on 95 samples from core CM92-43, in the distal MAD basin, were relayed on a laser Galai CIS 1 instrument with size interval of 1 µm and are discussed elsewhere in detail (Guerzoni et al., 1996).

HIGH-RESOLUTION SEISMIC STRATIGRAPHIC FRAMEWORK

New very high-resolution stratigraphic data in the Central Adriatic basin help define the late Quaternary depositional sequence (LQDS) based on reflection geometries and regional correlation. Within the LQDS, high-resolution seismic data allow the definition and regional correlation of four late Quaternary systems tracts (hereinafter listed in descending order) that developed during the last glacial-interglacial cycle (Fig. 3): the High-stand Systems Tract (HST), which rests above the maximum flooding surface (mfs; Correggiari et al., 1992); the Transgressive Systems Tract (TST), which rests above the Transgressive surface (ES1) and below the mfs (Trincardi et al., 1996b); the Low-stand Systems Tract (LST), which rests above a regional unconformity that originated during an interval of subaerial exposure of the shelf (sequence boundary); and a Falling sea-level Systems Tract (FST) that recorded the relative sea-level fall that occurred from oxygen-isotope Stage 5e through Stage 3 (Trincardi and Correggiari, 1998). All systems tracts vary laterally in thickness and facies assemblage and exhibit localized stratigraphic expansions with sediment accumulation rates exceeding 2 mm/yr (Trincardi et al., 1996b).

High-Stand Systems Tract

The HST of the late Quaternary depositional sequence in the Adriatic was deposited after the time of maximum marine transgression, which was about 5,500 years BP (Trincardi et al., 1996b; Asioli et al., 1996). The HST is a mud wedge up to 30 m thick (Figs. 4, 5) that correlates distally to an extensive mud drape less than 2 m thick. The depocenter of the HST is confined to the western side of the basin (Fig. 5); two factors control this sediment distribution: 1) most of the sediment entry points are located on the western side of the basin, where they drain the uplifting Apennine chain and the Po Plain to the north; and 2) the time-averaged effect of the southwestward baroclinic flow of bottom waters along this side of the basin (Malanotte-Rizzoli and Bergamasco, 1983; Artegiani et al., 1997; Trincardi et al., 1996b). North of the

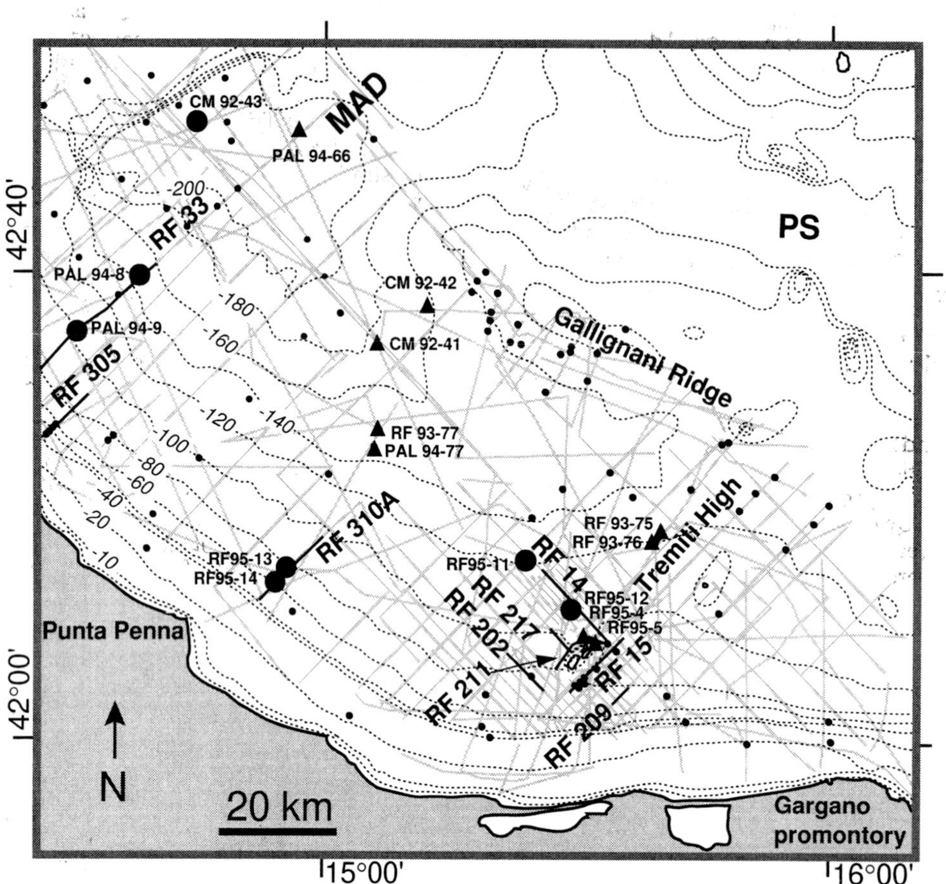

Fig. 2.—Data base of high-resolution seismic profiles and sediment cores in the Central Adriatic region. Seismic data discussed in this paper are shown as bold lines. Heavy dots indicate the key cores used in Figure 9 for dating and stratigraphic correlation; triangles refer to other cores discussed in the article. Bathymetric contours are shown for reference.

modern Po delta complex, the HST is undersupplied and consists of barrier-lagoon coastal systems rimming a sediment-starved continental shelf (Trincardi et al., 1994; Correggiari et al., 1996a, b).

In the shore-parallel depocenters of the HST the average sediment accumulation rate was on the order of 0.6 cm/year for the last 5,500 years; this is an underestimate that does not take into account the amount of time expressed by condensed deposition on the mfs. On seismic records, the HST consists of several stacked sigmoids that show decreasing thicknesses both seaward and landward; ^{210}Pb activity profiles across the youngest of these sigmoids show sediment accumulation rates up to 12 mm/year in the depocenter (Langone et al., 1996; Trincardi et al., 1996b).

Transgressive Systems Tract

The late Quaternary transgressive systems tract (TST) is bounded by a regional erosional surface (ES1; Fig. 6) at the base and by a regional downlap—the maximum flooding surface (mfs)—at the top (Trincardi et al., 1996a,b). The former surface is erosional on the shelf and becomes conformable in deeper waters (Fig. 4), and the latter marks an interval of condensed deposition that coincides with a turnover of planktonic foraminifera associations (Asioli, 1996), and represents the base of the overlying prograding HST (Correggiari et al., 1992; Trincardi et al., 1996b).

The late Quaternary TST in the Adriatic Sea consists of three complementary records: within the MAD, a muddy section forms a continuous marine succession (Jorissen et al., 1993; Asioli et al., 1996; Asioli, 1996); on the shelf west and south of MAD, transgressive deposits are up to 25 m thick and constitute a series of sedimentary bodies with an overall backstepping arrangement and a complex internal geometry (Trincardi et al., 1996a, b); on the low-gradient shelf north of the MAD, remnants of barrier-lagoon-estuary systems are separated by large horizontal gaps where a thin (typically 5 - 20 cm) lag of reworked shells caps alluvial plain deposits of the LST (Trincardi et al., 1994; Correggiari et al., 1996b). These differentiated records of the same interval allow a complementary reconstruction of the transgressive events: the most precise stratigraphic constraints come from the continuous deeper-water record (Asioli et al., 1996; Asioli, 1996; Langone et al., 1996; Lowe et al., 1996a,b); sedimentologic information on the lateral variability of sedimentary bodies comes from the composite shelf records (Trincardi et al., 1996a; Cattaneo et al., 1997; Correggiari et al., 1996a, b).

Lowstand Systems Tract

The LST was deposited during the last glacial maximum and includes a lower chaotic unit and an upper progradational wedge that filled the MAD from the north-

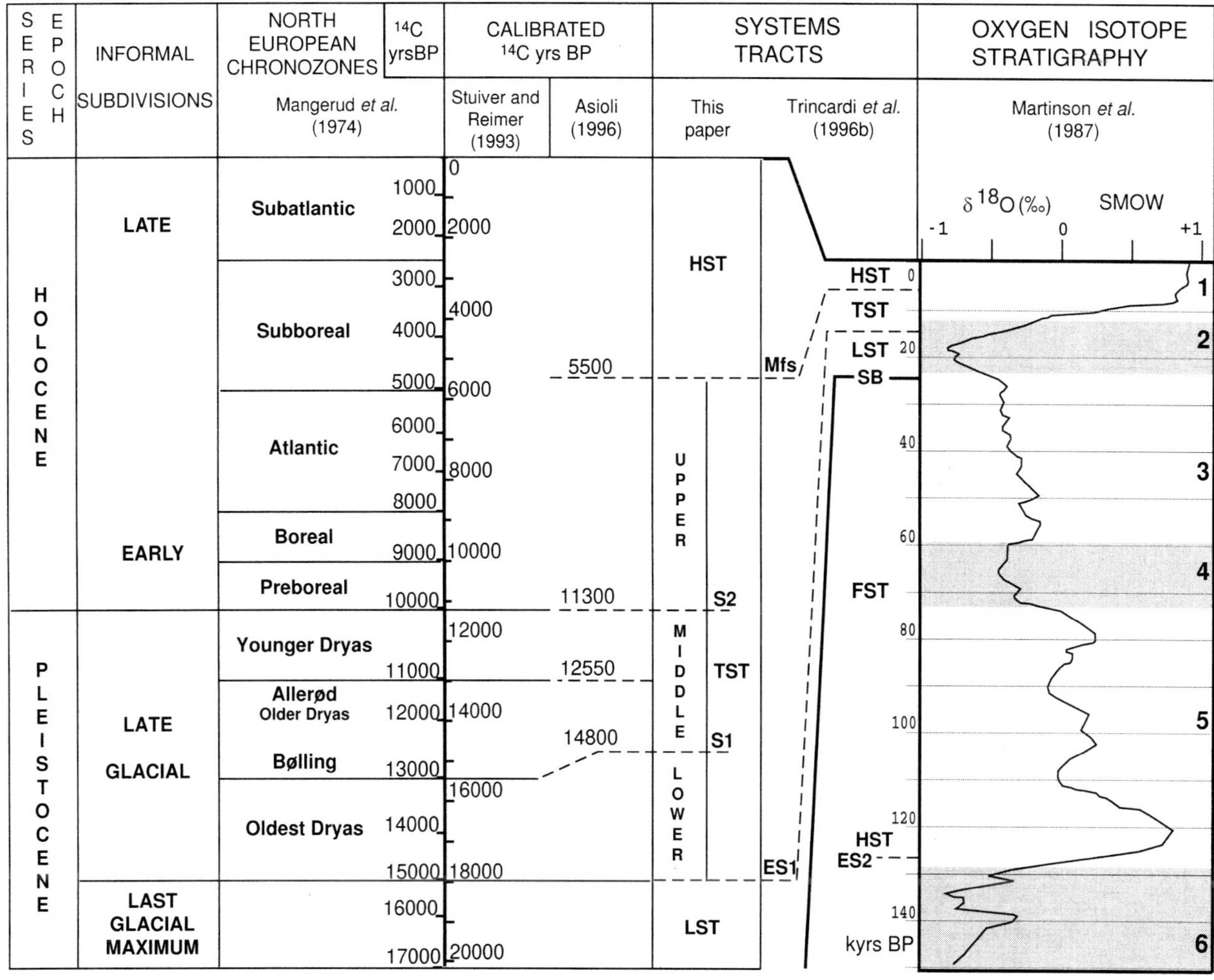

Fig. 3.—Stratigraphic scheme relating the systems tracts of the late Quaternary depositional sequence to the North European chronozones (in both calibrated and uncalibrated ^{14}C years) and to the standard marine oxygen isotope curve (cold stages are indicated by even numbers and shown as shaded areas).

west (Trincardi et al., 1996b). The lower chaotic unit is extensive, as much as 30 m thick, and composed of several stacked mass-failure deposits "ponded" within the MAD; sediment cores could not reach this unit because of the thickness of the overlying section (Trincardi et al., 1996b). The upper unit records 150 km of shelf progradation from the northwest under the combined supplies of the Po and other smaller Apennine and Alpine rivers. The total thickness of the LST is about 250 m in its depocenter, which is immediately north of the MAD (Fig. 5). The LST wedges out to a minimum thickness of a few meters on the southern flank of the MAD (Fig. 5). Farther to the south, on the shelf, the LST is replaced by the erosional surface ES1, which tops older deposits of forced regression (Trincardi and Correggiari, 1998). Cores reaching below surface ES1 and under the base of the LST in this area of slower accumulation rates prove that the entire progradational unit (lowstand wedge) is younger than oxygen isotope stage 3 (Trincardi and Correggiari, 1998).

SUBDIVISONS WITHIN THE TST ON THE CENTRAL ADRIATIC SHELF

The central Adriatic margin is the best site at which to introduce consistent internal subdivisions within the late Quaternary TST; these subdivisions are possible because of the high rates of sediment accumulation that accompanied the relative sea-level rise in this region and because of the fine-grained nature of the deposits; three seismic-stratigraphic units compose the TST and are separated by two regional surfaces: S1 and S2 (Figs. 4 , 5; Trincardi et al., 1996a). The lower and upper TST units record an abrupt landward shift of the shoreline, while the middle unit is prograding seaward and represents a regressive sedimentary body within the TST.

The three TST units record successive intervals of relative sea-level rise. Each of the three units progressively onlaps landward and encompasses a finite interval of landward migration of the paleoshoreline. In particular, the lower TST

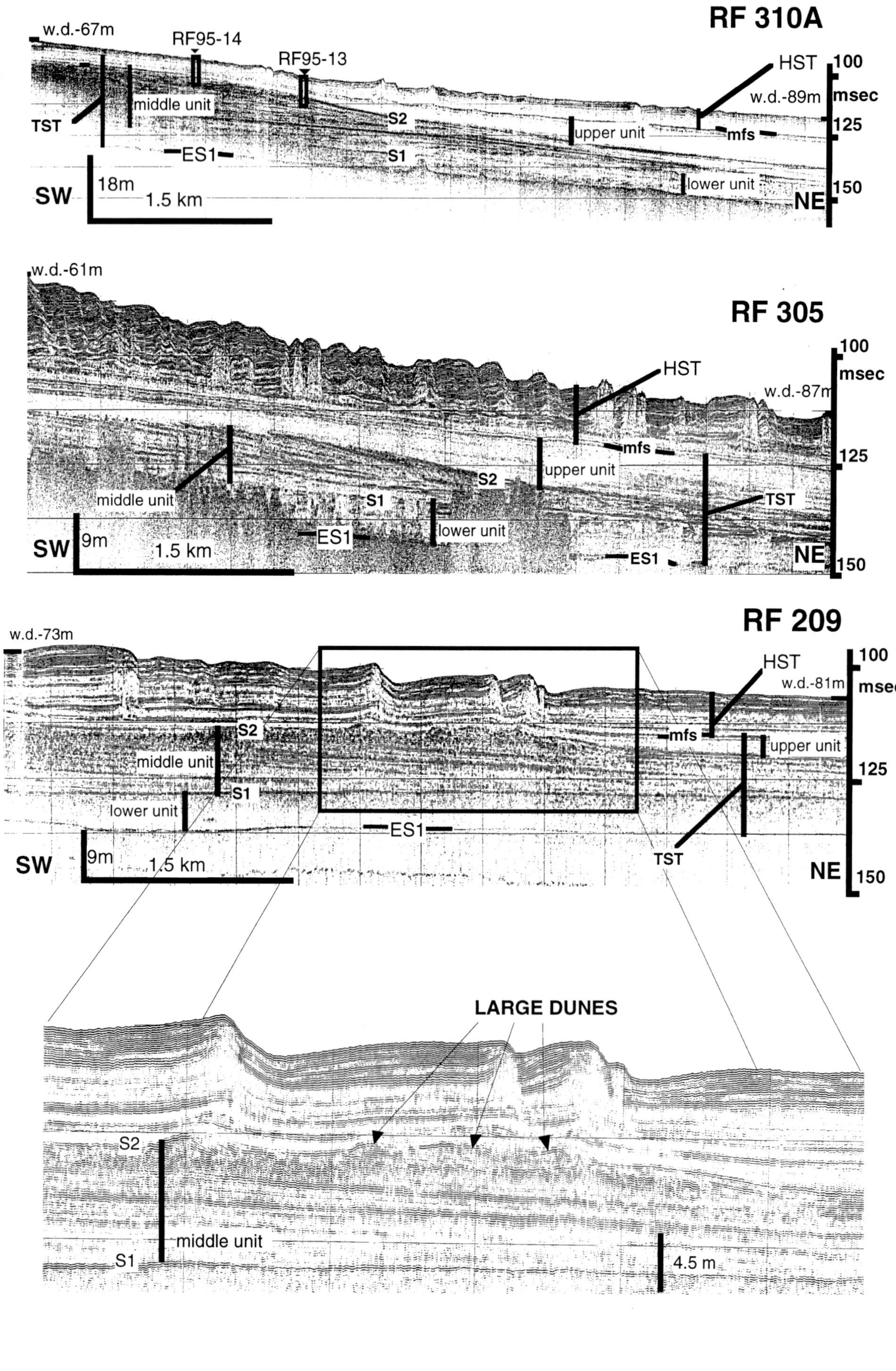

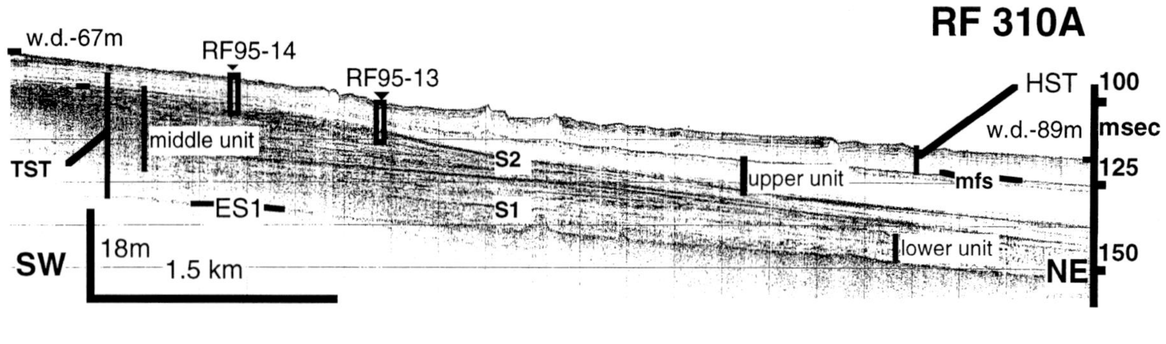
RF 310A
w.d.-67m
RF95-14
RF95-13
HST
100
msec
w.d.-89m
middle unit
TST
S2
upper unit
125
mfs
ES1
S1
18m
1.5 km
SW
lower unit
150
NE

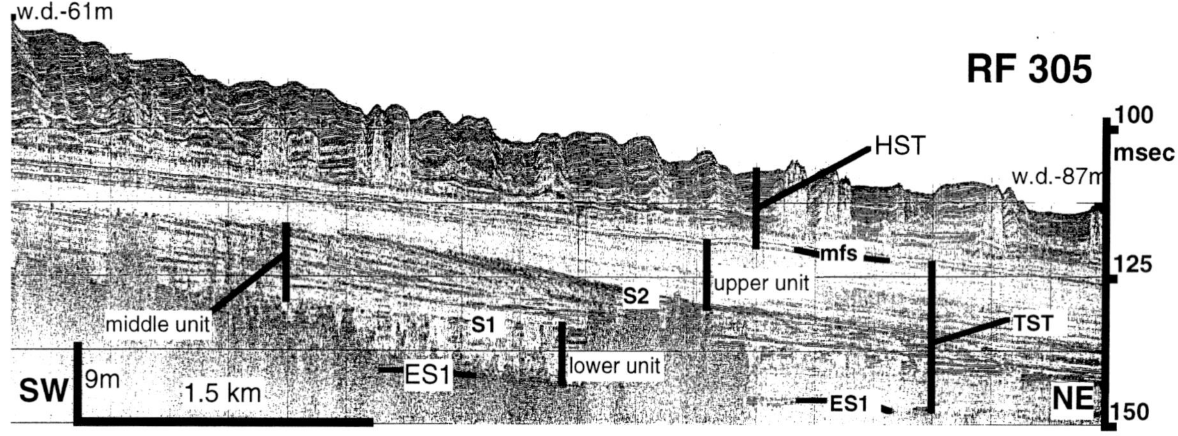
w.d.-61m
RF 305
100
msec
HST
w.d.-87m
mfs
125
upper unit
S2
TST
middle unit
S1
9m
1.5 km
SW
ES1
lower unit
ES1
NE
150

RF 209
w.d.-73m
100
HST
w.d.-81m
msec
S2
mfs
upper unit
middle unit
125
S1
lower unit
ES1
9m
1.5 km
SW
TST
NE
150
LARGE DUNES
S2
middle unit
S1
4.5 m

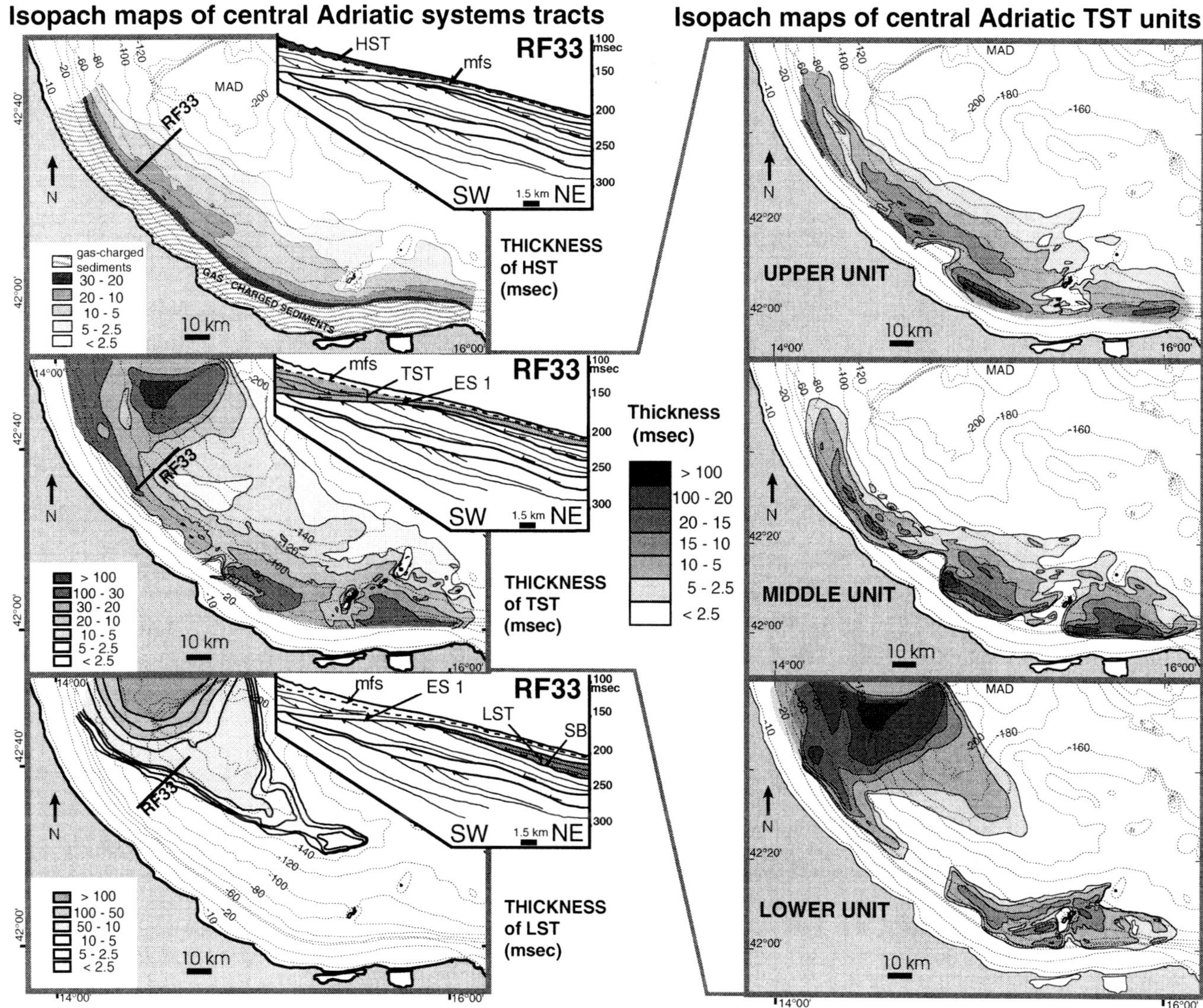

Fig. 5.—Isopach maps of the systems tracts that compose the late Quaternary depositional sequence in the central Adriatic basin (left). The LST depocenter reflects the location of the Po River north of the MAD during the last glacial maximum (bottom left); the HST depocenter reflects the dominant longshore dispersal of muddy sediments (upper left); the TST isopach map shows a combination of the above elements with a main depocenter north of the MAD and a secondary depocenter parallel to the coast (center left). By splitting the TST isopach map into those of its composing units (right), it is clear that the depocenters of both the middle (center right) and upper (upper right) TST units are elongated parallel to the coast. The depocenter of the lower TST unit (lower right) is more similar to that of the LST. All maps are expressed in msec two-way time. A simplified bathymetric contour is provided every 20 m.

←
Fig. 4.—Annotated 3.5-kHz reflection seismic profiles RF95-310A (with locations of sediment cores RF95-14 and RF95-13), RF95-305, and RF95-209 (see Fig. 2 for profile locations). The marine onlap of the upper TST unit is clear in two upper records seaward of the depocenter of the middle TST unit. The shingled reflectors of the middle TST unit are truncated at their tops (surface S2). Seismic profile RF209, south of the Tremiti High, shows more complex erosional/depositional features on top of the middle TST unit (reflector S2); in particular, the close-up shows asymmetric dunes at the top of laterally-continuous shingled reflectors of the middle unit.

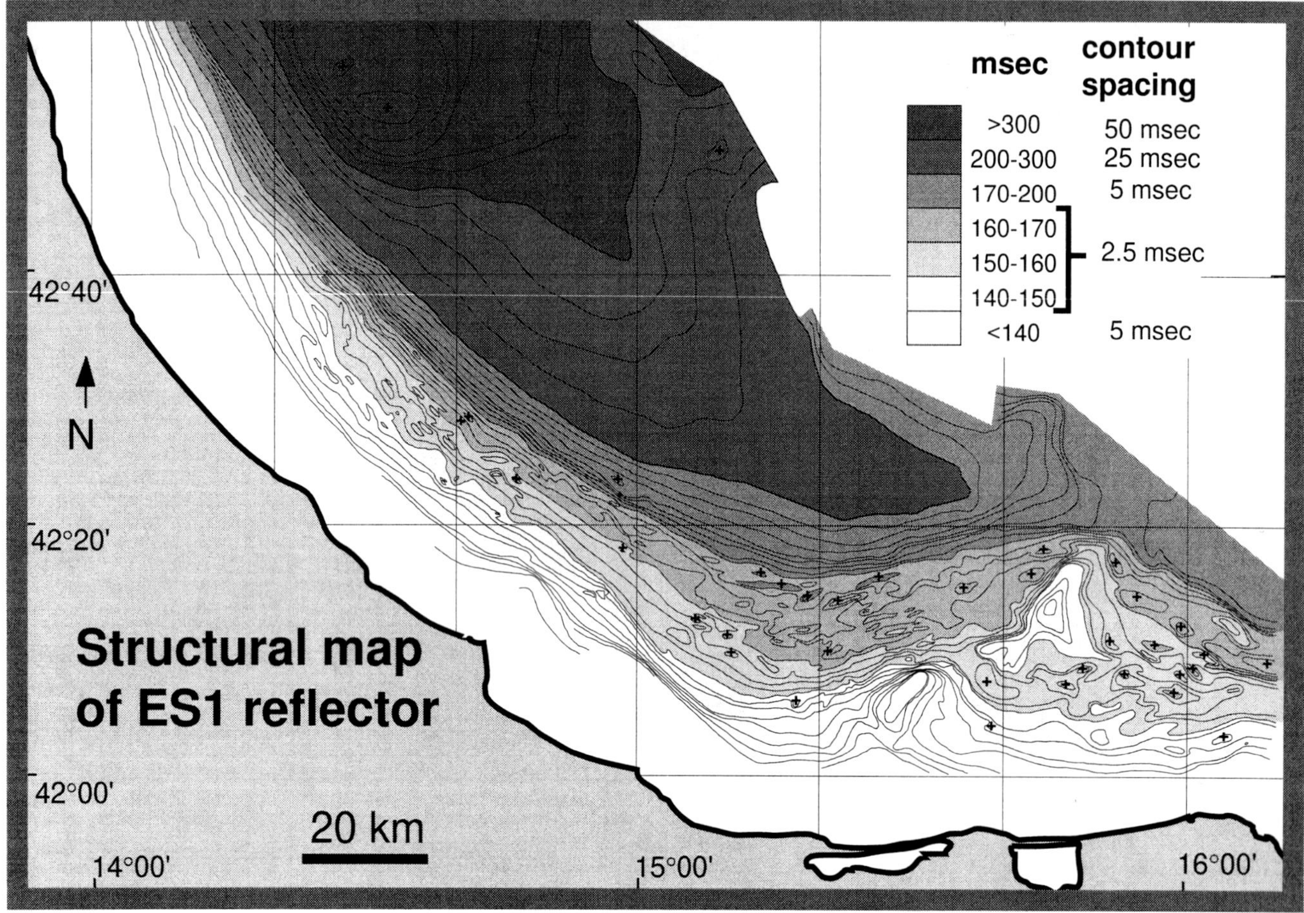

Fig. 6.—Structural map of the transgressive surface of the late Quaternary Depositional Sequence. This surface corresponds to seismic reflector ES1 and is characterized by a shore-parallel ridge-and-swale morphology of erosional origin; plus signs indicate relative lows on the topography of the surface. No evidence of subaerial exposure was found in the few sediment cores that reached this surface.

unit records the sea-level rise from its lowest position (ca. -120 m with respect to present sea-level) to ca -80 m, the middle TST unit from -80 m to -40 m, and the upper TST unit from -40 m to its present position reached at the time of maximum flooding (about 5,500 years BP). Therefore, the position attained by the shoreline at the end of the deposition of each of the three units of the TST was located respectively in the middle shelf, in the inner shelf, and landward of the modern shoreline (Fig. 5).

The Lower TST Unit

The lower TST unit, below reflector S1, is composed of laminated dark muds that are characterized by relatively high total organic carbon (TOC) fluxes and scarce faunal content (Langone et al., 1996; Cattaneo et al., 1997). This lithology dominates both the MAD basin and the outer shelf and slope areas. The unit records low-energy processes compared to the rest of the transgressive record and is influenced by freshwater input as testified by benthic faunas sampled in its upper part (*Ammonia tepida* and *A. perlucida*, Cattaneo et al., 1997); very scarce planktic forams were recovered from this unit on the shelf and in the MAD basin because of the dominance of fresh-water input, the still reduced water depth at time of deposition, and the poor connection to the rest of the Mediterranean. In the southern shelf area, around the Tremiti High, the lower TST unit is thicker in areas of greater erosion on the underlying transgressive surface (ES1). In this area, erosional features a few meters deep and few tens of meters wide incise the otherwise flat plane-parallel reflectors (Trincardi and Correggiari, 1998).

The Middle TST Unit

The middle unit of the late Quaternary TST is contained between reflectors S1 (at the base) and the S2 (at the top). In the central Adriatic basin, the middle TST unit consists of a set of low-angle, seaward-dipping reflectors that are truncated at their tops in water depths shallower than 100 m (reflector S2; Fig. 4). Compared to the other units, the middle TST unit is more variable laterally and includes coarser-grained sediment (Cattaneo et al., 1997). The unit backsteps with respect to the lower TST unit, showing that accommodation space was still being created during the deposition of this unit.

Three main depocenters correspond to progradational bodies with shingled medium to high amplitude reflections on the western coast of the basin (Fig. 5). The direction of progradation is oblique to the coast due to the interaction of the marine currents with structural highs on the shelf and complexieties in the coastline trend; along the direction of progradation, seismic facies and reflector geometries show a longshelf evolution to horizontal, faintly laminated acoustic returns with marine onlap against structural reliefs (Fig. 7).

The internal geometry of the middle TST unit is more complex near the Tremiti High, where the lap out termination of the middle TST unit against this relief changes systematically from west to east: in the western sector, the unit is several meters thick with subhorizontal and laterally continuous reflectors with marine onlap against the relief (Fig. 7, RF202). In the central area an erosional moat parallels the Tremiti High (Fig. 7, RF217). To the east, a steep erosional scour has a relief of 10-15 m (Fig. 7, RF211). Further east, where the Tremiti High loses its bathymetric expression, the erosional relief decreases progressively and is replaced by depositional mounds (Fig. 7, RF14; Fig. 8).

Mound deposits in the middle TST unit are clearly expressed on the modern bathymetry, although they are draped by younger deposits (Fig. 8). On seismic profiles these mounds are several kilometers wide with a relief of less than 10 m, gently convex in shape, and their internal geometry is characterized by repeated truncations and low-angle convergence of seismic reflectors toward areas of preferential erosion (Fig. 8). This geometry suggests that these mounds originated from alternating erosion and deposition caused by changes in intensity of bottom-hugging shelf currents. The tops of these mounds, even when erosional, do not show evidence of subaerial exposure in cores, such as roots or soil deposits. Based on the benthic faunas and seismic correlation with the inferred time-equivalent shoreline deposits, these mounds must have formed in water depths on the order of 20 to 40 m (Cattaneo et al., 1997).

The Upper TST Unit

The upper TST unit lies between surface S2 and mfs (Fig.4), and shows acoustically transparent to faintly laminated seismic facies with low amplitude and high-continuity reflectors. This unit is composed of marine mud with well-developed planktic faunas in the deep basin and the outer shelf (Asioli, 1996). In the central Adriatic Sea, the upper TST

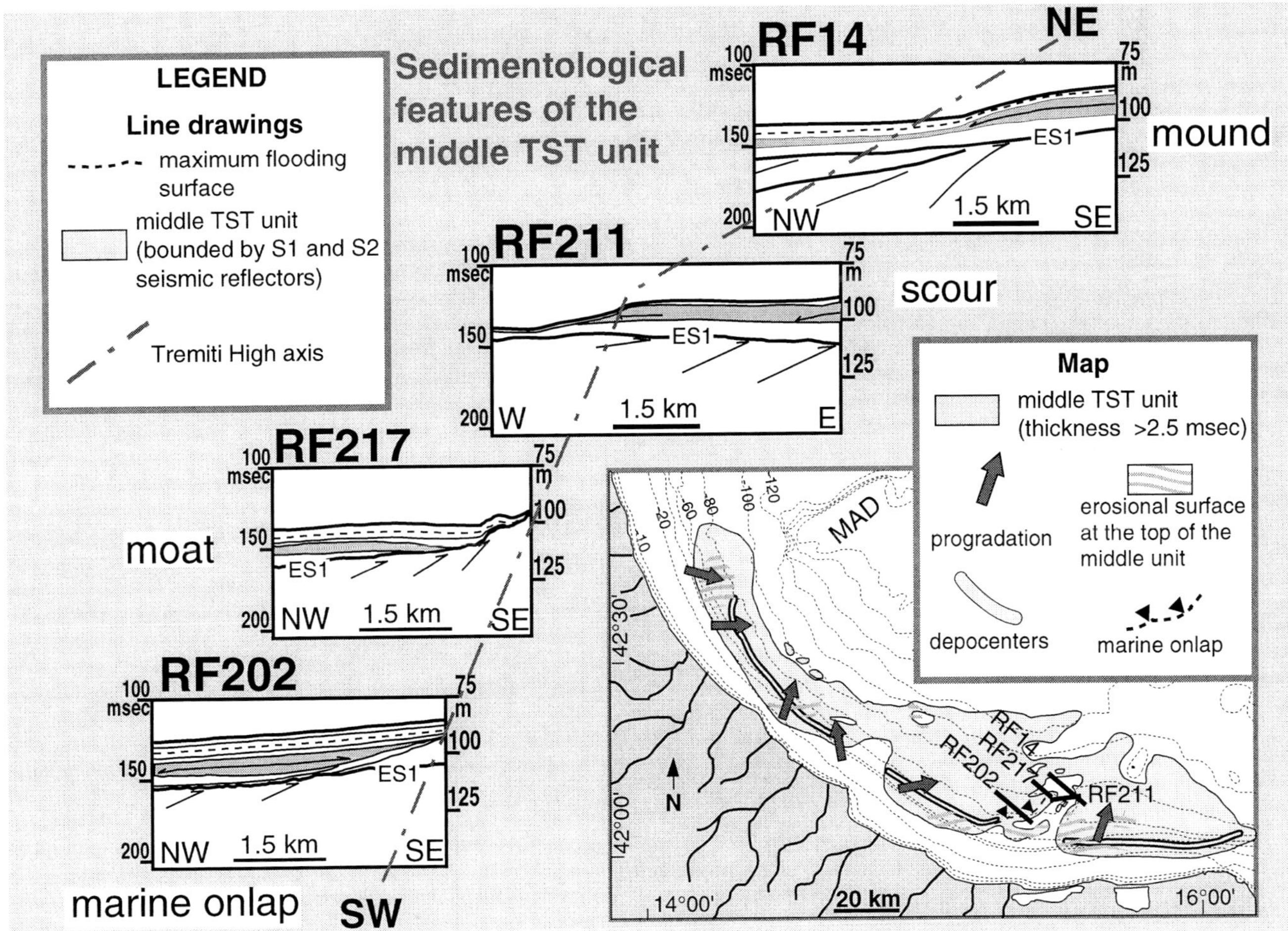

Fig. 7.—Line drawings showing the lateral changes in lap-out patterns of the middle TST unit (gray shading) against the Tremiti High (from Cattaneo et al., 1997): marine onlap (RF202); erosional moat (RF217); erosional scour (RF211); depositional mound (RF14). See text for explanation. The reference map shows the lateral relationships among these sedimentologic features of the middle TST unit.

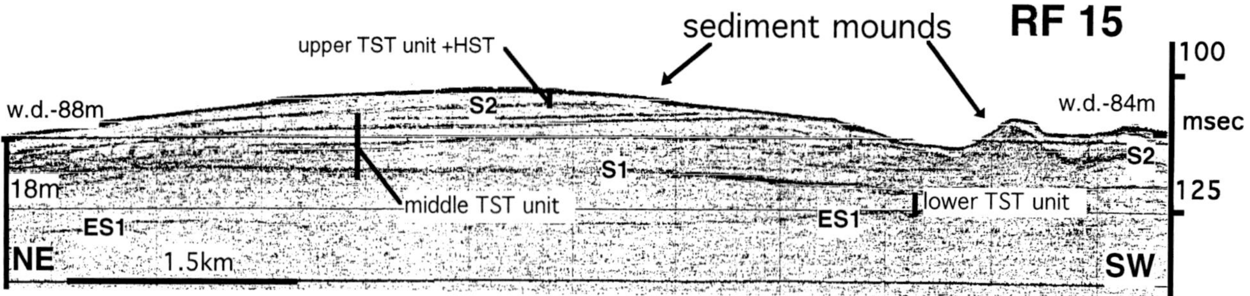

Fig. 8.—Seismic profile RF15 (3.5-kHz) showing the positions of S1 and S2 reflectors bounding the middle TST unit east of the Tremiti High. In this area the convergence and low-angle pinch out of seismic reflectors above S2 indicate that bottom currents may have been active during the deposition of the upper TST unit. In the rough area on the right portion of this record a coarse, pebbly lag coincides with reflector S2 .

unit shows main depocenters parallel to the coast and influenced by the presence of local tectonic highs; a secondary coast-parallel depocenter is elongated in a seaward position, where offshore muds onlap onto the depositional reliefs created by the underlying middle TST unit (Fig. 5).

Isopach Maps of the TST Units

The lower TST unit shows two distinct depocenters in the central Adriatic basin. The northern depocenter corresponds to an interval of progradation and aggradation that occurred during the earlier portion of the sea-level rise, when the rate of rise was slower (Fairbanks, 1989). At this time, the same fluvial apparatus that fed the LST were still capable of entirely filling the accommodation space created on the outer shelf, reaching the upper slope on the northwest flank of the MAD. The southern depocenter appears instead much thinner and is ponded in areas of bathymetric lows on the outer shelf (Fig. 5). In this area, the lower TST unit has a flat top and fills local lows and irregularities on the underlying erosional surface (ES1; Fig. 6) on the outer shelf. Isopach maps of the middle and upper TST units show a dominant shore-parallel elongation of depocenters (Fig. 5). This kind of thickness distribution clearly mimics that of the HST and hence may have resulted from an oceanographic dispersal regime similar to the modern one. In detail, each depocenter of the middle TST unit appears composite and shows a complex internal geometry, which indicates repeated erosional events and short-term changes in sediment flux (Fig. 4).

Sedimentologic Expression of TST Units and Chronostratigraphic Control

The stacking pattern of the late Quaternary transgressive record made it possible to selectively sample all of the TST units through standard coring techniques in complementary settings: deep basin (core CM92-43), outer shelf (PAL94-9, PAL94-8, RF95-11, RF95-12, RF95-4 and RF95-5), and inner shelf (RF95-13 and RF95-14). A scheme of the stratigraphic framework resulting from seven key cores (Fig. 9) shows that the lower TST unit is more than 14,800 calibrated years (Trincardi et al., 1996b) and therefore was deposited before the meltwater Ia defined in global $\delta^{18}O$ curves (Fairbanks, 1989; Bard et al., 1996). The middle TST unit records the late glacial interval including the Bølling/Allerød and Younger Dryas short-term climatic events, as indicated by pollen associations and ^{14}C dates (Fig. 3). This reconstruction is confirmed by physical correlation to the Younger Dryas interval defined by micropaleontologic ecozonations, $\delta^{18}O$ records, and pollen spectra in the deep-basin section (Asioli, 1996; Ariztegui et al., 1996; Lowe et al., 1996a, b). This shelf-basin correlation relies on the interpretation of high-resolution seismic profiles, tephrochronology, and sediment magnetic properties (Trincardi et al., 1996a, b).

The middle TST unit shows differentiated sedimentary facies: cores RF95-13 and RF95-14 (Fig. 9) penetrated the progradational depocenter of the middle TST unit on the inner-shelf and recovered alternating thin, sharp-based, silty layers in a muddy marine succession. Benthic foraminifera and shelf molluscs indicate that the cored unit was deposited in 20-40 m water depth (Cattaneo et al., 1997). One AMS ^{14}C determination yielded an age of 11,890 calibrated years within this succession, about 2 m below surface S2 (Trincardi et al., 1996a). Several dates from above reflector S2 confirm that this reflector is about 11,300 calibrated years BP (Asioli et al., 1996; Trincardi et al., 1996b). Cores PAL94-9 and PAL94-8 on the outer shelf reached below the distal equivalent of the middle TST unit to recover mud rich in organic carbon of the lower TST unit (Fig. 9). In both cores, the middle TST unit is replaced by a thin interval of sandy (PAL94-9) or bioclastic (PAL94-8) sediment with reworked, mixed forams (Fig. 9; Trincardi et al., 1996b). On the shelf north of the Tremiti High, the correlation can be traced through cores RF95-12 and RF95-11. The latter penetrated the entire TST above the transgressive surface ES1 (Trincardi and Correggiari, 1998).

Evidence of Shelf-wide Erosional Surfaces Within the Transgressive Record

A chronostratigraphic diagram summarizes the stratigraphic information from key cores in the central Adriatic basin (Fig. 10). Dated or interpolated ecozone limits, tephra layers, and erosional hiatuses are referred to time lines expressed in thousands of calibrated years. We report detailed subdivisons only for the late Quaternary TST although, on the southern shelf, several cores reached older units deposited during the falling stage of the late Quaternary sea-level cycle (Trincardi and Correggiari, 1998). The scheme shows the landward increase of the hiatus associated with subaerial exposure of the shelf and subsequent transgressive reworking (surface ES 1). The key surfaces S1 and S2, within the TST,

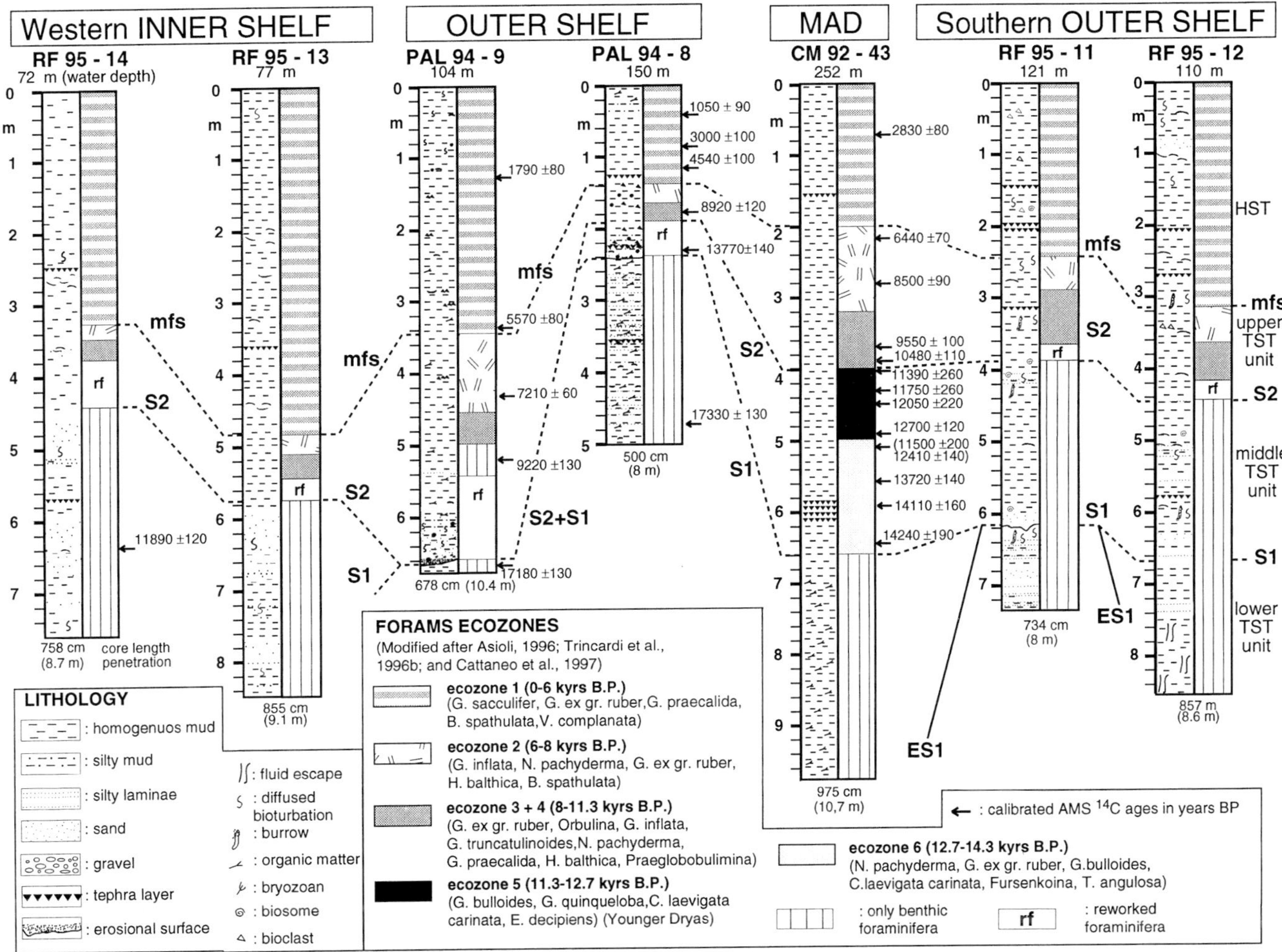

Fig. 9.—Stratigraphic correlation of sediment cores reaching the middle and lower TST units on the shelf and in the MAD (see Fig. 2 for core location). Core logs show lithology and ecozonations based on foraminifera assemblages (modified from Asioli, 1996). Note that planktic foraminifera are abundant in the middle unit only where waters were deeper (MAD). As a consequence, the upper limit of the interval characterized by lack of planktic forams (vertically ruled area) is time-transgressive and should not be used for correlation.

record short-lived intervals of laterally extensive erosion. Surface S1, at the top of the lower TST unit, marks an interval of rapid drowning and reworking of shallow marine deposits on the outer shelf and upper slope. Cores show contacts between marine muds above and below surface S1, separated by a lag of coarser sediment and reworked faunas (Trincardi et al., 1996b; Asioli, 1996).

All the shelf cores show that surface S2 corresponds to a reworked interval at the top of the middle TST unit. The amount of erosion associated with surface S2 is irregular along the margin. Erosional relief is highest between the Tremiti High and the Gargano promontory, where it is occasionally related to the occurrence of large bedforms that show high-angle internal clinoforms (Fig. 4, seismic line RF209). South of the Tremiti High, reflector S2 corresponds to a coarse, gravelly lag associated with reworked benthic forams from back-barrier, coastal, and inner-shelf environments (Cattaneo et al., 1997). Therefore, in this area, surface S2 is a ravinement surface, but seismic profiles show evidence of marine erosion down to 100 m deep; in these areas surface S2 must have a submarine erosional origin as expected in some models (Thorne and Swift, 1991).

Sediment Accumulation Rates in the MAD Basin

The stacking pattern and internal geometry of the three TST units on the Adriatic shelf clearly indicate an unsteady sediment flux during the late Quaternary sea-level rise. The erosional nature of the key surfaces on extensive shelf sectors also indirectly provides evidence of intervals of reduced sediment supply and increased marine reworking. In the basin, core CM92-43 provides the most continuous record for the late glacial and Holocene and is constrained by 16 AMS ^{14}C dates (Trincardi et al., 1996b; Langone et al., 1996). This core, which spans the last ca. 17,000 years, defines intervals of reduced or increased sediment flux associated with a number of short-term climatic and oceanographic changes during the late glacial and the Holocene (Fig. 11). Linear interpolation is made assuming constant sediment accumulation rates between successive dated

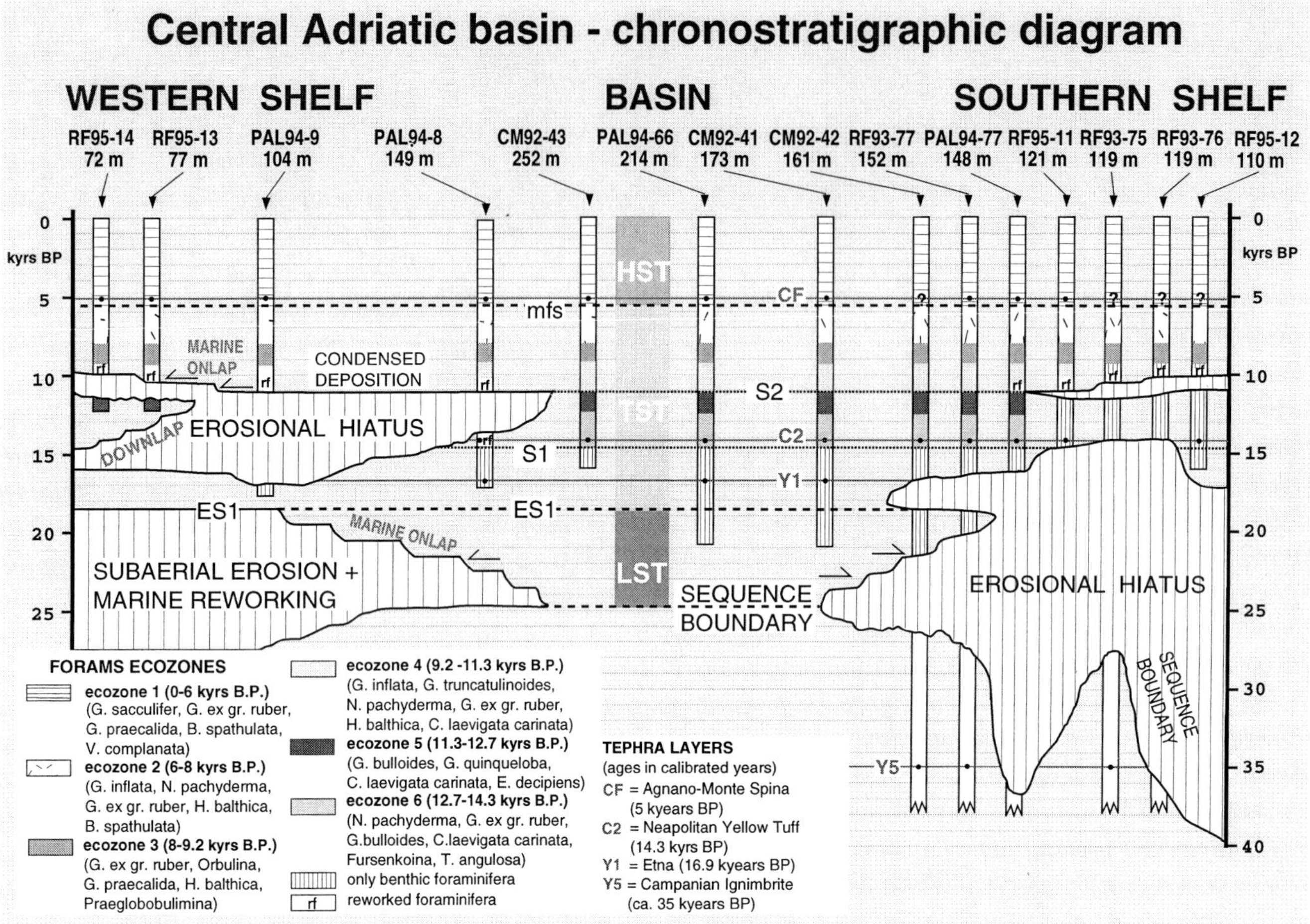

Fig. 10.—Chronostratigraphic correlation scheme defining major hiatuses within the late Quaternary deposits emplaced during sea-level low stand (LST), rise (TST), and high stand (HST) in the central Adriatic Sea. Core CM92-43, in the MAD basin, shows a continuous marine record of the last deglaciation. On the shelf, the late glacial record is far more complex because of the short-term pulses in sediment supply and changes in oceanographic regime discussed in the text.

points; this raw estimate shows that intervals of reduced sediment accumulation rates correspond to the distal equivalent of the key surfaces S1 and S2, respectively, at the base and top of the middle TST unit. Average sediment accumulation rates within each TST unit decrease overall from the lower unit of the TST to the HST. However, the deposition of the middle TST unit reflects a higher degree of variability in sediment flux compared to the other units. Maximum rates of sediment accumulation are documented in the upper part of the Younger Dryas and within the Bølling/Allerød warm intervals (Fig. 3, 11). Changes in dinocyst assemblages in the southern Adriatic also point to spikes of water runoff that are possibly related to changes in rain pattern or episodes of rapid melting of Alpine glaciers (Zonneveld, 1995).

Sediment Grain Size Analyses

We analyzed grains from the main lithologies within each of the TST units on the shelf and in particular from the coarser beds of the middle TST unit (Table 1; Fig. 12). Core CM92-43 in the MAD basin, described by 95 equally spaced samples, was used for reference (Guerzoni et al., 1996). Core RF95-12 reaches the lower TST unit and documents an overall fining-up trend in a continuous outer-shelf mud from the middle TST unit to the HST. In core RF95-14, between 302 cm and 402

Table 1.—Number of granulometric samples per stratigraphic unit in each shelf core.

seismic unit	core RF95-4	core RF95-5	core RF95-12	core RF95-14
HST	-	-	3	5
upper TST	1	-	1	10
middle TST	11	11	10	11
lower TST	-	-	4	-

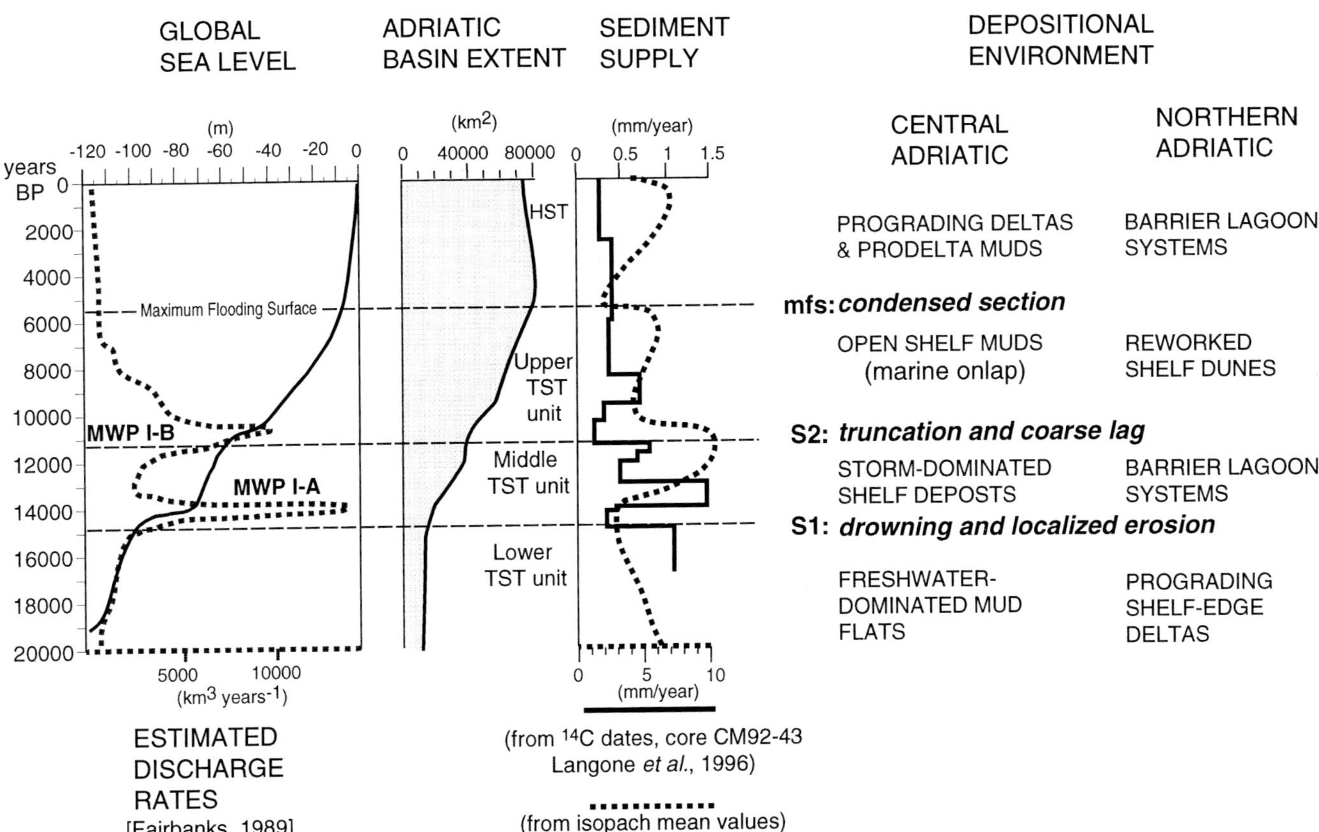

Fig. 11.—Qualitative diagram summarizing the interplay of sea-level (referred to the curve of Fairbanks, 1989), short-term changes in sediment flux to the margin, the areal extent of the basin, and the intensity of reworking and erosional processes. On the right are synthetic correlations of main depositional environments based on geometry of seismic reflectors and sedimentary facies in cores (capital letters), in the northern low-gradient and central-Adriatic high-gradient shelf sectors.

cm, the upper TST unit shows subtle changes in grain size. Samples in cores RF95-14, RF95-4, and RF95-5 define the coarser beds of the middle TST unit. These analyses help in comparing the coarser (silty to sandy) layers of the middle TST unit on the inner shelf in a progradational setting (core RF95-14) with those located more seaward within the sediment mound (cores RF95-4 and RF95-5).

In the basin, core CM92-43 includes the HST and most of the TST (Trincardi et al., 1996b, Asioli et al., 1996). The lower 3.55 m, below surface S1, consists of a dark laminated mud in which only benthic foraminifera are present (Asioli, 1996); this section corresponds to the lower TST unit, while the upper part of the core is composed of homogeneous mud. The fraction > 63 µm is around 1% of the total sediment, with a median sediment diameter of 3.8 mm and a mean of 5 mm in the lower TST unit and 3 µm in both the upper TST unit and the HST (Guerzoni et al., 1996). The main change in grain size, foraminifera ecozones, oxygen isotope, and sediment composition (calcite-dolomite ratio and clay mineral content) corresponds to surface S2 at 400 cm, which marks the boundary between the middle and upper TST units (Asioli, 1996; Calanchi et al., 1996). The mean clay mineral fraction (< 2 µm) is 25% with a range of 20-35% and an increasing trend toward the top (Guerzoni et al., 1996).

In the outer shelf north of the Tremiti High, core RF95-12 (Fig. 12) reached the lower TST unit in 110 m water depth. The lower TST unit is composed of muddy sediment with fluid escape structures (Trincardi and Correggiari, 1998); samples are silty clay with an average silt fraction of 47% by weight. The middle TST unit represents a peak in grain size with unimodal, poorly sorted, clayey silt and an increase of ~10% by weight of the silt fraction with respect to the underlying lower TST unit. An appreciable increase in grain size occurs within the middle TST unit from 625 cm to 525 cm (silt fraction from 52 to 66%), followed by a decrease upsection (silt fraction ~45%). Three samples in the HST indicate an upward increase in clay content, as observed in core CM92-43.

On the shelf north of the Tremiti High, cores RF95-4 and RF95-5 (Fig. 12) are in 104 and 99 m water depth, respectively, at the toe of a sediment mound of the middle TST unit. These cores show a lower section with alternating sharp-based sandy beds and marine muds that belong to the middle TST unit and a thin muddy record in the upper part that encompasses both the upper TST unit and the HST. The only moderately or well-sorted sand layers within the transgressive record were sampled in the middle TST unit of cores RF95-4 and RF95-5. Sand fraction values ranged from 56% to 98%. Sand layers show evidence of an upward increase in grain size and sorting when sampled at the base and top (Fig. 12).

On the inner shelf, core RF95-14 (Fig. 12) reached the middle TST unit and penetrated the shingled prograding reflectors below surface S2 in 72 m water depth. In the middle TST unit the coarser beds consist of sandy to clayey silt and

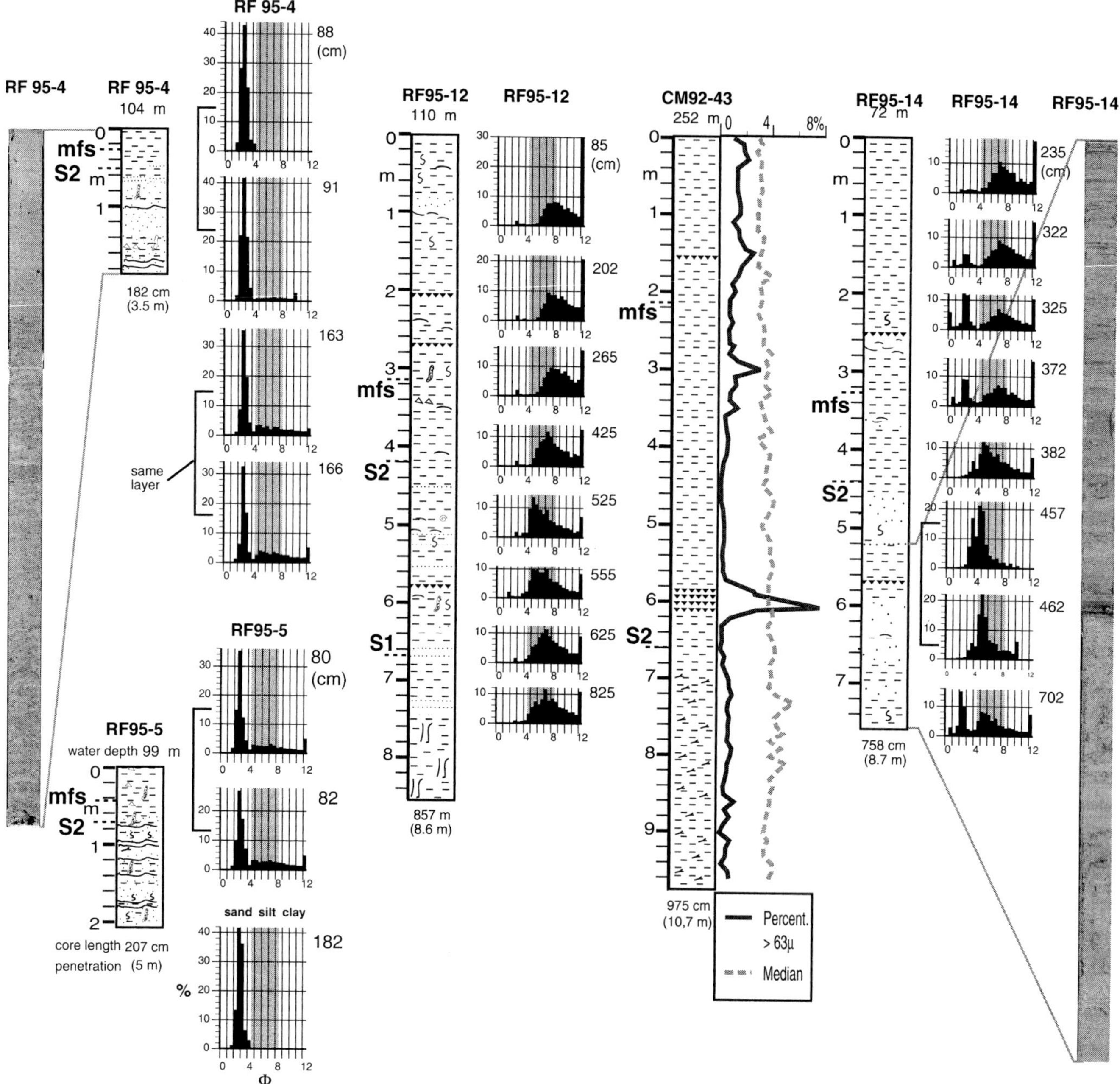

Fig. 12.—Lithology and grain-size analyses of cores RF95-4, RF95-5, RF95-12, CM92-43, and RF95-14. Sediment cores show changes in lithology within the TST record. Within the middle TST unit, in particular, cores RF95-4, RF95-5, and RF95-14 show sharp-based, thin, silty or sandy beds with high-angle erosional surfaces and mud drapes. Bioturbation, although present (large burrows), did not entirely destroy the physical structures. Small, asymmetric current ripples are occasionally detectable at core scale (6-cm diameter; Core RF95-4). The sand is lithic; marine molluscs and forams are present but not abundant. Core CM92-43 in the MAD basin is shown for reference.

are very poorly sorted; the sand fraction ranges from 10% to 35%. Silty beds show better sorting at the top; in a bed sampled at the base (457-459 cm) and at the top (462-463 cm) sand ranges from 9% to 25%, silt from 75% to 70%, and clay from 18% to 7%. In the upper TST unit, we noted an appreciable bimodal distribution of grain size, very poor sorting, and a sudden decrease in silt content (from ~60% below 382 cm to ~30% in the section above 372 cm). There is a concurrent increase in clay content from the base of upper TST unit to HST. The five HST samples in core RF95-14 include a sand fraction up to 8% because of the proximal location of this core.

Changes in sediment grain size of the coarser beds of the inner shelf (core RF95-14) and the southern shelf near the Tremiti High (cores RF95-4 and RF95-5) reflect differences in location and caliber of sediment sources and/or in the intensity of the oceanographic processes that disperse sediments. Samples collected closer to the coast (RF95-14) directly reflect changes in sediment source; samples taken more seaward record the effect of winnowing induced by bottom currents that were possibly amplified by flow restriction near structural reliefs. This is consistent with the observation of upward increase in grain size and better sorting documented in single layers.

CORRELATION OF THE LATE QUATERNARY TRANSGRESSIVE RECORD OF THE REST OF THE ADRIATIC BASIN

In this section we briefly extend the stratigraphic subdivisions of the transgressive record to the remaining part of the Adriatic semienclosed basin, based on published stratigraphic and sedimentologic data (Trincardi et al., 1994; Correggiari et al., 1996a, b). A significant landward shift in the location of sediment entry points is evident due to the reworking of the topset region of the underlying lower unit in areas previously characterized by a progradational configuration on the northern flank of the MAD (Fig. 11); this trend is accompanied by a deepening depositional environment testified by faunal assemblages (Asioli, 1996; Cattaneo et al., 1997).

The middle TST unit records the deposition and partial preservation of barrier-lagoon-estuary systems in the low-gradient shelf to the north in waters between 80 and 45 m (Trincardi et al., 1994; Correggiari et al., 1996b). These systems do not show time-equivalent units composed of offshore mud on the northern shelf (Figs. 11, 13). In the south, at the same time, accommodation was greater and time-equivalent offshore mud accumulated on the outer shelf, with enhanced erosional/depositional responses (marine onlaps, moats, depositional shelf mounds) due to flow restriction of shelf currents in the vicinity of drowned structural highs (Fig. 7).

Key surface S2 at the top of the middle TST unit marks an interval of intense and widespread submarine erosion (Figs. 11, 13). Above this surface, coastal systems similar to those of the underlying unit were deposited in water depths less than 40 m in the north Adriatic. Drowned coastal systems show the effects of marine reworking with increasing intensity upward (Correggiari et al., 1996a). During deposition of the upper TST unit, the horizontal translation induced by sea-level rise brought the coastline several tens of kilometers inland of its modern position. The upper TST unit displays a marked facies partitioning at basin scale with drowned coastal sand bodies reworked into large bedforms on the northern shelf; also present are accumulations of coeval offshore mud that show evidence of marine onlap on the midshelf in the central Adriatic (Figs. 11, 13).

DISCUSSION

Three main elements characterize the late Quaternary TST record in the Adriatic basin: 1) the stacking pattern of TST

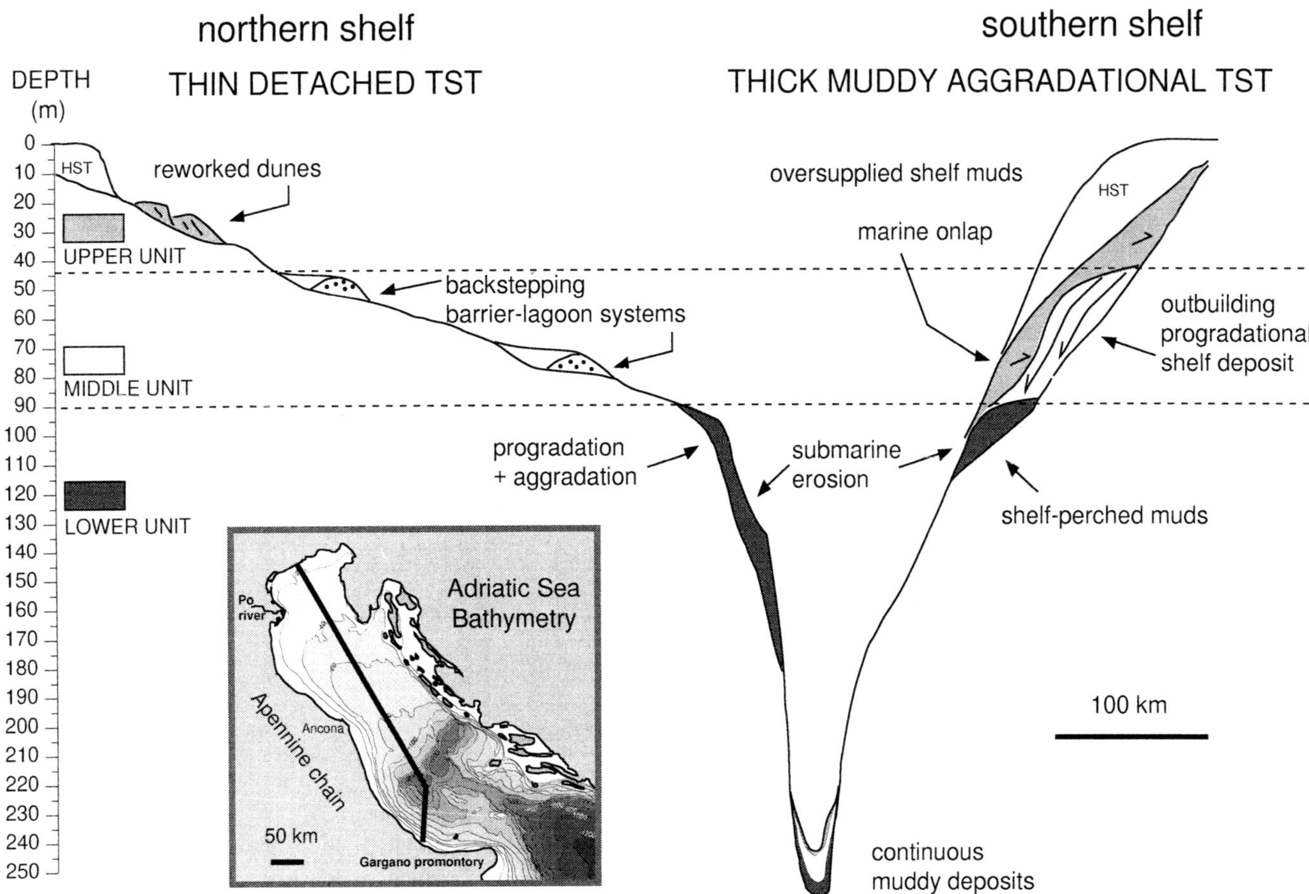

Fig. 13.—Simplified correlation scheme of transgressive deposits in the northern and central Adriatic shelves, respectively, based on published data and new data discussed in this paper. Facies partitioning is evident along the margin where barrier lagoon systems are the only TST records in the North and muddy shoreface to offshore deposits dominate to the south.

units, which points to an interval of increased sediment flux accompanied by a concurrent increase in grain size during the deposition of the middle TST unit; 2) the age of the middle TST unit, between 14,800 and 11,300 calibrated years, which encompasses an interval of extreme and short-term climate instability; and 3) the occurrence of facies partitioning along the basin axis (starved barrier-lagoon systems in the north passing into prograding shoreface deposits and muddy shelf systems in the south) and the elongation of isopach trends of the middle and upper TST units due to the longshore component in sediment dispersal. The role of preferential sediment routing toward the southeast is also supported by the distal marine onlap against the relief of the Tremiti High as well as by the location of depositional mounds on the "lee side" of main structural reliefs (Fig. 7). In this section we discuss the above evidence by comparing the Adriatic record with other late Quaternary systems and with sequence stratigraphic concepts originally developed for ancient deposits.

TST Internal Architecture and Supply Fluctuations

Sediment supply and sea level exert a dual control on the development of depositional sequences, although in most cases it is exceedingly difficult to disentangle these factors, and the choice of sea-level change as a primary control is therefore difficult to prove (Boyd et al., 1989; Schlager, 1993). The importance of changes in sediment supply has been addressed by several authors studying third-order depositional sequences (Galloway, 1989; Thorne and Swift, 1991; Posamentier and Allen, 1993; Schlager, 1993). Cant (1995) criticized the tendency to assume that sediment supply is invariable and adequate to fill all accommodation space in any depositional environment. Data from modern continental margins indicate that even during an interval dominated by high-rate and high-amplitude sea-level changes, sediment supply can alter the geometry of systems tracts (Morton and Price, 1987) or even outpace the eustatic rise (Kolla and Perlmutter, 1993). An additional example of the importance of supply flux in shaping depositional sequences comes from the distal reaches of the Bengal Fan, where turbidite deposition accompanied a significant portion of the late Quaternary sea-level rise and resulted from climatic forcing (Weber et al., 1997). In particular, increased precipitation and intensified monsoonal activity in the Himalayan area occurred during two humid pulses at ca. 12,500 and 10,000 ^{14}C years BP and coincided with times of rapid rise of global sea level (Weber et al., 1997). On the Adriatic shelf, intervals of rapid sea-level rise instead resulted in the reworking and drowning of extensive shelf sectors, and peak sediment fluxes characterized the middle TST unit that records a prolonged interval of extreme climatic instability (Asioli, 1996; Langone et al., 1996; Lowe et al., 1996a, b; Cattaneo et al., 1997). The complex geometry of the progradational middle TST unit signals significant change in supply regime compared to that of the underlying and overlying TST units.

Evidence of complex transgressive shelf records comes from other margins of the Mediterranean region. On the middle shelf of the Northern Alboran Sea, a progradational unit within the late Quaternary TST is capped by an erosional surface and occurs between two intervals of more condensed deposition in 60-m water depth (Hernandez-Molina et al., 1996). This unit was tentatively atttributed to a minor sea-level fall induced by the Younger Dryas cold interval (Hernandez-Molina et al., 1996). The Rôhne shelf is another example of composite late Quaternary TST, where backstepping parasequences include prograding shelf deposits that were emplaced during intervals of increased supply (Gensous et al., 1993).

Climate, Fluvial Evolution and Sediment Flux

Changes in fluvial pattern, runoff, amount, and type of sediment transported depend on a number of driving forces, like tectonism and climate, together with sea-level fluctuations (affecting mainly the lower trunks of rivers; Blum et al., 1994), and intrinsic mechanisms (presence of local thresholds, confluences, captures; Wescott, 1993). In fact, rather than just incising or aggrading, fluvial systems can adjust their equilibrium profiles during base-level changes through more complex responses such as modifying stream pattern, sinuosity, and channel geometry (Shumm, 1993).

The role of climate control on fluvial systems is clear even during intervals of large base-level changes such as those observed during the Quaternary (Blum et al., 1994). Changes in climate, hydrography, and soil vegetation cover within drainage basins are the major factors that control fluctuations in sediment supply. Lowland areas, where the typical longitudinal gradients of the rivers are on the order of some tens of centimeters per kilometer, are ideal sites for monitoring the effects of short-term climatic changes on river behavior (Huisink, 1997). At Maas River (Netherlands), a gradual transiton from braided to meandering occurred during the late glacial, followed by a temporary shift to braided conditions during the Younger Dryas period (Huisink, 1997). The return to braided river conditions could depend on a decrease in vegetation cover accompanied by the onset of a more irregular discharge regime. According to Vandenberghe (1995), at the scale of a full glacial-interglacial cycle (100,000 years, fourth order) river development follows climatic evolution; if shorter time scales can be resolved (100 years), the delay in vegetation response to short-term climate change causes a lag in sedimentary response. In general, major episodes of sediment discharge correspond to intervals of increased climatic variability (Vandenberghe, 1995).

For all these reasons, climate change during the deposition of the late Quaternary middle TST unit in the Adriatic may represent a key control on sediment supply to the shelf and basin. During meltwater pulses Ia and Ib (Fairbanks, 1989), the rates of base-level rise outpaced sediment flux to the basin, and river bedload was trapped in aggrading alluvial plains. Channel aggradation and flood-plain construction are likely at the origin of the drop in sediment supply observed on the shelf (Fig. 11). The occurrence of short-term climatic instability during the deposition of the middle TST unit which culminated in the Younger Dryas cold episode, may have determined a reduction in soil vegetation cover, increased denudation rates in mountain areas and a seaward shift of neutral point, expanding the erosional domain even during sea-level still stands. A decrease in vegetation cover enhanced soil erosion relative to melt-water times, allowing the outbuilding of the middle TST unit on the shelf and the increase in accumulation rates in the MAD slope basin (Fig. 11).

Long-shore Sediment Transport

Accommodation space defines the limits of potential sedimentation (Jervey, 1988; Thorne, 1992). A common assumption using the accommodation concept is that no deposition is expected in a given area until all the available space toward the source has been filled completely (Thorne, 1992; Cant, 1995). Shelf currents can reduce the effective accommodation; examples of the effects of such currents come from several modern pericontinental margins where bedform fields may even form on the outer shelf (Flemming, 1980; Ikehara and Kinoshita, 1994). In semienclosed basins the direct impact of oceanic currents is lower, but the combined action of storm events and cyclonic circulation affects areas of complex physiography where flow restriction may occur (Alexander et al., 1991). Under these conditions the effects of such currents should be observed within the architectural arrangement of HST and TST deposits.

The interaction of shelf currents and storm flows on a shelf in a microtidal setting was able to transport sand on the western Mediterranean shelf and deposit submarine transgressive sand ridges (Diaz and Maldonado, 1990). The distribution of the late Quaternary TST in the Gulf of Papua documents the role of an ocean current in dispersing sediment during the transgression (Harris et al., 1996). Evidence of a clockwise geostrophic system comes from budget calculations (inferred sediment supply versus TST thickness), the elongation of the TST depocenter, and the displacement of transgressive deltaic lobes with respect to the axis of the underlying incised valleys (Harris et al., 1996). On the Rhône outer shelf, high-resolution seismic reflection data show complex muddy deposits elongated parallel to the regional bathymetric contour (Gensous et al., 1993). The acoustically transparent to faintly laminated uppermost unit of the late Quaternary TST on the Rhône shelf records the deposition of muddy deposits parallel to the regional contour, which indicates that longshore transport plays a role in distributing silty clay sediments on this outer shelf setting (Gensous et al., 1993).

The overall stacking pattern and the complex internal geometry of the three late Quaternary TST units in the central Adriatic basin record sedimentary processes that occurred on two complementary time scales: on a shorter scale, storm action resulted in thin sharp-based sandy beds and mud drapes, likely deposited in a lower-shoreface/offshore transition environment (Figs. 9, 11); on a longer scale, seismic geometry and isopach maps are the time-averaged expressions of a shore-parallel current similar to that observed in the modern circulation. The suite of erosional/depositional geometries within the middle TST unit further indicates a complex interaction of marine current and morphologic reliefs (Figs. 7, 8, and 11). This kind of dispersal was not active during the deposition of the LST and lower TST and was likely triggered by basin widening (Trincardi et al., 1994) and sill drowning to the south (Asioli et al., 1996).

During the deposition of the middle and upper units of the late Quaternary TST in the Adriatic, shelf dynamics resulted in a southeastward and offshore sediment transport, which incompletely exploited the space available more landward. This interpretation is supported by the geometry of large-scale sedimentary bodies, the spatial distribution of their depocenters, and the internal lapout patterns rather than by sedimentologic paleocurrent indicators (Leckie and Krystinick, 1989; Hart et al., 1990; Duke, 1990). Changes in oceanographic regime similar to those observed in the Adriatic basin may have been relevant also in ancient stratigraphic records even if controlled by rapid accommodation-dominated cyclicity, although straightforward comparisons among modern and ancient shelf deposits is hampered by the different investigation techniques and the diverse scales of observation. Within ancient shelf successions, in particular, features like the moats, erosional scours, marine onlap, and depositional mounds recognized in the late Quaternary record of the Adriatic may be difficult to recognize. The lack of lateral continuity among TST deposits on low-gradient shelf affected by shore-parallel facies partitioning makes it difficult to determine whether or not shelf deposits are coeval, unless exceptional outcrop control is available (Cross, 1988). The possibility of demonstrating that sedimentary bodies not connected are coeval is poorer, expecially if sediment partitioning occurs.

CONCLUSIONS

The late Quaternary transgression occurred in a very brief interval of about 12,000 years. Despite the short duration of this interval, high-resolution seismic- and chrono-stratigraphic data from the central Adriatic shelf document the complexity of the late Quaternary TST. Several controlling factors are at the origin of the complexity in stacking pattern of the transgressive units and their lateral changes in facies association and depositional environment: 1) the stepwise nature of the sea-level rise; 2) the short-term climatic instability that resulted in fluctuations of sediment flux to the shelf; 3) the contrasting basin morphology and gradients of the shelf north and southwest of the MAD basin; and 4) the transgressive basin widening and consequent change in oceanographic regime.

The stepwise nature of the sea-level rise defined intervals of increased or reduced rates of accommodation space creation on the shelf (Anderson and Thomas, 1991; Thomas and Anderson, 1994); on the Adriatic shelf, the deposition of progradational coastal deposits during intervals of slower rise was separated by intervals of drowning and partial reworking during intervals of increased sea-level rise (Trincardi et al., 1994; Correggiari et al., 1996a).

The unsteady sediment flux to the basin resulted in an out-of-sequence stacking pattern of units within the TST. Biostratigraphic and chronologic control indicates that a peak in sediment flux to the shelf and basin records an interval during which climatic instability was extreme (between 14,800 and 11,300 calibrated years). The eustatic curve includes intervals of slower rise compared to the preceding and following intervals (Fairbanks, 1989).

The low-gradient of the north Adriatic shelf caused maximized landward shifts of shoreline facies for any given increment of sea-level rise. Sedimentation was not able to fill all the laterally extensive accommodation space created by the sea-level rise. Both the middle and upper TST units are expressed, in this area, by sets of drowned and partially preserved barrier lagoon systems, similar to the modern coastal lagoon systems that characterize the HST north of the modern Po delta. These undersupplied coastal systems are of equal age as the offshore-mud units stacked on the narrower and steeper shelf to the south. This change in transgressive facies along the strike of the basin is an example of lateral facies partitioning (Cross, 1988; Cant, 1995). This evidence, accompanied by the coast-parallel elongation of isopach trends and the occurrence of complex erosional/depositional features around structural highs on the southern shelf, implies a component of longshore sediment dispersal and the lateral advection of fine-grained sediments by shelf currents during the deposition of the middle and upper TST units in a fashion similar to that affecting the modern HST. The occurrence of marine onlap on outer shelf regions confirms that the preferential routing of shelf muds to this region occurred through submarine processes.

The transgressive widening of the shelf, accompanied by the drowning of the southern sill and the landward shift of fresh-water entry points, brought about the onset of the modern oceanographic regime. This onset occurred during deposition of the middle TST unit after the Pelagosa sill to the south drowned and the basin grew to a critical size. The role played by shelf currents (either thermohaline or storm-driven) in shaping the architecture of transgressive units is most evident around drowned structural highs (e.g., the Tremiti High), where a suite of erosional/depositional features developed in response to flow restriction and sediment redistribution.

ACKNOWLEDGMENTS

The stratigraphic data used for this work was derived from PALICLAS, a project of the European Union (EU5V-CT93-0267) that investigated paleoenvironmental change in the Mediterranean region. Biostratigraphic data, in particular, were kindly provided by and discussed with Alessandra Asioli whom we thank greatly. We thank Alessandra Asioli, Katherine Bergman, Bruce Power, and Gary Yeo for their careful review of the manuscript. We also acknowledge the help of our IGM colleagues during several cruises onboard the Urania research vessel. This is IGM contribution number 1113.

REFERENCES

ALEXANDER, C.R., DE MASTER, D.J. AND NITTROUER, C.A., 1991, Sediment accumulation in a modern epicontinental-shelf setting: The Yellow Sea: Marine Geology, v. 98, 51-72.

ANDERSON, J.B. AND THOMAS, M.A., 1991, Marine ice-sheet decoupling as a mechanism for rapid, episodic sea-level change: The record of such events and their influence on sedimentation: Sedimentary Geology, v. 70, 87-104.

ANDERSON, J.B., ABDULLAH, K., SARZALEJO, S., SIRINGAN, F. AND THOMAS, M.A., 1995, Late Quaternary sedimentation and high-resolution sequence stratigraphy of the Texas shelf, *in* de Batist, M. and Jacobs, P., eds., Geology of Siliciclastic Shelf Seas: London, Geological Society Special Publication 117, 155-169.

ARIZTEGUI, D., CHONDROGIANNI, C., WOLFF, G., ASIOLI, A., TERANES, J., BERNASCONI, S.M. AND MCKENZIE, J.A., 1996, Paleotemperature and paleosalinity history of the Meso Adriatic Depression (MAD) during the Late Quaternary: A stable isotopes and alkenones study, *in* Guilizzoni, P. and Oldfield, F., eds., Palaeoenvironmental Analysis of Italian Crater Lake and Adriatic Sediments (PALICLAS): Verbania Pallanza, Memorie Istituto Italiano di Idrobiologia 55, 219-230.

ARTEGIANI, A., PASCHINI, E., RUSSO, A., BREGANT, D., RAICICH, F. AND PINARDI, N., 1997, The Adriatic Sea general circulation. Part II: Baroclinic circulation structure: Journal of Physical Oceanography, v. 27, 1492-1514.

ASIOLI, A., 1996, High resolution foraminifera biostratigraphy in the central Adriatic basin during the last deglaciation: A contribution to the PALICLAS project, *in* Guilizzoni, P. and Oldfield, F., eds., Palaeoenvironmental Analysis of Italian Crater Lake and Adriatic Sediments (PALICLAS): Verbania Pallanza, Memorie Istituto Italiano di Idrobiologia 55, 197-217.

ASIOLI, A., TRINCARDI, F., CORREGGIARI, A., LANGONE, L., VIGLIOTTI, L., VAN DER KAARS, S. AND LOWE, J.J., 1996, The late Quaternary deglaciation in the Central Adriatic basin: Il Quaternario Italian Journal of Quaternary Sciences, v. 9, 763-770.

BARD, E., HAMELIN, B., ARNOLD, M., MONTAGGIONI, L., CABIOCH, G., FAURE, G. AND ROUGERIE, F., 1996, Deglacial sea-level record from Tahiti corals and the timing of global meltwater discharge: Nature, v. 382, 241-244.

BARTOLINI, C., CAPUTO, R. AND PIERI, M., 1996, Pliocene-Quaternary sedimentation in the Northern Apennine Foredeep and related denudation: Geological Magazine, v. 133, 244-273.

BLUM, M.D., TOOMEY, R.S. AND VALASTRO, S. JR., 1994, Fluvial response to Late Quaternary climatic and environmental change, Edward Plateau, Texas: Palaeoceanography, Palaeoclimatology, Palaeoecology, v. 108, 1-21.

BONDESAN, M., CASTIGLIONI, G.B., ELMI, C., GABBIANELLI, G., MAROCCO, R., PIRAZZOLI, P.A. AND TOMASIN, A., 1995, Coastal areas at risk from storm surges and sea-level rise in northeastern Italy: Journal of Coastal Research, v. 11, 1354-1379.

BOYD, R., SUTER, J. AND PENLAND, S., 1989, Relation of sequence stratigraphy to modern sedimentary environments: Geology, v. 17, 926-929.

CALANCHI, N., DINELLI, E., LUCCHINI, F. AND MORDENTI, A., 1996, "Paliclas" multi-proxy investigation of paleoenvironmental history of Central Italy: chemostratigraphy of late Quaternary sediments from Lago Albano and Central Adriatic Sea cores, *in* Guilizzoni, P. and Oldfield, F., eds., Palaeoenvironmental Analysis of Italian Crater Lake and Adriatic Sediments (PALICLAS): Verbania Pallanza, Memorie Istituto Italiano di Idrobiologia 55, 247-263.

CANT, D.J., 1995, Sequence stratigraphic analysis of individual depositional successions: Effects of marine/nonmarine sediment partitioning and longitudinal sediment transport, Mannville Group, Alberta Foreland Basin, Canada: American Association of Petroleum Geologists Bulletin, v. 79, 749-762.

CATTANEO, A., TRINCARDI, F. AND ASIOLI, A., 1997, Shelf sediment dispersal in the late Quaternary transgressive record around the Tremiti High (Adriatic sea): Giornale di Geologia, v. 59, 217-244.

CORREGGIARI, A., ROVERI, M. AND TRINCARDI, F., 1992, Regressioni "forzate," regressioni "deposizionali" e fenomeni di instabilita' in unita' progradazionali tardo-quaternarie: Giornale di Geologia Bologna, v. 54, 19-36.

CORREGGIARI, A., FIELD, M.E. AND TRINCARDI, F., 1996a, Late Quaternary transgressive large dunes on the sediment-starved Adriatic shelf, *in* de Batist, M. and Jacobs, P., eds., Geology of Siliciclastic Shelf Seas: London, Geological Society Special Publication 117, 155-169.

CORREGGIARI, A., ROVERI, M. AND TRINCARDI, F., 1996b, Late Pleistocene and Holocene evolution on the north Adriatic sea: Il Quaternario Italian Journal of Quaternary Sciences, v. 9, 697-704.

CROSS, T.A., 1988, Control on coal distribution in transgresive-regressive cycles, Upper Cretaceous, Western Interior, U.S.A., *in* Wilgus, C.K., Hastings, B.S., Kendall, C.G. St. C., Posamentier, H.W., Ross, C.A., Van Wagoner, J.C., eds., Sea Level Change: An Integrated Approach, Tulsa: Society for Sedimentary Geology (SEPM) Special Publication 42, 371-380.

DIAZ, J.I. AND MALDONADO, A., 1990, Transgressive Sand Bodies on the Maresme Continental Shelf, Western Mediterranean Sea: Marine Geology, v. 91, 53-72.

DUKE, W.L., 1990, Geostrophic circulation or shallow marine turbidity currents? The dilemma of paleoflow patterns in storm-influenced prograding shoreline systems: Journal of Sedimentary Petrology, v. 60, 870-883.

FAIRBANKS, R.G., 1989, A 17,000 year glacio-eustatic sea level record: Influence of glacial melting rates on the Younger Dryas event and deep-ocean circulation: Nature, v. 342, 637-642.

FLEMMING, B.W., 1980, Sand transport and bedform patterns on the continental shelf between Durban and Port Elisabeth (Southeast African continental margin): Sedimentary Geology, v. 26, 179-205.

FRANCO, P., JEFTIC, L., MALANOTTE-RIZZOLI, P., MICHELATO, A. AND ORLIC, M., 1982, Descriptive model of the North Adriatic: Oceanologica Acta, v. 5, 379-389.

GALLOWAY, W.E., 1989, Genetic stratigraphic sequences in basin analysis: Architecture and genesis of flooding surface bounded depositional units: American Association of Petroleum Geologists Bulletin, v. 73, 125-142.

GENSOUS, B., WILLIAMSON, D. AND TESSON, M., 1993, Late Quaternary transgressive and highstand deposits of a deltaic shelf (Rhône delta, France) *in* Posamentier, H.W., Summerhayes, C.P., Haq, B.U., Allen, G.P., eds., Sequence stratigraphy and facies associations: Oxford, International Association of Sedimentology Special Publication 18, 197-211.

GUERZONI, S., PORTARO, R., TRINCARDI, F., MOLINAROLI, E., LANGONE, L., CORREGGIARI, A., VIGLIOTTI, L., PISTOLATO, M., DE FALCO, G. AND BOCCINI, V., 1996, Statistical analyses of grain-size, geochemical and mineralogical data in core CM92-43, Central Adriatic basin, *in* Guilizzoni, P. and Oldfield, F., eds., Palaeoenvironmental Analysis of Italian Crater Lake and Adriatic Sediments (PALICLAS): Verbania Pallanza, Memorie Istituto Italiano di Idrobiologia 55, 231-246.

HARRIS, P.T., PATTIRATCHI, C.B., KEENE, J.B., DALRYMPLE, R.W., GARDNER, J.V., BAKER, E.K., COLE, A.R., MITCHELL, D., GIBBS, P. AND SCHROEDER, W.W., 1996, Late Quaternary deltaic and carbonate sedimentation in the Gulf of Papua foreland basin: Response to sea-level change: Journal of Sedimentary Research, v. 66, 801-819.

HART, B.S., VANTFOORT, R.M. AND PLINT, A.G., 1990, Is there evidence for geostrophic curents preserved in the sedimentary record of inner to middle-shelf deposits? – Discussion: Journal of Sedimentary Petrology, v.60, 633-635.

HERNANDEZ-MOLINA, F.J., SOMOZA, L. AND REY, J., 1996, Late Pleistocene-Holocene high-resolution sequence analysis on the Alboran Sea continental shelf, *in* de Batist, M. and Jacobs, P., eds., Geology of Siliciclastic Shelf Seas: London, Geological Society Special Publication 117, 1139-154.

HUISINK, M., 1997, Late glacial sedimentological and morphological changes in a lowland river in response to climatic change: The Maas, southern Netherlands: Journal of Quaternary Science, v. 12, 209-233.

IKEHARA, K. AND KINOSHITA, Y., 1994, Distribution and origin of subaqueous dunes on the shelf of Japan: Marine Geology, v. 120, 75-87.

JERVEY, M.T., 1988, Quantitative Geological Modelling of Siliciclastic Rock Sequences and their Seismic Expression, *in* Wilgus, C.K., Hastings, B.S., Kendall, C.G. St. C., Posamentier, H.W., Ross, C.A., Van Wagoner, J.C., eds., Sea Level Change: An Integrated Approach, Tulsa: Society for Sedimentary Geology (SEPM) Special Publication 42, 47-69.

JORISSEN, F.J., ASIOLI, A., BORSETTI, A.M., CAPOTONDI, L., DE VISSER, J.P., HILGEN, F.J., ROHLING, E.J., VAN DER BORG, K., VERGNAUD GRAZZINI, C. AND ZACHARIASSE, W.J., 1993, Late quaternary central Mediterranean biochronology: Marine Micropaleontology, v. 21, 169-189.

KOLLA, V. AND PERLMUTTER, M.A., 1993, Timing of turbidite sedimentation on the Mississippi fan: American Association of Petroleum Geologists Bulletin, v. 77, 1129-1141.

LANGONE, L., ASIOLI, A., CORREGGIARI, A. AND TRINCARDI, F., 1996, Age-depth modelling through the late Quaternary deposits of the central Adriatic basin, *in* Guilizzoni, P. and Oldfield, F., eds., Palaeoenvironmental Analysis of Italian Crater Lake and Adriatic Sediments (PALICLAS): Verbania Pallanza, Memorie Istituto Italiano di Idrobiologia 55, 177-196.

LECKIE, D.A. AND KRYSTINIK, L.F., 1989, Is there evidence for geostrophic curents preserved in the sedimentary record of inner to middle-shelf deposits?: Journal of Sedimentary Petrology, v. 59, 862-870.

LOWE, J.J., ACCORSI, C.A., ASIOLI, A., VAN DER KAARS, S. AND TRINCARDI, F., 1996a, Pollen-stratigraphical records of the last glacial-interglacial transition (ca. 14-9 ka BP) from Italy: A contribution to the PALICLAS project: Il Quaternario Italian Journal of Quaternary Sciences, v. 9, 627-642.

LOWE, J.J., ACCORSI, C.A., BANDINI MAZZANTI, M., BISHOP, A., VAN DER KAARS, S., FORLANI, L., MERCURI, A.M., RIVALENTI, C., TORRI, P. AND WATSON, C., 1996b, Pollen stratigraphy of sediment sequences from lakes Albano and Nemi (near Rome) and from the central Adriatic, spanning the interval from oxygen isotope Stage 2 to the present day, *in* Guilizzoni, P. and Oldfield, F., eds., Palaeoenvironmental Analysis of Italian Crater Lake and Adriatic Sediments (PALICLAS): Verbania Pallanza, Memorie Istituto Italiano di Idrobiologia 55, 71-98.

MALANOTTE RIZZOLI, P. AND BERGAMASCO, A., 1983, The dynamics of the coastal region of the northern Adriatic Sea: Journal of Physical Oceanography, v. 13, 1105-1130.

MILLIMAN, J.D. AND SYVITSKI, J.M.P., 1992, Geomorphic/tectonic control of sediment discharge to the ocean: the importance of small mountainous rivers: Journal of Geology, v. 100, 525-544.

MORTON, R.A. AND PRICE, W.A., 1987, Late Quaternary sea-level fluctuations and sedimentary phases of the Texas coastal plain and shelf, *in* Nummendal, D., Pilkey, O.H., Howard, J.D., eds., Sea-level fluctuation and coastal evolution, Tulsa: Society for Sedimentary Geology (SEPM) Special Publication 41, 181-198.

ORLIC, M., GACIC, M. AND LA VIOLETTE, P.E., 1992, The currents and circulation of the Adriatic Sea: Oceanologica Acta, v. 15, 109-124.

PASCHINI, E., ARTEGIANI, A. AND PINARDI, N., 1993, The mesoscale eddy field of the Middle Adriatic Sea during fall 1988: Deep-Sea Research, v. 20, 1365-1377.

PENLAND, S., BOYD, R. AND SUTER, J.R., 1988, Transgressive depositional systems of the Mississippi delta plain: A model for barrier shoreline and shelf sand development: Journal of Sedimentary Petrology, v. 58, 932-949.

POSAMENTIER, H.W. AND ALLEN G.P., 1993, Variability of the sequence stratigraphic model: Effects of local basin factors: Sedimentary Geology, v. 86, 91-109.

SCHLAGER, W., 1993, Accomodation and supply—a dual control on stratigraphic sequences: Sedimentary Geology, v. 86, 111-136.

SHUMM, S.A., 1993, Fluvial response to baselevel change: Implication for sequence stratigraphy: Journal of Geology, v. 101, 279-294.

SNEDDEN, J.W., TILLMAN, R.W., KRESIA, R.D., SCHWELLER, W.J., CULVER, S.J. AND WINN, R.D., 1994, Stratigraphy of a modern shoreface-attached sand ridge, Pehala ridge, New Jersey: Journal of Sedimentary Research, v. 64, 560-581.

STENNI, B., NICHETTO, P., BREGANT, D., SCARAZZATO, P. AND LONGINELLI, A., 1995, The $\delta^{18}O$ signal of the northward flow of Mediterranean waters in the Adriatic Sea: Oceanologica Acta, v. 18, 319-328.

STUIVER, M. AND REIMER, P.J., 1993, Extended ^{14}C data base and revised CALIB 3.0 ^{14}C age calibration program: Radiocarbon, v. 35, 215-230.

THOMAS, M.A. AND ANDERSON, J.B., 1994, Sea-level controls on the facies architecture of the Trinity/Sabine incised-valley system, Texas, *in* Darlymple, R.W., Boyd, R. and Zatlin, B.A., eds., Incised-valley systems: Origin and sedimentary sequences, Tulsa: Society for Sedimentary Geology (SEPM) Special Publication 51, 63-82.

THORNE, J.A., 1992, An Analysis of the Implicit Assumptions of the Methodology of Seismic Sequence Stratigraphy: American Association of Petroleum Geologists Memoire 53, 375-396.

THORNE, J.A. AND SWIFT ,D.J.P., 1991, Sedimentation on continental margins, VI: A regime model for depositional sequences, their component systems tracts, and bounding surfaces, *in* Swift, D.J.P., Oertel, G.F., Tillman, R.W. and Thorne, J.A., eds., Shelf Sand and Sandstone Bodies: Oxford, International Association of Sedimentology Special Publication 14, 189-225.

TRINCARDI, F. AND FIELD, M.E., 1991, Geometry, lateral variability, and preservation of downlapped regressive shelf deposits: Eastern Tyrrhenian margin, Italy: Journal of Sedimentary Petrology, v. 49, 775-790.

TRINCARDI, F., ASIOLI, A., CATTANEO, A., CORREGGIARI, A, VIGLIOTTI, L. AND ACCORSI, C.A., 1996a, Transgressive offshore deposits on the central Adriatic shelf: Architecture complexity and the record of the Younger Dryas short-term event: Il Quaternario Italian Journal of Quaternary Sciences, v. 9, 753-762.

TRINCARDI, F., CATTANEO, A., ASIOLI, A., CORREGGIARI, A. AND LANGONE, L., 1996b, Stratigraphy of the late Quaternary deposits in the central Adriatic basin and the record of short-term climatic events, *in* Guilizzoni, P. and Oldfield, F., eds., Palaeoenvironmental Analysis of Italian Crater Lake and Adriatic Sediments (PALICLAS): Verbania Pallanza, Memorie Istituto Italiano di Idrobiologia 55, 39-70.

TRINCARDI, F. AND CORREGGIARI, A., 1998, Quaternary forced-regression deposits in the Adriatic basin and the record of composite sea-level cycles, *in* D. Hunt, R., Gawthorpe, eds., Sedimentary Responses to Forced Regressions: Geological Society Special Publication 140, 118-134.

TRINCARDI, F., CORREGGIARI, A. AND ROVERI, M., 1994, Late Quaternary transgressive erosion and deposition in a modern epicontinental shelf: The Adriatic Semienclosed Basin: Geo-Marine Letters, v. 14, 41-51.

VANDENBERGHE, J., 1995, Timescales, climate and river development: Quaternary Science Reviews, v. 14, 631-638.

WEBER, M.E., WIEDICKE, M.H., KUDRASS, H.R., HUBSHER, C. AND ERLENKEUSER, H., 1997, Active growth of the Bengal Fan during sea-level rise and highstand: Geology, v. 25, 315-318.

WESCOTT, W.A., 1993, Geomorphic thresholds and complex response of fluvial systems - Some implications for sequence stratigraphy: American Association of Petroleum Geologists Bulletin, v. 77, 1208-1218.

ZONNEVELD, K.A.F., 1995, Paleoclimatic and palaeo-ecological changes during the last deglaciation in the Eastern Mediterranean; implications for dinoflagellate ecology: Review of Paleobotany and Palynology, v. 84, 221-253.

AN EXPANDED MODEL FOR MODERN SHELF SAND RIDGE GENESIS AND EVOLUTION ON THE NEW JERSEY ATLANTIC SHELF*

JOHN W. SNEDDEN
Mobil Exploration and Production Technical Center, P.O. Box 650232, Dallas, Texas 75265-0232 U.S.A.
RONALD D. KREISA
Mobil Exploration and Production Technical Center, P.O. Box 819047, Dallas, Texas 75281-9047 U.S.A.
RODERICK W. TILLMAN
Consultant, 2121 East 51st St., Suite 112, Tulsa, Oklahoma 74105 U.S.A.
STEPHEN J. CULVER
Department of Paleontology, The Natural History Museum, Cromwell Rd., London SW7 5BD, U.K.
AND
WILLIAM J. SCHWELLER
Chevron Oil Field Research Co., 1300 Beach Blvd., La Habra, California 90631 U.S.A.

ABSTRACT: The stepwise genesis and evolution of modern shelf sand ridges are investigated through chronostratigraphic analysis of four separate study sites on the New Jersey Atlantic shelf, ranging in depth from less than 4 meters (shoreface-attached ridge) to over 45 meters (detached ridge). Radiocarbon age-dated vibracores and high-resolution seismic surveys facilitated construction of a series of chronostratigraphic cross-sections of these 1 by 5 km-scale and greater ridges, which can range in vertical thickness to 10 meters. These data provide compelling support to an earlier morphodynamic model suggesting that shoreface-attached ridges may originate from ebb-tidal deltas and eventually detach from the shoreline during coastal transgression. A key element of the morphodynamic model involves cutting of the adjacent swale by an obliquely migrating tidal-inlet channel.

However, notable yet transitional differences exist between shoreface, nearshore, and offshore ridges in terms of age, microfaunal content, bathymetric profile, and cross-sectional area. This implies that ridges change considerably following coastal detachment. During this final phase, which we term ridge evolution, the ridges may migrate, change orientations, and possibly cannibalize earlier ridge and inlet-channel-fill complexes. This stage may be the most important in terms of what is preserved in the sedimentary record. Understanding the dynamic nature of these long-lived, transgressive shelf sand ridges may help resolve some of the debate regarding analogous ancient, stratigraphically-isolated marine sand bodies.

INTRODUCTION

Many of the world's continental shelves today are veneered by a series of bathymetric highs referred to as "shelf sand ridges". This ridge and swale topography is most common on wide, low sediment-supply shelves, covering the area from the shoreline (shoreface-attached) to detached sand bodies in water depths of 40 meters or more (Swift et al., 1972; Fig. 1). Oceanographic studies demonstrate that these sand ridges are not moribund, but are in dynamic equilibrium with the modern shelf wave and current flow regime (Gadd et al., 1978; Swift and Field, 1981). Present thinking regarding the origin of these bathymetric features generally falls into two categories: 1) those favoring formation at or near present-day water depths by modern dynamic processes (Swift et al., 1984; Boczar-Karakiewicz and Bona, 1991; Rine et al., 1991; Antia, 1994); and 2) those who propose reworking of earlier barrier islands (McClellen, 1973a; Stubblefield et al., 1984). This debate parallels the controversy regarding ancient shelf sand bodies, which are variously interpreted as having formed in-place as mid-shelf sand ridges (Tillman and Martinsen, 1984) or as lowstand shoreface deposits which were later transgressed (Bergman, 1994).

To address these questions and to develop a predictive model for the subsurface occurrence of these reservoir-scale sand bodies, a large-scale research project was formulated among several oil companies and academic institutions. Ridges in four separate sites on the New Jersey Atlantic shelf, ranging from 4 m to over 40 meters water depth, were thoroughly analyzed with vibracores, boxcores, grab samples and high resolution seismic data (Fig. 1). Two sites, Area 1B (which includes a single shoreface attached ridge) and Area 3, (a single mid-shelf ridge) yielded particularly rich datasets. The approach used here was to analyze stratigraphic units in terms of their sedimentology, radiocarbon ages, seismic geometry, and stratal relationships, thus establishing the chronostratigraphy. This time stratigraphic framework facilitated comparison between the four study sites and thus new insight into the evolutionary pathway defined by these long-lived, transgressive sand bodies. Building upon earlier morphodynamic models (Figueiredo, 1984; McBride and Moslow, 1991), we discuss ridge genesis and ridge-shoreline detachment stages and then present a new phase emphasizing post-detachment evolution of sand ridges.

MODERN SAND RIDGE GENESIS

Analysis of the sedimentary character and age relationships of stratigraphic units in Area 1B (Fig. 2) provides the basic model for ridge genesis. Key chronostratigraphic units observed in vibracores in Area 1B include modern upper ridge sand (0 to 0.9 ka), lower ridge sand (0.9 to 3.1 ka), swale/inlet fill (0.36 to 2.9 ka), and back-barrier/lagoon (3.7 to 5.9 ka; Figs. 3, 6A). Detailed sedimentological description of these units is given in Snedden et al. (1994). The occurrence of back-barrier/lagoonal deposits implies that a barrier island existed in the area of present-day Peahala ridge as late as 3.5 ky. This barrier island probably resembled those of the present day New Jersey coastline: microtidal, wave-dominated barriers

* based upon paper which won the SEPM Excellence in Oral Presentation Award, 1995 AAPG-SEPM Annual Meeting, Houston, Texas.

Isolated Shallow Marine Sand Bodies: Sequence Stratigraphic Analysis and Sedimentologic Interpretation.
SEPM Special Publication No. 64, Copyright © 1999
SEPM (Society for Sedimentary Geology), ISBN 1-56576-057-3, p. 147-163.

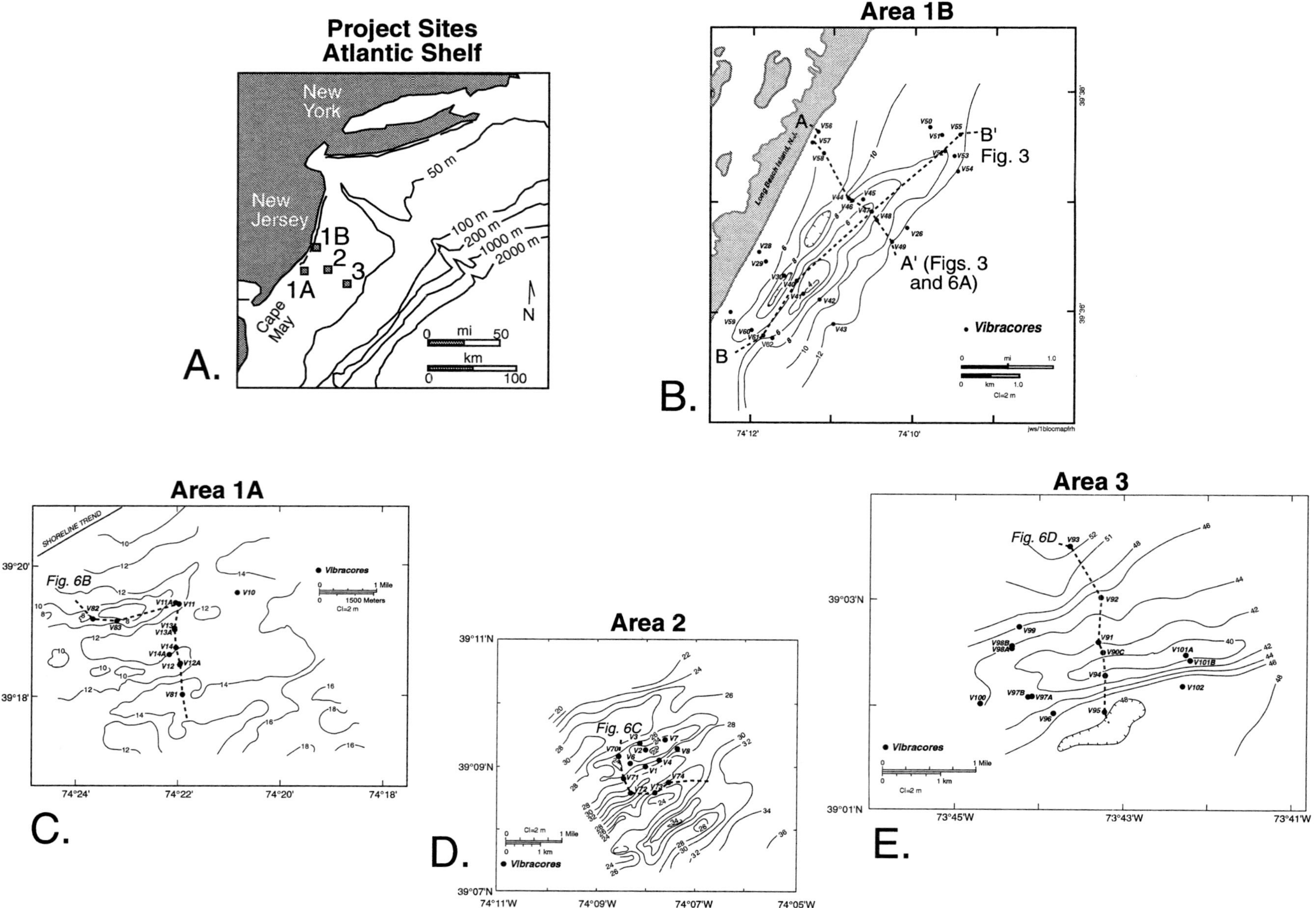

Fig. 1.—Location and bathymetry of project sites, with vibracores and chronostratigraphic cross-sections indicated. A) New Jersey shelf; B) Area 1B; C) Area 1A; D) Area 2; E) Area 3.

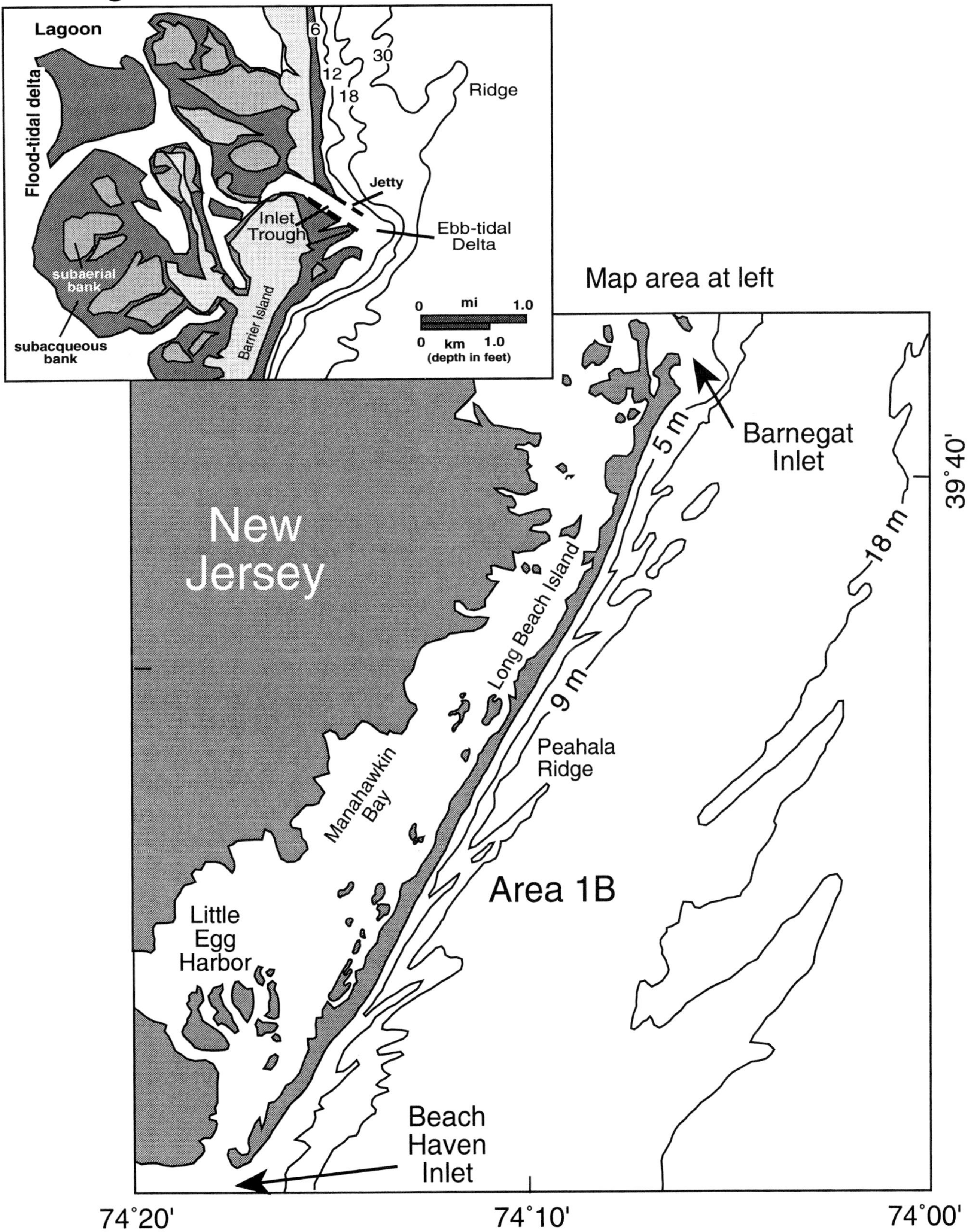

Fig. 2.—Location of Area 1B and adjacent localities, including Barnegat Inlet (upper inset).

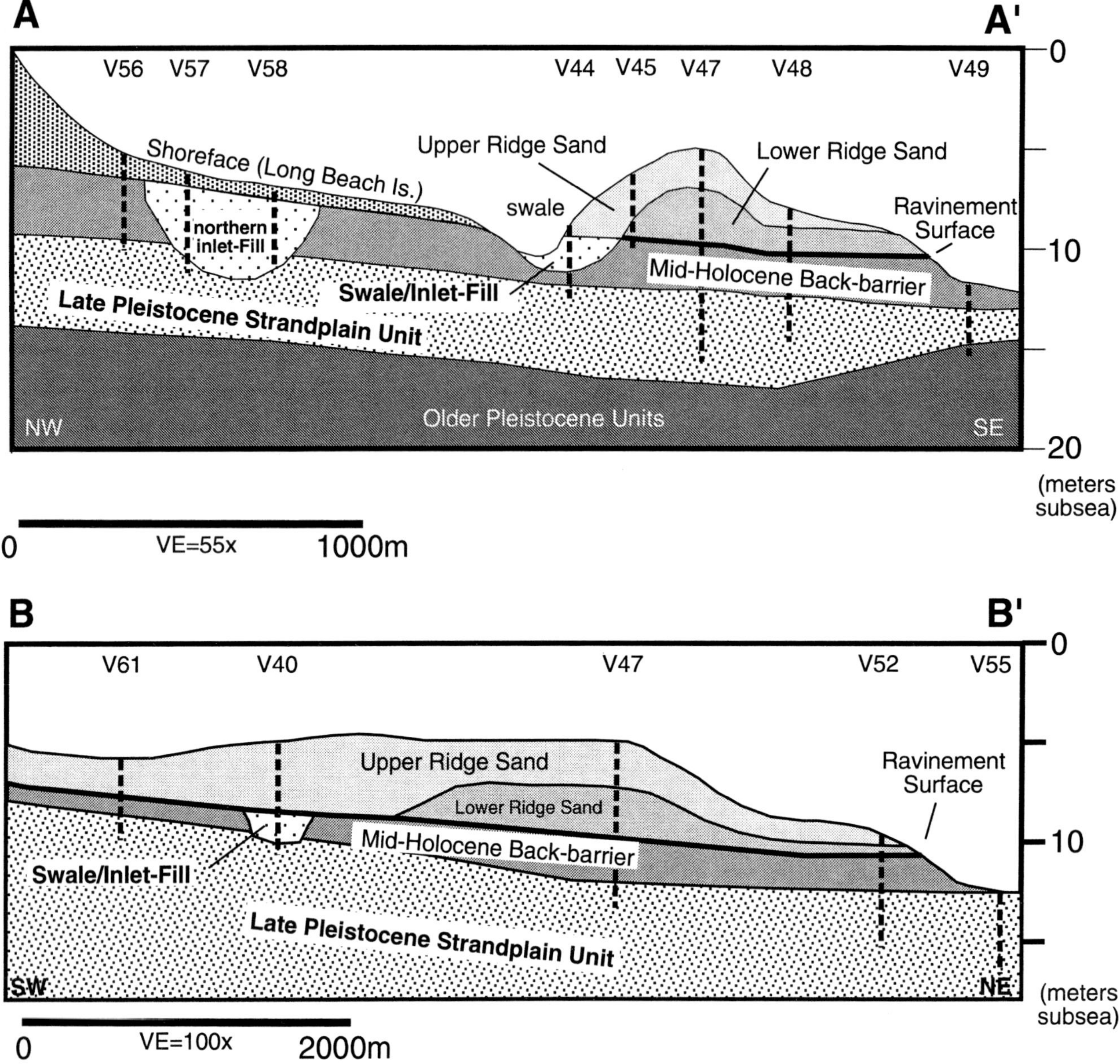

Fig. 3.—Cross-sections A-A' (across ridge) and B-B' (along ridge) based on seismic, vibracore, and bathymetric data. See Figure 1 for cross-section locations. Core interval indicated by dashed line at vibracore location.

separated by inlet-channels (Fig. 2). Even though the tidal range is small, back-barrier tidal prisms are appreciable (Nordstrom, 1987) and ebb-tidal deltas are developed seaward of the inlet mouth, as seen at modern Barnegat Inlet (Fig. 2, inset). In fact, ridge-like protuberances are known to develop on the outer fringes of these ebb-tidal deltas (Figueiredo, 1984; McBride and Moslow, 1991). The exact genetic process is still unknown, but Huthnance (1982) provides a conceptual explanation of how, once formed, these features will accrete and stabilize.

Vestiges of the ebb-tidal delta precursor to Peahala Ridge are observed in the lower ridge sand unit, which contains a mixture of shelf, bay, and marsh micro- and macrofaunal assemblages (Henderson, 1986; Culver and Snedden, 1996). Mixing of different ecological elements and relatively high organic content imply close proximity to an inlet mouth or an area of tidal exchange. It is notable, also, that the lower ridge sands occupy only the northern end of Peahala Ridge (Fig. 3), suggesting more recent accretion near the ridge attachment point.

Absence of shoreface/beach deposits equivalent in age to the back-barrier silts and clays beneath Peahala Ridge implies truncation and removal of the barrier island as the landward migrating shoreface passed through the area, beginning at about 3.5 ka. Accompanying the sea level rise was the likely southwestward migration of an inlet-channel, cutting the feature which today is the swale separating Peahala ridge from Long Beach Island. The progressive migration of the inlet-channel is documented in the sequence of radiocarbon dates from unaltered, *in situ* bivalve

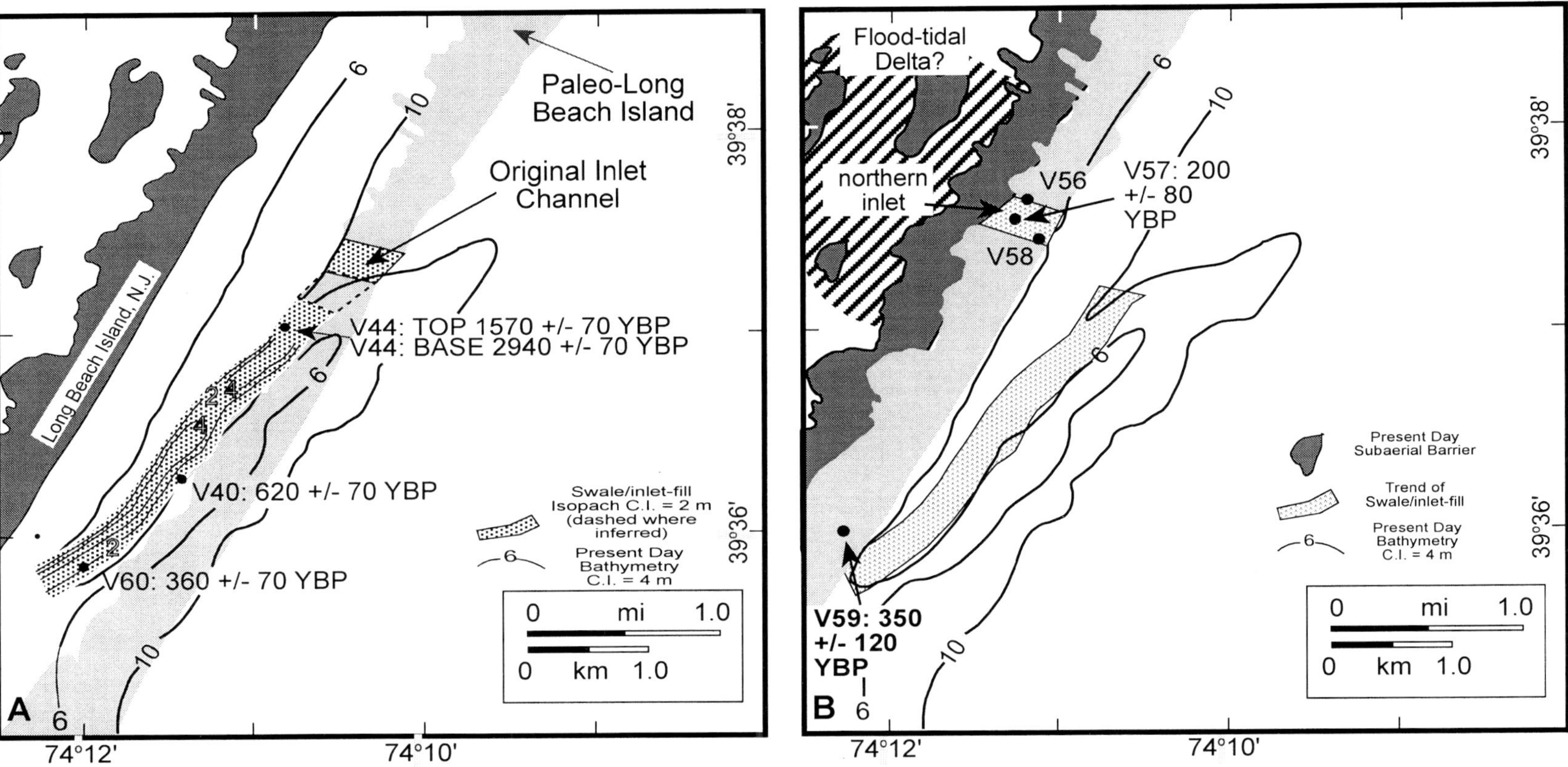

Fig. 4.—Paleogeographic reconstruction of Area 1B: A) 3500 yr B.P.—just prior to ravinement; B) 300 yr B.P.—following closure of southern inlet. Paleogeography superimposed upon present-day bathymetry and geography. Pertinent cores and radiocarbon dates are indicated.

shells taken from the unit (Fig. 4A). This shoreline-oblique motion is a probable vectoral result of westerly (landward) shoreline retreat and southerly longshore sediment motion, as suggested in the model offered by McBride and Moslow (1991).

Sometime prior to 0.30 ka, the southerly migrating inlet closed, leaving the swale and trailing body of sand today called Peahala Ridge (Fig. 4B). Obstruction by the trailing ridge precursor may have been one cause of the inlet closure. In addition, a second inlet opened to the north, capturing its tidal prism and causing closure (c.f. Fig. 3A). Regardless of the exact cause, it is evident that by 0.30 ka Peahala Ridge had acquired its present form as an oblique-trending, shoreface-attached ridge of sand. Closure of the northern inlet eventually ensued, although clues to its more recent existence are evident in the possible flood tidal delta-like arrangement of back-barrier islands behind Long Beach Island (Fig. 4B).

The active nature of Peahala Ridge today is clear from examination of vibracore sedimentology and radiocarbon dates, oceanographic measurements, and historical maps. The steeper landward margin of the ridge typically is underlain by coarse to medium-grained, cross-bedded to flat-bedded sand, implying frequent sediment transport. Ridge sands here yield the youngest ages, including several "modern" (Post-AD-1950) dates from shells recovered several meters below the sea surface. By contrast, finer lithologies, intact bivalves and older radiocarbon dates are more common on the seaward flank (Snedden et al., 1994).

Oceanographic measurements (Snedden et al., 1994) and radiogeochemical analyses of successive box core surveys (Oertel and Wong, 1987) confirm that intense storm-generated currents regularly impinge on the landward flank, with some flows veering over the ridge obliquely and others passing down the swale. This undoubtedly is a pattern repeated over the winter season, with storms passing across the area on a monthly, sometimes weekly basis.

However, comparison of historical bathymetric maps suggests that long-term vertical and lateral accretion of the landward margin of Peahala Ridge is also occurring (Fig. 5). Previous work on a similar shoreface-attached ridge in the False Cape area of Virginia indicates that crestal and landward accretion of the ridge is probably accomplished in summer months, as fair weather wave motion (Stokes drift) provides momentum for onshore migration of sediment (McHone, 1973). Erosion of the landward margin does take place during winter and spring storms but seems to be outweighed by the summer (fairweather) accretion (Fig. 5). Sidescan sonar records from Peahala Ridge clearly show onshore bedform migration during summer months (Snedden et al., 1994). Alternation of summer and winter flow patterns may explain the presence of a steep (erosional) but accreting landward flank, a motif not observed in offshore ridges, as discussed below.

To summarize, study of Peahala Ridge (Area 1B) documents how a shoreface-attached sand ridge can develop from an ebb-tidal delta, thus confirming an earlier morphological model (Figueiredo, 1984; McBride and Moslow, 1991). While the ebb-tidal delta model may not be the only means of ridge genesis (e.g., R. Dalrymple and E. Hoogendoorn, in press), it is an extremely attractive one as it requires no unique current flows (e.g., helicoidal flow, Stubblefield et al., 1984) and fits with the observed historical and geographic association of ridges with tidal inlets (McBride and Moslow, 1991).

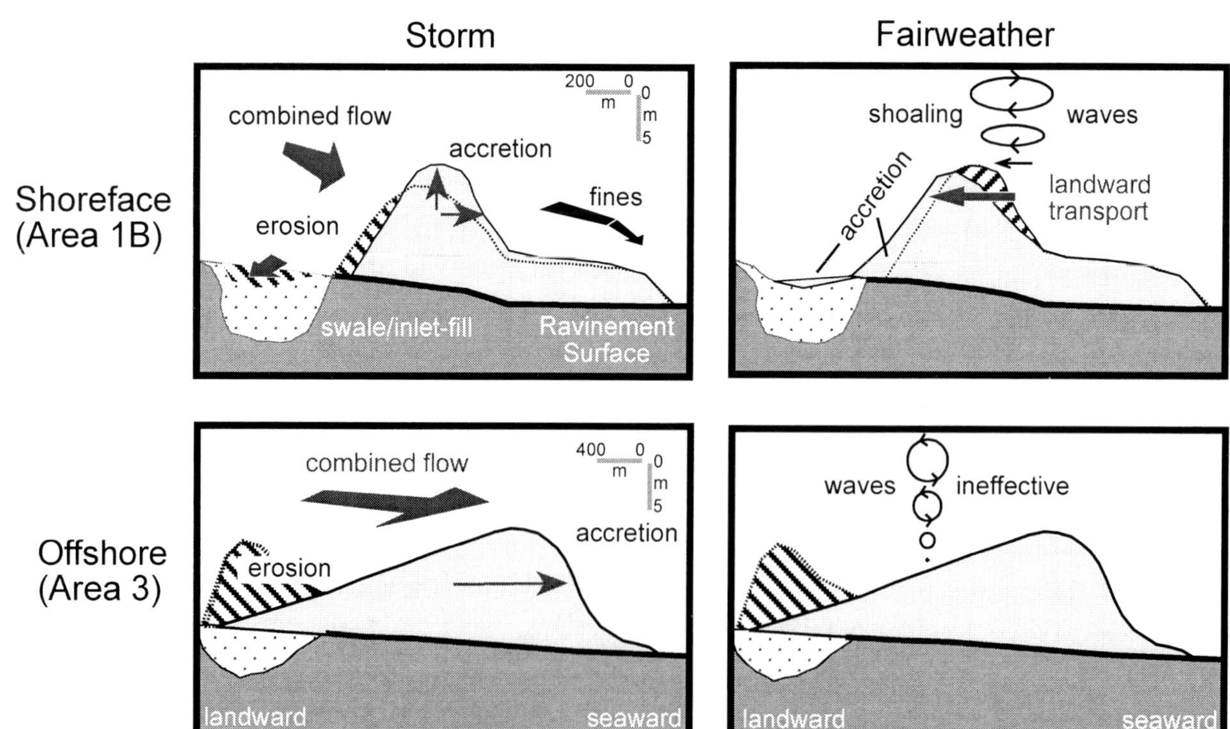

Fig. 5.—Storm and fairweather dynamics and ridge migration in nearshore and offshore areas. Based on current meter studies reported in Snedden et al. (1994) and McClelland (1973b) and bathymetric surveys of McHone (1973).

MODERN SAND RIDGE DETACHMENT

Ridges located further offshore, in water depths of 7 to 15 m (Area 1A), 22 to 33 m (Area 2) and 40 to 47 m (Area 3) exhibit a surprisingly similar stratigraphic architecture to that of Area 1B (Fig. 6). The uppermost unit in each area typically is relatively young, well-sorted, fine to coarse-grained, shell-rich sand containing microfauna characteristic of the water depth at which the ridges are presently located (Culver and Snedden, 1996). These are termed the upper ridge sands, following the terminology used in earlier studies (Rine et al., 1991; Snedden et al., 1994). Older ridge sands can also be identified in each area and are analogous in terms of geometry, sedimentary structures, and lithology to the lower ridge sand of Area 1B. Muds and silts comparable to the back-barrier lagoonal unit and coarse-grained, gravelly sands similar to the coastal-plain unit of Area 1B are also present in cores from these other areas.

These units do, however, differ significantly in terms of their respective ages, determined dominantly from radiocarbon dating of in-place macrofauna. There is a general increase in age of each unit moving into deeper water, although this is most apparent in the age of the youngest shells recovered (Table 1). The trend of older ages shows less of this pattern in the upper and lower ridge sands, due to reworking which tends to remove or destroy older shells.

Differences in microfaunal ecology are minor in these three units (Culver and Snedden, 1996) suggesting that the stepwise shift in radiocarbon ages reflects the landward migration of the New Jersey regional coastline from its Late Pleistocene lowstand position at approximately 90 meters subsea to its present location (Dillon and Oldale, 1978; Fig. 7). This view is supported by comparing the age and depth of the transgressive ravinement surface in each of the project sites to the local relative sea level curve for the last 20 ky (Fig. 7A) While there is some uncertainty in radiocarbon ages in each area, the transgressive surface in each of the project sites falls on or near the sea level curve. The Holocene sea level rise was punctuated by a series of shoreline stillstands, many marked by a wave cut declivity (Fig. 7B). Areas 2 and 3 are associated with two of the more pronounced scarps, called the Atlantis and Fortune/Tiger scarps. The geographic proximity is not likely to be coincidental, as the ridges of Areas 2 and 3 are just part of a larger field of ridges located just seaward of these former shorelines (Stubblefield et al., 1984).

This chronostratigraphic evidence provides confirmation of earlier models for ridge genesis and subsequent detachment based upon historical and morphological data (Figuereido, 1984; McBride and Moslow, 1991). McBride and Moslow (1991), in particular, provide a well-illustrated explanation of how shoreface-attached ridges can arise from ebb-tidal delta precursors, eventually detaching with continuing sea level rise to become shelf sand ridges (Fig. 8). The model is compelling as it explains several observations from the four project sites: 1) progressive shift in ages of similar sedimentary units from one area to another; 2) the lack of coeval barrier island deposits, probably removed during the erosional transgression; and 3) cutting of the immediately adjacent swale by an obliquely-migrating tidal inlet; and 4) the mixed faunal signature in the lower ridge sand of areas 1A and 2, which implies development in an earlier, lower sea level position.

POST-DETACHMENT EVOLUTION

Notable differences in ridge character exist between the four project areas, variations which may relate to processes of post-detachment ridge evolution (Fig. 8). This stage of ridge development is probably the most critical in determining the final stratigraphic architecture, that which is most likely to be preserved in the ancient sedimentary record. Three differences are most evident: 1) variations in the ridge bathymetric profile or symmetry from area to area; 2) stepwise changes in measurable dimensions of the ridges and; 3) variations in age, microfauna, and geometry of underlying channel-fill units, interpreted to be tidal inlet channel-fill successions.

Ridge Bathymetric Profile

Several workers have noted that a significant difference in ridge bathymetric profile or symmetry exists between shoreface-attached ridges and ridges located further offshore (Swift and Field, 1981; Stubblefield et al., 1984). In Area 1B, the steep side of Peahala ridge faces landward, in contrast to Area 3 where the ridge is strongly asymmetrical toward the southeast, seaward (Fig. 6). When viewed in isolation from intermediate depth ridges (Areas 1A and 2), this difference could be misconstrued as an indication of dissimilar origins. However, when the nearly symmetrical profiles of Areas 1A and 2 ridges are considered, the transitional nature of the ridge profile becomes apparent. Thus, some dynamic process must be causing a shift in ridge profile from landward to seaward-facing.

One explanation might be the offshore decline in wave/current ratios as a function of increasing water depths during the Holocene transgression. As mentioned earlier, shoreface-attached ridges such as those in Area 1B display an actively accreting landward margin due to fair-weather wave-gener-

Table 1.—Radiocarbon ages in the four project sites.

Area	Depth of Ravinement Surface (subsea)	Age of Pre-Ravinement Unit*	Age of Lower Ridge Sand Unit	Age of Upper Ridge Sand Unit
1B	8 - 10 m	3.7 - 5.9 ka	0.9 - 3.1 ka	0.0 - 0.9 ka
1A	14 - 15 m	6.5-8 ka**	3.6 - 6.4 ka	0.14 - 3.4 ka
2	27 - 28 m	8-10 ka**	4.9 - 7.5 ka	0.6 - 3.5 ka
3	42 - 47 m	10 - 12 ka	3.3 - 4.4 ka	0.7 - 1.8 ka

*interpreted back-barrier/lagoon or intertidal flats.

**age is estimated as there is insufficient material for radiocarbon dating.

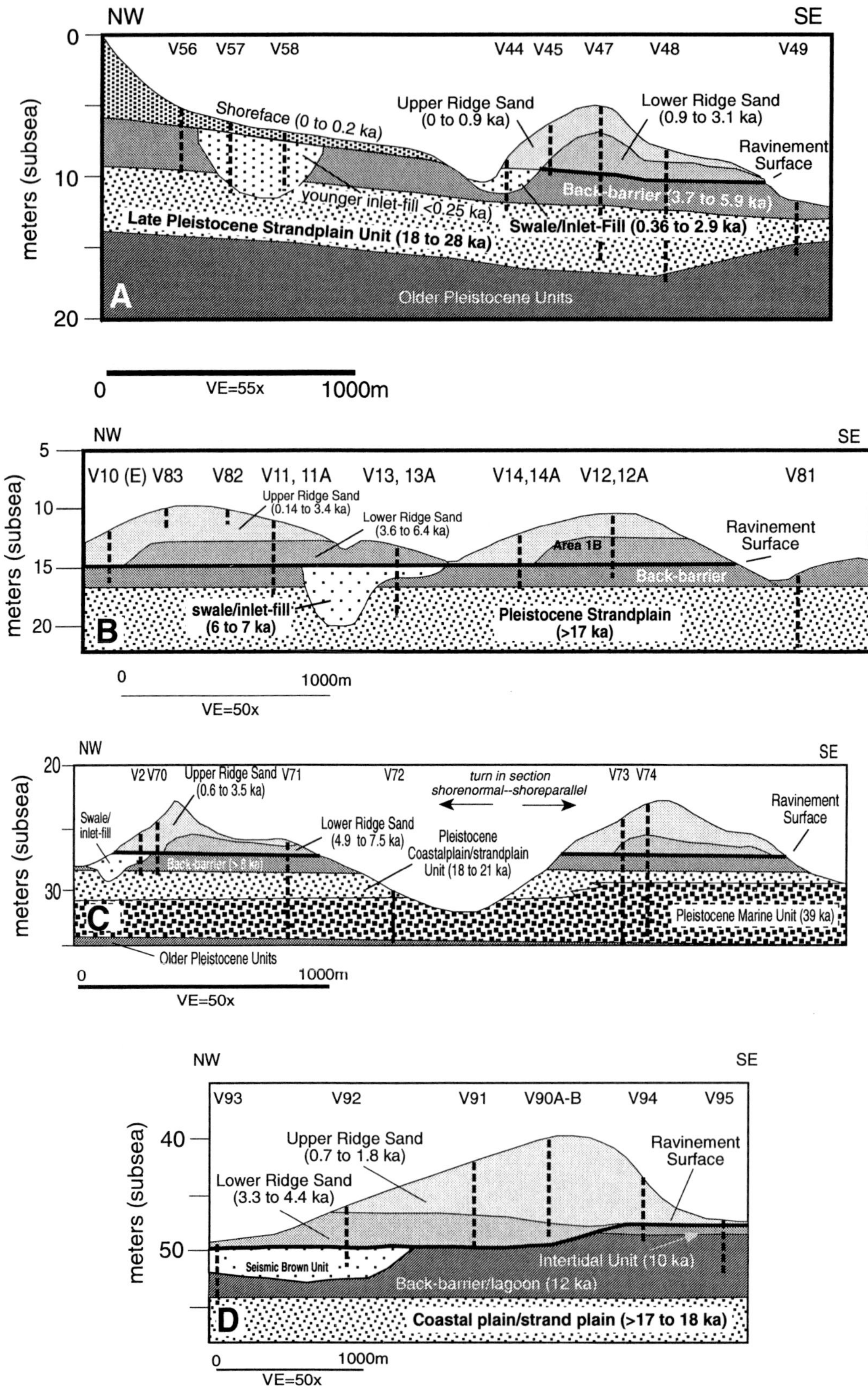

Fig. 6.—Chronostratigraphic cross-sections for A) Area 1B; B) Area 1A; C) Area 2; and D) Area 3. For locations see Figure 1.

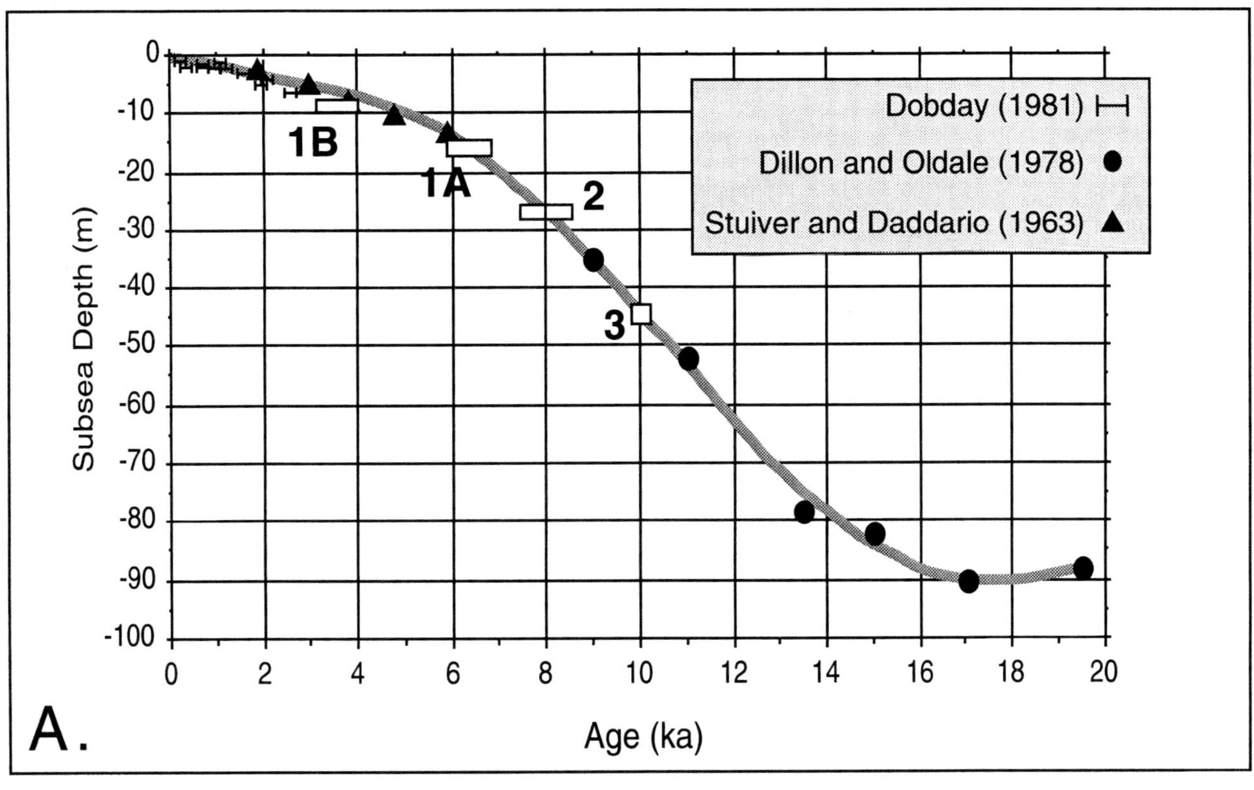

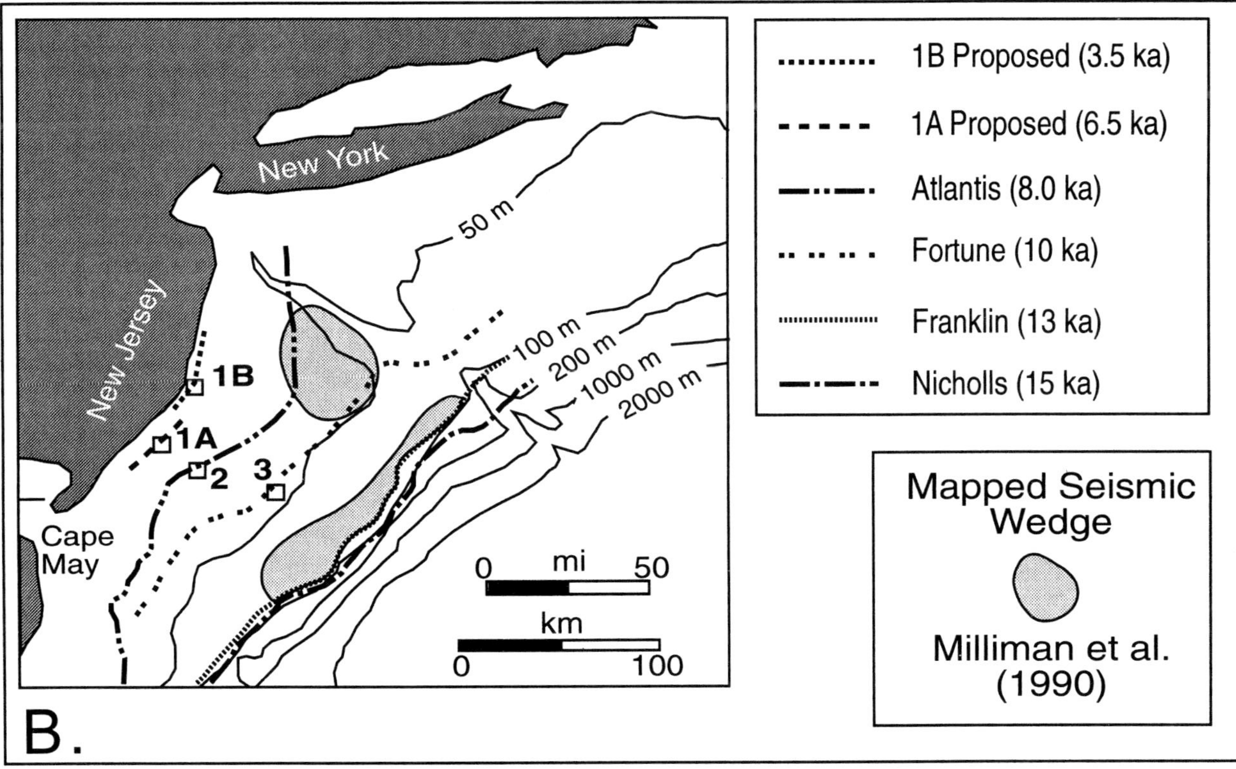

Fig. 7.—A) Age and depth of ravinement surface of all four study sites superimposed upon a relative sea-level curve for New Jersey Shelf. Curve drawn through data points of Dillon and Oldale (1978) from New Jersey Shelf; Dobday (1981) from Great Egg Bay, New Jersey; and Stuiver and Daddario (1963), from Brigatine City, New Jersey. Width of white box indicates corresponding uncertainty of age dating of the ravinement in each of the study sites; B) Project study sites superimposed upon four post-late Pleistocene lowstand shoreline positions of Dillon and Oldale (1978) and two other possible sites.

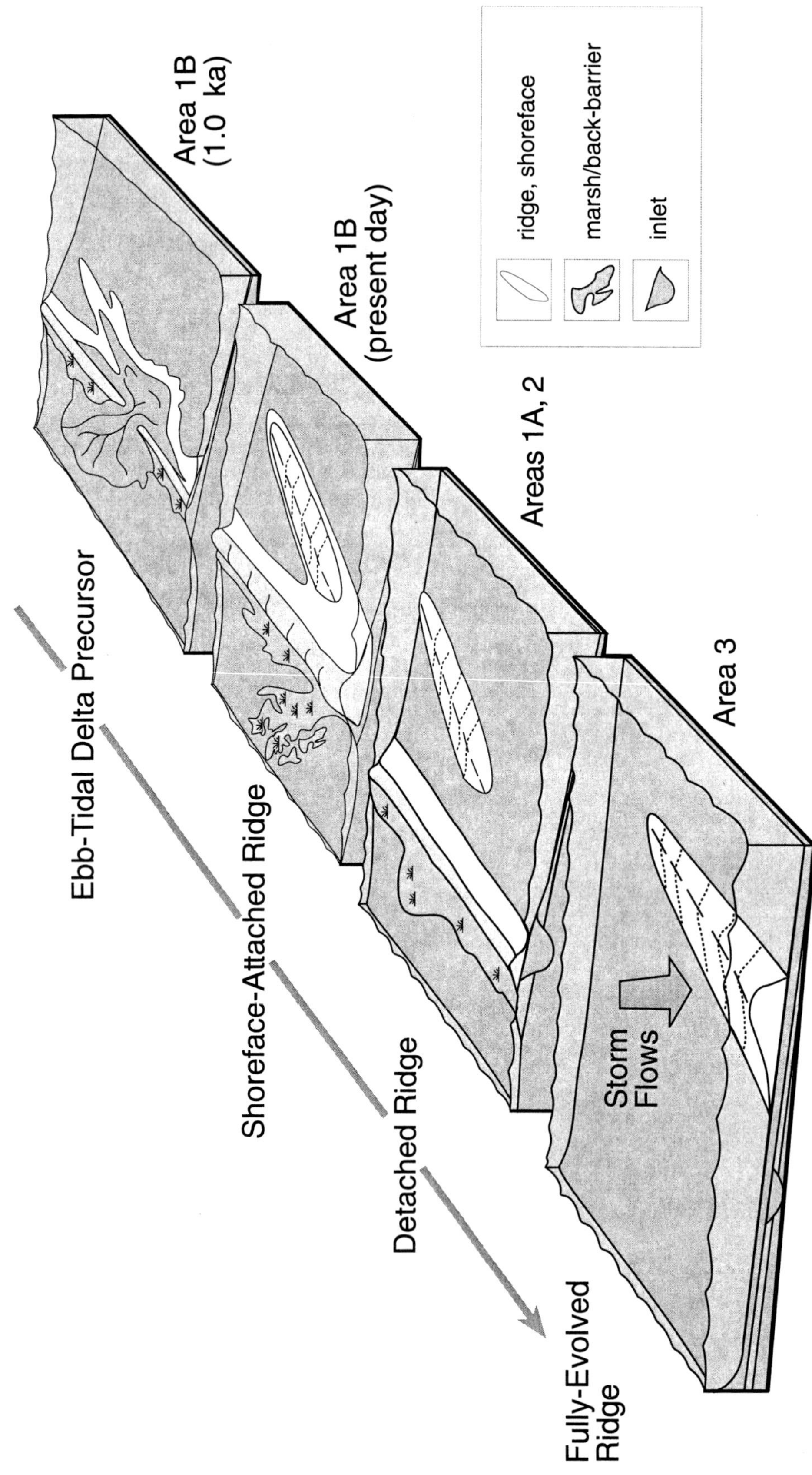

Fig. 8.—Evolution of ridges on the New Jersey Atlantic Shelf, USA. The first three evolutionary phases depicted here (ebb-tidal precursor, attached-, and detached-ridge) are modified after McBride and Moslow (1991).

ated landward transport (Fig. 5). By contrast, fair-weather waves are largely ineffective in the deeper water over the mid-shelf ridges as in Area 3 (McClellen, 1973b). The dominant sediment transport direction for the mid-shelf ridges is to the south and southeast, as geostrophically-balanced storm currents pass obliquely across the ridge (Gadd et al., 1978). If indeed the Area 3 ridge developed initially as a shoreface-attached ridge as in Area 1B, it has presumably experienced a slow but steady decrease in fair-weather wave energy since its origin at approximately 10 ky. As fair-weather wave energy declined, the sand body probably began to act much like a bedform, re-orienting itself with its steep or leeside down current to the south and southeast (Fig. 5).

In fact, we believe that these storm flows have enough strength and frequency to induce seaward migration of the entire ridge sand body. Such large-scale bedform migration has been observed in tide-dominated settings (Jones et al., 1965) and under high-velocity geostrophic currents (Ramsey et al., 1996) and is generally in the direction of the steep face (McCave, 1971) as suggested here. Although migration rates of the ridges are probably low on a yearly basis, migration distances are probably not inconsequential in light of the long Holocene transgressive time frame.

Facilitating this migration is the transport across and accretion of sand upon the crest of the ridge. Huthnance (1982) provides an dynamic model for sand ridge growth as sediment-laden flows pass obliquely across the ridge crest (see also Snedden and Dalrymple, this volume). In the model, the oblique orientation of the ridge crest enhances bottom friction, reducing shear stress and causing sand deposition. This may explain why ridges on the Atlantic shelf (and worldwide) tend to be oriented obliquely to the prevailing shore-parallel currents.

In Area 3, the sand body appears to have migrated to a position directly overlying the candidate swale/inlet-fill unit, the seismic brown unit (Fig. 6). It is possible that after shoreline detachment, the ridge may have actually migrated landward for a period of time as result of aforementioned fair-weather wave action, later reversing direction to move to its present position as currents began to dominate sand transport (c.f. Fig. 5). Truncation of the older ridge sand unit on the northern flank of the ridge is possible evidence for this reversal of ridge migration directions.

One obvious consequence of slow but steady ridge migration is overturning or reworking of the older ridge sands, a process which apparently removed many of the more obvious signs of the initial ridge genesis as well as introducing younger age shells. Shallow shelf, marsh, and lagoonal microfauna are present in only trace quantities in the older ridge sand unit of Area 3, the dominant forms reflecting present water depths (Table 2; Culver and Snedden, 1996). However, one would not expect these delicate foram tests to survive extensive reworking during 10,000 years of storms. Experimental and physical observations demonstrate that agglutinated foraminifera are particularly susceptible to mechanical breakage during high energy events (Miller and Ellison, 1982; Culver and Snedden., 1996). However, indications of the earlier shallow water phase in this unit are evident from examination of the macrofaunal assemblage, which includes marsh grass snails (*Littorina irrorata*), razor clams (*Ensis directus*), oysters (*Crassostrea virginica*), bay scallops (*Argopecten irradians*), and surf clams (*Spisula solidissima*) (Henderson, 1986; Table 2). While upward reworking of shell material is always a concern, it is unlikely in this case as older units (Late Pleistocene fluvial-coastal plain, Middle to Early Holocene back-barrier) are generally non-fossiliferous.

Ridge Dimensions.—

Another important difference among ridges relates to the actual amount of ridge sand developed at a given site (Fig. 9). Earlier studies noted an offshore increase in ridge cross-sectional area (Swift and Field, 1981). The observed increase is accomplished largely through a "broadening" of the ridges, as the length/width ratio declines, perhaps as a function of the superimposition of early landward and later seaward ridge migration phases (c.f. Fig. 5). Grain size of the ridge sand unit also increases in the offshore direction, although local variations appear to be important as well (e.g., relatively fine-grain size of Area 2 ridge sand). Possibly during ridge reworking and migration, sediment is added to the ridge from erosion of adjacent swales, while finer sizes are moved offshore through progressive sorting (Swift et al., 1972). Huthnance (1982) provides a theoretical explanation of how once developed, ridges such as this can accrete or enlarge as a consequence of topographic "feedback" to the fluid flow regime.

Channel-fill Units.—

One of the more enigmatic units in the study sites is a sand and mud succession filling a 2 to 5 m deep channel-cut (Fig. 6). Initially called the seismic brown unit, it was first recognized in early studies of Area 1B (Figueiredo, 1984). Later work in Area 1B, as discussed, indicated that cutting and filling of this unit occurred during shoreline oblique migration of a tidal inlet channel (Snedden et al., 1994). Similar channel-fills were later identified in Areas 1A, 2, and 3 with better high resolution seismic data acquired in 1984. Understanding the origin and

Table 2.—Micro- and macrofaunal content of the upper and lower ridge sand units.

Area and Water Depth	Microfaunal Habitat*	Macrofaunal Habitat**
Area 1B (4 - 12 m)	Very Inner Shelf	Very Inner Shelf/Intertidal/Inlet-influenced
Area 1A (7 - 15 m)	Inner Shelf	Similar to Area 1B
Area 2 (22 - 33 m)	Inner to Middle Shelf	Offshore Marine and Inlet/Bay
Area 3 (40 - 47 m)	Middle Shelf	Offshore Marine and Bay/Estuary/Inlet

*Culver and Snedden (1996); **Henderson (1986)

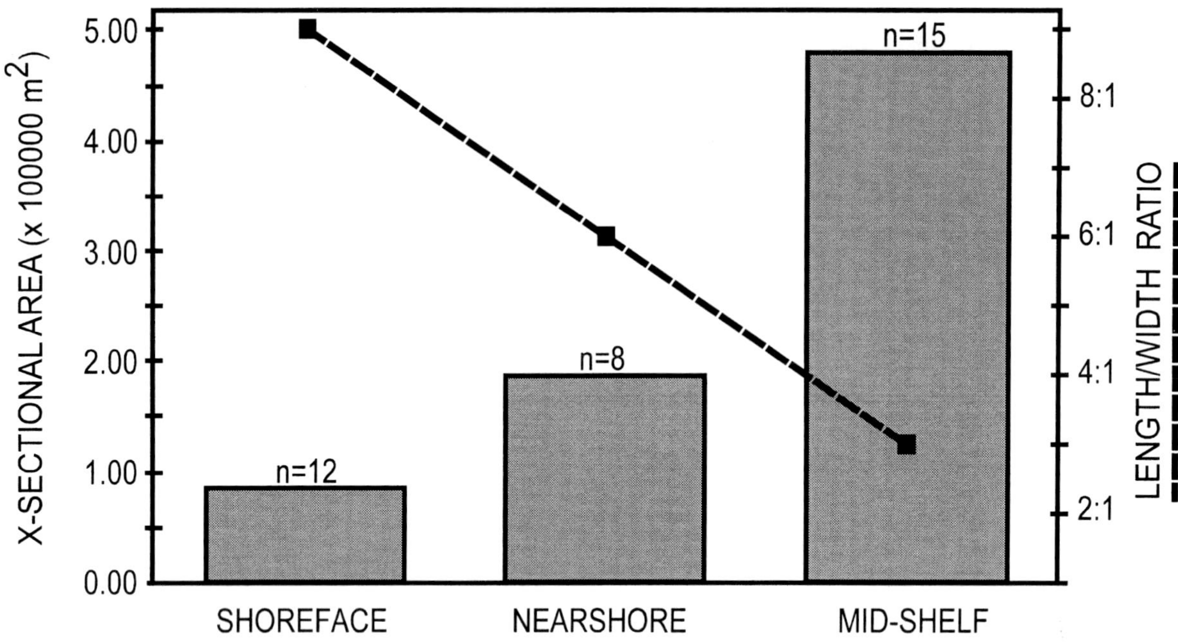

Fig. 9.—Comparison of average cross-sectional area (shaded squares) and length/width ratios (dashed line) of shoreface-attached, nearshore, and mid-shelf ridges, Atlantic Shelf. From tabular data provided in Swift and Field (1981).

evolution of this unit is important in explaining the genesis of the detached shelf ridges (Areas 1A, 2 and 3) and thus the general model for ridge development on the Atlantic shelf. Two alternative models for the origin of this feature can be considered: 1) post-transgressive shelf swale incision and fill; and 2) coastal tidal inlet-channel migration.

Like in Area 1B, the channel-fill unit in Areas 1A, 2 and 3 lies beneath or near the present day swale, paralleling the trend of the upper ridge sand body. Most cores from this unit are quite similar in geometry, lithology, sedimentary structures, and microfaunal content. The mixed (back-barrier and inner shelf) micro- and macro-faunal content supports the argument that the channel-fill unit was formed by a tidal inlet and filled-in to become a swale between ridges (Fig. 6). Age dating from this unit in Areas 1B and 1A fits with the estimated time frame of shoreline transgression in this area (Fig. 7A, Table 1). There is insufficient material for radiocarbon dating in cores from Area 2.

The channel-fill (seismic brown) unit in Area 3 is more problematic (Fig. 6D). Like the swale/inlet-fill of other areas, it is incised into back/barrier/lagoonal deposits. It also is directly overlain by ridge sands, similar to that observed in the other areas. However, the seismic brown unit of Area 3 trends more at an angle to the present-day ridge, generally north to south, in contrast to the ridge-parallel orientation apparent in other areas (Fig. 4). Few cores penetrate the central portion of the unit here but one core that does (V97) yields marine microfauna and relatively young radiocarbon dates (6.35 ka) in light of the probable timing of transgressive ravinement and shoreline detachment (c.f. Fig. 7A; Table 1).

One explanation for the relatively young radiocarbon dates and mid-shelf microfauna in the channel-fill unit of Area 3 is that the channel-like feature was cut and filled by shelf currents or wave action. Radiocarbon dates from the unit suggest that filling would have had to occur within the last 6500 years, when water depths probably exceeded 20 meters and the adjacent ridge was fully detached from the shoreline (Fig. 7A). An example of such deepwater erosion is observed in Area 2 where the swale separating the two ridges was excavated four meters downward, exposing Pleistocene strata (Fig. 6C).

The problem with such an explanation for the seismic brown unit in Area 3 is that processes promoting swale erosion are decidedly non-depositional. Comparative boxcore studies of Peahala ridge indicate that sediments settling in the adjacent swale are removed during passage of winter storms (Oertel and Wong, 1987). Recently deposited sediments are found in none of the present swales of Areas 1B, 1A, 2 or 3. Nor is deep swale erosion very common; in the study sites and elsewhere, it usually is measured at a meter or less (c.f. McHone, 1973; Swift and Field, 1981). The unusually deep swale erosion of Area 2 is probably related to confinement of storm flows between the two ridges (Fig. 6) and consequent enhanced bottom shear. In fact, the presence of a deep swale may have something to do with the fine-grain size of the Area 2 ridges. By isolating the ridge sands on bathymetric highs and preventing significant ridge migration, little overturning of the original sediment volume has occurred, a process which tends to increase the overall grain size through progressive sorting.

One well-established process for cutting deep, channel-like features in coastal settings is that of tidal inlet migration. Migrating tidal inlet channels are particularly common in the microtidal, wave-dominated settings (like that of the New Jersey coastline) and are know to incise as much as five meters or more into the underlying substrate (Moslow and Tye, 1985). The channel-fill feature in the four project sites shows an average thalweg depth of four to five meters, comparable in size to that observed in modern tidal inlet channels. In fact, the only other process known to incise such a deep channel is river entrenchment during lowstands (Posamentier et al., 1992). However, this implies a significant gap in time between incision (15 to 18 ka; see Fig. 7A) and infilling (6.35 ka), with little or no supporting evidence to suggest this in cores (e.g., paleosols, exposure surfaces).

While it is probable that the seismic brown unit in Area 3 was cut by a migrating tidal inlet, it is unlikely that the present infilling sediment was produced at the same time, given its age and microfaunal content. Rather, later excavation and infilling by younger sediments must have occurred, although the specific means by which that occurred can only be a matter of speculation. It should be noted, however, that intertidal/marsh and shallow inner shelf microfauna are still present in trace quantities (Culver and Snedden, 1996), suggesting reworking of earlier tidal channel-infill sediment.

The unusual map geometry of the seismic brown unit in Area 3, trending in a N-S direction rather than the NE-SW orientation of the channel-fill units in other areas is still puzzling. Such a configuration could, however, be the result of local progradation and southwestern tidal inlet migration, such as during a brief regressive episode (c.f. McBride and Moslow, 1991). An abrupt glacial advance in the 9 to 10 ka time frame has been documented in the North Atlantic (Kaufman et al., 1993) and may have been the cause of a seaward progradation of a paired inlet/ridge complex originally present in Area 3.

MODEL SUMMARY

Based on geomorphic associations and regional observations of modern coastlines, McBride and Moslow (1991) present a graphic model showing progression of ridges from initial genesis as inlet-associated ebb-tidal deltas to shoreline-detached sand bodies (Fig. 8). Data from the Atlantic Shelf project support that model.

However, we see the need to add a significant post-detachment evolutionary phase, when ridges migrate, change profile symmetry, increase their volume, modify their texture, age and paleontological content (Fig. 9). When viewed in a time/space domain, this phase actually spans the longest period of time in the ridge history (Fig. 10). The possibility of intra-ridge hiati may exist, although we cannot imagine long time spans with little or no storm flow modification. The offshore increase in ridge cross-sectional area (Fig. 9) observed by Swift and Field (1981) implies that some sediment volume is added during this phase, thus differentiating the ridge evolutionary phase from condensed intervals *sensu stricto*. The hiatus developed below the ridges is also partly a function of the continuous turnover of ridge sands and destruction of older (shallower) macro- and micro-fauna (Fig. 10).

We should also add that, during the post-detachment evolution phase, considerable erosion of adjacent swales occurs, thus accentuating the ridge topography (Rine et al., 1991) and possibly removing signs of the earlier genetic sequence. The transfer of significant amounts of sediment from swales to ridges is also possible during this time, although faunal and textural studies would suggest otherwise (Miller and Ellison, 1982). Much of the sand added to ridges during this relatively long time frame is probably derived from shoreface erosion and obliquely offshore-trending geostrophic flows generated during major storm events (Snedden et al., 1988; Snedden and Nummedal, 1991).

IMPLICATIONS FOR STUDY OF ANCIENT MARINE SANDBODIES

Study of the New Jersey Atlantic Shelf sand-ridges has taken place against a backdrop of rapidly evolving interpretations on possible ancient analogs (see Bergman and Walker, Tillman, this volume). Presently, two divergent viewpoints exist regarding the genesis of isolated sand bodies found on ancient marine shelves: those advocating relative sea level falls, and others proposing more dynamic factors (i.e., storm currents, infragravity waves). For example, sand bodies in the Shannon Sandstone member of Wyoming were originally proposed as storm-built sand ridges formed largely in place on a wide Cretaceous seaway (Tillman and Martinsen, 1984; Fig. 11A). More recently, however, Shannon sand bodies have been interpreted as lowstand shoreface deposits (Bergman, 1994). The latter explanation is particularly appealing in light of the forced regression model of Posamentier et al. (1992), which demonstrates how isolated marine sand bodies can be formed through a rapid relative sea level fall (Fig. 11B).

Studies at Peahala Ridge and other areas of the New Jersey shelf demonstrate that both eustatic and dynamic factors are at work in creating modern shelf sand ridges (Fig. 11C). Ridges can originate from a shoreline-associated precursor like an ebb-tidal delta, requiring that sea level fall (or rise) to near or at the position of the future ridge. Dynamic reworking and growth of the sand body following transgressive ravinement then occurs. Ridges continue their evolution until final burial by sediments of the prograding (highstand) shoreline system.

We would argue that a similar interaction of eustatic and dynamic events could be responsible for formation of ancient sand bodies like those found in the Shannon of Wyoming. The effect of imposing short term, high energy storm flows on a relatively long-lived marine sand body has clearly been underestimated. Study of shelf sand body evolution during transgressive episodes is obviously needed.

Implications for sequence stratigraphic models are also evident. Current models tend to underemphasize post-depositional modification of transgressive sand bodies. By contrast, studies of modern sediments like those of the New Jersey shelf demonstrate the active nature of these sand bodies initially formed several thousands of years ago. The sand bodies evolve during the long transgressive periods, sometimes taking on a form very different from the initial sand body geometry.

Preservation and eventual incorporation into the stratigraphic record is not an issue for shelf sand ridges, particularly when one considers that shelf sand ridges are the byproducts of transgression, having survived vigorous erosional shoreface retreat. We would expect that any return to progradational conditions following their development would tend to facilitate final preservation as sediments of subsequent highstand shoreline/delta systems would progressively bury these sand bodies (Snedden and Dalrymple, this volume).

CONCLUSIONS

Our study of the New Jersey Atlantic shelf indicates that both eustatic and dynamic factors are at work creating and modifying sand ridges. Eustatic factors determine whether a shoreline passed through a site, providing in succession an ebb-tidal delta precursor, a shoreface-attached ridge, and finally, a detached sand ridge (the ebb-tidal delta model of McBride and Moslow, 1991). However, post-detachment ridge evolution causes shift in position and form, as well as the observed increase in sand volume and enhancement of texture. Thus, our expanded model builds upon previous work on modern processes (e.g., Swift et al., 1984; Antia, 1993) and

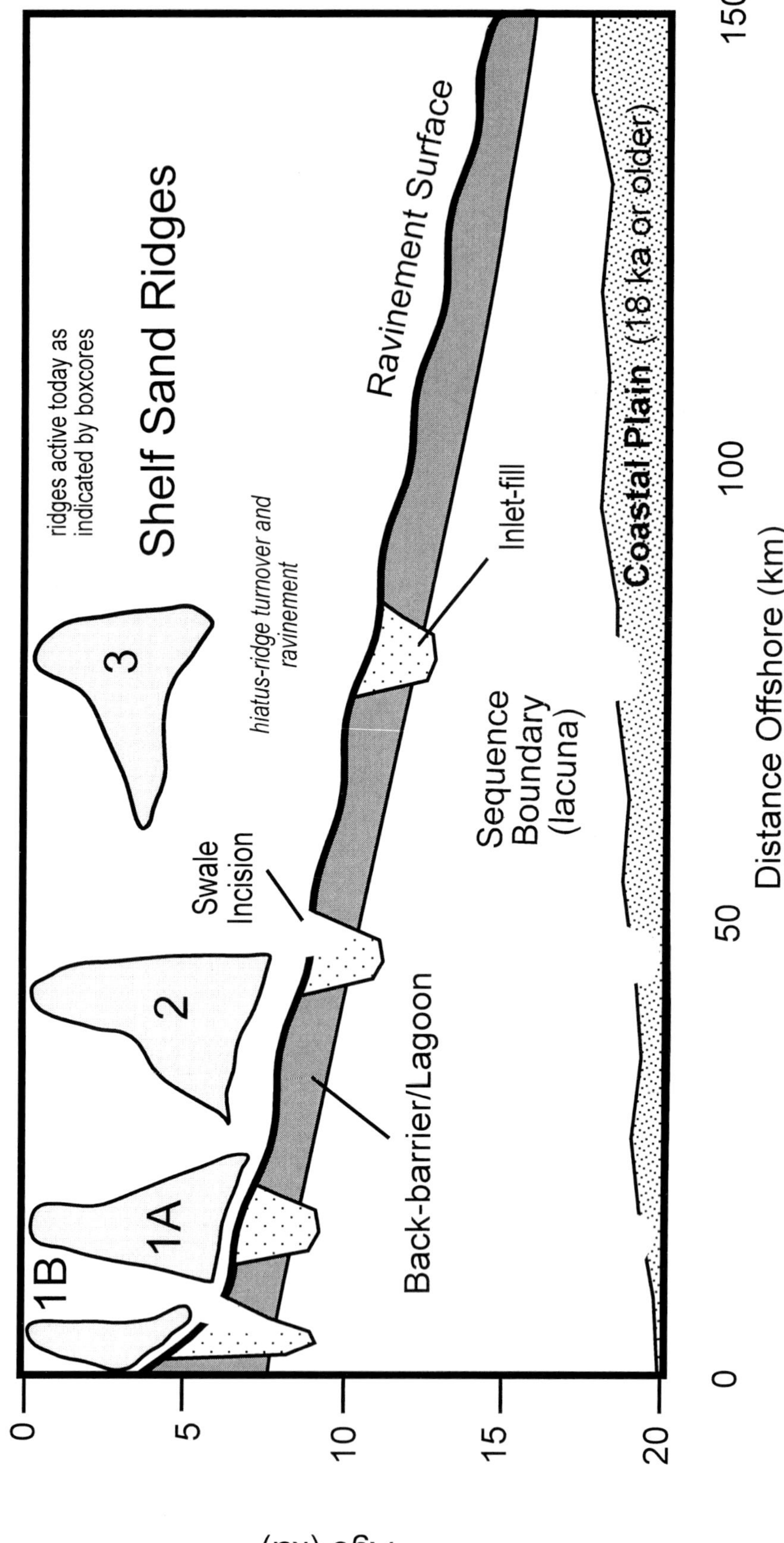

Fig. 10.—Wheeler diagram (time-space plot) of stratigraphic units for New Jersey Atlantic Shelf, 20 ka to present. Hiatus and lacuna for sequence boundary, ravinement surface, and ridge reworking are indicated by white gaps.

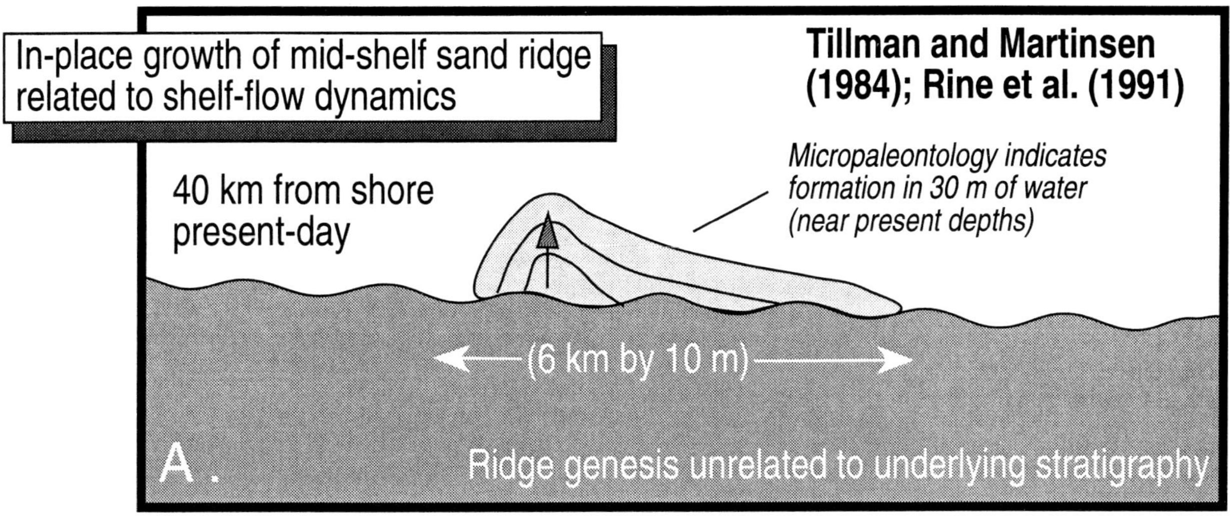

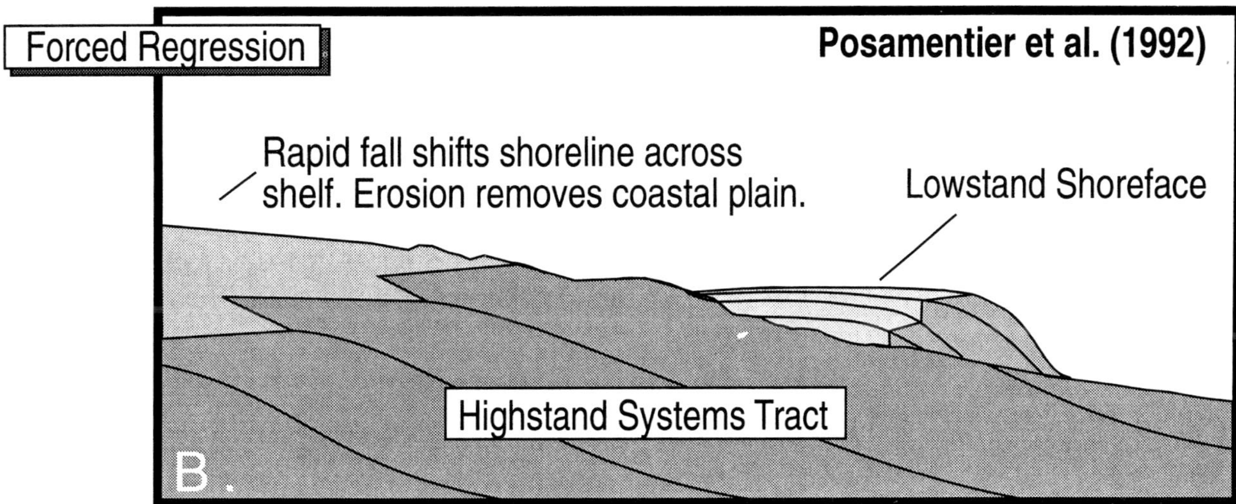

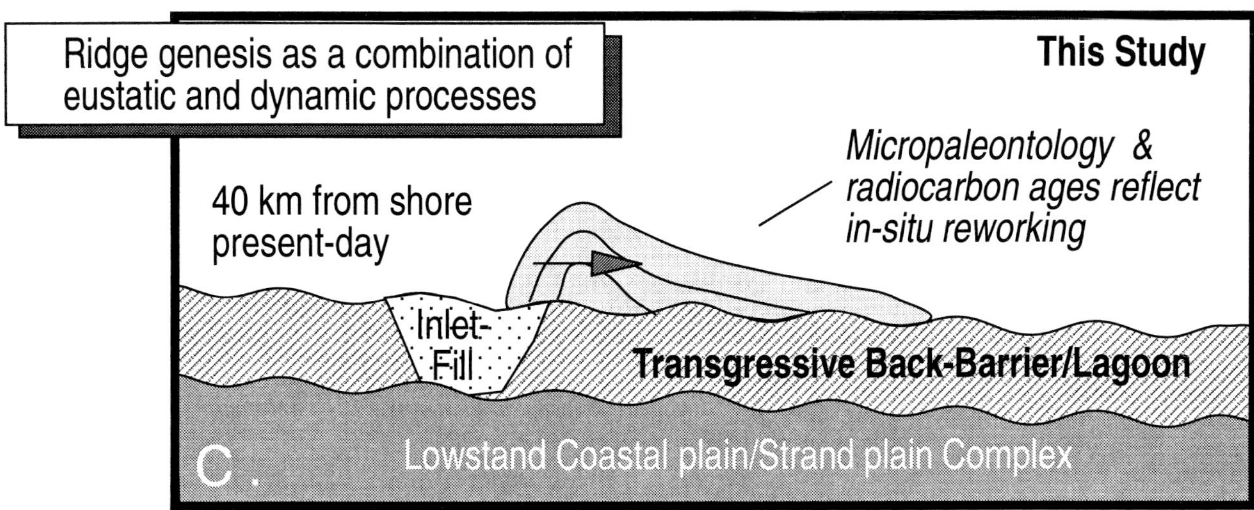

Fig. 11.—Three schematic models for genesis of isolated marine sand bodies: A) in-place growth; B) forced regression; C) combination of eustatic and dynamic factors (this study). No scale intended, but some dimensions indicated.

longer-term eustatic forces (Stubblefield et al., 1984; Rine et al., 1991) for shelf sand ridge genesis.

We view the New Jersey Atlantic shelf as an analog for ancient continental margins where tides are small, sediment supply is limited, transgressive episodes are long-lived, and rates of sea level rise are relatively high. Possible ancient analogs do exist, for example, Mesozoic strata of Siberia (CIS), Bathonian sediments of the North Sea, Carboniferous rocks of the Mid-continent U.S.A., and portions of the Rocky Mountain Cretaceous section (e.g., Tillman, this volume; Bergman and Walker, this volume).

ACKNOWLEDGMENTS

The model presented by Randolph McBride and Tom Moslow in 1991, though not one we originally advocated, proved ultimately to best fit the Atlantic Shelf data and we appreciate their kind understanding and comments. We gratefully acknowledge the contributions of Kirby Rodgers. This manuscript also benefited from reviews by Rick Sarg, Tim Hurley, Keith Conrad, John Armentrout, Randolph McBride, John Anderson, Martin Gibling, and Robert Dalrymple. Discussions with Dag Nummedal are also greatly appreciated.

REFERENCES

Antia, E.E., 1993, Surficial grain-size statistical parameters of a North Sea shoreface-ridge:patterns and process implications: Geo-Mar. Letters, p. 172-181.

Antia, E.E., 1994, The ebb-tidal delta model of shoreface ridge origin and evolution: appraisal and applicability along the southern North Sea barrier island coast: a discussion: Geo-Mar. Letters, v. 14, p. 59-64.

Bergman, K.M., 1994, Shannon Sandstone in Hartzog Draw-Heldt Draw Fields (Cretaceous, Wyoming, USA) reinterpreted as lowstand shoreface deposits: Journal of Sedimentary Research, v. B64, p. 184-201.

Boczar-Karakiewicz, B. and Bona, J.L., 1991, Wave-dominated shelves: a model of sand ridge formation by progressive, infragravity waves: *in* Knight, R.J., and McLean, J.R., eds., Shelf Sands and Sandstones, Can. Soc. of Petrol. Geol. Mem., v. 11, p. 163-179.

Culver, S.J. and Snedden, J.W. 1996 Foraminiferal implications for the formation of New Jersey shelf sand ridges: Palaios, v. 11, p. 161-175.

Dillon, W.P. and Oldale, R.N., 1978, Late Quaternary sea-level curve: reinterpretation based upon glaciotectonic influence: Geology, v. 6, p. 56-60.

Dobday, M.P., 1981, The Holocene geologic history of the Great Egg Harbor river estuary: unpublished M. S. thesis, Temple University, 200 pp.

Figueiredo, A.G., 1984, Submarine sand ridges: geology and development, New Jersey, USA: unpub. Ph.D. dissertation, Univ. of Miami, 385 pp.

Gadd, P.E., Lavelle, J.W. and Swift, D.J.P., 1978, Calculations of sand transport on the New York shelf using near bottom current meter observations: Jour. of Sedim. Petrology, v. 48, p. 239-252.

Henderson, S.W., 1986, Report on mollusc shells used for carbon dating from Areas 1A, 1B, 2, and 3, Altantic Shelf (New Jersey) Project Vibracores: unpubl. report, Atlantic Shelf Project, 12 pp.

Huthnance, J.M., 1982, On one mechanism forming linear sand banks: Estuar. Mar. Coast. Science, v. 14, p. 79-99.

Jones, N.S., Kain, J.M. and Stride, A.H., 1965, The movement of sand waves on Warts Bank, Isle of Man: Mar. Geology, v. 3, p. 329-336.

Kaufmann, D.S., Miller, G.H., Stravers, J.A. and Andrews, J.T., 1993, Abrupt early Holocene (9.9 to 9.6 ka) ice-stream advance at the mouth of Hudson Strait, Arctic Canada: Geology, v. 21, p. 1063-1066.

McBride, R.A. and Moslow, T.F., 1991, Origin, evolution, and distribution of shoreface sand ridges, Atlantic inner shelf, U.S.A.: Mar. Geology, v. 97, p. 57-85.

McCave, I.N., 1971, Sand waves in the North Sea off the coast of Holland: Mar. Geology, v. 10, p. 199-225.

McClelland, C.E., 1973a, Nature and origin of the New Jersey continental shelf topographic ridges and depressions: unpubl. Ph.D. dissertation, Univ. of Rhode Island, 94 pp.

McClelland, C.E., 1973b, New Jersey shelf near bottom current meter records and recent sediment activity: Jour. Sed. Petrology, v. 43, no. 2, p. 371-380.

McHone, J.F., Jr., 1973, Morphologic time series from a submarine sand ridge on the southern Virginia coast: unpublished M.S. thesis, Old Dominion University, Norfolk, Virginia, 59 pp.

Miller, D.J. and Ellison, R.L., 1982, The relationship of foraminifera and submarine topography on the New Jersey-Delaware shelf: Geol. Soc. of America Bull., v. 93, p. 239-245.

Milliman, J.D., Jiezao, Z. and Ewing, J.I., 1990, Late Quaternary sedimentation on the outer and middle shelf: result of two local deglaciations?: Jour. of Geology, v. 98, p. 966-976.

Moslow, T.F. and Tye, R.S., 1985, Recognition and characterization of Holocene tidal inlet sequences: Mar. Geology, v. 63, p. 129-151.

Nordstrom, K.F., 1987, Management of tidal inlets on barrier island shorelines: Jour. Shoreline Management, v. 3, p. 169-190.

Oertel, G.T. and Wong, G.T., 1987, Sea bed response to storm events at Peahala Ridge, New Jersey: unpubl. report, Atlantic Shelf Project, 18 pp.

Posamentier, H.W., Allen, G.P., James, D.P. and Tesson, M., 1992, Forced regressions in a sequence stratigraphic framework: concepts, examples, and exploration significance: A.A.P.G. Bull., v. 76, p. 1678-1709.

Ramsey, P.J., Smith, A.M. and Mason, T.R., 1996, Geostrophic sand ridge, dune fields, and associated bedforms from the northern KwaZulu-Natal shelf, south-east Africa: Sedimentology, v. 43, p. 407-419.

Rine, J.M., Tillman, R.W., Culver, S.J. and Swift, D.J.P., 1991, Generation of Late Holocene ridges on the middle continental shelf of New Jersey, USA—Evidence for formation in a mid-shelf setting based upon comparison with a nearshore ridge: *in* Swift, D.J.P., Oertel, G.F., Tillman, R.W. and Thorne, J.A., eds., Shelf Sand and Sandstone, Geometry, Facies, and Sequence Stratigraphy, Inter. Assoc. Sediment. Spec. Public. No. 14, p. 395-426.

Snedden, J.W., Nummedal, D. and Amos, A.F., 1988, Storm and Fairweather combined-flow on the central Texas continental shelf: Jour. Sed. Pet., vol. 58, no. 4, p. 580-595.

Snedden, J.W. and Nummedal, D., 1991, Origin and geometry of storm-deposited sand beds in modern sediments of the Texas continental shelf: *in* Swift, D.J.P., Tillman, R.W. and Oertel, G.F., eds., Spec. Public. Int. Assoc. Sediment. No. 14, p. 283-308.

Snedden, J.W., Kreisa, R.D., Tillman, R.W., Schweller, W.J., Culver, S.J. and Winn, R.D., Jr., 1994, Stratigraphy and genesis of a modern shoreface-attached sand ridge, Peahala Ridge, New Jersey: Jour. of Sedim. Research, v. B64, p. 560-581.

Stubblefield, W.L., McGrail, D.W. and Kersey, D.G., 1984, Recognition of transgressive and post-transgressive sand ridges on the New Jersey continental shelf—a reply: *in* Tillman, R.W. and Siemers, C.T., eds., Siliciclastic Shelf Sediments: SEPM Spec. Publ. No. 34, p. 37-41

Stuiver, R.M. and Daddario, J.J., 1963, Submergence of the New Jersey coast: Science, v. 142, p. 951.

Swift, D.J.P., Kofoed, J.W., Saulsbury, F.P. and Sears, P., 1972, Holocene evolution of shelf surface, central and southern Atlantic Shelf of North America: *in* Swift, D.J.P., Duane, D.B. and Pilkey, O.H., eds., Shelf Sediment Transport, Process and Pattern, Stroudsberg, Pennsylvania, Dowden, Hutchison and Ross, p. 447-498.

Swift, D.G. and Field, M.E., 1981, Evolution of a classic sand ridge field: Maryland Sector, North American inner shelf: Sedimentology, v. 28, p. 461-482.

Swift, D.J.P., McKinney, T.F. and Stahl, L., 1984, Recognition of transgressive and post-transgressive sand ridges on the New Jersey continental shelf—discussion: *in* Tillman, R.W. and Siemers, C.T., eds., Siliciclastic Shelf Sediments: SEPM Spec. Publ., No. 34, p. 25-36.

Tillman, R.W. and Martinsen, R.S., 1984, The Shannon shelf ridge sandstone complex, Salt Creek Anticline area, Powder River basin, Wyoming: *in* Tillman, R.W. and Siemers, C.T., eds., Siliciclastic Shelf Sediments: SEPM Spec. Publ., No. 34, p. 85-142.

SEDIMENTARY FACIES AND GENESIS OF HOLOCENE SAND BANKS ON THE EAST TEXAS INNER CONTINENTAL SHELF

ANTONIO B. RODRIGUEZ AND JOHN B. ANDERSON
Department of Geology and Geophysics, Rice University, Houston, TX 77005, U.S.A.
FERNANDO P. SIRINGAN
University of the Philippines, Quezon City, Philippines
AND
MARCO TAVIANI
Istituto per la Geologia Marina, Via Zamboni 65, 40127 Bologna, Italy

ABSTRACT: Sediment cores and high-resolution seismic and side-scan sonar data were collected from four shelf banks on the east Texas inner continental shelf. Sabine, Heald, Shepard, and Thomas banks all have similar sediment facies, structure, and genesis. The banks are composed of three facies (top to bottom): A) an interbedded shell hash and sand unit; B) a muddy-sand unit characterized by a seaward-prograding and chaotic seismic facies, and C) an interbedded sand and mud unit characterized by landward-dipping seismic reflectors. These three sediment facies represent amalgamated storm beds, lower-shoreface or ebb-tidal delta, and back-barrier/flood-tidal delta environments respectively. Facies B and C were deposited during a time of relatively slow sea-level rise and were stranded on the shelf during a rapid sea-level rise. Facies A is the result of storms and wind-driven currents reworking the paleoshoreline deposits on the shelf. The banks are drowned paleo-shorelines restricted to the area above and immediately adjacent to the Trinity and Sabine incised fluvial valleys. This association is explained by the greater thickness of sands within the valleys (larger sources of sands), greater accommodation space, and greater subsidence rate within the valleys (greater preservation potential).

INTRODUCTION

Previous studies of modern and ancient shelf sand bodies have produced two rather different schools of thought concerning the origins of these bodies. One school argues that formation is the result of hydrodynamic processes that occur on the continental shelf (Swift et al., 1974; Huthnance, 1982; Swift et al., 1984; Swift and Rice, 1984; Boczar-Karakiewicz and Bonas, 1986; Tillman and Martinsen, 1984, 1987; Rine et al., 1991); and the second school contends that formation is the result of coastal lithosomes being overstepped and stranded on the shelf during transgression (Curray, 1960; Sanders and Kumar, 1975; Stubblefield et al., 1984; Penland et al., 1988; Bergman, 1994; Walker and Bergman, 1993; Wagle and Veerayya, 1996). These two schools probably do not explain the origin of all shelf sand ridges and banks; in fact, they represent end-member type models.

Debate about the origin of the lower Campanian Shannon Sandstone in the Powder River Basin is an example of a recent controversy concerning the origin of shelf sand bodies (Tillman and Martinsen, 1984, 1987; Gaynor and Swift, 1988; Walker and Bergman, 1993; Bergman, 1994). The Shannon Sandstone consists of a series of elongated sand bodies contained within marine shales. One group of investigators believes that these sand bodies were deposited as a shelf ridge complex 160 km from the shoreline (Tillman and Martinsen, 1984, 1987; Gaynor and Swift, 1988; Rine et al., 1991). Another group of investigators argues that the Shannon sandbodies are shoreface deposits stranded and preserved on the shelf due to eustatic fluctuations (Walker and Bergman, 1993; Bergman, 1994).

Sand ridges on the New Jersey continental shelf have been studied as possible modern analogs to ancient shelf sand bodies. However, interpretations are mixed and indicate that the ridges were either deposited as a result of hydrodynamic processes occurring on the shelf (Swift et al., 1978; Swift et al., 1984; Rine et al., 1991) or as barrier islands that were "drowned" and modified during transgression (Stubblefield et al., 1984).

Snedden et al. (1994) conducted a detailed study of Peahala ridge located on the New Jersey continental shelf. The formation of this ridge was interpreted as the result of overstepping of a coastal lithosome (ebb-tidal delta) during transgression and subsequent reworking of this feature by hydrodynamic processes (Snedden et al., 1994, Snedden et al., this volume).

The east Texas inner continental shelf is a mud-dominated shelf. The only sands on the shelf are in the form of banks or are located within the Trinity/Sabine incised valley. This type of setting is very different from that at any other modern bank study site, but it is similar to the lower Campanian Shannon Sandstone in the Powder River Basin, for which east Texas banks are considered a reasonable modern analog.

The origin and evolution of four shelf sand banks (Thomas, Shepard, Heald, and Sabine banks) on the east Texas continental shelf were studied to better understand the mechanisms for shelf sand body formation (Fig. 1). These sand banks currently exist at water depths of -29 m, -17 m, -15 m, and -12 m, respectively, and are surrounded by marine muds. The shallower banks (Sabine, Heald, and Shepard) exhibit signs of active sediment transport and deposition, while it is unclear whether the outer, deeper bank (Thomas Bank) is currently active or is being buried in shelf muds.

Curray (1960) was the first to study the four shelf sand banks and speculated that they were formed during the Holocene transgression. Nelson and Bray (1970) performed a subsequent study using high-resolution seismic data, sediment cores, and surface grab samples from Sabine and Heald banks. They concluded that these banks formed in a shallow marine environment due to the reworking of shoreline deposits. Our investigation focused on the facies architecture, stratigraphic occurrence relative to key seismic surfaces, distribution relative to the Trinity/Sabine incised fluvial valley, and origin of the banks.

Isolated Shallow Marine Sand Bodies: Sequence Stratigraphic Analysis and Sedimentologic Interpretation.
SEPM Special Publication No. 64, Copyright © 1999
SEPM (Society for Sedimentary Geology), ISBN 1-56576-057-3, p. 165-178.

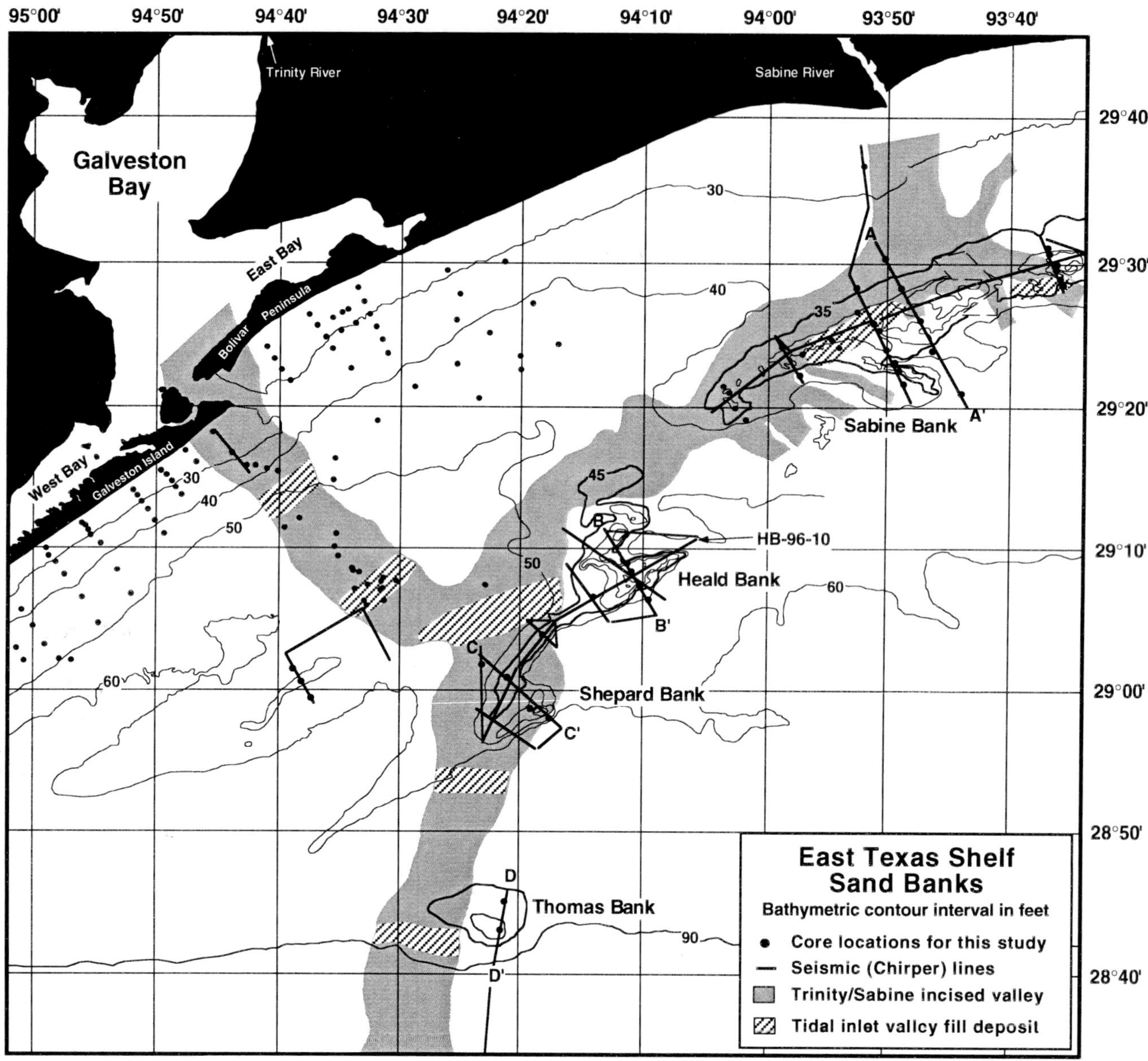

Fig. 1.—Bathymetric map of the east Texas shelf showing the locations of seismic lines, cores, and cross-sections. The Trinity/ Sabine incised river valley (gray) and tidal inlet deposits (striped) preserved within it are also shown (Thomas and Anderson, 1994; Siringan, 1993).

STUDY AREA

The Texas coast is characterized by fair weather astronomical tides ranging from 45 to 60 cm and relatively low-amplitude waves with periods commonly within 4-6 s (Morton and McGowen, 1980). The maximum mean significant wave height reported for this area is 2.1 m, and the maximum significant wave period is 8.3 s. Historical wave data indicate that wave heights can reach up to 7 m (Armstrong, 1980).

Unidirectional bottom flows produced by wind forcing are probably the most important mechanism for shelf sediment transport in the Gulf of Mexico (Morton, 1981). Fair weather wind- and storm-generated currents on the Texas shelf generally flow along-shore to the southwest (Morton, 1977; Snedden et al., 1988). However, Snedden et al. (1988) observed that on the shallow portions of the central Texas shelf (southwest of the study area) the southwest current switches to the northeast during summer months, coinciding with a southerly shift in prevailing winds. Average current speeds reported for the east Texas shelf range from about 10 to 30 cm/s. Current speeds measured during the passage of tropical cyclones ranged from 53 cm/s to 180 cm/s, with speeds farther offshore reaching in excess of 250 cm/s (Armstrong, 1980). During tropical storm Delia, bottom currents up to 200 cm/s at -18 m water depth were recorded approximately 30 km west-southwest of Sabine Bank (Forristall et al., 1977). The center of tropical storm Delia passed within 5 km west of the current meter.

Late Pleistocene to Holocene History

A brief overview of the response of depositional systems in the study area to the rise of sea level during the last glacial eustatic cycle (18,000 yr BP to present) is necessary before bank formation and origin can be understood. A more detailed description of the late Quaternary stratigraphy of shelf sediments in the study area can be found in Thomas and Anderson (1994) and Anderson et al. (1992, 1996).

During the Stage 2 sea-level lowstand, the Trinity and Sabine rivers incised into older highstand fluvial-deltaic deposits on the continental shelf; these valleys converge approximately 50 km offshore (Fig. 1). During the subsequent rise of sea-level, the incised valleys backfilled with fluvial, estuarine, and coastal (flood and ebb tidal delta) deposits. Seismic and lithologic data used to map lithofacies within the incised valley indicates incomplete valley-fill sequences bounded by intermediate flooding surfaces (Thomas and Anderson, 1994). The valley-fill facies architecture is characterized by backstepping parasequences that have been attributed to an episodic or step-like rise of sea level (Anderson et al., 1991; Thomas and Anderson, 1994). Periods of relatively slow sea-level rise are manifested by tidal inlet and associated flood and ebb tidal delta deposits. The banks lie adjacent to or directly over these tidal inlet deposits (Fig. 1). The interfluve portions of the study area are characterized by a thin (< 1m) layer of bioturbated marine mud resting directly on stiff, varigated Pleistocene (Beaumont Formation) clay (Fig. 2). The Beaumont formation consists of mainly coastal and delta plain silts and clays deposited on the subaerially exposed shelf during the previous highstand. The surface separating these units is an amalgamated stage 2 sequence boundary and transgressive ravinement surface (Siringan and Anderson, 1994).

METHODS

Approximately 4200 km of high-resolution seismic data, 15 km of side-scan sonar data, and 57 sediment cores were collected aboard the R/V *Lone Star* from the east Texas shelf (Fig. 1). A Datasonics CAP 6000 Chirper system using a 10 ms linear sweep from 1.5-7 kHz was used to collect the seismic data, which were then processed using ProMAX software. The processing consisted of applying a bandpass filter (500-800-2000-4000 Hz), automatic gain control (3 ms operator length), and trace mixing (over 7 traces). The method used for seismic interpretation centered around the recognition of important bounding surfaces and the identification of variations in seismic character. The acoustic travel time to depth conversion was computed under the assumption that the velocity of the upper few meters of sediment is the same as that of salt water (1525 m/s).

The side-scan sonar data were collected using a Datasonics SIS 1000 system. Sediment cores provide ground truth for the seismic data. They were collected using a pneumatic hammer coring device and a piston coring device, both capable of recovering cores up to 5 m in length. Radiocarbon dating (radiometric and accelerated mass spectrometry (AMS) techniques) of shell and peat material sampled from the cores was performed by Beta Analytic Inc.

RESULTS

Seismic Data Analysis

The banks are large bathymetric features. Sabine Bank, the largest feature (600 km^2), rests 7.5 m above the surrounding seafloor. Echo-sounding profiles oriented SE-NW across and perpendicular to the long axis of the banks

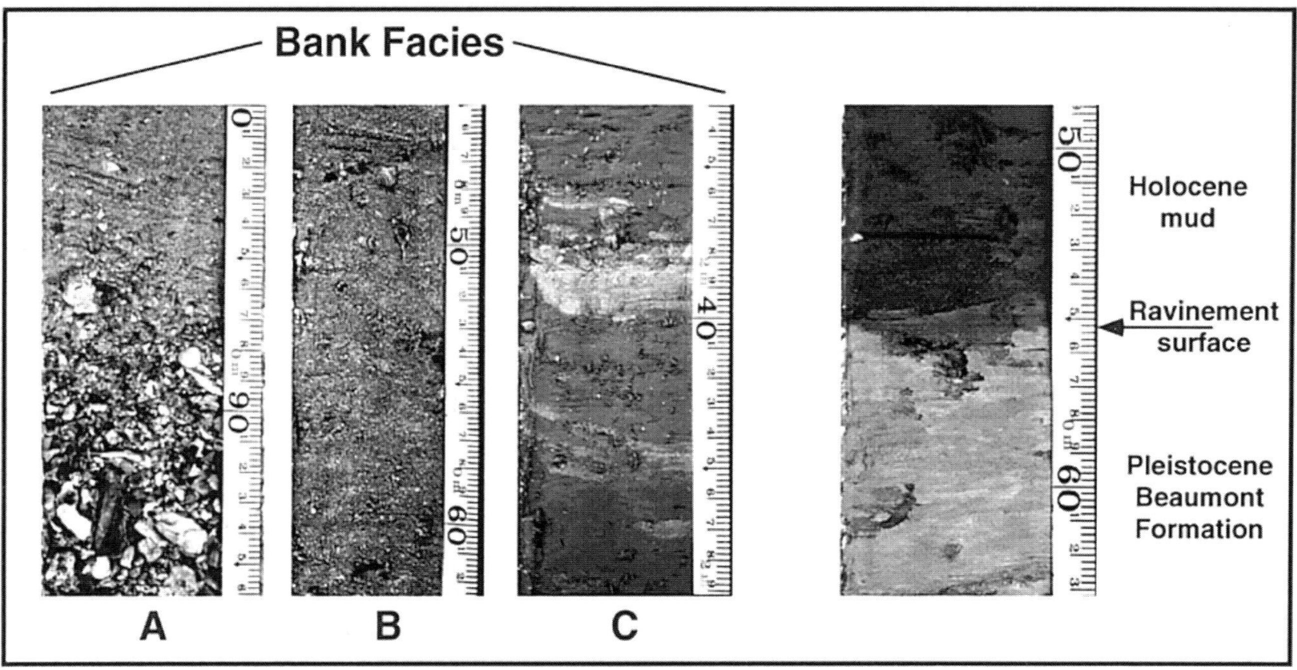

Fig. 2.—Representative core photographs that characterize the sedimentary units (lithologic units A, B and C) that make up the banks, and the interfluvial portions of the study area.

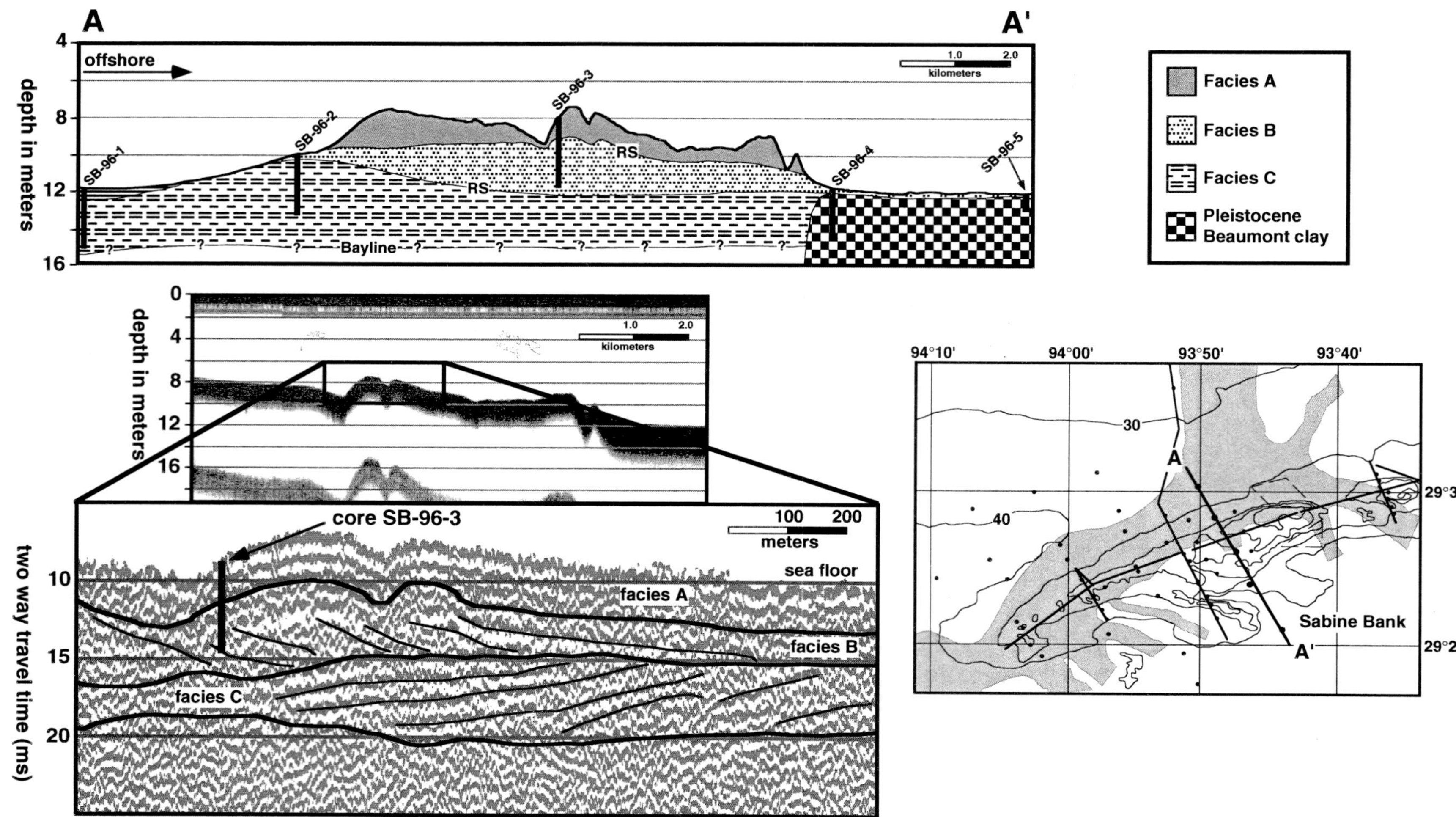

Fig. 3.—Cross-section A-A', echo-sounding profile, and a portion of a seismic (chirper) line through Sabine Bank. Five cores collected along the transect are indicated on the cross-section.

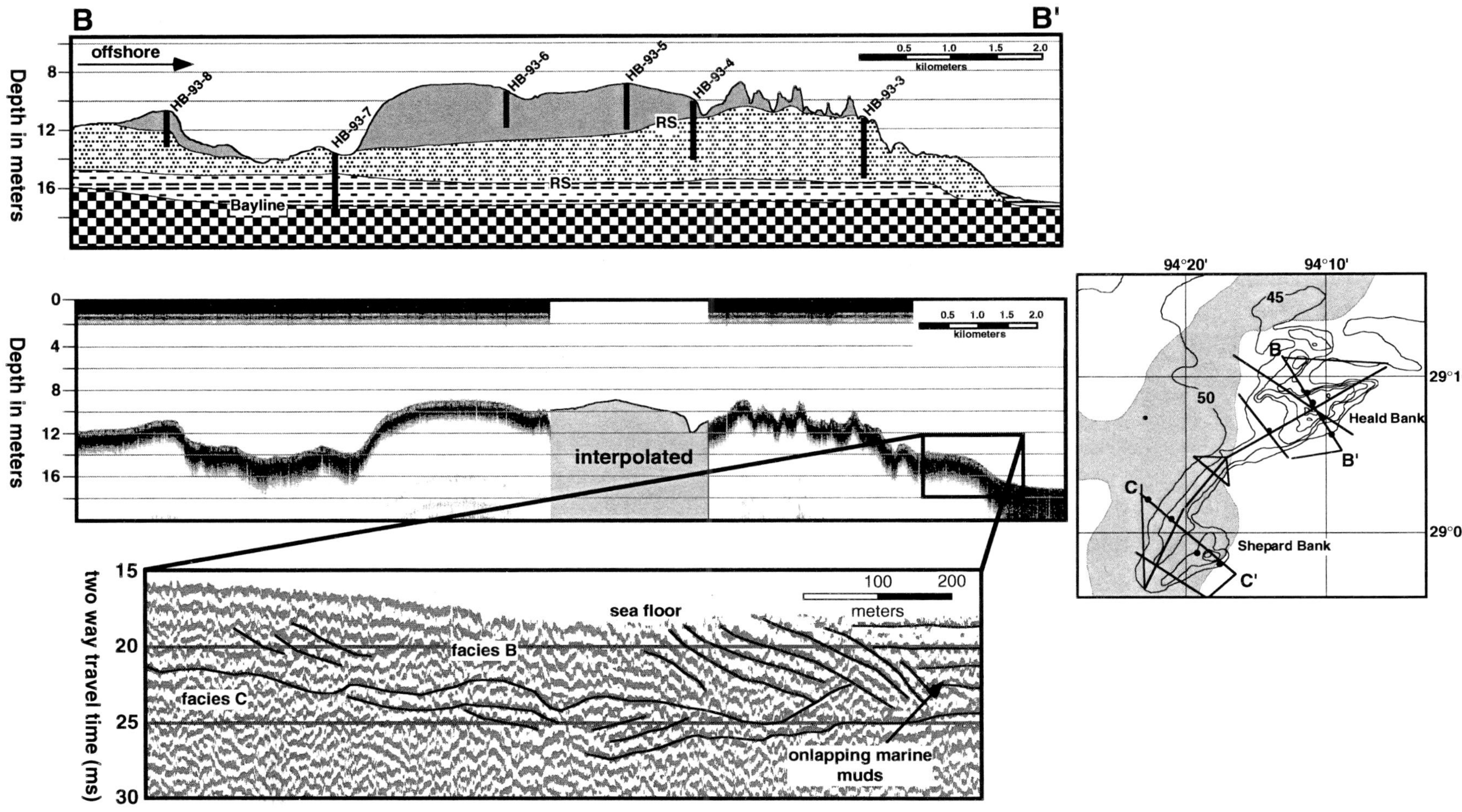

Fig. 4.—Cross-section B-B', echo-sounding profile, and a portion of a seismic (chirper) line through Heald Bank. Six cores collected along the transect are indicated on the cross-section.

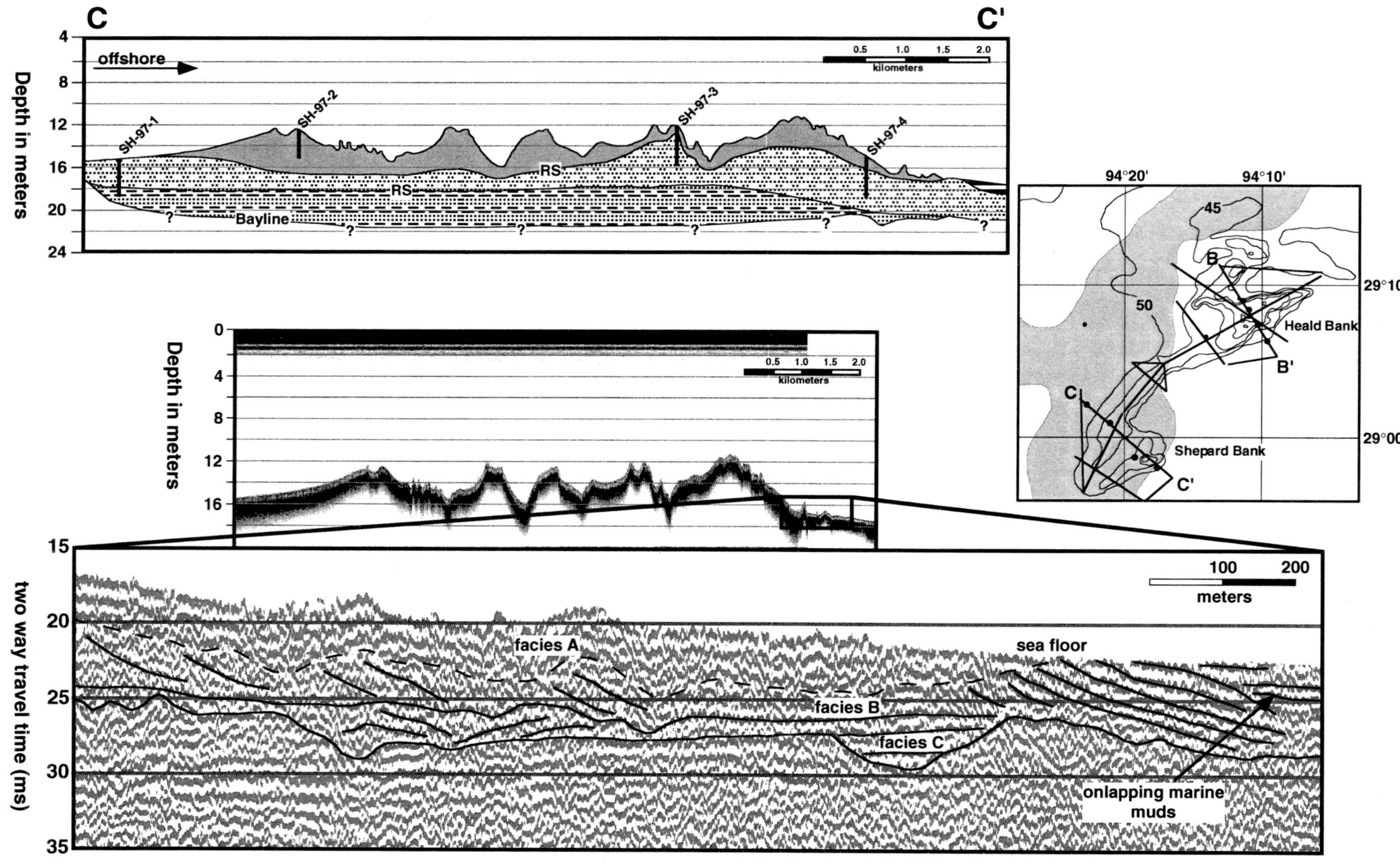

Fig. 5.—Cross-section C-C', echo-sounding profile, and a portion of a seismic (chirper) line through Heald Bank. Four cores collected along the transect are indicated on the cross-section.

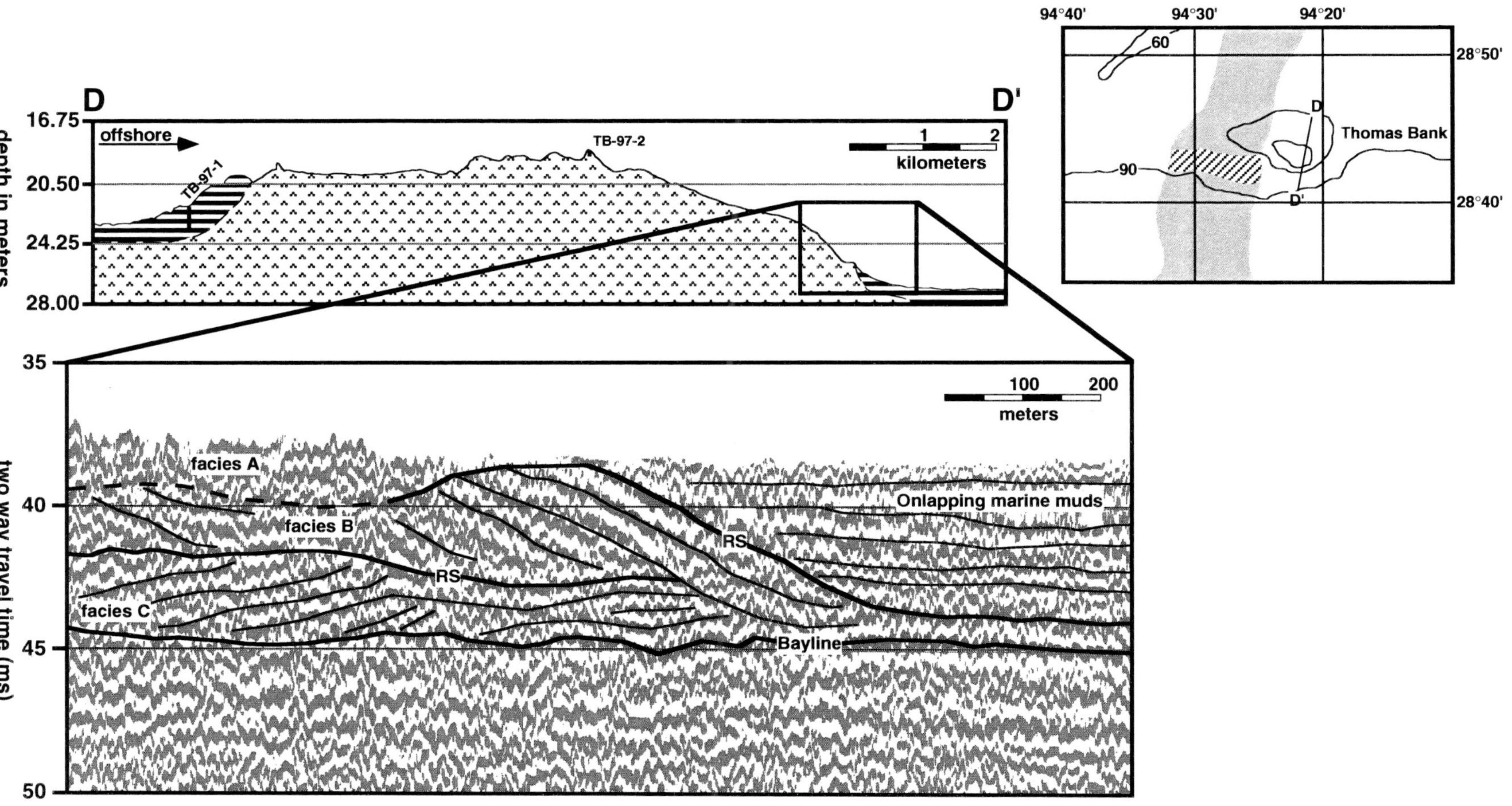

Fig. 6.—Echo-sounding profile trace D-D', and a portion of a seismic (Chirper) line through Thomas Bank. Only two cores were collected along this transect; therefore, no cross-section is shown.

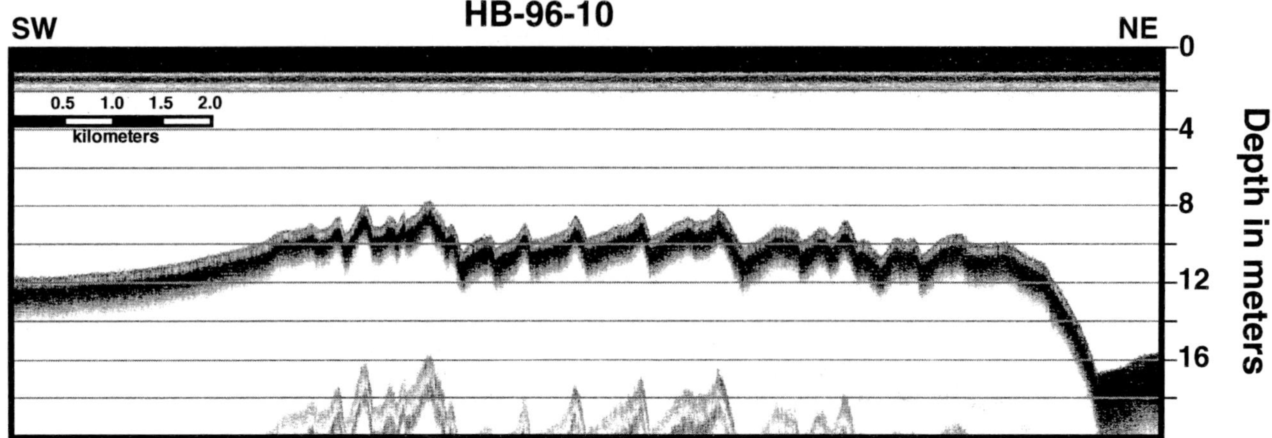

Fig. 7.—Echo-sounding profile across Heald Bank showing large-scale (2-m-high and 700-m-long) sand waves migrating toward the northeast.

show steeply dipping seaward flanks, irregular tops, and gently dipping landward flanks (Figs. 3, 4, 5, and 6). Figure 7 shows an echo-sounding axial profile oriented SW-NE across Heald Bank. Directly on top of the bank, very large (2-m-high and 700-m-long) two-dimensional or three-dimensional subaqueous dunes (Ashley, 1990) are shown migrating toward the NE. Side-scan sonar data acquired along the same transect did not image any smaller-scale bedforms. The NE flank of Heald Bank is the steepest feature in the study area (Fig. 7).

The banks contain three seismic facies units (SFU), A through C. A is the uppermost unit capping each bank (Figs. 3, 4, 5, and 6) and is characterized by a chaotic to acoustically reverberating seismic reflection pattern. Imaging the internal stratal geometries within SFU A is difficult due to the high amplitude sea-bottom reflection. This facies either shows a sharp erosional contact with, or is difficult to distinguish from, SFU B. SFU B is characterized by seaward-dipping reflectors. SFU B is most discernible on the seaward side of the banks, where it is exposed at the seafloor. This facies generally has a sharp erosional contact with SFU C, which is characterized by landward-dipping reflectors and large-scale cross-bedding. This facies is thickest directly over the incised valley.

Lithologic and Paleontologic Data Analysis

Only cores collected from the banks or the Trinity/Sabine incised valley contain any significant amount of transgressive deposits. Cores from interfluvial areas of the shelf penetrated a thin layer (tens of centimeters thick) of Holocene bioturbated, olive-gray sandy mud with basal clay rip-ups lying on mottled green and red, very stiff Pleistocene (Beaumont Formation) clay (Fig. 2). The shear strength of the clay is >1.0 kg/cm^2.

The correlation of seismic facies units with lithologic units is straightforward; the banks are composed of three general lithologic units, each of which corresponds with one of the seismic facies units. From top to bottom these are: A) an interbedded shell hash and sand unit; B) a muddy sand unit, and C) an interbedded sand and mud unit (Fig. 2). Unit A, the shelly sand facies, is approximately 2-5 m thick at the crest of the banks. The shell hash beds reach up to 25 cm in thickness while the sand beds are up to 45 cm thick. Contacts between the beds are generally sub-horizontal. Shells from unit A were sourced from a variety of environments, including bay-lagoonal (*Crassostrea virginica*), back-barrier/nearshore (*Crassinella lunulata, Mulinia lateralis, Natica pusilla, Anchis obesa*), and open marine (*Plicatula gibbosa, Strigilla mirabilis, Semele bellastriata*; Parker, 1960; Andrews, 1992). The boundary between units A and B (the muddy sand facies) varies from sharp to gradational.

Unit B is 1-3 m thick and composed of a clean to <10% mud-bioturbated sand. Grain size of the sand ranges from 2.5 to 2.0 phi. This facies subcrops on the bank flanks. Most of the shells from unit B (*Natica pusilla, Olivella dealbata, Ervilia concentrica, Caecum johnsoni,* and *Anadara transversa*) represent shoreface/inlet environments (Parker, 1960; Andrews, 1992). The boundary between unit B and unit C, the interbedded sand and clay facies, is gradational and difficult to identify from cores alone. The seismic data help to resolve this problem.

Unit C is slightly burrowed with thin (<2 cm), intercalated layers of sand and silt. Shells from this unit include *Ostrea equestris, Nassarius acutus, Nuculana concentrica,* and *Mulinia lateralis.* Although all the units contain *Mulinia lateralis,* this species is most abundant in unit C, where some layers are monospecific. This shell assemblage is typical of bay/inlet environments (Parker, 1960; Andrews, 1992).

Nelson and Bray (1970) obtained radiocarbon dates for 15 samples from Sabine Bank and for 20 samples from Heald Bank. An additional three radiocarbon dates from Heald Bank and two from Sabine Bank were obtained for the recent study. Marine shells collected from Sabine Bank unit A yielded ages ranging from 215 ± 150 yr BP to 383 ± 210 yr BP. Back-barrier shells from Sabine Bank unit A yielded ages ranging from 2,210 ± 243 yr BP to 7,500 ± 330 yr BP (Nelson and Bray, 1970) indicating that the shells are relict material. Nelson and Bray (1970) reported a radiocarbon date of 4,630 ± 215 yr BP from an *Anadara transversa* shell acquired from Sabine Bank at the base of unit B. An AMS date of 4490 ± 50 yr BP was measured from an articulated *Ostrea equestris* shell sampled from Sabine Bank at the top of unit C. Freshwater peats sampled from below Sabine Bank at the base of unit C yielded a radiocarbon age of 7,800 ± 70 yr BP. *Mulinia lateralis* shells from Heald Bank in unit B yielded an age of 7065 ± 275 yr BP. Freshwater peats sampled from below unit C in Heald Bank yielded an age of 8570 ± 70 yr BP.

DISCUSSION

The sand banks are presently being modified by currents and storms as indicated by bedforms on Heald Bank (Fig. 7). High gradients associated with seaward (southern) sloping bank flanks are attributed to the fact that storms typically approach the east Texas coast from the SW (Snedden et al., 1988). Unit A is mostly composed of reworked sand and shells. Radiocarbon dating of marine and back-barrier shells from Sabine Bank unit A yielded quite variable ages, 215 ± 150 yr BP to 7,500 ± 330 yr BP (Nelson and Bray, 1970), indicating a contribution of modern and reworked material. The gradational nature of the contact between units A and B, as observed from core and seismic data, indicates reworking of unit B into unit A. The older shells from unit A in both Sabine Bank and Heald Bank are always back-barrier or shoreface/inlet species, which indicates that they were sourced from excavation of earlier deposits. The presence of both back-barrier and marine mollusc shells indicates reworking and bed-load transport at current water depths. Thus, unit A represents amalgamated storm beds presently being modified and deposited at water depths of up to -24 m.

Unit B is only exposed at the seafloor on the banks' seaward flanks, which are most heavily affected during storms. SFU B shows seaward-dipping reflectors and an inlet or shoreface faunal assemblage, which indicates that this unit represents either an ebb-tidal delta or lower shoreface environment. Unit B lacks distinct sand mud interbeds and laminations, characteristic of the modern Bolivar Roads ebb-tidal delta (Siringan and Anderson, 1993). The lithology instead resembles the bioturbated shoreface deposits offshore of Galveston Island (Siringan and Anderson, 1994).

SFU C shows landward-dipping reflectors and a bay or tidal inlet faunal assemblage, which indicates that this unit represents either a flood-tidal delta or other back-barrier environment. The seismic expression of the modern Bolivar Roads flood-tidal delta shows channel stacking and cut-and-fill geometries near the inlet, and landward-dipping reflectors near the bay (Siringan and Anderson, 1993). This compares well to the geometries observed in SFU C. Siringan and Anderson (1993) collected 18 cores from the modern Bolivar Roads flood-tidal delta. The distal portions of the flood-tidal delta were described as mud-dominated (interlaminated clay and fine sand) with few shell beds (Siringan and Anderson, 1993). Unit C has similar lithologies.

The examination of Thomas Bank was intended to provide an example of a bank situated below storm influence. Thomas Bank was first studied by Thomas and Anderson (1994), although they referred to it as -29 m bank. Thomas and Anderson (1994) suspected that preserved barrier island/lagoon or strandplain facies directly underlie Thomas Bank, similar to Sabine, Heald, and Shepard banks. Our hypothesis was that Thomas Bank would be draped in marine muds because of its depth. Although the burial process has clearly begun, Thomas Bank is not currently draped in marine muds. Two piston cores were collected from Thomas Bank. The first core location was on the northern bank flank and sampled 133 cm of olive-gray marine mud and the second location was on the top of the bank and sampled 20 cm of well-sorted, shelly sand 1.5-2.0 phi (Facies A; Fig. 6). The thick marine mud unit sampled from the northern bank flank is evidence that Thomas Bank may be in the process of becoming completely buried.

Sabine Bank appears to be the youngest reworked paleoshoreline deposit in the study area. Radiocarbon dates from fresh water peats below unit C and an *Ostrea equestris* shell at the top of unit C indicates that Sabine Bank is between 7,800 ± 70 yr BP and 4,490 ± 50 yr BP in age. The gradational contact between units B and C makes it difficult to ascertain whether the dates from the *Anadara transversa* shell (4,630 ± 215 yr BP) in the base of unit B and the *Ostrea equestris* shell (4,490 ± 50 yr BP) in the top of unit C are actually from those units or from the gradational contact. Because the *Ostrea equestris* shell was articulated, it was preserved *in situ* and therefore the 4,490 ± 50 yr BP date measured from it is the best youngest age estimate for this paleoshoreline. Modern and 7,500 ± 330 yr BP ages from unit A also indicate active reworking at present water depths. Fresh water peats below unit C show the age of the Heald Bank paleoshoreline to be between 8570 ± 70 yr BP and 7065 ± 275 yr BP (from *Mulinia lateralis* shells in unit B). The timing of Thomas Bank and Shepard Bank formation is unknown but by inference from the Holocene sea level curve (Bard et al., 1990), the paleoshoreline associated with Thomas Bank is approximately 9,500 years old. Shepard Bank either formed before or simultaneous with Heald Bank.

Bank Formation and Evolution

Bank formation is closely tied to changing rates of sea-level rise during transgression, the Trinity/Sabine incised valley, and the hydrodynamic setting of the shelf. During the last lowstand, an irregular surface—a sequence boundary—formed on the shelf. This surface marks the top of the Beaumont Formation. During the subsequent transgression this surface was reshaped by wave erosion and the incised valleys were infilled with back-stepping parasequences (Thomas and Anderson, 1994). The back-stepping parasequences within the Trinity/Sabine incised valley record flooding events during rapid sea-level rises. Tidal inlet deposits within the valley and sand banks above and adjacent to the valley represent episodes when sea-level rose slowly, allowing barriers and associated inlets to evolve.

Figure 8 illustrates the four general stages of bank evolution in the study area. During stage one, sea-level rose relatively slowly, enabling stabilization of the shoreline and construction of barrier islands and tidal inlets. Units B and C were deposited at this time. Unit C was deposited in back-barrier bays situated within and adjacent to the Trinity/Sabine incised valley. Hence, these deposits were formed in an environment very similar to modern East Bay and West Bay (Fig. 1). The base of unit C represents a flooding surface—a bayline—as evidenced by peat near the contact, back-barrier muds above the contact, and Pleistocene clay below the contact. During stage 2, the rate of sea-level rise accelerated. The barrier island responded by retreating shoreward, depositing unit B over unit C as the shoreface reestablished itself. The contact between unit C and unit B is a ravinement surface. Unit B represents the preserved shoreface deposits of the prograding shoreline.

Subsidence rates are much faster over the incised valley (0.62 cm/yr based on tide gauge records from Galveston Bay) relative to the interfluvial areas of the shelf (0.01 cm/yr; Paine, 1993); this ultimately helped to preserve unit C along the flanks of the valley. Stage 3 was a time of rapid sea-level rise, which caused the coastal lithosomes to be "drowned," or overstepped, leaving them isolated on the shelf. This final flooding produced a second ravinement surface that truncated the shoreface (unit B) and is located within or at the base of unit A.

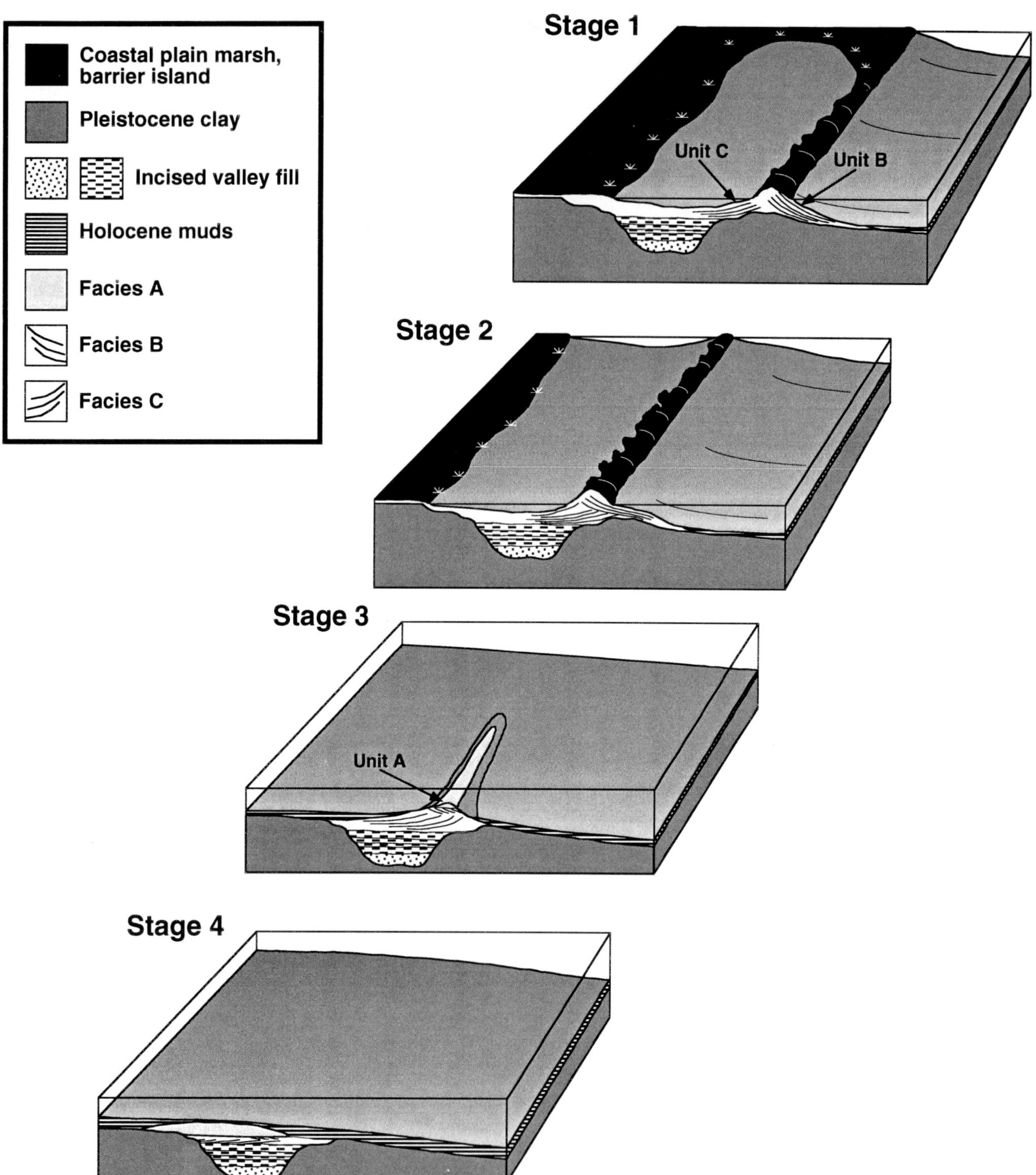

Fig. 8.—Block diagram illustrating barrier island formation, retreat, overstepping, and reworking, which are the four stages of bank genesis and evolution.

An examination of how Galveston Island and Bolivar Peninsula would respond to a rapid rise in sea level illustrates the above sequence of events. The Bolivar Peninsula and Galveston Island sand bodies both thicken toward the Trinity River incised valley (Fig. 9). This is due to Bolivar Roads tidal inlet, a stable, mixed-energy, tide-dominated tidal inlet/delta complex that sits directly over the incised valley and the easily compacted/eroded soft estuarine valley fill deposits (Siringan and Anderson, 1993). The depth to the modern ravinement surface is marked by the onlap of offshore marine

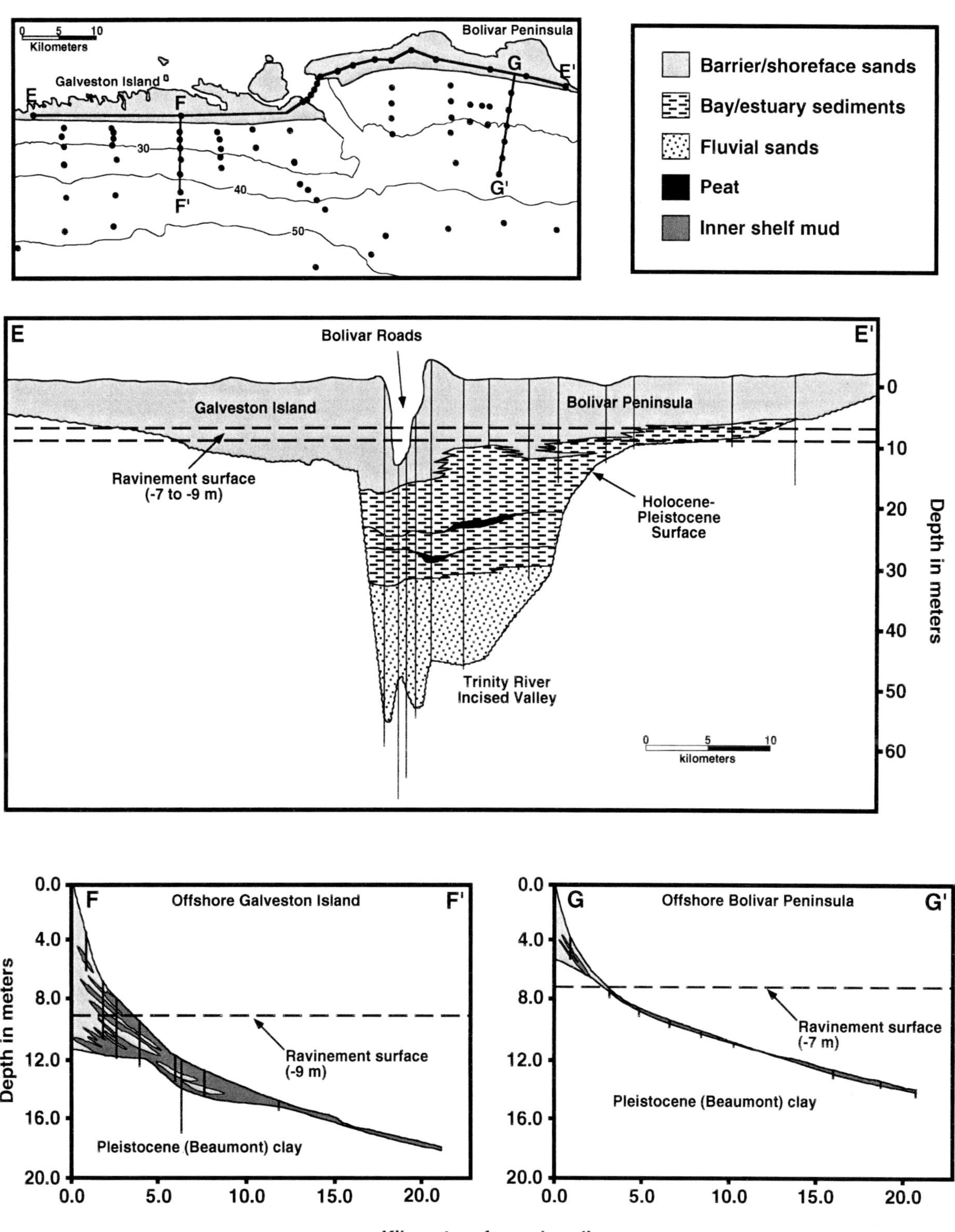

Fig. 9.—Cross-section E-E' through Galveston Island and Bolivar Peninsula shows thick barrier island and shoreface sands located within and adjacent to the Trinity River incised valley. Profiles G-G' and F-F' are representative core transects that illustrate how shoreface ravinement has removed virtually all coastal deposits on the inner shelf. The ravinement surface is marked by offshore marine muds onlapping either Pleistocene deposits (offshore Bolivar Peninsula) or lower shoreface deposits (offshore Galveston Island) and occurs at -7m to -9 m water depths. With continued sea-level rise and current depths of ravinement, coastal lithosomes will be eroded everywhere except in the incised valley (modified from Cole and Anderson, 1982, Siringan and Anderson, 1993, and Siringan and Anderson, 1994).

muds onto shoreface deposits and corresponds closely to the toe of the shoreface (abrupt change in the shoreface profile). The ravinement surface increases from -7 m off Bolivar Peninsula to -9 m off Galveston Island (Siringan and Anderson, 1994; Fig. 9). Offshore Bolivar Peninsula the ravinement surface cuts deep into the shoreface profile, leaving only a thin layer of Holocene marine mud resting on Pleistocene clay (Fig. 9). Offshore Galveston Island the ravinement surface has cut the shoreface at a higher level, preserving shoreface deposits beneath marine muds (Fig. 9). When the ravinement surface cuts through the expanded barrier island/inlet complex; the shoreline deposits located adjacent to and directly over the Trinity River incised valley—a relatively thick section of the coastal lithosome—will be preserved (Fig. 9). Bolivar Roads tidal inlet will continue to redistribute sediment to the tidal delta complex and the adjacent shoreline during the initial sea-level rise. Hence, the tidal delta complex and the shoreline adjacent to it will remain in place longest before the entire barrier island complex is removed or truncated and finally submerged. It is this portion of the barrier island complex that will be stranded on the shelf as a bank. In the case of offshore banks, initial flooding created the lower ravinement surface, and then progradation and aggradation of coastal lithosomes created the bank topography. Galveston Island sits atop a ravinement surface and has a similar history (Bernard et al., 1970).

The overstepping events correspond to episodes of significant flooding, as indicated by backstepping (tens of kilometers) of the valley fill facies in the Trinity/Sabine incised valley (Thomas and Anderson, 1994). Subsequent reworking of units B and C by storms and wind-driven currents led to the formation of unit A. With time, unit A continues to thicken at the expense of the lower units. Heald, Shepard, and Thomas banks are older and have thicker unit A's than Sabine Bank, indicating that the longer these features are stranded on the shelf, the sandier they become. The banks also migrate up-dip (landward) through time as the lower units are reworked into unit A. Thomas Bank has migrated up-dip approximately 5 km from its associated tidal inlet deposit (Thomas and Anderson, 1994). The banks may evolve to a point where only unit A is preserved as a shelf sand body encapsulated within marine shelf muds; however, none of the banks studied have reached that point. The 133 cm of marine mud sampled from the northern flank of Thomas Bank are evidence that it has started to become encapsulated. Holocene sea-level rise was very rapid, certainly in comparison with nonglacial intervals of time, such as the Mesozoic. Given a slower rate of sea-level rise, bank isolation and encapsulation in marine muds on the shelf are more likely.

This sand bank formation and evolution model is similar to what Berne et al. (1994) evoke in the southern North Sea for the formation of Middelkerke Bank, and to what Snedden et al. (in this volume) evoke for the formation and evolution of sand ridges on the Atlantic shelf. There are important differences though. Snedden et al. (in this volume) interpret a very rigorous final stage of ridge evolution where a ridge loses almost all of its original characteristics through extensive migration and sediment volume increases. Thomas Bank is reaching its final stage of evolution by becoming encapsulated in marine muds; however, its seismic facies have remained similar in character to those of the younger banks, and it is still situated adjacent to the Trinity/Sabine incised valley. Thomas Bank may still evolve to a point where only facies A is present; however, given the present shelf setting, it will most likely be entirely encapsulated in marine muds before it is completely reworked. Hence, these banks on the east Texas shelf will always remain in a predictable position: adjacent to the Trinity/Sabine incised valley and a tidal delta complex preserved within the valley (Thomas and Anderson, 1994).

CONCLUSIONS

Holocene sand banks of the east Texas shelf are composed of three facies (top to bottom): A) an interbedded shell hash and sand unit; B) a muddy-sand unit characterized by a seaward-prograding and chaotic seismic facies, and C) an interbedded sand and mud unit characterized by landward-dipping seismic reflectors. Unit C represents a back-barrier/flood-tidal delta environment and unit B represents a lower-shoreface or ebb-tidal delta environment. These coastal lithosomes were stranded on the shelf during rapid transgression. Unit A was deposited as storms and wind-driven currents reworked ancestral paleoshoreline deposits on the shelf. It is possible that the lower units (B and C) will become completely reworked into unit A, resulting in a shelf-sand body encapsulated within shelf muds. Thus, sand banks on the east Texas shelf are formed through a combination of hydrodynamic processes occurring on the shelf and overstepping and stranding of coastal lithosomes during transgression.

ACKNOWLEDGMENTS

We thank Captain Mark Herring and the crew of the R/V *Lone Star*, who helped collect the data for this project. We also acknowledge John Bradford, who greatly aided in the collection and processing of the seismic data. A critical review by John Snedden that improved this manuscript is appreciated. Funding for this project was provided by a consortium of oil companies including Agip, Amoco, BP, Conoco, Exxon, Marathon, Mobil, Pan Canadian, Shell, and Union Pacific Resources and The American Chemical Society-Petroleum Research Fund. This is Istituto di Geologia Marina scientific contribution #1098.

REFERENCES

ANDERSON, J.B., ABDULAH, K., SARZALEJO, S., SIRINGAN, F. AND THOMAS, M.A., 1996, Late Quaternary sedimentation and high-resolution sequence stratigraphy of the east Texas shelf, *in* DeBatist, M. and Jacobs, P., eds., Geology of Siliciclastic Shelf Seas: Boulder, Geological Society of America Special Publication 117, p. 95-124.

ANDERSON, J.B., SIRINGAN, F.P., SMYTH, W.C. AND THOMAS, M.A., 1991, Episodic nature of Holocene sea-level rise and the evolution of Galveston Bay (abs.): Gulf Coast Section, Society of Economic Paleontologists and Mineralogists Foundation 12th Annual Research Conference, Program and Abstacts, p. 8-14.

ANDERSON, J.B., THOMAS, M.A., SIRINGAN, F.P. AND SMYTH, W.C., 1992, Quaternary evolution of the east Texas coast and continental shelf, *in* Fletcher, C.H., and Wehmiller, J.F., eds., Quaternary Coasts of the United States: Marine and Lacustrine Systems: Tulsa, Society of Economic Paleontologists and Mineralogists Special Publication 48, p. 253-263.

ANDREWS, J., 1992, A Field Guide to Shells of the Texas Coast: Texas Monthly Field Guide Series, Gulf Publishing Company, Houston, 176 p.

ARMSTRONG, R.S., 1980, Current patterns and hydrography, vol. iv, *in* Jackson, W.B., and Wilkens, E.P., eds., Environmental assessment of Buccaneer gas and oil field in the north western Gulf of

Mexico: Springfield, National Oceanic Atmospheric Administration Technical Memorandum, NMFS-SEFC-40, 33 p.

ASHLEY, G.M., 1990, Classification of large-scale subaqueous bedforms: A new look at an old problem: Journal of Sedimentary Petrology, v. 60, p. 160-172.

BARD, E., HAMELIN, B. AND FAIRBANKS, R.G., 1990, U-Th ages obtained by mass spectrometry in corals from Barbados: Sea level during the past 130,000 years: Nature, v. 346, p. 456-458.

BERGMAN, K.M., 1994, Shannon sandstone in Hartzog Draw-Heldt Draw fields (Cretaceous, Wyoming, USA) reinterpreted as lowstand shoreface deposits: Journal of Sedimentary Research, v. B64, p. 184-201.

BERNARD, H.A., MAJOR, C.F., JR., PARROT, B.S. AND LEBLANC, R.J., 1970, Recent sediments of southeast Texas—A field guide to the Brazos alluvial and deltaic plains and the Galveston barrier island complex, *in* Bureau of Economic Geology Guidebook 11: Austin, The University of Texas, 132 p.

BERNE, S., TRENTASAUX, A., STOLK, A., MISSIAEN, T. AND DE BATIST, M., 1994, Architecture and long term evolution of a tidal sandbank: The Middelkerke Bank (southern North Sea): Marine Geology, v. 121, p. 57-72.

BOZCAR-KARAKIEWICZ, B. AND BONA, J.L., 1986, Wave-dominated shelves: A model of sand ridge formation by progressive, infragravity waves, *in* Knight, R.J. and Mclean, J.R., eds., Shelf Sands and Sandstones: Calgary, Canadian Society of Petroleum Geologists Memoir, v. 11, p. 163-179.

COLE, M.L. AND ANDERSON, J.B., 1982, Detailed grain size and heavy mineralogy of sands of the northeastern Texas gulf coast: Implications with regard to coastal barrier development: Transactions—Gulf Coast Association of Geological Societies, v. 32, p. 555-563.

CURRAY, J.P., 1960, Sediments and history of the Holocene transgresion, continental shelf, northwest Gulf of Mexico, *in* Shepard, F.P., Phleger, F.B. and Van Andel, T.H., eds., Recent Sediments, Northwest Gulf of Mexico: Tulsa, American Association of Petroleum Geologists, p. 221-266.

FORRISTALL, G.C., HAMILTON, R.C. AND CARDONE, V.J., 1977, Continental shelf currents in Tropical Storm Delia: Observation and theory: Journal of Physical Oceanography, v. 7, p. 532-546.

GAYNOR, G.G. AND SWIFT, D.J.P., 1988, Shannon Sandstone depositional model: Sand ridge dynamics on the Campanian western interior shelf: Journal of Sedimentary Petrology, v. 58, p. 868-880.

HUTHNANCE, J.M., 1982, On the formation of sand banks of finite extent: Estuarine, Coastal and Shelf Science, v. 15, p. 277-299.

MORTON, R.A., 1977, Historical shoreline changes and their causes, Texas Gulf Coast: Gulf Coast Association of Geological Societies Transactions, v. 27, p. 352-364.

MORTON, R.A., 1981, Formation of storm deposits by wind-forced currents in the Gulf of Mexico and the North Sea, *in* Nio, S.D., Shuttenhelm, R.T.E. and van Weering, C.E., eds., Holocene Marine Sedimentation in the North Sea Basin: Oxford, International Association of Sedimentologists Special Publication 5, p. 385-396.

MORTON, R.A. AND MCGOWEN, J.H., 1980, Modern depositional environments of the Texas Coast, *in* Bureau of Economic Geology Guidebook 20: Austin, The University of Texas, 167 p.

NELSON, H.F. AND BRAY, E.E., 1970, Stratigraphy and history of the Holocene sediments in the Sabine-High Island area, Gulf of Mexico, *in* Morgan, J.P., ed., Deltaic Sedimentation Modern and Ancient: Tulsa, Society of Economic Paleontologists and Mineralogists Special Publication 15, p. 48-77.

PAINE, J.G., 1993, Subsidence of the Texas coast: Inferences from historical and late Pleistocene sea levels: Tectonophysics, v. 222, p. 445-458.

PARKER, R.H., 1960, Ecology and distributional patterns of marine macro-invertebrates, nothern Gulf of Mexico, *in* Phleger, F.B. and Van Andel, T.H., eds., Recent Sediments, Northwest Gulf of Mexico: Menasha, American Association of Petroleum Geologists, p. 302-337.

PENLAND, S., BOYD, R. AND SUTER, J.R., 1988, Transgressive depositional systems of the Mississippi delta plain: A model for barrier shoreline and shelf sand development: Journal of Sedimentary Petrology, v. 58, p. 932-949.

RINE, J.M., TILLMAN, R.W., CULVER, S.J. AND SWIFT, D.J.P., 1991, Generation of Late Holocene ridges on the middle continental shelf of New Jersey, USA—Evidence for formation in a mid-shelf setting based upon comparison with a nearshore ridge, *in* Swift, D.J.P., Oertel, G.F., Tillman, R.W. and Thorne, J.A., eds., Shelf Sand and Sandstone Geometry, Facies and, Sequence Stratigraphy: Boston, International Association of Sedimentologists Special Publication 14, p. 395-426.

SANDERS, J.E. AND KUMAR, N., 1975, Evidence of shoreface retreat and in-place "drowning" during Holocene submergence of barriers, shelf off Fire Island, New York: Geological Society of America Bulletin, v. 86, p. 65-76.

SIRINGAN, F.P., 1993, Coastal lithosome evolution and preservation during an overall rising sea-level: east Texas Gulf Coast and continental shelf: unpublished Ph. D. Dissertation, Rice University, Houston, 226 p.

SIRINGAN, F.P. AND ANDERSON, J.B., 1993, Seismic facies, architecture, and evolution of the Bolivar Roads tidal inlet/delta complex, east Texas Gulf Coast: Journal of Sedimentary Petrology, v. 63, p. 794-808.

SIRINGAN, F.P. AND ANDERSON, J.B., 1994, Modern shoreface and inner-shelf storm deposits off the east Texas coast, Gulf of Mexico: Journal of Sedimentary Research., v. B64, p. 99-110.

SNEDDEN, J.W., NUMMEDAL, D. AND AMOS, A.F., 1988, Storm- and fair-weather combined flow on the central Texas continental shelf: Journal of Sedimentary Petrology, v. 58, p. 580-595.

SNEDDEN, J.W., KREISA, R.D., TILLMAN, R.W., SCHWELLER, W.J., CULVER, S.J. AND WINN, R.D., 1994, Stratigraphy and genesis of a modern shoreface-attached sand ridge, Peahala Ridge, New Jersey: Journal of Sedimentary Research, v. B64, p. 560-581.

STUBBLEFIELD, W.L., MCGRAIL, D.W. AND KERSEY, D.G., 1984, Recognition of transgressive and post-transgressive sand ridges on the New Jersey continental shelf—a reply, *in* Tillman, R.W. and Siemers, C.T., eds., Siliciclasitic Shelf Sediments: Tulsa, Society of Economic Paleontologists and Mineralogists Special Publication 34, p. 37-41.

SWIFT, D.J.P., DUANE, D.B. AND MCKINNEY, T.F., 1974, Ridge and swale topography of the middle Atlantic bight: Secular response to Holocene hydraulic regime: Marine Geology, v. 14, p. 1-43.

SWIFT, D.J.P, PARKER, G., LANFREDI, N.W., PERILLO, G. AND FIGGE, K., 1978, Shoreface-connected sand ridges on American and European shelves: A comparison: Estuarine and Coastal Marine Science, v. 7, p. 257-273.

SWIFT, D.J.P., MCKINNEY, T.F. AND STAHL, L., 1984, Recognition of transgressive and post-transgressive sand ridges on the New Jersey continental shelf—disscussion, *in* Tillman, R.W. and Siemers, C.T., eds., Siliciclasitic Shelf Sediments: Tulsa, Society of Economic Paleontologists and Mineralogists Special Publication 34, p. 25-36.

SWIFT, D.J.P. AND RICE, D.D., 1984, Sand bodies on muddy shelves: A model for sedimentation in the western interior Cretaceous seaway, North America, *in* Tillman, R.W. and Siemers, C.T., eds., Siliciclastic Shelf Sediments: Tulsa, Society of Economic Paleontologists and Mineralogists Special Publication 34, p. 43-62.

THOMAS, M.A. AND ANDERSON, J.B., 1994, Sea-level controls on the facies architecture of the Trinity/Sabine incised-valley system, Texas continental shelf, *in* Boyd, R., Zaitlin, B.A. and Dalrymple, R., eds., Incised-valley systems: Origin and sedimentary sequences: Tulsa, Society of Economic Paleontologists and Mineralogists Special Publication 51, p. 63-82.

TILLMAN, R.W. AND MARTINSEN, R.S., 1984, The Shannon shelf ridge sandstone complex, Salt Creek Anticline area, Powder River basin, Wyoming, *in* Tillman, R.W. and Siemers, C.T., eds., Siliciclasitic Shelf Sediments: Tulsa, Society of Economic Paleontologists and Mineralogists Special Publication 34, p. 1-34.

TILLMAN, R.W. AND MARTINSEN, R.S., 1987, Sedimentologic model and production characteristics of Hartzog Draw Field, Wyoming, a Shannon shelf ridge sandstone, *in* Tillman, R.W. and Weber, K.J., eds., Reservoir Sedimentology: Tulsa, Society of Economic Paleontologists and Mineralogists Special Publication 40, p. 15-112.

WAGLE, B.G. AND VEERAYYA, M., 1996, Submerged sand ridges on the western continental shelf off Bombay, India: Evidence for Late Pleistocene-Holocene sea-level changes: Marine Geology, v. 136, p. 79-95.

WALKER, R.G. AND BERGMAN, K.M., 1993, Shannon sandstone in Wyoming: A shelf-ridge complex reinterpreted as lowstand shoreface deposits: Journal of Sedimentary Petrology, v. 63, p. 839-851.

SEDIMENT TRANSPORT IN THE WESTERN INTERIOR SEAWAY OF NORTH AMERICA: PREDICTIONS FROM A CLIMATE-OCEAN-SEDIMENT MODEL

RUDY SLINGERLAND

Department of Geosciences, The Pennsylvania State University, University Park, Pennsylvania 16802 U.S.A.

AND

TIMOTHY R. KEEN

Naval Research Laboratory, Oceanography Division, Stennis Space Center, Mississippi 39529 U.S.A.

ABSTRACT: Whether from the foreshore, shoreface, shelf, or incised estuarine valleys, sedimentary deposits along the western edge of the Western Interior seaway quite uniformly record southerly directed paleoflows. Cardium Formation shoreface gravels at Willesden Green, Alberta, decrease in clast size to the southeast. Isoliths outlining clastic wedges, such as the Chalk Creek, are recurved to the south. Large-scale cross-strata in rocks considered to be either shelf sand ridges or detached shorefaces, such as the Kakwa and Musreau Members of the Cardium Formation and the Straight Cliffs Formation of southwestern Utah, indicate southerly directed paleocurrents. Estuarine incised valley-fills trend south or southeast, reflecting a high-stand shelf topography inherited by rivers as they cut across the inner shelf in response to a high-order sea-level drop. To explain this uniformity we conducted two numerical experiments that predict circulation and sediment transport paths in the seaway in response to 1) mean annual atmospheric forcing and 2) the passage of a mid-latitude winter storm. The mean annual forcing for the early Turonian is computed by GENESIS, an NCAR global climate model; the cyclone is computed using an idealized hurricane model. For the mean annual experiment, circulation of the seaway is computed using a three-dimensional, turbulent flow, coastal ocean model under the following initial and boundary conditions: 1) paleobathymetry according to a new interpretation of the litho- and bio-stratigraphy for the early Turonian; 2) fresh water runoff and precipitation-evaporation magnitudes as computed by GENESIS; 3) temperatures and salinities of the Boreal and Tethys Oceans based on GENESIS atmospheric temperatures; and 4) mean annual and daily wind stresses computed by GENESIS. For the storm experiment, circulation is forced solely by wind stresses.

Results show that these boundary conditions combine to produce a basin-wide counterclockwise gyre that arises from Coriolis acceleration acting on runoff jets trapped along the coast, abetted by latitudinal temperature and salinity gradients and a north-south, wind shear-couple. Individual storm events reverse the general circulation locally, but summed over a storm's duration, the storm-driven fluid and sediment transport augments the mean-annual transport of the gyre. Thus, net sediment transport directions along the western margin of the seaway, both on the shelf and in the wave-driven littoral zone, were southerly because the mean annual wind field, latitudinal temperature gradient, and fresh water runoff from land created a background circulation consisting of southerly geostrophic flow there. In addition, counterclockwise rotating, mid-latitude cyclonic storms passing southeastward over the central seaway drove a net southerly littoral drift in the foreshore and shore-parallel geostrophic flows whose net sediment transport was southerly. Along the eastern margin the computed shelf and littoral transport was to the north.

INTRODUCTION

The origin of isolated shallow-marine sandbodies is unknown, and nowhere is answering this question more vexing than along the western margin of the Cretaceous interior seaway. Examples there include the Shannon Sandstone (Spearing, 1976; Seeling, 1978; Shurr, 1984; Swift and Rice, 1984; Gaynor and Swift, 1988; Tillman and Martinsen, 1984, 1987; Walker and Bergman, 1993; Bergman, 1994), Eagle Sandstone (Rice, 1976, 1980; Rice and Shurr, 1983; this volume), Tocito Sandstone, (Snedden and Nummedal, 1990; Nummedal and Riley, 1991; Valasek, 1995; Jennette and Jones, 1995), and Viking Formation (Evans, 1970; Hein et al.,1986; Leckie, 1986; Raddysh, 1988; Power, 1988; Downing and Walker, 1988; Posamentier and Chamberlain, 1993; Walker, 1995; Walker and Wiseman, 1995), all of which have been variously interpreted as offshore sand ridges, incised estuarine valley fills, or incised shoreface deposits. In each case, sedimentological paleoflow indicators, such as the type, scale, and orientation of bedforms and sediment textures, have played an important role in the interpretation. Whether these sedimentological indicators can be used to differentiate among the different interpretations depends on what bedforms, textures, and transport directions we expect along the western margin of the seaway in these various depositional settings. These expectations are usually derived from modern analogs (e.g., Part 2, this volume), but this is always risky because argument by analogy only succeeds if all other factors forcing circulation in the modern and ancient basins, such as the regional climate, are similar.

The purpose of this paper is to provide an alternative basis for interpreting paleoflow indicators and sedimentary textures observed in coastal and shelf sediments of the Western Interior seaway. Here we deduce the expected circulation and sediment transport paths in the seaway from first principles using a climate/ocean/sedimentation model. Admittedly, this method too has its pitfalls. While based on the conservation laws, climate/ocean/sedimentation models parameterize some important processes and by computational necessity must be run at coarse resolution. More importantly, the accuracy of the results is strongly dependent upon the initial conditions and the boundary conditions. Nevertheless, given our present best estimates of the seaway's paleogeography, paleobathymetry, and paleoclimate, these computations should provide the best estimate to date of the seaway's circulation and sediment transport because they constrain the complex sedimentary processes using fundamental principles of mass, energy, and momentum conservation.

Early conjectures on the circulation of the seaway relied upon intuition and analogy with modern oceans (see Parrish, Gaynor, and Swift, 1984; or Hay, Eicher, and Diner, 1993). Attempts to compute the circulation of the seaway from first principles started with Parrish et al. (1984), who computed the wind-driven circulation in present-day Colorado and Wyoming. Slater (1985) calculated the independent tides of the seaway using a primitive equation model and concluded that the tides were microtidal everywhere. Subsequent calcu-

Isolated Shallow Marine Sand Bodies: Sequence Stratigraphic Analysis and Sedimentologic Interpretation.
SEPM Special Publication No. 64, Copyright © 1999
SEPM (Society for Sedimentary Geology), ISBN 1-56576-057-3, p. 179-190.

lations of the co-oscillating tides by Ericksen and Slingerland (1990) suggested that they were larger but still microtidal, with the exception of the southeastern coast. Ericksen and Slingerland also computed the wind-driven circulation of the entire seaway in response to mean annual winds and a mid-latitude extratropical storm. More recent circulation hindcasts (Slingerland et al., 1996; Jewell, 1996) have included thermohaline forcing as well as mean annual winds. None of these studies has attempted to compute sediment transport in the seaway.

Here we present results from the first climate/ocean/sedimentation computations for the seaway. They show that the net sediment transport direction along the western margin of the seaway in offshore sand ridges, incised estuarine valley fills, and incised shoreface deposits should have been to the south. These predictions are consistent with paleoflow observations there, which indicate a spatially and temporally uniform net sediment transport to the south.

METHODOLOGY

To calculate the sediment transport rates and directions in the Western Interior seaway, one must compute the expected atmospheric forcing and resulting wind-driven and thermohaline flows, as well as oscillatory currents associated with surface waves. The steady and oscillatory currents must then be combined using an appropriate bottom boundary layer model (BBLM) to compute bottom shear stresses. Given the history of bed stresses, sediment transport rates and directions must then be calculated using a sediment transport and bed conservation model. Two cases are presented: Case I) sediment transport due to mean annual thermohaline and wind-driven circulation, and Case II) sediment transport during an 8-day storm.

Sediment Transport Due to Thermohaline and Wind-driven Circulation

Atmospheric forcing for this case comes from the climate for the early Turonian as computed by GENESIS, a global climate model developed by Pollard and Thompson (1992) at the National Center for Atmospheric Research and reported in earlier studies by Barron, Fawcett, Pollard, and Thompson (1993) and Slingerland et al. (1996). GENESIS consists of an atmospheric general circulation model (AGCM) coupled to surface models of soil, snow, sea-ice and a slab ocean, and includes a Land-Surface-Transfer Model that computes near-surface fluxes of heat, moisture, and momentum in the presence of vegetation. The AGCM is an extensively modified version of the National Center for Atmospheric Research Community Climate Model version 1 (CCM1). Each of the seasonal cycle experiments is executed for 12 years to bring the model into dynamic equilibrium; values of atmospheric variables for the subsequent three years are stored for later analysis. These experiments are based on the continental paleogeography and topography of Barron (1987) with atmospheric CO_2 concentrations four times the present day value.

The wind-driven and thermohaline circulation of the seaway for the mean annual case is taken from Slingerland et al. (1996), who calculated it using CIRC, a coastal ocean model derived by Leendertse (Leendertse and Liu, 1977; Keen and Slingerland, 1993). Shoreline positions and depth contours were based on the relationships between lithofacies and depths observed in modern marine settings, and supported by biofacies data from selected sites within the basin (see Slingerland et al., 1996, for a discussion). Precipitation or evaporation over the seaway in that study was simulated by adding or subtracting fresh water from the upper layer of the seaway in proportion to the net precipitation hindcast by GENESIS. Fresh-water runoff from the adjacent continent entered the model seaway through 18 rivers spaced roughly every 4° latitude along the seaway's eastern and western shorelines. River discharges into the seaway were set equal to the precipitation minus evaporation for each river's drainage basin. Surface shear stresses arising from the mean annual wind field were applied to each wet node in CIRC. Output of the model consisted of U, V, and W velocities, temperature, salinity, and water surface elevations. After 7 years of spin-up, the model system approached dynamic equilibrium. At that point the mass of water entering the seaway through precipitation, runoff, and counterflow at the entrances was balanced by the mass leaving by evaporation and surface flows to the world ocean.

Because the resulting circulation is relatively weak, we do not compute the surface wave field, bottom bed stresses, and bed conservation for this case. Instead we qualitatively discuss the implications for fine-grained sediment transport.

Sediment Transport During an 8-Day Storm

The experiment that simulated sediment transport during the passage of a storm over the seaway was forced by an ideal, extratropical winter cyclone. We defined an average storm track across the seaway using the standard deviation of the geopotential height field as computed by GENESIS (Ericksen and Slingerland, 1990). Along the track we passed an ideal cyclone using the cyclone model discussed in Keen and Slingerland (1993). The forward storm speed is 10 m/s, the pressure difference is 53 mm of mercury, and the maximum radius is 1000 km. The simulation starts at day zero when the storm center is northwest of the seaway and ends 8 days later after it has passed to the southeast.

The resulting wind-driven circulation in the seaway was computed using the Princeton Ocean Model (POM) (Blumberg and Mellor, 1987; Mellor, 1993; for its previous application to the Western Interior seaway, see Jewell, 1996). POM is a primitive equation ocean model that uses a free surface and sigma coordinate system. Vertical mixing is computed using the turbulence closure sub-model of Mellor and Yamada (1982). The model grid consists of 112 x 260 nodes in the horizontal, spaced 30 km apart, and 11 sigma coordinates; model external time-step is 100 seconds. POM is initialized from an experiment in which the circulation was driven solely by mean annual winds as computed by GENESIS. The northern and southern open boundaries of the basin are treated using a radiation boundary condition with water surface elevations relaxed to interior values.

The bathymetry and boundary geography for Case II are given in Figure 1. Minimum water depth is 10 m; maximum is 950 m. Sediment in the seaway consists of two types. In water depths shallower than 30 m the bed consists of five size classes of very fine sand with a mean grain size of 2.5 ϕ and a standard deviation equal to 1 ϕ. In waters deeper than 30 m the bed consists of 5 size classes with a mean size of 7.5 ϕ and a standard deviation of 2 ϕ.

Storm waves are computed from the wind field using the forecasting equations for fetch-limited conditions (Corps of Engineers, 1984):

$$H_s = 5.112x10^{-4} U_a \sqrt{f} \qquad \text{(EQ 1)}$$

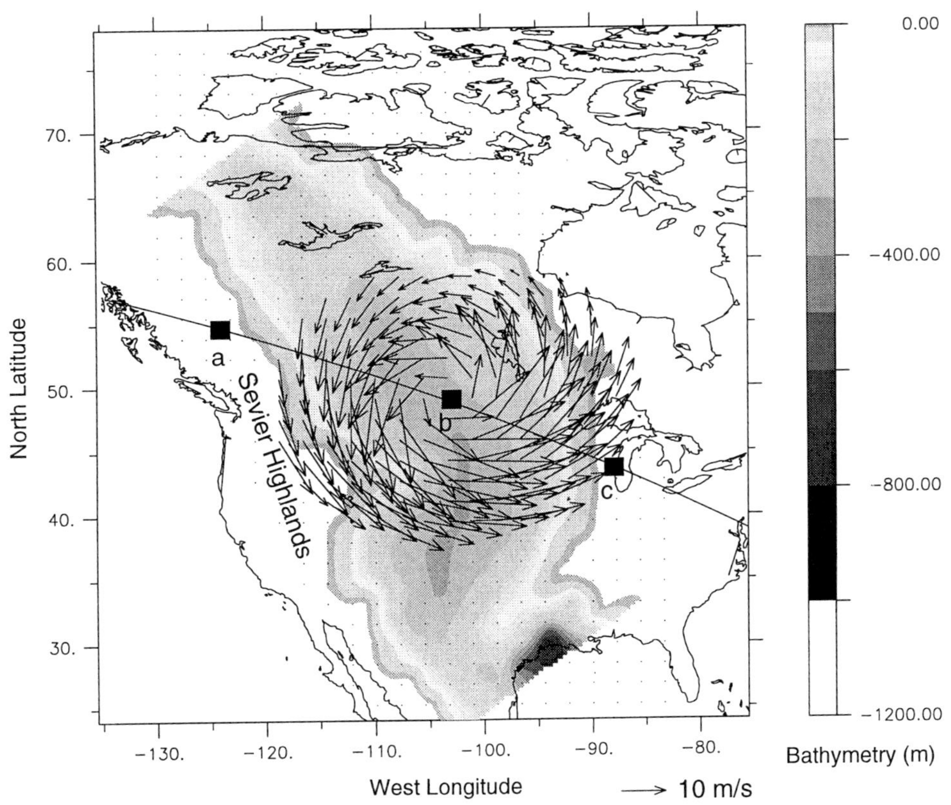

Fig. 1.—Paleobathymetric map of the Early Turonian *W. coloradoense* Biozone in the western interior basin (redrawn from Sageman and Arthur, 1994). In one numerical experiment an idealized extra-tropical cyclone passes from west to east along the solid line. Solid squares represent location of storm eye at a) 2.17, b) 4.14, and c) 6.13 days after the start of computations. Wind speed magnitudes indicated by arrow at bottom right.

$$T_m = 6.238x10^{-2}\sqrt[3]{U_a f} \quad \text{(EQ 2)}$$

where H_s = significant wave height (m); U_a = adjusted wind speed (m/s), given by:

$$U_a = 0.71U^{1.23} \quad \text{(EQ 3)}$$

where U = wind speed (m/s) at 10 m above the water surface, f = fetch (m), and T_m = period (s) of the peak of the wave spectrum. The wave propagation direction is set equal to the wind direction. Wave height and period are then used to compute the wave orbital amplitude and speed in the wave bottom boundary layer using linear wave theory.

Suspended sediment profiles are computed using the Glenn and Grant (1987) suspended-sediment, stratified BBLM with modifications discussed by Keen and Glenn (1994). The unidirectional current driving the BBLM is taken from POM's lowest layer. The bottom roughness is computed by the BBLM and is dependent on sediment characteristics as well as the combined wave-current flow. An active layer is defined as the height of ripples plus the thickness of the near-bed sediment transport layer (Grant and Madsen, 1982). This layer represents interactions between the bed and the flow during a model time step. Sediment resuspension and erosion cannot exceed the active layer depth, thereby greatly reducing the erosion of fine material reported previously (e.g., Keen and Slingerland, 1993). Suspended sediment transport rates are computed from the sediment and current profiles computed by the BBLM at each point. A bed conservation equation is solved at each model grid point to calculate regions of erosion or deposition for each sediment size.

RESULTS

Case I: Mean Annual Circulation and Sediment Transport

As reported in Slingerland et al. (1996), the steady-state surface circulation of the seaway (Fig. 2) consists of a basin-scale counterclockwise gyre. This flow extends to the bed in all water depths less than about 100 m. On the eastern shelf currents flow to the north; on the western shelf currents flow to the south. Surface currents are on the order of a few centimeters per second. Below 100 m, waters collect in the core of the seaway through weak caballing and flow along its thalweg, exiting to the north and south.

This simple estuarine circulation owes its existence to a relatively complex forcing. Sensitivity studies in which the circulation was computed independently for each forcing factor indicate that the density and wind-driven flows are additive, with the thermohaline forcing being the strongest. Fresh and therefore buoyant river water enters the seaway from its eastern and western margins and creates offshore-dipping water-surface slopes, down which this buoyant water subsequently flows. In the process it is deflected to the right by the Coriolis force and piles up along each coast, until the offshore pressure force arising from the surface slope just balances the Coriolis force. This water then moves along

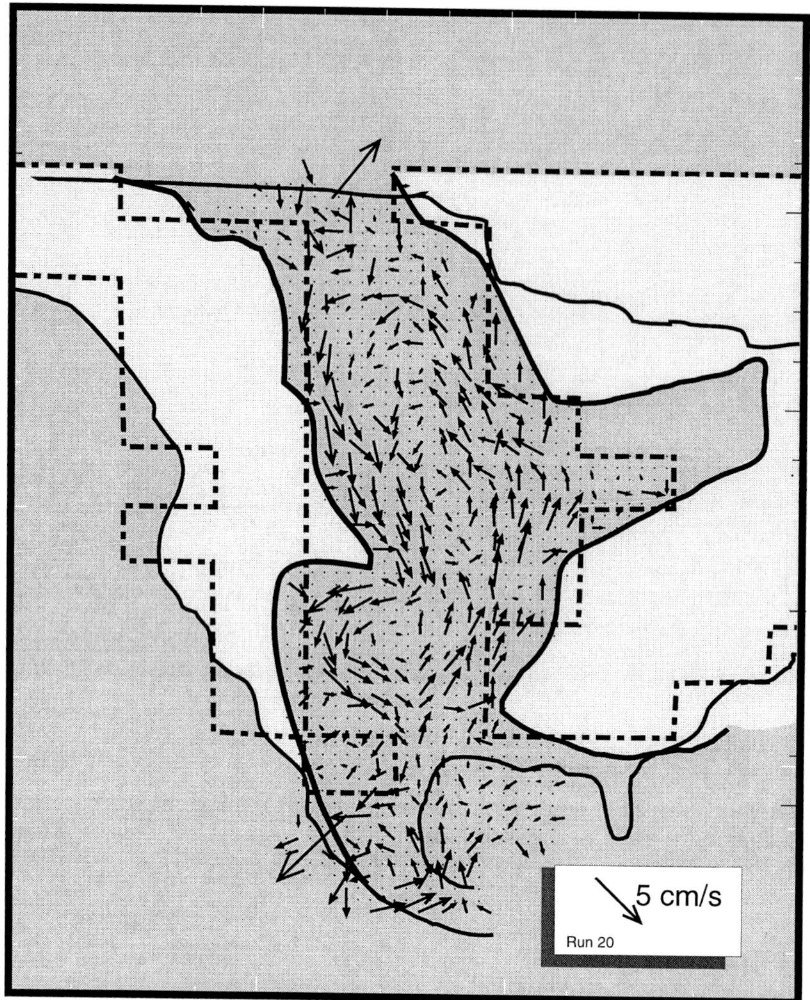

Fig. 2.—Mean annual steady-state circulation in top 10 meters of water column as hindcast by a coastal ocean model subject to atmospheric forcing computed by a global climate model (from Slingerland et al., 1996). Dash line represents paleogeography of North America seen by GENESIS. Circulation in the seaway consists of a large cyclonic gyre.

isobaths as geostrophically confined jets, avoiding the center of the seaway.

The shear couple arising from the coastal jets, and water surface slopes arising from contraction of the water column as it densifies toward the basin center, draw Boreal and Tethyan surface waters into the seaway where they mix as they shear past one another. The mixed waters, being denser than either component, downwell, split into two flows, and return to the global ocean, with 60% of the flux into the Boreal Ocean.

No sediment transport was computed for the mean annual circulation in Slingerland et al. (1996). We can conjecture however, that during fair weather this background circulation would advect suspended sediment delivered to the shelf by river plumes, much as muds of the Amazon are carried north along the Guiana coast by the Guiana Current. Consequently, fair weather, fine-grained sediment transport in the seaway should follow the trajectories shown in Figure 2.

This isn't the whole story of course, because sediment transport rates are proportional to a high power of flow velocity. Consequently, the bulk of sediment transport on the shelf, shoreface, and in the nearshore of the seaway must have occurred during storms, much as occurs on today's continental shelves (Swift et al., 1979; Madsen et al., 1993; Vincent, Young, Swift, 1983; Wright, Xu, and Madsen, 1994; Green et al., 1995).

Case II: Storm-driven Circulation and Sediment Transport

Superimposed on the mean annual circulation of the seaway must have been the stronger but shorter duration flows caused by storms. If modern shelves are the key to the past, then these flows consisted of two parts, an oscillatory component due to water surface waves and a wind and pressure gradient-driven, quasi-steady component. As discussed above, the time integral of these two components determines the net sediment transport magnitude and direction.

Here the time integral is computed during a model storm's transit across the seaway. Initially, the storm eye is located over the Sevier Highlands (Fig. 1) and weaker winds at the front of the storm blow predominantly to the north, parallel to shore. By day 2 (Fig. 1, location a), winds blow

a)

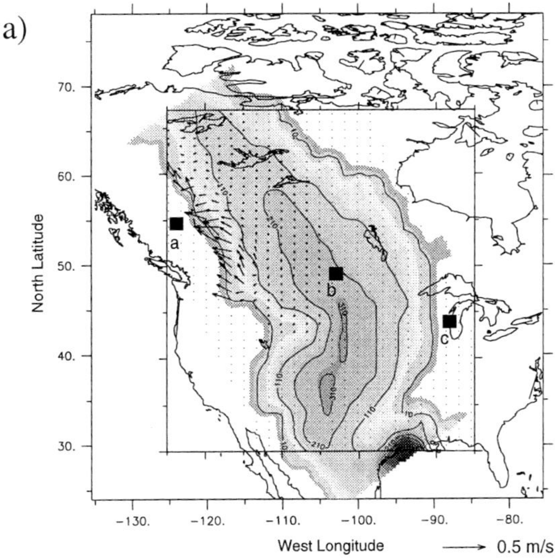

b)

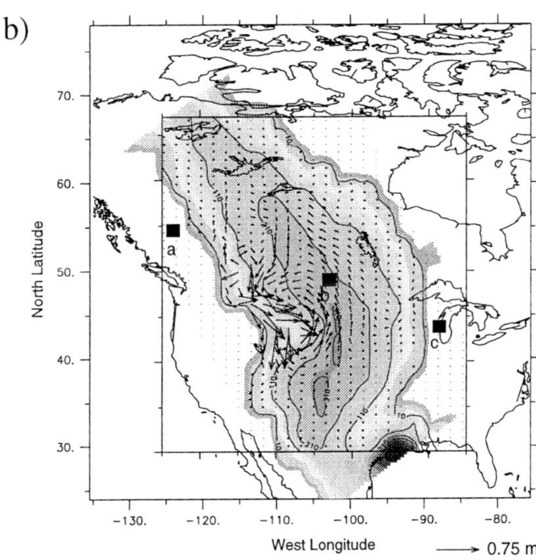

c)

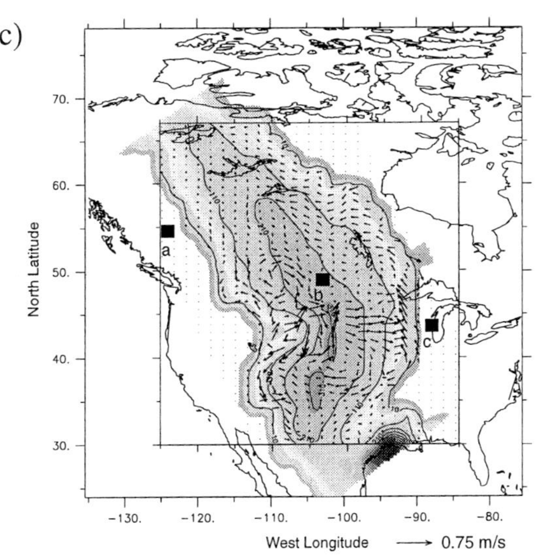

predominantly offshore along the U.S. portion of the seaway and onshore in northern Alberta. Bottom currents computed by POM at this time are northerly and onshore (Fig. 3a) along the Canadian portion of the seaway's west coast. This pattern arises because of geostrophy and because the water motion reflects the history of wind stresses. As the wind stress accelerates surface waters downwind (northward), a cross-stream Coriolis force is created that deflects the surface waters to the right (offshore). Sea levels fall along the coast until a cross-shelf water surface slope creates a shoreward-directed pressure force of sufficient magnitude to balance the Coriolis force. Bottom waters, being loosely decoupled from the wind stress, sense the water surface slope and flow down it. Thus, their net motion is northward and shoreward.

The wind sea (shown for day 4 in Fig. 4) consists of asymmetric distributions of wave heights and periods that decrease away from the storm center. Higher values occur to the south due to the higher winds there (c.f., Fig. 1 and Fig. 2). This pattern is translated self-similarly as the storm eye advances, and therefore the wind sea on day 2 can be reconstructed by recentering the distributions shown in Figure 4 on site a. The direction of wave advance is assumed to be coparallel with the local wind vectors.

As noted above, early in the storm's transit, winds blow predominantly alongshore to the north. The wind sea then should consist of waves between zero and 6 m high with periods of less than 9 s approaching the coast from the south. These waves should create a littoral drift to the north. South of the storm track strong winds blow offshore, creating a wave field that propagates offshore at a right angle to the wind-driven currents.

The nonlinear interaction between the quasi-steady and oscillatory currents in the bottom boundary layer creates the bed stress that entrains and transports sediment. The magnitude of the bed stress as computed in the BBLM is not only a function of the magnitudes of the waves and currents but also of water depth—because bottom wave orbital parameters vary with location in the water column—and sediment type—because turbulence damping due to suspended sediment and friction due to bedforms varies with grain size. Regions of significant bed stress on day 2 (Fig. 5a) are restricted to two regions shallower than 30 m: a region spanning about 10° latitude immediately south of the storm track and a much smaller region to the north. These are regions of both significant waves and wind-driven currents. Sediment in transport moves northward and shoreward at this time (Fig. 3).

By day 4, the storm has progressed to the center of the seaway (Fig. 1) such that winds are predominantly north-directed on the east coast and south-directed on the west coast. In response to this evolving wind field, the storm currents are significantly intensified and reoriented (Fig. 3b). A vast region of the western shelf from the U.S.-Canadian border to southern Colorado is swept by currents that average 0.5 m/s. In waters shallower than about 100 m, the currents are roughly coast parallel and southerly with a small offshore component. North

←

Fig. 3.—Bottom currents as computed by POM at 5 m above the bed after: a) 2.17 days of simulation when the strongest winds are restricted to the western margin of the basin; b) 4.14 days when the storm is in the center of the basin; and c) 6.13 days as the storm approaches the eastern margin.

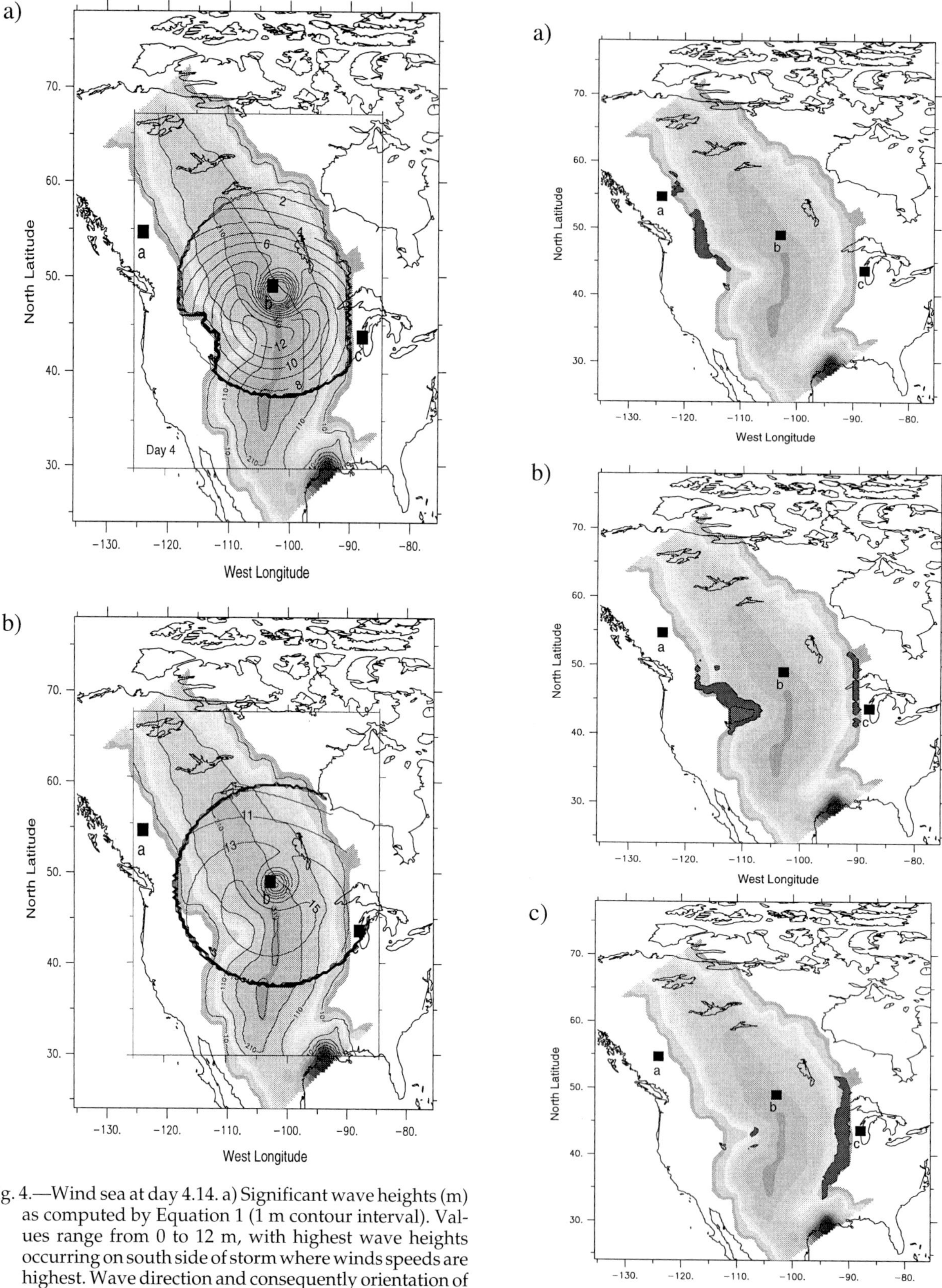

Fig. 4.—Wind sea at day 4.14. a) Significant wave heights (m) as computed by Equation 1 (1 m contour interval). Values range from 0 to 12 m, with highest wave heights occurring on south side of storm where winds speeds are highest. Wave direction and consequently orientation of oscillatory stroke are assumed to follow local wind vectors (Fig. 1); b) Significant wave periods (seconds) computed by Equation 2 (2 second contour interval).

of the border the shelf currents diminish and turn north. Currents on the east coast are still weak. The currents flow onshore in the north and offshore in the south in deeper waters. These bottom currents are isobathyal geostrophic currents, similar to currents on the east coast of the United States during a nor'easter (Lee et al., 1985).

Along the western shelf in the region of strong bottom flow, the wind sea (Fig. 4) increases in height and period from 5 m and 11 second waves in the north to a maximum of 12 m, 15 second waves in the south. As expected, this combination of currents and waves leads to a region of high bed-shear stress (Fig. 5b) centered on the Montana-Wyoming promontory. Along the east coast an extensive region of shallow water experiences modest shear stresses, mainly due to the wind sea.

By day 6 the storm eye has passed over the eastern coast (Fig. 1). Consequently winds blow predominantly to the south along the eastern shelf. Bottom currents (Fig. 3c) now comprise a melange of forced and relaxation flows. Bottom currents along the western shelf are primarily driven by a barotropic Kelvin wave propagating counterclockwise along the western shelf. Bottom currents along the eastern shelf are a result of downwelling associated with coastal setup and relaxation flows as the wind stress decreases. The wind sea is now confined to the eastern shelf. Although wind-driven flows are still relatively strong on the western shelf, significant bottom shear stresses (Fig. 5c) are principally confined to shallow waters of the eastern shelf, reflecting the importance of the storm waves there.

The net sediment transport arising from this complicated history of bed-shear stress magnitudes and directions is given in Figure 6. In this plot, the direction of sediment transport has been taken into account in integrating the sediment flux so that if sediment at a site first moves north and then south at the same rate and for the same duration, then the net transport rate at the site will be zero. The net sediment transport direction on the western shelf is roughly isobathyal to the south. Highest magnitudes occur on the Montana-Wyoming promontory, where sediment is transported in water depths of up to 100 m. On the eastern shelf, net transport is to the north with lower magnitudes. The small magnitude arises partly because southerly transport during the latter stages of the storm's passage cancel northerly transport earlier in the storm's transit.

The bulk of the material transported is sand from the inner shelf, where the greatest shear stresses are calculated. Limited transport occurs offshore because the shear stresses are weaker there and the active layer is quite thin.

Two storm beds remain after the storm passes, one on the western shelf and one on the eastern shelf (Fig. 7). Both span about 15° of latitude and are confined to waters shallower than 100 m. Thickness ranges from a seaward feather-edge to 54 cm in about 20 m of water. The bed consists predominantly of very fine sand.

In summary, although the quasi-steady and oscillatory currents arising from the passage of an extratropical storm are multidirectional, the computed net sediment transport on the western margin of the seaway is southerly, reinforcing finer-grained sediment transport to the south due to the mean annual circulation. Consequently, we also expect the observed paleoflow indicators and sedimentary textures from foreshore, shoreface, and offshore settings in the region of the Cretaceous storm track to record this southward transport.

←

Fig. 5.—Dark patches denote regions of significant wave-current shear velocities ($U_{*cw} > 1$ cm/s) computed from the benthic boundary layer model. (a) High shear velocities occur along the western margin as the storm enters the seaway. (b) High shear velocities occur on both the western and eastern shelves as the storm crosses midway. (c) High shear velocities are limited to the eastern margin. Maximum shear velocities remained at about 8 cm/s throughout the storm's passage but changed location. For reference, bed shear stress is proportional to U^2_{*cw}.

EVIDENCE FOR SOUTHERLY SEDIMENT TRANSPORT

Evidence for southerly sediment transport along the western margin of the Cretaceous Interior seaway has been obtained by numerous field studies of nearshore marine units. Table 1 summarizes units described in the literature that contain paleoflow indicators and were deposited in the western half of the seaway during the Cretaceous, when the seaway was through-going.

The majority record southerly directed paleoflows. Where the indicators are cross-strata of dunes, it seems reasonable that the net sediment transport direction also was southerly. Quite remarkably, all depositional environments sampled, whether the foreshore, shoreface, shelf, or incised estuarine valleys, show this southerly transport. Even estuarine valley-fills cut at low-stand—such as suggested for the Shannon, the Tocito, the Sego sandstones—all contain south-directed paleoflow indicators. Why rivers feeding a roughly north-south shoreline should have turned south, paralleling the shore as they extended seaward during low-stand remains a puzzle. Jennette et al. (1995) call upon tectonic control, but the evidence is circumstantial. Here we propose that the rivers inherited a shelf bathymetry of southerly recurved, subaqueous spits and shoals created by deflection of delta plumes at highstand. Therefore, as sea level fell, the rivers were steered south to debouch on local east-west trending shorelines.

CONCLUSIONS

Our modeling results indicate that net sediment transport along the western margin of the seaway—whether in offshore sand ridges, incised estuarine valley fills, or incised shoreface deposits—would have been to the south. This occurred because the mean annual wind field, latitudinal temperature gradient, and fresh water runoff from land created a background circulation consisting of southerly geostrophic flow on the western shelf. In addition, counterclockwise-rotating, mid-latitude cyclonic storms passing southeastward over the central seaway drive a net southerly littoral drift in the foreshore and shore-parallel geostrophic flows whose net suspended load transport is southerly. Finally, one can conjecture that in response to these mean and event-generated southerly flows, sediment plumes at river mouths constructed submarine topographies at highstand that recurved southward. When sea level fell, river courses may have been steered by this subtle topography to incise shore-parallel or oblique valleys filled with southerly directed, estuarine valley-fill sandbodies

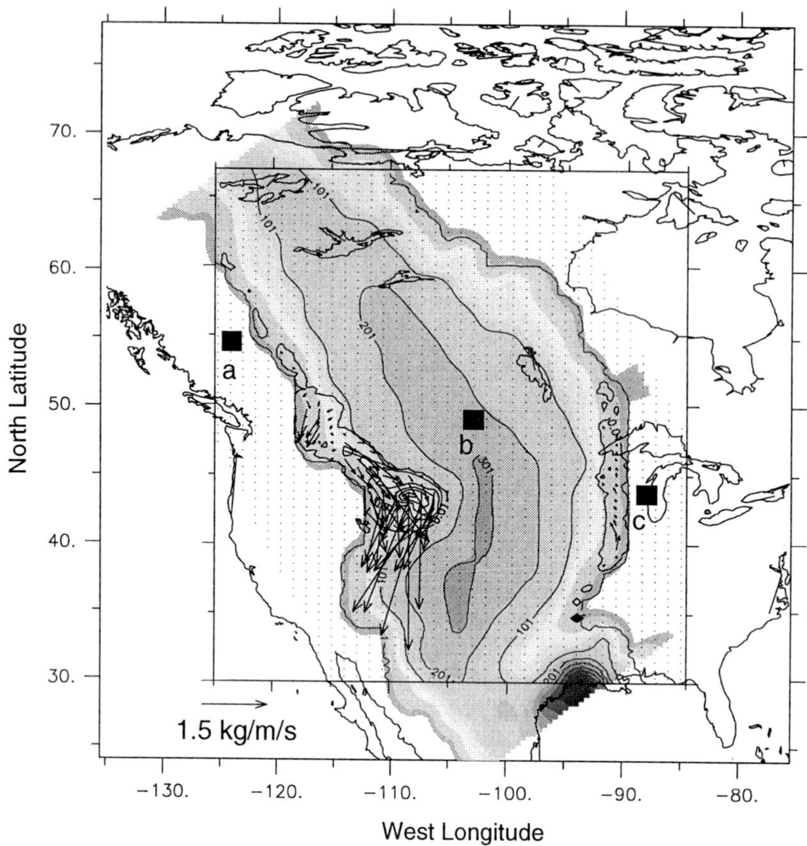

Fig. 6.—Net direction and magnitude of transport of suspended load (kg/m/s) during the eight days of the storm. Values reflect integration of sediment concentration over the lower 10 m of the water column times wind-driven current speed in lowest layer of POM expressed as kg per unit width per unit time.

ACKNOWLEDGMENTS

R. Slingerland was supported in part by National Science Foundation grant EAR-9117398; T. R. Keen was supported by Program Element 62435N of the Office of Naval Research. We thank J. T. Parrish and D. A. Leckie for their comments on the manuscript,

REFERENCES

Barron, E.J., 1987, Cretaceous paleogeography: Palaeogeography, Palaeoclimatology, Palaeoecology, v. 40, p. 103-133.

Barron, E.J., Fawcett, P.J., Pollard, D. and Thompson, S.L., 1993, Model simulations of Cretaceous climates: The rope of geography and carbon dioxide: Philosophical Transactions of the Royal Society of London B, v. 341, p. 307-316.

Bergman, K.M., 1994, Shannon Sandstone in Hartzog Draw-Heldt Draw fields (Cretaceous, Wyoming, USA) reinterpreted as lowstand shoreface deposits, Journal of Sedimentary Research, Section B: Stratigraphy and Global Studies, v. 64, p. 184-201.

Bhattacharya, J., Walker, R.G., 1991, River- and wave-dominated depositional systems of the Upper Cretaceous Dunvegan Formation, northwestern Alberta, Bulletin of Canadian Petroleum Geology, v. 39, p. 165-191.

Blumberg, A.F. and Mellor, G.L., 1987, A description of a three-dimensional coastal ocean circulation model, *in* Heaps, N.S., ed., Three Dimensional Coastal Ocean Models: Washington, D.C., American Geophysical Union Coastal and Estuarine Sciences Series, v. 4, p. 1-16.

Boyles, J.M. and Scott, A.J., 1982, A model for migrating shelf bar sandstones in Upper Mancos Shale (Campanian), northwestern Colorado: American Association of Petroleum Geologists Bulletin, v. 66, p. 491-508.

Campbell, C.V., 1971, Depositional model Upper Cretaceous Gallup beach shoreline, Ship Rock area, northwestern New Mexico: Journal of Sedimentary Petrology, v. 41, p. 395-409.

Corps of Engineers, 1984, Shore Protection Manual, 4th ed. Vicksburg, Miss., Dept. of the Army, Waterways Experiment Station, Coastal Engineering Research Center, Washington, D.C.

Cotter, E., 1975, Late Cretaceous sedimentation in a low-energy coastal zone: The Ferron Sandstone of Utah: Journal of Sedimentary Petrology, v. 45, p. 669-685.

Downing, K.P., Walker, R.G., 1988, Viking Formation, Joffre Field, Alberta; shoreface origin of long, narrow sand body encased in marine mudstones: American Association of Petroleum Geologists Bulletin, v. 72, p. 1212-1228.

Ericksen, M.C. and Slingerland, R., 1990, Numerical simulations of tidal and wind-driven circulation in the Cretaceous Interior seaway of North America: Geological Society of America Bulletin, v. 102, p. 1499-1516.

Evans, W.E., 1970, Imbricate linear sandstone bodies of Viking Formation in Dodsland-Hoosier area of southwestern Saskatchewan, Canada: The American Association of Petroleum Geologists Bulletin, v. 54, p. 469-486.

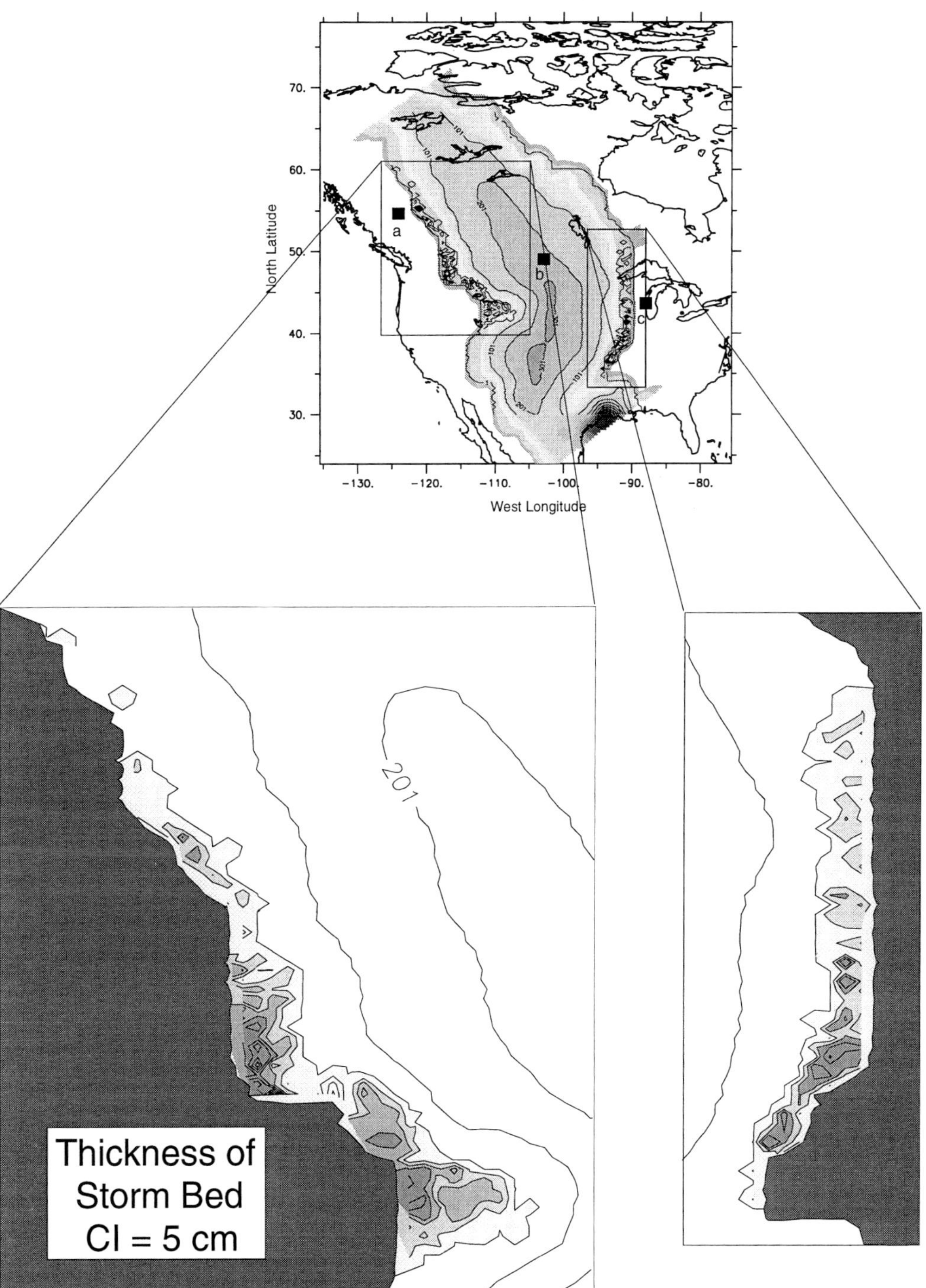

Fig. 7.—Thickness (cm) of storm bed deposited during the 8-day simulation. Lightest shading denotes region of bed thickness between 0 and 5 cm. A bed extends to depths of 75 m along the western margin where flow was strongest. The bed is limited to depths of about 20 m along the eastern margin.

FITZSIMMONS, R.O., 1995, High resolution sequence stratigraphy of the Upper Cretaceous Eagle Formation, Wyoming, U.S.A., *in* Fitzsimmons, R.O., Parsons, B. and Swift, D.J.P., eds., Tongues, Ridges, and Wedges, Society of Economic Paleontologists and Mineralogists Research Conference Field Guide, June 24-29, 1995, p. 29.

GAYNOR, G.C., SWIFT, D.J.P., 1988, Shannon Sandstone depositional model; sand ridge dynamics on the Campanian Western Interior Shelf, Journal of Sedimentary Petrology, v. 58, p. 868-880.

GLENN, S.M. AND GRANT, W.D., 1987, A suspended sediment stratification correction for combined wave and current flows: Journal Geophysical Research, v. 92, p. 8244-8264.

Table 1.—Paleocurrent data from nearshore marine sandbodies along the western margin of the Cretaceous Western Interior seaway.

Unit	Age	Depositional Environments and Paleocurrent Data
1. Tocito Ss.; northwestern New Mexico (Van Wagoner, et al., 1991)	Coniacian	Transgressive shelf sand ridge or incised estuarine valley: offshore unimodal large-scale cross-strata dip to SE; large wave-ripple crests trend NW-SE; shore parallel sediment transport to the SE on a transgressive restricted shelf or SE transport in a shore-parallel estuary
2. Kenilworth Mbr., Blackhawk Fm.; Book Cliffs, Utah (Taylor and Lovell, 1995)	Campanian	Upper shoreface: "Trough cross-bed orientations...in the upper shoreface deposits...have a dominant southerly component and suggest that there were strong shore-parallel currents."
3. Shannon Ss., east-central Wyoming (Tillman and Martinsen, 1984)	Campanian	Shelf ridge complex: "Transport directions determined from high-angle cross-beds indicate a southwest transport direction."
4. Dunvegan Fm.; northwestern Alberta (Bhattacharya and Walker, 1991)	Cenomanian	Barrier bar: "A major distributary channel probably fed this barrier bar at its northeastern end and sand was transported to the southwest by longshore drift."
5. Burnstick Mbr., Cardium Fm.; Alberta (Pattison and Walker, 1992)	Turonian	Incised shoreface: "We interpret the alongstrike trends as the result of ... southwestward sediment transport in the shoreface."
6. Virgelle Mbr, Eagle Fm.; Bighorn Basin, Wyoming (Fitzsimmons, 1995)	Campanian	Incised estuarine valley: bipolar N-S dipping trough cross-strata fill a N-S trending, shore parallel, incised valley
7. Baytree Mbr., Cardium Fm.; northwestern Alberta (Hart and Plint, 1989)	Turonian	Shoreface: "The dominant southeast (i.e. shore-parallel) cross-bedding, in both sandstones and conglomerates indicates a strong longshore component to sediment transport."
8. Rusty Mbr., Ericson Ss.; SW Wyoming (Martinsen et al., 1997)	Campanian	Valley-fill sandstones: "In contrast to surrounding delta plain sediments, where paleoflow was to the ESE, the paleocurrents within valley-fill sandstones are directed to the SSW."
9. Duffy Mtn. Ss., Mancos Shale; Northwestern Colorado (Boyles and Scott, 1982)	Campanian	Shore-parallel shelf bars: " The sediment was probably derived from....southwestern Wyoming."
10. Eagle Ss.; north-central Montana (Rice, 1980)	Upper Cretaceous	Delta-front: "The sand was probably deposited as river-mouth bars that were redistributed by dominately southward-flowing marine longshore currents..."
11. Gallop Ss.; northwestern New Mexico (Campbell, 1971)	Upper Cretaceous	Elongate, shore-parallel, northwest-southeast trending, offshore bar: Cross-laminae dip to the southeast.
12. Ferron Ss.; Castle Valley, Utah (Cotter, 1975)	Upper Cretaceous	Low energy coast: "Sediment....was derived from the large Vernal Delta, located north and west of the Castle Valley outcrops, and was transported generally southwestward, parallel with the coast."
13. Hygiene Ss.; northern Colorado (Kitely and Field, 1984)	Late Campanian	Mid-outer shelf sand ridges: "The sand was derived from the west, transported eastward, and then redistributed by southward-flowing storm and oceanic currents."

GRANT, W.D. AND MADSEN, O.S., 1982, Movable boundary roughness in unsteady oscillatory flow: Journal Geophysical Research, v. 87, p. 1797-1808.

GREEN, M.O., VINCENT, C.E., MCCAVE, I.N., DICKSON, R.R., REES, J.M. AND PEARSON, N.D., 1995, Storm sediment transport; observations from the British North Sea shelf: Continental Shelf Research, v. 15, p. 889-912.

HART, B.S., PLINT, A.G., 1989, Gravelly shoreface deposits; a comparison of modern and ancient facies sequences: Sedimentology, v. 36, p. 551-557.

HAY, W.W., EICHER, D.L. AND DINER, R., 1993, Physical oceanography and water masses in the Cretaceous Western Interior seaway, *in* Caldwell, W.G.E. and Kauffman, E.G., eds., Evolution of the Western Interior Basin: St. John's, Geological Association of Canada Special Paper 39, p. 297-318.

HEIN, F.J., DEAN, M.E., DEIURE, A.M., GRANT, S.K., ROBB, G.A., LONGSTAFFE, F.J., 1986, The Viking Formation in the Caroline, Garrington and Harmattan East fields, western south-central Alberta; sedimentology and paleogeography: Bulletin of Canadian Petroleum Geology, v. 34, p. 91-110.

JENNETTE, D.C., JONES, C.R., 1995, Sequence stratigraphy of the Upper Cretaceous Tocito Sandstone; a model for tidally influenced incised valleys, San Juan Basin, New Mexico, *in* Van Wagoner, J.C., Bertram, G.T., Sequence stratigraphy of foreland basin deposits; outcrop and subsurface examples from the Cretaceous of North America: American Association of Petroleum Geology Memoir, v. 64, p. 311-347.

JEWELL, P.W., 1996, Circulation, salinity, and dissolved oxygen in the Cretaceous North American seaway: American Journal of Science, v. 296, p. 1093-1125.

KEEN, T.R. AND GLENN, S.M., 1994, A coupled hydrodynamic-bottom boundary layer model of Ekman flow on stratified continental shelves: Journal of Physical Oceanography, v. 24, p. 1732-1749.

KEEN, T.R. AND SLINGERLAND, R.L., 1993, A numerical study of sediment transport and event bed genesis during Tropical Storm Delia: Journal of Geophysical Research, v. 98, p. 4775-4791.

KITELEY, L.W., FIELD, M.E., 1984, Shallow marine depositional environments in the Upper Cretaceous of northern Colorado, *in* Tillman, R.W. and Siemers, C.T., Siliciclastic shelf sediments: Tulsa, Society of Economic Paleontologists and Mineralogists Special Publication, 34, p. 179-204.

LECKIE, D.A., 1986, Tidally influenced, transgressive shelf sediments in the Viking Formation, Caroline, Alberta: Bulletin of Canadian Petroleum Geology, v. 34, p. 111-125.

LEE, T.N., KOURAFALOU, V., WNAG, J.D. AND HO, W.J., 1985, Shelf circulation from Cape Canaveral to Cape Fear during Winter; *in* Atkinson, L.P., Menzel, D.W. and Bush, K.A., Oceanography of the Southeastern U.S. Continental Shelf: Washington, D.C., American Geophysical Union Coastal and Estuarine Sciences, v. 2, p. 33-62.

LEENDERTSE, L.J. AND LIU, S.K., 1977, A three-dimensional model for estuaries and coastal seas: v. IV, turbulent energy computation: Santa Monica, Rand Report Number R-2187-OWRT.

MADSEN, O.S., WRIGHT, L.D., BOON, J.D. AND CHISHOLM, T.A., 1993, Wind stress, bed roughness and sediment suspension on the inner shelf during an extreme storm event, *in* Huntley, D.A., Nearshore and coastal oceanography: Continental Shelf Research, v. 13, p. 1303-1324.

MARTINSEN, O.J., RYSETH, A. AND HELLAND-HANSEN, W., 1997, Sandrich, estuarine valley-fill sandstones, Rust Member, Ericson Sandstone (Campanian), SW Wyoming: Sedimentology, stratigraphy, and tectonic significance (abs.): Society of Economic Paleontologists and Mineralogists—Canadian Society of Petroleum Geologists Joint Convention, p. 184.

MELLOR, G.L., 1993, Users Guide for a Three-dimensional, Primitive Equation, Numerical Ocean Model. Report to the Institute of Naval Oceanography: Princeton, New Jersey, Princeton University, 35 p.

MELLOR, G.L. AND YAMADA, T., 1982, Development of a turbulence closure model for geophysical fluid problems: Review of Geophysics and Space Physics, v. 20, p. 851-875.

NUMMEDAL, D. AND RILEY, G.W., 1991, Origin of late Turonian and Coniacian unconformities in the San Juan Basin, *in* Van Wagoner, J.C., Jones, C.R., Taylor, D.R., Nummedal, D., Jennette, D.C. and Riley, G.W., Sequence stratigraphy applications to shelf sandstone reservoirs; outcrop to subsurface examples: American Association of Petroleum Geologists Field Conference, p. 1-6.

PARRISH, J.T., GAYNOR, G.C. AND SWIFT, D.J.P., 1984, Circulation in the Cretaceous Western Interior seaway of North America, a review, *in* Stott, D.F. and Glass, D.J., (eds), The Mesozoic of Middle North America: Canadian Society of Petroleum Geologists Memoir, v. 9, p. 221-231.

PATTISON, S.A.J., WALKER, R.G., 1992, Deposition and interpretation of long, narrow sandbodies underlain by a basinwide erosion surface; Cardium Formation, Cretaceous Western Interior seaway, Alberta, Canada: Journal of Sedimentary Petrology, v. 62, p. 292-309.

POLLARD, D. AND THOMPSON, S.L., 1992, User's Guide to the GENESIS Global Climate Model Version 1.02: Boulder, National Center for Atmospheric Research ICS, 58 p.

POSAMENTIER, H.W. AND CHAMBERLAIN, C.J., 1993, Sequence-stratigraphic analysis of Viking Formation lowstand beach deposits at Joarcam Field, Alberta, Canada, *in* Posamentier, H.W., Summerhayes, C.P., Haq, B.U. and Allen, G.P., Sequence stratigraphy and facies associations: International Association of Sedimentologists Special Publication, 18, p. 469-485.

POWER, B.A., 1988, Coarsening-upwards shoreface and shelf sequences; examples from the Lower Cretaceous Viking Formation at Joarcam, Alberta, Canada, *in* James, D.P. and Leckie, D.A., Sequences, stratigraphy, sedimentology; surface and subsurface: Canadian Society of Petroleum Geologists Memoir, v. 15, p. 185-194.

RADDYSH, H.K., 1988, Sedimentology and "geometry" of the Lower Cretaceous Viking Formation, Gilby A and B fields, Alberta, *in* James, D.P. and Leckie, D.A., Sequences, stratigraphy, sedimentology; surface and subsurface: Canadian Society of Petroleum Geologists Memoir, v. 15, p. 417-429.

RICE, D.D., 1976, Depositional environments of Eagle Sandstone, North-central Montana; aid for hydrocarbon exploration, American Association of Petroleum Geologists Bulletin, v. 60, p. 1408.

RICE, D.D., 1980, Coastal and deltaic sedimentation of Upper Cretaceous Eagle Sandstone; relation to shallow gas accumulations, North-central Montana: American Association of Petroleum Geologists Bulletin, v. 64, p. 316-338.

RICE, D.D., SHURR, G.W., 1983, Patterns of sedimentation and paleogeography across the Western Interior seaway during time of deposition of Upper Cretaceous Eagle Sandstone and equivalent rocks, Northern Great Plains, *in* Reynolds, M.W., Dolly, E.D., Mesozoic paleogeography of the West-Central United States: Rocky Mountain Paleogeography Symposium, v. 2, p. 337-358.

SAGEMAN, B.B. AND ARTHUR, M.A., 1994, Early Turonian paleogeographic/paleobathymetric map, Western Interior, US, *in* Caputo, M.V., Peterson, A. and Franczyk, K.J., eds., Mesozoic Systems of the Rocky Mountain Region, USA: Tulsa, Society of Economic Paleontologists and Mineralogists, Rocky Mountain Section, p. 457-469.

SEELING, A., 1978, The Shannon Sandstone, a further look at the environment of deposition at Heldt Draw Field, Wyoming, The Mountain Geologist, v. 15, p. 133-144.

SHURR, G.W., 1984, Geometry of shelf-sandstone bodies in the Shannon Sandstone of southeastern Montana, *in* Tillman, R.W. and Siemers, C.T., Siliciclastic shelf sediments: Tulsa, Society of Economic Paleontologists and Mineralogists Special Publication, 34, p. 63-83.

SLATER, R.D., 1985, A numerical model of tides in the Cretaceous seaway of North America: Journal of Geology, v. 93, p. 333-345.

SLINGERLAND, R., KUMP, L.R., ARTHUR, M.A., FAWCETT, P.J., SAGEMAN, B.B. AND BARRON, E.J., 1996, Estuarine circulation in the Turonian-Western Interior seaway of North America: Geological Society of America Bulletin, v. 108, p. 941-952.

SNEDDEN, J.W. AND NUMMEDAL, D., 1990, Coherence of surf zone and shelf current flow on the Texas (U.S.A.) coastal margin; implications for interpretation of paleo-current measurements in ancient coastal sequences: Sedimentary Geology, v. 67, p. 221-236.

SPEARING, D.R., 1976, Upper Cretaceous Shannon Sandstone; an offshore, shallow-marine sand body: Casper, Wyoming Geological Association Guidebook, v. 28, p. 65-72.

SWIFT, D.J.P., YOUNG, R.A., CLARKE, T.L., VINCENT, C.E., NIEDORODA, A. AND LESHT, B., 1979, Sediment transport in the Middle Atlantic Bight of North America; synopsis of recent observations, *in* Nio, S.D., Shuettenhelm, R.T.E., van Weering, T.C.E., Holocene marine sedimentation in the North Sea basin: International Association of Sedimentologists Special Publication, v. 5, p. 361-383.

SWIFT, D.J.P., RICE, D.D., 1984, Sand bodies on muddy shelves; a model for sedimentation in the Western Interior seaway, North America, *in* Tillman, R.W. and Siemers, C.T., Siliciclastic shelf sediments: Tulsa, Society of Economic Paleontologists and Mineralogists Special Publication, 34, p. 43-62.

TAYLOR, D.R. AND LOVELL, R.W.W., 1995, High-frequency sequence stratigraphy and paleogeography of the Kenilworth Member, Blackhawk Formation, Book Cliffs, Utah, U.S.A., *in* Van Wagoner, J.C. and Bertram, G.T., Sequence stratigraphy of foreland basin deposits; outcrop and subsurface examples from the Cretaceous of North America: American Association of Petroleum Geologists Memoir, v. 64, p. 257-275.

TILLMAN, R.W. AND MARTINSEN, R.S., 1984, The Shannon shelf-ridge sandstone complex, Salt Creek Anticline area, Powder River basin, Wyoming, *in* Tillman, R.W., Siemers, C.T., Siliciclastic shelf sediments: Tulsa, Society of Economic Paleontologists and Mineralogists Special Publication, 34, p. 85- 142.

TILLMAN, R.W., MARTINSEN, R.S., 1987, Sedimentologic model and production characteristics of Hartzog Draw Field, Wyoming, a Shannon shelf-ridge sandstone, *in* Tillman, R.W. and Weber, K.J., Reservoir sedimentology: Tulsa, Society of Economic Paleontologists and Mineralogists Special Publication, 40, p. 15-112

VALASEK, D., 1995, The Tocito Sandstone in a sequence stratigraphic framework; an example of landward- stepping small-scale genetic sequences, *in* Van Wagoner, J.C. and Bertram, G.T., Sequence stratigraphy of foreland basin deposits; outcrop and subsurface examples from the Cretaceous of North America: American Association of Petroleum Geologists Memoir, v. 64, p. 349-369.

VAN WAGONER, J.C., JONES, C.R., TAYLOR, D.R., NUMMEDAL, D., JENNETTE, D.C. AND RILEY, G.W., 1991, Sequence stratigraphy applications to shelf sandstone reservoirs; outcrop to subsurface examples: Tulsa, American Association of Petroleum Geologists Field Conference Guidebook.

VINCENT, C.E., YOUNG, R.A. AND SWIFT, D.J.P., 1983, Sediment transport on the of Long Island shoreface, North American Atlantic shelf; role of wave and currents in shoreface maintenance: Continental Shelf Research, v. 2, p. 163-181.

WALKER, R.G., 1995, Sedimentary and tectonic origin of a transgressive surface of erosion; Viking Formation, Alberta, Canada: Journal of Sedimentary Research, Section B: Stratigraphy and Global Studies, v. 65, p. 209-221.

WALKER, R.G. AND WISEMAN, T.R., 1995, Lowstand shorefaces, transgressive incised shorefaces, and forced regressions; examples from the Viking Formation, Joarcam area, Alberta: Journal of Sedimentary Research, Section B: Stratigraphy and Global Studies, v. 65, p. 132-141.

WALKER, R.G. AND BERGMAN, K.M., 1993, Shannon Sandstone in Wyoming; a shelf-ridge complex reinterpreted as lowstand shoreface deposits: Journal of Sedimentary Petrology, v. 63, p. 839-851.

WRIGHT, L.D., XU, J.P. AND MADSEN, O.S., 1994, Across-shelf benthic transports on the inner shelf of the Middle Atlantic Bight during the "Halloween storm" of 1991: Marine Geology, v. 118, p. 61-77.

GENETIC CHARACTERISTICS OF GLAUCONITE AND SIDERITE: IMPLICATIONS FOR THE ORIGIN OF AMBIGUOUS ISOLATED MARINE SANDBODIES

SHARON A. STONECIPHER

Marathon Oil Company, P.O. Box 269, Littleton, CO 80160 U.S.A.

ABSTRACT: Authigenic glauconite and siderite form under a limited range of well-documented geological and geochemical conditions. Because glauconite precursors need to remain at or near the sediment water surface for long periods of time in a setting where they can be repeatedly exhumed and shallowly buried, glauconite typically develops on the outer margins of continental shelves in areas of low sediment input. Based on these requirements, glauconite has traditionally been used as an indicator for transgressive sequences because transgressions tend to trap sediment on the continents. However, from a sequence stratigraphic standpoint, glaucony may be present in virtually any part of a depositional sequence due to remobilization. Glaucony can provide useful information for sequence stratigraphy only if variations in its abundance, physicochemical properties, and spatial/temporal characteristics are carefully documented.

Siderite typically forms in one of two distinct environments: one characterized by strongly reducing conditions (methanogenic zone), and one under slightly reducing conditions (post-oxic zone). Methanogenic siderite is more common in continental and fresh-water lacustrine than marine deposits. Post-oxic conditions are commonly associated with marine environments exhibiting moderately low concentrations of organic matter and low sedimentation rates. Siderite is also frequently found in association with sequence boundaries where it occurs as a secondary cement below the lowstand surface of erosion (LSE).

These restrictions on environment of origin provide information on the hydrologic regime and, by inference, the depositional and sequence stratigraphic setting of the host sediment in which these minerals are found. By examining the genetic significance of minerals such as glauconite and siderite, the origin of ambiguous, controversial, isolated marine sand bodies such as those discussed elsewhere in this volume may be clarified. This paper summarizes what is currently known about the chemical characteristics of these minerals and discusses generalized models of their distribution in a variety of sequence stratigraphic settings.

INTRODUCTION

Authigenic glauconite and siderite form under a limited range of well-documented geological and geochemical conditions. For example, glauconite typically develops on the outer margins of continental shelves in areas of low sediment input. Glauconite precursors need to remain at or near the sediment water surface for long periods of time in a setting where they can be repeatedly exhumed and shallowly buried. Based on these requirements, glauconite has traditionally been used as an indicator for transgressive sequences because transgressions tend to trap sediment on the continents. Similarly, the presence of siderite connotes precipitation either under fresh-water or marine reducing conditions in sediments where the available iron content exceeds the amount of sulfate entrained in depositional pore waters. The difference between these two environments is reflected in the stable isotope and trace element geochemistry of the siderites.

These restrictions on environments of origin have been used as geochemical fingerprints to provide information on the hydrologic regime and, by inference, the depositional and sequence stratigraphic setting of the host sediment. By examining the genetic significance of minerals such as glauconite and siderite, as well as other very early authigenic cements, the hope is that evidence will be found to help resolve the origin of ambiguous, controversial, isolated marine sand bodies such as those discussed elsewhere in this volume. This paper presents a summary of what is currently known about the chemical characteristics of these minerals and discusses generalized models of their distribution in a variety of sequence stratigraphic settings.

ORIGIN OF GLAUCONITE

Glaucony Versus Verdine Facies — Locus of Formation

In the past, the term glauconite has been used interchangeably to designate both a morphological form—that is, just about any green pellet—as well as a specific mineral. However, many green grains are composed not of the mineral glauconite, but rather of various other components. Odin and Matter (1981) argued successfully to replace the generic term glauconite by several more specific descriptors. They proposed the term "glaucony" to represent the depositional as well as diagenetic facies in which glauconitic minerals consisting of 2:1 layer, iron-rich, dioctahedral micas with varying interstratifications of expansible layers are likely to occur. The glaucony facies is distinct from the verdine facies, which includes other green-pelleted minerals, berthierine and chamosite. Berthierine is a 1:1 layered, trioctahedral, 7Å material (essentially an iron-rich kaolinite) that is sometimes confused with the 2:1:1 layered, trioctahedral, 14Å chamosite, which is essentially an iron-rich chlorite. The verdine facies is typically found on the inner shelf and is particularly abundant in estuarine environments or immediately offshore from fluvial deltas (Fig. 1).

According to Odin and Matter (1981), the glaucony facies typically forms in two different oceanographic environments. First, it most commonly occurs on the shelves of stable continental margins in water depths between 60 m and 550 m. It is particularly abundant on the outer shelf and upper slope between 200 m and 300 m water depth. Second, it may also accumulate at or near the top of isolated mid-ocean tectonic

Isolated Shallow Marine Sand Bodies: Sequence Stratigraphic Analysis and Sedimentologic Interpretation.
SEPM Special Publication No. 64, Copyright © 1999
SEPM (Society for Sedimentary Geology), ISBN 1-56576-057-3, p. 191-204.

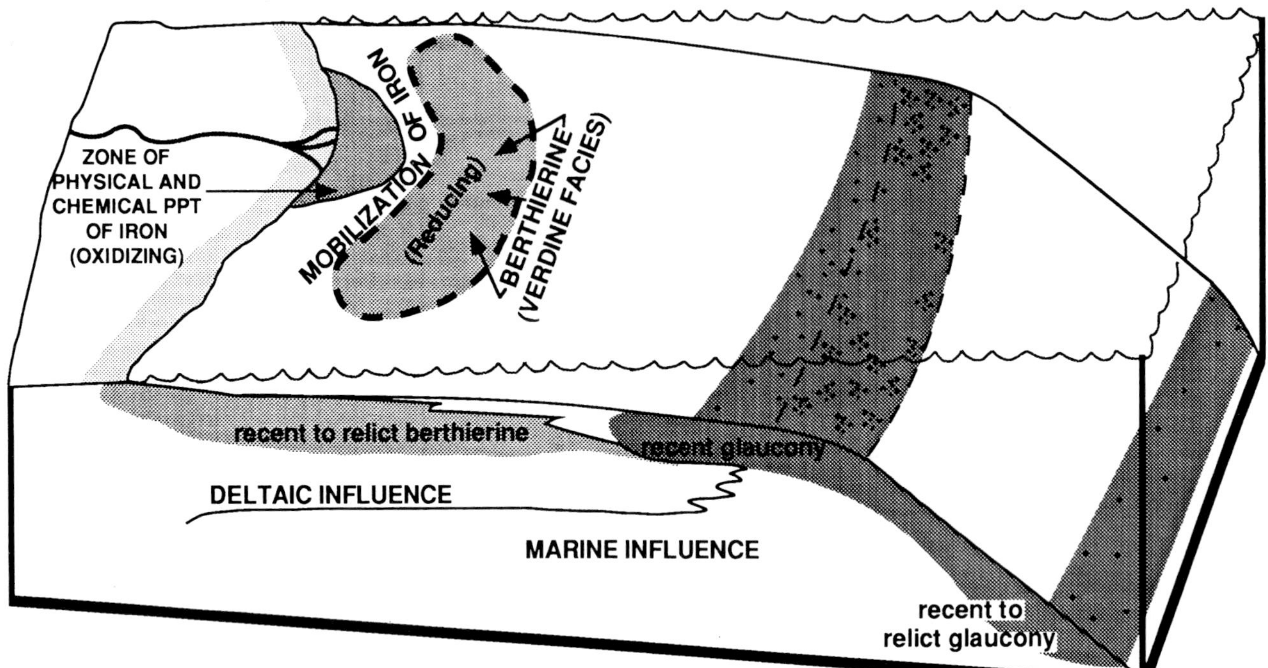

Fig. 1.—Schematic diagram showing relative geographic and geochemical relationships between the verdine and glaucony facies.

highs ranging between 400-1000 m depth. Both of these settings occur in open marine environments at some distance from any source of sediment input in an area where bottom currents are likely to be active. Close to the shoreline, detrital influx commonly exceeds the winnowing ability of currents, resulting in high accumulation rates and rapid burial of sediment. This prevents glauconite formation. Where water is deeper and detrital input lower, winnowing causes reworking of the sediment, resulting in repeated burial and reexposure of detrital grains at the sediment/water boundary where glauconitization takes place. Below 300 m, energy is commonly so low that detrital grains are simply buried instead of being reworked. Much of the glauconitic material found at these depths may have been transported from shallower depths. It should be noted, however, that contour currents may rework the seafloor at depths greater than 1000-2000 m, so some of the glauconite found in deep-water settings could be authigenic.

Mechanism of Formation and Evolution of Glaucony

A widely accepted theory that explains the process of glauconitization is the one brought forward by Odin and Matter (1981). It involves a delicate balance between access to and semi-isolation from sea water in a porous and typically granular substrate. Glauconitization occurs at the boundary between oxidizing sea water and reducing interstitial waters. Isolated pores are required to provide the microenvironments in which the earliest glauconitic minerals form (Fig. 2a). This porosity may be in the form of borings and natural or solution cavities in biogenic particles, as well as surface fractures and fissures in fecal pellets, mineral grains, and rock fragments. The semi-restricted microenvironment protects both initial crystal growth and subsequent recrystallization processes from the aggressive diluting and oxidizing influence of typical sea water. Conditions with temperatures below 15°C and pH around 8 are favorable. Eh conditions are at the oxidation-reduction boundary which allows iron to be mobilized as Fe^{+2} from the sediment, and stabilized as Fe^{+3} in the crystal structure of the glauconitic mineral. At this earliest, or nascent stage of glauconitization, the mineral precipitates typically consist of iron-rich glauconitic smectites with 2-4% K_2O (Figs. 3 and 4). These authigenic smectites gradually fill available pore space and begin to impart the first pale green coloration to the substrate grains. At this point, the initial substrate begins to disappear, often by dissolution. The more stable the substrate, the longer it takes to disappear. Calcareous substrates are the least stable, and are easily and rapidly replaced by authigenic clays. Conversely, silicate substrates are more resistant to alteration, and remnants of these starting materials are often found in ancient glauconies, even in highly evolved grains.

As the substrate is destroyed, it leaves a new system of pores that, in turn, become filled with new authigenic clays. In addition, the earlier formed glauconitic smectites begin to recrystallize to more illitic forms. At this stage, individual grains are said to be slightly evolved (Fig. 2b), and consist largely of light green glauconitic illite/smectite mixed-layer clays (see Figs. 3 and 4) with K_2O contents of 4-6%.

With continued evolution, recrystallization toward more illitic forms with less expansible layers continues and the initial texture and crystal structure of the substrate becomes almost completely obscured. Extensive recrystallization of earlier formed glauconitic materials is accompanied by a significant amount of new illitic glauconite precipitation, causing an increase in volume. The volume increase, more pronounced in the interior than in the exterior of substrate grains, leads to the formation of the characteristic bulbous, cracked habit of the glauconite grains (Fig. 2c), and the well developed green color. At this stage, K_2O contents range from 6-8%, and the grains are said to be evolved.

If environmental conditions remain suitable, the cracks created in the previous stage are filled with new precipitates reestablishing a smooth surface on the grains. These highly evolved, dark green, well-rounded grains contain K_2O contents in excess of 8% (Fig. 2d) and consist almost entirely of glauconitic illite (see Fig. 3).

Because glauconitization requires prolonged contact with sea-water, the presence of glaucony in a sediment can suggest a hiatus in sediment accumulation. In addition, according to Odin and Fullagar (1988), the stage of evolution of the glauconite grains can be used as a rough measure of the length of the hiatus. Smectitic glaucony can be found in a sediment even if there was no significant hiatus. Glaucony with a mean potassium content of 7% K_2O suggests a significant break in sedimentation (10^4-10^5 yrs). A lengthier interruption in sediment accumulation (about 10^6 yrs) would be suggested by the presence of highly evolved glaucony, which is dark green in color, contains K_2O contents in excess of 8%, and could possibly be goethitized or phosphatized.

The effects of burial diagenesis on the continued evolution of glauconite have not as yet been documented. Glauconitic minerals presumably could continue to evolve toward more illitic forms with time and continued burial, but only if a suitable source of potassium was available. However, the rate at which this process would take place has not been examined, nor whether the process would take place at the same rate for the range of glauconitic minerals. Thus, it is not clear whether such an evolutionary process would merely shift all of the glauconitic minerals present to a slightly more mature form, or whether the process could increase or decrease the range of glauconite forms present by causing some of them to mature more rapidly than others.

The glauconite evolution process may be halted at any stage if the environment becomes unsuitable (Odin and Fullagar, 1988). Two main factors appear to be involved: sea-level change and burial. A drop in sea-level will likely expose the grains to a more oxidizing environment, while a transgressive event may cause phosphatization. In addition, although glauconitization may still take place at depths within the sediment of 1 m or more, burial below this depth typically halts the process. Consequently, a rapid increase in detrital influx is likely to inhibit or entirely prevent glauconitization.

GEOLOGIC SIGNIFICANCE OF GLAUCONITE

Glaucony-bearing horizons are commonly associated with transgressions; they typically occur at the bases of transgressive/regressive cycles. Development of sequence stratigraphic principles (Vail et al., 1984; Posamentier et al., 1988; Loutit et al., 1988) has emphasized the possible stratigraphic significance of glaucony; however, the distribution of glaucony within the different parts of a depositional sequence has never been well determined. Glaucony is most frequently described as being present within rocks that are interpreted to be part of the condensed section (CS) of a depositional

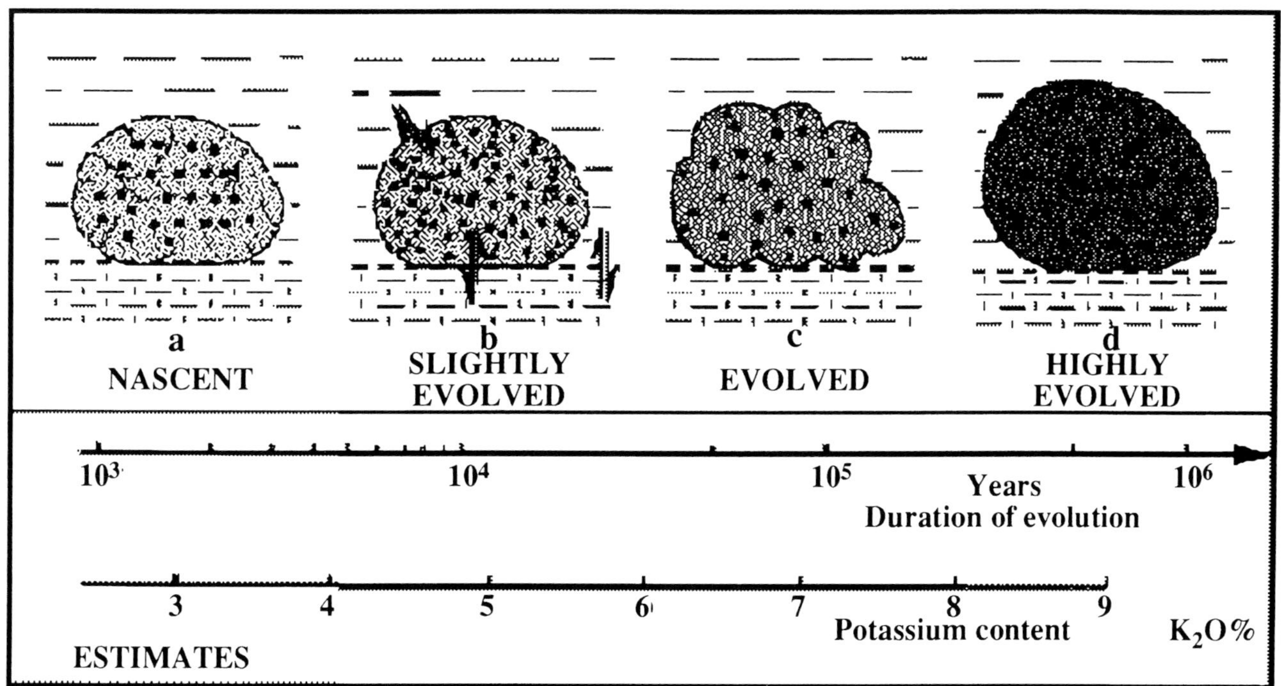

Fig. 2.—Stages of development in the glauconitization of a granular substrate: a) nascent; b) slightly evolved; c) evolved; d) highly evolved (Odin and Fullagar, 1988).

sequence with abundant planktonic and benthic microfossils, phosphorites, and organic debris (Loutit et al., 1988; Haq, 1991; Swift et al., 1991). However, glaucony has also been described as being present in a variety of other settings that include the base of the transgressive systems tract (TST) (Baum and Vail, 1988; Van Wagoner et al., 1990), throughout the entire TST marking the base of each parasequence (Mitchum and Van Wagoner, 1991), in the lowstand systems

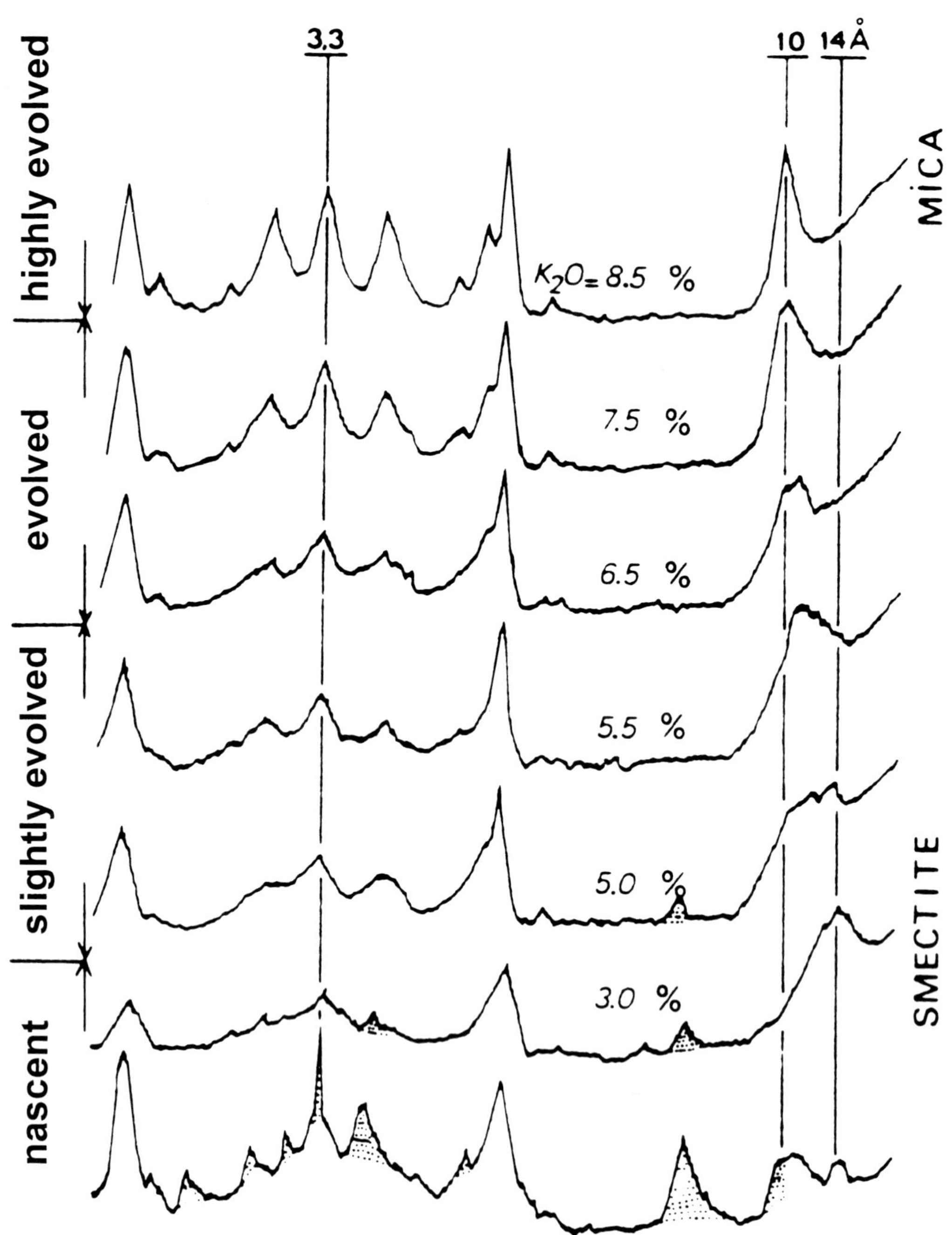

Fig. 3.—X-ray diffraction patterns depicting the mineralogical evolution of glauconitic minerals from smectite to illite. The substrate (stippled pattern) is still visible in the patterns of nascent and slightly evolved glaucony. Note the change in the position and shape of the (001) peak (between 14 and 10 Å) and the increase in definition of the (hkl) peaks of the authigenic component with increased evolution (after Odin and Matter, 1981).

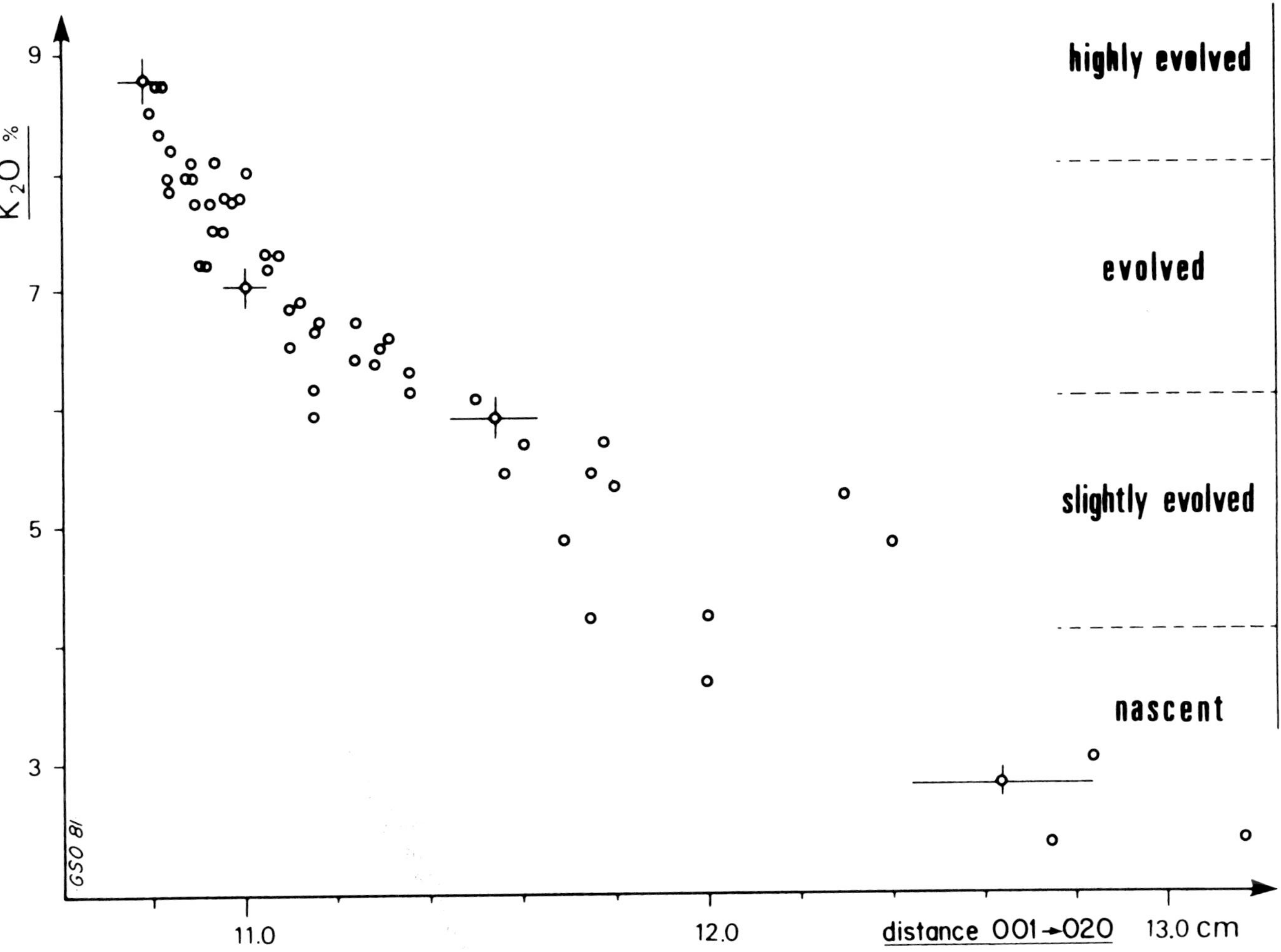

Fig. 4.—Variation of K_2O content and distance between the (001) and (020) peaks as a function of increasing evolution of glauconitic minerals. Distances were measured on diffraction patterns using KαCu radiation with a goniometer running at 1° per minute and a recorder at 1 cm per minute (Odin and Fullagar, 1988).

tract/shelf margin wedge (LST/SMW) (Baum and Vail, 1988; Glenn and Arthur, 1990; Walker and Bergman, 1993), and, rarely, in the highstand systems tract (HST) (Surlyk, 1991; Amorosi, 1993; Amorosi and Centineo, 1997). Thus, it would seem that the presence of glaucony alone is not suggestive of any specific systems tract or position within that tract. However, it should be noted that there is considerable debate about the interpretations of some of these depositional environments in which glauconite has been found, so the situation may not be as confused as it seems.

Amorosi (1995) showed that a reliable sequence stratigraphic-interpretation of glaucony-bearing units requires information in addition to just the presence or absence of glaucony, including:

1. distribution of glaucony within the glaucony-bearing unit. Information on the relative abundance of the glaucony, the thickness of the glaucony-bearing unit, and vertical variations (increasing/decreasing upward trends) is necessary to make a reasonable stratigraphic interpretation.
2. maturity of the glaucony (Table 1). As indicated above, glauconitization begins with the precipitation of K-poor glauconitic smectite that evolves toward K-rich glauconitic mica. Identification of the evolutionary stage of the glauconite requires the use of X-ray diffraction analyses to determine the mineralogic structure and chemical analyses to determine potassium content. Approximations based on grain color and paramagnetic susceptibility may also be used as a rough estimate of maturity (Odin and Matter, 1981; Amorosi, 1993).
3. spatial and temporal characteristics of the glaucony (Table 2). One of the most important pieces of information required to interpret the distribution of glaucony is the genetic relationship between the glaucony and its host sediment, that is, was the glaucony formed *in situ* or transported in from elsewhere. Facies analyses must be integrated with information on factors such as maturity, chemical composition, mineralogy, distribution, and in some cases, even the radiometric age of the glaucony to differentiate whether it is autochthonous versus

allochthonous, and intrasequential (parautochthonous) versus extrasequential (detrital/reworked). For detailed information on the criteria for performing such a differentiation, the reader is referred to Amorosi (1993).

DISTRIBUTION OF GLAUCONY IN SILICICLASTIC SEQUENCES

As indicated above, glaucony can be present throughout a variety of depositional successions. In addition, processes such as penecontemporaneous remobilization of autochthonous glaucony during transgression and reworking of glaucony from older deposits during sea-level lowstands are likely to result in significant concentrations of allochthonous glaucony in anomalous stratigraphic settings. This can result in a mixture of autochthonous and allochthonous and, in some cases, parautochthonous and detrital glaucony within the same deposit (Amorosi, 1993, 1995). In addition, the distribution and physicochemical characteristic of glaucony can undergo more or less pronounced modifications due to local changes in characteristics such as subsidence, sediment supply, availability of iron, Eh/pH, and the size and composition of available substrates. Consequently, the distribution of glaucony is controlled by factors much more complex than would be suggested by most recent sequence stratigraphic models (Table 3).

In the following discussions, a number of documented examples describing the distribution of glaucony in various depositional systems will be examined.

Glaucony in the Lowstand Systems Tract (LST)

In classic Type 1 depositional sequences (Vail et al., 1991), the subaerial exposure of the shelf caused by the fall in relative sea level results in conditions very unfavorable for the formation of autochthonous glaucony. However, extensive erosion of these previously submerged shelf deposits, together with older underlying or onshore formations, makes it possible to produce significant deposits of detrital reworked glaucony. These deposits (Fig. 5) are typically associated with deep water turbidites of the lowstand fan complex, but may also be found in coastal-belt and estuarine sand deposits of the proximal lowstand wedge (Baum and Vail, 1988; Glenn and Arthur, 1990; Amorosi, 1993).

When glaucony is present in the shelf-margin wedge of a Type 2 sequence (Fig. 6), the distribution of glaucony is complicated by the fact that the outer part of the shelf is not subaerially exposed. Consequently, if sediment supply is relatively low, the glaucony in such a sequence may be both autochthonous and allochthonous. In this setting, it could be possible that paralic/deltaic deposits such as lagoonal or foreshore/shoreface sediments could contain detrital glaucony. For example, Walker and Bergman (1993) interpreted the lower Cretaceous Shannon Sandstone of Wyoming as being a glauconitic shoreline sand. It should be noted, however, that the great abundance of glauconite in these sands, questions about the stability of glauconite in a high-energy, oxidizing shoreline setting, and the serious debate over the interpretation of this sand as a shoreline deposit (see Tillman, Swift and Parsons, and Bergman and Walker in this volume, as well as Tillman and Martinsen, 1987) cast doubt about whether or not glauconite has been proven to occur within shoreface facies. Within offshore marine deposits, at the point where the Type 2 sequence boundary becomes conformable, it would still be possible for autochthonous glaucony to form along omission surfaces at the bases of parasequences. Glaucony may also be present within accreting sand-waves and bars within the shelf-margin wedge, however, there the glauconite probably has a parautochthonous or detrital origin (Amorosi, 1993).

Glaucony in the Transgressive Systems Tract (TST) and Condensed Section (CS)

As discussed above, a transgression, or rise in sea level, typically favors the development of glaucony primarily because sediment is trapped on the continent. However, glaucony can develop in a variety of settings within this transgressive systems tract. Autochthonous glaucony (Fig. 7) is typically present at the tops of burrowed, shoaling-upward, coarse-grained sequences, marking marine flooding surfaces within backstepping, retrogradational parasequence sets (Van Wagoner et al., 1990; Bhattacharya and Walker, 1991). Maturity and concentration of glaucony within these units vary as a function of the type of flooding surface, which could be a ravinement/TSE, minor flooding surface, or maximum flooding surface, but generally increase from the base to the top of the TST.

Table 1.—Classification of glaucony at different stages of evolution. From Amorosi, 1995.

Glaucony	Maturity	K_2O Content	Mineralogical Structure	Paramagnetic Susceptibility	Color
nascent	low	<4%	smectite	low	pale green
slightly evolved	moderate	4-6%	↓	moderate	light green
evolved	high	6-8%	↓	high	green
highly evolved	very high	>8%	mica (Glauconite)	very high	dark green

Table 2.—Classification of glaucony by spatial vs. temporal characteristics. From Amorosi, 1995.

	Spatial Characteristics	Temporal Characteristics	Proposed Terminology	Synonyms
Glaucony	Autochthonous (in place)	Intrasequential	Autochthonous	Authigenic, *in situ*
	Allochthonous (transported)	Intrasequential	Parautochthonous	Perigenic
		Extrasequential	Detrital	Allogenic, reworked

Parautochthonous glaucony is commonly concentrated in the upper, high-energy portions of transgressive shelf deposits. Because this glaucony has been winnowed either landward or seaward by waves, tides, and storms before its evolution is complete, it is likely to have low to moderate maturity. Parautochthonous glaucony may also be associated with detrital glaucony in estuarine/incised-valley fills common to Type 1 sequence boundaries, especially if older glaucony-rich units were exposed during a previous lowstand. Transgressive sheet sands, formed during shoreline retreat, may contain virtually any type of glaucony, including autochthonous grains (Swift et al., 1971).

Thus, the interpretation of glauconies in the TST, particularly in its lower portions, is often equivocal because many types of glaucony are commonly found within this systems tract.

The association of very mature, glaucony-rich sediments with the condensed section (CS) of a depositional sequence is one of the clearest and most consistent relationships. The maximum rise in sea level traps sediment in more landward areas, resulting in the conditions most favorable for both the development and full evolution of glaucony, especially in the outer shelf and slope areas (see Fig. 7). Glaucony in the CS is often associated with phosphate grains and abundant fossil debris and may be associated with hardgrounds or burrowed omission surfaces. The landward extension of the CS, the maximum flooding surface (MFS), is also a site for concentration of glaucony. In this setting, glaucony can form in the uppermost part of the coarse-grained sand facies, at the boundary between marginal-marine sands and outer-shelf deposits.

Although condensed sections commonly contain the greatest abundance of highly evolved glaucony, extended depositional hiatuses are not exclusively related to condensed sections. Each omission surface at the base of a parasequence could, in theory, be marked by the presence of evolved glaucony. Likewise, depending on local conditions, not all condensed sections contain highly evolved glaucony.

Glaucony in the Highstand Systems Tract (HST)

The highstand systems tract is characterized by a basinward-shift of facies and a situation in which sediment accumulates faster than accommodation space is created. This situation is highly unfavorable for verdissement, which explains the relatively rare occurrence of glaucony in this setting.

The lower HST may contain minor, slightly evolved, autochthonous glaucony in the lower portions of parasequences (Fig. 8), either within fine-grained outer-shelf deposits or at the tops of shoaling-upward sequences. These deposits mark the last stages of glaucony production in areas that are still protected from clastic input. Abundance and maturity of glaucony in these deposits typically decrease upward, showing a trend opposite to that in the TST. Parautochthonous glaucony, of low- to moderate maturity, may also be found in the upper portions of parasequences, which are associated with cross-bedded bars and sand waves. Glaucony is extremely rare in the upper portion of the HST.

ORIGIN OF SIDERITE

Chemical Environment — Locus of Formation

The conditions under which siderite can form are well known and fairly limited. Berner (1981) showed that the formation of siderite requires reducing conditions to permit mobilization of Fe^{+2} and low sulfate activity. Waters rich in sulfate tend to have high sulfide activities where reducing conditions occur. Under these conditions any free, ferrous ion would react with sulfide to produce pyrite rather than siderite.

Berner showed that siderite typically forms in one of two distinct geochemical environments, one characterized by strongly reducing condition (methanogenic zone) and one under slightly reducing conditions (suboxic or post-oxic zone). Because methanogenic zones are favored in environments characterized by low sulfate concentrations, high sedimentation rates, and high organic carbon concentrations (Coleman, 1985), methanogenic formation of siderite is more common in continental sediments than in marine deposits. Accordingly, siderite forms a common early cement in settings such as anoxic fresh-water lakes, and swamp and marsh deposits, where the siderite commonly seen replaces and forms nodules around plant roots. Siderite nodules are also common in coal measures (Curtis, 1967) and oil shale deposits (Weber and Hofmann, 1982; Mason and Kirchner, 1993; Gibson et al., 1994). Methanic condi-

Table 3.—Conceptual framework of glaucony distribution in depositional sequences and expected features of glaucony*. From Amorosi, 1995.

	Sequence Stratigraphic Location	Host Deposit	Origin	Maturity
HST	Upper part of parasequence	High-energy shelf deposits	P	Low to very high
	Lower part of parasequence	Low-energy shelf deposits	A	Low to moderate
CS	Maximum flooding surface	Top of marginal-marine sand-stones	A	High to very high
	Uppermost TST-lowermost HST	Hardgrounds	A	High to very high
	Uppermost TST-lowermost HST	Outer-shelf to slope deposits	A	High to very high
TST	Ravinement surface	Transgressive sheet sandstones	P, A, D	Variable
	Incised valley fill	Estuarine deposits	P,D	Variable
	Upper part of parasequence	High-energy shelf deposits	P	Low to moderate
	Lower part of parasequence	Low-energy shelf deposits	A	Low to high
LST	Shelf margin wedge	Offshore-marine deposits	A,P	Low to moderate
	Shelf margin wedge	Coastal-belt deposits	D	Variable
	Proximal lowstand wedge	Coastal-belt deposits	D	Variable
	Lowstand fan complex	Deep-water turbidites	D	Variable

* P = Parautochthonous; A= Autochthouous; D=Detrital

tions can also develop in marine pore waters if organic contents are sufficient to consume all of the oxygen and sulfate, but these deposits are generally rare. Matsumoto (1989) described an unusual occurrence of siderite forming as a result of decomposition of methane hydrate in continental margin deposits along the Blake Outer Ridge.

Post-oxic conditions are commonly associated with marine environments that display relatively low concentrations of organic matter and low sedimentation rates. Here, sufficient organic matter is deposited so that all dissolved oxygen is consumed during organic decomposition by aerobic micro-organisms. However, the amount of organic matter is not sufficient to bring about sulfidic conditions and, as a result, further organic matter decomposition takes place by first nitrate, then manganese, and then iron reduction, finally resulting in siderite formation. Siderite has also been shown

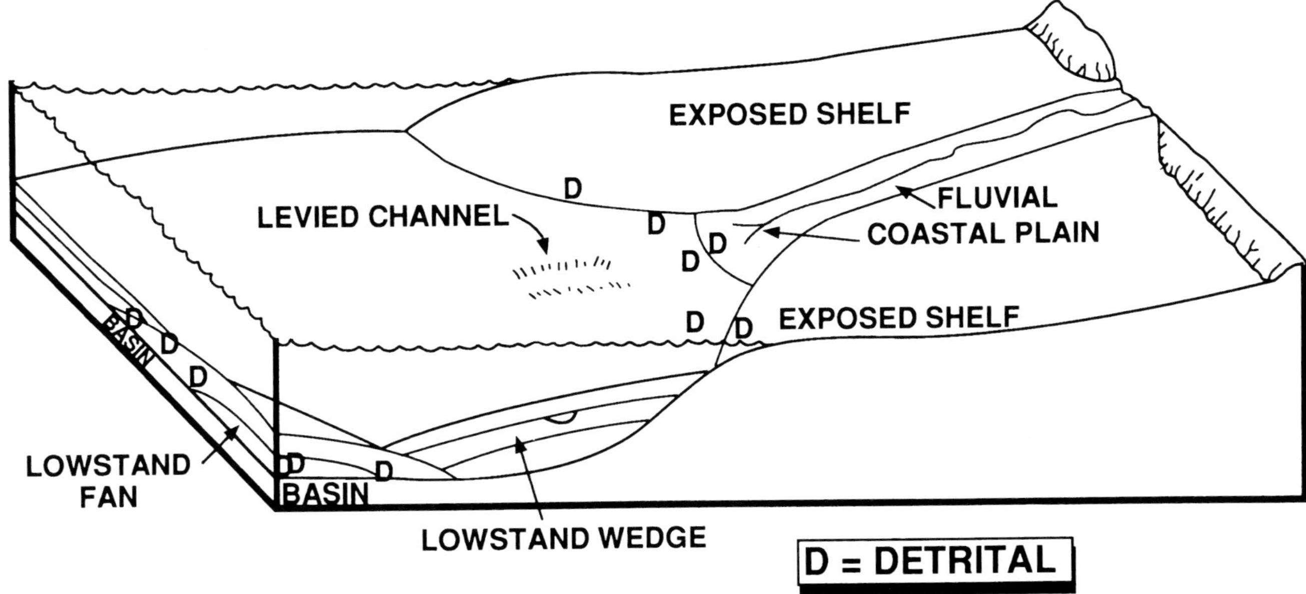

Fig. 5.—Distribution of glaucony in the lowstand systems tract.

to form in post-oxic conditions in sediments that are subjected to alternating anoxic and oxic pore waters such as nearshore and intertidal deposits (Coleman, 1985), and burrowed hardgrounds (Sellwood, 1970). Siderite has also been observed in lacustrine deposits that form at the suboxic boundary, where rising CO_2 and reduced iron intermingle (Bahrig, 1988).

Thus, siderite appears to form in a variety of marine as well as nonmarine environments. The next section will explore whether the siderite precipitated in these various environments can be differentiated on the basis of chemical or isotopic characteristics.

Chemical and Isotopic Differentiation Between Marine- and Fresh-Water Siderites

As discussed above, because siderite is often one of the earliest minerals to precipitate in sediments, there is considerable interest in using siderite as an indicator of depositional environment. As a consequence, a number of workers have looked at the possibility of using elemental and/or isotopic compositions to differentiate between siderites precipitated in marine versus fresh-water systems.

Mozley (1989) used data from a variety of sources to show that the elemental composition of early diagenetic siderite is

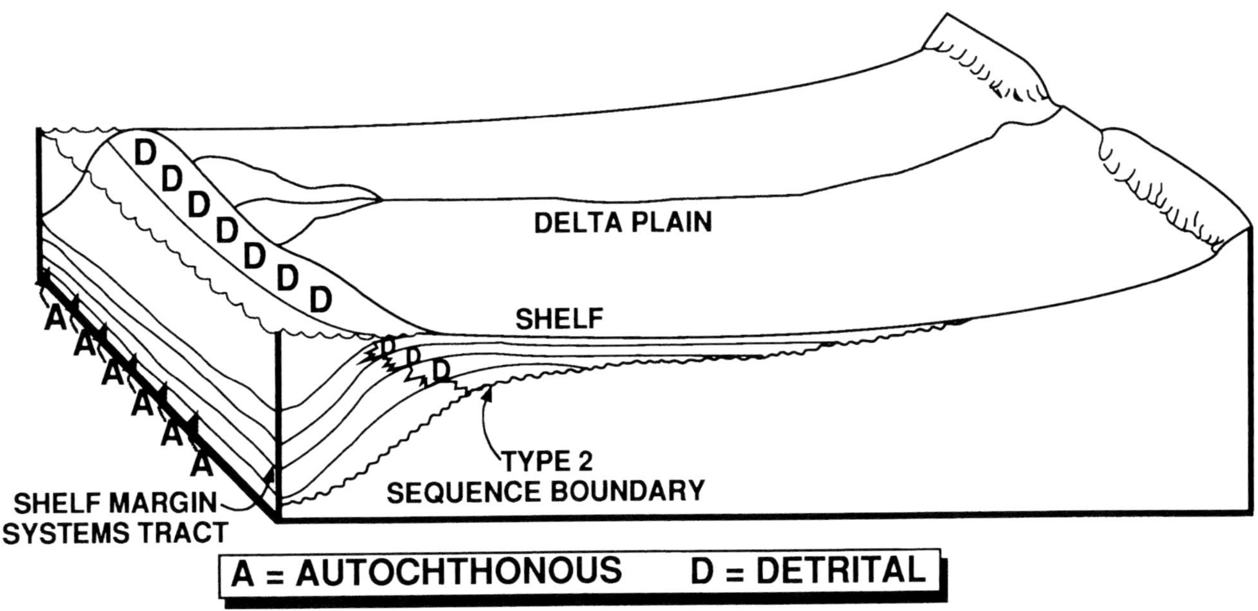

Fig. 6.—Distribution of glaucony in the lowstand shelf-margin wedge.

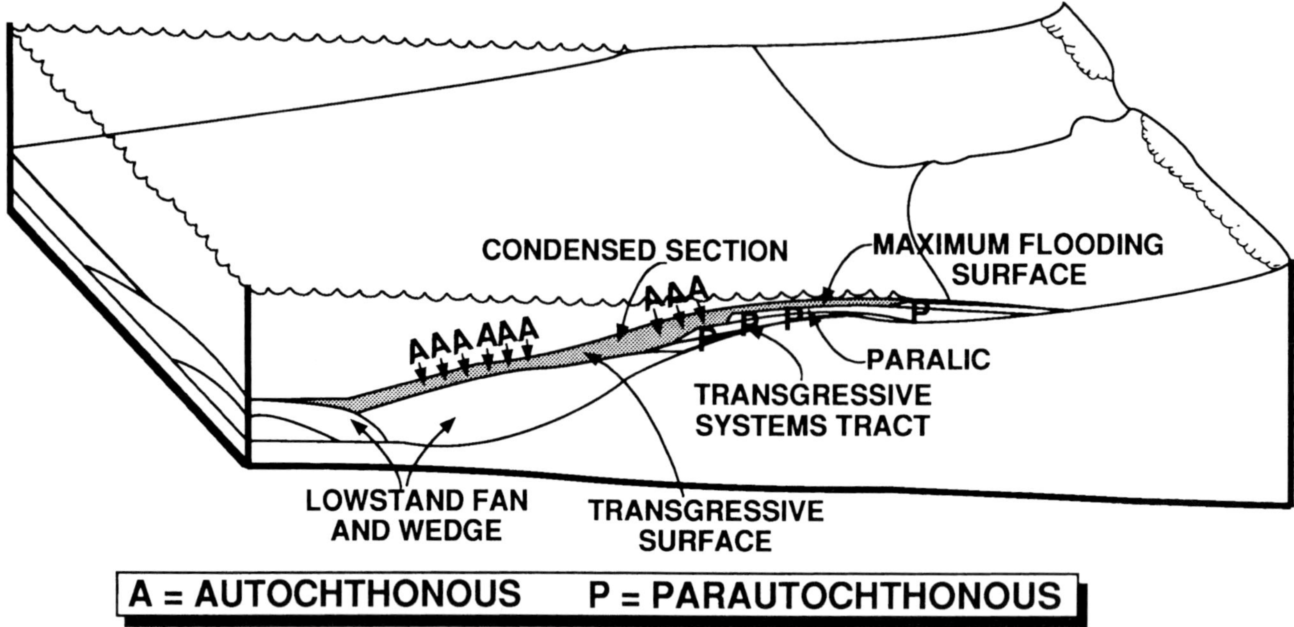

Fig. 7.—Distribution of glaucony in the transgressive systems tract.

highly variable and that there is considerable overlap in the data for siderite from marine and fresh water. Nevertheless, marine and fresh-water siderites are characterized by distinctive compositional trends (Fig. 9). Marine siderite is always very impure and exhibits extensive substitution of Mg (up to 41 mol% $MgCO_3$), and minor Ca (up to 15 mol%). Marine siderites also generally contain very low Mn contents (<1 mol% $MnCO_3$).

In contrast, siderite from fresh water is frequently very pure (>90 mol% $FeCO_3$), often attaining end-member composition. In addition, fresh-water siderites frequently exhibit high concentrations of Mn (>2 mol% $MnCO_3$), and high Ca/Mg ratios relative to marine siderites.

Similarly, Mozley and Wersin (1992) examined the relationship between siderite isotopic compositions and depositional environment based on a compilation of isotopic data

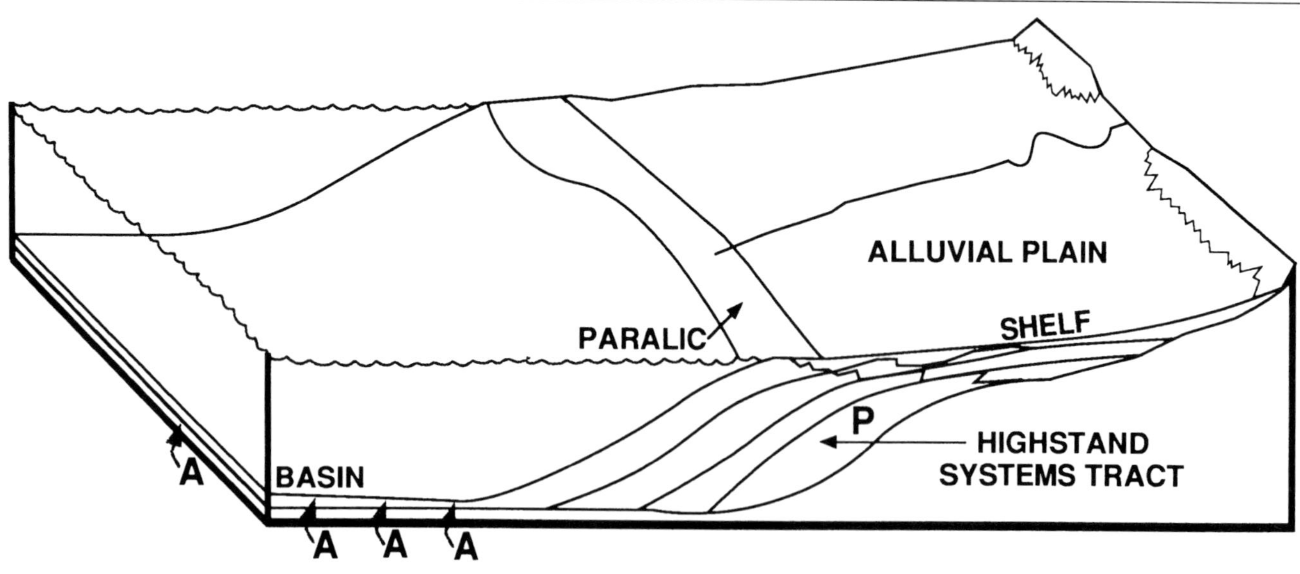

Fig. 8.—Distribution of glaucony in the highstand systems tract.

from the literature. The data (Fig. 10) show that, although there is significant overlap in the stable isotopic compositions for marine versus fresh-water siderites, there are also large areas in which no overlap occurs. The most notable difference is the generally higher $\delta^{13}C$ values of continental versus marine forms. These values suggest that siderites with $\delta^{13}C$ values less than -8 ‰ are most likely of marine origin, while those with $\delta^{18}O$ values less than -13 ‰ and with positive $\delta^{13}C$ values are most likely of nonmarine origin.

Mozley and Wersin suggest that the most reliable way of inferring the depositional environment of siderite host sediment is to combine compositional and isotopic data, thereby resolving some of the ambiguity in both methods. Mozley and Wersin point out, however, that a complicating factor in the use of siderite composition to infer depositional environment is the early introduction of waters from a different depositional environment into the sediment. Siderite from complex depositional sequences with intercalated marine and non-marine strata (e.g., coal-measure siderite) can show very confused compositional patterns. Similarly, siderite formed in association with an unconformity is more likely to have a composition reflecting the depositional environment of the rocks overlying the unconformity than that of the host rock in which it occurs. Thus, great care should be used in the interpretation of compositional data.

SUMMARY

The conditions required for the formation of glaucony suggest that if it is truly authigenic, it should be a reliable indicator of slow sedimentation in a marine setting. However, from a sequence stratigraphic standpoint, glaucony may be present in virtually any part of a depositional sequence due to remobilization. Glaucony can provide useful information for sequence stratigraphy only if variations in its abundance, physicochemical properties (degree of evolution), and spatial/temporal characteristics (autochthonous versus allochthonous; intrasequential versus extrasequential) are carefully documented.

Although there are many exceptions, the following generalizations may be made:

1. Autochthonous glaucony is associated with open-marine deposits throughout the TST and in the lower part of the HST. It is typically distributed with an upward increase and then a decrease in abundance and maturity. The CS and MFS represent intervals where the maximum concentration of the most highly evolved glaucony can be expected.

2. Allochthonous intrasequential (parautochthonous) glaucony is most common in wave-, tide-, and storm-dominated deposits of the TST and HST, but minor occurrences may also be found in the LST. This type of glaucony typically shows low concentrations and low to moderate maturity.

3. Allochthonous extrasequential (detrital) glaucony is typically present in the LST and the lower TST. Abundance and maturity vary widely depending on the character of the sediment source.

The formation of siderite requires reducing conditions and low sulfate activity. Siderite typically forms in one of two distinct environments: one characterized by strongly reducing conditions (methanogenic zone), and one under slightly reducing conditions (post-oxic zone). Methanogenic siderite is more common in continental and fresh-water lacustrine than marine deposits. Post-oxic conditions are commonly associated with marine environments exhibiting moderately low concentrations of organic matter and low sedimentation rates. Siderite is also frequently found in association with sequence boundaries where it occurs as a secondary cement below the LSE.

Marine versus fresh-water siderites may be differentiated on the basis of their chemical and isotopic compositions; however, because of common overlap in the respective compositional fields, both types of data should be compared.

Marine siderite is always very impure and exhibits extensive substitution of Mg and minor Ca. In contrast, siderite from fresh water is frequently very pure, but may contain minor Mn and high Ca/Mg ratios relative to marine siderites.

Siderites with $\delta^{13}C$ values less than -8 ‰ are most likely of marine origin, while those with $\delta^{18}O$ values of less than -13 ‰ and with positive $\delta^{13}C$ values are most likely of nonmarine origin.

The chemical restrictions on environment of origin discussed above have been used to provide information on the hydrologic regime and, by inference, on the depositional and sequence stratigraphic setting of the host sediment. By examining the genetic significance and distribution of minerals such as glauconite and siderite, as well as other very early authigenic cements, evidence may be found that will help resolve the origin of ambiguous, controversial, isolated marine sand bodies such as those discussed elsewhere in this volume. To test the models discussed in this paper, a project has been started to collect data on glauconite distribution in several formations whose depositional origin is currently under debate by sedimentologists and sequence stratigraphers. Data from this project will be presented in a future series of papers as it becomes available.

REFERENCES

Amorosi, A., 1993, Intèrêt des niveaux glauconieux et volcano-sédimentaires en stratigraphie: Exemple de dépôts de bassins tectoniques miocènes des Alpinnins et comparison avec quelque dèpôts de plate-forme stable, Ph.D. Dissertation: Mémoires des Sciences de la Terre no. 93-12, Université Pierre et Marie Curie, Paris, 194 p.

Amorosi, A., 1995, Glaucony and sequence stratigraphy: A conceptual framework of distribution in siliciclastic sequences: Journal of Sedimentary Research, v. B65, p. 419-425.

Amorosi, A. and Centineo, M.C., 1997, Glaucony from the Eocene of the Isle of Wight (southern UK); implications for basin analysis and sequence-stratigraphic interpretation: Journal of the Geological Society of London, v. 154, p. 887-896.

Baum, G.R. and Vail, P.R., 1988, Sequence stratigraphy concepts applied to Paleogene outcrops, Gulf and Atlantic basins, *in* Wilgus, C.K., Hastings, B.S., Kendal, C.G. St. C., Posamentier, H.W., Ross, C.A. and Van Wagoner, J.C., eds., Sea Level Changes: An Integrated Approach: Tulsa, Society for Sedimentary Geology Special Publication 42, p. 309-328.

Berner, R.A., 1981, A new geochemical classification of sedimentary environments: Journal of Sedimentary Petrology, v. 51, p. 359-365.

Bhattacharya, J. and Walker, R.G., 1991, Allostratigraphic subdivision of the Upper Cretaceous Dunvegan, Shaftesbury, and Kaskapau Formations in the subsurface of northwestern Alberta: Bulletin of Canadian Petroleum Geology, v. 39, p. 145-164.

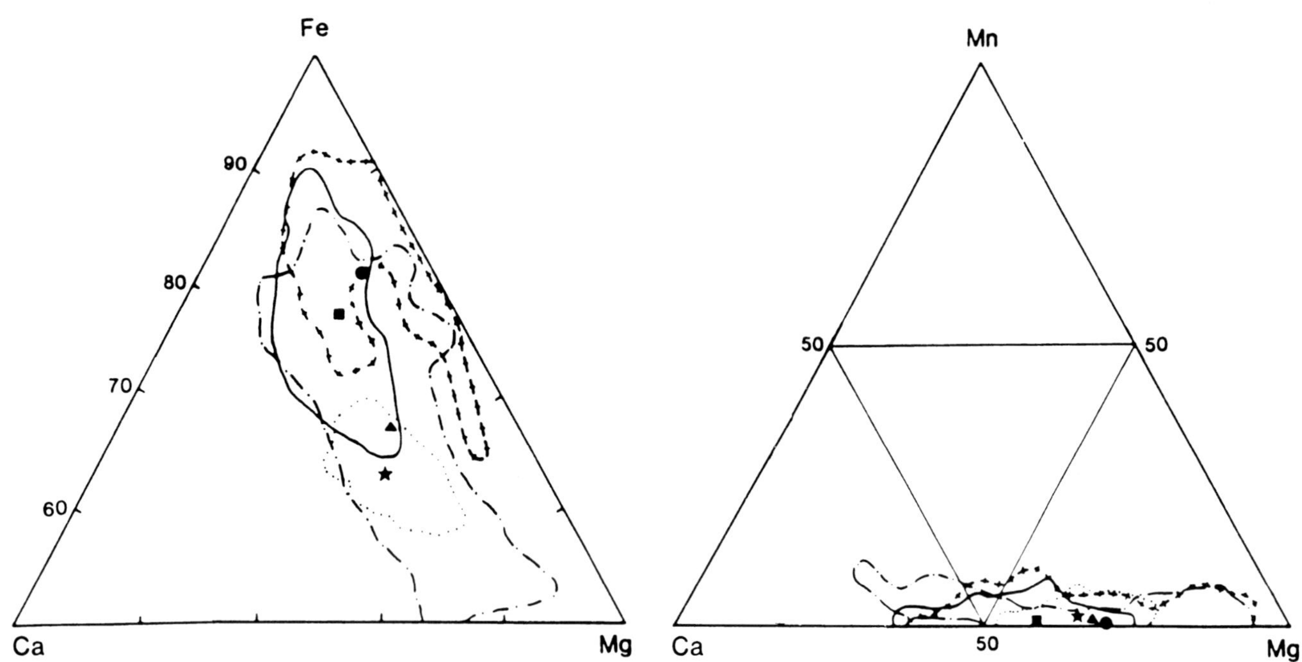

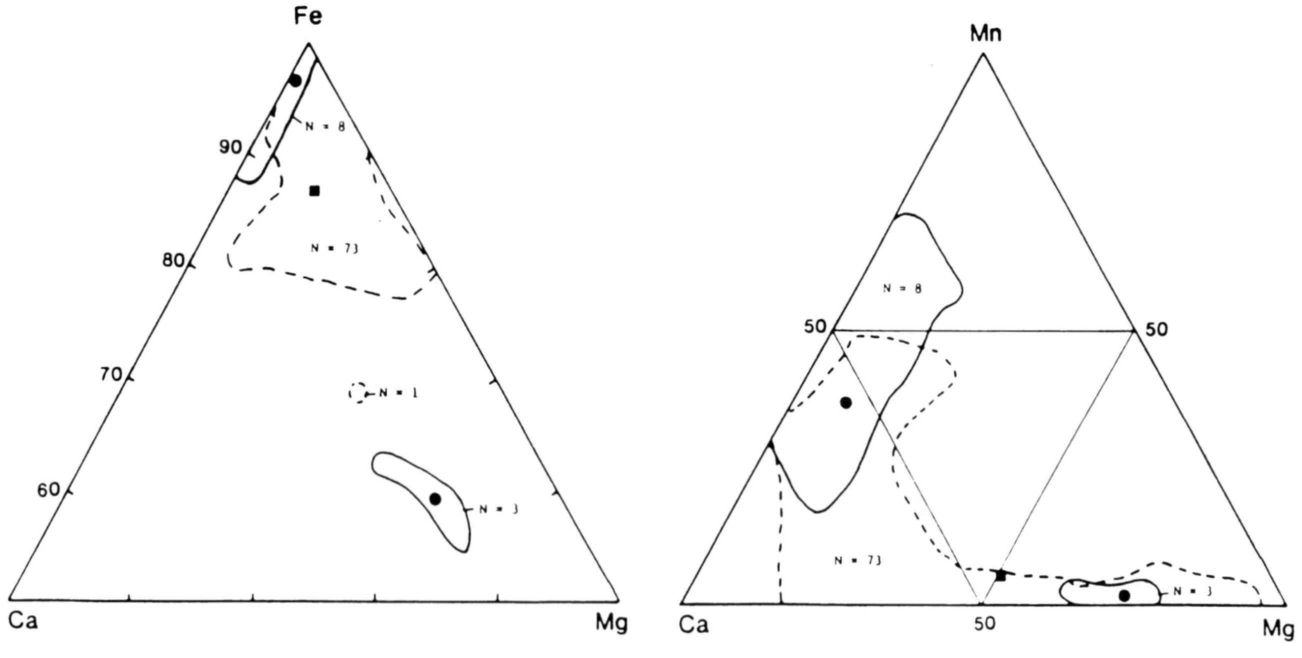

Fig. 9.—Elemental compositions of marine versus fresh-water siderites. Ternary $CaCO_3$-$MgCO_3$-$FeCO_3$ (left) and $CaCO_3$-$MgCO_3$-$MnCO_3$ (right) plots for electron microprobe data (Mozley, 1989).

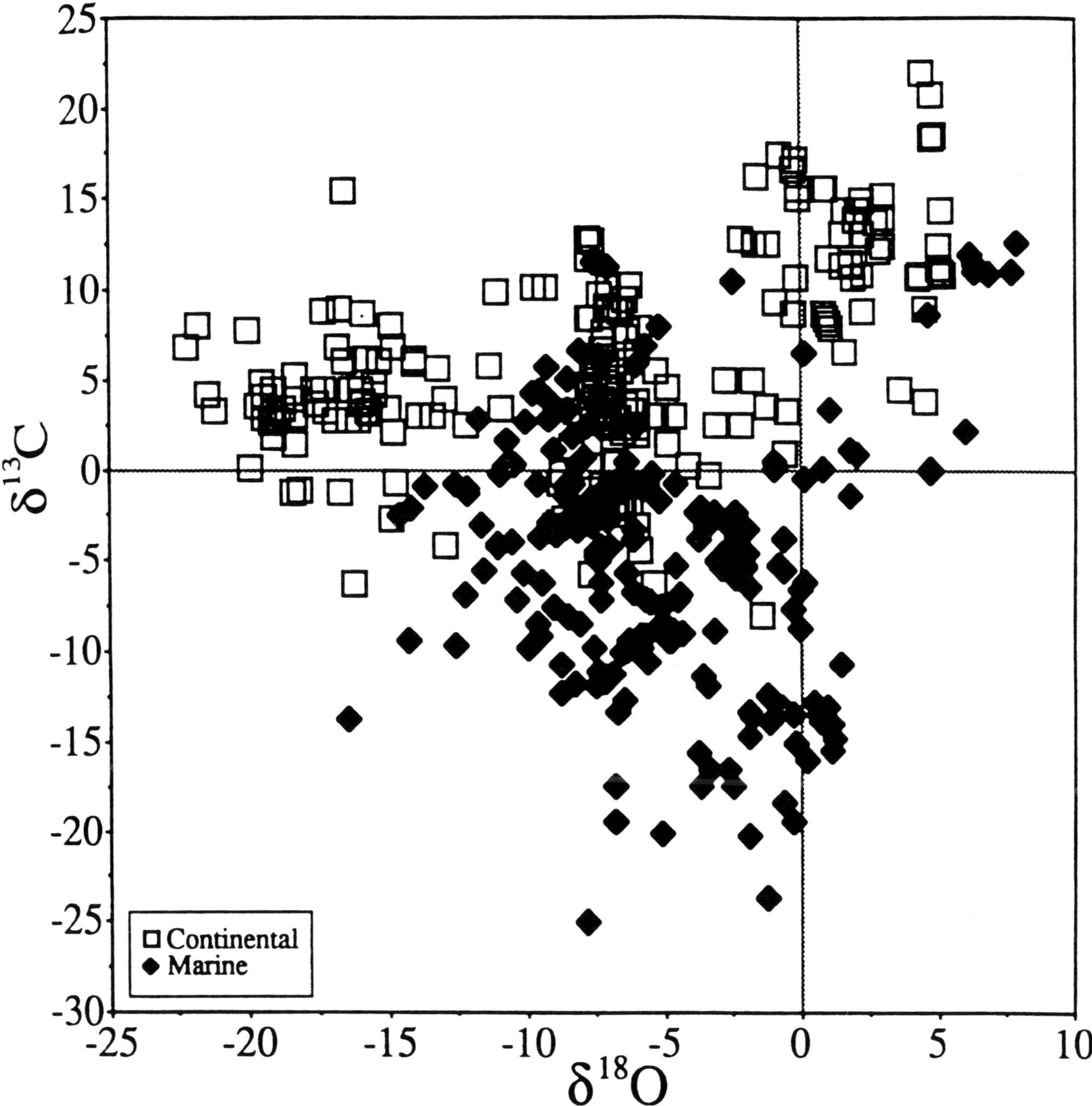

Fig. 10.—Isotopic compositions of marine versus continental siderites (Mozley and Wersin, 1992).

COLEMAN, M.L., 1985, Geochemistry of diagenetic non-silicate minerals; kinetic considerations: Philosophical Transactions of the Royal Society of London, v. A315, p. 39-56.

CURTIS, C.D., 1967, Diagenetic iron minerals in some British carboniferous sediments: Geochimica et Cosmochimica Acta, v. 31, p. 2109-2123.

GIBSON, P.J., SHAW, H.F. AND SPIRO, B., 1994, The nature and origin of sideritic ironstone bands in the Tertiary Lowmead and Duaringa Basins, Queensland: Australian Journal of Earth Sciences, v. 41, p. 255-263.

GLENN, C.R. AND ARTHUR, M.A., 1990, Anatomy and origin of a Cretaceous phosphorite-greensand giant, Egypt: Sedimentology, v. 37, p. 123-154.

HAQ, B.U., 1991, Sequence stratigraphy, sea-level change, and significance for the deep sea, *in* Macdonald, D.I.M., ed., Sedimentation, Tectonics, and Eustacy: Sea-level Changes at Active Margins: Oxford, International Association of Geologists Special Publication 12, p. 3-39.

LOUTIT, T.S., HARDENBOL, J., VAIL, P.R. AND BAUM, G.R., 1988, Condensed sections: The key to age determination and correlation of continental margin sequences, *in* Wilgus, C.K., Hastings, B.S., Kendal, C.G. St. C., Posamentier, H.W., Ross, C.A. and Van Wagoner, J.C., eds., Sea Level Changes: An Integrated Approach: Tulsa, Society for Sedimentary Geology Special Publication 42, p. 183-213.

Mason, G.M. and Kirchner, G., 1993, Bacterial influence on the deposition of siderite, Green River Formation, Wyoming: University of Kentucky and U.S. DOE 1993 Eastern Oil Shale Symposium Proceedings, p. 381-389.

Matsumoto, R., 1989, Isotopically heavy oxygen-containing siderite derived from the decomposition of methane hydrate: Geology, v. 17, p. 707-710.

Mitchum, R.M., Jr. and Van Wagoner, J.C., 1991, High-frequency sequences and their stacking patterns: Sequence-stratigraphic evidence of high frequency eustatic cycles: Sedimentary Geology, v. 70, p. 131-160.

Mozley, P.S., 1989, Relation between depositional environment and the elemental composition of early diagenetic siderite: Geology, v. 17, p. 704-706.

Mozley, P.S. and Wersin, P., 1992, Isotopic composition of siderite as an indicator of depositional environment, Geology, v. 20, p. 817-820.

Odin, G.S. and Fullagar, P.D., 1988, Geological significance of the glaucony facies, *in* Odin, G.S., ed., Green Marine Clays: Amsterdam, Elsevier, Developments in Sedimentology 45, p. 295-332.

Odin, G.S. and Matter, A., 1981, De glauconarium origine: Sedimentology, v. 28, p. 611- 641.

Posamentier, H.W., Jervey, M.T. and Vail, P.R., 1988, Eustatic controls on clastic deposition I: Conceptual framework, *in* Wilgus, C.K., Hastings, B.S., Kendal, C.G. St. C., Posamentier, H.W., Ross, C.A. and Van Wagoner, J.C., eds., Sea Level Changes: An Integrated Approach: Tulsa, Society for Sedimentary Geology Special Publication 42, p. 109-124.

Sellwood, B.W.,1970, The genesis of some sideritic beds in the Yorkshire Lias (England): Journal of Sedimentary Petrology, v. 41, p. 854-858.

Surlyk, F., 1991, Sequence stratigraphy of the Jurassic-lowermost Cretaceous of East Greenland: American Association of Petroleum Geologists Bulletin, v. 75, p. 1468-1488.

Swift, D.J.P., Stanley, D.J. and Curray, J.R., 1971, Relict sediments on continental shelves: A reconstruction: Journal of Geology, v. 79, p. 322-346.

Swift, D.J.P., Phillips, S. and Thorne, J.A., 1991, Sedimentation on continental margins, IV: Lithofacies and depositional systems, *in* Swift, D.J.P., Oertel, G.F., Tillman, R.W. and Thorne, J.S., eds., Shelf Sand and Sandstone Bodies: Geometry, Facies, and Sequence Stratigraphy: Oxford, International Association of Geologists Special Publication 14, p. 89-152.

Tillman, R.W. and Martinsen, R.S., 1987, Sedimentological characteristics and production model of Hartzog Draw Field, Wyoming, a Shannon shelf ridge sandstone, *in* Tillman, R.W. and Weber, K.J., eds., Reservoir Sedimentology: Tulsa, Society for Sedimentary Geology Special Publication 40, p. 15-112.

Vail, P.R., Audemard, F., Bowman, S.A., Eisner, P.N. and Perez-Cruz, C., 1991, The stratigraphic signatures of tectonics, eustacy, and sedimentology - an overview, *in* Einsele, G., Ricken, W. and Seilacher, A., eds., Cycles and Events in Stratigraphy: Berlin, Springer-Verlag, p. 617-659.

Vail, P.R., Hardenbol, J. and Todd, R.G., 1984, Jurassic unconformities, chronostratigraphy, and sea-level changes from seismic stratigraphy and biostratigraphy, *in* Schlee, J.S., ed., Interregional Unconformities and Hydrocarbon Accumulation: Tulsa, American Association of Petroleum Geologists Memoir 36, p. 1129-1144.

Van Wagoner, J.C., Mitchum, R.M., Jr., Campion, K.M. and Rahmanian, V.D., 1990, Siliciclastic Sequence Stratigraphy in Well Logs, Cores, and Outcrops: Tulsa, American Association of Petroleum Geologists Methods in Exploration Series No. 7, 55 p.

Weber, J. and Hofmann, U., 1982, Kernbohrungen in der eozënen Fossillagerstätte Grube Messel bei Darmstadt: Geologische Abhandlung Hessen, v. 83, 58 p.

Walker, R.G. and Bergman, K.M., 1993, Shannon Sandstone in Wyoming: A shelf-ridge complex reinterpreted as lowstand shoreface deposits: Journal of Sedimentary Petrology, v. 63, p. 839-851.

MARINE AND MARGINAL MARINE MUDSTONE DEPOSITION: PALEOENVIRONMENTAL INTERPRETATIONS BASED ON THE INTEGRATION OF ICHNOLOGY, PALYNOLOGY AND FORAMINIFERAL PALEOECOLOGY

JAMES A. MACEACHERN
Earth Sciences, Simon Fraser University, Burnaby, British Columbia V5A 1S6, Canada
CHARLES R. STELCK AND S. GEORGE PEMBERTON
Department of Earth and Atmospheric Sciences, University of Alberta, Edmonton, Alberta T6G 2E3, Canada

ABSTRACT: This study integrates ichnological, palynological and foraminiferal assemblages of selected mudstones, principally from the late Albian Viking Formation of central Alberta, to characterize their paleoecology. Ichnological assemblages have been based on the analysis of more than 300 cored intervals. Twenty-two palynological samples were analyzed from 9 wells, and 14 foraminiferal samples were analyzed from 7 wells. The selected samples comprise 5 depositional "types", namely: 1) transgressive systems tract shelf deposits; 2) highstand systems tract lower offshore deposits; 3) transgressive systems tract upper offshore deposits; 4) transgressive systems tract embayment/lagoon deposits; and 5) transgressive systems tract estuarine incised valley central basin deposits.

The palynological assemblages were most useful in separating fully marine from strongly brackish-water (estuarine) environments within the Viking Formation. Peridinioid dinoflagellate concentrations are anomalously high in samples from the central basin mudstone of estuarine incised valley complexes. Foraminiferal assemblages were sufficient to differentiate most of the subenvironments, based on changes in the diversity and abundance of genera and species, as well as the presence of environmentally tolerant forms. The paucity of foraminifera within the estuarine incised valley mudstone limited their paleoecologic utility in these deposits.

Ichnological analysis was able to differentiate between shelf, lower offshore, upper offshore, embayment/lagoon and estuarine incised valley central basin mudstones. In addition, this data can be collected during initial facies descriptions without the need for laboratory analysis, making it an essential paleontologic discipline in facies analysis.

INTRODUCTION

Delineating the depositional position of many sandstone bodies within the marine and marginal marine realm can be problematic, and has been a hotly debated issue in the sedimentological community. The widespread use of genetic stratigraphic techniques and the concomitant recognition of bounding stratigraphic discontinuities has, however, facilitated the interpretation of depositional environments. Nevertheless, the sequence stratigraphic interpretation of the bounding discontinuities, which fundamentally affects the ultimate interpretation of the succession in question, is not always agreed upon (e.g., the Shannon Sandstone; cf., Walker and Bergman, 1993; Bergman, 1994; Bergman and Walker, 1995; Sullivan et al., 1995).

The difficulties in interpreting the paleoenvironmental setting of marine-influenced sandstone stems from the fact that deposition of coarse-grained bodies tends to occur rapidly and under relatively high energy conditions. The result is that many sandstones are dominated by primary physical sedimentary structures that reflect short-lived, high energy traction transport. Such deposits, however, commonly lack unequivocal evidence of their depositional environment. Trough cross-stratified sandstone, for example, can and does occur in many depositional settings, ranging from fluvial systems, deltaic/strandline complexes, nearshore bars, shelf sand ridges and submarine fan channels.

The integration of ichnology with classical sedimentology has improved the interpretation of sedimentary successions. Unfortunately, direct application of ichnology to many sandstone bodies is limited, due to the predominance of primary physical sedimentary structures and subordinance or even total absence of paleontologic assemblages. In many instances, therefore, the interpretation of sandstone and conglomerate lies largely with the character of intercalated, genetically related muddy facies. Muddy facies typically accumulate slowly and reflect lower energy settings, favouring the development of a paleontologic component. Unlike most physical sedimentary features, these biological elements are strongly controlled by the attendant environment. By analyzing the trace fossil, palynological and foraminiferal components of the genetically related mudstone, an understanding of the more economically important coarse-grained deposits is possible.

Samples were collected from depositional successions that have been described in several previous papers and for which detailed paleoenvironmental and sequence stratigraphic interpretations have been outlined (see below). This paper integrates the sedimentology and ichnology of selected mudstones within the Joli Fou, Viking and Westgate formations of Alberta, with palynological and/or foraminiferal fossil assemblages, to derive paleontologic criteria for interpretation of the paleoenvironments. The selected samples comprise 5 depositional "types", namely: 1) transgressive systems tract shelf deposits; 2) highstand systems tract lower offshore deposits; 3) transgressive systems tract upper offshore deposits; 4) transgressive systems tract embayment/lagoon deposits; and 5) transgressive systems tract estuarine incised valley central basin deposits. The paper also gauges the reliability of ichnological interpretations, by testing them against the historically accepted accuracy of foraminiferal and palynological paleoecology.

GENERAL GEOLOGIC SETTING

The study area centres mainly around the Joffre Field, with the addition of two wells from the Willesden Green Field and one from the Crystal Field (Fig. 1). Comparisons are also drawn from the Giroux Lake/Kaybob North area. Most samples were collected from the Viking Formation,

Isolated Shallow Marine Sand Bodies: Sequence Stratigraphic Analysis and Sedimentologic Interpretation.
SEPM Special Publication No. 64, Copyright © 1999
SEPM (Society for Sedimentary Geology), ISBN 1-56576-057-3, p. 205-225.

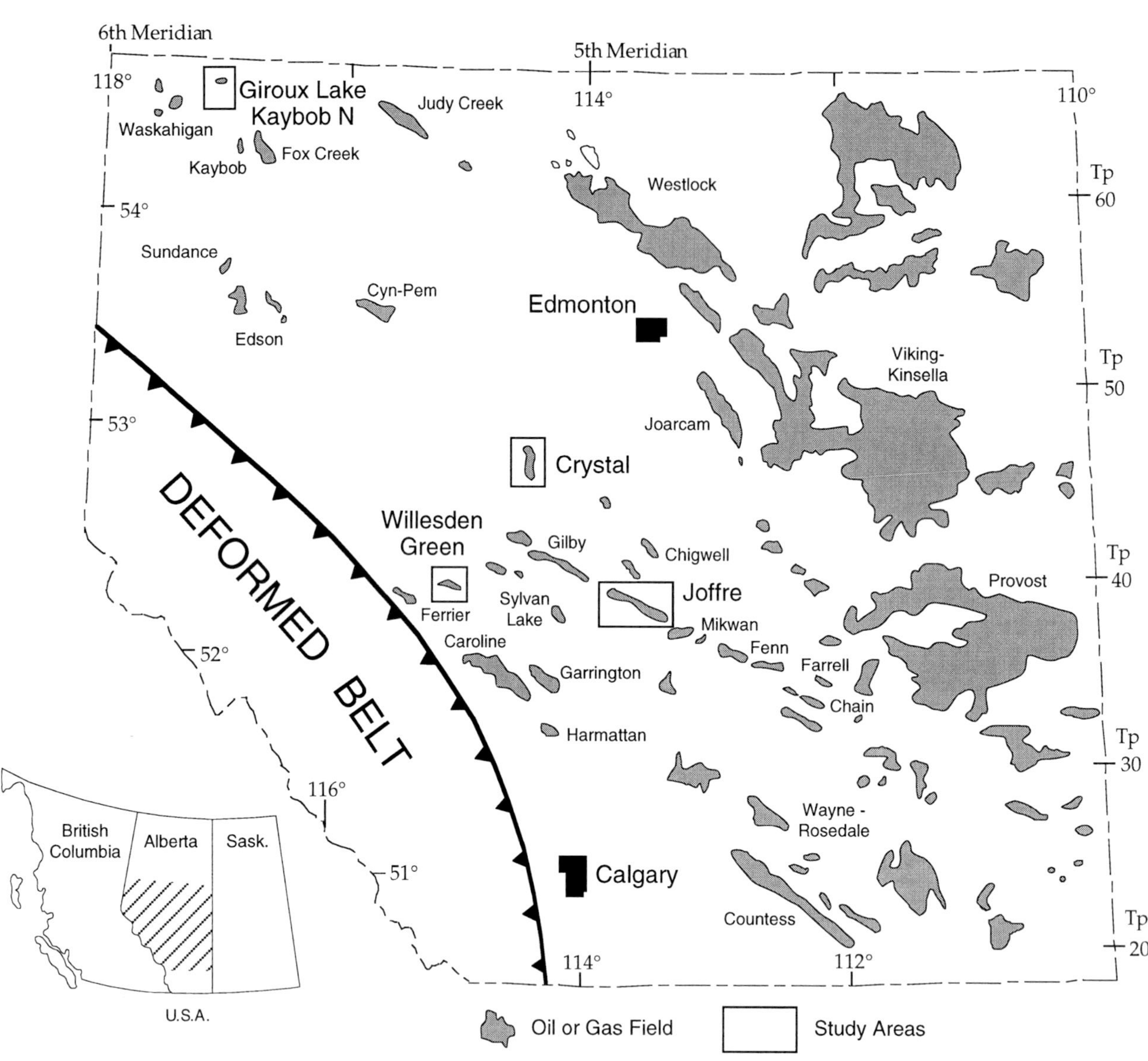

Fig. 1.—Study area. Modified after J. MacEachern et al. (1998).

with the exception of a single sample collected near the base of the Westgate Formation directly overlying the Viking. Comparison is made to a sample analyzed from the underlying Joli Fou Formation in the Giroux Lake area of central Alberta (Fig. 2). The Late Albian Viking Formation passes upward from marine shales of the Joli Fou Formation and is overlain by the transgressive marine shales of the Westgate Formation (Fig. 2). The Joli Fou Formation unconformably overlies the Mannville Group and is roughly equivalent to the Skull Creek shale of the Colorado Group in Montana and the Thermopolis shale in Wyoming (McGookey et al., 1972; Weimer, 1984). The Viking Formation is roughly equivalent to the Paddy Member of the Peace River Formation (Stelck and Leckie, 1990), the upper part of the Bow Island Formation (Glaister, 1959), as well as the Muddy Sandstone, Newcastle Formation and J-Sandstone in Montana, Wyoming and Colorado, respectively (McGookey et al., 1972; Weimer, 1984). The shales of the Westgate Formation are stratigraphically equivalent to the lower part of the Shaftesbury Formation (Stelck and Leckie, 1990; Bloch et al., 1993), and to part of the Hasler Formation in N.E. British Columbia (Stelck and Leckie, 1990). In the United States, these shales are equivalent to the Mowry shale in Montana and North Dakota (McGookey et al., 1972).

DATABASE AND METHODS

Twenty-two 75 gram samples of mudstone were collected from the Viking and Westgate formations of central Alberta and sent for sample preparation, as well as preliminary palynological and foraminiferal analysis by G. Dolby and Associates. The consultants returned a written report, the analyzed samples, and the sample residues upon project completion. The foraminiferal sample residues of 14 of these samples were then picked completely, to analyze the entire assemblage. The assemblage was considered in terms of

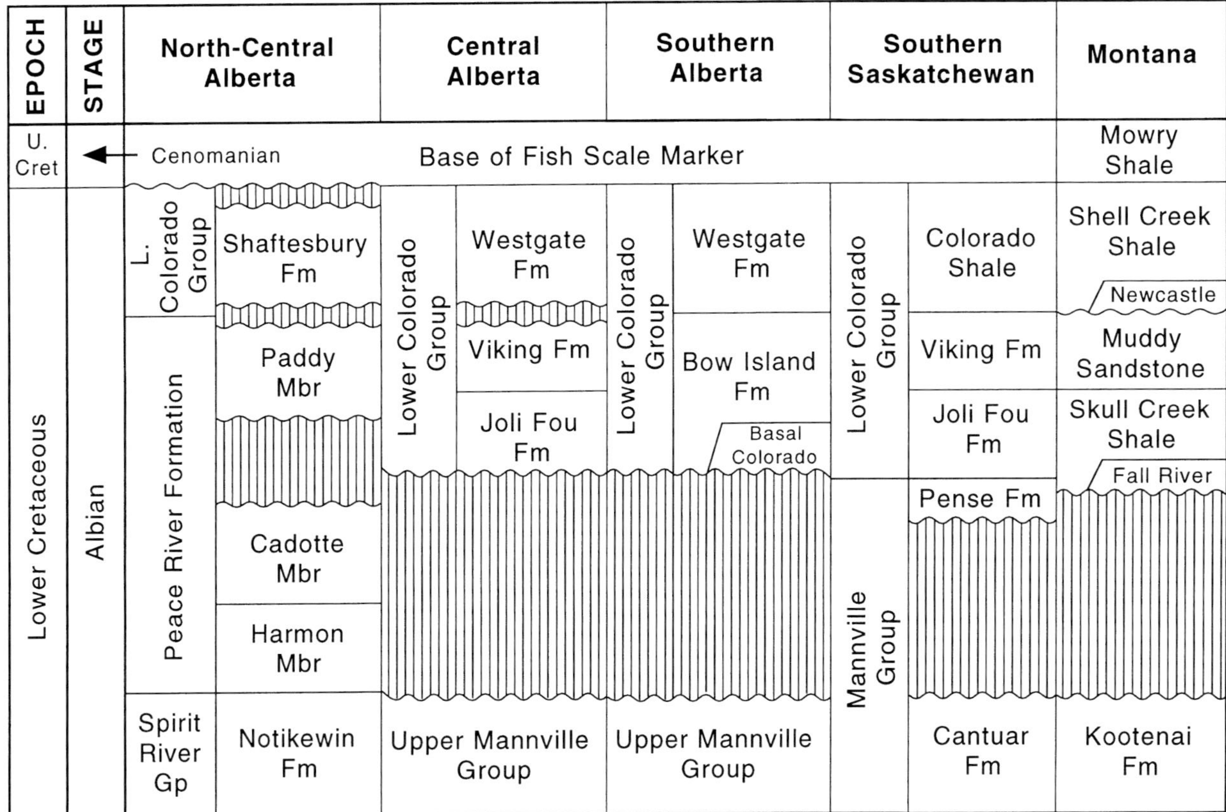

Fig. 2.—Stratigraphic correlation chart. Modified after J. MacEachern et al. (1998).

abundance, diversity of genera and species, and comparison of environmentally sensitive genera to environmentally tolerant genera. Sensitivity was gauged in terms of salinity conditions (cf., Koke and Stelck, 1985; Stelck and Koke, 1987). Larger (150 gram) samples from the Giroux Lake/Kaybob North area were used for comparison purposes and form part of a biostratigraphic analysis of the Viking in this area (J. MacEachern et al., in prep). The palynological data were compiled from the consultant's lists and are expressed in percentage by numbers.

The cored intervals, from which the samples were collected, were described in detail, including a thorough sedimentological and ichnological analysis. These cores constituted parts of previous studies in the Viking Formation, in which ichnology, sedimentology and high-resolution sequence stratigraphy were conducted (MacEachern et al., 1992; 1995; 1998; MacEachern and Pemberton, 1994) and which built upon the previous work of a large number of other researchers.

ICHNOLOGY, PALYNOLOGY AND FORAMINIFERAL PALEOECOLOGY OF MUDSTONES

The mudstones were collected from 5 principal depositional types. These are arranged from most marine to least marine in character; namely: transgressive systems tract shelf deposits, highstand systems tract lower offshore deposits, transgressive systems tract upper offshore deposits, transgressive systems tract embayment/lagoon deposits, and transgressive systems tract estuarine incised valley central basin deposits.

This paper employs the terminology of "upper offshore" and "lower offshore" as originally presented by Howard (1971; 1972), Howard and Reineck (1981), Howard and Frey (1984), Vossler and Pemberton (1989), and Frey (1990). This has subsequently been modified by MacEachern and Pemberton (1992), and Pemberton and MacEachern (1995; 1997). This usage reflects the difficulty of reconciling modern physiographic terminology with that used for the preserved depositional record. In our usage, the upper offshore is regarded to lie below fairweather wave base, but adjacent to (and grading from) the lower shoreface of a prograding shoreline. This proximity to the lower shoreface results in significant amounts of intercalated sand and silt with clay during ambient (non-storm) conditions, generating a sandy mudstone facies. The upper offshore is characterized by a highly diverse mixture of grazing/foraging organisms, deposit feeding organisms, with rarer suspension feeding organisms and carnivores, whose structures generate the archetypal *Cruziana* ichnofacies. The lower offshore grades basinward from the upper offshore, and passes basinward into the "shelf" as used in this paper. The lower offshore receives significant amounts of silt and lesser sand, with clay during fairweather (non-storm) conditions, generating a silty mudstone facies. The lower offshore is dominated by a mixture of grazing/foraging organisms and deposit feeding organisms, whose structures produce assemblages reflecting the distal *Cruziana* ichnofacies. The shelf receives little silt and sand during fairweather conditions and is characterized by dark shales. This setting is characterized by a dominance of foraging/grazing organisms with rarer deposit feeding and farming organisms, producing trace fossil suites reflecting the *Zoophycos* ichnofacies.

We recognize that upper and lower offshore, as used here, constitutes what some workers regard as the inner or proximal portion of the shelf (e.g., Walker and Plint, 1992; Johnson and Baldwin, 1996). We have found it useful, however, to distinguish these facies tracts or subenvironments lying intermediate between the mudstone-dominated portion of the shelf and the sand-dominated shoreface. In more strongly storm influenced settings, the introduction of hummocky stratified sandstones with intervening muddy facies obscures the distinction between shoreface and shelf environments, and is commonly regarded as the "offshore transition" (e.g., Hunter et al., 1979; Ainsworth and Pattison, 1994), "heterolithic offshore" (Hadley and Elliott, 1993), "offshore-to-shoreface transition" (Walker and Wiseman, 1995; Mellere and Steel, 1996) or "lower shoreface-inner shelf transition" (Walker and Plint, 1992). We suggest that our "offshore" corresponds to this transitional position. As our usage is based upon ambient or fairweather deposition rather than storm event deposition, this "offshore" setting can be subdivided effectively into an "upper" and "lower" offshore reflecting slightly more proximal and distal positions, respectively. We reserve "shelf" for the silt-poor, suspension deposition dominated environment lying more distally, similar to that employed by others within more storm-dominated successions.

Transgressive Systems Tract (TST) Shelf Deposits

A single sample for quantitative analysis of foraminiferal and palynological suites was collected from mudstone of the Westgate Formation at 1415.6 m in the 12-07-38-24W4 well (Fig. 3). Mudstones of the Westgate Formation (previously the unnamed Colorado Shales) have been interpreted to reflect open marine shelf deposition within a transgressive systems tract (Leckie et al., 1990; Davies and Walker, 1993). The ichnological assemblage is based on the analysis of more than 200 cored intervals of the lower Westgate Formation (MacEachern et al., 1992; MacEachern, 1994).

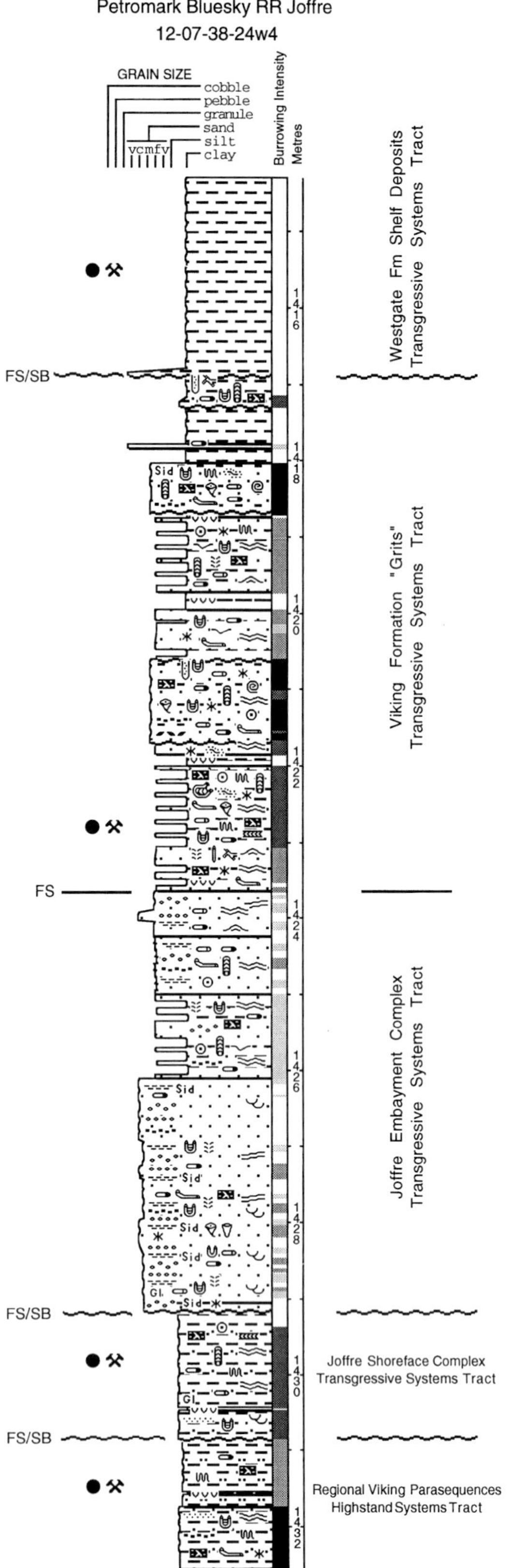

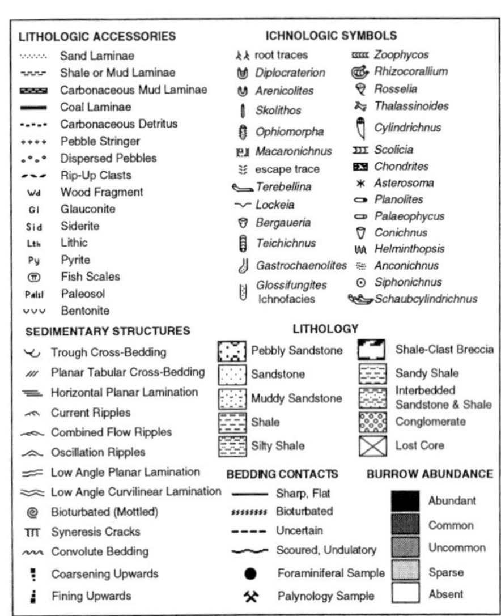

Fig. 3.—Interpreted graphic litholog for the facies succession in the 12-07-38-24W4 well. Sample types and locations are designated to the left of the litholog. The legend included with the figure applies to all other lithologs in the paper. FS = flooding surface; FS/SB = amalgamated flooding surface and sequence boundary.

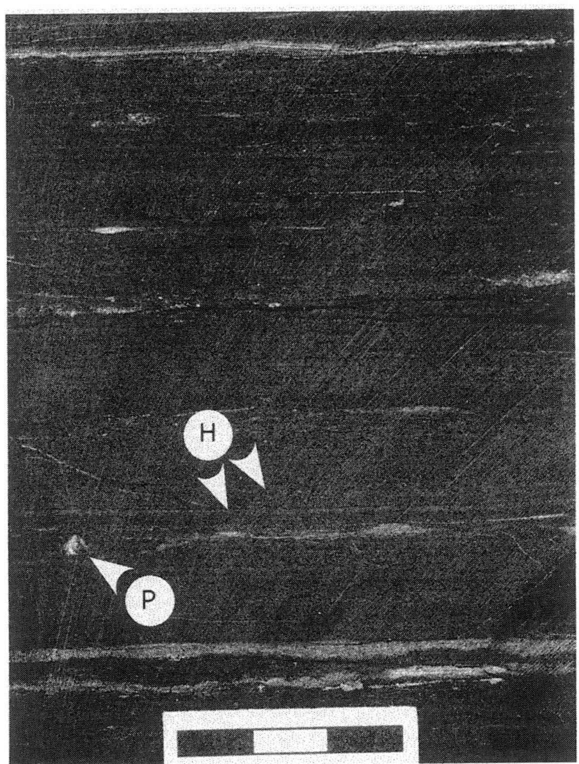

Fig. 4.—Photo of mudstone facies from the transgressive systems tract shelf deposits of the Westgate Formation. Rare numbers of *Helminthopsis* (H) and *Planolites* (P) are sporadically present in the unit. Scale is 3 cm long. Well 12-35-42-07W5; 2244.3 m.

The mudstone is black, moderately carbonaceous, and weakly fissile (Fig. 4). Foraminiferal sample residues show an absence of interstitial silt or sand. Westgate Formation mudstone typically contains rare siderite-cemented horizons and nodules, pyrite nodules, pyritized plant fragments, and coalified carbonaceous detritus. Thin, parallel laminated to combined flow ripple laminated, very fine to lower fine-grained sandstone beds, 1-3 cm thick, are intercalated and correspond to distal tempestites. Disseminated fish scales are locally present on bedding planes. Trace fossils appear uncommon and include *Planolites*, *Helminthopsis*, *Terebellina*, *Lockeia*, *Anconichnus/Phycosiphon*, and very rare *Chondrites* and *Zoophycos* (Fig. 5), typically visible near the tops of the laminated sandstone interbeds. Within the thin sandstone interbeds, rare fugichnia (escape structures) as well as very rare, diminutive *Skolithos* and *Diplocraterion* are present. The trace fossil suite corresponds to a poorly developed *Zoophycos* to distal *Cruziana* ichnofacies, with sporadic development of an opportunistic *Skolithos* ichnofacies colonizing the distal tempestite sandstones.

The foraminiferal assemblage is the most diverse of all samples collected, and consists of 16 genera and 23 species (Table 1). The sample also contains a high number of environmentally sensitive foraminifers, notably *Miliammina*, *Ammodiscus*, *Glomospirella*, *Ammobaculites*, *Bulbobaculites* and *Verneuilina*. The assemblage is also abundant, containing 223 specimens collected from a 75 gram sample. Associated fossils consist of coal and lignite fragments, vertebrate remains, spheroid bodies (locally pyritized) and *Inoceramus* shell prisms. A qualitatively analyzed sample of the Westgate from the Giroux Lakes/Kaybob North area also displays high numbers of foraminifera, comprising >14 genera and 20 species, with similar environmentally sensitive forms, and the addition of *Scherochorella*, *Verneuilinoides* and *Gravellina*. A sample collected from transgressive shelf deposits of the Joli Fou

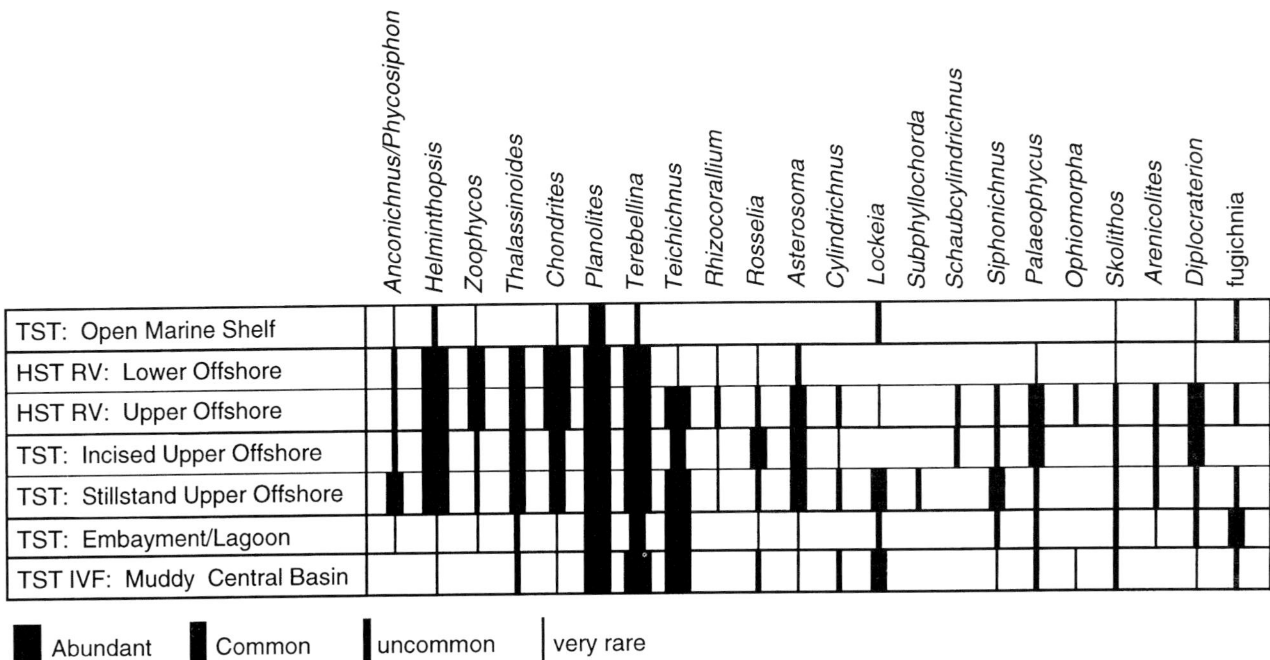

Fig. 5.—Ichnological assemblages typical of various mudstone types within the Joli Fou, Viking and Westgate formations. Assemblages are based on analyses of more than 300 cored intervals. TST = transgressive systems tract; HST = highstand systems tract; IVF = incised valley fill; RV = regional Viking parasequences.

Formation in the same area contains 11 genera and 19 species, with abundant environmentally sensitive forms as well (J. MacEachern et al., in prep).

The palynological sample comprises an assemblage characterized by a reduced proportion of terrestrial plant material (36%) and a high marine microplankton (64%) content (Table 2). The terrestrial plant component is dominated by small spores and bisaccate pollen, which are largely transported by wind and have wide dispersal distances from land. The marine microplankton population is mainly dominated by simple indeterminant dinoflagellate cysts, which may be of ceratioid affinity.

Highstand Systems Tract (HST) Lower Offshore Deposits

The silty mudstone of the regional Viking Formation parasequences is regarded as lower offshore deposits (MacEachern and Pemberton, 1992; 1994) of a highstand systems tract (Pattison, 1991; MacEachern et al., 1992; Posamentier and Chamberlain, 1993; Pemberton and MacEachern, 1995). A single quantitatively analyzed foraminiferal/palynological sample was collected from lower offshore silty mudstone of the Viking Formation in the 12-07-38-24W4 well, at 1431.4 m (Fig. 3). The ichnological assemblage is based on the analysis of more than 300 cored intervals (cf., Pemberton et al., 1992; MacEachern, 1994; MacEachern and Pemberton, 1994; Pemberton and MacEachern, 1995; MacEachern et al., 1998). The silty mudstone typically comprises the basal facies of the informally named "Regional Viking" parasequences that prograded toward the northeast during early to late highstand conditions following transgression of the Joli Fou Sea.

The silty mudstone typically shows intense bioturbation (Fig. 6). Silt is dispersed biogenically throughout the facies, though it may be present locally as discontinuous laminae. The facies commonly contains pyrite nodules and framboids, as well as small amounts of glaucony, carbonaceous detritus, fish bones and teeth, and siderite cemented bands. Rare, very fine-grained sandstone interbeds (<2 cm thick) are locally intercalated. These sandstone beds contain low angle wavy parallel lamination or, more rarely, combined flow ripple lamination, and are interpreted to reflect distal tempestites.

Trace fossils are abundant and uniformly distributed throughout the facies (Fig. 6). *Helminthopsis*, *Chondrites*, *Planolites* and *Terebellina* comprise the dominant elements of the suite, and are present in moderate to abundant numbers in most intervals (Fig. 5). *Zoophycos* and *Thalassinoides* are present in most of the intervals, in rare to moderate numbers, and constitute the secondary elements. *Asterosoma*, *Teichichnus*, *Rhizocorallium*, *Palaeophycus*, *Anconichnus*/*Phycosiphon*, *Rosselia*, *Skolithos* and *Diplocraterion* occur in few of the intervals, and when present, occur in very rare to rare amounts. The suite is characterized by a diverse and abundant trace fossil assemblage reflecting mainly specialized (equilibrium) grazing, and deposit feeding behaviours. This constitutes the distal *Cruziana* ichnofacies. *Palaeophycus*, *Teichichnus*, *Skolithos*, *Diplocraterion* as well as some of the *Rosselia* and *Asterosoma* are associated with colonization of the distal tempestites and do not reflect faunal behaviours associated with ambient or fairweather conditions.

The foraminiferal assemblage is quite diverse, and consists of 11 genera and 15 species (Table 1). In addition, the sample contains a high number of environmentally sensitive taxa, notably *Miliammina*, *Ammobaculites*, *Gravellina*, *Gaudryina*, and *Verneuilina*. The foraminifera are abundant, with 145 specimens collected from a 75 gram sample. Samples from the Giroux Lake area similarly display large numbers of foraminifera, comprising 10 genera and 16 species, many of which are environmentally sensitive. Accessory fossils consist of plant debris (cuticles and megaspores) coal and lignite fragments, fish remains, pyritized diatoms, prolate algal spore bodies and rare *Inoceramus* shell prisms.

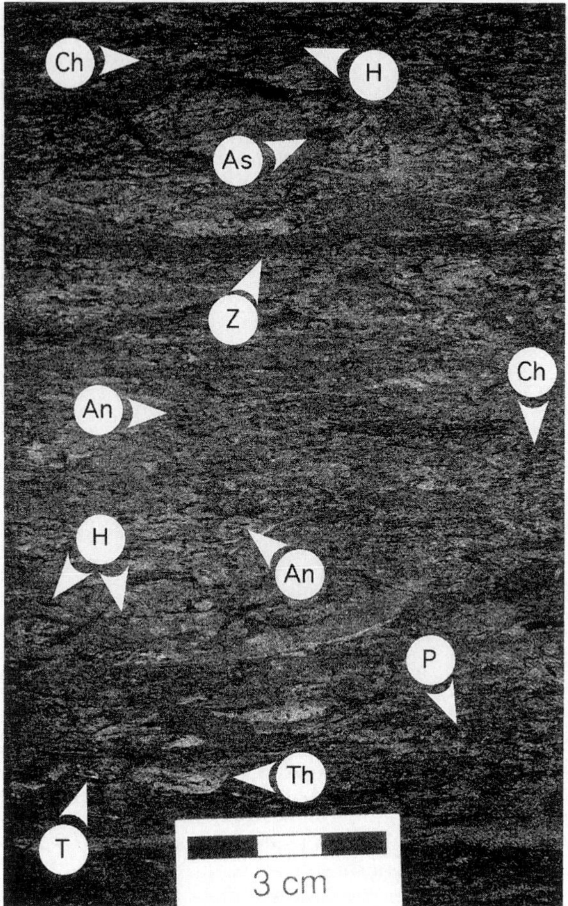

Fig. 6.—Photo of mudstone facies from the highstand systems tract lower offshore deposits of the Viking Formation. The unit is thoroughly bioturbated with a distal *Cruziana* assemblage, including *Chondrites* (Ch), *Helminthopsis* (H), *Zoophycos* (Z), *Anconichnus* (An), *Planolites* (P), *Thalassinoides* (Th), *Asterosoma* (As) and *Terebellina* (T). Well 14-20-40-06W5; 2252.4 m.

The palynological sample comprises an assemblage characterized by roughly equal numbers of terrestrial plant (53%) and marine microplankton (47%) material (Table 2). The terrestrial plant component is dominated by wind transported small spores and bisaccate pollen, which have wide dispersal distances from land, although some megaspores are present, which were probably transported fluvially to the coast and drifted basinward. The marine microplankton population is mainly dominated by simple indeterminant dinoflagellate cysts, which may be of ceratioid affinity.

Transgressive Systems Tract (TST) Upper Offshore Deposits

Two types of transgressive systems tract upper offshore deposits were quantitatively analyzed. The first type corresponds to upper offshore deposits overlying a transgressive

Table 1.—Foraminiferal data collected from 14 samples in 7 wells. IVF= incised valley fill; Bay = lagoonal/embayment.

Legend: s single; ○ rare (2-8); ◐ common (9-19); ● many (20-199)

Well	Sample	*Hippocrepina barksdalei*	*Hippocrepina sp?*	*Hyperammina emacerata*	*Hyperammina emacerata var A*	*Hyperammina sp. wide*	*Thuramminoides septagonalis*	*Thuramminoides sp. B*	*Psammosphaera sp. A*	*Psammosphaera sp. B*	*Saccammina alexanderi s.s.*	*Saccammina lathrami*	*Saccammina D?*	*Ammodiscus anthosatus?*	*Glomospirella reata*	*Psammonyx ? sp.*	*Reophax deckeri ssp B*
12-07-38-24W4	1415.6m		s	○			○		○	◐				s	◐	○	
12-07-38-24W4	1422.8m				○		○		○								
12-07-38-24W4	1429.8m				s												
12-07-38-24W4	1431.4m						○		◐	○	◐						○
06-30-38-25W4	1521.3m						○										
06-30-38-25W4	1522.3m							○	○	◐							
11-07-39-26W4	1545.5m										○						
11-07-39-26W4	1549.3m						○	○	◐	○							
03-24-38-25W4	1435.5m								◐	○		○					
03-24-38-25W4	1442.6m	s							○	○							
16-11-40-04W5	1720.3m								○				○				
16-11-40-04W5	1728.2m								○								
11-15-41-07W5	2263.5m			○				s									
06-27-41-07W5	2294.1m					s			○								

Well	Sample	*Reophax fuscus*	*Reophax incompta*	*Reophax troyeri*	*Reophax tundraensis*	*Reophax vasiformis*	*Reophax vasiformis var A*	*Miliammina awunensis*	*Miliammina inflata*	*Miliammina subelliptica*	*Psamminopelta bowsheri*	*Haplophragmoides linki*	*Haplophragmoides yukonensis?*	*Ammobaculites erectus*	*Ammobaculites fragmentarius*	*Ammobaculites tyrrelli*	*Ammobaculites jolifouensis*	*Bulbobaculites engleri*	*Bulbobaculites sp C*	*Pseudobolivina sp. A?*
12-07-38-24W4	1415.6m			○		s	◐	◐	○		○	○			◐	●			●	○
12-07-38-24W4	1422.8m	○	○	◐								○		●	○	◐	●		○	
12-07-38-24W4	1429.8m		◐	○								○			○	○			○	
12-07-38-24W4	1431.4m			○		○	○	○				○					●			
06-30-38-25W4	1521.3m				○					○		○				●	○		◐	
06-30-38-25W4	1522.3m		○	○	○					s		◐		◐	○	◐			○	
11-07-39-26W4	1545.5m			○							s	○								
11-07-39-26W4	1549.3m			○								○			○	○				
03-24-38-25W4	1435.5m			◐								○				◐		◐	●	
03-24-38-25W4	1442.6m	○					◐					○			◐					
16-11-40-04W5	1720.3m											○						○		
16-11-40-04W5	1728.2m											○								
11-15-41-07W5	2263.5m	○										◐	○							
06-27-41-07W5	2294.1m											○								

Well	Sample	*Trochammina alcanensis*	*Trochammina depressa*	*Trochammina exigua*	*Trochammina gatesensis*	*Trochammina rutherfordi?*	*Trochammina umiatensis*	*Gaudryina canadensis*	*Verneuilina canadensis*	*Verneuilinoides borealis*	*Verneuilinoides sp.*	*Gravellina chamneyi*	diatoms	fungal spores (round)	prolate algal spore bodies	pyrite spheroid bodies	spheroid bodies (rough/glauc)	ovoid bodies (lt. brown)	clear glassy spheres	fish remains	shell fragments (clams)	Facies Interpretation
12-07-38-24W4	1415.6m	○			○		○		○							○		○	s	○		Shelf
12-07-38-24W4	1422.8m					○		○		◐			○			○		○				Offshore
12-07-38-24W4	1429.8m			○						○			◐		○	○	●	○	●	○		Offshore
12-07-38-24W4	1431.4m		○					○	s			●			s			○		◐		Offshore
06-30-38-25W4	1521.3m										○				○	◐		○				Bay
06-30-38-25W4	1522.3m									○			○	○	○	○		●			○	Bay
11-07-39-26W4	1545.5m														s		○	○				Bay
11-07-39-26W4	1549.3m												○		○	○		◐		○		Bay
03-24-38-25W4	1435.5m												○			○		◐	●			Bay
03-24-38-25W4	1442.6m									○			s		○							Bay
16-11-40-04W5	1720.3m														○	○		s				IVF
16-11-40-04W5	1728.2m														○							IVF
11-15-41-07W5	2263.5m														○	◐			s			IVF
06-27-41-07W5	2294.1m		○											s	s	s		○				IVF

Table 2.—Palynological data collected from 22 samples in 9 wells. Values in % by numbers. TST = transgressive systems tract; HST = highstand systems tract; IVF = incised valley fill; RV = regional Viking parasequences.

Well	Sample Values in Percent / Depth	Terrestrial Assemblage	Spores and Small Pollen	*Taxodiaceae*	Bisaccate Pollen	*Schizaceae*	Fungi	*Classopollis*	Microplankton Assemblage	Simple Indeterminate cysts	Ceratioid	Peridinioid	Proximate	Chorates	Acritarchs	Facies Interpretation
12-07-38-24W4	1415.6m	**(36.0)**	26.0	3.0	7.0	0.0	0.0	0.0	**(64.0)**	53.0	2.5	3.5	3.5	1.0	0.5	TST: Open Marine Shelf
	1422.8m	**(55.5)**	38.5	5.5	9.0	1.0	0.0	1.5	**(44.5)**	34.5	4.0	1.5	1.0	3.0	0.5	TST: Stillstand Upper Offshore
	1429.8m	**(44.0)**	27.0	4.5	10.5	2.0	0.0	0.0	**(56.0)**	46.5	2.5	2.5	1.5	2.5	0.5	TST: Incised Upper Offshore
	1431.4m	**(53.0)**	28.5	3.0	18.0	2.5	0.0	1.0	**(47.0)**	32.5	2.0	9.0	1.0	2.0	0.5	HST (RV): Lower Offshore
06-30-38-25W4	1520.4m	**(76.5)**	48.5	11.5	14.0	0.0	0.0	0.0	**(24.5)**	14.5	3.5	1.5	0.5	4.5	1.5	TST: Embayment/Lagoon
	1521.3m	**(55.5)**	30.0	7.5	18.0	0.0	0.0	0.0	**(44.5)**	18.0	5.5	2.0	23.5	4.5	1.0	TST: Embayment/Lagoon
	1522.0m	**(49.5)**	24.0	5.5	19.0	0.5	0.5	0.0	**(50.5)**	36.0	3.0	4.5	1.5	5.0	0.5	TST: Embayment/Lagoon
	1522.3m	**(56.5)**	40.0	4.5	12.0	0.0	0.0	0.0	**(43.5)**	25.5	5.0	7.0	2.0	3.5	0.5	TST: Embayment/Lagoon
11-07-39-26W4	1545.5m	**(65.0)**	49.0	1.5	14.0	0.5	0.0	0.0	**(35.0)**	22.0	6.0	2.0	4.5	0.5	0.0	TST: Embayment/Lagoon
	1547.3m	**(38.0)**	30.5	3.5	3.5	0.0	0.5	0.0	**(62.0)**	34.5	1.5	7.0	17.5	0.5	0.5	TST: Embayment/Lagoon
	1549.3m	**(48.0)**	32.5	5.5	8.5	1.0	0.5	0.0	**(52.0)**	44.5	1.5	3.5	0.0	1.5	1.0	TST: Embayment/Lagoon
	1550.3m	**(40.0)**	30.0	4.0	5.0	1.0	0.0	0.0	**(60.0)**	52.5	1.0	4.5	0.0	0.5	1.5	TST: Embayment/Lagoon
03-24-38-25W4	1435.5m	**(63.0)**	41.5	2.0	17.0	2.5	0.0	0.0	**(37.0)**	18.5	6.0	8.5	0.5	2.0	0.5	TST: Embayment/Lagoon
	1440.4m	**(41.5)**	29.5	4.0	7.0	1.0	0.0	0.0	**(58.5)**	29.0	11.5	15.0	1.0	1.5	0.5	TST: Embayment/Lagoon
	1442.6m	**(43.5)**	29.0	5.5	7.0	2.0	0.0	0.0	**(56.5)**	20.5	13.0	14.5	3.0	5.0	0.5	TST: Embayment/Lagoon
13-13-38-25W4	1438.0m	**(52.5)**	33.0	9.0	10.5	0.0	0.0	0.0	**(47.5)**	28.0	8.0	6.0	1.5	2.5	1.5	TST: Embayment/Lagoon
	1441.6m	**(52.5)**	26.0	13.0	13.5	0.0	0.0	0.0	**(47.5)**	36.0	2.0	7.0	2.0	0.5	0.0	TST: Embayment/Lagoon
08-14-38-25W4	1433.8m	**(44.0)**	29.5	2.5	11.5	0.5	0.0	0.0	**(56.0)**	30.0	7.0	14.0	3.0	2.0	0.0	TST: Incised Upper Offshore
16-11-46-04W5	1720.3m	**(32.5)**	19.0	0.0	11.0	2.0	0.5	0.0	**(67.5)**	13.0	2.5	50.0	0.0	0.0	2.0	TST IVF: Muddy Central Basin
	1728.2m	**(31.0)**	19.5	0.5	10.0	1.0	0.0	0.0	**(69.0)**	12.5	5.0	47.5	2.5	1.5	0.0	TST IVF: Muddy Central Basin
11-15-42-07W5	2263.5m	**(35.0)**	26.5	2.0	6.5	0.0	0.0	0.0	**(65.0)**	21.0	1.5	42.0	0.5	0.0	0.0	TST IVF: Muddy Central Basin
06-27-41-07W5	2294.1m	**(34.5)**	24.5	4.5	5.5	0.0	0.0	0.0	**(65.5)**	31.0	2.0	31.5	0.5	0.0	0.5	TST IVF: Muddy Central Basin

discontinuity and forming the distal and basal facies of a transgressively incised shoreface. The second reflects upper offshore deposition within the informally named upper Viking Formation "Grits" (cf., Stelck, 1958).

The offshore mudstone samples from the transgressively incised shoreface complex were collected from the 12-07-38-24W4 well at 1429.8 m and from the 08-14-38-25W4 well at 1433.8 m, both from the Joffre field. Only the former sample was analyzed quantitatively for foraminifera (Fig. 3) although both were quantitatively analyzed for palynomorphs (Table 2). The ichnological assemblage is based upon analysis of more than 50 cored intervals. The original sedimentological and stratigraphic work of Downing and Walker (1988) recognized that the facies overlies a discontinuity they termed E1, subsequently redefined as BD-1 (Burton and Walker, this volume). They and MacEachern et al. (1995; 1998) agree that the facies constitutes the basal offshore portion of a transgressively incised shoreface complex (Fig. 8).

The succession to which it belongs consists of three facies, constituting an overall coarsening-upward succession. A complete facies succession consists of a thin granule- to pebble-sized lag mantling BD-1, an amalgamated sequence boundary and transgressive surface of erosion (FS/SB), grading upward into 1) granular sandy mudstone, through 2) muddy sandstone, and into 3) interstratified wavy parallel laminated-to-burrowed sandstone. The samples were collected from upper offshore granular sandy mudstone deposits.

The sandy mudstone near the base of the succession contains dispersed chert pebbles and granules, as well as very coarse-grained chert and quartz sand (Fig. 7). Glaucony, carbonaceous detritus, pyritized wood fragments, siderite mudstone clasts and pyrite nodules are locally common. Discontinuous lenses of sharp-based, parallel laminated, fine-grained sandstone are commonly intercalated, and are interpreted to reflect distal storm beds.

The facies is moderately to thoroughly burrowed with a diverse, uniformly distributed, equilibrium trace fossil assemblage dominated by *Helminthopsis*, *Chondrites*, *Terebellina*, *Planolites*, *Asterosoma*, *Thalassinoides*, and *Teichichnus*, with subordinate *Zoophycos*, *Anconichnus*/*Phycosiphon*, *Rosselia*, *Cylindrichnus*, *Schaubcylindrichnus*, *Palaeophycus*, *Skolithos*, *Siphonichnus*, *Rhizocorallium*, *Arenicolites*, *Diplocraterion* and fugichnia (Fig. 5). The trace fossil suite corresponds to a fully marine, unstressed, archetypal *Cruziana* ichnofacies, that reflects a dominance of deposit feeding behaviour interspersed with subordinate grazing, passive carnivore and suspension feeding behaviours.

The foraminiferal assemblage is fairly diverse, consisting of 8 genera and 9 species (Table 1). The sample contains some environmentally sensitive foraminifers, notably *Ammobaculites*, *Bulbobaculites* and *Verneuilinoides*. The suite consists mainly of coarsely agglutinated fauna. The abundance of foraminifers is relatively low, containing only about 46 specimens recovered from a 75 gram sample. This suggests that sedimentation rates were considerably higher than in the previously described facies. Accessory fossils consist of coal and lignite fragments, bivalve shell fragments, fish remains, diatoms and prolate algal spore bodies. The assemblage compares favourably with 2 additional foraminiferal suites collected from transgressively incised offshore deposits within the Giroux Lake area. These suites contain 9-11 genera and 10-14 species, including environmentally sensitive forms such as *Miliammina*, *Gravellina*, *Scherochorella*, *Verneuilina* and *Ammodiscus*, in addition to the genera listed above. Suites contain up to 364 specimens from a 150 gram sample (~182 specimens/75 gram).

The palynological assemblage is characterized by terrestrial plant contents of 44% and marine microplankton contents of 56% (Table 2). The terrestrial plant component is dominated by small spores and bisaccate pollen, probably transported long distances by wind. The marine microplankton population is dominated mainly by simple indeterminant

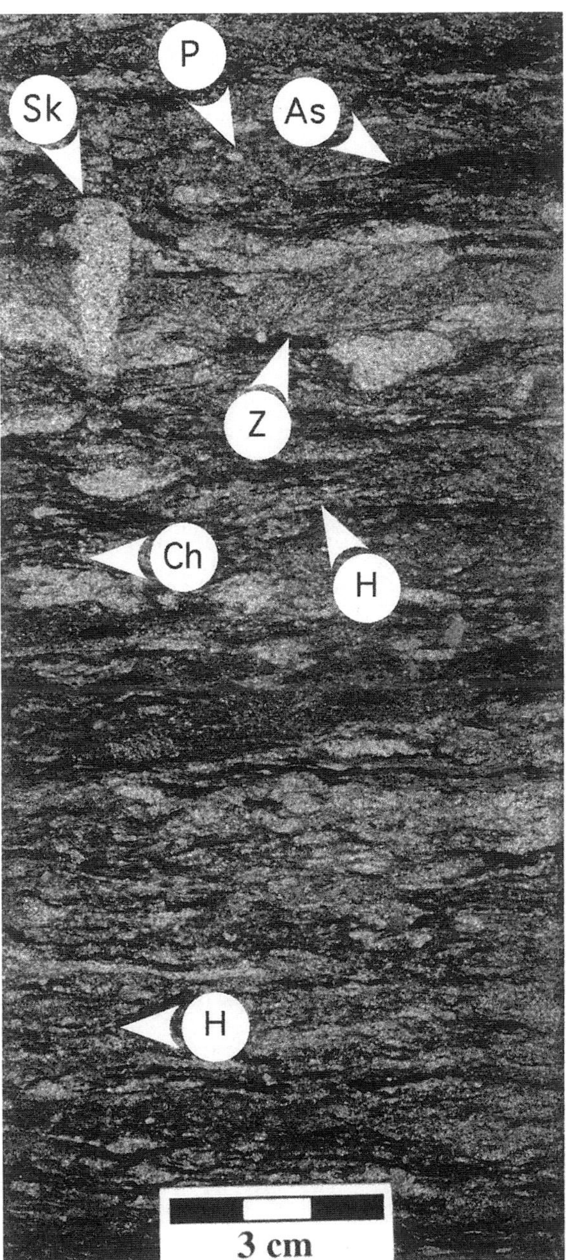

Fig. 7.—Photo of granular sandy mudstone facies from transgressive systems tract upper offshore deposits of the Viking Formation. The facies constitutes the basal unit of an incised shoreface complex in the Joffre Field area. The unit is moderately to thoroughly burrowed with an archetypal *Cruziana* assemblage of *Planolites* (P), *Asterosoma* (As), *Skolithos* (Sk), *Zoophycos* (Z), *Chondrites* (Ch), and *Helminthopsis* (H). Well 08-14-38-25W4; 1433.2 m.

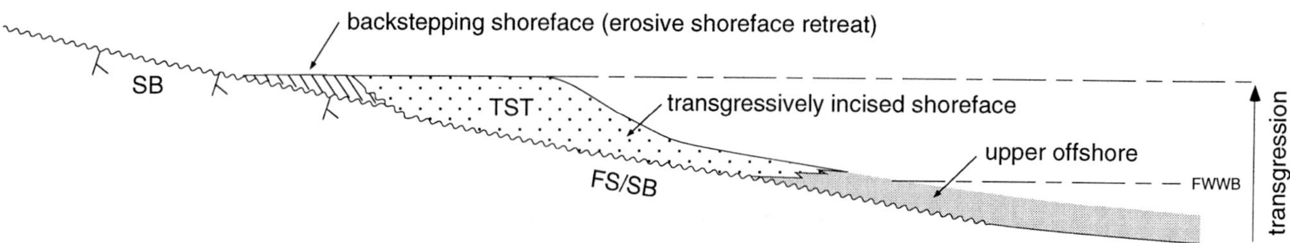

Fig. 8.—Model of a transgressively incised shoreface and associated upper offshore facies. In the model, relative sea level has fallen far to the right of the diagram, developing a subaerial sequence boundary (SB). A subsequent rise in relative sea level results in transgressive erosion that modifies the subaerial portion of the sequence boundary (FS/SB) as it passes progressively landward. As sea level rises, the position of fairweather wave base (FWWB) also rises. When sedimentation exceeds the rate of relative sea level rise, a shoreface progrades across this transgressively incised surface. Lower shoreface deposits persist seaward to the position of FWWB associated with the transgression. Upper offshore mudstone forms seaward of the lower shoreface, but may still overlie the transgressively modified erosion surface, because the surface was cut prior to progradation and while sea level lay at a lower position.

dinoflagellate cysts, which may be of ceratioid affinity, and with highly variable concentrations of peridinioids (2.5-14%).

A single sample of the upper offshore deposits within the informally named upper Viking "Grits" was quantitatively analyzed for foraminifera and palynology, from the 12-07-38-24W4 well at 1422.8 m (Fig. 3). Several samples have also been analyzed qualitatively from these upper Viking transgressive systems tract deposits in the Giroux Lake and Kaybob North area (MacEachern et al., 1998.). Ichnological assemblages are based upon the analysis of more than 300 cored intervals across central Alberta (e.g., MacEachern et al., 1992; in prep; MacEachern, 1994).

These deposits have been studied in a number of field locations (e.g., Leckie, 1986; Downing and Walker, 1988; Boreen and Walker, 1991; MacEachern et al., 1992, 1998; Davies and Walker, 1993; Posamentier and Chamberlain, 1993; Walker and Wiseman, 1995), and are regarded as offshore marine deposits of a complex transgressive systems tract. The facies consists of moderately to thoroughly bioturbated, interstratified sandstone and mudstone, containing dispersed granules and stringers of pebbles (Fig. 9). The facies overlies transgressive surfaces that vary from nonerosional flooding surfaces (FS) to transgressive surfaces of erosion (TSE). The facies contains dispersed lower fine to upper medium grained sand, with carbonaceous detritus, pyritized wood fragments, siderite nodules and cemented bands, pyrite nodules, and pyrite framboids. Thin beds of bentonite and bentonitic shale are locally intercalated within the facies. Sandstone beds are erosionally based, with wavy parallel lamination and lesser oscillation to combined flow ripple lamination, 1-7 cm in thickness. Mudstone is weakly fissile, dark grey in colour, locally contains significant interstitial silt and sand, and ranges from 1-10 cm in thickness.

The primary stratification and bedding styles are typically disrupted by biogenic reworking. Trace fossils comprise numerous *Helminthopsis*, *Chondrites*, *Terebellina*, *Teichichnus*, *Planolites*, *Asterosoma*, *Rosselia*, *Siphonichnus*, *Palaeophycus* and fugichnia, with associated *Anconichnus/Phycosiphon*, *Zoophycos*, *Lockeia*, *Thalassinoides*, *Ophiomorpha*, *Skolithos*, *Arenicolites*, *Scolicia*, *Rhizocorallium*, and *Diplocraterion* (Fig. 5). The suite reflects a mixture of deposit feeding, grazing, passive carnivore and suspension feeding behaviours. The assemblage comprises a fully marine, generally unstressed (equilibrium) archetypal *Cruziana* ichnofacies, within a moderately storm-influenced, upper offshore setting. Some intervals display sporadically distributed zones of diminished bioturbation and reduced diversity, suggesting episodic rates of deposition and/or proximity to deltaic or estuarine depositional settings.

The foraminiferal assemblage from the 12-07-38-24W4 sample is fairly diverse, consisting of 10 genera and 15 species (Table 1), and contains some environmentally sensitive foraminifers, notably *Ammobaculites*, *Bulbobaculites*, *Gaudryina* and *Verneuilinoides*. The suite consists largely of coarsely agglutinated fauna. The foraminiferal assemblage is abundant, containing 155 specimens collected from a 75 gram sample. Accessory fossils consist of coal and lignite fragments, fish remains, diatoms and prolate algal spore bodies. This is consistent with observations of samples from the Giroux Lake and Kaybob North field areas, which are characterized by assemblages containing between 7-11 genera and 9-17 species, with up to 239 specimens in 150 gram samples (~120/75 grams). Environmentally sensitive genera include those listed above, with the addition of *Scherochorella*, *Miliammina*, *Gravellina* and *Ammodiscus*.

The palynological assemblage is characterized by terrestrial plant contents of 55.5% and marine microplankton contents of 44.5% (Table 2). As in the previous samples, the terrestrial plant component is dominated by small spores and bisaccate pollen, probably transported long distances by wind. The marine microplankton population is likewise dominated by simple indeterminant dinoflagellate cysts, which may be of ceratioid affinity. Peridinioid dinoflagellates constitute minor constituents of the assemblage.

Transgressive Systems Tract (TST) *Embayment/Lagoonal Deposits*

A total of 6 foraminiferal samples and 13 palynological samples were collected from several mudstone intervals in the Joffre Field area. Ichnological assemblages are based on the analysis of more than 50 cored intervals. Although the depositional environment of this mudstone remains a subject of some controversy, researchers generally agree that the facies accumulated within a transgressive systems tract. Downing and Walker (1988) originally regarded the mudstone as shelf or offshore, an interpretation endorsed by Burton and Walker (this volume). In contrast, MacEachern et al. (1995; 1998; this volume) regard the mudstone as lagoonal in origin. This mudstone comprises the intervening fine-grained facies interstratified with low angle planar and trough cross-stratified conglomerate, pebbly sandstone

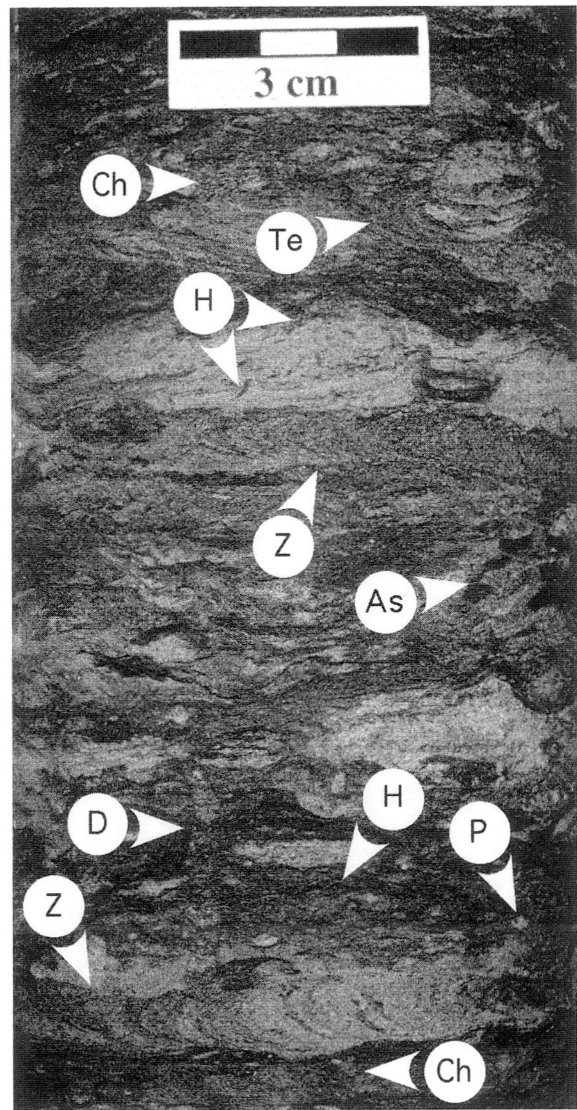

Fig. 9.—Photo of sandy mudstone facies from transgressive systems tract upper offshore deposits of the upper Viking Formation (Viking "Grits"). The unit is moderately to thoroughly burrowed with an archetypal *Cruziana* ichnofacies, characterized by *Chondrites* (Ch), *Teichichnus* (Te), *Helminthopsis* (H), *Zoophycos* (Z), *Asterosoma* (As), *Planolites* (P) and *Diplocraterion* (D). Well 10-13-38-25W4; 1432.6 m.

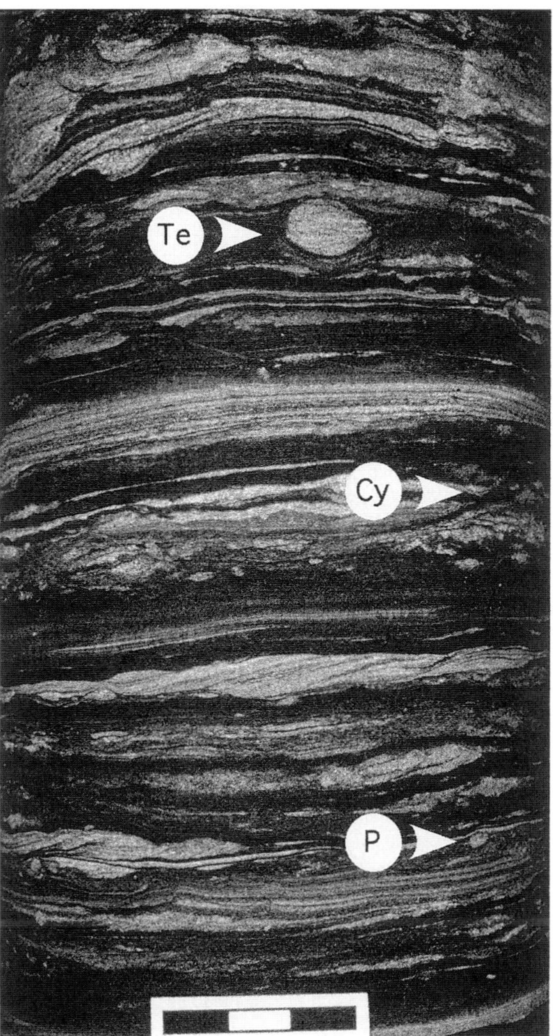

Fig. 10.—Photo of interstratified mudstone and sandstone facies from the transgressive systems tract embayment/ lagoonal deposits of the Viking Formation. The facies constitutes the basal units of parasequences within the Joffre embayment complex. The unit is burrowed by a low diversity opportunistic assemblage consisting of *Teichichnus* (Te), *Planolites* (P) and *Cylindrichnus* (Cy). Scale is 3 cm long. Well 02-18-38-24W4; 1459.2 m.

and sandstone that constitutes the reservoir of the Joffre field.

The embayment/lagoonal deposits are characterized by interbedded mudstone and sandstone (Fig. 10), which comprise the basal units of individual parasequences within the Joffre embayment complex (MacEachern et al., this volume; Fig. 11). Sandstone beds constitute as little as 5% of the succession near the base of a parasequence, to as much as 75% toward the top, and range from 1.0-15.0 cm in thickness. These beds are well sorted, lower fine in grain size, and sharp based, consisting of oscillation ripples, combined flow ripples, and wavy parallel lamination. Glaucony is common in the mudstone, particularly within the basal parasequence. Mudstone beds range from 1.0-20.0 cm in thickness, are silt and sand poor, dark grey to black in colour, and contain carbonaceous detritus. Mudstone beds are locally siderite cemented or display displacive siderite nodule development. Pyrite content within the mudstone is variable.

The facies is weakly burrowed with a sporadically distributed and low diversity trace fossil suite (Fig. 5). Bioturbation intensity and diversity of ichnogenera tend to increase toward the top of the Joffre embayment complex, as well as toward the top of each individual parasequence. In addition, muddier portions of the succession possess the lowest trace fossil diversity, lowest degree of burrowing, and the most diminutive trace fossil genera. As sand content increases, trace fossils become slightly more diverse, more abundant, and more robust. The facies is wholly dominated by

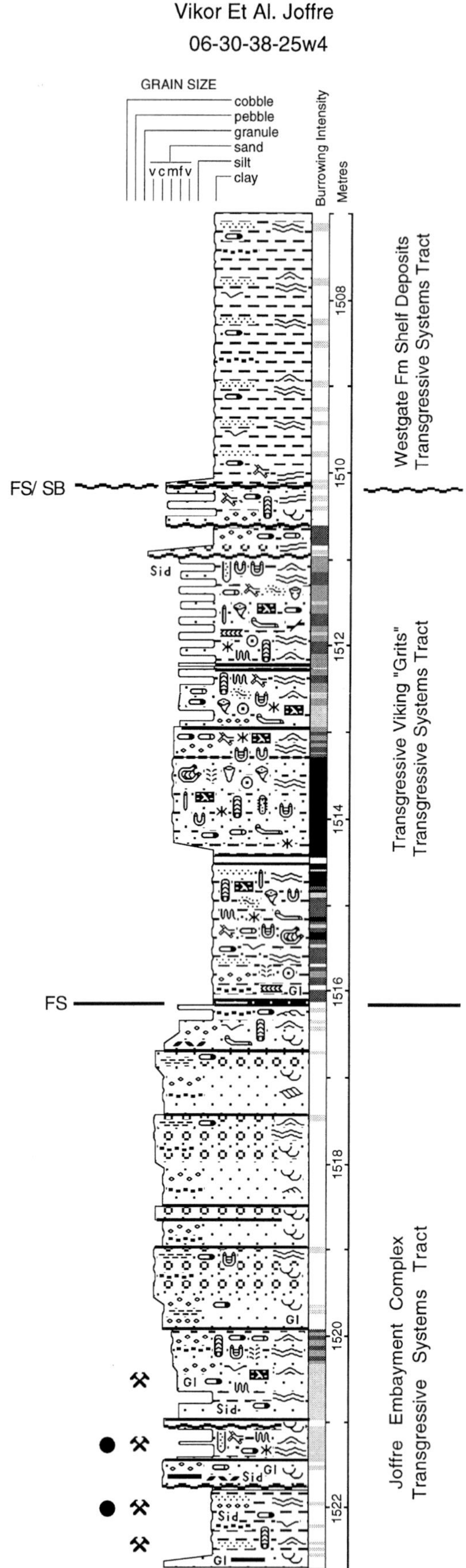

Teichichnus, *Planolites*, and *Terebellina*, with subordinate but moderate numbers of *Palaeophycus*, *Lockeia*, *Skolithos*, and *Thalassinoides*. Near the top of the Joffre embayment complex, *Siphonichnus*, *Arenicolites*, *Diplocraterion*, and fugichnia become significant elements of the suite. *Chondrites*, *Asterosoma*, *Rosselia*, *Anconichnus/Phycosiphon*, *Helminthopsis*, and *Zoophycos*, though present, are restricted to near the top of the Joffre embayment complex and are exceedingly rare. The predominant trace fossil suite corresponds to a low diversity (stressed) mixed *Skolithos-Cruziana* assemblage, typical of environments subject to reduced and/or fluctuating salinity, episodic deposition and variable substrate consistency (Pemberton and Wightman, 1992; Ranger and Pemberton, 1992; Pemberton et al., 1992; MacEachern and Pemberton, 1994). The localized presence and sporadic distribution of specialized, salinity sensitive ichnogenera within the suite attests to periodic pulses of more normal marine conditions within the embayment. The upward increase in trace fossil diversity heralds the onset of marine transgression which ultimately terminated marginal marine conditions within the Joffre area.

The foraminiferal assemblages within the mudstone facies are quite variable throughout the succession (Table 1). Near the base of the Joffre embayment complex, the mudstone directly overlies an amalgamated sequence boundary and flooding surface (FS/SB), or a highly glauconitic pebbly sandstone reflecting a transgressive sand sheet developed on the FS/SB. In both cases, the samples are closely associated with initial transgression of the Joffre area and display a somewhat stronger marine character. Toward the top of the Joffre embayment complex, samples also reflect an increase in marine influence, associated with transgressive flooding of the Joffre Field area. Samples associated with the middle portion of the Joffre embayment complex display a reduced marine influence.

The two samples from the 06-30-38-25W4 well (at 1521.3 m and 1522.3 m) and one from the 03-24-38-25W4 well (at 1442.6 m) were collected near the base of the succession and contain between 6-8 genera and 8-13 species of foraminifera (Figs. 11 and 12). The samples also contain a few environmentally sensitive foraminifers, notably *Miliammina*, *Ammobaculites*, and *Verneuilinoides*. The foraminifers are generally abundant, though variable, ranging from 47-105 specimens collected from a 75 gram sample. Accessory fossils consist of plant debris (cuticles and megaspores) coal and lignite fragments, pyritized diatoms, prolate algal spore bodies, fungal spores, and rare bivalve shell fragments.

The two samples from the 11-07-39-26W4 well (at 1545.5 m and 1549.3 m) were collected from mudstone within the middle parasequences of the Joffre embayment complex (Fig. 13). The samples contain between 4-5 genera and 4-8 species of foraminifers, and possess only environmentally tolerant forms. Abundances of specimens are also quite low, ranging from 13-40 individuals. Coalified (lignitic) material, pyritized diatoms, rare fish fragments and prolate algal spore bodies are present as accessory elements.

A single 75 gram sample from the 03-24-38-25W4 well at 1435.5 m was collected from mudstone within the upper-

←
Fig. 11.—Interpreted graphic litholog for the facies succession in the 06-30-38-25W4 well. Sample types and locations are designated to the left of the litholog. See Figure 3 for the legend of symbols. FS = flooding surface; FS/SB = amalgamated flooding surface and sequence boundary.

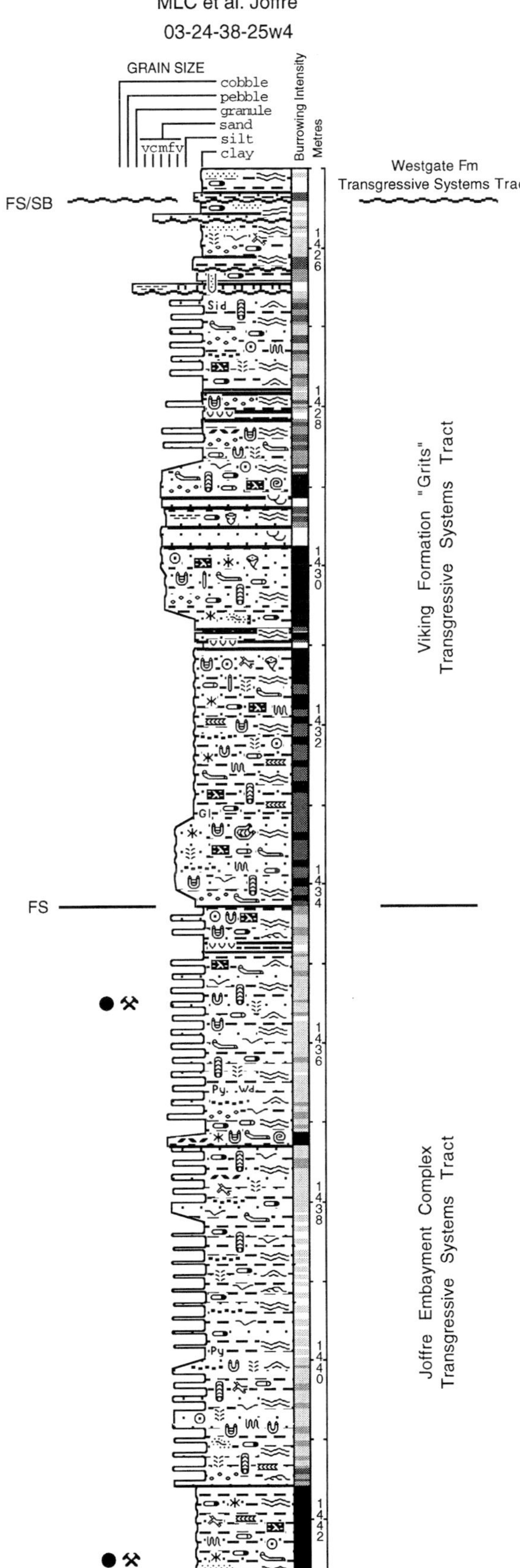

most parasequence of Joffre embayment complex (Fig. 12). The sample contains a more abundant and diverse foraminiferal assemblage, totaling 94 specimens that comprise 6 genera and 7 species. Environmentally sensitive foraminifers are restricted to *Bulbobaculites* and *Ammobaculites*.

The Joffre embayment complex was sampled from 13 intervals in 4 wells for palynological analysis (Table 2). Unlike the ichnological and foraminiferal assemblages, no significant variation between individual parasequences could be detected. In addition, the palynological samples, though varied, are surprisingly similar to the open marine assemblages discussed above. The suites contain terrestrial plant concentrations ranging from 38.0-76.5% (averaging 52.4%) and marine microplankton concentrations varying from 24.5-62.0% (averaging 47.6%). The terrestrial plant component is dominated by wind transported small spores and bisaccate pollen, which have wide dispersal distances from land. The marine microplankton population is dominated by simple indeterminant dinoflagellate cysts, which may be of ceratioid affinity. Peridinioid dinoflagellates vary from 1.5-15.0% and average 6.4%.

Transgressive Systems Tract (TST) *Estuarine Incised Valley Fill Complexes*

A total of 4 foraminiferal/palynological samples were collected from 3 wells, within transgressive systems tract estuarine incised valley fill deposits (Tables 1 and 2). Ichnological assemblages are based on the study of more than 75 cored intervals from 5 incised valley complexes (cf., MacEachern, 1994; MacEachern and Pemberton, 1994). The mudstone constitutes the muddy central basin complexes of wave-dominated estuarine incised valley systems in the Viking Formation, well studied from both the Crystal Field (Reinson et al., 1988; Pattison, 1991, 1992; MacEachern and Pemberton, 1994) and the Willesden Green Field (Boreen and Walker, 1991; MacEachern and Pemberton, 1994).

The central basin deposits consist of the regular interbedding of two facies. The dominant facies consists of the interstratification of moderately to intensely burrowed sandy mudstone, weakly burrowed sand- and silt-poor dark mudstone, and thin sandstone beds and interlaminae (Fig. 14). Sandy mudstone beds range from 2-50 cm, though generally less than 25 cm, and locally contain dispersed pebbles, carbonaceous detritus, and coalified wood fragments. Dark, sand- and silt-poor, organic-rich mudstone beds and interlaminae are commonly intercalated. Discrete sandstone interlaminae and interbeds are upper very fine to upper fine in grain size, are regularly interstratified with the sandy mudstone beds, and are typically sharp based with combined flow ripples, oscillation ripples, rare current ripples and wavy parallel laminae. Syneresis cracks are locally present. The facies displays wavy, lenticular and pinstripe composite bedsets and is interpreted to reflect fairweather deposition of sand and mud within the central basin.

The second facies consists of 5-25 cm thick sandstone beds, only rarely erosionally amalgamated into simple bedsets thicker

←
Fig. 12.—Interpreted graphic litholog for the facies succession in the 03-24-38-25W4 well. Sample types and locations are designated to the left of the litholog. See Figure 3 for the legend of symbols. FS = flooding surface; FS/SB = amalgamated flooding surface and sequence boundary.

than 50 cm. The bulk of the sandstone is upper very fine to upper fine grained, though very rare lower medium grained beds are present. The beds are well sorted with rare interlaminae of mud toward the tops of some beds. Organic detritus is variable in abundance but low overall. Stratification dominantly consists of wavy parallel laminae, locally grading into combined flow or oscillation ripple lamination. Less commonly, beds consist of combined flow, oscillation and very rare current ripple lamination. Thin conglomerate beds are very rarely present, ranging from 2-25 cm in thickness.

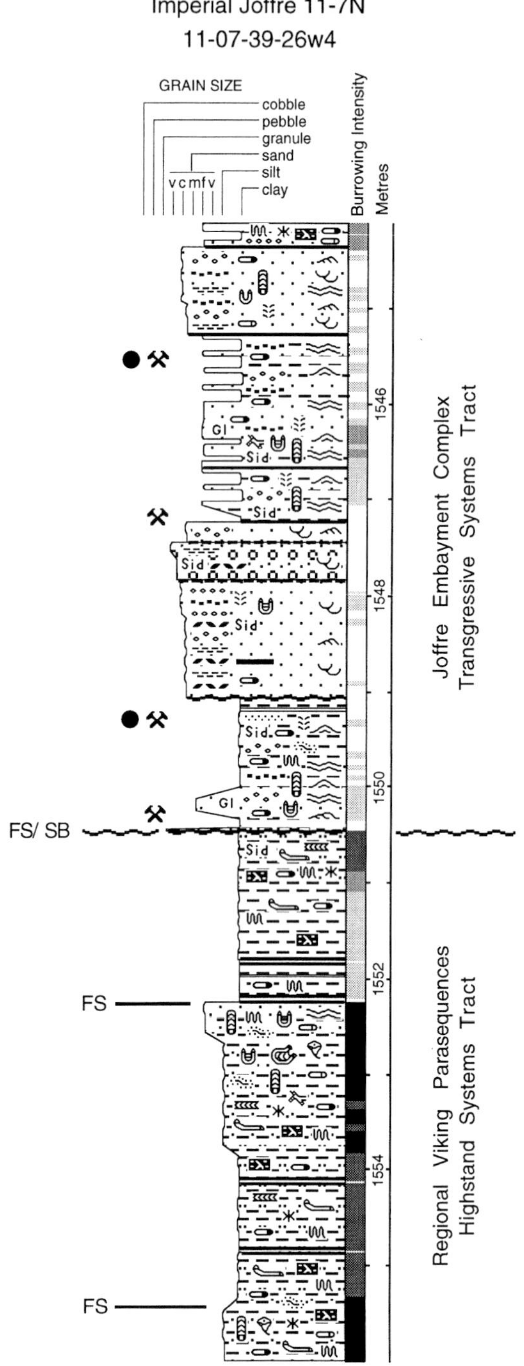

Fig. 13.—Interpreted graphic litholog for the facies succession in the 11-07-39-26W4 well. Sample types and locations are designated to the left of the litholog. See Figure 3 for the legend of symbols. FS = flooding surface; FS/SB = amalgamated flooding surface and sequence boundary.

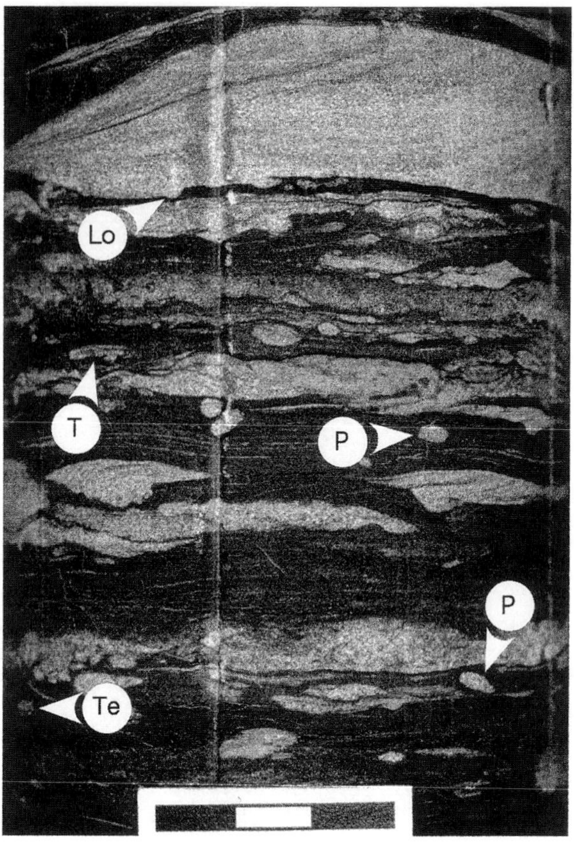

Fig. 14.—Photo of interstratified mudstone and sandstone facies from the transgressive systems tract estuarine incised valley complex in the Viking Formation. The facies constitutes the central basin deposits of the Willesden Green valley and is weakly burrowed with a low diversity, opportunistic suite of *Lockeia* (Lo), *Terebellina* (T), *Planolites* (P) and *Teichichnus* (Te). Scale is 3 cm long. Well 11-15-42-07W5; 2263.1 m.

The central basin deposits contain an ichnological assemblage characterized by reduced diversity and sporadic distribution of ichnogenera. The suite is dominated by *Planolites, Teichichnus* and *Terebellina*, with *Palaeophycus, Skolithos, Lockeia, Rosselia, Thalassinoides, Cylindrichnus* and fugichnia comprising the secondary elements (Fig. 5). *Asterosoma, Ophiomorpha, Diplocraterion, Chondrites, Siphonichnus*, and *Helminthopsis* are exceedingly rare constituents. Dominant ichnogenera in the fairweather sandy shales correspond to structures produced by trophic generalists (i.e., opportunistic organisms; cf., Pianka, 1970; Grassle and Grassle, 1974; Jumars, 1993). These assemblages are characteristic of highly stressed settings, such as those subject to salinity fluctuations. Those traces dominating the tempestite sandstone also reflect rapid opportunistic colonization of the event bed. These assemblages, however, are characteristic of stresses imposed by episodic deposition and variable substrate consistency. The regular alternation between these facies produces a low diversity,

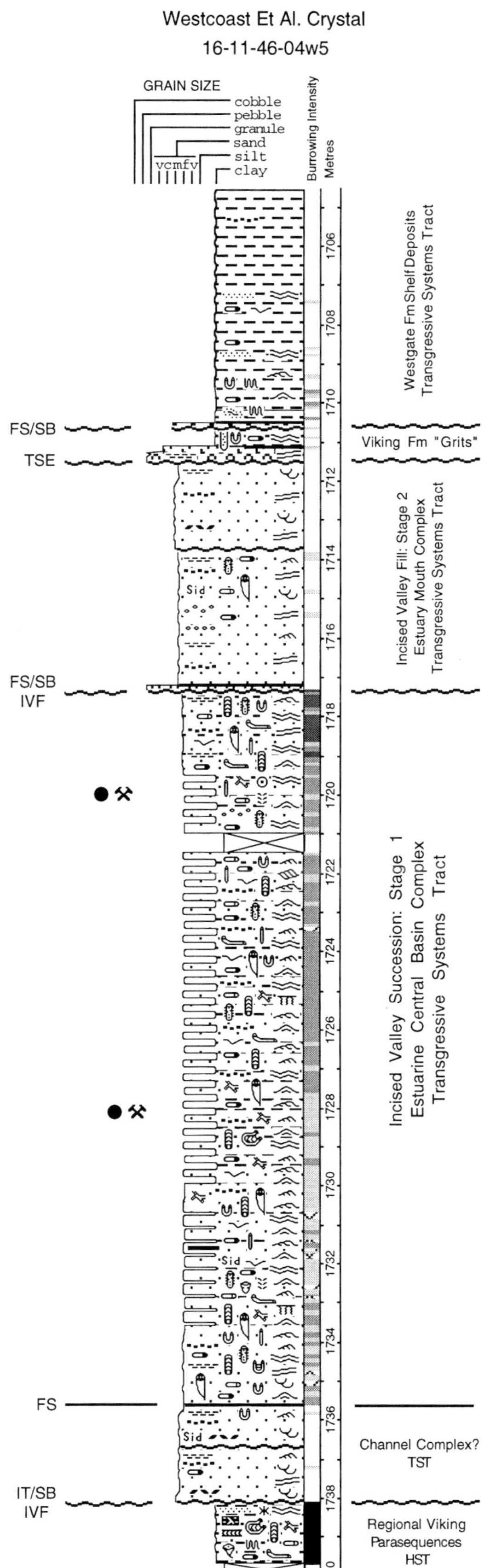

mixed *Skolithos-Cruziana* ichnofacies that reflects all three stresses typical of brackish bays and lagoons (Pemberton and Wightman, 1992; Pemberton et al., 1992; Ranger and Pemberton, 1992; MacEachern and Pemberton, 1994). Most structures tend to be small, corresponding partly to the diminutive size of tracemakers adapted to the rigours of brackish water habitation (Remane and Schlieper, 1971; Croghan, 1983), and partly to the predominance of juvenile animals over adults. The presence of syneresis cracks supports salinity reduction and/or fluctuations. The sporadic distribution of the secondary and accessory elements, which reflects more elaborate and specialized feeding and grazing behaviours, indicates that central basin salinities probably ranged from predominantly brackish to short periods of nearly fully marine conditions (MacEachern, 1994; MacEachern and Pemberton, 1994).

The 4 foraminiferal samples were collected from three wells (Table 1). The 16-11-46-04W5 well occurs in the Crystal valley complex (Fig. 15) and the 11-15-42-07W5 (Fig. 16) and 06-27-41-07W5 wells are from the Willesden Green valley complex. The diversity of the foraminiferal suites are highly reduced, characterized by 2-4 genera and 2-5 species. Environmentally sensitive forms are exceedingly rare, manifest by *Bulbobaculites* occurring in only one of the four samples. All samples contain low numbers of foraminifers, ranging from 10-22 specimens in 75 gram samples. Low abundances may reflect a combination of salinity stress and high sedimentation rates. Accessory fossils include rare prolate algal spores and vitrinitic debris.

The palynological assemblages from the four samples contrast markedly with those of the previous samples (Table 2). Terrestrial plant material ranges from 31.0-35.0% (averaging 33.3%) and microplankton material ranges from 65.0-69.0% (averaging 66.7%). Peridinioid dinoflagellates, however, comprise 31.5-50.0% of the assemblage (averaging 42.8%), and accounts for 48.1-74% (averaging 63.9%) of the microplanktonic component. This high concentration of peridinioid dinoflagellates distinguishes the incised valley deposits from all other mudstones sampled (Fig. 17).

DISCUSSION

The interpretation of marine and marginal marine sandstones as shore-attached bodies, shore-detached bodies, or estuarine incised valley complexes is typically hampered by the seemingly ambiguous nature of many of their primary physical sedimentary structures. Most physical sedimentary structures within sandstone reflect short-lived depositional processes that rarely reflect ambient or "typical" depositional conditions. In the case of sandbodies containing trace fossil or body fossil information, a better understanding of the sandstone's genesis is possible. Unfortunately, most sandstones possess limited biogenic constituents and are dominated by primary physical sedimentary structures. Consequently, in many cases, the paleodepositional interpretation

←
Fig. 15.—Interpreted graphic litholog for the facies succession in the 16-11-46-04W5 well. Sample types and locations are designated to the left of the litholog. See Figure 3 for the legend of symbols. IVF = incised valley fill; FS = flooding surface; FS/SB = amalgamated flooding surface and sequence boundary; IT/SB = amalgamated initial transgressive surface and sequence boundary; TSE = transgressive surface of erosion.

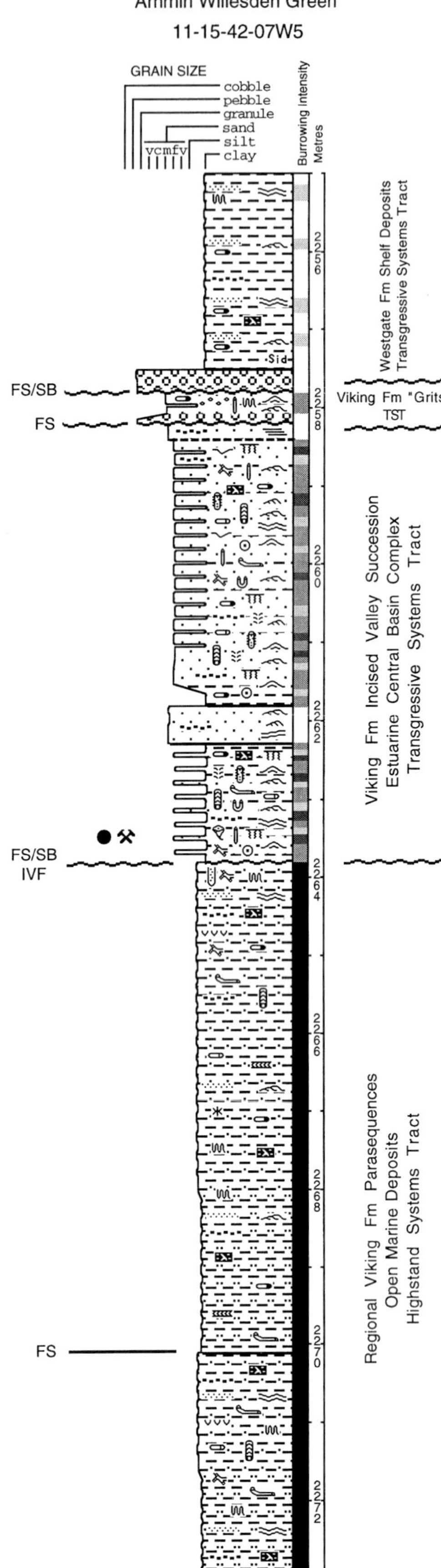

of such sandbodies is not based on direct analysis of the sandstone, but rather, from the study of interstratified muddy facies.

Paleoenvironmental analysis of mudstone deposits has traditionally been limited to palynological and microfossil analysis. More recently, the integration of ethology-based ichnology with conventional sedimentology has improved paleoenvironmental analysis. Ichnology is well suited to paleoenvironmental analysis, without time consuming and typically expensive sample processing and microfossil collecting methods. This is not to downplay the value of foraminiferal or palynological analysis. Not only are these disciplines essential for biostratigraphic analysis, they constitute the only viable paleontologic technique for the study of facies that, for whatever reasons, lack trace fossil and/or body fossils.

This is particularly apparent from the study of transgressive shelf deposits of the Westgate Formation. MacEachern et al. (1992) and MacEachern (1994) outlined the problems in the paleoenvironmental interpretation of this black, fissile, and largely unburrowed mudstone. Trace fossils, where visible, are typically associated with thin interbeds and interlaminae of fine-grained, parallel laminated sandstone, interpreted as distal tempestites. The apparent absence of burrowing within the shelf mudstone appears to be a contradiction; conventional ichnologic knowledge suggests that such mudstone should contain the *Nereites* or *Zoophycos* ichnofacies (Seilacher, 1967; Ekdale et al., 1984; Pemberton et al., 1992; cf., Fig. 5). The abundance of a diverse and environmentally sensitive benthic foraminiferal assemblage (Table 1), and the abundance of microplankton (Table 2) within this mudstone indicates that the absence of burrowing is not a function of a stressed depositional environment, such as oxygen deficiency at the sediment-water interface. As well, the large numbers of fossil specimens indicates that high sedimentation rates are unlikely to be a significant factor in the reduced intensity of bioturbation. Rather, the lack of ichnogenera appears to be a taphonomic problem. It is likely that the assemblage was dominated by bedding plane-parallel (non-penetrative), dark, fecal-infilled grazing and foraging structures, typical of shelf environments. These structures were produced on dark, soupy mud substrates. During burial, dewatering and compaction, these biogenic structures effectively "disappear" because of a lack of lithologic contrast and pronounced re-ordering of clay particles. Hence, the mudstone appears unburrowed but was probably thoroughly reworked by epifaunal and some infaunal organisms. It is not surprising, therefore, that ichnogenera preservation is typically restricted to the tops of distal storm beds where lithologic contrast exists and more rarely, as casts along the bases of the tempestites. If only the ichnogenera that are unrelated to opportunistic colonization of the storm beds are considered, then a *Zoophycos* suite is recognized. This relationship could not easily be demonstrated except through the integration of ichnology, palynology and foraminiferal paleoecology. A similar interpretation can be applied to the seemingly

←
Fig. 16.—Interpreted graphic litholog for the facies succession in the 11-15-42-07W5 well. Sample types and locations are designated to the left of the litholog. See Figure 3 for the legend of symbols. IVF = incised valley fill; FS = flooding surface; FS/SB = amalgamated flooding surface and sequence boundary.

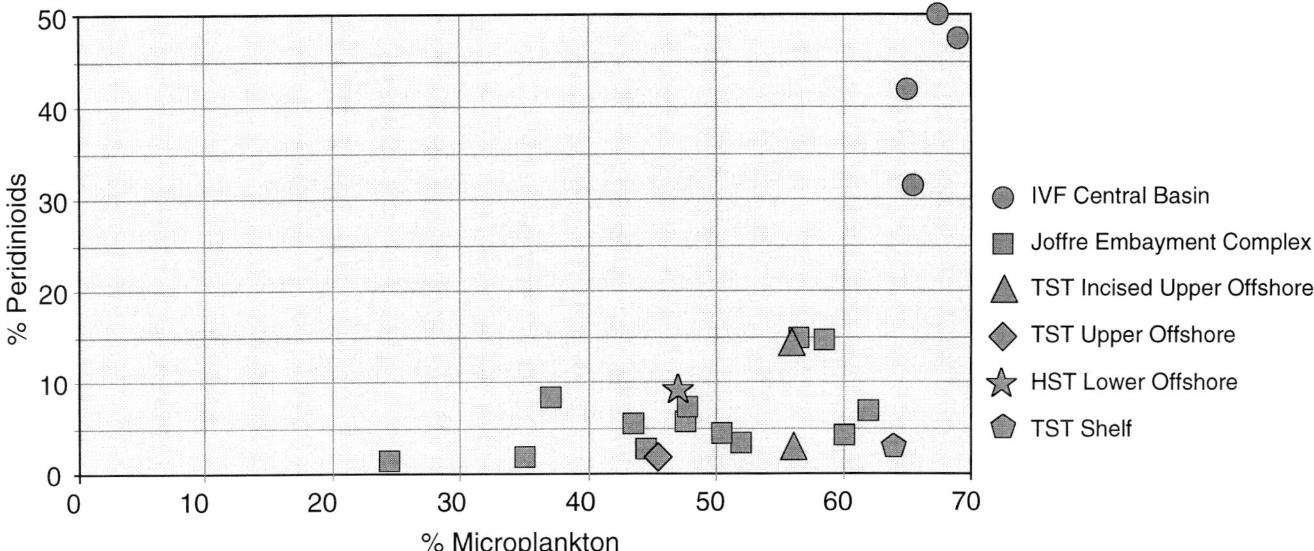

Fig. 17.—Palynological graph displaying concentration of peridinioid dinoflagellates versus concentration of microplankton for the different mudstone types. Note that only the estuarine incised valley central basin mudstone is easily differentiated. Percentage values recorded by numbers. TST = transgressive systems tract; HST = highstand systems tract; IVF = incised valley fill.

unburrowed Joli Fou Formation shelf mudstone, as well as the basal mudstone facies in some of the highstand systems tract parasequences of the Viking Formation.

With increasing lithologic variation within mudstone facies, however, trace fossils become increasingly more easily identified and yield depositional information directly from core analysis. For each mudstone type (with the exception of the shelf mudstone discussed above), the foraminiferal assemblages agree with paleoenvironmental interpretations derived from ethology-based ichnological analysis. Unfortunately, the palynological data collected are considerably more ambiguous, due to the abundance of wind transported terrestrial material and the planktonic character of the dinoflagellates. These factors make refined subenvironmental delineations difficult, and therefore, differentiation between shelf, lower offshore, upper offshore and embayment/lagoonal environments was not possible using palynology alone (Fig. 17). The foraminiferal assemblages afforded much better discrimination between shelf, offshore and embayment/lagoonal deposits (Table 1). Only the ichnological assemblages, however, permitted the segregation of shelf, lower offshore, upper offshore, and embayment/lagoonal mudstone deposits. All three techniques were able to differentiate the estuarine incised valley central basin deposits from the embayment/lagoon and open marine mudstones.

None of the paleontologic techniques were able to clearly differentiate between highstand and transgressive systems tract deposits. As well, in the case of the upper offshore deposits underlying the transgressively-incised Joffre shoreface complex, no paleontologic characteristics could be identified as discretely different from those of the offshore mudstone in the Viking "Grits". Variations between these samples can be explained by differing sedimentation rates and differentiated using physical sedimentology coupled with genetic stratigraphy.

As previously described, the transgressive shelf mudstone is characterized by a taphonomically controlled paucity of burrowing reflecting a *Zoophycos* ichnofacies. Foraminiferal suites reflect a highly diverse, abundant assemblage, consisting of numerous environmentally sensitive genera. This demonstrates a marine environment, at least as fully marine as encountered within the Albian of the Western Interior Seaway. The palynological suite merely indicates marine to marginal marine conditions, well removed from the shoreline.

Lower offshore silty mudstone is characterized by a diverse, uniformly distributed, distal *Cruziana* ichnofacies. The upper offshore sandy mudstone contains a diverse archetypal *Cruziana* ichnofacies. In both the highstand systems tract and the transgressive systems tract, the suites comprise K-selected (equilibrium) assemblages, reflecting open marine, unstressed settings. In the case of the highstand deposits, grain sizes are finer and trace fossils tend to occur somewhat more uniformly, indicating slow, continuous sedimentation, which permitted thorough biogenic reworking of the substrate by infaunal and epifaunal organisms. In contrast, the transgressive systems tract offshore deposits contain dispersed chert granules and small pebbles. Trace fossils are similarly diverse, but tend to be less uniformly distributed and typified by slightly lower overall bioturbation intensities than in highstand equivalents. This indicates less uniform and somewhat higher sedimentation rates compared to the highstand deposits. These offshore deposits probably reflect pulses of sedimentation during gradual transgression rather than widespread rapid transgression, which would be characterized by reduced sedimentation rates due to sediment trapping in shallower water settings. The foraminiferal suites of all these offshore deposits are typified by diverse and environmentally sensitive genera (Table 1). Abundances of foraminifera within the transgressive offshore deposits are slightly lower than in the highstand systems tract sample, supportive of higher sedimentation rates. The database is clearly too small, however, to be considered representative. Foraminiferal diversities and abundances in the granular upper offshore sandy mudstone underlying the transgressively incised Joffre shoreface complex is also somewhat reduced. Given the diverse trace fossil suite and the presence of environmentally sensitive foraminifera, this reduction in numbers of specimens is also interpreted to reflect higher sedimentation rates compared to highstand equivalents.

The mudstone of the Joffre embayment complex shows pronounced ichnological differences from the suites within offshore and shelf deposits. Although varied with respect to bioturbation intensity, distribution and diversity of ichnogenera, and uniformity of burrowing, the overall suite is dominated by the simple structures of trophic generalists. These structures reflect r-selected (opportunistic) behaviours, characteristic of environmentally stressful environments. The most common stress reflected by an assemblage dominated by *Teichichnus*, *Planolites* and *Terebellina* is a reduction and/or fluctuation in water salinity (Pemberton and Wightman, 1992; Ranger and Pemberton, 1992; MacEachern and Pemberton, 1994). The rare and sporadic occurrences of specialized structures such as *Helminthopsis*, *Anconichnus/Phycosiphon* and *Zoophycos* suggests that nearly fully marine conditions were achieved in the embayment/lagoon for short periods. The abundance of interstitial silt and sand, and the presence of oscillation, combined flow, and current ripple lamination within sandstone beds indicates that the lowered intensity of burrowing and low diversity of trace fossils within these deposits is not analogous to the taphonomic control experienced by the shelf mudstone of the Westgate Formation. There is ample lithological contrast within the facies to permit preservation of a variety of ichnological structures. As well, the trace fossils present within the Joffre embayment complex mudstone are penetrative, not surface grazing structures, demonstrating a reduced diversity *Cruziana* to mixed *Skolithos-Cruziana* ichnofacies, typical of more energetic, shallower water environments (MacEachern and Pemberton, 1992; Pemberton et al., 1992; MacEachern et al., 1998). Some paucity of burrowing can be attributed to sporadically higher sedimentation rates, suggested by the low abundances of foraminifera recovered from a few of the samples.

The foraminiferal suites support a marginal marine environment, characterized by reduced though variable salinities. The presence of only subnormal rather than strongly reduced salinities are reflected by the presence of several environmentally sensitive genera (e.g., *Ammobaculites*, *Bulbobaculites*, and *Verneuilinoides*). The suites do show, however, an overall reduction in foraminiferal diversity, as well as a reduction in the numbers and diversity of environmentally sensitive forms compared to the offshore and shelf deposits (Table 1), attesting to subnormal salinities. Some embayment samples are strongly stressed, such as the 11-07-39-26W4 sample at 1545.5 m, which contains only 4 genera and 4 species and yielded only 13 specimens. Others are quite marine, such as the 06-30-38-25W4 sample at 1522.3 m, which contains 8 genera and 13 species, and yielded 105 specimens. These variations probably reflect a combination of variable salinities and variable sedimentation rates. Most samples, however, simply display marine signals and sedimentation rates intermediate between those of the offshore/shelf samples and the estuarine incised valley samples. The character of the ichnological suites and the presence of benthic foraminifera indicate that reduced oxygenation and water stratification are unlikely to be controlling paleoecologic factors in these deposits. The common occurrence of traction transported sandstone interbeds also supports a mixed water column. The palynological assemblages within these samples, however, show no significant differences from those recovered from the shelf and offshore deposits (Table 2).

The ichnological suite of the estuarine incised valley central basin mudstone is similar in many regards to the mudstone of the Joffre embayment complex. The central basin mudstone, however, displays an overall lower trace fossil diversity, suggesting a more stressful depositional setting. Environmentally sensitive forms (i.e., those reflecting specialized behaviours of organisms unable to endure physiological stress, such as *Helminthopsis* and *Zoophycos*) are entirely absent from the suites, or occur only very rarely within discrete horizons, that reflect short-lived periods of reduced physico-chemical stress. The dominant ichnogenera comprise simple structures, reflecting r-selected dwelling/feeding behaviours of trophic generalists (opportunistic omnivores; cf., Pianka, 1970; Grassle and Grassle, 1974; Jumars, 1993), which are able to consume a variety of food resource types (e.g., *Teichichnus*, *Planolites*, ?*Terebellina*). Many of the ichnogenera tend to be smaller in size than their fully marine counterparts, reflecting a combination of tracemaker adaptation to diminutive size (facilitating osmotic regulation; cf., Remane and Schlieper, 1971; Croghan, 1983), as well as an abundance of juvenile rather than adult structures due to high mortality rates. These characteristics are typical of physico-chemically stressful environments, particularly those susceptible to salinity reductions and salinity fluctuations. The variability in burrowing intensities, sporadic distribution of individual ichnogenera, and the presence of discrete opportunistic event bed suites and ambient (fairweather) suites are similar to that encountered in the Joffre embayment deposits as well. This generates a low diversity (stressed) mixed *Skolithos-Cruziana* ichnofacies, which attests to fluctuations in energy, fluctuations in substrate consistency, fluctuations in sedimentation rate, fluctuations in salinity and general reduction of salinity (Pemberton et al., 1992; MacEachern, 1994; MacEachern and Pemberton, 1994; Pemberton and MacEachern, 1995). This is strongly supported by the foraminiferal suites which typically lack environmentally sensitive forms, and contain very low abundances of specimens (Table 1). The palynological assemblages of the central basin mudstones are uniquely different from the offshore/shelf and embayment/lagoon deposits, in that they contain high concentrations of peridinioid dinoflagellates (Table 2, Fig. 17). These are characteristic of highly stressed, brackish water conditions (G. Dolby, pers. commun., 1994), supportive of the ichnological interpretation. Rather surprisingly, the central basin deposits show a reduction in the terrestrial plant component of the assemblage, despite the interpreted proximity to land. This may be an artifact of the unusually high concentrations of peridinioid dinoflagellates, which comprise much of the microplankton assemblage, and comparatively diminish the proportion of terrestrial elements. Vitrinitic plant debris is generally more abundant within these samples, supporting close proximity to land. It is curious, however, that megaspores and other seed cases are not significant components of the palynological assemblage.

CONCLUSIONS

The integration of ichnology and sedimentology with foraminiferal micropaleontology and palynology affords the opportunity to test the paleoenvironmental utility of these different disciplines. Coarse-grained deposits typically lack or have markedly reduced concentrations of biogenic elements, hampering their depositional interpretation. Consequently, five fine-grained deposit types, namely: 1) transgressive systems tract shelf; 2) highstand systems tract lower offshore; 3) transgressive systems tract upper offshore; 4) transgressive systems tract embayment/lagoon; and 5) transgressive systems tract estuarine incised valley central basin deposits were analyzed from the Joli Fou, Viking and Westgate formations of central Alberta. The ability to recognize these subenvironments clearly has important implications for the depositional interpretation of interstratified and genetically related, coarse-grained deposits, many of which have economic significance.

In most regards, the fossil assemblages derived from each paleontologic discipline strongly support one another, strengthening the paleoecologic interpretation of the mudstone. Further, the integration of the disciplines was able to account for complexities which could not be resolved using only a single discipline. Although ideally each paleontological discipline would be employed in a paleoenvironmental study, this is, unfortunately, generally impractical.

The palynological assemblages appeared to be most useful in separating marine from strongly brackish-water (estuarine) conditions within the Viking Formation, but did not appear to be able to supply criteria by which shelf, offshore, and weakly to moderately brackish embayment settings could be differentiated. Foraminiferal assemblages, on the other hand, were able to differentiate most of these subenvironments and supplied essential data to explain the apparent absence of burrowing in the shelf mudstone. In addition, the mudstone of the Joffre embayment complex could be differentiated from open marine mudstone, confirming the conclusions reached from ichnological analyses of these deposits (MacEachern et al., 1995; 1998; this volume). The paucity of foraminifers within the estuarine incised valley mudstone, however, limited the paleoecologic utility of this discipline in these deposits.

Ichnology, a relative newcomer with respect to the paleoecologic interpretation of the rock record, is not only suited to direct visual identification of biogenic components, but is able to differentiate between shelf, lower offshore, upper offshore, embayment/lagoon and estuarine incised valley central basin mudstones. This ichnological data can be collected during initial description of the rocks and, therefore, facilitates rapid delineation of paleoenvironments without the need for costly and time consuming laboratory procedures.

ACKNOWLEDGEMENTS

This paper derives partly from a post-doctoral project undertaken by the first author, funded by a Natural Sciences and Engineering Research Council (NSERC) Collaborative Research and Development (CRD) grant 180563 awarded to S.G. Pemberton, as well as through NSERC operating grant 184293 and a Simon Fraser President's Research Grant awarded to J.A. MacEachern. Some of the data was collected while the first author was engaged at PanCanadian Petroleum Ltd. The authors would also like to thank PanCanadian Petroleum Ltd. for their financial and logistical support throughout the course of this study. Dr. G. Dolby and M.T. Gallagher of G. Dolby and Associates processed the samples and conducted preliminary fossil identifications. Dr. Dolby carried out all palynological analyses. Dr. Brian A. Zaitlin is gratefully acknowledged for imparting his insights into the paleoenvironments of the Viking Formation. The manuscript was greatly improved through the critical review of A.G. Plint and J. Dale.

REFERENCES

AINSWORTH, R.B. AND PATTISON, S.A.J., 1994, Where have all the lowstands gone? Evidence for attached lowstand systems tracts in the Western Interior of North America: Geology, v. 22, p. 415-418.

BERGMAN, K.M., 1994, Shannon Sandstone in Hartzog Draw-Heldt Draw fields reinterpreted as detached lowstand shoreface deposits: Journal of Sedimentary Research, v. B64, p. 184-201.

BERGMAN, K.M. AND WALKER, R.G., 1995, High resolution sequence stratigraphic analysis of the Shannon Sandstone in Wyoming, using a template for regional correlation, *in* Swift, D.J.P., Snedden, J.W. and Plint, A.G., eds., Tongues, ridges and wedges: highstand *versus* lowstand architecture in marine basins: Society of Economic Paleontologists and Mineralogists Research Conference, Powder River and Bighorn Basins, Wyoming, June 24-29, unpaginated.

BLOCH, J., SCHRÖDER-ADAMS, C., LECKIE, D.A., MCINTYRE, D.J., CRAIG, J. AND STANILAND, M., 1993, Revised stratigraphy of the lower Colorado Group (Albian to Turonian), Western Canada: Bulletin of Canadian Petroleum Geology, v. 41, p. 325-348.

CROGHAN, P.C., 1983, Osmotic regulation and the evolution of brackish- and fresh-water faunas: Journal of the Geological Society, v. 140, p. 39-46.

EKDALE, A.A., BROMLEY, R.G. AND PEMBERTON, S.G., 1984, Ichnology: Trace fossils in sedimentology and stratigraphy: Society of Economic Paleontologists and Mineralogists Short Course 15, 317 p.

FREY, R.W., 1990, Trace fossils and hummocky cross-stratification, Upper Cretaceous of Utah, Palaios, v. 5, p. 203-218.

GLAISTER, P., 1959, Lower Cretaceous of southern Alberta and adjoining areas: American Association of Petroleum Geologists Bulletin, v. 43, p. 590-640.

GRASSLE, J.F. AND GRASSLE J.P., 1974, Opportunistic life histories and genetic systems in marine benthic polychaetes: Journal of Marine Research, v. 32, p. 253-284.

HADLEY, D.F. AND ELLIOTT, T., 1993, The sequence-stratigraphic significance of erosive-based shoreface sequences in the Cretaceous Mesaverde Group of northwestern Colorado, *in* Posamentier, H.W., Summerhayes, C.P., Haq, B.U. and Allen, G.P., eds., Stratigraphy and facies associations in a sequence stratigraphic framework: International Association of Sedimentologists, Special Publication, v. 18, p. 521-535.

HOWARD, J.D., 1971, Comparison of the beach-to-offshore sequence in modern and ancient sediments, *in* Howard, J.D., Valentine, J.W. and Warme, J.E., eds., Recent advances in paleoecology and ichnology: American Geological Institute, Short Course Lecture Notes, p. 148-183.

HOWARD, J.D., 1972, Trace fossils as criteria for recognizing shorelines in stratigraphic record, *in* Rigby, J.K. and Hamblin, W.K., eds., Recognition of ancient sedimentary environments: Society of Economic Paleontologists and Mineralogists Special Publication 16, p. 215-225.

HOWARD, J.D. AND FREY, R.W., 1984, Characteristic trace fossils in nearshore to offshore sequences, Upper Cretaceous of east-central Utah: Canadian Journal of Earth Sciences, v. 21, p. 200-219.

HOWARD, J.D. AND REINECK, H.E., 1981, Depositional facies of a high energy beach-to-offshore sequence: comparison with low-energy sequence: American Association of Petroleum Geologists, Bulletin, v. 65, p. 807-830.

HUNTER, R.E., CLIFTON, H.E. AND PHILLIPS, R.L., 1979, Depositional processes, sedimentary structures, and predicted vertical sequences in barred nearshore systems, southern Oregon coast: Journal of Sedimentary Petrology, v. 49, p. 711-728.

JOHNSON, H.D. AND BALDWIN, C.T., 1996, Shallow siliciclastic seas, *in* Reading, H.G., ed., Sedimentary environments: processes, facies and stratigraphy: 3rd edition, Cambridge, Blackwell Science Ltd., p. 232-280.

JUMARS, P.A., 1993, Concepts in Biological Oceanography: New York, Oxford University Press, 348p.

KOKE, K.R. AND STELCK, C.R., 1985, Foraminifera of a Joli Fou Shale equivalent in the Lower Cretaceous (Albian) Hasler Formation, northeastern British Columbia: Canadian Journal of Earth Sciences, v. 22, p. 1299-1313.

LECKIE, D.A., 1986, Tidally influenced, transgressive shelf sediments in the Viking Formation, Caroline, Alberta: Bulletin of Canadian Petroleum Geology, v. 34, p. 111-125.

LECKIE, D.A., SINGH, C., GOODARZI, F. AND WALL, J.H., 1990., Organic-rich, radioactive marine shale: a case study of a shallow-water condensed section, Cretaceous Shaftesbury Formation, Alberta, Canada: Journal of Sedimentary Petrology, v. 60, p. 101-117.

MACEACHERN, J.A., 1994, Integrated ichnological-sedimentological models: applications to the sequence stratigraphic and paleoenvironmental interpretation of the Viking and Peace River formations, west-central Alberta: Ph.D. thesis, University of Alberta, Edmonton, 618p.

MACEACHERN, J.A. AND PEMBERTON, S.G., 1992, Ichnological aspects of Cretaceous shoreface successions and shoreface variability in the Western Interior Seaway of North America, *in* Pemberton, S.G., ed., Applications of Ichnology to Petroleum Exploration: Society of Economic Paleontologists and Mineralogists, Core Workshop 17, p. 57-84.

MACEACHERN, J.A. AND PEMBERTON, S.G., 1994, Ichnological aspects of incised valley fill systems from the Viking Formation of the Western Canada Sedimentary Basin, Alberta, Canada, *in* Dalrymple, R.W., Boyd, R. and Zaitlin, B.A., eds., Incised valley systems: origin and sedimentary sequences: Society of Economic Paleontologists and Mineralogists, Special Publication 51, p. 129-157.

MACEACHERN, J.A., BECHTEL, D.J. AND PEMBERTON, S.G., 1992, Ichnology and sedimentology of transgressive deposits, transgressively-related deposits and transgressive systems tracts in the Viking Formation of Alberta, *in* Pemberton, S.G., ed., Applications of Ichnology to Petroleum Exploration: Society of Economic Paleontologists and Mineralogists, Core Workshop 17, p. 251-290.

MACEACHERN, J.A., PEMBERTON, S.G. AND ZAITLIN, B.A., 1995, A late lowstand to early transgressive coarse-grained tongue from the Viking Formation of the Joffre Field, Alberta: embayment complex or shoreface wedge?, *in* Swift, D.J.P., Snedden, J.W. and Plint, A.G., eds., Tongues, ridges and wedges: highstand *versus* lowstand architecture in marine basins: Society of Economic Paleontologists and Mineralogists Research Conference, Powder River and Bighorn Basins, Wyoming, June 24-29, unpaginated.

MACEACHERN, J.A., ZAITLIN, B.A. AND PEMBERTON, S.G., 1998, High-resolution sequence stratigraphy of early transgressive deposits, Viking Formation, Joffre Field, Alberta, Canada: American Association of Petroleum Geologists Bulletin v. 82, p. 729-756.

MCGOOKEY, D.P., HAUN, J.D., HALE, L.A., GOODELL, H.G., MCCUBBIN, D.G., WEIMER, R.J. AND WULF, G.R., 1972, Cretaceous System, *in* Mallory, W.W., ed., Geologic Atlas of the Rocky Mountain Region, U.S.A.: Rocky Mountain Association of Geologists, Denver, Colorado, p. 190-228.

MELLERE, D. AND STEEL, R.J., 1995, Facies architecture and sequentiality of nearshore and 'shelf' sandbodies: Haystack Mountains Formation, Wyoming, USA: Sedimentology, v. 42, p. 551-574.

PATTISON, S.A.J., 1991, Sedimentology and allostratigraphy of regional, valley-fill, shoreface and transgressive deposits of the Viking Formation (Lower Cretaceous), Central Alberta: Unpublished Ph.D. thesis, McMaster University, 380 p.

PATTISON, S.A.J., 1992, Recognition and interpretation of estuarine mudstones (central basin mudstones) in the tripartite valley-fill deposits of the Viking Formation, Central Alberta, *in* Pemberton, S.G., ed., Applications of ichnology to petroleum exploration, a core workshop: Society of Economic Paleontologists and Mineralogists, Core Workshop 17, p. 223-249.

PEMBERTON, S.G. AND MACEACHERN, J.A., 1995, The sequence stratigraphic significance of trace fossils: examples from the Cretaceous foreland basin of Alberta, Canada, *in* Van Wagoner, J.C. and Bertram, G., eds., Sequence stratigraphy of foreland basin deposits- outcrop and subsurface examples from the Cretaceous of North America: American Association of Petroleum Geologists, Memoir 64, p. 429-475.

PEMBERTON, S.G. AND MACEACHERN, J.A., 1997, The ichnological signature of storm deposits: the use of trace fossils in event stratigraphy, *in* Brett, C.E., ed., Paleontological event horizons: ecological and evolutionary implications: Columbia University Press, p. 73-109.

PEMBERTON, S.G. AND WIGHTMAN, D.M, 1992, Ichnological characteristics of brackish water deposits, *in* Pemberton, S.G., ed., Applications of ichnology to petroleum exploration, a core workshop: Society of Economic Paleontologists and Mineralogists, Core Workshop 17, p. 141-167.

PEMBERTON, S.G., MACEACHERN, J.A. AND FREY, R.W., 1992, Trace fossil facies models: environmental and allostratigraphic significance, *in* Walker, R.G. and James, N., eds., Facies Models: Response to Sea Level Change, p. 47-72.

PIANKA, E.R., 1970, On r and k selection: American Naturalist, v. 104, p. 592-597.

POSAMENTIER, H.W. AND CHAMBERLAIN C.J., 1993, Sequence stratigraphic analysis of Viking Formation lowstand beach deposits at Joarcam field, Alberta, Canada, *in* Posamentier, H.W., Summerhayes, C.P., Haq, B.U. and Allen, G.P., eds., Stratigraphy and facies associations in a sequence stratigraphic framework: International Association of Sedimentologists, Special Publication, v. 18, p. 469-485.

POSAMENTIER, H.W., ALLEN, G.P., JAMES, D.P. AND TESSON, M., 1992, Forced regressions in a sequence stratigraphic framework: concepts, examples, and exploration significance: American Association of Petroleum Geologists, Bulletin v. 76, p. 1687-1709.

RANGER, M.J. AND PEMBERTON, S.G., 1992, The sedimentology and ichnology of estuarine point bars in the McMurray Formation of the Athabasca Oil Sands Deposit, northeastern Alberta, Canada, *in* Pemberton, S.G., ed., Applications of Ichnology to Petroleum Exploration: Society of Economic Paleontologists and Mineralogists, Core Workshop 17, p. 401-421.

REINSON, G.E., CLARK, J.E. AND FOSCOLOS, A.E., 1988, Reservoir geology of Crystal Viking Field, Lower Cretaceous estuarine tidal channel-bay complex, south-central Alberta: American Association of Petroleum Geologists, Bulletin, v. 72, p. 1270-1294.

REMANE, A. AND SCHLIEPER, C., 1971, Biology of Brackish Water: New York, Wiley, 372p.

SEILACHER, A., 1967, Bathymetry of trace fossils: Marine Geology, v. 5, p. 413-428.

STELCK, C.R., 1958, Stratigraphic position of the Viking sand: Journal of the Alberta Society of Petroleum Geologists, v. 6, p. 2-7.

STELCK, C.R. AND KOKE, K.R., 1987, Foraminiferal zonation of the Viking interval in the Hasler Shale (Albian), northeastern British Columbia: Canadian Journal of Earth Sciences, v. 24, p. 2254-2278.

STELCK, C.R. AND LECKIE, D.A., 1990, Biostratigraphy of the Albian Paddy Member (Lower Cretaceous Peace River Formation), Goodfare, Alberta: Canadian Journal of Earth Sciences, v. 27, p. 1159-1169.

SULLIVAN, M., VAN WAGONER, J., FOSTER, M., STUART, R., JENETTE, D., LOVELL, R. AND PEMBERTON, S.G., 1995, Lowstand architecture and sequence stratigraphic control on Shannon incised valley distribution, Hartzog Draw Field, Wyoming, *in* Swift, D.J.P., Snedden, J.W. and Plint, A.G., eds., Tongues, ridges and wedges: highstand *versus* lowstand architecture in marine basins: Society of Economic Paleontologists and Mineralogists Research Conference, Powder River and Bighorn Basins, Wyoming, June 24-29, unpaginated.

VOSSLER, S.M. AND PEMBERTON, S.G., 1989, Ichnology and paleoecology of offshore siliciclastic deposits in the Cardium Formation (Turonian, Alberta, Canada): Palaeogeography, Palaeoclimatology, Palaeoecology, v. 74, p. 217-229.

WALKER, R.G. AND BERGMAN, K.M., 1993, Shannon Sandstone in Wyoming: a shelf-ridge complex reinterpreted as lowstand shoreface deposits: Journal of Sedimentary Petrology, v. 63, p. 839-851.

WALKER, R.G. AND PLINT, A.G., 1992, Wave- and shallow-marine shallow marine systems, *in* Walker, R.G. and James, N., eds., Facies models: response to sea level change: Geological Association of Canada, St. John's, Newfoundland, p. 219-238.

WALKER, R.G. AND WISEMAN, T.R., 1995, Lowstand shorefaces, transgressive incised shorefaces, and forced regressions: examples from the Viking Formation, Joarcam area, Alberta: Journal of Sedimentary Research, v. B65, p. 132-141.

WEIMER, R.J., 1984, Relation of unconformities, tectonics, and sea level changes, Cretaceous of the Western Interior, U.S.A., *in* Schlee, J.S., ed., Interregional unconformities and hydrocarbon accumulation: American Association of Petroleum Geologists, Memoir 36, p. 7-35.

THE ORIGIN OF THE TOCITO SANDSTONE AND ITS SEQUENCE STRATIGRAPHIC LESSONS

DAG NUMMEDAL

Unocal Corporation, 14141 Southwest Freeway, Sugar Land, TX 77478, U.S.A.

AND

GREGORY W. RILEY

Amoco Production Company, 501 WestLake, Houston, TX 77079, U.S.A.

ABSTRACT: A detailed and comprehensive study of all Tocito outcrops along the margins of the San Juan basin, New Mexico, combined with an equally comprehensive study of the subsurface Tocito by David Jennette and Clive Jones, provides new insights into the origin of these enigmatic, oil-rich, shale-encased sand bodies. The Tocito Sandstone is entirely Coniacian in age. It is separated from the underlying Gallup Sandstone on the Four Corners Platform by a major unconformity that locally is associated with a lacuna up to 3 million years. In the southeast San Juan basin, however, there is no demonstrable biostratigraphic gap.

The formation of the Tocito is intimately connected with an episode of Coniacian tectonism that triggered the uplift of the Waterflow Anticline that trends northwest-southeast across the Four Corners platform and adjacent regions of the San Juan basin. Uplift of this anticline commenced immediately after Gallup Sandstone deposition and gave rise to an elevated seafloor region across which shallow marine tidal (and storm?) currents eroded several hundred feet of Mancos Shale. Concurrently, the stream system that previously had fed the wave-dominated shorelines of the Gallup, the Torrivio system, began delivering its load into the embayment (or strait?) formed behind the uplifted anticline. In that setting of strongly enhanced tides and reduced wave action, the lower Tocito formed as a tide-dominated delta complex. In late Coniacian time regional subsidence set in, probably by reversal of movement on the inferred fault below the flank of the anticline. The ensuing marine transgression truncated much of the tidal delta facies of the lower Tocito, as well as some of the fluvial Torrivio, and formed a set of shelf sand ridges across much of the Four Corners platform.

Studies on modern continental shelves have revealed that such juxtaposition of transgressive shelf sand ridges directly above an early Holocene ravinement surface that has truncated underlying regressive deltaics is quite common. Therefore, an accurate documentation of the differences in distribution, geometry, vertical textural trends, and component facies between the lower, regressive member and the upper, transgressive member of the Tocito Sandstone may provide a useful template for a better understanding of the many similar delta/estuary/shelf sand-stone complexes in the rock record. The Tocito Sandstone has excellent porosity and is the major oil reservoir rock in the San Juan basin. Documentation of the Tocito sand body and shale body geometries, therefore, will help us better understand fluid flow in such reservoirs, of which there probably are many more than currently recognized.

INTRODUCTION

The Tocito Sandstone has been the subject of numerous studies because of its prolific oil production (Sabins, 1972; McCubbin, 1969), its intriguing stratigraphic relationship to the Gallup Sandstone (Campbell, 1979; Molenaar, 1983; Nummedal and Molenaar, 1995; Jennette and Jones, 1995), and its controversial origin (Tillman, 1985; Jennette et al., 1991; Nummedal and Riley, 1991). Jennette and Jones (1995) published an elegant and comprehensive study of the Tocito Sandstone based on data from many outcrops and over 1300 well logs covering the Four Corners Platform and the adjacent area of the northwest corner of the San Juan basin. (Fig. 1) This paper, which is based on a Ph.D. dissertation by Greg Riley (1993), complements the Jennette and Jones study by placing more emphasis on the outcrop data. Together, the two extensive data sets form the basis for an internally consistent, yet complex, model of the origin and sequence stratigraphy of the Tocito Sandstone. Even with essentially similar facies interpretations, different authors still reach contrasting conclusions about the overall origin of the Tocito Sandstone and about the sequence stratigraphic lessons that this complex sandstone unit can teach us.

This paper addresses the three most intractable problems with the Tocito. For a review of these problems, refer to Tillman (1985) and Riley (1993). First, there is controversy about the basal Tocito unconformity. Is it subaerial or submarine? Is it localized on the Four Corners platform or regional in extent? What is the duration of the associated lacuna? New megafossil finds and the time scales of Obradovich (1993) and Gradstein et al. (1995), combined with sedimentological analysis of strata above and below this unconformity, now provide answers to these questions. Second, the role of tectonism versus eustasy has figured in many recent papers on the Tocito. Resolving this is a most crucial issue, because the value of the Tocito as a "case study" to test our emerging sequence stratigraphic models is critically dependent on clear differentiation of near-field (local) versus far-field (ups and downs of the whole continent) tectonic effects. Third, with all the current emphasis on the architecture of hydrocarbon reservoirs, the Tocito has great potential value as an analog because it is a major hydrocarbon reservoir immediately east of the outcrop belt on the Four Corners platform. However, to avoid the misapplication of ongoing architectural studies, the depositional systems of the Tocito as well as their temporal and stratigraphic changes must be correctly interpreted.

CHRONOSTRATIGRAPHY

The age of the Tocito Sandstone and its surrounding units has recently been better constrained as a result of the collection of several new species of age-diagnostic megafossils across the northwest San Juan basin (Nummedal and Molenaar, 1995; Molenaar et al., 1996; Riley, 1993). Concurrently, the absolute ages of Cretaceous biozones have been improved due to argon-argon analysis of biostratigraphically calibrated bentonites (Obradovich, 1993). Considering the crucial value of a correct chronostratigraphic framework in

Isolated Shallow Marine Sand Bodies: Sequence Stratigraphic Analysis and Sedimentologic Interpretation.
SEPM Special Publication No. 64,
SEPM (Society for Sedimentary Geology), ISBN 1-56576-057-3, p. 227-254.

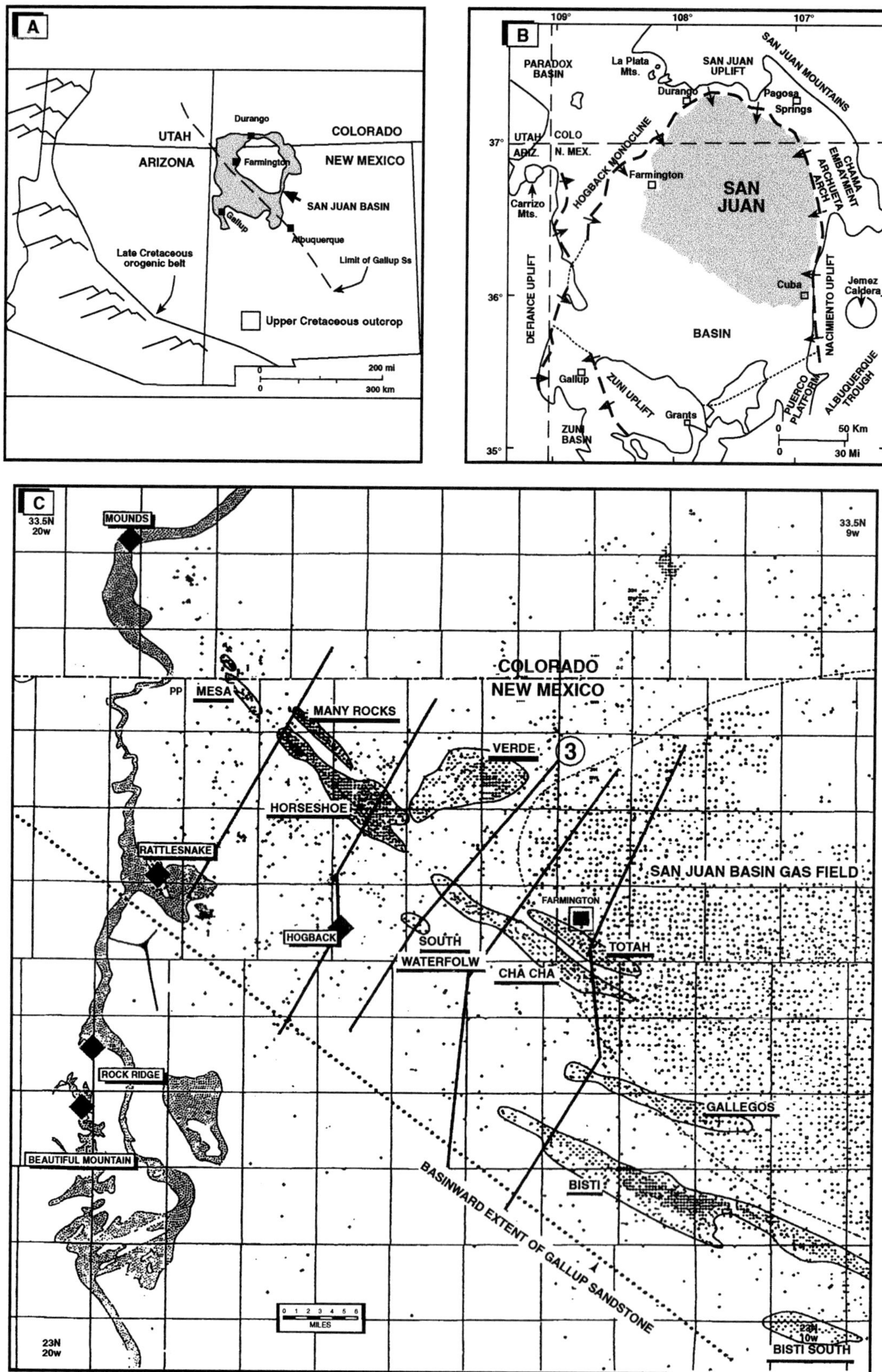

Fig. 1.—Location maps of A) the San Juan basin in northwest New Mexico, B) major structural elements around the margins of the San Juan basin, and C) the Four Corners Platform, where most of the outcrops discussed in this paper are located, and the adjacent western parts of the San Juan basin. Major oil fields producing from the Tocito Sandstone are indicated. Map C is modified from Jennette and Jones (1995), and their cross-sections are indicated.

any stratigraphic analysis, the revised biostratigraphy is presented first. Combined with the physical stratigraphic evidence to be discussed below, the new age data shed considerable light on the nature of the basal Tocito unconformity. Because this unconformity separates the Tocito from the underlying Gallup Sandstone, we will review the chronostratigraphy of the Gallup Sandstone first.

Gallup Sandstone

The Tocito Sandstone overlies the Gallup Sandstone and its basinward equivalent interval of the Mancos Shale. The Gallup Sandstone represents an episode of major delta progradation in the San Juan basin, spanning most of Late Turonian time, from about 89.7 Ma to 88.3 Ma (Nummedal and Molenaar, 1995). Several collections of *Prionocyclus germarii* document that the major phase of Gallup progradation (the C tongue) occurred within this zone, the youngest of the late Turonian ammonite zones (Cobban, 1984; Obradovich, 1993; Molenaar et al., 1996). In the southeastern parts of the San Juan basin (Figs. 1A ,1B), *Inoceramus erectus* (or "*Cremnoseramus erectus*") has been recovered from the uppermost Gallup (Maxwell et al., 1989). The wide age range of *I. erectus* (Fig. 2) does not provide conclusive evidence as to whether the youngest Gallup here is late Turonian or early Coniacian. On the Four Corners platform, collections of *Prionocyclus germarii* immediately below the B tongue of the Gallup Sandstone near its pinch-out yield a late Turonian age, wheras only the long-ranging *Inoceramus erectus* and *Lopha sannionis* have been recovered from the Gallup and the mudstone immediately above the B tongue, below the basal Tocito unconformity.

Age assignment for the top of the Gallup, therefore, is extrapolated from the biostratigraphic data after applying reasoning based on sedimentation rates. The three oldest tongues of the Gallup Sandstone (tongues F through D) record about 0.6 million years of time (89.5 - 88.9 Ma; Nummedal and Molenaar, 1995; Fig. 7). Assuming constant long-term sediment accumulation rates, tongues C through A probably represent another 0.6 m.y. Thus, the most plausible age for the top of the Gallup Sandstone is 88.3 Ma, at the end of the early Coniacian stage. Both *Lopha sannionis* and *Inoceramus erectus* are the only age-diagnostic fossils in the youngest Gallup Sandstone and range from latest Turonian to early Middle Coniacian. An age of 88.3 Ma for the top Gallup Sandstone agrees fully with existing biostratigraphic constraints.

The Gallup-equivalent Mancos Shale interval is deeply truncated along a northwest-southwest-trending zone basinward of the Gallup Sandstone pinch-out across the northwest part of the San Juan basin and adjacent Four Corners platform (Molenaar and Baird, 1992; Nummedal and Molenaar, 1995; Molenaar et al., 1996). At a particularly fossil-rich Tocito outcrop near the Colorado/New Mexico state line at the "Plunge Pool" (PP in Fig. 1C), clasts of reworked *Prionocyclus wyomingensis* were recovered within the basal Tocito lag, and *in situ* specimens of *Prionocylus macombi, Inoceramus dimidius,* and *Lopha lugubris* were found in the shales just below, confirming an early, Late Turonian age for this section (about 90 Ma). There is no evidence for similar truncation of the Lower Mancos Shale in the southeastern parts of the San Juan basin.

Tocito Sandstone

All megafossil collections support an entirely Coniacian age for the Tocito Sandstone. In the southeastern areas of the San Juan basin (e.g., near Guadalupe on the Rio Puerco; Molenaar et al., 1996, section C-C') a series of discontinuous Tocito Sandstone outcrops contain *Inoceramus erectus* and *I. browni*. The ammonite *Forresteria peruana* has been recovered from Tocito outcrops at the nearby Pipeline Road. These fossil recoveries place the basal Tocito in the southeastern parts of the San Juan basin within the earliest Coniacian ammonite zone. Thus, the age of the basal Tocito here is biostratigraphically indistinguishable from that of the youngest Gallup Sandstone (Fig. 3B).

On the Four Corners platform to the northwest of the San Juan basin (Fig. 1B), the basal Tocito Sandstone is much younger. The most updip Tocito sandbodies on this platform occur along the east flank of Beautiful Mountain (Fig. 1C). Here the Tocito overlies a thick section of Gallup Sandstone consisting of the C, B and A tongues. The only age-diagnostic megafossil recovered from these outcrops is *Inoceramus* (or *'Platyceramus') stantoni* of late Coniacian age (Fig. 2). This particular *Inoceramus* species is widespread and common, and is therefore thought to record a reliable age. At the Hogback oil field, about 8 km northeast of the Gallup pinch-out (Fig. 1C), W.A. Cobban, D. Nummedal, and G. Riley recovered a well-preserved specimen of *Peroniceras westphalicum*, identifying the lower body of this Tocito outcrop as late middle Coniacian in age. A specimen of *Inoceramus deformis* (late form) in the lower Tocito at the same location is also consistent with a middle Coniacian age. Finally at the "Plunge Pool" outcrop referred to above, the Tocito has yielded in-place specimens of *Platyceramus stantoni* and the ammonites *Gauthiericeras roquei* and *Peroniceras tridorsatum*, demonstrating a late Coniacian age (Fig. 3A).

The basal Tocito unconformity

The biostratigraphic data base for the Tocito and Gallup Sandstones, as well as the related Mancos Shale, permits the construction of a reliable chronostratigraphic framework, including the accurate extent of the lacuna associated with the basal Tocito unconformity (Fig. 3A). In the southeast San Juan basin there is no biostratigraphic evidence of any hiatus: both the top Gallup and entire (thin) Tocito Sandstone section fall within the early Coniacian stage. Across the Four Corners Platform, in contrast, there is a major basal Tocito unconformity, ranging in duration from about 1 m.y. in updip areas such as at Beautiful Mountain, to about 3 m.y. at the "Plunge Pool" outcrop locality. Riley (1993) presents further details regarding the lacuna on the basal Tocito unconformity. The lacuna between the lower and upper Tocito Sandstone depicted in Figure 3A cannot be resolved biostratigraphically, but it is clearly demonstrable in physical stratigraphic data to be discussed below.

STRUCTURE

Several regional markers, including the top Juana Lopez Member of the Lower Mancos Shale and four internal sequence boundaries in the Tocito, were mapped regionally across the northwest San Juan basin by Jennette and Jones (1995). Figure 4 shows the central region of the five sections they constructed across the subsurface Tocito Sandstone belt (see Fig. 1C for location of section). The section demonstrates that the Mancos Shale interval is uplifted about 75 m immediately seaward of the Gallup Sandstone pinch-out. As documented in Molenaar et al. (1996) this uplift is even greater (about 110 m) across the Four Corners Platform some 30 to 40 km farther to the northwest. As shown by Jennette and Jones

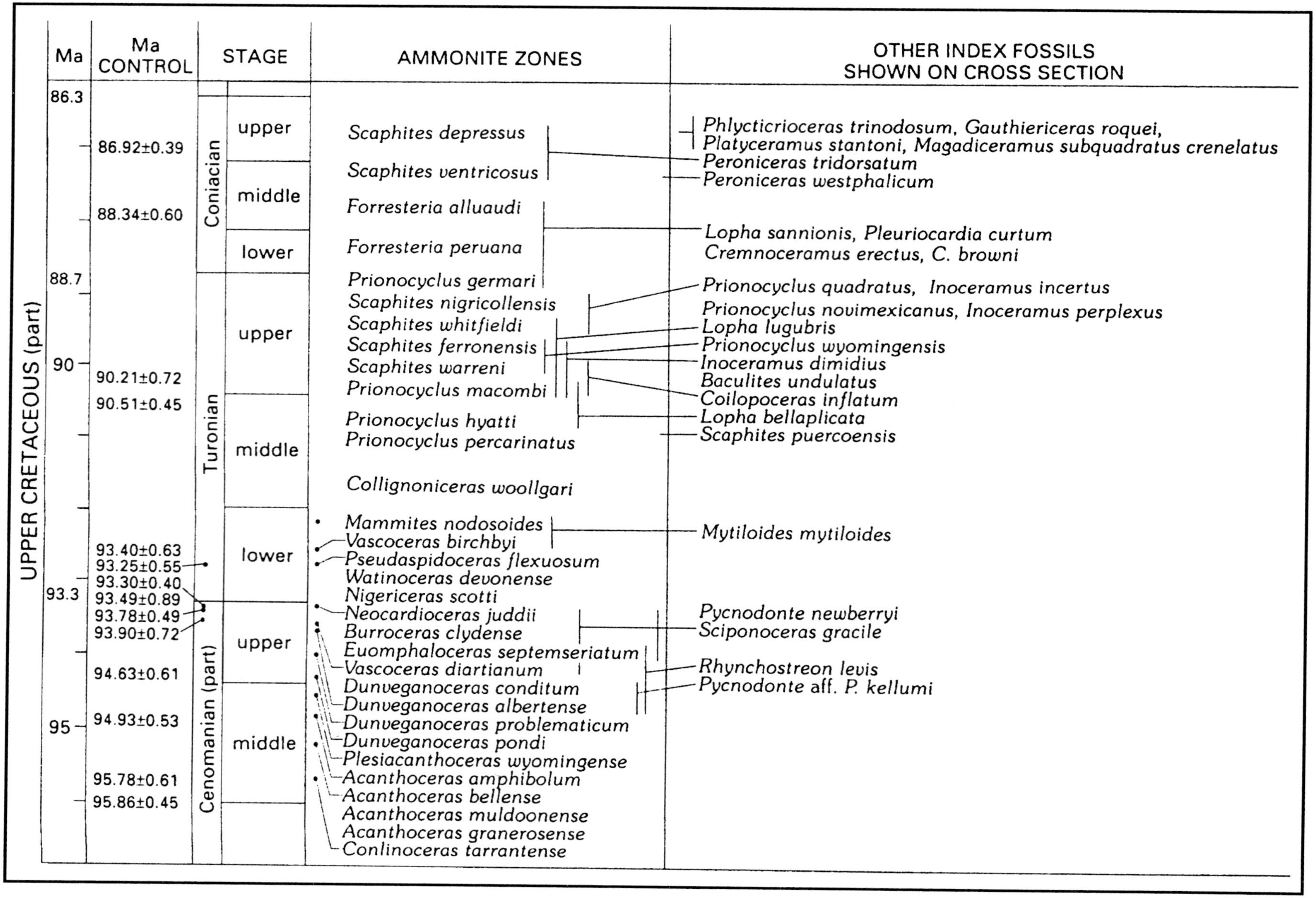

UPPER CRETACEOUS (part)

Ma	Ma CONTROL	STAGE		AMMONITE ZONES	OTHER INDEX FOSSILS SHOWN ON CROSS SECTION
86.3		Coniacian	upper	Scaphites depressus	Phlycticrioceras trinodosum, Gauthiericeras roquei,
	86.92±0.39				Platyceramus stantoni, Magadiceramus subquadratus crenelatus
				Scaphites ventricosus	Peroniceras tridorsatum
			middle		Peroniceras westphalicum
	88.34±0.60			Forresteria alluaudi	Lopha sannionis, Pleuriocardia curtum
			lower	Forresteria peruana	Cremnoceramus erectus, C. browni
88.7		Turonian	upper	Prionocyclus germari	Prionocyclus quadratus, Inoceramus incertus
				Scaphites nigricollensis	Prionocyclus novimexicanus, Inoceramus perplexus
				Scaphites whitfieldi	Lopha lugubris
				Scaphites ferronensis	Prionocyclus wyomingensis
90	90.21±0.72			Scaphites warreni	Inoceramus dimidius
	90.51±0.45			Prionocyclus macombi	Baculites undulatus
			middle		Coilopoceras inflatum
				Prionocyclus hyatti	Lopha bellaplicata
				Prionocyclus percarinatus	Scaphites puercoensis
				Collignoniceras woollgari	
			lower	Mammites nodosoides	Mytiloides mytiloides
	93.40±0.63			Vascoceras birchbyi	
	93.25±0.55			Pseudaspidoceras flexuosum	
	93.30±0.40			Watinoceras devonense	
93.3	93.49±0.89	Cenomanian (part)	upper	Nigericeras scotti	
	93.78±0.49			Neocardioceras juddii	Pycnodonte newberryi
	93.90±0.72			Burroceras clydense	Sciponoceras gracile
				Euomphaloceras septemseriatum	
	94.63±0.61			Vascoceras diartianum	Rhynchostreon levis
				Dunveganoceras conditum	Pycnodonte aff. P. kellumi
			middle	Dunveganoceras albertense	
95	94.93±0.53			Dunveganoceras problematicum	
				Dunveganoceras pondi	
				Plesiacanthoceras wyomingense	
	95.78±0.61			Acanthoceras amphibolum	
	95.86±0.45			Acanthoceras bellense	
				Acanthoceras muldoonense	
				Acanthoceras granerosense	
				Conlinoceras tarrantense	

Fig. 2.—Middle Cenomanian to Coniacian Western Interior ammonite fossil zones and other age-diagnostic molluscan fossils. Numerical time scale and radiometric age control from Obradovich (1993). Figure from Molenaar et al. (1996).

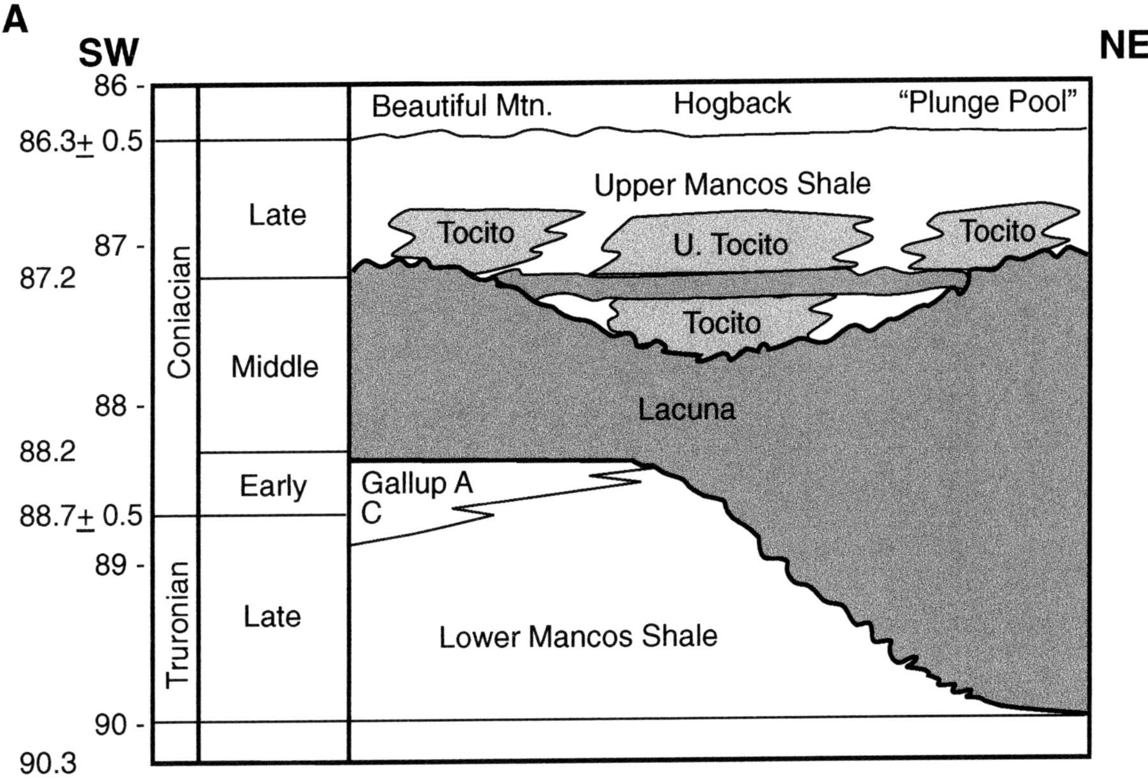

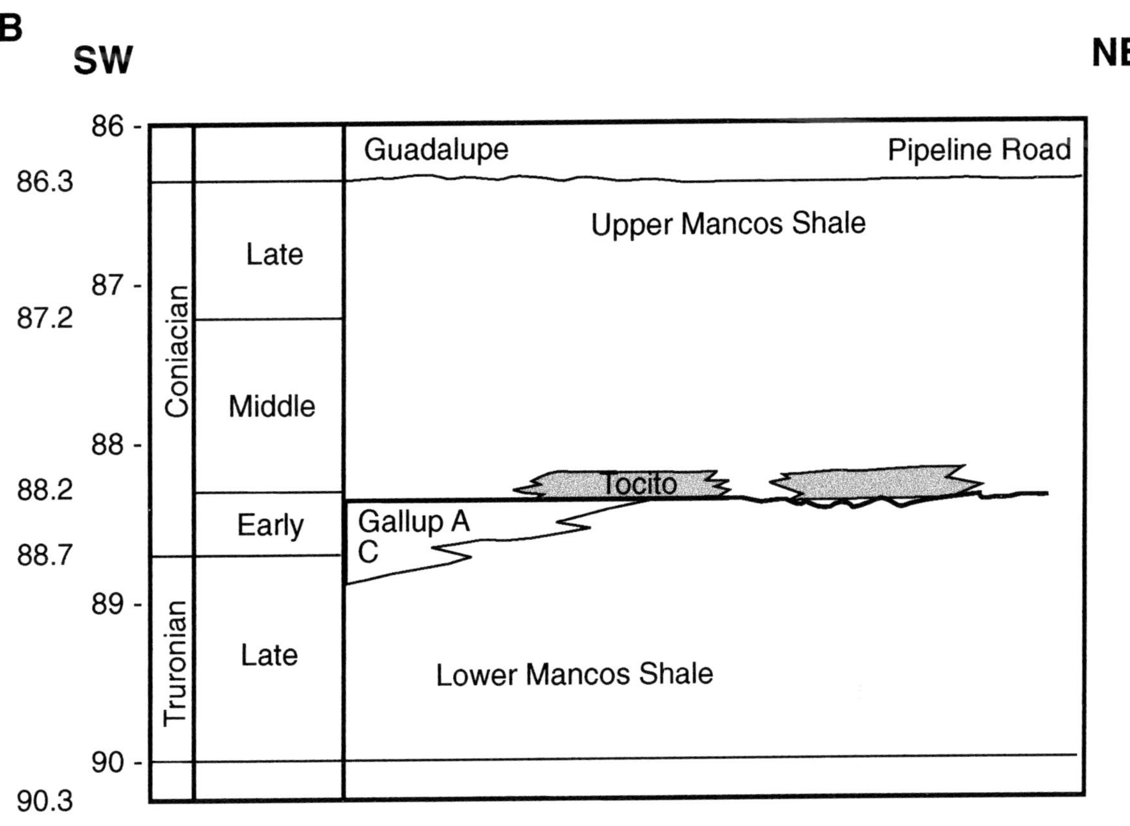

Fig. 3.—Chronostratigraphy of the Gallup and Tocito sandstones across A) the Four Corners Platform, and B) the outcrop belt along the southeast margin of the San Juan basin. Note the wide lacuna separating the two units on the Four Corners Platform, and the lack of a biostratigraphically resolvable break in the southeast. Also, the Tocito Sandstone in the southeast is older than across the Four Corners platform. Ages from Obradovich (1993) and Berggren et al. (1996).

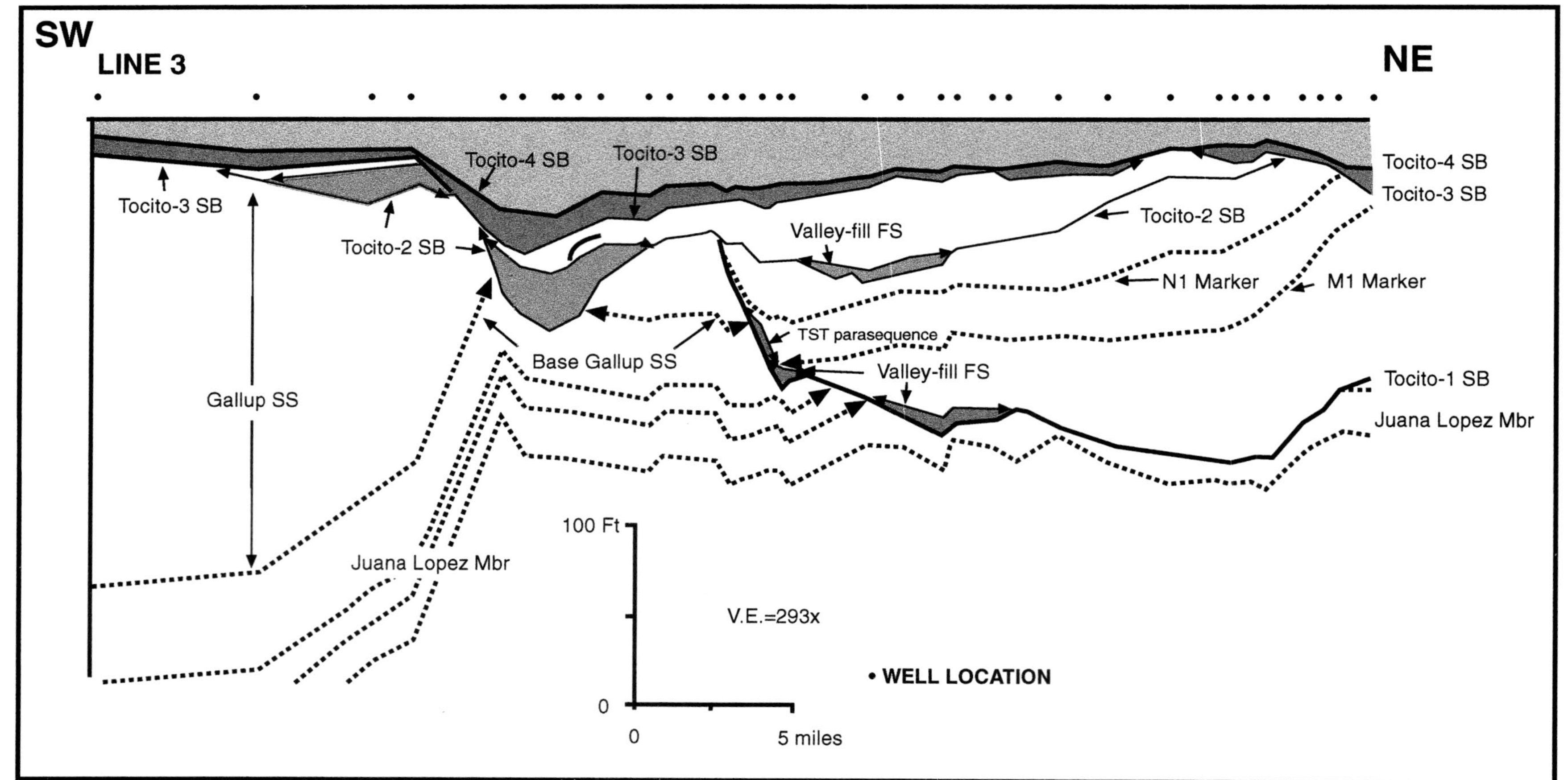

Fig. 4.—Detailed cross-section of the Tocito interval across the Cha-Cha and South Waterflow fields in the northwest corner of the San Juan basin (see Fig. 1 for location of section #3). The top of the Tocito interval is used as datum in the section. Note the structural high at the Juana Lopez and base Gallup levels at the location of the Waterflow oil field, the deep truncation of the Gallup-equivalent Mancos Shale interval toward the northeast at the "basal Tocito unconformity", and the complex pattern of onlap within the Tocito interval onto this surface toward the southwest. Figure modified from Jennette and Jones (1995). Jennette and Jones (1995) were able to differentiate four high-order sequences within the Tocito in the subsurface, only the upper two of which extend onto the outcrop belt of the Four Corners Platform.

(1995), as well as in earlier regional studies (e.g., McCubbin, 1969), the magnitude of this uplift decreases to the southeast away from the section shown in Figure 4. McCubbin (1969) gave the structure the name "Waterflow Anticline."

Basement faults

The regional structural relief documented in all the studies cited above reflects the basement structure on the Four Corners platform. Stevenson and Baars (1986) documented by means of deep well penetrations that the age of the basal Phanerozoic section above basement in the Four Corners region ranges from Cambrian to Pennsylvanian. Mapping these age distributions revealed a set of basement blocks bounded by northwest-southeast and southwest–northeast trending faults (Fig. 5A). Figure 5A shows only a small window of a similar pattern of structural grain characterizing much of the Colorado Plateau. A subsequent study of regional seismic profiles by Huffman and Taylor (1991) showed that the same fault pattern also offsets top of basement (Fig. 5B). There is abundant evidence (Stevenson and Baars, 1986) that these faults were repeatedly reactivated during the many orogenies that affected the Rocky Mountain region during the Phanerozoic. It is therefore not at all surprising, and hardly a coincidence, that movement on some of these faults and uplift of the Waterflow Anticline occurred during the Coniacian, which was a time of active thrusting on the Sevier orogenic belt some 500 km to the west (Pang and Nummedal, 1995).

The regional subsurface studies cited earlier clearly document that the Juana Lopez marker is uplifted across the Waterflow Anticline and areas farther to the northeast, relative to younger (presumably level) maximum flooding surfaces within the Upper Mancos Shale section above the Tocito/Gallup interval as well as with respect to the top of the Tocito (Fig. 4). The time of uplift was Early Coniacian because there is no evidence for thinning of Gallup or equivalent Mancos strata toward the Waterflow Anticline. The initial uplift appears to have extended beyond the area of the Waterflow Anticline proper, because the basal Tocito unconformity can be traced tens of kilometers farther landward. During the subsequent filling episode, subsidence also appears to have been initiated near the crest of the anticline and then gradually expanded to account for the regional southwestward onlap of internal Tocito strata (Fig. 4, and Molenaar and Baird, 1992).

The basement fault pattern, uplift, and subsidence history discussed above suggests to us that the Waterflow Anticline is a drape fold across a footwall block uplifted along fault "WA" in Figure 5A in response to movement in the Sevier thrust belt. The adjacent hanging wall block to the southwest gave rise to the bathymetric trough in which much of the lower Tocito Sandstone was deposited (see below). According to Stevenson and Baars (1986), the basal Phanerozoic section is of Pennsylvanian age south of fault WA and Devonian to the north. This suggests that the direction of movement on the WA fault in the Late Cretaceous was the inverse of its early Paleozoic movement.

STRATIGRAPHY

The Tocito Sandstone forms a series of discrete northwest-to-southeast elongate sandstone bodies encased in shale (Figs. 1, 4; McCubbin, 1969; Molenaar, 1983; Jennette and Jones, 1995). They all overlie the Gallup Sandstone as well as landward and seaward equivalent strata and are commonly but not always separated from these underlying units by an unconformity (Fig. 6). Based on more than 100 measured outcrop sections from Molenaar (1983), Nummedal and Molenaar (1995), Kofron (1987), Bergsohn (1988), and Riley (1993) and lateral tracing of contacts between critical sections, the stratigraphic pattern shown in Figs. 7 - 11 has emerged. Figure 7 emphasizes the regional pattern of the Tocito Sandstone on the Four Corners (NW) and Puerco (SE) platforms bordering the San Juan basin, while Figs. 8-11 present data both of the regional stratigraphy and the internal facies of the Tocito along the San Juan River and its tributaries across the Four Corners platform.

Regional

The Gallup Sandstone and equivalent offshore mudstone facies of the Mancos Shale overlie a distinct regional marine carbonate marker, referred to as the Juana Lopez Member of the Mancos Shale (Fig. 6). The Juana Lopez is very distinctive, both in outcrop and subsurface logs, and is commonly used as a regional reference marker. It is interpreted to be capped by a maximum flooding surface (Nummedal and Molenaar, 1995) that was probably a fairly level surface, representing a gently seaward-dipping ramp along the shallow western margin of the Cretaceous Western Interior Seaway at a time of very low clastic sediment accumulation rates. Where the outcrop extent permitted, all sections in Figs. 7-11 were measured from this regional marker.

Over a time period of about 1.2 m.y., just before the Early Coniacian episode of local, fault-related uplift in the Waterflow Anticline area, the Gallup Sandstone had formed a regional extensive delta complex across much of the southwestern and southern parts of what was to become the San Juan basin (McCubbin, 1982; Molenaar, 1983). This delta complex consists of at least six sequences, each one probably lasting only a few hundred thousand years. Nummedal and Molenaar (1995) interpreted each sequence to have formed as follows. The Gallup shoreface facies and equivalent offshore mudstone facies formed during successive episodes of relative sea level fall and lowstand. The associated Dilco coastal plain strata record dominantly estuarine facies of the transgressive systems tract and poorly preserved highstand fluvial facies. The wedge of Gallup Sandstone and equivalent Mancos Shale depicted in Figures 7A and B represent the distal, lowstand part of this Gallup sequence set.

The relationship of the coarse-grained Torrivio fluvial sandstone bodies to the rest of the Gallup/Dilco sequence set remains controversial. The presence of multiple, stacked Torrivio Sandstone bodies (Fig. 6), the existence of a coarse, "Torrivio-type" sandstone between the Gallup E and F shorefaces just west of the town of Gallup (control point 28A in Molenaar et al., 1996), and the petrographic similarities between Torrivio and Gallup Sandstones argues strongly for the interpretation that the Torrivio and Gallup sandstones are coeval. Paleocurrent data for the Torrivio Sandstone depict a consistent fluvial current flowing toward the northeast, perpendicular to the regional Gallup shorelines (Bergsohn, 1988). Several arguments, however, have also been presented for the view that the Torrivio post-dates the entire Gallup succession. If the first view is correct then the Torrivio is interpreted as a series of stacked, lowstand fluvial valley fills, recording episodes of valley fill during early relative sea level rise for each of the high-frequency Gallup sequences. If the latter view is correct, the Torrivio represents a tectonostratigraphic unit recording changes in the up-dip drainage basin in response to the same uplift that

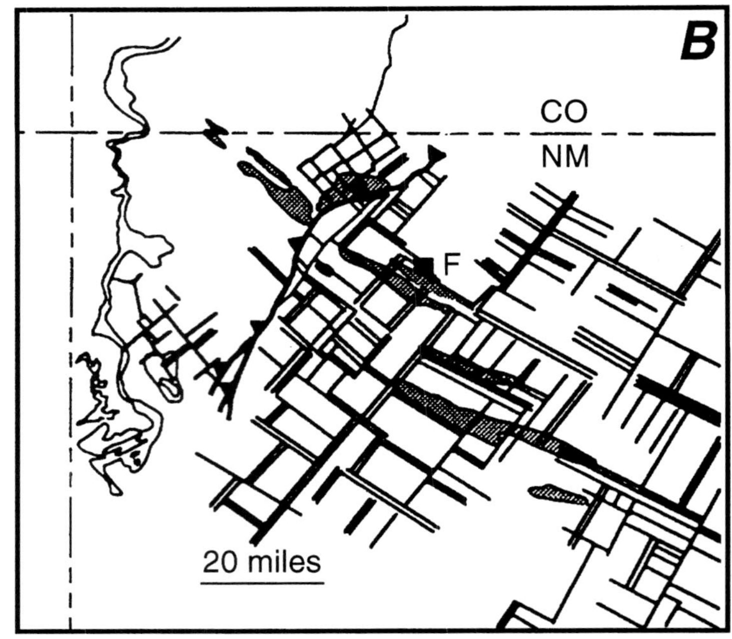

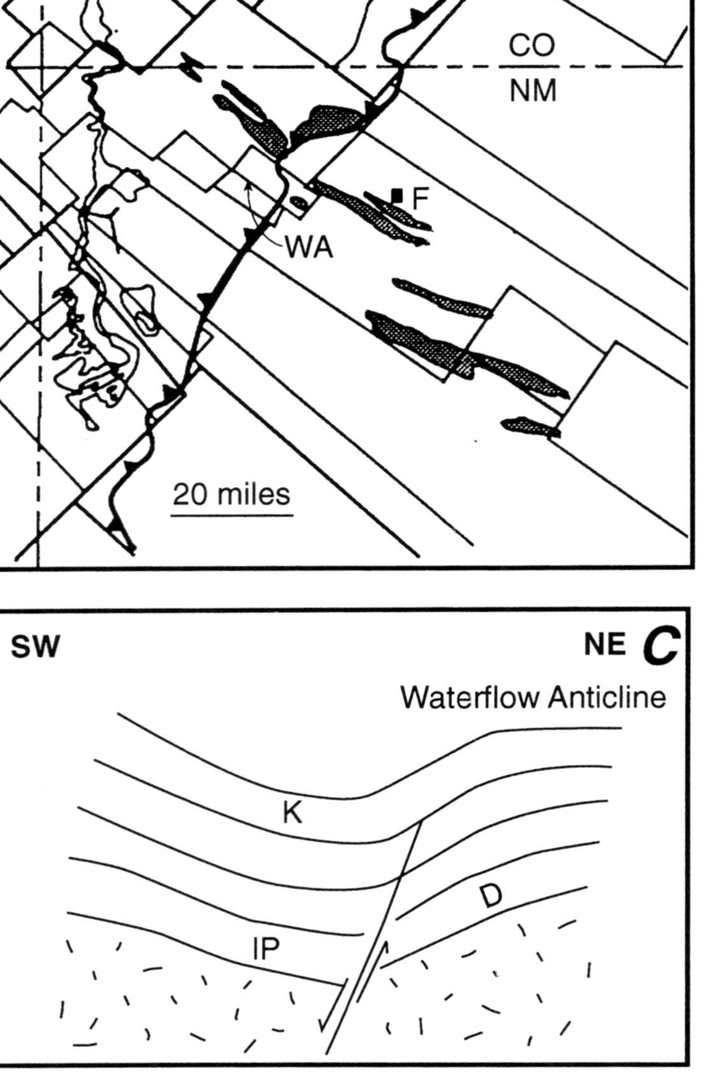

Fig. 5.—A) Map of basement terranes across the Four Corners region derived by mapping the age of the basal Phanerozoic section from deep wells (From Stevenson and Baars, 1986). The terrane map delineates a prominent northwest- and northeast-trending basement fault pattern. B) Top of basement fault pattern across the same region derived from reflection seismic data (From Huffman and Taylor, 1989). C) Interpretation of the Waterflow Anticline as a drape fold across a footwall block uplifted along fault "WA" in Figure 5A in response to movement in the Sevier thrust belt to the west during the Late Cretaceous. The adjacent hanging wall block to the southwest gave rise to the bathymetric trough in which much of the lower Tocito Sandstone was deposited (see below, and Figs. 17 and 18). F designates the location of Farmington, New Mexico.

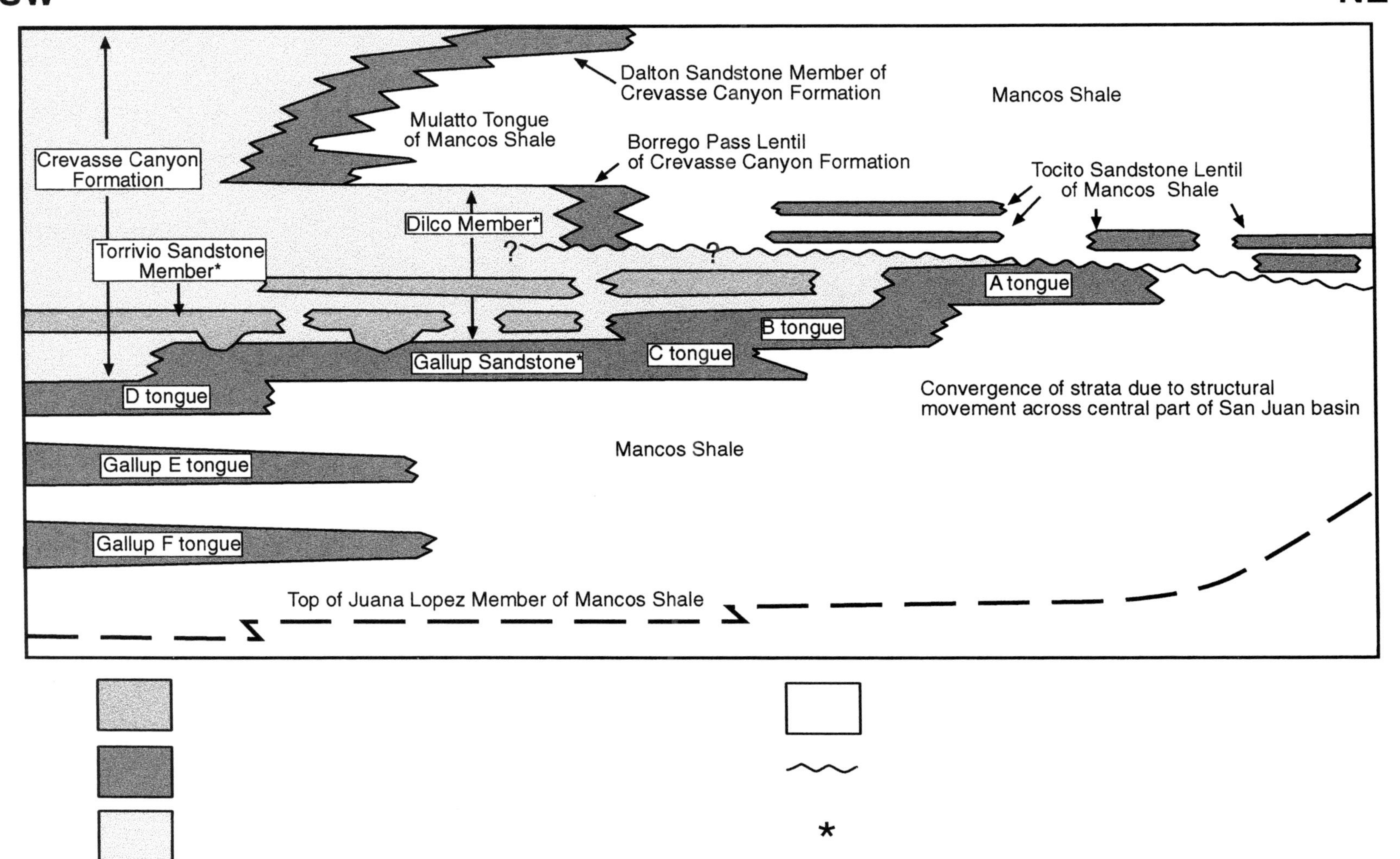

Fig. 6.—Schematic lithostratigraphic cross-section emphasizing the relationship between the Tocito Sandstone and adjacent units. The stratigraphic terminology in this figure was introduced by Nummedal and Molenaar (1995); the formal definitions of the Dilco and Torrivio Sandstone Members differ slightly from earlier usage in that both are now members of the Crevasse Canyon Formation. The basal Tocito unconformity overlies the Torrivio Sandstone Member and extends an unknown distance landward.

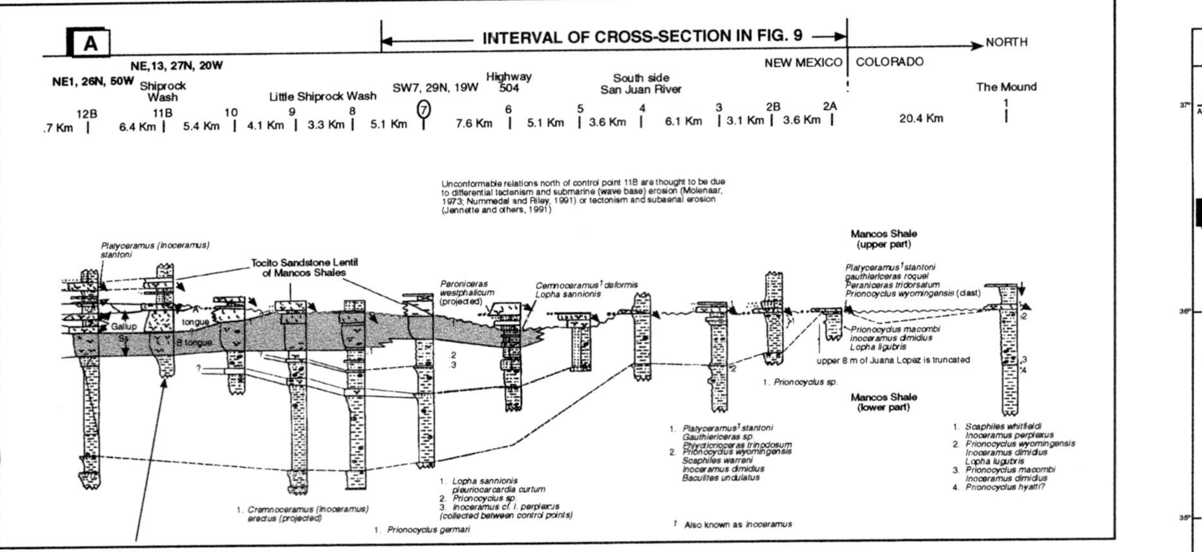

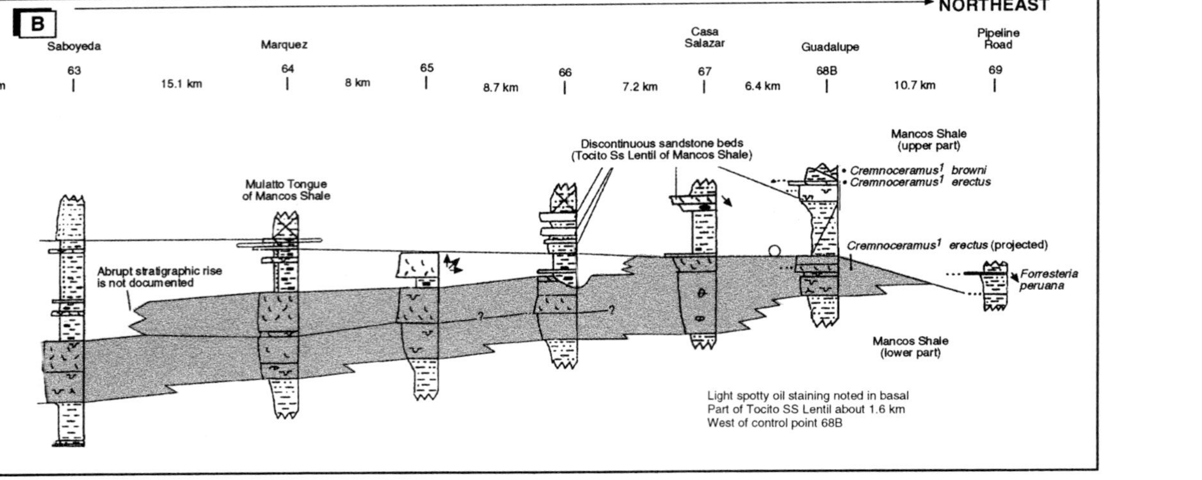

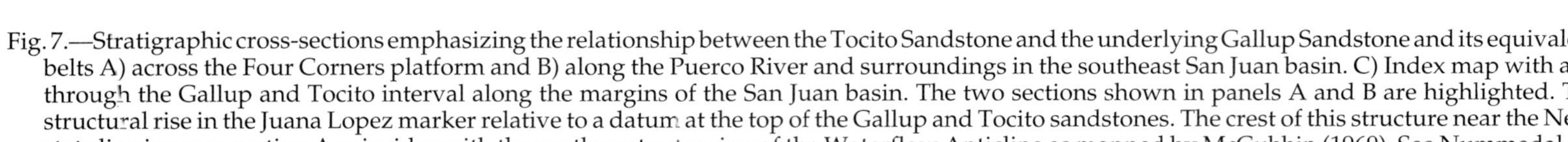

Fig. 7.—Stratigraphic cross-sections emphasizing the relationship between the Tocito Sandstone and the underlying Gallup Sandstone and its equivalent along the outcrop belts A) across the Four Corners platform and B) along the Puerco River and surroundings in the southeast San Juan basin. C) Index map with all measured sections through the Gallup and Tocito interval along the margins of the San Juan basin. The two sections shown in panels A and B are highlighted. There is a prominent structural rise in the Juana Lopez marker relative to a datum at the top of the Gallup and Tocito sandstones. The crest of this structure near the New Mexico-Colorado state line in cross-section A coincides with the northwest extension of the Waterflow Anticline as mapped by McCubbin (1969). See Nummedal and Molenaar (1995) or Molenaar et al. (1996) for complete cross-sections.

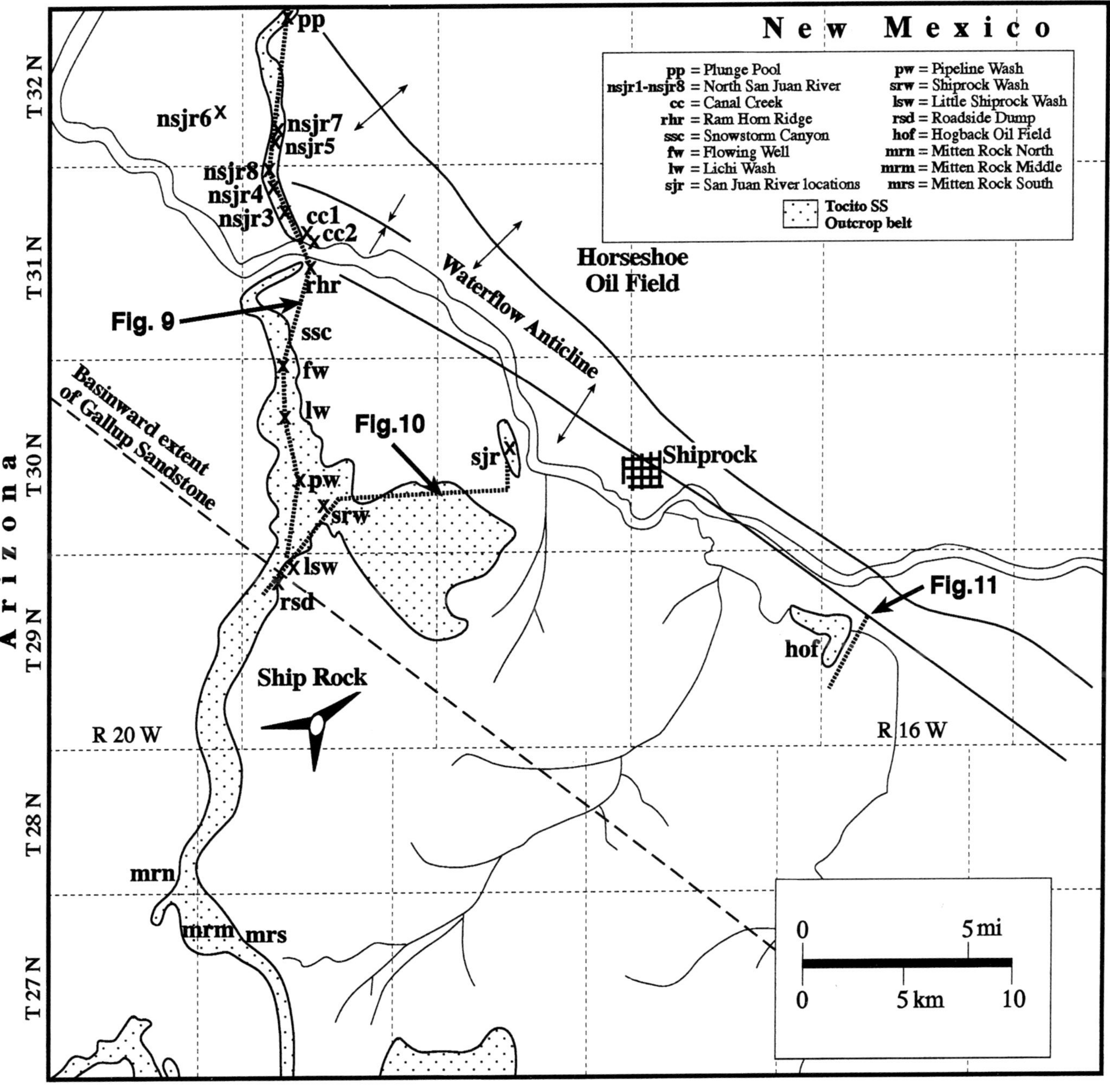

Fig. 8.—Location map for the principal outcrop control of the Tocito Sandstone across the Four Corners Platform. Outcrop belt is stripped. Abbreviations for measured sections used in Tocito cross-sections in this paper are listed below. All sections are in San Juan County, New Mexico. Sections shown in Figures 9, 10, and 11 are highlighted. Section locations: Mitten Rock South (mrs); Mitten Rock Middle (mrm); Mitten Rock North (mrn); Roadside Dump (rsd); Little Shiprock Wash (lsw); Shiprock Wash (srw); Pipeline Wash (pw); Lichi Wash (lw); Flowing Well (fw); Snowstorm Canyon (ssc); Ram Horn Ridge (rhr); Canal Creek (cc); N. San Juan River 3 (nsjr3); N. San Juan River 4 (nsjr4); N. San Juan River 8 (nsjr8); N. San Juan River 5 (nsjr5); N. San Juan River 7 (nsjr7); Plunge Pool (pp); San Juan River (sjr-large number of closely-spaced sections); Hogback Oil Field (hof-large number of closely-spaced sections).

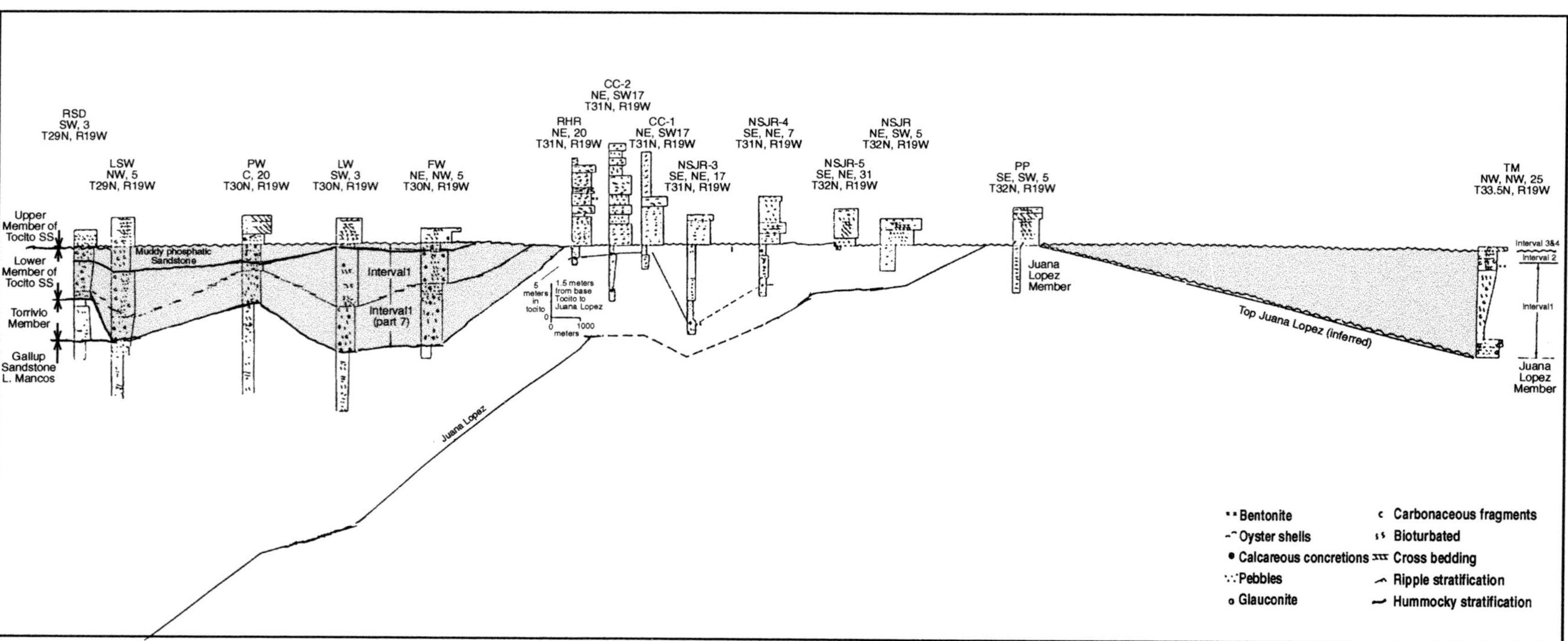

Fig. 9.—South-to-north cross-section of the Tocito Sandstone across the northwestern outcrop belt on the Four Corners platform. See Figure 8 for location. The section includes about half of the total number of measured outcrop sections and is compiled to emphasize the stratigraphic relationships rather than the details of the local facies geometry. The section is flattened on the inferred ravinement surface at the base of the upper member of the Tocito Sandstone. Note the northward truncation of the lower member of the Tocito Sandstone and its total absence across the crestal area of the Waterflow Anticline between the Ramhorn Ridge (rhr) and the Plunge Pool (pp) outcrops. Also note the ridge-shaped cross-section of the upper member. Note difference in vertical and horizontal scales.

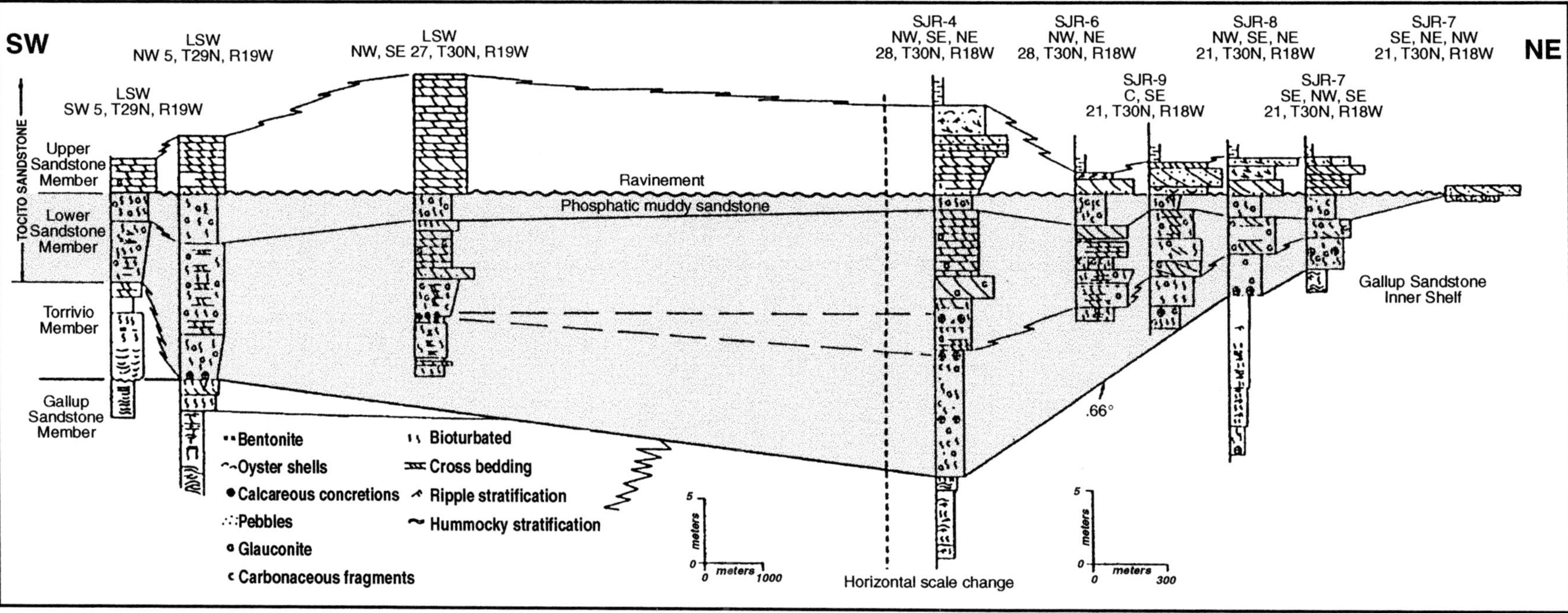

Fig. 10.—Southwest-to-northeast cross-section connecting the outcrops at Ship Rock Wash to the eastern outcrop belt along the San Juan River. See Figure 8 for location. The ravinement surface at the base of the upper member is used as a datum. The distribution of the lower member of the Tocito Sandstone shown in this and the previous figure correspond to the trough above the hanging wall block on the southwest side of the Waterflow Anticline fault depicted in Figure 5C. Note the coarsening-upward textural trend in the lower Tocito member in the deepest part of the trough at section SJR-4, and the presence of mostly coarse-grained (high-energy) facies against the flank of the anticline on the north side. Also note the ridge-shaped geometry of the upper member of the Tocito. Note difference in horizontal and vertical scales.

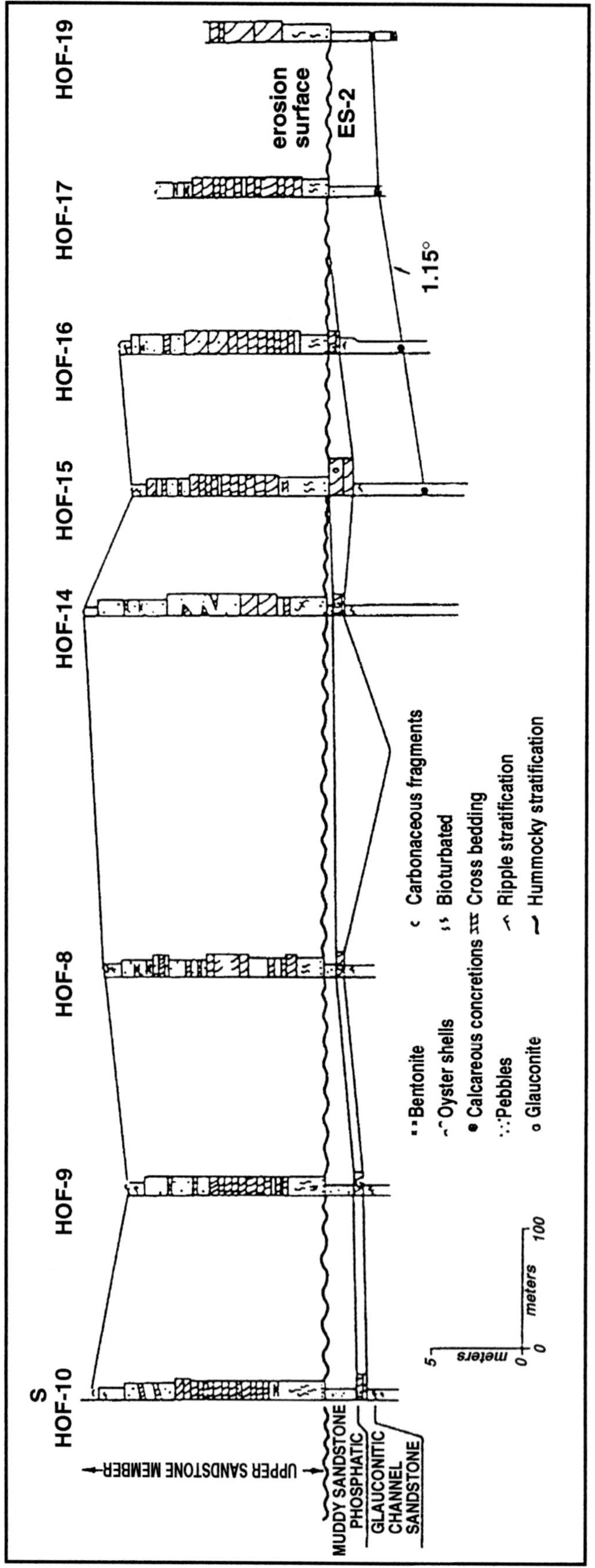

Fig. 11.—South-to-north cross-section along the eastern canyon wall at the Hogback Oil Field outcrop. See Figure 8 for location. The lower member of the Tocito Sandstone here is quite thin and dominated by a glauconitic, cross-bedded sandstone overlain by a muddy sandstone with phosphate pebbles. The ravinement surface at the base of the upper member is inferred to be the same as the one traced in the western outcrop belt (Figs. 9, 10) and is used as the datum. The ravinement truncates the muddy phosphatic sandstone toward the north, according to the same pattern observed along the western outcrop belt. The upper member of the Tocito is particularly well-developed in this outcrop and is generally characterized by a coarsening then fining upward textural trend. Note difference in horizontal and vertical scales.

created the Waterflow Anticline. In either case, coarse-grained Torrivio fluvial facies did reach the most distal Gallup shoreline and directly underlie the marine Tocito Sandstone across much of the northwestern outcrop belt from Beautiful Mountain to Shiprock Wash (Fig. 1C). At several locations in this outcrop belt, the top Torrivio was remobilized by wave action during the marine Tocito transgression, leaving behind large-scale wave ripples at the inferred ravinement surface. Alternatively, it is possible that both interpretations are correct. In this case the lower Torrivio Sandstone would represent updip fluvial valley fills associated with transgression of each Gallup Sandstone Tongue, whereas the uppermost and most extensive Torrivio represents a basinward shift in this facies belt and is coeval with the Tocito Sandstone.

From this evidence and the petrographic similarities between the Torrivio and Tocito sandstones (Riley, 1993) we infer that the Torrivio Sandstone, at least in part, represents proximal, fluvial facies age-equivalent to the Tocito Sandstone. We will return to the specific issue of transformation of this coarse fluvial sand to marine sandbodies below.

Local

Three regional outcrop sections across the Four Corners Platform document the detailed stratigraphy and facies of the Tocito Sandstone in the area between the seaward pinch-out of the Gallup Sandstone and the Waterflow Anticline (Fig. 8). The cross-sections are marked by the three following distinctive surfaces. 1) The basal Tocito unconformity, which everywhere is characterized by erosion (Fig. 12A, B), separates underlying muddy fine sandstone or sandy mudstone of the distal Gallup shoreface facies from overlying medium to coarse-grained sandstone. 2) An internal surface of erosion that overlies a distinctive band of phosphatic muddy sandstone (Fig. 12C, D) and truncates the Tocito section progressively deeper to the northeast such that the phosphatic muddy sandstone band is absent in the northeastern half of the region, north of section FW in Figure 9. This surface separates the informal lower and upper members of the Tocito Sandstone. 3) An abrupt contact between the Tocito Sandstone and overlying Upper Mancos Shale (for example, section SJR-4, Fig. 10) that is commonly characterized by dense patches of the oyster *Pseudoperna congesta,* which unfortunately has an age range that spans the entire Coniacian stage.

The lower Tocito Sandstone forms a northwest-to-southeast-trending sandbody with a maximum isopach thickness in excess of 15 m in the San Juan River outcrops just to the west of the town of Shiprock (Fig. 10). It is abruptly truncated north of the Flowing Well section (Fig. 9), the San Juan River outcrops (Fig. 10), and the north side of the Hogback Oil Field (Fig. 11), whereas it extends as a thin unit (less than 5 m) for an unknown distance to the south. The cross-sectional geometry of this sandbody depends strongly on the choice of datum. The erosion surface at the base of the upper Tocito Sandstone was chosen as datum for the following reasons. Over short distances, such as in the outcrop belt at the Hogback Oil Field (Fig. 11) or at the San Juan River outcrop belt (Fig. 10, right part) this surface is observed to be perfectly level. Also, as will be documented below, this surface is interpreted as a regional ravinement surface. Therefore, due to its inferred origin by wave planation the surface must have been relatively flat when it formed (Nummedal and Swift, 1987). Relative to this chosen datum, the lower Tocito Sandstone occupies an elongate trough that was filled by a generally upward-coarsening succession of marginal marine strata (for example the section at San Juan River 4, Fig. 10) and then capped by a phosphatic muddy sandstone band a few meters thick (Fig. 12C). At the Hogback Oil Field outcrop (Fig. 11), the lower Tocito Sandstone member also contains a sandy channel fill, which is locally enhanced by the fact that a channel-base diastem lies at or near the basal Tocito unconformity in this area. The isopach map of the lower Tocito Sandstone member (Fig. 13A) shows that the axis of maximum thickness parallels the pinch-out edge.

At the "Roadside Dump" (rsd, Fig. 13A) location on the landward flank of the lower Tocito trough (Figs. 9, 13A) one can directly observe the relationship between the Torrivio and Tocito Sandstones. From the base up (Fig. 9, left) one observes first a coarsening-upward section of interbedded HCS-stratified sandstone and mudstone with *Lopha sannionis* oysters, capped by a bed with a well-developed *Glossifungites* ichnofacies. This is the distal shoreface of the Gallup Sandstone. A fluvial (distributary) channel of the coarse-grained Torrivio Sandstone abruptly truncates this succession. The Torrivio, in turn, is overlain by a coarsening-upward, 7 m thick section of the lower Tocito Sandstone, capped by the phosphatic muddy sandstone band and a single bed of the upper Tocito Sandstone. This stratigraphic pattern further supports the inference made above, based on the regional relationships, that the most distal Torrivio postdates Gallup deposition and probably represents proximal (terrestrial) facies of the same age as the Tocito.

The upper Tocito Sandstone also forms northwest- to-southeast-trending sandbodies across the Four Corners platform (Fig. 13B). Two parallel sandbodies crop out in the area, the most landward one attains a maximum thickness of about 10 m east of Shiprock Wash (Fig. 10) and thins to less than 5 m at its basinward margin in the San Juan River outcrop belt and even less along its updip margin at the Roadside Dump locality (Fig. 10). This ridge-shaped, cross-sectional geometry was documented independently by Kofron (1987). The seaward sandbody attains a maximum thickness of 13 m where the western outcrop belt crosses the San Juan River in the "Canal Creek" area (Fig. 13B). A thin sheet of Tocito Sandstone, only a few meters thick, connects the two ridge-shaped isopach thicks to the north and south. The lower Tocito Sandstone member is truncated by erosion just at the south flank of the northern ridge in the upper sandstone, suggesting that erosion of the lower sand acted as a local source for the development of the upper sandstone ridge similar to the way that an ebb-tidal delta core is interpreted to have been the source for the upper, transgressive part of the Holocene Peahala Ridge off New Jersey (Snedden et al., 1995).

At the Hogback Oil Field outcrop the upper Tocito Sandstone member is very well developed, and attains a local maximum thickness of 14 m, thinning rapidly both to the north and the south. Due to inadequate control between the Hogback Oil Field exposures and the more continuous outcrops along the western outcrop belt (Fig. 8), the exact relationship between the Hogback Oil Field sand ridges and the two farther west remains unresolved.

FACIES A — THE LOWER TOCITO SANDSTONE

A rich fossil assemblage of ammonites, oysters, and inoceramids, accompanied by marine authigenic minerals (Riley, 1993) and a diverse suite of shallow marine trace fossils (Nummedal et al., 1993), documents a marine origin

Fig. 12A.—Basal Tocito unconformity dipping to the northeast, toward Waterflow Anticline, in Snowstorm Canyon (see Fig. 8). A thin interval of the lower Tocito Sandstone member onlaps the sequence boundary toward the southwest. The upper member caps the skyline. Bruce Kofron at boundary.

Fig. 12B.—Basal Tocito unconformity at Mark Pasley's hand truncating distal Gallup/Mancos transition facies on the south flank of the Waterflow Anticline.

Fig. 12C.—Phosphate pebbles in sandy mudstone at the top of the lower Tocito member in Shiprock Wash. Scale in cm.

Fig. 12D.—Lag of phosphate pebbles on the ravinement surface at the base of the upper member of the Tocito Sandstone at the Hogback oil field.

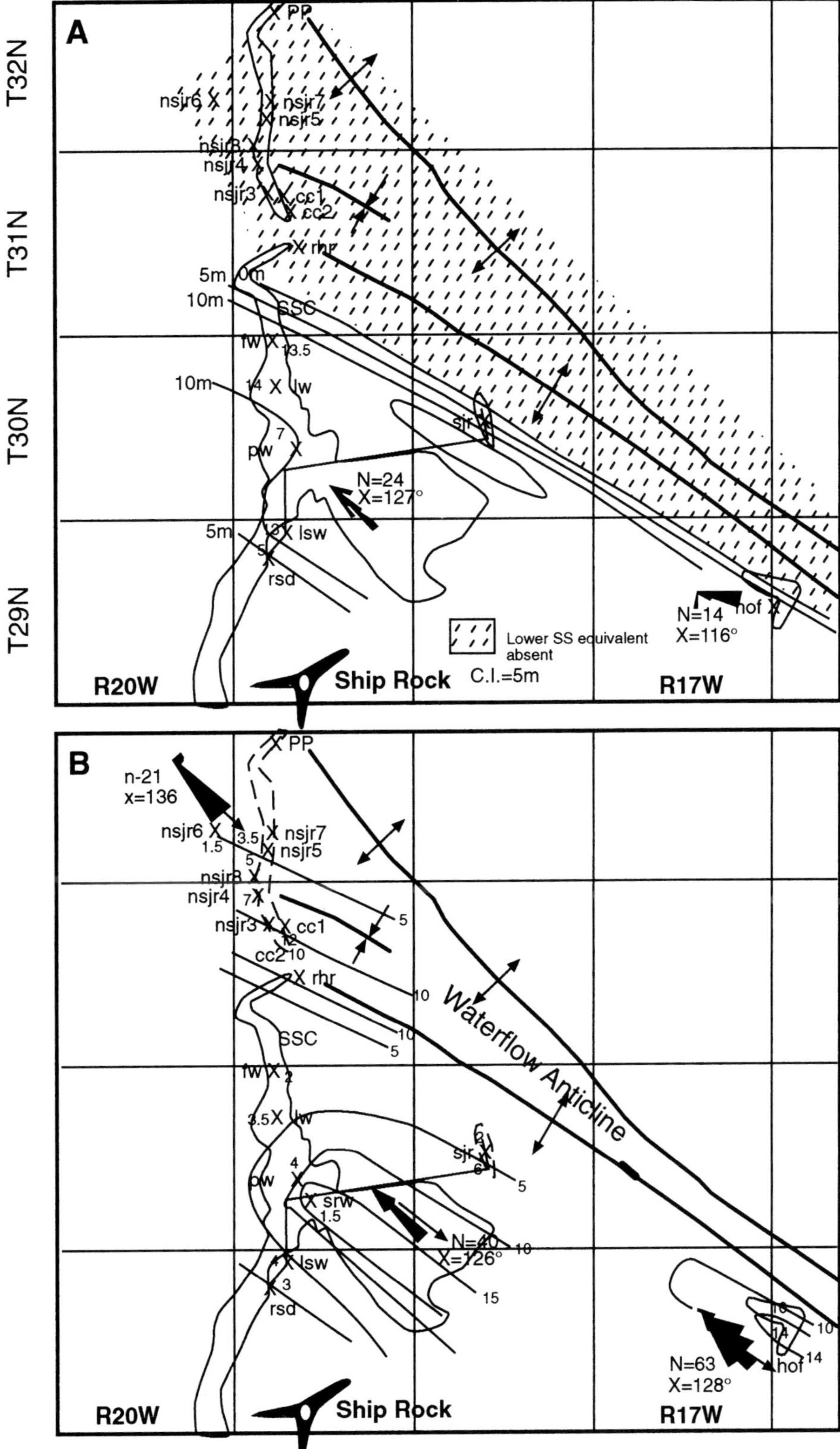

Fig. 13.—A) Isopach map of the lower member of the Tocito Sandstone compiled from outcrop control. The lower sandstone forms a northwest-to-southeast-trending isopach thick just landward (SW) of the Waterflow Anticline. It is entirely absent in a band across the crest of the anticline. Paleocurrents (black rose diagrams) were generally parallel to the axis of the isopach thick, trending a bit more easterly at the Hogback oil field (hof). B) Isopach map of the upper member of the Tocito Sandstone compiled from outcrop control. Underlined numbers record minimum values. The upper sandstone forms several northwest-to-southeast-trending "ridge-shaped" isopach thicks both above and landward of the Waterflow Anticline. Paleocurrents (black rose diagrams) were generally parallel to the axis of the isopach thick in the upper sandstone member.

for both the Tocito Sandstone and the intervening shales. Beyond their marine affinity, the Tocito Sandstone outcrops across the Four Corners platform are highly diverse in facies composition, thickness, and lateral extent.

The lower Tocito Sandstone consists of three lithofacies: 1) bioturbated muddy sandstone (a subfacies of which contains phosphate pebbles, referred to above), 2) burrowed to intensely bioturbated sandstone, and 3) cross-bedded sandstone with set thicknesses ranging from .15 to 2.0 m The cross-bedded sandstone typically contains abundant glauconite.

Bioturbated muddy sandstone

This facies is superficially similar to the Lower Mancos Shale, consisting of bioturbated, muddy sandstone with locally as much as 20% dispersed mud (base of Fig. 14A). The average grain size is less than 200 μm, but quite a few quartz grains up to 1-2 mm are dispersed throughout. Glauconite is ubiquitous in this facies and ranges from 2% to 7 %. Local concentrations of carbonate nodules also occur. Glauconite and dispersed coarse quartz grains differentiate this facies from the subjacent shelf facies of the Mancos Shale. One subfacies of the bioturbated muddy sandstone locally contains high concentrations of phosphate nodules and higher mud percentage; this is the phosphatic muddy sandstone that caps the lower Tocito Sandstone member. Trace fossils in the bioturbated muddy sandstone include *Teichichnus, Terrebellina, Chondrites,* and *Rhizocorallium,* all common members of the *Cruziana* ichnofacies that represent deposit feeding behavior near wave base (MacEachern and Pemberton, 1994).

Electron microprobe analysis of a polished thin section established that the ratio of Fe^{+3} to tetrahedral Al^{+3} of the Tocito grains corresponds to that of true glauconite rather than ferric illite. Microprobe analysis of the phosphate nodules and cements establish that these are composed of calcium fluorapatite (Sabins, 1972; Riley, 1993). The phosphatic nodules and cements also contain abundant quartz and feldspar grains as well as patches of Fe-rich carbonate.

Modern glauconites and phosphates occur mostly in shallow marine settings; locally phosphates also occur in lagoonal and estuarine settings (Odin and Matter, 1981; Cullen et al., 1990). Because sedimentary phosphate and glauconite are believed to form near the sediment-water interface, their presence is generally taken to indicate slow rates of sediment accumulation, which allows sufficient time for authigenesis. Therefore, these minerals are often (but not always) associated with condensed sections (Loutit et al., 1988). The phosphatic muddy sandstone subfacies probably represents a lagoonal or estuarine setting that formed after abandonment of the underlying sandstone distributary channels, landward of (stratigraphically below) the transgressive ravinement at the base of the upper Tocito Sandstone member.

The Gallup/Dilco system in the San Juan basin is about 160 m thick and was deposited in about 1.2 m.y.; its average sedimentation rate was 0.13 mm/yr. In contrast, the Tocito section on the Four Corners platform is generally 10-20 m thick, yet represents a time span of 1-2 m.y.; thus, the average sediment accumulation rates were about 0.01 mm/yr. Moreover, sedimentation rates during deposition of thick, cross-bedded Tocito facies must have been several orders of magnitude higher than that during deposition of some of the bioturbated mudstone. Thus, the sedimentation rate for the phosphatic muddy sandstone was probably much less than 0.01 mm/yr. Therefore, the phosphate- and glauconite-rich facies represents a highly condensed section.

Burrowed sandstone

This facies consists of beds typically 1 to 25 cm thick and separated by 2 to 5 cm thick muddy sandstone interbeds (Fig. 14D). Intense burrowing commonly destroys physical sedimentary structures, but sufficient "ghosts" of cross-bedding remain to infer that this facies originated by episodic migration of two-dimensional dunes. Burrows are concentrated on "pause planes" within cross-bed sets that demonstrate episodic bedform movement.

Cross-bedded sandstone

This facies consists of cross-bedded sandstone with set thicknesses ranging from .15 m to 2.0 m, grain size is typically from 350 μm to 500 μm and glauconite is abundant, locally condensed in cross laminae at more than 50%. Foresets dip to the southeast and show a repetitive pattern of alternating angular and tangential contacts with the toesets. Toesets in turn are generally heterolithic with wavy and lenticular bedding. In addition, these sandstone cross-beds commonly contain mud drapes, reactivation surfaces, and superimposed smaller cross-bed sets that formed in response to secondary(?) flow aligned parallel to the crest of the large two-dimensional dunes. Combined, these structures leave little doubt about deposition of the cross-bedded sandstone by tidal currents. The two sandstone facies contain abundant, primarily vertical burrows of *Skolithos, Thalassinoides, Ophiomorpha,* and rare *Rosselia,* an association that is characteristic of the *Skolithos* ichnofacies and indicative of suspension feeding behavior in higher energy settings (MacEachern and Pemberton, 1994).

Vertical and lateral facies relationships

The three facies of the lower Tocito Sandstone are generally stacked in a coarsening-upward succession, with the thick, glauconitic sandstone beds forming lenticular sandbodies near the top of the sequence (Fig. 15A). Laterally, the bioturbated muddy sandstone facies is best developed in the deepest part of the structural trough landward of the Waterflow Anticline (Fig. 10). Along the shallower flanks only the higher-energy facies are developed.

FACIES B - THE UPPER TOCITO SANDSTONE

The upper Tocito Sandstone is composed of five distinctive lithofacies: 1) burrowed sandstone, 2) interbedded sandstone and heterolithic strata, 3) cross-bedded sandstone with sets ranging from 30 cm to 3 m, 4) pebbly sandstone, and 5) ripple-bedded sandstone. The bioturbated muddy sandstone facies, so dominant in basal parts of the lower Tocito Sandstone, is absent from the upper Sandstone Member. Other facies differences between the two members include the presence of a pebbly sandstone facies, ripple-bedded sandstone facies, and thicker sets of cross-bedded sandstone in the upper member.

The burrowed sandstone facies in the upper member is identical to that in the lower member and is interpreted as a product of migrating straight to slightly sinuous dunes. The trace fossil assemblages are also the same in the burrowed sandstone facies of the upper and lower Tocito members.

Fig. 14A.—All photos on this page depict the facies of the lower member of the Tocito Sandstone. These facies are inferred to represent a tide-dominated delta confined within the trough landward of the Waterflow Anticline. This photo shows a 15-m-thick coarsening-upward succession of the lower Tocito at section SJR-4 (see Fig. 15A for measured section).

Fig. 14B.—Base of distributary channel near top of lower Tocito member at the Hogback Oil Field at Bruce Kofron's hand.

Fig. 14C.—Close-up of glauconitic, distributary channel in the lower Tocito Sandstone. This is unit 2 in section HOF-13 (Fig. 15B).

Fig. 14D.—Typical expression of the bioturbated, muddy sandstone facies ("grunge" facies) in the lower member of the Tocito. Photo taken near base of unit 2 in section SJR-4 (Fig. 15A).

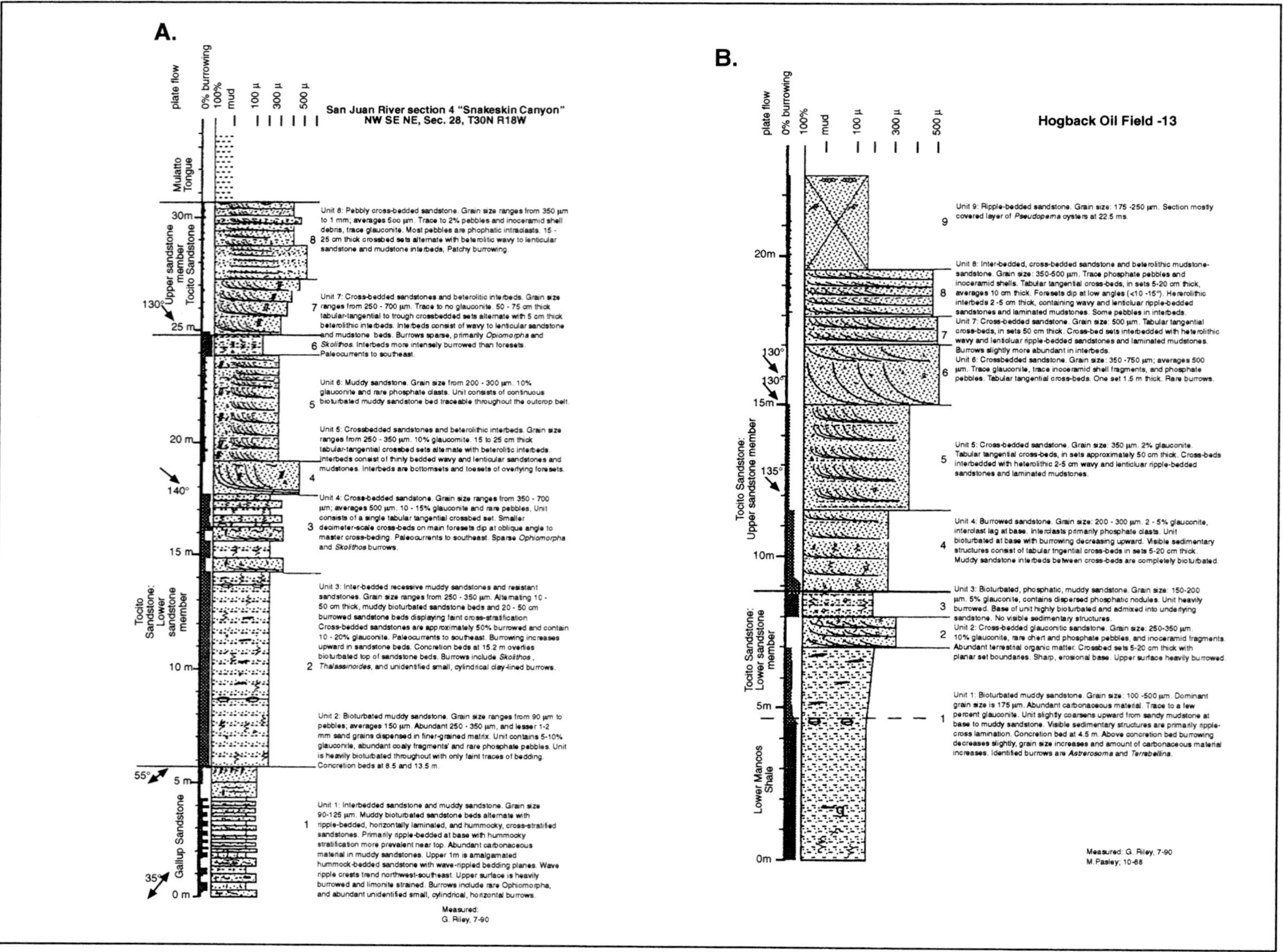

Fig. 15.—A) Measured outcrop section through the Tocito Sandstone at San Juan River, section 4 (See Fig. 8 for location). This is the thickest exposure of the lower Tocito member (19 m thick) and is located near the axis of the trough landward of the Waterflow Anticline. Note the coarsening-upward textural trend in the lower member, the sharp ravinement at the base of unit 7, and the upward-coarsening-then-fining trend in the upper member (which here is only 6 meters thick). B) Measured outcrop section through the Tocito Sandstone at the Hogback oil field, section 13. The contact between the lower Tocito and underlying Mancos Shale is texturally indistinct but easily identified by the presence of glauconite and abundant carbonaceous material in the Tocito. The glauconitic channel depicted in Figure 14 C is unit 2, the ravinement surface at the base of the upper member depicted in Figure 12 D is at the base of unit 4, and the contact with the overlying Mancos Shale is at the base of unit 9.

Horizontally laminated subunits of this facies probably represent preserved top- and bottom-sets of such dunes. Abundant burrows and carbonate cement suggest overall slow sedimentation rates for this facies.

Interbedded sandstone and heterolithic strata

Sandstone beds ranging from 5 cm to 25 cm in thickness are interbedded with heterolithic strata of similar thickness. The heterolithic strata are internally lenticular, wavy, or flaser-bedded (Fig. 16A). In most cases the heterolithic strata can be traced laterally into the toesets of the cross-bedded sandstone facies. Therefore, the interbedded facies is interpreted as distal toe- and bottom-sets of large, migrating dunes. Marine burrows are present, including common members of the *Cruziana* ichnofacies, such as *Thalassinoides, Paleophycus, Planolites*, and *Teichichnus*. These, however, occur in much lower concentrations than in the burrowed sandstone facies discussed above, suggesting that the interbedded facies represents a depositional system with higher sedimentation rates.

Cross-bedded sandstone

Sets of cross-bedded medium to coarse sandstone, ranging in thickness from .3 m to 3 m, are common within the upper Tocito Sandstone facies. They are particularly spectacular along the Chaco River just west of the Hogback oil field, where they include overturned foresets, reactivation surfaces, repetitive sets of bundles bounded by regularly spaced shale drapes, and alternating angular and tangential basal foreset contacts (Fig. 16B). Trace fossils in the cross-bedded facies belong to the *Skolithos* ichnofacies and include *Ophiomorpha, Rosselia, Thalassinoides*, and *Skolithos*. Overall, burrows in this facies are less common than in any other Tocito facies. This facies has been interpreted in several field guides to the area (Nummedal and Wright, 1989; Jennette et al., 1991; Nummedal and Riley, 1991) and papers (Nummedal et al., 1993; Jennette and Jones, 1995). The facies represents a combination of fields of migratory two-dimensional dunes and large-scale solitary three-dimensional dunes (tidal sand bars). The dominant tidal currents flowed toward the southeast, but about 10% of the cross-beds document the presence of a subordinate tidal current component toward the northwest. A particularly impressive perspective on the extent and lateral geometry of the tidal dune field is found near Lichi Wash, where large, gently arcuate dunes representing the upper part of the upper Tocito Sandstone member have been exhumed over an area of more than 3 km^2 (Fig. 16C). Internally, the dune field at Lichi Wash shows well-developed tidal bundles (Fig. 16D). The dominant paleocurrent direction in the Lichi Wash Dune Field was toward the southeast.

Pebbly sandstone

Only the upper Tocito Sandstone member contains sufficient pebbly strata to single them out in a separate facies. The matrix is composed of coarse quartz grains ranging from about 0.3 mm to 2 mm. These are mixed with millimeter- to centimeter-sized pebbles that consist of reworked phosphate clasts, extrabasinal chert pebbles, and abundant broken and abraded inoceramid shell fragments. Burrowing is variable. The facies locally truncates the cross-bedded and interbedded facies. It is interpreted as a "lag facies" associated with the focusing of tidal currents on the upcurrent margin(?) of the Hogback Oil Field sandbody.

Ripple-bedded sandstone

A ripple-bedded facies commonly caps the upper member of the Tocito Sandstone. This is a fine grained sandstone facies (grain size 150-200 µm), often mixed with a few pebbles, and locally carbonate-cemented or stabilized with colonies of *Pseudoperna congesta* oyster shells that are sometimes attached to large fragments of inoceramid shells. Ripples are dominantly wave- and combined-flow ripples. Immediately above this facies is the abrupt transition into the laminated dark gray shales of the Mulatto Tongue of the Mancos Shale. This facies is interpreted to represent the final stage ("moribund" stage, Kenyon et al., 1981) in the growth of the tidal sandbody, at a time when it occupied a water depth beyond the reach of the strong shallow-water tidal currents, or were laterally sheltered from their effects.

The overlying laminated dark shales suggest that the episode of sand ridge building recorded in the upper member of the Tocito Sandstone was followed by very slow hemipelagic sedimentation in an anaerobic or nearly anaerobic environment, probably on a fairly deep seafloor.

Vertical and lateral facies relationships

The details of the facies architecture of the upper Tocito member are complex (Kofron, 1987; Bergsohn, 1988; Riley, 1993) and will not be discussed in this paper. Along the crest of the Waterflow Anticline in outcrops between the Ram Horn Ridge to the North San Juan River region (Fig. 9), the upper Tocito Sandstone shows a fining-upward pattern and is internally composed mostly of medium-scale, cross-bedded, interbedded and ripple-bedded facies. In the thick, ridge-like sandbodies near Shiprock Wash (Fig. 12B) and Hogback Oil Field (Fig. 11-15B) there is a very complex distribution of all the described facies, but a subtle coarsening- then fining-upward pattern prevails.

The basal surface of the upper Tocito Sandstone member is regionally level and erosionally truncates underlying strata toward the northeast (toward the Waterflow Anticline; Fig. 9, 10). The topmost facies of the underlying lower Tocito Sandstone is the phosphatic pebble mudstone, inferred above to represent a low-energy, marginal-marine environment with slow rates of sediment accumulation such as an estuary or a lagoon. The basal facies of the upper Tocito Sandstone is generally a high-energy interbedded or cross-bedded marine facies. Therefore, the erosional surface is inferred to be a regional ravinement associated with shoreline transgression driven, perhaps, by inversion of movement on the fault below the Waterflow Anticline.

GENESIS OF THE TOCITO SANDSTONE

Lower member

Following the end of Gallup coastal plain progradation in the early Coniacian, tectonic movement in the Sevier orogenic belt in southern Utah triggered reactivation of the northwest-trending basement faults across the Four Corners platform. The area around the Waterflow Anticline, which had accumulated distal Gallup prodelta sediments in a water depth of about 100 m, was uplifted about 70-150 m and exposed to erosion by shallow marine tidal currents. Tides, which were regionally fairly strong along the southwestern seaway margin (Ericksen and Slingerland, 1990), were further enhanced by the formation of either an embayment or a strait(?) along this newly tectonically active margin of the

Fig. 16A.—All photos on this page depict the facies of the upper member of the Tocito Sandstone. These facies are inferred to represent a field of shelf sand ridges that formed both across and landward of the Waterflow Anticline during the transgression that accompanied the ultimate subsidence of the anticline. This photo shows a typical expression of interbedded sandstone and heterolithic strata in the Hogback oil field outcrop.

Fig. 16B.—Tidal bundle sets in the upper Tocito along the Chaco River west of the Hogback oil field. Note that crossbeds change rhythmically between angular and tangential basal set contacts. Paleocurrent flows to the southeast.

Fig. 16C.—Airphoto of the arcuate geometry of an exhumed sandwave field at the top of the upper Tocito near Lichi Wash (see Fig. 8 for location). Paleoflow (SE) was toward the right side of the photo. Photo reproduced from Campbell (1969).

Fig. 16D.—One of the tital bundles near Lichi Wash in cross-section. Thick foresets near Don Swift's hand are inferred spring tide bundles; the thinner, more bioturbated bundles on either side of the photo probably represent neap tides.

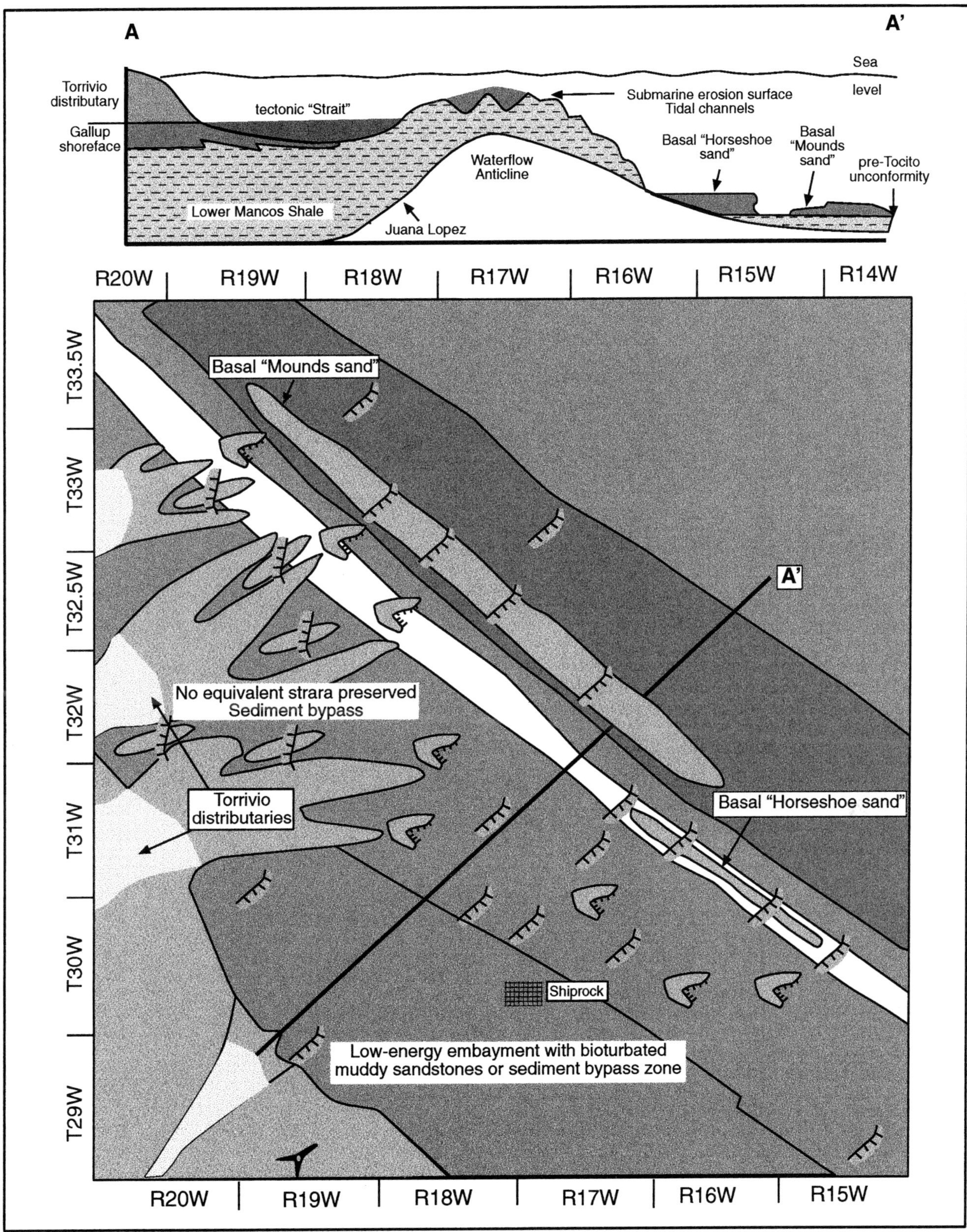

Fig. 17.—Model for the depositional system of the lower member of the Tocito Sandstone and contemporaneous tide-influenced lower reaches of the Torrivio alluvial valley. Note that the Torrivio fluvial channels entered the littoral zone essentially perpendicular to the trough that had formed landward of the Waterflow Anticline. The trough is inferred to overlie a subsiding hanging-wall block as depicted in Fig. 5C. Strong tidal currents in the trough maintained a mostly aggradational and locally progradational tide-dominated delta. Paleocurrents were structurally confined and flowed to the southeast.

seaway. Tidal currents were topographically constrained to flow parallel to the crest of the uplifted anticline as well as the shoreline; unidirectional flows to the southeast were dominant (Fig. 17).

The fine-grained distal shoreface to shelf facies of the Mancos Shale was removed by tidal erosion and redeposited as part of the fine-grained facies within the Tocito interval farther downcurrent to the southeast. The Torrivio fluvial channels that flowed across a regional coastal plain toward the northeast became tide-influenced distributaries along the seaway margin. In these distributaries, sand was remobilized into tidal sand bars across the high-energy front of a tide-dominated delta. Once the sand entered the marine domain, the paleocurrents turned to the southeast in response to the shoreline-parallel tidal currents. A near balance between rate of sediment supply from the Torrivio streams and rate of subsidence in the trough, combined with an efficient marine reworking of the fluvial supply, prevented the shoreline from prograding very far into the basin. Only locally, mostly in the axial zone of the tidal troughs, is there stratigraphic evidence for significant delta progradation. Generally, the lower Tocito Sandstone is characterized by an aggradational stacking pattern across most of the trough landward of the Waterflow Anticline.

The cross-sectional geometry of the lower Tocito Sandstone, therefore, reflects both the concave erosional bases of tidal-scoured troughs as well as the convex tops of the tidal sand bars that fill the top of the section within these troughs. The basal Tocito unconformity is a product of this regional tidal-current erosion.

The asymmetric thickness of the lower Tocito Sandstone across the Four Corners platform (Figs. 9, 10, 13A) is a measure of the differential subsidence that occurred across the hanging wall block landward of the Waterflow Anticline fault. The subsiding hanging wall created increased accommodation space that was filled with essentially aggradational facies of the distal, tide-dominated delta front. Because the deepest preserved part of this hanging wall basin must have subsided the fastest, this location initially filled with the deepest water environment and accumulated the thickest section of the bioturbated muddy sandstone facies. This stratigraphic pattern should be emphasized, because if the basal Tocito sequence boundary was formed by subaerial (fluvial) erosion, as has been suggested (Jennette et al., 1991), the deepest parts of the erosional trough would have been the preferred site for accumulation of coarse fluvial channel fill, not deepwater marine facies, as observed.

The early Coniacian tectonic event probably began with broad regional uplift, causing shallowing across much of the Four Corners platform, as suggested by the broad trend of uplift in the Juana Lopez marker (Fig. 7). Marine erosion would have been widespread. Late in the uplift phase, the Waterflow Anticline fault (Fig. 5C) had propagated upward nearly to the surface, causing differential subsidence across the abrupt boundary between the anticline and the hanging wall trough on its landward side. As the footwall rose, the region northeast of the fault experienced continued relative sea-level fall and a coarse lag accumulated on the sequence boundary in response to submarine tidal erosion. Concurrently, subsidence occurred above the hanging wall to the southwest while a rapid sediment influx from Torrivio streams kept the top of the sedimentary column nearly at sea level, building the aggradational stacking pattern of the tide-dominated delta. It follows from this model that erosion on the basal Tocito sequence boundary across the high Waterflow Anticline continued long after deposition of the lower Tocito Sandstone had begun within the adjacent trough.

Upper member

The upper Tocito Sandstone member is interpreted as a set of transgressive shelf sand ridges (Fig. 18) for the following reasons. The sandbodies trend northwest-to-southeast, parallel to the inferred paleoshoreline, their ridge-shaped, cross-sectional geometry merges into thin sand sheets, their body and trace-fossil assemblages are fully marine, their internal sedimentary structures are strongly indicative of tides, and their paleocurrents are consistent with shore-parallel flow to the southeast. Numerous detailed studies of modern shelf sand ridge systems document that all the listed attributes are expected in sand ridge fields on tide-dominated shelves (Swift, 1976; Stride, 1982; Belderson, 1986). Moreover, as documented by Berne et al. (1988), many of the sand ridge fields on modern shelves directly overlie estuarine or regressive, locally tide-dominated deltaics where the ridge sand was quarried from during the erosional retreat of the shoreface. Thus, the inferred association of transgressive shelf sand ridges of the upper Tocito Sandstone that erosionally overlie a tide-dominated delta succession of the lower member is fully consistent with the documented evolution of similar complexes on Holocene shelves.

SEQUENCE STRATIGRAPHIC LESSONS

The geologic interpretation of the Tocito Sandstone advanced in this paper was achieved by integration of local (Jennette and Jones, 1995) and regional (Pang and Nummedal, 1995) structural and tectonic understanding, precise megafossil-based biostratigraphy (Molenaar et al., 1996), extensive subsurface and outcrop stratigraphic studies (Jennette and Jones, 1995; Riley, 1993), and the collection of detailed sedimentologic information and its interpretation in the context both of hydrodynamics and models for modern depositional systems (Riley, 1993). The resulting interpretation, while consistent with stratigraphic principles, would have been difficult to achieve by starting with existing sequence stratigraphic models. Such models are based on many assumptions that are invalid in a setting of syn-depositional structural movement, such as what occurred in the Late Cretaceous across the Four Corners platform. Here we address the specifics of how some of these differences are expressed.

First, the basal Tocito sequence boundary deserves some attention. This surface forms a major regional erosional unconformity across a wide belt of the Four Corners platform and the adjacent San Juan basin (Molenaar and Baird, 1992; Jennette and Jones, 1995; Riley, 1993). Across the unconformity, shallow-water sandy marine sediments overlie deeper-water marine mudrock. The surface meets all the criteria of a sequence boundary, yet there is no evidence that it was formed by subaerial erosion. No fluvial sediments are found anywhere, no paleosols, nor "lowstand deltas" at the downcurrent terminus of the "incised valleys." This sequence boundary formed in response to regional tectonic uplift and scour by tidal and shallow-marine storm currents, and it may not have any direct relationship to a eustatic lowstand.

Second, as uplift of the Waterflow Anticline approached its climax in response to fault block movement, a subsiding trough developed on the adjacent hanging wall block. In such settings, continued footwall uplift further enhanced the sequence boundary there, while concurrent local sea level rise

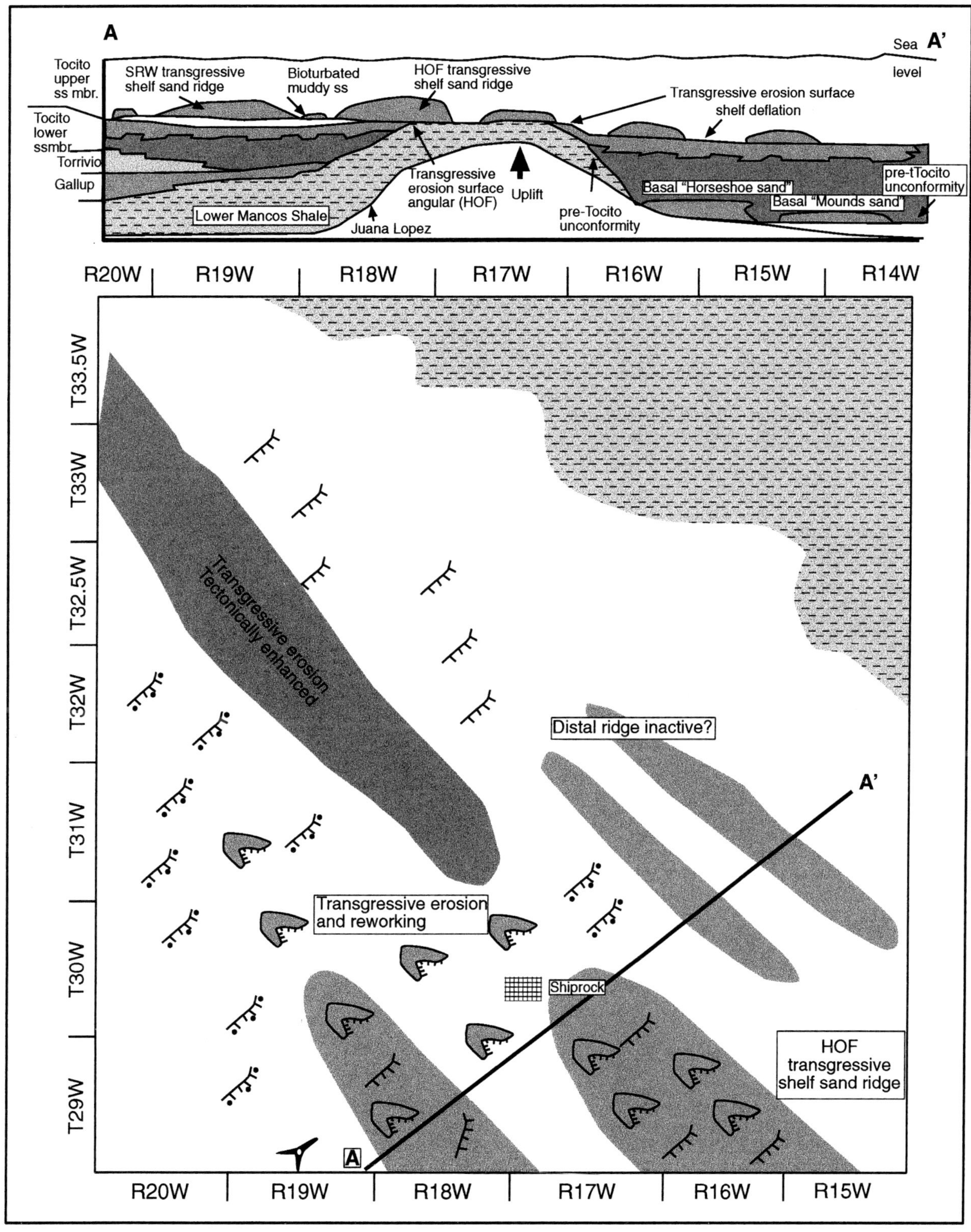

Fig. 18.—Model for the depositional system of the upper member of the Tocito Sandstone. At this stage (about one million years later than in Fig. 17), both the Waterflow Anticline and surrounding areas on the Four Corners Platform were all subsiding. A landward-encroaching ravinement surface (top panel) truncated the underlying tide-dominated delta and distal fluvial Torrivio deposits and redeposited a series of shelf sand ridges. The lower Tocito Sandstone, if it ever were deposited across the crest of the Waterflow Anticline, was preserved only in the structural low and deeper areas seaward of the anticline (not discussed in this paper), whereas the upper member of the Tocito forms a widespread transgressive sand across much of the Four Corners platform.

prevailed in the adjacent half graben. Such is the case landward of the Waterflow Anticline, where the lower Tocito Sandstone accumulated in such a shore-parallel half graben. The tide-dominated deltas that formed in this trough probably received sediments both from the landward hangingwall ramp in the form of Torrivio distributaries and from concurrent erosion due to sea level lowering across the still submerged(?) Waterflow Anticline ridge to the northeast (Fig. 18). If relative sea level was in fact falling across the seaward anticline while it was rising across the hangingwall block, it is a bit difficult to assign the proper systems tract to the lower member of the Tocito Sandstone. Here we use the systems tract definitions loosely and refer to the lower Tocito Sandstone member as a lowstand systems tract. The observed Tocito stacking pattern of aggradation or local progradation is geometrically consistent with this classification.

Third, the uplift phase ended during the middle Coniacian, and renewed subsidence again became prevalent across the region. This subsidence was accompanied by regional transgression and the deposition of a zone of shelf sand ridges that extended far beyond the more limited area of earlier Tocito sedimentation (Jennette and Jones, 1995). Sequence stratigraphic terminology causes no problem at this stage; the upper Tocito Sandstone member is clearly in the transgressive systems tract. On top of the upper Tocito is the condensed section at the base of the Mulatto Tongue of the Mancos Shale (Pasley et al., 1993). This condensed section probably contains the maximum flooding surface of this sequence and correlates lithostratigraphically to the base of the Dalton regressive sandstone near Gallup (Nummedal and Molenaar, 1995). Unfortunately, no age-diagnostic fossils have yet been retrieved near this landward pinch-out of the Mulatto Tongue.

Fourth, the relationship between the basal Tocito unconformity and unconformities close in age elsewhere in the Western Interior presents a few intriguing problems. Because erosion of the basal Tocito unconformity was structurally driven there is no physical reason why unconformities of the same age should exist elsewhere along the margin of the same seaway. For example, the "basal Niobrara unconformity," which is biostratigraphically very well constrained at Pueblo, Colorado, is Late Turonian in age (Kauffman et al., 1993), is clearly older. That unconformity appears to correlate to a relative sea-level fall along the southwestern San Juan basin that triggered the most significant regression of the Gallup shoreline, the Gallup C-tongue (Nummedal and Molenaar, 1995). On the other hand, because the Early Coniacian uplift across the Four Corners platform probably was driven by movement in the southern parts of the Sevier thrust belt (Pang and Nummedal, 1995), regional unconformities of correlative age and origin may exist within Coniacian strata elsewhere in the Rocky Mountains. Perhaps the unconformity at the base of the Calico beds in the Kaiparowits Plateau, reported by Shanley et al. (1992), is of related age and origin.

CONCLUSIONS

The Tocito Sandstone is entirely Coniacian in age, and it is separated from the underlying Gallup Sandstone on the Four Corners Platform by a major unconformity that locally is associated with a lacuna of as much as 3 m.y. In the southeast San Juan basin there is no demonstrable biostratigraphic gap.

Formation of the basal Tocito sequence boundary and deposition of the coarse-grained Tocito Sandstone itself are both intimately associated with far-field tectonism in the Sevier thrust belt to the west and near-field reactivation of basement fault blocks across the Four Corners platform. The sequence boundary formed by shallow marine erosion during uplift of the Waterflow Anticline block. Delivery of coarse-grained sediment to the shoreline was possibly accentuated by tectonism in the upper Torrivio drainage basins and was certainly facilitated by the enhanced tidal currents in straits and bays along the tectonically reactivated shoreline.

Most of the lower member of the Tocito Sandstone was deposited in a tide-dominated delta within a bay or strait sheltered from oceanic waves by the Waterflow Anticline high. At the very top of the lower member lies a thin wedge of phosphatic, pebbly mudstone that was deposited in an estuary. The upper member of the Tocito was deposited as a series of sand ridges on a tide-dominated shelf during a regional transgression driven by reversal of movement on the Waterflow Anticline fault. A distinct ravinement surface separates the lower and upper members across the entire Four Corners platform. Much of the lower member has been truncated by this ravinement surface across the Waterflow Anticline high.

The Tocito Sandstone has excellent porosity and is the major oil reservoir rock in the San Juan basin. Documentation of the Tocito sand body and shale body geometries, therefore, will help us better understand fluid flow in such reservoirs, which are probably more numerous than is currently recognized.

ACKNOWLEDGMENTS

Field work on the Tocito Sandstone was supported over many years by Amoco, BP, Conoco, Mobil and Unocal. We owe our many stratigraphy colleagues at those companies our sincere gratitude for their support. Funding was also received from National Science Foundation Grant EAR - 9205811 to D. Nummedal. G. W. Riley particularly appreciates a dissertation grant from Mobil. Donald Swift, Roderick Tillman, and the late K. Molenaar greatly stimulated our interest in the Tocito problem and shared many keen insights into shallow marine sedimentation and Cretaceous stratigraphy of the Rocky Mountains. We also thank the many students at Louisiana State University who contributed in numerous ways to the completion of this project, especially Mark Pasley, Paul Templet, Ming Pang, Bruce Kofron, and Ivo Bergsohn.

REFERENCES

Belderson, R.H., 1986, Offshore tidal and non-tidal sand ridges and sheets: Differences in morphology and hydrodynamic setting, *in* Knight, R.J., and McLean, J.R., eds., Shelf Sands and Sandstones: Calgary, Canadian Society of Petroleum Geologists Memoir II, p. 293-301.

Bergsohn, I., 1988, Lithofacies architecture of the Tocito Sandstone, northwest New Mexico: Unpublished M. Sc. Thesis, Baton Rouge, Louisiana State University, 170 p.

Berne, S., Auffret, J.P. and Walker, P., 1988, Internal structure of subtidal sandwaves revealed by high resolution seismic: Sedimentology, v. 35, p. 5-20.

Campbell, C.V., 1979, Model for Beach Shoreline of Gallup Sandstone (Upper Cretaceous) of Northwestern New Mexico: Socorro, New Mexico Bureau of Mines and Mineral Resources Circular 164, 32 p.

Cobban, W.A., 1984, Mid-Cretaceous ammonite zones, Western Interior, United States: Bulletin of the Geological Society of Denmark, v. 33, p. 71-89.

CULLEN, D.J., CHALLIS, G.A. AND DRUMMOND, G.W., 1990, Late Holocene estuarine phosphogenesis in Raglan Harbour, New Zealand: Sedimentology, v. 37, p. 847-857.

ERICKSEN, M.C. AND SLINGERLAND R., 1990, Numerical simulation of tidal and wind-driven circulation in the Cretaceous Interior seaway of North America: Geological Society of America Bulletin, v. 102, p. 1499-1516.

GRADSTEIN, F.M., AGTERBERG, F.P., OGG, J.G., HARDENBOL, P., VAN VEEN, P., THIERRY, J. AND HUANG, Z. 1995, A Triassic, Jurassic and Cretaceous time scale; *in* Berggren, W.A., Kent, D.V., Aubry, M.P. and Hardenbol, J., eds., Geochronology, Time Scales and Global Stratigraphic Correlation: Tulsa, Society For Sedimentary Geology (SEPM) Special Publication 54, p. 95-126.

HUFFMAN, A.C., JR. AND TAYLOR, D.J. 1991, Basement fault control on the occurrence and development of San Juan basin energy resources: (abstract), Geology Society of America, Rocky Mountain Section meeting, Abstracts with Programs, v. 23, p. 34.

JENNETTE, D.C. AND JONES, C.R., 1995, Sequence stratigraphy of the Upper Cretaceous Tocito Sandstone: A model for tidally influenced incised valleys, San Juan basin, New Mexico, *in* Van Wagoner, J.C. and Bertram, G.B., eds., Sequence Stratigraphy in Foreland Basins: Tulsa, American Association of Petroleum Geologists Memoir 64, p. 311-347.

JENNETTE, D.C., JONES, C.R., VAN WAGONER, J.C. AND LARSEN, J.E., 1991, High-resolution sequence stratigraphy of the Upper Cretaceous Tocito Sandstone: The relationship between incised valleys and hydrocarbon accumulation, San Juan basin, New Mexico, *in* Van Wagoner, J.C., Nummedal, D., Jones, C.R., Taylor, D.R., Jennette, D.C. and Riley, G.W., eds., Sequence Stratigraphy Applications to Shelf Sandstone Reservoirs - outcrop to subsurface examples: Tulsa, American Association Petroleum Geologist Field Conference Guidebook. Not consecutively paginated.

KAUFFMAN, E.G., SAGEMAN, B.B., KIRKLAND, J.I., ELDER, W.P., HARRIES, P.J. AND VILLAMIL, T., 1993, Molluscan biostratigraphy of the Cretaceous Western Interior basin, North America, *in* Caldwell, W.G. and Kauffman, E.G. eds., Evolution of the Western Interior Basin: St. John's, Geological Association of Canada Special Paper 39, p. 397-434.

KENYON, N.H., BELDERSON, R.H., STRIDE, A.H., JOHNSON, A.H., 1981, Offshore tidal sandbanks as indicators of net sand transport and as potential deposits: Oxford, International Association of Sedimentologists Special Publication 5, p. 257-268.

KOFRON, B.M., 1987, Facies characteristics of the Upper Cretaceous Tocito Sandstone, San Juan basin, New Mexico, Unpublished, M. Sc. Thesis, Baton Rouge, Louisiana State University, 104 p.

LOUTIT, T.S., HARDENBOL, J., VAIL, P.R. AND BAUM, G.R., 1988, Condensed sections: The key to age determination and correlation of continental margin sequences, *in* Wilgus, C.K., Hastings, B.S., Kendall, St. C., Posamentier, H.W., Ross, C.A. and Van Wagoner, J.C., eds., Sea-Level Changes: An Integrated Approach: Tulsa, Society for Sedimentary Geology (SEPM) Special Publication 42, p. 184-213.

MACEACHERN, J.A. AND PEMBERTON, S.G., 1994, Ichnological aspects of incised valley-fill systems from the Viking Formation on the western Canada sedimentary basin, Alberta, Canada, *in* Dalrymple, R. W., Boyd, R. and Zaitlin, B.A., eds., Incised-valley Systems: Origin and Sedimentary Sequences: Tulsa, Society for Sedimentary Geology (SEPM) Special Publication 51, p. 129-157.

MAXWELL, C.H. ANDERSON, O.J., LUCAS, S.J., CHAMBERLIN, R.M. AND LOVE, D.W., 1989, First-day road log, from Albuquerque to Mesita, Laguna, Acoma, McCarthys and Grants, *in* Anderson, O.J., Lucas, S.G., Love, D.W. and Cather, S.M., eds., Southeast Colorado Plateau: Socorro, New Mexico Geological Society, 40th Annual Field Conference Guidebook p. 1-24.

MCCUBBIN, D.G., 1969, Cretaceous strike valley sandstone reservoirs, northwestern New Mexico: Tulsa, American Association of Petroleum Geologists Bulletin, v. 53, p. 2114-2140.

MCCUBBIN, D.G., 1982, Barrier island and strandplain facies, *in* Scholle, P.A. and Spearing, D., eds., Sandstone Depositional Environments: Tulsa, American Association of Petroleum Geologists Memoir 31, p. 247-280.

MOLENAAR, C.M AND BAIRD, J.K., 1992, Regional stratigraphic cross sections of Upper Cretaceous rocks across the San Juan basin, northwestern New Mexico and southwestern Colorado: Washington, D.C., U.S. Geological Survey Open File Report 92-257.

MOLENAAR, C.M., 1983, Principal reference section and correlation of Gallup Sandstone, northwestern New Mexico, *in* Hook, S.C., ed., Contributions to Mid-Cretaceous Paleontology and Stratigraphy of New Mexico: Socorro, New Mexico Bureau of Mines and Mineral Resources, Circular 185, p. 29-40.

MOLENAAR, C.M., NUMMEDAL, D. AND COBBAN, W.A., 1996, Regional stratigraphic cross sections of the Gallup Sandstone and associated strata around the San Juan Basin, New Mexico, and parts of adjoining Arizona and Colorado: Washington, D.C., U.S. Geological Survey, Oil and Gas Investigations Chart, OC-143.

NUMMEDAL, D. AND MOLENAAR, C.M., 1995, Sequence stratigraphy of a ramp-setting deltaic succession: the Gallup Sandstone, New Mexico, *in* Van Wagoner J.C. and Bertram, G.B., eds., Sequence Stratigraphy in Foreland Basins: Tulsa, American Association of Petroleum Geologists Memoir 64, p. 277-310.

NUMMEDAL, D. AND RILEY, G.W., 1991, Origin of Late Turonian and Coniacian unconformities in the San Juan basin, *in* Van Wagoner, J.C., Nummedal, D., Jones, C.R., Taylor, D.R., Jennette, D.C. and Riley, G.W., eds., Sequence Stratigraphy Applications to Shelf Sandstone Reservoirs outcrop to subsurface examples: Tulsa, American Association of Petroleum Geology Field Conference Guidebook, not consecutively paginated.

NUMMEDAL, D. AND SWIFT, D.J.P., 1987, Transgressive stratigraphy at sequence-bounding unconformities: some principles derived from Holocene and Cretaceous examples, *in* Nummedal, D., Pilkey, O.H. and Howard, J.D., eds., Sea-Level Fluctuation and Coastal Evolution: Tulsa, Society for Sedimentary Geology (SEPM) Special Publication 41, p. 241-260.

NUMMEDAL, D. AND WRIGHT, R., 1989, Cretaceous shelf sandstones and shelf depositional sequences, Western Interior Basin, Utah, Colorado and New Mexico: Washington, D.C. 28th International Geological Congress Field Trip Guidebook T119, 87 p.

NUMMEDAL, D., WOLTER, N.R., FLEMING, T.F. AND BERGSOHN, I., 1993, Lowstand and transgressive shelf sandstones in Upper Cretaceous strata of the San Juan basin, New Mexico, *in* Caldwell, W.G. and Kauffman, E.G., eds., Evolution of the Western Interior Basin: St. John's, Geological Association of Canada Special Paper 39, p. 199-218.

OBRADOVICH, J.D., 1993, A Cretaceous time scale, *in* Caldwell, W.G.E. and Kauffman, E.G., eds., Evolution of the Western Interior Basin, St. John's, Geological Association of Canada Special Paper 39, p. 379-396.

ODIN, G.S. AND MATTER, A., 1981, De glauconarium origine: Sedimentology, v. 28, p. 611-641.

PANG, M. AND NUMMEDAL, D., 1995, Flexural subsidence and basement tectonics, the Cretaceous western interior basin, USA: Geology, v. 23, p. 173-176.

PASLEY, M.A., RILEY, G.W. AND NUMMEDAL, D., 1993, Sequence stratigraphic significance of organic matter variations: Examples from the Upper Cretaceous Mancos Shale of the San Juan basin, New Mexico, *in* Katz, B. and Pratt, L., eds., Organic facies and sequence stratigraphy: Tulsa, American Association of Petroleum Geology Studies in Geology, v. 29, p. 221-241.

RILEY, G.W., 1993, Origin of a Coarse-Grained Shallow Marine Sandstone Complex: The Coniacian Tocito Sandstone, Northwestern New Mexico: Unpublished Ph.D. Dissertation, Baton Rouge, Louisiana State University, p. 251.

SABINS, F.F., JR., 1972, Comparison of Bisti and Horseshoe Canyon stratigraphic traps, San Juan basin, New Mexico, *in* King, R.E., ed., Stratigraphic Oil and Gas Fields, American Association of Petroleum Geologists Memoir 16, p. 610-622.

SHANLEY, K.W., MCCABE, P.J. AND HETTINGER,R.D., 1992, Tidal influence in Cretaceous fluvial strata from Utah, USA: A key to sequence stratigraphic interpretation: Sedimentology, v. 39, p. 905-930.

SNEDDEN, J.W., TILLMAN, R.W., KREISA, R.D., SCHWELLER, W.J., CULVER, S.J. AND WINN, JR., R.D., 1995, Stratigraphy and genesis of a modern shoreface-attached sand ridge, Peahala Ridge, New Jersey: Journal of Sedimentary Research, v. B64, p. 560-581.

STEVENSON, G.M. AND BAARS, D.L., 1986, The Paradox: a pull-apart basin of Pennsylvanian age, *in* Peterson, J.A., ed., Paleotectonics and Sedimentation: Tulsa, American Association of Petroleum Geologists Memoir 41, p. 513-540.

STRIDE, A.H., 1982, ed., Offshore Tidal Sands: London, Chapman & Hall, p. 222.

SWIFT, D.J.P., 1976, Continental shelf sedimentation, *in* Stanley, D.J. and Swift, D.J.P., eds., Marine Sediment Transport and Environmental Management: New York, John Wiley, p. 311-350.

TILLMAN, R.W., 1985, The Tocito and Gallup sandstones, New Mexico, a comparison, *in* Tillman, R.W., Swift, D.W.J. and Walker, R.G. eds., Shelf Sands and Sandstone Reservoirs: Tulsa, Society for Sedimentary Geology (SEPM) Short Course Notes 13, p. 403-464.

LINEAR TRANSGRESSIVE SHOREFACE SANDBODIES CONTROLLED BY FLUCTUATIONS OF RELATIVE SEA LEVEL: LOWER CRETACEOUS VIKING FORMATION IN THE JOFFRE-MIKWAN-FENN AREA, ALBERTA, CANADA

JAMIE BURTON* AND ROGER G. WALKER

Department of Geology, McMaster University, Hamilton, Ontario L8S 4M1, Canada

ABSTRACT: The Viking Formation in the Joffre-Fenn area consists of 1) a lower succession of regionally extensive, stacked, sandier-upward successions of open marine to shoreface bioturbated sandy mudstone and muddy sandstone, overlain by 2) a series of long, narrow, linear sandbodies with controversial depositional environments. The lower extensive successions are cut by several bounding discontinuities (BD surfaces). Following the first major drop of relative sea level, BD1 represents a surface of marine transgression, and during a pause in transgression, a shoreface succession prograded northeastward. The succession between BD1 and the next surface BD2 IT (IT = initial transgression) or BD2 RT (RT = resumed transgression) comprises the oil-producing sandbody in the Fenn area. A second major drop in relative sea level was followed by a second transgression on surface BD2 IT. In this paper we contrast our interpretation of prograding shoreface deposits with that of J. MacEachern et al. (2, this volume). They have interpreted the main Joffre sandbody as a series of northeast-prograding bayhead deltas that built into a lagoon, barred to the northeast. Whatever its environment, the sandbody was truncated by a resumed rise of relative sea level, forming surface BD2 RT. The linear sandbody at Joffre (a shoreface in our interpretation) dies out southeastward along-strike, and it appears to be a coincidence that Fenn (producing from the BD1-BD2 IT or RT interval) is exactly along-strike from Joffre, where oil production is from the BD2 IT-BD2 RT interval. We suggest that the linear sandbodies in the Joffre-Fenn area are neither ridges, wedges, nor tongues. Exact depositional environments are controversial. This paper examines the reasoning behind the different interpretations, and shows that sandbodies exactly along-strike from each other are not necessarily at the same stratigraphic horizon.

INTRODUCTION

The phrase "ridges, wedges, and tongues" includes those controversial long, narrow sandbodies that occur at many stratigraphic horizons in the Western Interior Seaway of North America. The sandbodies are encased in marine mudstone and therefore appear to rest in offshore marine settings kilometers or tens of kilometers from their time-equivalent shorelines. The classic example is the Shannon Sandstone of Wyoming, which is interpreted to lie about 160 km from the time-equivalent shoreline (Tillman and Martinsen, 1984, 1987). In this interpretation, sand was moved by shelf currents to the depositional site, where unspecified processes focused the sand into linear shelf ridges. Fluctuations of relative sea level play no part in this interpretation.

Examination of similar sandbodies in the subsurface of Alberta has led to an alternative interpretation, namely that the sandbodies encased in marine mudstone are lowstand and transgressive shoreface deposits, abandoned in apparently offshore positions by continued transgression. This was first suggested for the Viking Formation by Beaumont (1984) and developed for the Cardium Formation by Bergman and Walker (1987, 1988) and Pattison and Walker (1992). Specific shoreface interpretations were applied to sandbodies in the Viking Formation by Downing and Walker (1988), Power (1988), Posamentier and Chamberlain (1993), and Walker and Wiseman (1995). Even the classic "shelf ridges" of the Shannon Sandstone have recently been reinterpreted as lowstand shoreface deposits (Bergman, 1994; Bergman and Walker, 1995; Walker and Bergman, 1993). In this paper, we will accept as a working hypothesis the interpretation of Downing and Walker (1988), namely that the Joffre sandbody was deposited in a transgressively incised shoreface that prograded a short distance during a slight pause in the overall transgression.

The purpose of this paper is to reexamine some of the linear sandbodies in the Viking Formation (Fig. 1), particularly Joffre and its apparent extension into the Mikwan and Fenn areas (Figs. 2, 3, 4). At Joffre (Fig. 3), the sandbody interfingers seaward (northeastward) with black mudstone, and our first objective is to describe this relationship and interpret it in light of the existing shoreface interpretation of Downing and Walker (1988).

This leads directly to our second objective, which is to contrast this interpretation with that of MacEachern et al. (2, this volume). We particularly thank James MacEachern for agreeing to a built-in discussion and reply in both this paper and his own in this volume, and for giving us preprints of his work in press. In their interpretation (J. MacEachern et al., pers. commun., 1998) they suggest that the sandbody consists of a series of embayment-filling, bay-head deltas, and that the black mudstone is lagoonal. This lagoonal interpretation implies that there was originally some form of barrier to the northeast, although no evidence of this remains—as J. McEachern et al. (pers. commun., 1998) state, it is a barrier "whose deposits are conspicuously absent from the depositional record."

Our third objective is to describe the along-strike changes from Joffre southeastward, showing that the shoreface sand dies out rapidly into a mudstone drape on the shoreface incision. Finally, by careful mapping of key bounding discontinuities, we will show that the sandbodies at Mikwan and Fenn, apparently alongstrike from Joffre (Fig. 2), are at a different stratigraphic horizon within the Viking Formation.

PREVIOUS WORK

Gas was first discovered in the Viking Formation in the Viking-Kinsella area in 1917, and the name Viking was given by Slipper (1918). Classic interpretations of Viking depositional environments fall into two main categories—nearshore and barrier settings (DeWiel, 1956; Shelton, 1973; Tizzard and Lerbekmo, 1975) and offshore bars or shelf ridges (Evans, 1970; Koldijk, 1976; Boethling, 1977; Beaumont, 1984; Hein et

* Present address; Imperial Oil Resources Ltd., P.O. Box 2480 Station M, Calgary, AB T2P 3M9, Canada.

Isolated Shallow Marine Sand Bodies: Sequence Stratigraphic Analysis and Sedimentologic Interpretation.
SEPM Special Publication No. 64, Copyright © 1999
SEPM (Society for Sedimentary Geology), ISBN 1-56576-057-3, p. 255-272.

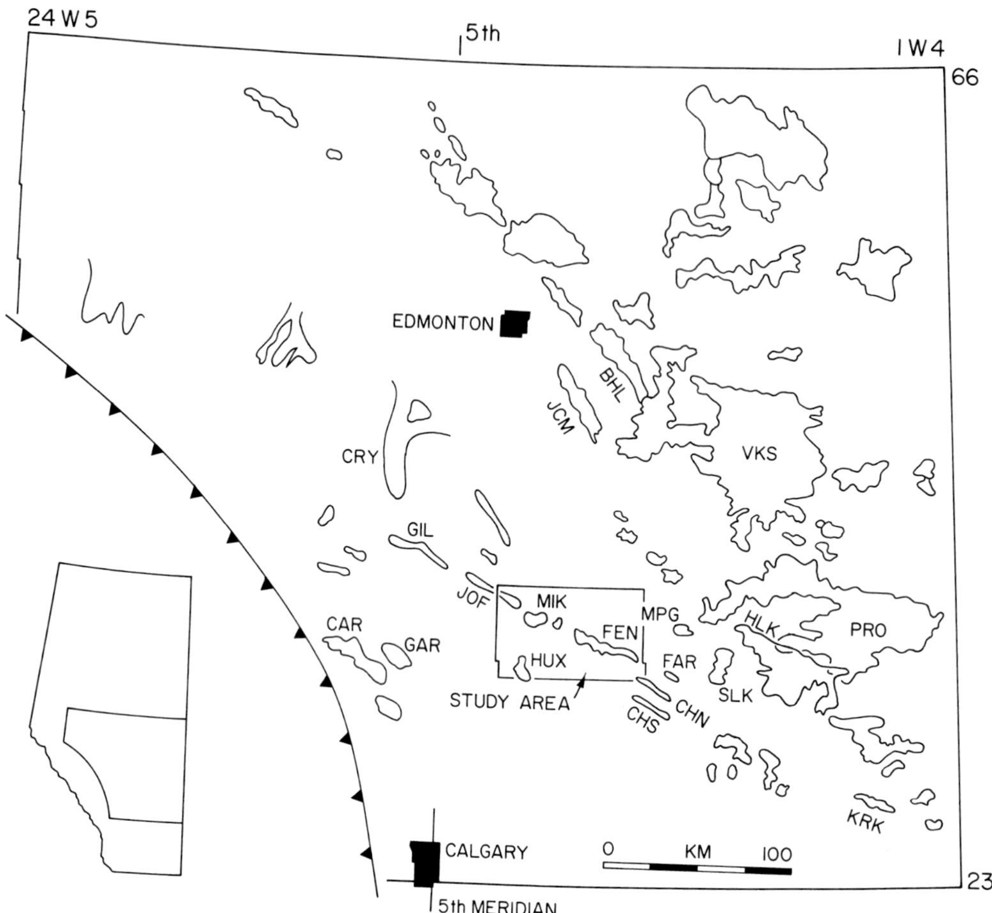

Fig. 1.—Location of Viking oil and gas fields, with study area superimposed. Note that the entire study area lies west of the fourth meridian. The fields identified are BHL = Beaverhill lake; CAR = Caroline; CHN = Chain North; CHS = Chain South; CRY = Crystal; FAR = Farrell; FEN = Fenn; GAR = Garrington; GIL = Gilby; HLK = Hamilton Lake; HUX = Huxley; JCM = Joarcam; JOF = Joffre; KRK = Kirkwall; MIK = Mikwan: MPG = Maple Glen; PRO = Provost; SLK = Sullivan Lake; VKS = Viking/Kinsella. Inset lower left shows area of main map within Province of Alberta.

al., 1986; Amajor and Lerbekmo, 1990). Beaumont (1984) was the first to suggest the importance of relative sea-level fluctuations in influencing the position and nature of Viking sandbodies.

The work of Bergman (Bergman and Walker, 1986, 1987, 1988) on linear sandbodies in the Cardium Formation showed that the coarse-grained Cardium sediments (coarse, very coarse and granule sandstone, and conglomerate) rested on asymmetrical bounding discontinuities (shown schematically in Fig. 5) that are interpreted as incised shorefaces formed during pauses in an overall transgression. Similar ideas involving transgressive, incised shorefaces were first applied to Viking sandbodies by Downing and Walker (1988) at Joffre. Subsequently, similar interpretations were suggested by Power (1988) at Joarcam, Raddysh (1988) at Gilby, Posamentier and Chamberlain (1993) at Joarcam, and Walker and Wiseman (1995) at Joarcam and Beaverhill Lake (Fig. 1). In each of these studies, asymmetrical bounding discontinuities were identified beneath the sandbodies and incised shoreface interpretations were proposed. The most recent general review of the entire Viking Formation is that of Reinson et al. (1994).

Relatively little work has been published regarding sand bodies in the present study area. The first suggestion of relative sea-level changes in the Joarcam to Joffre area was made by Beaumont (1984). He suggested that during a regressive phase, "sea level fell rapidly, subaerially exposing most of the shelf...[with] delta complexes to the east." Ensuing transgression "occurred so quickly that the original irregular shape of the deltaic sandbodies...was not altered appreciably by shoreface erosion...as the transgressive phase continued, shoreface erosion of sediment deposited during minor regressions and transgressions supplied sand to the shelf that was subsequently restructured into linear sand bodies, such as the Joffre and Joarcam Fields, by the shelf hydraulic regime." Beaumont's ideas were modified by Raddysh (1988) at Gilby and Downing and Walker (1988) at Joffre. In both of these studies, it was suggested that there was a major fall of relative sea level followed by transgression. Shoreface incision took place at times of stillstand during the overall rise of sea level, and the coarse sandbodies formed in the shoreface from sediment supplied by rivers and reworked along-shore by wave-generated currents. This eliminates the problem of exactly how sediment was

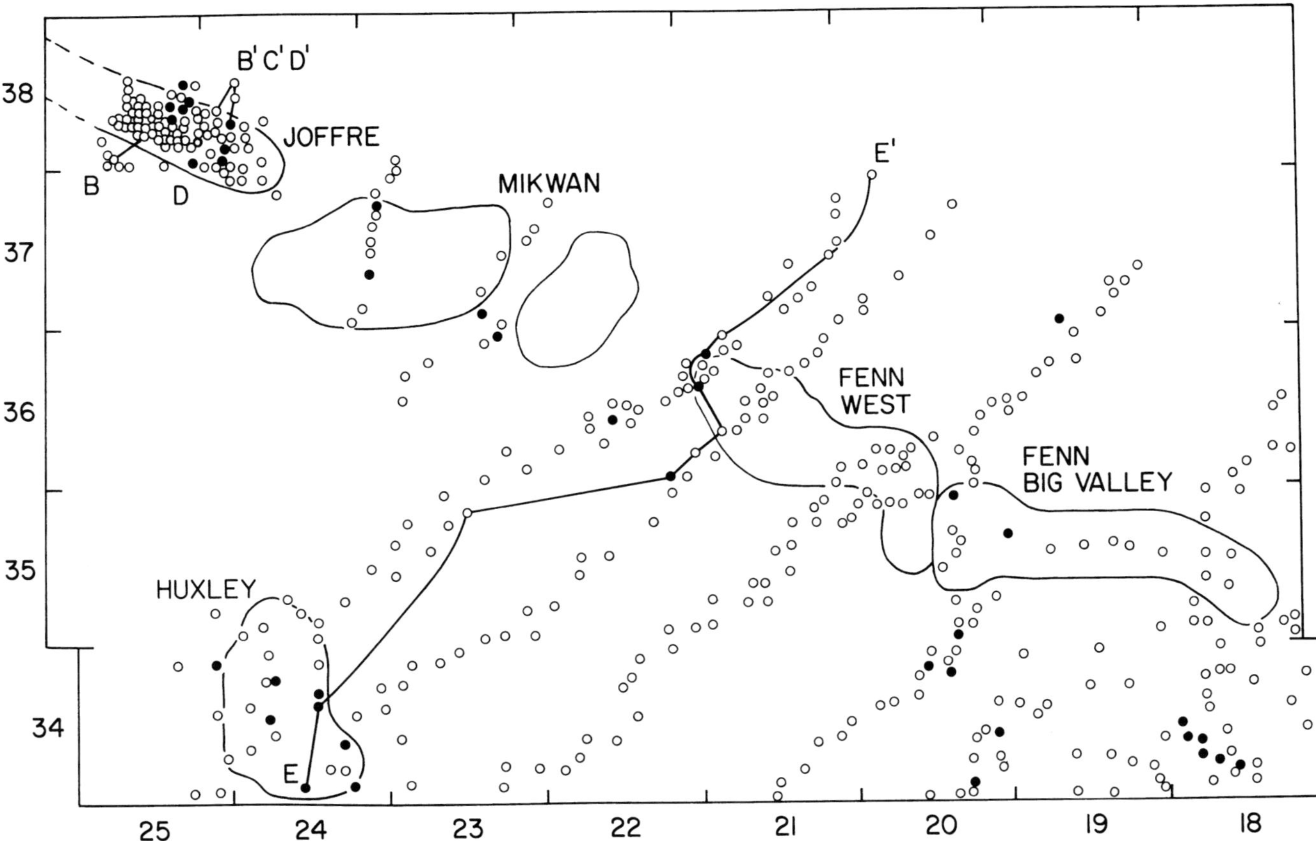

Fig. 2.—Study area showing well logs (open circles) and cores (black circles) used in this study. Locations of cross-sections also shown; see Figure 3 for details at Joffre. Scale: each Township and Range is 6 mi, or roughly 10 km. All ranges are west of the fourth meridian.

transported across the shelf and formed into linear bars by the "shelf hydraulic regime."

The most recent studies are those of J. MacEachern et al. (1, 2, this volume), who has identified a "stressed" trace fauna in mudstone tongues that interfinger with the Joffre sandstones. As an alternative to the incised shoreface interpretation, they suggested that the coarse sandbody was supplied directly into a lagoon by rivers flowing from the southwest, with some form of barrier (not preserved) lying parallel to Joffre to the northeast. We will comment more on this interpretation below.

In summary, the offshore bar interpretations (Evans, 1970; Koldijk, 1976; Boethling, 1977; Beaumont, 1984; Hein et al., 1986; Amajor and Lerbekmo, 1990) involved sand transport from a distant shoreline out across the shelf by storm currents, with focusing of the sand into linear bars by open marine processes. By contrast, the interpretations invoking fluctuations of relative sea level (Beaumont, 1984; Downing and Walker, 1988) suggest that the linear sandbodies represent lowstand or transgressive shoreline or lagoonal deposits abandoned in apparently offshore positions by continued transgression.

STUDY AREA, STRATIGRAPHY AND DATA BASE

This study emphasizes the seaward and along-strike changes in the architecture of the Joffre sandbody (Figs. 2, 3), as far as the eastern end of Fenn Big Valley. In this area, the Late Albian Viking Formation can be divided lithostratigraphically or allostratigraphically (Fig. 4). Lithostratigraphically, the Viking Formation is defined by the appearance of the first sandstone above the shales of the Joli Fou Formation, and the disappearance of the last sandstone below the Westgate Formation. The advantage of an allostratigraphic scheme is that bounding discontinuities can be used to define a high-resolution internal Viking stratigraphy. Core control shows that the first sandstone appears gradationally toward the top of an overall sandier-upward succession, here termed Regional Viking succession 1 (RV1 in Fig. 4). This is the producing horizon in the Hamilton Lake Field, southwest of the study area, and is termed the Hamilton Lake sandstone (HLS in Fig. 4). Because the base of the HLS is gradational and does not represent an important change in style or environment of deposition, we suggest that the most appropriate bounding discontinuity for definition of the base of the Viking Alloformation is the maximum flooding surface below RV1. We tentatively pick this at the point where the gamma- ray log shows the highest API values and the resistivity log shows the lowest values—the two logs thus appear to form a narrow "neck" within what is lithostratigraphically regarded as the Joli Fou Formation. No cores penetrate this horizon within the study area; it is tentatively picked as the maximum flooding surface because the gamma-ray log suggests the most organic-rich, and perhaps most slowly deposited shale. RV1 appears to become progressively sandier upward from this point.

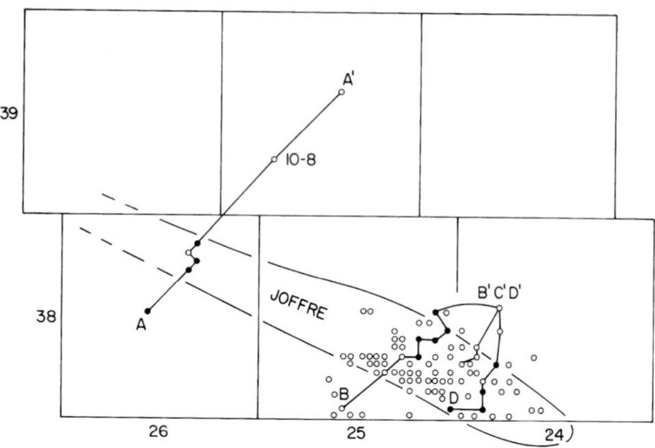

Fig. 3.—Detail of the Joffre area, showing well locations used and locations of cross- sections A through D. AA' is one of the cross-sections used by Downing and Walker (1988), and BB', CC,' and DD' are newly constructed for this paper. Core locations are shown in black, and only the southeastern end of Joffre (some of the wells in T 38, R 24, and R25) has been reexamined in this paper. All ranges are west of the fourth meridian.

The top of the Viking Formation varies across the basin, as shown by Davies and Walker (1993). The top of the Viking Alloformation is therefore taken at the next major bounding discontinuity, the condensed horizon known as Base of Fish Scales (BFS in Fig. 4). Bounding discontinuities BD1 through BD4 have been determined from core control and are shown in Figure 4. The relationship between the BD terminology and the original terminology of Downing and Walker (1988) is explained in the caption of Figure 6.

Well log and core control is shown in Figure 2. We have reexamined all of the eight available cores in the southeasternmost 10 km of the Joffre Field itself, along with 80 adjacent well logs (Fig. 3). In the rest of the study area (Fig. 2), we have used 180 well logs and have examined 29 selected cores. Correlations of key surfaces are based on the calibration of well logs with cores and the construction of cross-sections.

TERMINOLOGY OF BOUNDING DISCONTINUITIES IN A SHOREFACE SETTING

In general, correlations are made with some guiding principle in mind, be it that units are all layer cake and tabular, or that units all downlap onto a lower surface; there are many other possibilities. We believe that there is good

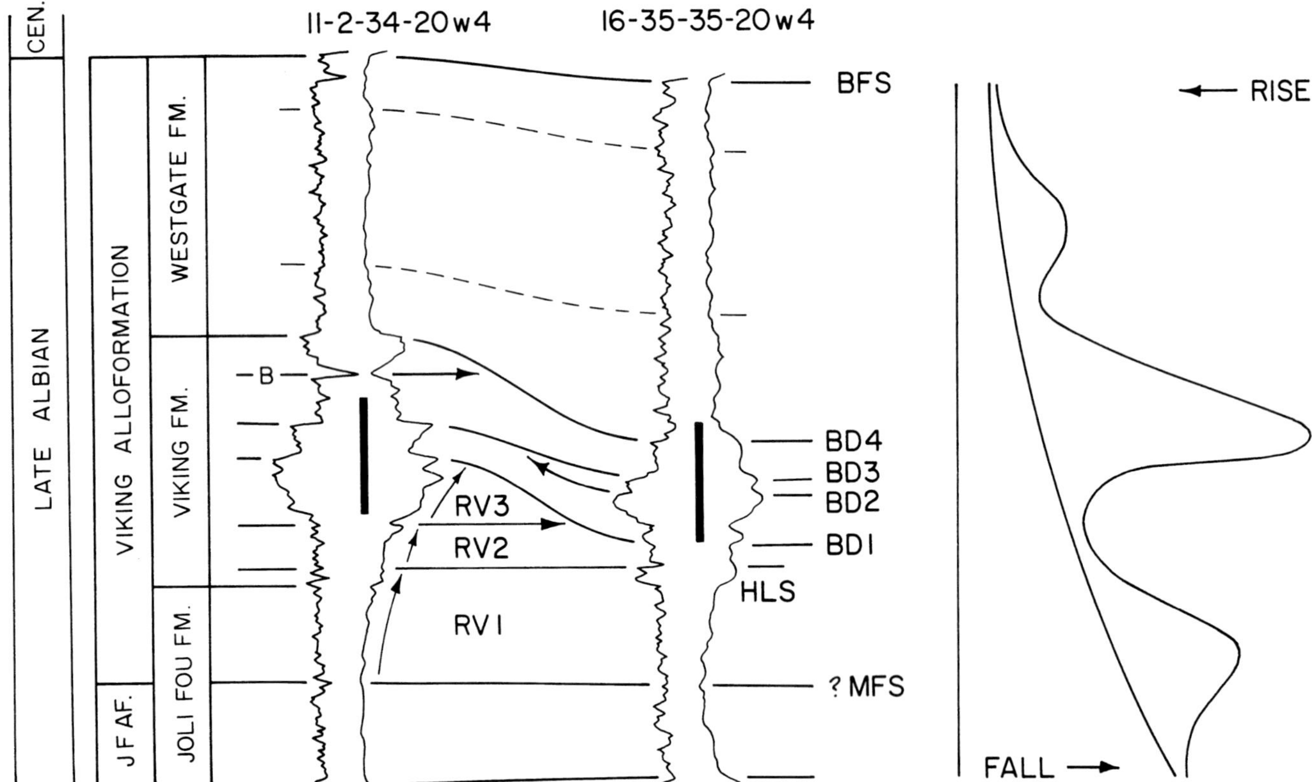

Fig. 4.—Stratigraphy of the Viking and adjacent formations. Lithostratigraphy includes the Joli Fou, Viking, and Westgate formations. Allostratigraphy includes the Joli Fou and Viking alloformations. Biostratigraphically, the succession is Late Albian. The stratigraphy is illustrated by two well logs from south of Fenn (Fig. 2), with RV2 used as a datum. RV = Regional Viking; MFS indicates a possible maximum flooding surface within the Joli Fou Formation (and used allostratigraphically as the lower bounding discontinuity of the Viking Alloformation). BD = bounding discontinuity; B = bentonite; and BFS = Base of Fish Scales marker; HLS = Hamilton Lake sandstone. Part of the long-term and short-term eustatic sea level curves are shown to the right (from Haq et al., 1988). Note erosion of RV2 and RV3 by BD1, erosion of BD2 by BD3, and truncation of the bentonite by BD4. Black bars indicate core control. For scale, the total section shown in well 11-2-34-20W4 is about 100 m.

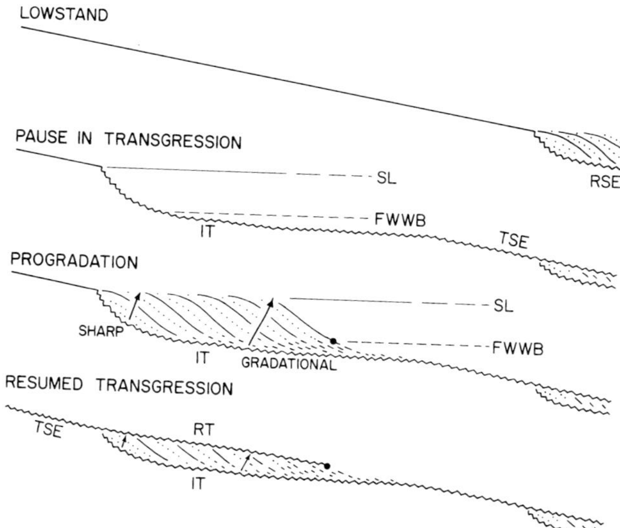

Fig. 5.—Terminology and concepts used to describe incised shorefaces. RSE = regressive surface of erosion; TSE = transgressive surface of erosion; IT = initial transgression; RT = resumed transgression. Arrows indicate sharp and gradationally based sandier-upward prograding shoreface successions. SL = sea level; FWWB = fair-weather wave base. The four stages, lowstand, pause in transgression, progradation and resumed transgression are described in the text. The black dot indicates fair-weather wave base at the time of maximum progradtion. It is also the seaward limit of the bounding discontinuity made during resumed transgression. Seaward of the black dot, the RT surface passes into a correlative conformity.

evidence for lowstand and incised shoreface settings in the Viking (Downing and Walker, 1988; Raddysh, 1988; Power, 1988; Posamentier and Chamberlain, 1993; Walker and Wiseman, 1995), and we introduce a terminology for initial transgressive (IT) and resumed transgressive (RT) bounding discontinuities in transgressive shoreface settings.

At lowstand of relative sea level (RSL), a regressive surface of erosion (RSE) lies beneath a lowstand shoreline sandbody (Fig. 5, lowstand). Subsequent transgression forms a transgressive surface of erosion (TSE), which rises steeply if the rise of RSL is rapid and gently if the rise of RSL is slow. During a pause in the rise of RSL (Fig. 5) the TSE is cut horizontally by wave action, principally at fair-weather wave base (FWWB, Fig. 5). This surface is described as an IT surface in Figure 5. At incised shorefaces, there is commonly a lower IT surface and a higher RT surface. Thus the terms IT and RT can only be used in places (e.g., incised shorefaces) where a regional transgressive surface of erosion locally splits into two surfaces.

During a pause in the rise of RSL, there is a possibility for sediment progradation if there is sufficient supply of sediment. Sediment will prograde and downlap onto the IT surface. Above FWWB, the shoreface succession will be sharp-based and dominantly sandy. Below FWWB, the shoreface succession will begin with offshore mudstone and will grade up into lower shoreface sandstone. At the time of maximum progradation, FWWB is marked by a black dot in Figure 5 (progradation). When the rise of RSL resumes, erosion will again take place down to FWWB, and a surface of RT will develop from the black dot in Figure 5. The surface will climb relatively steeply if the rate of RSL rise is rapid, or gently if the rate is slow. Thus, more or less of the progradational shoreface package will be preserved, but in either case, the foreshore and beach, and most of the upper shoreface, stand little chance of preservation. Thus, the transgressive shoreface deposits will be preserved between the IT and RT surfaces, and the shoreface sandbody itself will be sharply or gradationally based. Seaward of the black dot, transgressive mudstone will lie gradationally on progradational offshore mudstone, and the RT surface will pass into a correlative conformity that would be very difficult to identify in core.

The conceptual ideas of Figure 5 have been applied to Viking shorefaces in this paper. To avoid confusion, we emphasize that the BD2 surface is a regional surface that can be traced across the entire study area (Fig. 2). In most places it is a single transgressive surface of erosion, except that in the area of Joffre, Mikwan, and Fenn fields, it splits into BD2 IT and BD2 RT, as explained below.

These ideas will be used, as appropriate, in trying to correlate the various bounding discontinuities recognized in cores and calibrated with the well logs.

CROSS-SECTIONS

We present five cross sections in Figs. 6-10. Figure 6 shows one of the original cross-sections through Joffre from Downing and Walker (1988). Figs. 7-9 show three new cross-sections from the southeastern end of Joffre, and Figure 10 shows a regionally extensive section from Huxley to Fenn. Cored wells from the southeastern end of Joffre are correlated in Figs. 11 and 12.

The Joli Fou Formation consists of marine mudstone that transgress the underlying nonmarine Mannville Group rocks. The maximum flooding surface (?MFS) is taken at the narrowest neck in the gamma-ray and induction logs, but there is no core control to support this pick. Three Regional Viking sandier upward successions, designated RV1 through 3 on the cross-sections, lie above the ?MFS. The Hamilton Lake sandstone (HLS) occurs at the top of RV1. There is also a thick sandstone at the top of RV3 in the Huxley area (Fig. 10), which forms the producing sandstone in that field (Fig. 2).

The RV successions are truncated progressively northeastward by BD1 and by BD2 IT, which truncates BD1 as well as the RV successions (Figs. 7-12). In the study area, BD2 IT is not truncated by BD2 RT (despite the indications in Fig. 5). We assume that this truncation did take place, but that the evidence has been removed by erosion of BD2 RT by BD3. BD4 can be traced over the entire area and truncates BD3 at the northernmost end of cross-section E (Fig. 10).

The relationships discussed have also been mapped (Fig. 13). The northern erosional edge of RV3 generally lies to the south of the fields, and the edge of RV2 lies to the north of the fields. The sandstone at the top of RV1 is the producing sandstone in Hamilton Lake Field (HLK in Fig. 1). Where BD1 cuts down through RV2, it exposes the top of RV1 and the Hamilton Lake sandstone (Fig. 13). Note also that BD1 rises again stratigraphically northward, as explained in the description of BD1 (below).

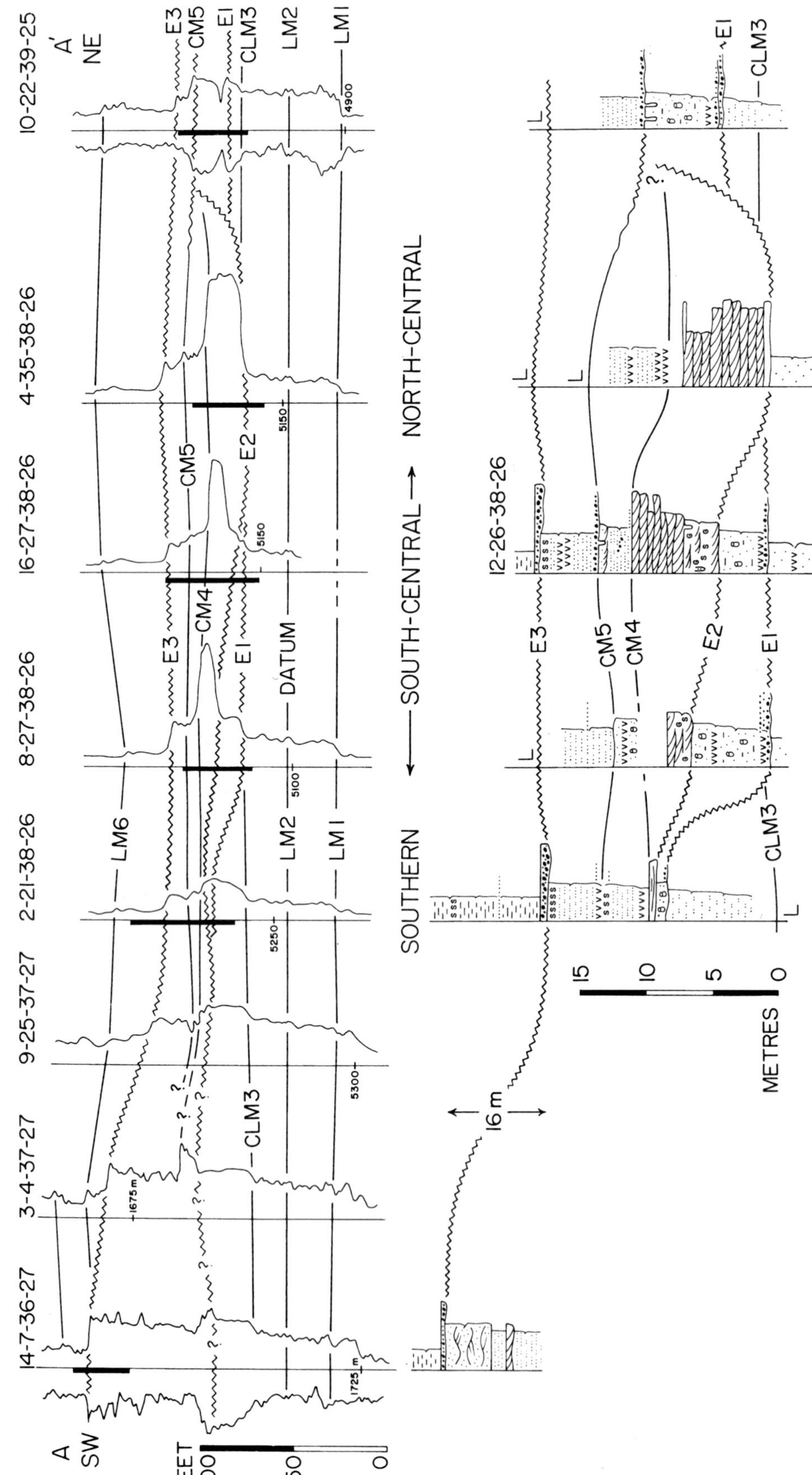

Fig. 6.—Cross-section AA′, from Downing and Walker (1988). The old and new terminologies correspond in this manner; E1 = BD1; E2 = BD2 IT; CM4 = BD2 RT; CM5 = BD3; and E3 = BD4. L indicates a log "pick", LM indicates a log marker, CM a core marker, and CLM is a core and log marker. Note that E2 (BD2 IT) rises stratigraphically northeastward as well as southwestward.

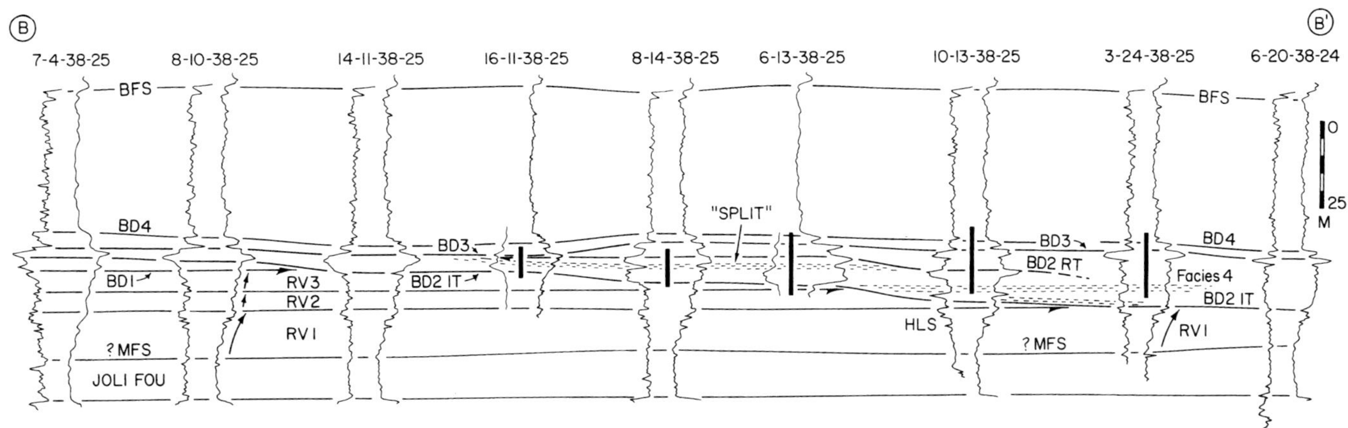

Fig. 7.—Cross-section BB' from the southeastern end of Joffre (located in Fig. 3). Black bars indicate locations of cores, which are correlated in Figure 11. "Split" refers to a muddier interval within the shoreface succession between BD2 IT and BD2 RT. HLS = Hamilton Lake sandstone; other abbreviations are explained in the caption of Figure 4.

Differences in Correlation

Our cross-section (Fig. 11) presents a slightly different correlation from that of J. MacEachern et al. (pers. commun., 1998), where they suggest that the surface that we term the top of RV2 in well 6-13-38-25 is actually FS/SB1 (BD1). This disagreement does not affect the overall interpretation of either group. However, we correlate the sandstone immediately below BD2 RT in 10-13 with the Facies 1 sandstone in 6-13, while J. MacEachern et al. (pers. commun., 1998) correlate the sandstone in 10-13 with the Facies 3 sandstone in 6-13. This changes the number of parasequences that can be recognized in the main Viking sandbody.

DESCRIPTIONS OF ALLOSTRATIGRAPHIC UNITS AND BOUINDING DISCONTINUITIES

Joli Fou MFS to BD1

The section between the postulated maximum flooding surface within the Joli Fou shales and the first major bounding discontinuity (BD1) consists of three "Regional Viking" sandier-upward successions with a maximum combined thickness of 43 m. All thicknesses tend to be reduced northeastward by erosion (Figs. 7-10).

Succession RV1 has a maximum thickness of 21 m in the southeastern part of the study area. The lower part consists of black to gray mudstone, grading up into sandy mudstone and muddy sandstone (very-fine to fine-grained). The upper part is almost totally bioturbated by a marine assemblage of traces including *Helminthopsis*, *Terebellina*, *Planolites*, *Teichichnus*, *Arenicolites*, *Zoophycos*, *Paleophycus*, and *Skolithos*. Glauconite is present but not abundant.

Succession RV2 is up to 9 m thick. The overall sandier-upward succession begins with a bentonitic mudstone with lenticular sandstone beds commonly a few centimeters in thickness. The commonest trace is *Helminthopsis*. Very-fine grained sandstone beds become more common upward, and the trace fossil suite becomes more diverse, including *Planolites*, *Helminthopsis*, *Terebellina*, *Chondrites*, *Teichnichnus*, *Zoophycos*, and *Rosselia*.

Succession RV3 is up to 25 m thick. It begins with bentonitic mudstone with very little sandstone, but with abundant and diverse traces including *Terebellina*, *Schaubcylindrichnus*, *Teichichnus*, *Chondrites*, *Helminthopsis*, *Anconichnus*, *Skolithos*, *Asterosoma*, *Zoophycos*, *Cylindrichnus*, *Siphonichnus*, and *Rosselia*. The succession becomes sandier upward, and the upper part consists dominantly of very-fine to fine-grained sandstone beds up to 25 cm thick. Both HCS and angle-of-repose cross stratification are present. The trace fauna decrease in abundance and diversity, to include *Ophiomorpha*, *Paleophycus*, *Asterosoma*, and *Macaronichnus simplicatus*.

Bounding Discontinuity 1

BD1 is typically very sharp, although in places the contact is blurred by bioturbation. It can be correlated and mapped over the entire study area, but is locally cut out by BD2. *Skolithos* burrows up to 50 cm deep penetrate down from BD1 along with small, unlined *Thalassinoides* and rare *Rhizocorallium*. BD1 is overlain by a lag of coarse material, including coarse sand grains and granules and rare pebbles of chert up to 1.5 cm in diameter. The coarse grains passively fill

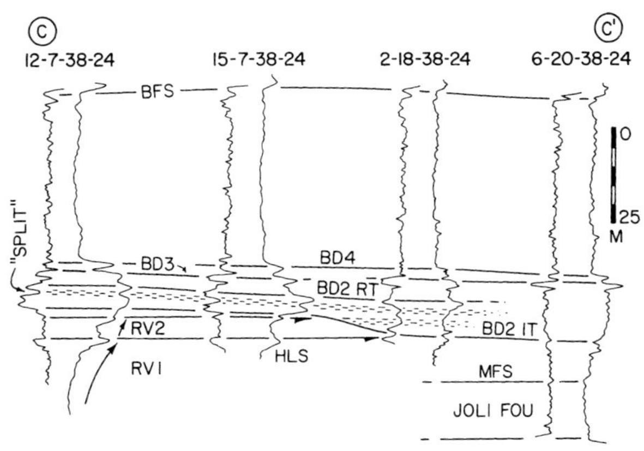

Fig. 8.—Cross-section CC', located in Figure 3. There is no core control. The section shows the interpreted onlap of a lower muddy horizon against BD2 IT, and a higher muddy horizon (the "split") that continues across the entire cross-section.

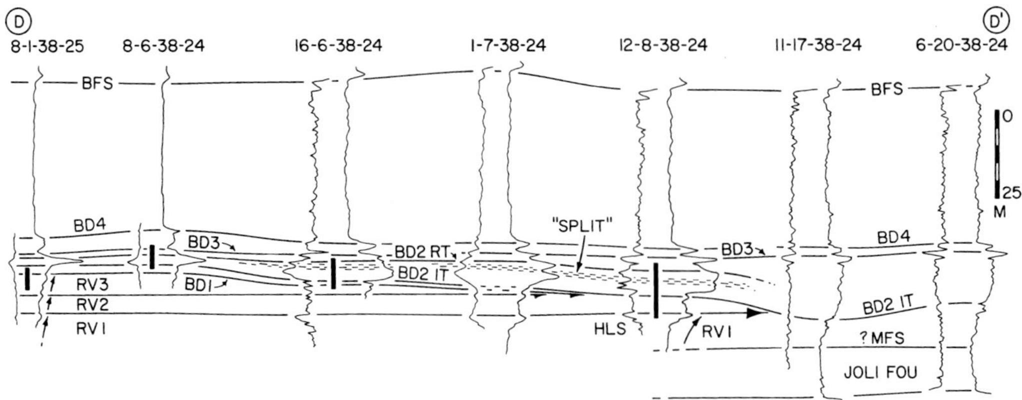

Fig. 9.—Cross-section DD', located in Figure 3. Note the location of the "split," and the erosion of BD1 by BD2 IT. BD2 IT also appears to rise stratigraphically slightly northeastward (toward D'). Three of the cores are shown in Figure 12.

the *Skolithos* and *Thalassinoides* burrows. This firm-ground *Glossifungites* suite of traces supports the idea that BD1 is a transgressive surface of erosion (TSE).

The relationship of BD1 to the underlying regional Viking successions RV2 and RV3 is shown in Figure 13. Note that the BD1 surface dips southward in the northern part of the study area, cutting into the Hamilton Lake sandstone (diagonal ruling, Fig. 13). The angle of dip, compared with a datum drawn on top of RV2, is between 0.08 and 0.015° (estimated from relationships and correlations shown in Fig. 10). South of the diagonally ruled area, the northern limits of RV2 and RV3 are shown. The northeastward dip of the BD1 surface averages about 0.03° southwest of the Fenn West field. This northeastward dip flattens northeastward and begins to rise again stratigraphically, giving a southwestward (landward) dip to BD1. In our section E (Figs. 2, 10), BD1 rises about 8 m stratigraphically in 6 km between 16-13-37-21 and 15-32-37-20, implying a gradient of 0.0013 or an angle of 0.07°. These angles hardly seem steep enough to propose a valley wall or topographic high to the northeast of the linear fields, and we will propose an alternative interpretation below.

Allomember Between BD1 and BD2

This allomember is preserved over much of the southwestern part of the study area, but is cut out by BD2 in the central and northeastern parts of the area. In the southwest, the allomember is up to 11 m thick and is characterized by a sandier- and coarsening-upward trend. The lower part consists of bioturbated, sandy mudstone with open marine traces including abundant *Helminthopsis*, *Planolites*, *Terebellina*, and *Zoophycos*. Less common traces include *Siphonichnus*, *Teichichnus*, *Rosselia*, *Schaubclyindrichnus*, *Asterosoma*, *Paleophycus*, and *Cylindrichnus*. *Chondrites* is rare. In this lower part of the allomember, coal clasts, and granules and pebbles of chert up to 1 cm in diameter are common and tend to be concentrated toward the base. Where the upper part of the succession is preserved (typically within the area of Fenn field), it consists of fine sandstone with angle-of-repose cross-stratification in sets up to 20 cm thick. Bioturbation is scarce, but includes large *Skolithos* and *Diplocraterion* burrows up to 13 cm deep.

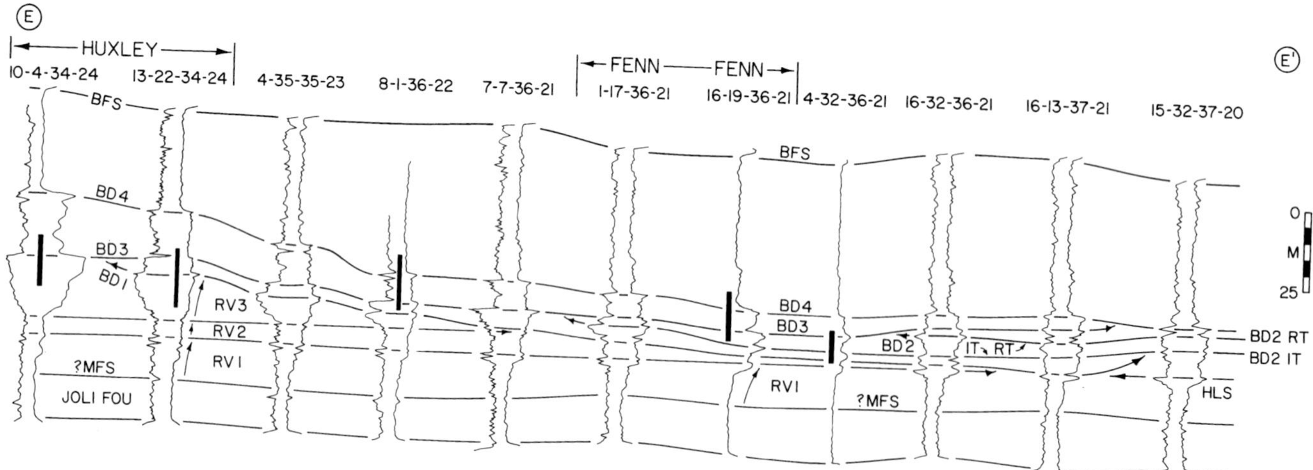

Fig. 10.—Cross-section EE', located in Figure 2. Note that the main producing sandbody at Huxley is at the top of the RV3 succession. Note also that BD1 cuts down into the Hamilton Lake Sandstone (HLS) in 16-13-37-21, and that BD2 RT is truncated by BD3 (between 4-32 and 16-32). Before erosion by BD3, BD2 RT presumably truncated BD2 IT somewhere southwest of well 1-17-36-21.

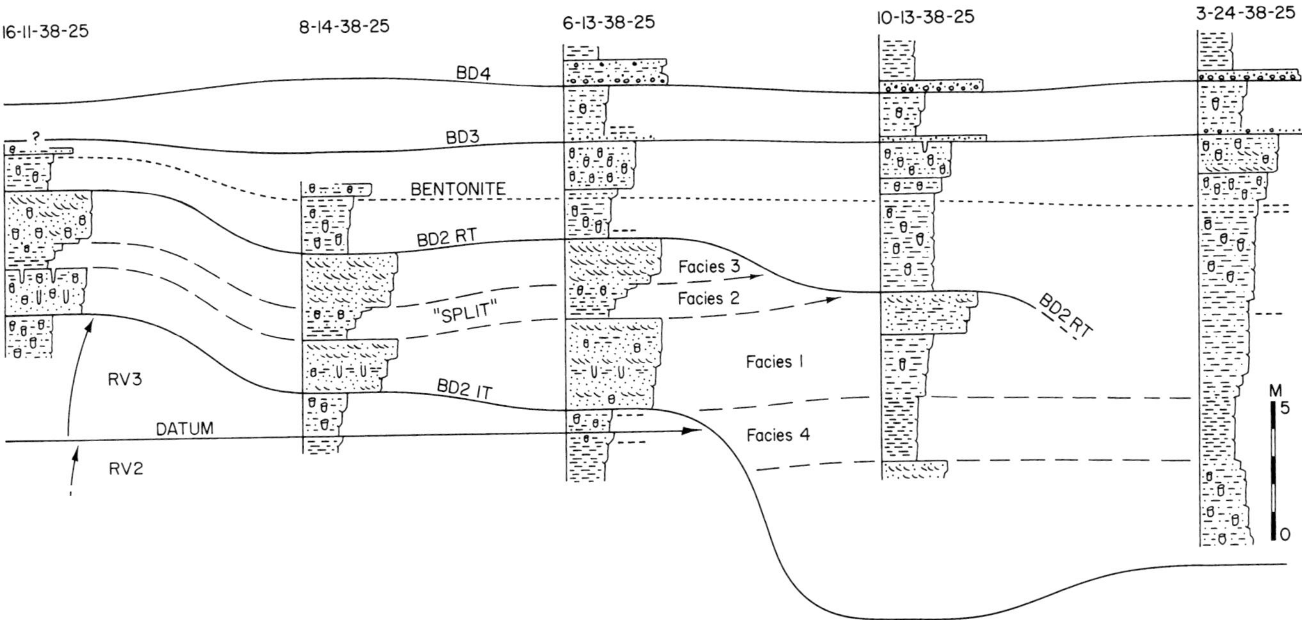

Fig. 11.—Core cross-section from BB' (Fig. 7), located in Figure 3. Legend is shown in Figure 12. Note that the black mudstone of Facies 4 lies stratigraphically below the Facies 1 sandstone in 6-13. We interpret this mudstone to pinch out against BD2 IT. We interpret the sandstone of Facies 1 in 6-13 as shoreface deposits; they appear to pass seaward into 10-13, where we interpret a sandier-upward prograding shoreface succession. Facies 2 (the split) and Facies 3 are interpreted as having been truncated by BD2 RT, which grades seaward (into 3-24) into a correlative conformity. These relationships are explained diagrammatically in Figure 5. The black dot in Figure 5 that represents the limit of progradation and fair-weather wave base is interpreted to lie slightly seaward (toward 3-24) of the lettering "BD2 RT" in the figure above.

Description of Bounding Discontinuity 2

BD2 is a single transgressive surface of erosion over most of the study area. Locally, it splits into two parts, BD2 IT and BD2 RT (as explained in Fig. 5). In the Joffre, Mikwan, and Fenn areas the BD2 IT surface is normally very sharp. It is mantled by a lag no thicker than 2 cm, with granules and small pebbles of chert up to 6 mm in diameter. In places, *Skolithos* and *Diplocraterion* burrows penetrate at least 20 cm down from the contact and are passively filled with medium- to coarse-grained sand. BD2 IT is interpreted as a firmground *Glossifungites* surface.

In the Joffre and Mikwan areas, BD2 RT is also sharp. In most cores there is no transgressive lag, but locally a lag up to 16 cm thick is present, with chert pebbles up to 1 cm in diameter. In the Fenn area, the BD2 RT surface is commonly indistinct and there is no coarse lag. The geometrical relationships of these surfaces are described below.

Geometry of Bounding Discontinuity 2

Downing and Walker (1988) noted that the BD2 IT (their surface E2; Fig. 6) surface rose stratigraphically between wells 4-35-38-26 and 10-22-39-25 (Fig. 6). Unfortunately, there is no well control immediately north of sections B, C, and D (Fig. 3), and the morphology of BD2 cannot be determined there. Superficially, in section AA' (Fig. 6) the northeastward stratigraphic rise of BD2 suggests a channel-like morphology, with the axis lying at the northeastern edge of the field and trending parallel to the field. The vertical exaggeration and lack of horizontal scaling of the wells in Figure 6 emphasize this channel-like relationship. However, it is important to estimate the true dip of the BD2 surface northeast of Joffre. We have added one more control well to the cross-section of Downing and Walker (1988), 10-8-39-25 (Fig. 3). Between wells 4-35-38-26 and 10-8-39-25, the BD2 surface rises about 6 m stratigraphically in a horizontal distance of 4.7 km, giving a gradient of 0.0013, or an angle of 0.07°. If the stratigraphic rise were achieved in less than 4.7 km, the angle would be steeper, but there is no control between these two wells. The gradient of 0.0013 is the same as that calculated for BD1, presented above.

Step in Bounding Discontinuity 2 IT Along Northern Edge of Joffre

A prominent step can be seen in BD2 IT between wells 6-13-38-25 and 10-13-38-25 in Figure 7; between wells 15-7-38-24 and 2-18-38-24 in Figure 8; and between wells 12-8-38-24 and 11-17-38-24 in Figure 9. The step can be traced along the northern edge of Joffre almost to the northwestern end of the field (Downing and Walker, 1988, Fig. 3, between wells 6-7-39-26 and 11-7-39-26). It can also be correlated southeastward, but becomes subdued and hard to recognize at Fenn (Fig. 10).

Allostratigraphic Unit Between BD2 IT and BD2 RT

This unit comprises the main producing sandbody in the Joffre field. To the south, BD2 RT truncates BD2 IT (cross-section in Fig. 6, where CM4 [BD2 RT] truncates E2 [BD2 IT] between wells 9-25-37-27 and 2-21-38-26). Sections B, C, and D (Figs. 7, 8, 9) do not extend far enough south to encounter this point of truncation. In cross-section E, the point of truncation of BD2 IT by BD2 RT is not preserved, due to subsequent erosion of BD2 RT by BD3 between wells 7-7-36-21 and 1-17-36-21 (Fig. 10).

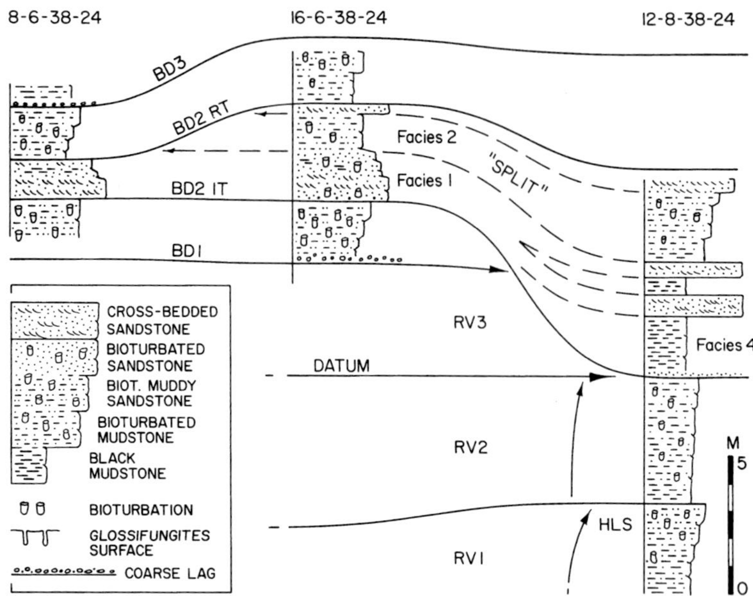

Fig. 12.—Core cross-section from DD' (Fig. 9), located in Figure 3. Note again that Facies 4 lies stratigraphically below the Facies 1 sandstone interpreted here as shoreface. We suggest that the Facies 4 mudstone pinches out against the BD2 IT surface.

There are four main facies within this unit. Facies 1 lies sharply on BD2 IT and consists of fine- to medium-grained glauconitic sandstone with some interbedded mudstone. The rocks are mostly bioturbated but contain some preserved sets of cross bedding 3-20 cm thick. One of the most prominent trace fossils is *Skolithos*, which occurs both within the facies and in places (16-11-38-25, Fig. 12) at the top of the facies, demarcating a *Glossifungites* surface. The top of this facies is normally very sharp.

Facies 2 is finer grained and splits the main Joffre sandbody into two parts ("split" in Figs. 7-9, 11, 12). It rests abruptly on Facies 1 and consists of a sandier-upward succession of mudstone and bioturbated muddy sandstone. The basal mudstone is black, with few thin silty laminations and very little macroscopic bioturbation. Upward, the number and thickness of very-fine sandstone beds increase, as does the amount of bioturbation (dominated by *Planolites* and a few *Skolithos*). The split can be mapped through the northern part of Joffre using prominent signatures in the gamma-ray and resistivity logs (Figs. 7, 8, 9). Southward, the split appears to be truncated by BD2 RT (between wells 8-6-38-24 and 16-6-38-24; Figs. 9, 11) or BD3 (between wells 14-11-38-25 and 16-11-38-25; Fig. 8). Northeastward, the split grades laterally into muddier facies and cannot be recognized (Figs. 7-9).

The split grades upward into Facies 3, medium- to coarse-grained sandstone. Facies 3 forms the upper part of the main sand body at Joffre, and is dominated by cross-bedding in sets about 4-20 cm thick. The top of this facies (and the top of the Joffre sandbody) is defined by BD2 RT. This surface is sharp, but there are no rip-up clasts and no pebble lag. Facies 3 is well-preserved in the west (Fig. 11), but is truncated by BD2 RT in the east (Fig. 12). Together, the sandier Facies 1 and 3 show "a well-developed trend toward finer sizes southeastward along the length of Joffre field. The pebbly and granular sandstone is prominent in the northwest, but the pebbles and granules gradually disappear southeastward, leaving only the cross-bedded glauconitic and nonglauconitic sandstones" (Downing and Walker, 1988). Northeastward, the sandstone of Facies 3 is truncated by BD2 RT (between wells 6-13-38-25 and 10-13-38-25 in Fig. 7). In Figs. 8 and 9, the Facies 3 sandstone either grades laterally into mudstone or is truncated by BD2 RT; there is not enough well and core control to distinguish between these two alternatives. However, if truncation by BD2 RT can be seen in the cross-section of Figure 7, it is likely to have occurred in the cross-sections of Figs. 8 and 9 also.

Facies 4 consists of black mudstone with thin very-fine to fine-grained sandstone beds up to about 1 cm thick. There is some indication of cross-lamination and a lenticularity suggestive of symmetrical ripples. The most characteristic feature of the mudstone is the restriction of the trace fauna to small examples of *Planolites*. There are two main layers of mudstone each about 2 m thick, and they occur exclusively along the northern edge of the Joffre field. The lower layer abruptly overlies the BD2 IT surface (or overlies the lag on this surface), as shown in Figure 12. It is also shown in Figure 7. It is not present in Figure 6, and probably lies northeast of well 4-35-38-26. The second layer is shown in Figs. 7 and 8, and in the core cross-section of Figure 12. The black mudstone layers have abrupt contacts with adjacent sandstone facies in well 12-8-38-24 (Fig. 12), but have more gradational contacts with bioturbated muddy siltstone in wells 10-13-38-25 and 3-24-38-25 (Fig. 11).

The topographic position of the Facies 4 mudstone is extremely important. They occur within the BD2 IT step described above, and always occur topographically below the sandstone of Facies 1. This is particularly clear in Figure 11, where the top of the Facies 4 mudstone layer lies below the base of the Facies 1 sandstone (where it rests on BD2 IT). It is also clear in Figure 12, where Facies 4 mudstone (in well 12-8-34-24) lies below the base of the Facies 1 sandstone in well 16-6-38-24. Similar relationships can be seen in the cross-sections of Figs. 7 and 8. We do not believe that there is sufficient core and well log evidence to correlate the black mudstone layers with any horizons above the top of the BD2 IT step, and hence we show the mudstone onlapping the step and pinching out against it. The juxtaposition of me-

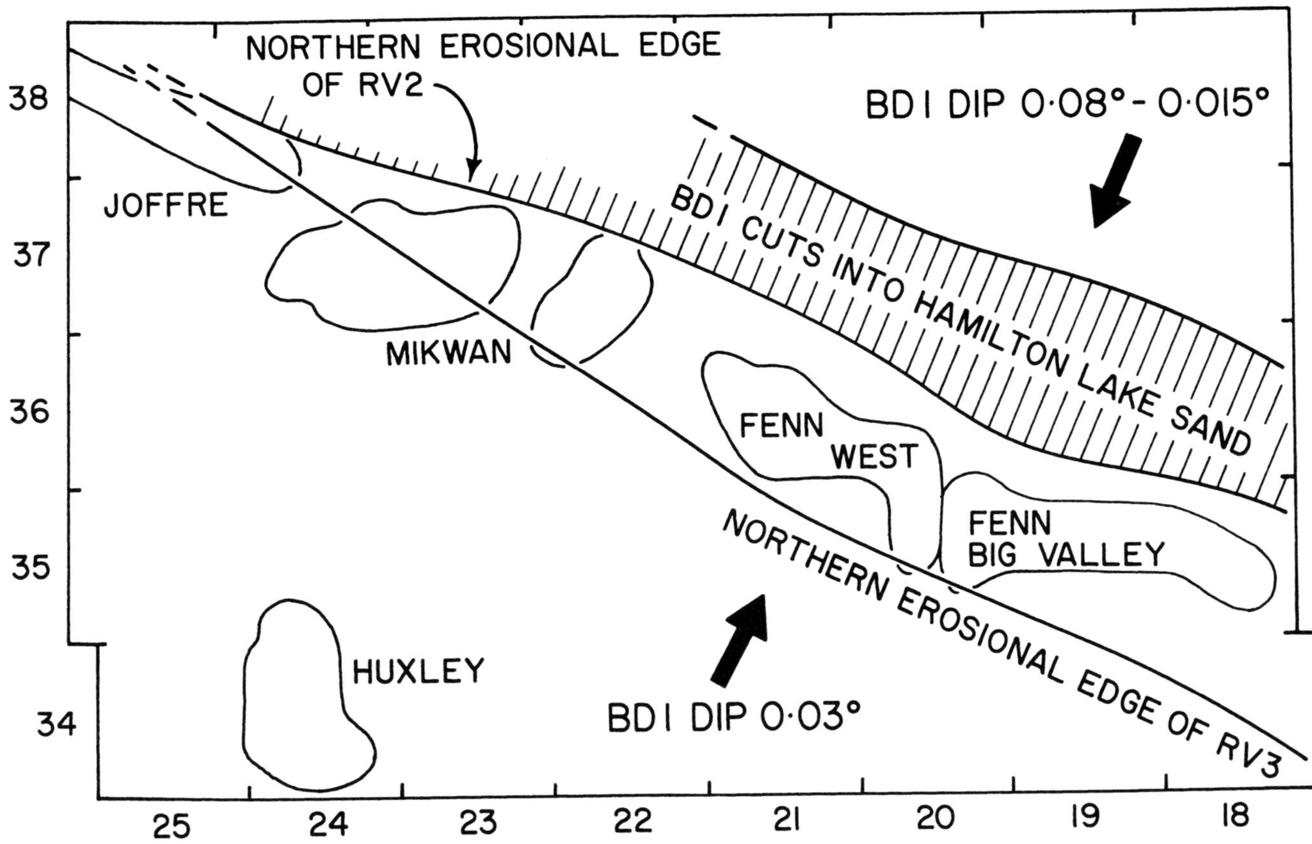

Fig. 13.—Map showing the northern erosional edges of RV2 and RV3 as they have been truncated by BD1. In the cross-shaded area, BD1 cuts down into the Hamilton Lake sandstone, and then rises again stratigraphically northeastward. The calculated dips on BD1 are about 0.03° northeastward, and 0.08-0.015° southwestward. All ranges are west of the fourth meridian.

dium- to coarse-grained, cross-bedded sandstone with mudstone horizons is unusual, and our interpretation, discussed below, differs from that of J. MacEachern et al. (pers. commun., 1998).

Allostratigraphic Unit Between BD2 IT to BD2 RT Traced Southeastward

Southeast of the tip of Joffre there is little core control at this stratigraphic interval. In the Mikwan area (Fig. 2, and McIntosh, 1995), there are about five cores that may span the interval. BD1 is easily picked, but BD2 IT is more difficult. The best two cores are 10-8-37-23 and 7-13-37-24; they both show a 1-2.5 m thick succession of thoroughly bioturbated sandstone but no evidence of the mudstone "split" that is prominent in the Joffre area. Farther southeastward, in the Fenn area, two cores (16-19-36-21 and 4-32-36-21; Fig. 10) span the interval and show a gradational sandier-upward succession 5.75 m thick. It begins with black mudstone interbedded with discontinuous laminations of very fine sandstone up to about 1 cm thick. The only traces are very small *Planolites*. Upward, the succession becomes sandier, but remains thoroughly bioturbated (mostly slightly larger *Planolites*). There are no coarse, cross-bedded facies similar to those at Joffre.

Bounding Discontinuity BD2 RT

BD2 RT is commonly a sharp surface across the study area, but it does not have a lag of coarser material. It can be correlated southwestward until it is truncated by BD3 (between wells 14-11-38-25 and 16-11-38-25 in Fig. 7; and between wells 4-32-36-21 and 16-32-36-21 in Fig. 10). Northward, BD2 RT cannot be traced far beyond the edge of the field (for example, northeast of well 10-13-38-25 in Fig. 7; northeast of well 12-8-38-24 in Fig. 9). Conceptually, RT surfaces pass seaward into correlative conformities (Fig. 5), and this is implied by the dashed lines in all of the cross-sections.

Allomember Between BD2 RT and BD3

BD2 RT is overlain by a sandier-upward succession of bioturbated mudstone, muddy sandstone, and sandstone displayed in cores from wells 6-13-38-25 and 10-13-38-25 (Fig. 12). The basal mudstone is interlaminated with sandstone on a centimeter scale, and contain variable amounts of bioturbation. The traces include *Planolites*, *Terebellina*, and *Rhizocorallium*. The sandy mudstone and sandstone contain lenticular, combined-flow ripples and some low-angle to flat lamination. Traces include *Terebellina* and *Skolithos*. There is a prominent bentonite just below the sandstone at the top of this sandier-upward succession (Fig. 11). Its absence in Figure 12 suggests erosion of the BD2 RT to BD3 allomember by BD3.

Bounding Discontinuity 3

BD3 can be traced across the entire study area. It truncates various underlying facies and discontinuities, and in the southwest it is overlain by a lag up to 19 cm thick. The lag consists of a clast-supported chert pebble conglomerate (maximum pebble size about 2 cm) or medium-grained sandstone with scattered pebbles. The sediment below the contact is in places sideritized to a thickness of about 30 cm, and unlined Skolithos burrows up to 30 cm long penetrate down from the contact.

Relationship Between BD2 IT, BD2 RT, and BD3

In the eastern part of the study area, BD2 IT is truncated by BD2 RT, locating the preserved shoreface incision north of Fenn Big Valley and through the middle of Fenn West. This incision forms the southern edge of the main productive shoreface sandbody at Joffre. It is clear that neither Fenn West nor Fenn Big Valley are lateral continuations of the Joffre shoreface sandbody. South of Fenn Big Valley, BD2 RT is truncated by BD3 (Fig. 14).

Relationships are reversed in the Mikwan area, where BD3 truncates BD2 RT in the north and BD2 IT farther south. Thus, the preserved position of the shoreface incision at Mikwan and at the southeastern end of Joffre is due to truncation of BD2 IT by BD3, not by BD2 RT. The original position of the shoreface incision (BD2 IT truncated by BD2 RT) must have lain to the south of the line marking the truncation of BD2 IT by BD3 in Figure 14.

In the transgressive shoreface model (Fig. 5, bottom panel), the RT surface truncates the IT surface as transgression resumes. However, in the cross section of Figure 10, the point of incision of BD2 IT by BD2 RT is not preserved because BD3 truncates BD2 RT between wells 4-32-36-21 and 16-32-36-21. The various points of truncation can be mapped and are shown in Figure 14. In the Fenn Big Valley area, BD2 IT is truncated by BD2 RT along the northern edge of the field, and BD2 RT is truncated by BD3 about one township south of the field. In the Fenn West area, the two erosional edges cross, such that BD2 RT is truncated by BD3 (Fig. 10 between wells 4-32 and 16-32) and BD2 IT continues farther to the south before being truncated by BD3 (Fig. 10, between wells 7-7 and 1-17). The same relationship of surfaces can be seen in the Joffre cross-sections (Fig. 7).

Allomember Between BD3 and BD4

In the Joffre-Mikwan-Fenn area, the BD3-BD4 allomember is generally about 4-6 m thick. It thickens southwestward to about 20 m at Huxley (well 10-4-34-24 in Fig. 10). The characteristic succession alternates from intervals of well-preserved, interbedded sandstone and mudstone to intervals dominated by bioturbation. In the well-preserved intervals, sandstone beds are up to 7 cm thick, with undulating laminations or unidirectional (current) ripple cross-lamination. Bioturbation is scarce, but includes *Planolites* and *Cylindrichnus*. In the bioturbated intervals, bedding is almost totally destroyed by *Terebellina, Planolites, Cylindrichnus, Teichichnus, Helminthopsis, Schaubcylindrichnus, Siphonichnus, Zoophycos*, and small mud-lined *Skolithos*.

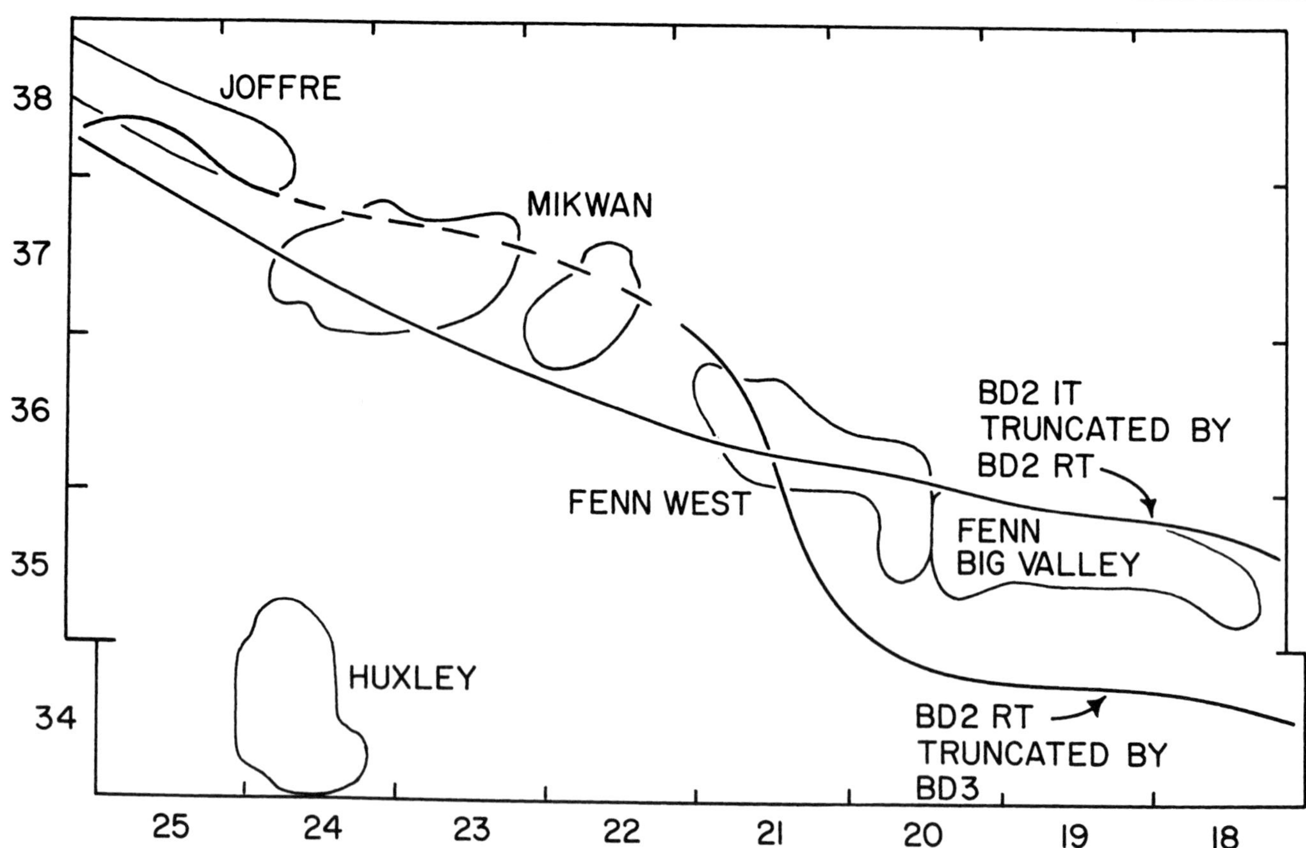

Fig. 14.—Map showing where BD2 IT is truncated by BD2 RT, and where BD2 RT is truncated by BD3. The two lines cross at Fenn West, showing that farther to the west the "BD2 IT truncated by BD2 RT" line becomes "BD2 IT truncated by BD3." All ranges are west of the fourth meridian.

Bounding Discontinuity 4

BD4 (VE4 of previous publications, see Walker, 1995) can be traced across the entire extent of the Viking in the subsurface, and is one of the most continuous and extensive bounding discontinuities to have been mapped in Alberta (Walker, 1995). In the study area, it sharply truncates underlying strata and is overlain by a lag up to 30 cm thick. This lag consists of alternating beds of conglomerate (pebbles up to 3.5 cm in diameter), pebbly sandstone, and mudstone. The bases of the coarser beds are sharp and loaded into underlying mudstone. The conglomerate is structureless and ranges from clast- to matrix-supported. Bioturbation is commonly intense, forming a churned mixture of mud, sand, and pebbles. The trace fauna include abundant *Teichichnus*, *Helminthopsis*, *Terebellina*, and *Planolites*, with less common *Skolithos*, *Diplocraterion*, *Thalassinoides*, *Zoophycos*, *Asterosoma*, and *Schaubcylindrichnus*. Shafts penetrate downward from BD4 for as much as 11 cm, and include *Skolithos*, *Diplocraterion*, and *Thalassinoides*.

Allomember Between BD4 and Base of Fish Scales

In the study area, this allomember is equivalent to the Westgate Formation of Bloch et al. (1993). It is 28-60 m thick, and consists mostly of very black mudstone with thin (2-3 cm) beds of very fine-grained sandstone. These beds are normally sharp-based and contain gently undulating lamination and ripple cross-lamination. Bioturbation is rare (*Planolites* and *Helminthopsis* are the most common traces), but fish scales are common on bedding planes.

Within the mudstone there are numerous siderite horizons 2-30 cm thick. Above several of the siderites are thin (1-19 cm) fine- to medium-grained sandstone beds with scattered chert granules and pebbles (maximum diameter is 2 cm). These sandstone beds are sharp-based and totally bioturbated by *Teichichnus*, *Planolites*, *Diplocraterion*, *Skolithos*, *Terebellina*, and *Helminthopsis*. The siderite layers and associated sandstone beds become less common upward from BD4 into the Westgate Formation.

INTERPRETATIONS OF ALLOMEMBERS BELOW BD2

Successions RV1, RV2, and RV3

In the study area, successions RV1 and RV2 appear to have formed by aggradation in an open marine setting. We also interpret the RV3 succession to have formed by aggradation in open marine conditions, but with progradation of lower to middle shoreface environments suggested by the sandiness and trace fauna (particularly *M. simplicatus*) of the upper part of the succession. This interpretation is in full agreement with the microfaunal, palynological, and ichnological evidence published by MacEachern et al. (1, this volume).

Bounding Discontinuity 1

BD1 erosionally truncates regional Viking successions RV3 and RV2 and cuts down into the top of RV1 (the Hamilton Lake sand). The geometry of the surface is shown in Figure 13. The stratigraphic rise of the surface from the diagonally ruled area (Fig. 13) southwestward suggests that BD1 is a transgressive surface of erosion. There is no line of incision within the study area that suggests a pause in transgression and local shoreface incision and progradation (as there is associated with BD2).

The implication of this interpretation is that after deposition of RV3, relative sea level lowered, such that the shoreline moved northeastward out of the study area, which was entirely subaerially exposed. There are no incised valleys associated with the BD1 surface, nor any other direct evidence of subaerial exposure, but apart from a deep incised valley, we would not expect features of subaerial exposure (root traces, paleosols) to survive transgressive erosion. The minimum distance of shoreline movement would have been about 65 km. If the gradient of the seafloor on top of RV3 had been about 1:1000, the fall and rise of relative sea level would have been 65 m. For a gradient of 1:2000, the fluctuation would have been 32.5 m. These gradients will be further discussed with respect to BD2.

Allostratigraphic Unit Between BD1 and BD2

This sandier-upward succession is interpreted as an open marine to shoreface succession that prograded into the accommodation space created during the BD1 transgression. No indications of beach or high-energy upper shoreface (cross-bedded sandstone) are preserved, presumably because of erosion by BD2 IT. The microfaunal, palynological, and ichnological data published by MacEachern et al. (1, this volume) indicate a "fully marine, unstressed, archetypal *Cruziana* ichnofacies," in agreement with our interpretation of an open marine to shoreface succession.

Bounding Discontinuity BD2 IT

Following the progradation of the BD1-BD2 allostratigraphic unit, the study area was subaerially exposed as the shoreline advanced northeastward. A second major transgression moved the shoreline southwestward again, with an important pause in transgression when a shoreface was incised (Fig. 14). The preserved position of this incision lies just southwest of Joffre and can be correlated continuously to a position immediately north of Fenn Big Valley.

It was originally recognized by Downing and Walker (1988) that the BD2 IT surface rises stratigraphically northeastward from the northern edge of Joffre (surface E2 in Fig. 6). Indeed, in Figure 6, the vertical exaggeration makes the BD2 IT (E2) surface resemble an incised valley. However, the southwestward dip of the BD2 IT surface northeast of Joffre has an angle of about 0.07°, or a gradient of 0.0013, as calculated from geometrical relationships shown in Figure 10. The southwestward-dipping geometry is not indicative of a channel, but may be interpreted as a function of the rate of transgression and the fact that datums are conventionally drawn horizontally (Fig. 15). We argue that a transgressive surface of erosion cannot be formed with a landward-dipping (southwestward-dipping) gradient—this would imply increasing wave scour and depth of erosion in a landward direction. The minimum condition is that a transgressive surface of erosion (TSE) is cut horizontally at fair-weather wave base during a pause in relative sea-level rise. The normal condition is that a TSE rises gently stratigraphically during rise of relative sea level. The northeastward-rising TSE northeast of the linear fields (Figs. 6, 10) is probably an artifact due to the fact that our datums have been drawn horizontally (LM2 in Fig. 6, and top of RV2 in Figs. 7-9). The only reason for doing this is convention—geologically, the top of RV2 probably dipped gently northeastward into the basin at the time it was deposited. Therefore, if the top of RV2 is given a gradient of about 1 in 1000 (0.0013 or 0.07° ; Fig. 15) into the basin, the northeastern portion of BD2 IT becomes

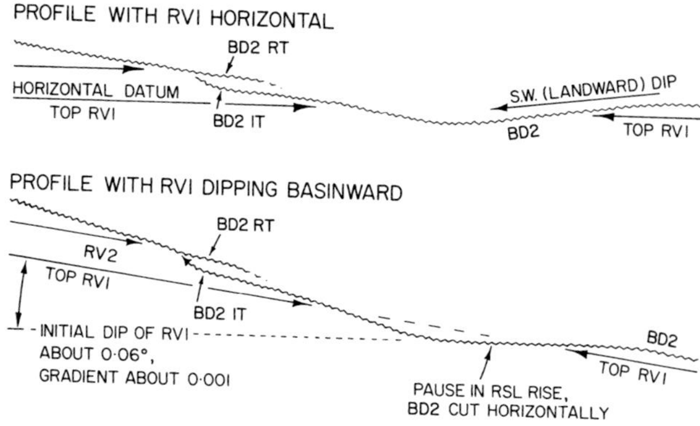

Fig. 15.—Diagram to show (above) that if the top of RV1 is drawn horizontally, BD2 has a landward dip on the northeastern side of the profile (to the right). This is not considered likely, or even possible. The section below has therefore been rotated to make BD2 horizontal on the right. This implies that the top of RV1 would have had a basinward dip of about 0.001 (1 m per kilometer) at the time that the BD2 surface was cut horizontally during a pause in overall sea-level rise.

horizontal. This would imply that it was cut during a pause in relative sea level rise, and the geometry of the rest of the surface across the fields and to the southwest is a function of variable rates of relative sea-level rise.

If this analysis is correct, we conclude that there is no evidence of a strike-parallel channel or embayment of any sort, and no evidence for any form of topographic high (past or present) northeast of Joffre, as suggested by McEachern et al. (2, this volume).

We may also infer that the shoreline associated with BD2 moved across the width of the study area (southwest to northeast and back). If the original gradient of the substrate was 0.0013, the absolute fluctuation of sea level would have been 1.3 m per kilometer of shoreline transgression. The seaward limit of BD2 is difficult to estimate due to poor control in the northeast (Fig. 14), but the minimum movement was probably about 50 km, implying a rise of relative sea level of about 65 m. Although this appears to be a large figure, at today's rate of relative sea-level rise (about 1 mm/yr) it would have taken place in a mere 65,000 years. At Pleistocene rates, closer to 10 mm/yr, the rise would be achieved in 6,500 years.

Our BD2 surface is the same as that termed FS/SB2 by J. MacEachern et al. (pers. commun., 1998). We agree about the identification and geometry of the surface, but disagree about the interpretation of the succession between BD2 IT and BD2 RT (their succession between FS/SB2 and FS, termed an "embayment complex").

INTERPRETATIONS OF THE ALLOMEMBER BETWEEN BD2 IT RT—IS JOFFRE A SHOREFACE DEPOSIT?

This allomember contains the main sandbody in Joffre Field and represents the unit where our interpretation differs from that of J. MacEachern et al. (pers. commun., 1998). To the south of Joffre, it can be shown that the sandbody rests within an asymmetrical incision, interpreted by Downing and Walker (1988) as a shoreface cut during a pause in an overall transgression. We have found no evidence that compels a change in this interpretation. In contrast, J. MacEachern et al. (pers. commun., 1998) interpret the coarse-grained facies as representing tidal creeks and channels feeding bay-head deltas that prograded into a barred embayment.

The first line of evidence in favor of an incised shoreface interpretation is the geometry of the lower discontinuity, BD2 IT. This is asymmetrical, steeper on the southwestern (landward) side, and much flatter on the seaward side. The incision is essentially linear, as is the seaward limit of coarse facies (Figs. 14, 18, 22 of MacEachern et al., 2, this volume). This seaward limit forms a very smooth and rather straight line, suggesting alongshore movement of sediment by wave-driven currents. This suggestion is reinforced by the observation of Downing and Walker (1988) that there is a "well-developed trend toward finer sizes southeastward along the length of Joffre field. The pebbly and granular sandstone is prominent in the northwest, but the pebbles and granules gradually disappear southeastward, leaving only the cross-bedded glauconitic and nonglauconitic sandstones." The implication is a single fluvial point of input of sediment to the shoreface in the northwest, with about 35 km of alongstrike sediment dispersal in the shoreface to the southeast.

The cross-bedded sandstone is compatible with a shoreface interpretation. Facies 1 (Fig. 7) is up to about 4 m in thickness, and is sharply overlain by mudstone at the base of the "split." The fact that only 4 m of section is preserved suggests that the facies represents lower shoreface (higher parts of the shoreface may have been eroded, as shown in Figure 5, bottom panel). The sandstone in 6-13 (Fig. 7) grades seaward into a sandier-upward, prograding succession of mudstone and sandstone in well 10-13, suggesting that 1) in 6-13 the sandstone was deposited above FWWB on the bounding discontinuity (left arrow, bottom panel, Fig. 5), while 2) in 10-13 the mudstone was deposited below FWWB (right arrow, bottom panel, Fig. 5) and grades upward into lower shoreface sandstone. Even farther seaward, the succession passes into mudstone in 3-24, and somewhere in this succession is the correlative conformity that grades landward into BD2 RT.

In an alternative scenario, it can be argued that progradation may have been accompanied by a drop of relative sea level (a forced regression). If this were the case, the entire shoreface succession would have been deposited in very little accommodation space, giving a complete (lower, middle, and upper) succession in a thickness of only a few meters. Thus the preserved 4 m succession may represent both lower and middle (and possibly part of the upper) shoreface, with only the beach and foreshore removed during the resumed transgression.

Facies 2, in the "split" (Figs. 11, 12), consists of a sandier-upward succession that begins with black mudstone. The contact of Facies 2 black mudstone on Facies 1 sandstone can be interpreted as a flooding surface, implying a pause in shoreface progradation and a minor deepening of relative sea level. Progradation quickly resumed (sandier upward Facies 2 succession), and the bioturbated muddy sandstone of Facies 2 grades rapidly into coarse, cross-bedded sandstone of Facies 3. These probably represent lower to middle shoreface, with the cross-bedding representing dunes moving southeastward in the longshore drift system.

There is a prominent "step" in the BD2 IT erosion surface that can be traced across the entire study area. It occurs between 6-13 and 10-13 in Figure 7, and between 16-6 and 12-8 in Figure 9. It can be seen that the black mudstone of Facies 4 onlap this stepped topography, suggesting that they were deposited transgressively before progradation of the overlying Facies 1 shoreface sandstone. This relationship does not show in Figure 17 of J. MacEachern et al. (pers. commun.,

1998)—in their cross-section, the black mudstone interfingers with the lowermost Joffre sandstone between wells 10-13 and 6-13. These topographic differences may be partly due to choice of datum. We have assumed that the top of regional Viking succession RV2 was essentially flat when deposited, and that shoreface incision can be compared with this horizon. J. MacEachern et al. (pers. commun., 1998) have used a flooding surface above BD2 RT as a datum. This may have distorted topographic relationships below the datum if the datum itself has erosional topographic relief (like BD2 RT).

We interpret the black mudstone onlapping the BD2 IT surface in the same way that Davies and Walker (1993) interpreted black mudstone onlapping the VE4 (BD4) surface at Caroline and Garrington. The mudstone represents deposition during transgression of the BD2 IT surface, but as soon as there is a pause in transgression, progradation of shoreface deposits begins. A resumption of transgression would lead to cutoff of sediment supply and deposition of transgressive mudstone abruptly on top of shoreface deposits (the lower mudstone part of Facies 2 in the "split," Figs. 7 and 9, and particularly the base of Facies 2 in well 6-13 in Fig. 11). Facies 3 represents another stillstand and progradation. After these minor oscillations of relative sea level, the main transgression resumed and surface BD2 RT was cut into Facies 1, 2, and 3 (Figs. 7, 9).

Embayment Interpretation of J. MacEachern et al. (pers. commun., 1998)

J. MacEachern et al. (pers. commun., 1998) present three reasons why the BD2 IT-BD2 RT succession "is not consistent with the previously proposed transgressively incised shoreface interpretation of Downing and Walker (1988)." They state that the "trough cross-stratified coarse clastics must correspond to upper shoreface (nearshore) conditions," and note that they are interstratified with "marginal marine fine-grained deposits." However, Downing and Walker (1988) did not claim that the coarse deposits were "upper shoreface"; indeed, they stated (p. 1225) that the cross-bedded glauconitic sandstone and granule sandstone represent "middle to lower shoreface deposits." We suggest here that they probably represent lower shoreface, simply because upper shoreface (and to a lesser extent middle shoreface) sediments are eroded during the resumed transgression. If the coarse layers are lower shoreface, perhaps moved by longshore currents during storms, they can easily be interstratified with thin mudstone layers representing much quieter conditions. The mudstone rip-up clasts are not restricted to the upper shoreface, as implied by J. MacEachern et al. (pers. commun., 1998), but may have formed as storm conditions increased and coarse sediment was transported across a muddy substrate.

If the coarse sediments reflect "nearshore deposition, they should grade seaward into contemporaneous middle- and lower-shoreface burrowed to hummocky/swaley stratified sandstones" (J. MacEachern et al., pers. commun., 1998). We do not necessarily agree with this supposition, and we are not aware of any example in the Western Interior Seaway where coarse cross-bedded facies are associated with swaley cross-stratification (which commonly develops in fine-grained sandstone). We agree that the coarse facies passes seaward within 400 m into mudstone, and we agree that this mudstone contains a low diversity, stressed trace fossil suite. We also accept that this facies contains fewer foraminifers than mudstone below BD2 IT and above BD2 RT (see next section). However, we suggest that the stress could have been related to high concentrations of suspended sediment, high rates of deposition on the bed, unusually soft substrates, and times of freshwater flooding into the sea. The stress does not necessarily require an embayment with a barrier on the seaward side (see next section).

The Embayment Interpretation of J. MacEachern et al. (2, this volume)

The succession between BD2 IT and BD2 RT is interpreted by MacEachern et al. (2, this volume) as an "embayment complex." They propose a barrier to the northeast, and instead of shoreface deposition along the Viking shoreline, they suggest a series of bayhead deltas supplied by many rivers flowing from the southwest.

We believe that there are several problems with this interpretation. First, their proposed model "requires the existence of a barrier complex lying northeastward of the zero edge of the Joffre Embayment Complex; a barrier whose deposits are conspicuously absent from the depositional record" (MacEachern et al., 2, this volume). We agree that these deposits are conspicuously absent. The suggestion of a barrier is based on the faunal characteristics of the mudstone in the BD2 IT-BD2 RT succession. The foraminiferal evidence is based on six samples, three of which are from the base of the succession and contain a "generally abundant, though variable" number of foraminifers, including environmentally sensitive forms such as *Miliammina*, *Ammobaculites*, and *Verneuilinoides*. This suggests relatively unstressed, more-or-less marine conditions. Another one of the six samples comes from near the top of the interval, and also contains "a more abundant and diverse foraminiferal assemblage." Only two of the six samples, both from the same core (11-7-39-26), contain low abundances of foraminifers, and only environmentally tolerant forms. MacEachern et al. (2, this volume) then comment that the "palynological samples, though varied, are surprisingly similar to the open marine assemblages" found above BD2 RT and below BD2 IT. Ichnologically, the BD2 IT - BD2 RT succession corresponds to a "low diversity (stressed) mixed *Skolithos-Cruziana* assemblage, typical of environments subject to reduced and/or fluctuating salinity, episodic deposition, and variable substrate consistency." The problem remains as to whether a barrier (not preserved) must be proposed to enclose an embayment, or whether the existing evidence is compatible with mudstone being deposited rapidly immediately seaward of a shoreface. In such an environment, stress might be related to rapid deposition, soft to soupy substrates, and freshwater input from the rivers supplying sediment to the shoreface. Because the evidence for the embayment is not overwhelming, and the barrier itself is not preserved, we prefer the simpler shoreface interpretation, as originally suggested by Downing and Walker (1988).

The coarse cross-bedded sandstone along the Joffre shoreline is difficult to interpret as high-energy shoreface deposits if they are located on the landward side of a protected embayment. Consequently, MacEachern et al. (2, this volume) reinterpret them as a succession of "tidal channels, tidal creeks, and marine-influenced distributary channels of bay-head delta systems," admitting at the same time that "the absence of cyclic, tidally-bundled mud couplets marking the foresets makes this interpretation equivocal" (MacEachern et al., 2, this volume). If the concept of an embayment is rejected, there is no need to propose a series of tidal channels and bay-head deltas.

The geometry of the proposed bay-head deltas is also problematic. The seaward edge of the coarse-grained facies at Joffre is remarkably straight (MacEachern et al., 2, this volume, Figs. 14, 18, 22), implying that each of the bay-head deltas prograded exactly the same distance, and that no "lobate" shoreline geometries were developed. It is even more remarkable that the shoreline is so straight, given that the deltas would have been prograding into a very shallow embayment. Even slight differences in discharge of sediment would have led to major differences in the distance of progradation. The total preserved thickness of the BD2 IT-BD2 RT succession is only 6-8 m; three different successions prograded into this accommodation space in the MacEachern et al. (2, this volume) interpretation.

Finally, J. MacEachern et al. (pers. commun., 1998) note that an "upward increase in sandstone content reflects shallowing of the bay during fill, and locally may indicate close proximity to a bay head delta." However, in the next sentence, they indicate that the trace fossil assemblage is "consistent with a salinity-stressed setting....coupled with increasingly more normal-marine conditions upward." The increasingly more normal-marine conditions upward do not seem to be consistent with bay shallowing and close proximity to a bay head delta.

CONTINUATION OF THE JOFFRE TREND TOWARD MIKWAN AND FENN

On the maps (Figs. 1, 2), the Viking fields from Joffre to Fenn Big Valley appear to fall in a straight line, superficially suggesting that they all lie within the same shoreface incision, and that the reservoir rocks all lie in the BD2 IT to BD2 RT stratigraphic interval. This is not the case, as can be seen in a detailed examination of the cross-section in Figure 10.

This cross-section shows that at the southern end, the producing sandstone at Huxley belongs to the Regional Viking succession RV3. This succession coarsens and becomes sandier-upward, passing from offshore marine bioturbated mudstone and sandstone into shoreface deposits containing *Macaronichnus segregatis*. In 10-04-34-24, the shoreface is truncated by BD3 and the section remains muddy and bioturbated in the lower part of the BD3-BD4 interval where there is core control. In the adjacent well, 04-27-22-34-24, the BD1-BD3 interval consists of a sandier-upward succession of well-bioturbated mudstone and sandstone, with a sharp lower contact (BD1) and a concentration of glauconite grains on that surface. The sand content increases from about 40-60% upward, but remains in the very-fine (upper part) to fine (lower part) size range. Some sharp-based event beds up to 8 cm thick are preserved in the succession, and contain flat to undulating laminations within the beds. Mudstone layers up to 1 cm thick are preserved. The trace fauna is characterized by abundant *Helminthopsis*, with *Planolites*, *Terebellina*, *Siphonichnus*, *Paleophycus*, *Rosselia*, *Zoophycos*, and small *Skolithos* also present.

The succession described above lies above BD1. Both BD2 surfaces have been truncated farther to the north (near wells 1-17 and 16-32, Fig. 10).

In the Fenn area (Fig. 10), all of the bounding discontinuities are present, including BD2 IT and BD2 RT. However, the succession between BD2 IT and RT tends to be muddy and bioturbated in the Fenn area, and the best sand development (and all of the production) is in the BD1-BD2 IT interval. This is best seen in well 1-17-36-21. Similar relationships can be seen in cross-sections through the western part of Fenn West, and the eastern and western ends of Fenn Big Valley (Fig. 10, and cross-sections in Burton, 1997). Note in Figure 2 that there is very little core control in the Fenn area; well 16-19 fails to penetrate the main producing horizon, and well 4-32 lies at the northern edge of the field, where the sandbody has essentially passed seaward into much finer grained facies.

The BD1-BD2 IT interval appears to become sandier-upward in wells 7-7-36-21 and 1-17-36-21 (Fig. 10). We interpret this to indicate a prograding offshore to shoreface succession, as at Joffre. The important difference is that at Joffre, the upper part of this succession was truncated by BD2 IT, while at Fenn, more of the succession has been preserved. In 4-32-36-21, the BD1-BD2 IT succession is about 2.2 m thick. It begins with a pebble lag on BD1 and then consists of a totally bioturbated, fully marine muddy sandstone. The BD2 IT surface is associated with a 10 cm siderite layer, above which there is a very abrupt facies change into overlying silt-laminated dark mudstone.

At Fenn West, the productive sandbody lies between BD1 and BD2 IT, as in well 1-17-36-21 (Fig. 10). Note that this sandbody is progressively cut out by BD3 to the southwest. By contrast, in Fenn Big Valley, BD2 IT is truncated by BD2 RT exactly along the northern edge of the field. The producing sandbody therefore lies between BD1 and BD3, and the updip trap (along the northern edge of the field) is due to fine-grained, transgressive mudstone that lies on the BD2 IT surface and extends from BD2 IT to BD2 RT.

CONCLUSIONS

1. The Regional Viking successions were deposited in open marine settings. They coarsen and become sandier-upward, reaching shoreface environments in the Huxley area. Production in Huxley Field is from the shoreface sandstone in RV3.

2. After deposition of the RV successions, there was a major lowering of relative sea level, followed by transgression. Pauses in the transgression, with shoreface progradation, gave rise to a sandier-upward succession on surface BD1, which is productive in the Fenn area.

3. There was another major drop of relative sea level and another transgression, forming surface BD2 IT (Initial Transgression). The succession between BD2 IT and BD2 RT (Resumed Transgression) also coarsens and becomes sandier-upward, but its origin is controversial (see below).

4. The geometry of both BD1 and BD2 IT with respect to a horizontally drawn datum shows that the surfaces dip southwestward (landward) north of the fields, but the dip is northeastward (seaward, as would be expected) south of the fields. The change of dip of the surfaces suggests an east-southeast-west-northwest trending "valley," but the dips of the valley walls are very gentle, with angles from 0.015° to 0.08°. A rotation of the datum downward to the northeast by 0.015° to 0.08° would make the presently southwestward-dipping erosion surface horizontal; it could then be interpreted as a surface cut during a stillstand of relative sea level. This apparent valley may have first given rise to the concept of a high, or barrier, to the north of Joffre.

5. We interpret the succession between BD2 IT and BD2 RT at Joffre to have been deposited in a series of prograding shorefaces during a pause in overall transgression.

6. We disagree with MacEachern et al. (this volume) that the coarse deposits at Joffre represent bayhead deltas that prograded into a brackish (stressed) lagoon with a barrier to the north.

7. The Joffre shoreface incision and the overlying coarse-grained facies cannot be traced along-strike to the east-southeast. The Fenn West and Fenn Big Valley Fields lie along-strike from Joffre, but the productive interval lies between BD1 and BD2 IT (and not between BD2 IT and BD2 RT, as at Joffre).

ACKNOWLEDGMENTS

We have enjoyed many discussions with James MacEachern and thank him for freely sharing his ideas and data, and for suggesting many core locations at the eastern end of Joffre for our examination. We have tried to structure this paper so that readers can appreciate how the two different interpretations of Joffre were developed. We also thank David James and Wascana Energy, Inc., for logistical help in making maps and giving access to well logs. The work was funded by the Natural Sciences and Engineering Research Council of Canada.

REFERENCES

AMAJOR, L.C. AND LERBEKMO, J.F., 1990, The Viking (Albian) reservoir sandstones of cental and south-central Alberta, Canada. Part II—Lithofacies analysis, depositional environments and paleogeographical setting: Journal of Petroleum Geology, v. 13, p. 421-436.

BEAUMONT, E.A., 1984, Retrogradational shelf sedimentation: Lower Cretaceous Viking Formation, central Alberta, *in* Tillman, R.W. and Siemers, C.T., (eds.), Siliciclastic Shelf Sediments: Tulsa, Society for Sedimentary Geology (SEPM) Special Publication 34, p. 163-177.

BERGMAN, K.M., 1994, Shannon Sandstone in Hartzog Draw-Heldt Draw fields (Cretaceous, Wyoming, USA) reinterpreted as lowstand shoreface deposits: Journal of Sedimentary Research, v. B64, p. 184-201.

BERGMAN, K.M. AND WALKER, R.G., 1986, Cardium Formation conglomerates at Carrot Creek field: Offshore linear ridges or shoreface deposits?, *in* Moslow, T.F. and Rhodes, E.G., (eds.) Modern and Ancient Shelf Clastics: A Core workshop: Tulsa, Society for Sedimentary Geology (SEPM) Core Workshop 9, p. 217-268.

BERGMAN, K.M. AND WALKER, R.G., 1987, The importance of sea level fluctuations in the formation of linear conglomerate bodies: Carrot Creek Member, Cretaceous Western Interior Seaway, Alberta, Canada: Journal of Sedimentary Petrology, v. 57, p. 651-665.

BERGMAN, K.M. AND WALKER, R.G., 1988, Formation of Cardium erosion surface E5, and associated deposition of conglomerate; Carrot Creek Field, Cretaceous Western Interior Seaway, Alberta, *in* James, D.P. and Leckie, D.A., (eds.), Sequences, Stratigraphy, Sedimentology: Surface and Subsurface: Calgary, Canadian Society of Petroleum Geologists Memoir 15, p. 15-24.

BERGMAN, K.M. AND WALKER, R.G., 1995, High-resolution sequence stratigraphic analysis of the Shannon Sandstone in Wyoming, using a template for regional correlation: Journal of Sedimentary Research, v. B. 65, p. 255-264.

BLOCH, J., SCHRODER-ADAMS, C., LECKIE, D.A., MCINTYRE, D.J., CRAIG., J. AND STANILAND, M., 1993, Revised stratigraphy of the lower Colorado Group (Albian to Turonian), western Canada: Bulletin of Canadian Petroleum Geology, v. 41, p. 325-348.

BOETHLING, F.C., 1977, Typical Viking sequence; a marine sand enclosed with marine shales: Oil and Gas Journal, March 28, 1977, p. 172-176.

BURTON, J.A., 1997, Sedimentology, ichnology and high-resolution allostratigraphy of the Lower Cretaceous Viking Formation, central Alberta, Canada: Unpublished M. Sc. Thesis, McMaster University, Hamilton, Canada, 145 p.

DAVIES, S.D. AND WALKER, R.G., 1993, Reservoir geometry influenced by high-frequency forced regressions within an overall transgression; Caroline and Garrington fields, Viking Formation (Lower Cretaceous), Alberta: Bulletin of Canadian Petroleum Geology, v. 41, p. 407-421.

DEWIEL, J.E.F., 1956, Viking and Cardium not turbidity current deposits: Journal of the Alberta Society of Petroleum Geologists, v. 4, p. 173-177.

DOWNING, K.P. AND WALKER, R.G., 1988, Viking Formation, Joffre field, Alberta: Shoreface origin of long, narrow sand body encased in marine mudstones: American Association of Petroleum Geologists Bulletin, v. 72, p. 1212-1228.

EVANS, W.E., 1970, Imbricate linear sandstone bodies of Viking Formation in Dodsland-Hoosier area of southwestern Saskatchewan, Canada: American Association of Petroleum Geologists Bulletin, v. 54, p. 469-486.

HEIN, F.J., DEAN, M.E., DEIURE, A.M., GRANT, S.K., ROBB, G.A. AND LONGSTAFFE, F.J., 1986, The Viking Formation in the Caroline, Garrington and Harmattan East fields, western south-central Alberta: Sedimentology and paleogeography: Bulletin of Canadian Petroleum Geology, v. 34, p. 91-110.

KOLDIJK, W.S., 1976, Gilby Viking "B"; a storm deposit, *in* Lerand, M.M., (ed.), The sedimentology of selected oil and gas reservoirs in Alberta: Calgary, Canadian Society of Petroleum Geologists, Core Conference Proceedings, p. 62-77.

MCINTOSH, L., 1995, The sedimentology and stratigraphy of the Lower Cretaceous (Albian) Viking Formation, Joffre/Mikwan fields, Alberta, Canada: Unpublished B. Sc. Thesis, McMaster University, Hamilton, Canada, 71 p.

PATTISON, S.A.J. AND WALKER, R.G., 1992, Deposition and interpretation of long, narrow sand bodies underlain by a basinwide erosion surface: Cardium Formation, Cretaceous Western Interior Seaway, Alberta, Canada: Journal of Sedimentary Petrology, v. 62, p. 292-309.

POSAMENTIER, H.W. AND CHAMBERLAIN, C.J., 1993, Sequence stratigraphic analysis of Viking Formation lowstand beach deposits at Joarcam field, Alberta, Canada, *in* Posamentier, H.W., Summerhayes, C.P., Haq, B.U. and Allen, G.P., (eds.), Stratigraphy and Facies Associations in a Sequence Stratigraphic Framework: Oxford, International Association of Sedimentologists Special Publication 18, p. 469-485.

POWER, B.A., 1988, Coarsening-upwards shoreface and shelf sequences: Examples from the Lower Cretaceous Viking Formation at Joarcam, Alberta, Canada, *in* James, D.P. and Leckie, D.A., (eds.), Sequences, Stratigraphy, Sedimentology: Surface and Subsurface: Calgary, Canadian Society of Petroleum Geologists Memoir 15, p. 185-194.

RADDYSH, H.K., 1988, Sedimentology and "geometry" of the Lower Cretaceous Viking Formation, Gilby A and B fields, Alberta, *in* James, D.P. and Leckie, D.A., (eds.), Sequences, Stratigraphy, Sedimentology: Surface and Subsurface: Calgary, Canadian Society of Petroleum Geologists Memoir 15, p. 417-429.

REINSON, G.E., WATERS, W.J., COX, J. AND PRICE, P.R., 1994, Cretaceous Viking Formation of the Western Canadian Sedimentary Basin, *in* Mossop, G. and Shetsen, I., (eds.), Geological Atlas of the Western Canadian Sedimentary Basin: Calgary, Canadian Society of Petroleum Geologists / Alberta Research Council, p. 353-363.

SHELTON, J.S., 1973, Models of sandstone deposits; a methodology for determining sand genesis and trend: Oklahoma Geological Survey Bulletin, v. 118, p. 91-94.

SLIPPER, S.E., 1918, Viking gas field: Structure of the area: Calgary, Geological Survey of Canada, Summary Report 1917, Part C, p. 6-7.

TILLMAN, R.W. AND MARTINSEN, R.S., 1984, The Shannon shelf-ridge sandstone complex, Salt Creek anticline area, Powder River Basin, Wyoming, *in* Tillman, R.W. and Siemers, C.T., (eds.), Siliciclastic Shelf Sediments: Tulsa, Society for Sedimentary Geology (SEPM) Special Publication 34, p. 85-142.

TILLMAN, R.W. AND MARTINSEN, R.S., 1987, Sedimentologic model and production characteristics of Hartzog Draw field, Wyoming, a Shannon shelf-ridge sandstone, *in* Tillman, R.W. and Weber, K.J., (eds.), Reservoir Sedimentology: Tulsa, Society for Sedimentary Geology (SEPM) Special Publication 40, p. 15-112.

TIZZARD, P.G. AND LERBEKMO, J.F., 1975, Depositional history of the Viking Formation, Suffield area, Alberta, Canada: Bulletin of Canadian Petroleum Geology, v. 23, p. 715-752.

WALKER, R.G., 1995, Sedimentary and tectonic origin of a transgressive surface of erosion: Viking Formation, Alberta, Canada: Journal of Sedimentary Research, v. B65, p. 209-221.

WALKER, R.G. AND BERGMAN, K.M., 1993, Shannon Sandstone in Wyoming: A shelf-ridge complex reinterpreted as lowstand shoreface deposits: Journal of Sedimentary Petrology, v. 63, p. 839-851.

WALKER, R.G. AND WISEMAN, T.R., 1995, Lowstand shorefaces, transgressive incised shorefaces, and forced regressions: Examples from the Viking Formation, Joarcam area, Alberta: Journal of Sedimentary Research, v. B65, p. 132-141.

COARSE-GRAINED, SHORELINE-ATTACHED, MARGINAL MARINE PARASEQUENCES OF THE VIKING FORMATION, JOFFRE FIELD, ALBERTA, CANADA

JAMES A. MACEACHERN

Earth Sciences, Simon Fraser University, Burnaby, British Columbia V5A 1S6, Canada

BRIAN A. ZAITLIN

PanCanadian Petroleum Ltd.

AND

S. GEORGE PEMBERTON

Department of Earth and Atmospheric Sciences, University of Alberta, Edmonton, Alberta T6G 2E3, Canada

ABSTRACT: The Viking Formation of the Joffre Field area comprises parts of three discrete sequences. The reservoir facies lies within the lower part of Sequence 3, interpreted to reflect shoreline-attached, marginal marine deposits. Sequence 3 was initiated by a fall of relative sea level with associated subaerial exposure and erosional scour, generating a sequence boundary. This sequence boundary was erosionally modified by ravinement during the ensuing transgression, to form a broad NW-SE trending asymmetric incision at Joffre, referred to as basal discontinuity 2 (BD-2).

BD-2 is mantled by conglomeratic lags and bioturbated, glauconitic transgressive sand sheets (Facies A). These are overlain by moderately burrowed, trough and low angle planar cross-stratified sandstone, pebbly sandstone and conglomerate, concentrated along the southern (landward) margin of BD-2, and interpreted to reflect distributary channels (Facies B/C/D). These channels fed sediment to prograding bay-head delta fronts, which coalesced to form broad, NW-SE trending coarse clastic aprons (Facies B/C/D and F). Each bay-head delta apron progressively interfingers with, and ultimately passes into weakly burrowed interbedded sandstone and mudstone (Facies E). These fine-grained deposits contain oscillation ripples, storm-generated wavy parallel laminations, and low diversity trace fossil suites, and are interpreted as brackish-water bay deposits.

Fluctuations in the rate of transgression resulted in the shifting of brackish-water bay deposits over channel/bay-head delta complexes, delineating three discrete marginal marine parasequences. Each parasequence trends northwest-southeast, onlaps relief on BD-2 along its landward (southwestern) margin, and offlaps/downlaps to the northeast. Resumed regional transgression generated a flooding surface (FS) with associated ravinement, that terminated brackish-water deposition and returned the study area to fully marine, offshore conditions.

INTRODUCTION

The Viking Joffre Field (Lower Cretaceous) is part of an elongate trend of fields, including Gilby, Mikwan, Fenn and Chain, extending NW-SE for approximately 250 km in central Alberta (Fig. 1). Joffre is the southeastern-most oil field in a trend that becomes gas-prone to the south. MacEachern et al. (1998) summarized the Viking Formation succession in the Joffre area, integrating ichnology with sedimentology and high resolution sequence stratigraphy to identify three discrete sequences. The reservoir interval corresponds to the lower portion of Sequence 3 and consists of three stacked, anomalously coarse-grained, NW-SE trending, narrow, linear conglomeratic sandstone bodies, interstratified at their distal (NE) edges with dark, weakly burrowed mudstone. The mapped distribution of Lower Sequence 3 and the basinward limits of the reservoir facies are shown in Figure 2. This paper concentrates on the characteristics, paleoenvironments, stacking, and distribution of the three parasequences that comprise the reservoir interval of Sequence 3.

The reservoir succession at Joffre was previously interpreted as a single incised conglomeratic shoreface, stranded during transgression in an offshore to shelf setting (Downing and Walker, 1988), an interpretation now widely accepted. Further work by Burton and Walker (this volume) refines this model to account for the interstratification of mudstone-dominated facies with conglomerate, by introducing two high-order lowstand events during an overall transgression. Their model now proposes two stacked lowstand or forced regressive conglomeratic shorefaces generated by high frequency relative falls in sea level during an overall transgression. Many characteristics of the reservoir succession, however, are incompatible with a shoreface interpretation in our opinion. These characteristics have been discussed in detail in MacEachern et al. (1998), and are outlined here in the "Interpretation of Lower Sequence 3" section of the present paper.

Additionally, the succession superficially resembles an estuarine incised valley complex, due to the apparent shape of the basal contact, the dominance of trough cross-bedded coarse clastics, and the interstratification of conglomeratic sandstones with brackish-water mudstones. Again, MacEachern et al. (1998) have outlined the incompatibilities with this model, which are also summarized here in the "Interpretation of Lower Sequence 3" section of the present paper.

STUDY AREA AND DATA BASE

The Joffre Field, located in Townships 37-39, Ranges 24-27W4 in central Alberta, Canada, was discovered in 1953, and extends for some 35 km along a NW-SE trend (Fig. 2). The Viking Formation within the Joffre Field contained some 14831 x 10^3m^3 (93 x 10^6 barrels US) of original oil in place (ERCB, 1994). Established reserves constituted 6451 x 10^3m^3, reflecting 2481 x 10^3m^3 from primary and 3970 x 10^3m^3 from enhanced (3087 x 10^3m^3 water flood and 883 x 10^3m^3 solvent flood) recovery techniques. To date, cumulative production totals approximately 5982.5 x 10^3m^3 (37.6 x 10^6 barrels US).

The Lower Cretaceous Viking Formation consists predominantly of westerly derived siliciclastics, and reflects northeastward progradation of sediment into the developing Alberta foreland basin in response to the progressive uplift of the Cordillera. Subsidence within the basin has since resulted in a southwesterly dip for the Viking Formation. Closure in

Isolated Shallow Marine Sand Bodies: Sequence Stratigraphic Analysis and Sedimentologic Interpretation.
SEPM Special Publication No. 64, Copyright © 1999
SEPM (Society for Sedimentary Geology), ISBN 1-56576-057-3, p. 273-296.

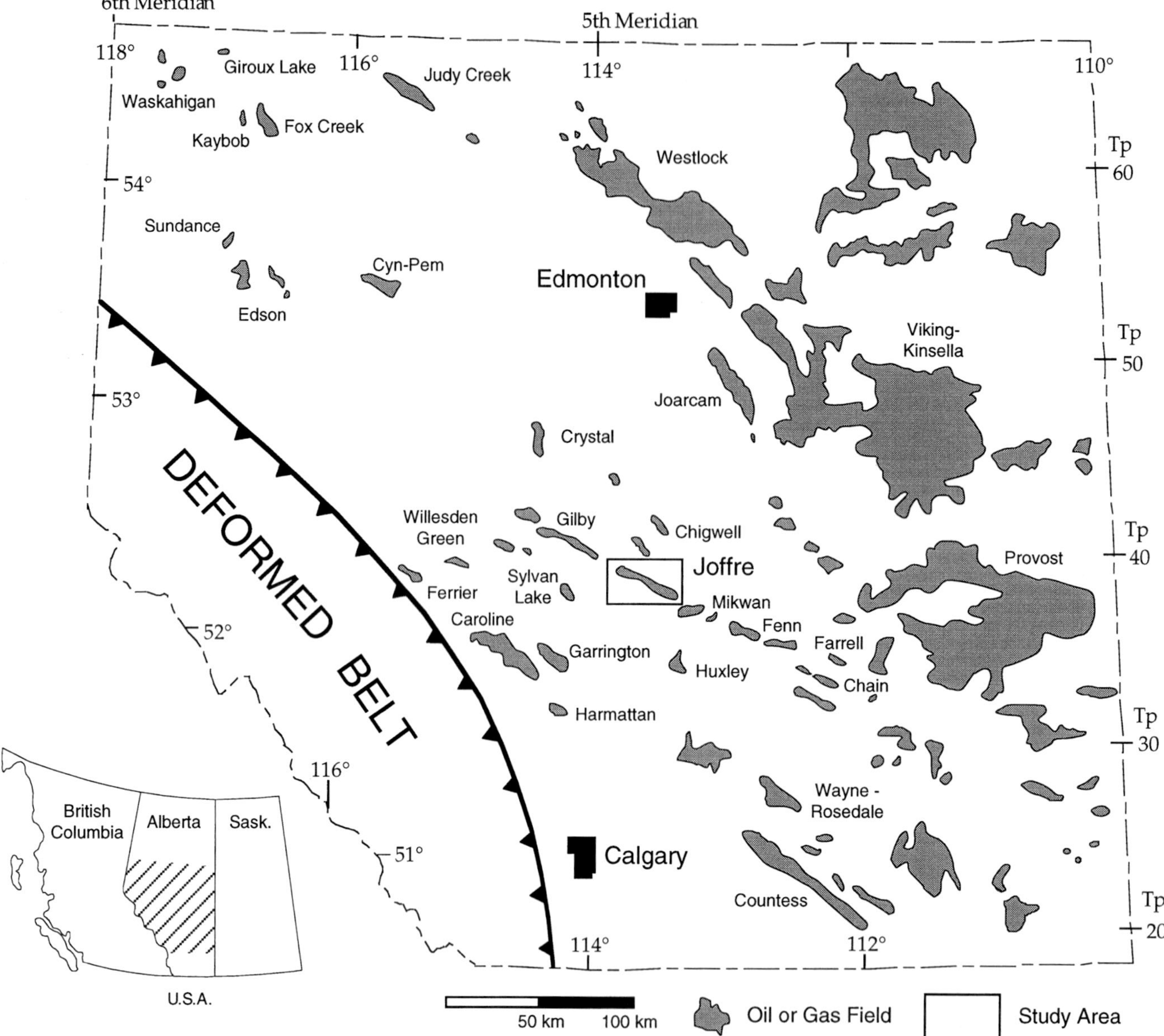

Fig. 1.—Major Viking Formation hydrocarbon field locations in Alberta, Canada.

the field is therefore a combination structural/stratigraphic trap, reflecting structurally high distal edges of coarse-grained parasequences that interfinger basinward with mudstone.

The study area contains approximately 950 wells that penetrate the Viking Formation, of which approximately 280 contain core from the interval. This study used data from 110 cores (Fig. 2), which were logged in detail to integrate physical sedimentological, ichnological, and sequence stratigraphic data. Selected core lithologs were used in the construction of 9 (6 dip-oriented and 3 strike-oriented) stratigraphic facies cross sections.

In addition, 700 of the 950 geophysical well-log suites were analyzed to delineate the internal stratigraphic discontinuities. "Picks" from these wells were incorporated into a database used for all mapping. Selected gamma-ray and resistivity geophysical well log responses were used to construct 8 regional stratigraphic cross-sections and 17 local (field-scale) stratigraphic cross-sections. The regional cross-sections and large-scale mapping delineated the distributions, geometries and thickness variations of the discontinuity-bound sequences employed in development of the general model (MacEachern et al., 1998). The field-scale well log and litholog facies cross-sections were used to determine the orientation, distribution, and geometry of the three parasequences comprising the lower part of Sequence 3.

REGIONAL STRATIGRAPHIC RELATIONSHIPS

The Viking Formation is Late Albian in age. It passes upward from marine shale of the Joli Fou Formation and is overlain by the transgressive marine shale of the Westgate Formation (Fig. 3). The Joli Fou Formation unconformably overlies the Mannville Group and is roughly equivalent to the Skull Creek shale of the Colorado Group in Montana and the Thermopolis shale in Wyoming (McGookey et al., 1972; Weimer, 1984). The Viking Formation is roughly equivalent

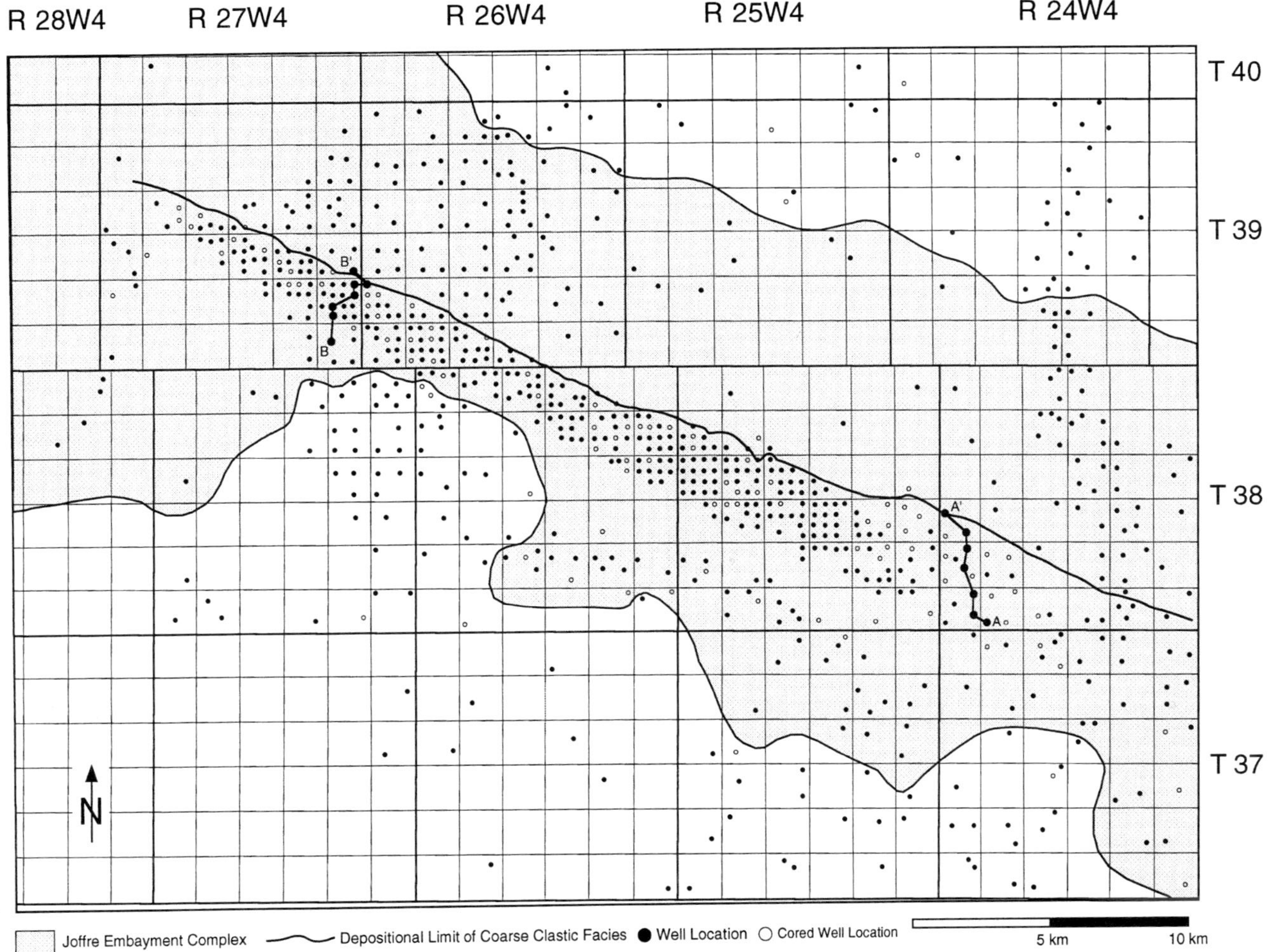

Fig. 2.—Joffre Field study area. The map shows the depositional limit of coarse clastic reservoir facies as well as cross-section lines A-A' and B-B'.

to the Paddy Member of the Peace River Formation (Stelck and Leckie, 1990), the upper part of the Bow Island Formation (Glaister, 1959), as well as the Muddy Sandstone, Newcastle Formation and J-Sandstone in Montana, Wyoming and Colorado, respectively (McGookey et al., 1972; Weimer, 1984). The shales of the Westgate Formation are stratigraphically equivalent to the lower part of the Shaftesbury Formation (Stelck and Leckie, 1990; Bloch et al., 1993), and to part of the Hasler Formation in N.E. British Columbia (Stelck and Leckie, 1990). In the United States, these shales are equivalent to the Shell Creek (Caldwell et al., 1993; Obradovich, 1993). The Mowry shale in Montana and Wyoming is regarded to be Cenomanian in age (Cobban and Kennedy, 1989) and is equivalent to the Base of Fish Scales Marker and the overlying shale.

The Viking Formation is highly complex and contains numerous stratigraphic discontinuities. Attempts to subdivide the interval into regionally correlative genetic units have been undertaken by Boreen and Walker (1991), Pattison (1991), Davies and Walker (1993) and most recently by Burton and Walker (this volume). These attempts have sought to establish a formal allostratigraphic framework for the Viking Formation that conforms to the rules of the North American Code of Stratigraphic Nomenclature (NACSN, 1983). Others have taken a sequence stratigraphic approach to the subdivision of the interval (e.g., Posamentier and Chamberlain, 1993; Leckie and Reinson, 1993). To date, a paucity of good internal markers and lack of a precise biostratigraphic framework for the interval in central Alberta have limited the ability of researchers to carry their correlations reliably across the Western Canada Sedimentary Basin.

VIKING STRATIGRAPHY OF THE JOFFRE AREA

The Viking Formation, as preserved within the Joffre Field area (Fig. 4), contains parts of three discrete sequences separated by two regionally extensive, transgressively modified sequence boundaries (MacEachern et al., 1998). These major stratigraphic breaks were accurately delineated by Downing and Walker, (1988) and constituted the fundamental bounding discontinuities of their allostratigraphic units. Boreen and Walker (1991) correlated the lower of the two surfaces to their VE3a surface. These discontinuities are referred to as BD-1 and BD-2, employing the terminology of Burton and Walker (this volume), although our interpretations regarding the specific details of their genesis differ somewhat.

The basal sequence (Sequence 1) consists of stacked, NW-SE trending, regionally extensive, coarsening-upward shelf

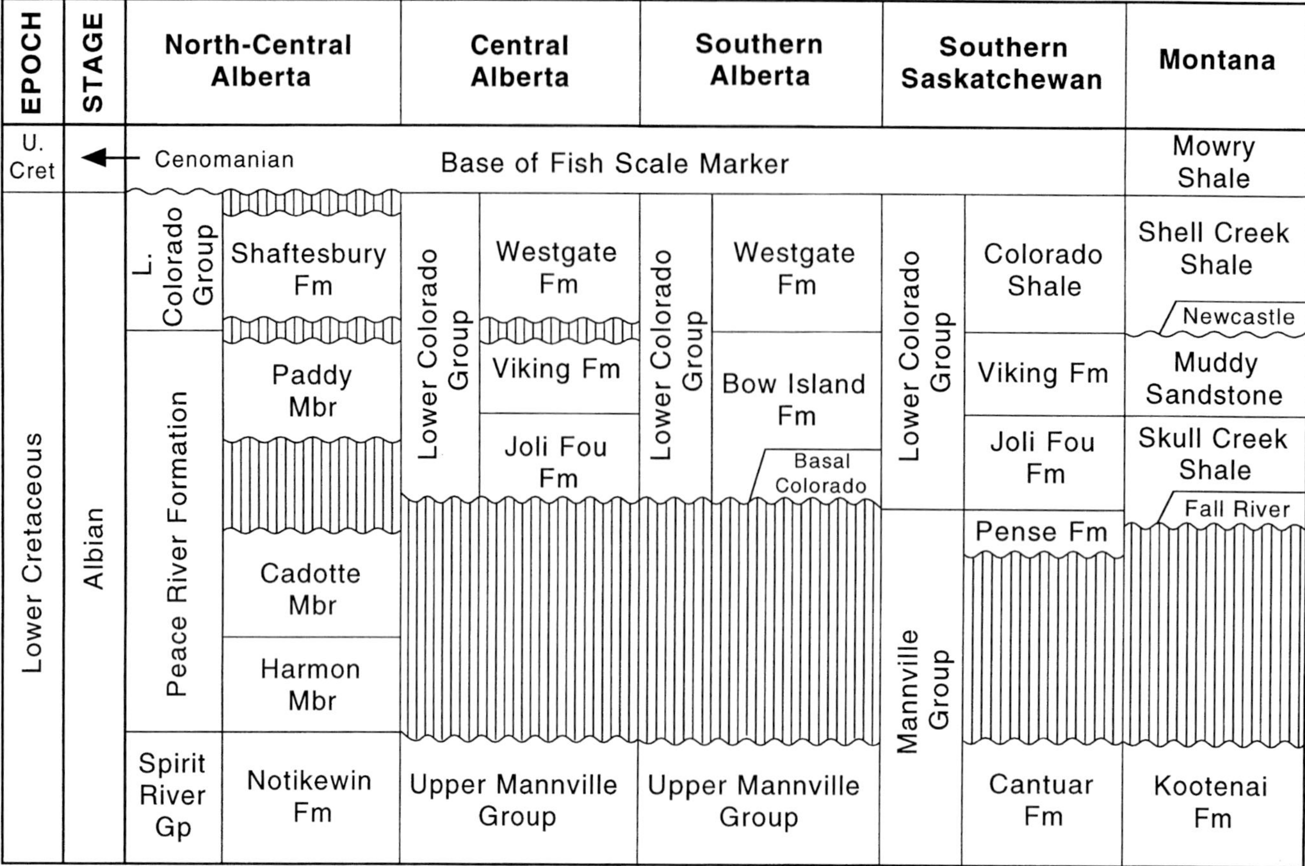

Fig. 3.—Stratigraphic correlation chart for the Viking Formation.

to lower shoreface parasequences informally referred to as the Regional Viking. The succession records progressive development of distal, through archetypal, and into proximal *Cruziana* ichnofacies. The facies are characterized by intense and uniformly distributed burrowing, as well as a diverse assemblage of ichnogenera (Fig. 5), reflecting slow, continuous, fully marine deposition with minimal influence of storm events. These parasequences downlap onto transgressive marine shales of the Joli Fou Formation, and constitute a progradational parasequence set of a highstand systems tract.

This marine parasequence set is erosionally truncated by an amalgamated sequence boundary and flooding surface (BD-1) typically demarcated by the *Glossifungites* ichnofacies. The discontinuity slopes steeply along the southwestern (landward) edge, and flattens out to the northeast, forming an asymmetric, scarp-like geometry. BD-1 is overlain by thoroughly bioturbated, fully marine, gritty sandy mudstone and muddy sandstone of Sequence 2, and is interpreted as an incised, early transgressive shoreface that prograded northward during a pause in the rate of relative sea level rise. Sequence 2 is truncated by the overlying BD-2 discontinuity, and near the southeastern end of Joffre, this succession is preserved as an erosional remnant. Toward the northwestern end, Sequence 2 is largely removed, and BD-2 typically rests directly on the highstand marine parasequences of Sequence 1.

The deposits of Sequence 2 are incised by BD-2, which forms a NW-SE trending trough, locally demarcated by the *Glossifungites* ichnofacies. The deposits overlying BD-2 constitute part of Sequence 3 and contrast markedly with the fully marine deposits of Sequence 1 and Sequence 2. The lower part of Sequence 3 consists of three stacked, coarse-grained parasequences, separated by marginal marine flooding surfaces (Fig. 4). The lower part of Sequence 3 is overlain and locally truncated by a regionally extensive flooding surface, designated as FS. Facies deposited above FS reflect a return to fully marine offshore conditions in the study area.

LOWER SEQUENCE 3

The lower part of Sequence 3, lying between BD-2 and FS, constitutes the Viking Formation reservoir interval in the Joffre Field. The map of lower Sequence 3 (Fig. 2) shows two erosional "zero" edges, related to erosional truncation by the overlying regional flooding surface FS. The map also demarcates the northern limit of lower Sequence 3 conglomeratic facies in the Joffre area, and constitutes the maximum progradational limits of the three parasequences. Northward of these coarser deposits, the succession is dominated by muddy facies, and internal parasequences cannot easily be delineated in core or on geophysical well logs. For this reason, detailed cross-sections are restricted to the southern (landward) margin of the deposit.

Lower Sequence 3 consists of three stacked parasequences (Fig. 4). Each parasequence overlies an erosional surface in the landward direction that passes northward and northeastward into a nonerosional and locally gradational downlap surface. The basal surface is regarded as a progradational/ depositional surface of autocyclic origin, based on the facies relations discussed below. Coarse grained facies of Parasequence 1 rest on P1, those of Parasequence 2 rest on P2, and equivalent facies of Parasequence 3 rest on P3. Each parasequence is terminated by a *marginal marine* flooding surface overlain by weakly burrowed mudstone-dominated facies. Parasequence 1 is terminated by F2, Parasequence 2 is capped by F3, and Parasequence 3 is truncated by the regional flooding surface FS that marks the return to fully marine conditions in the study area. F1 constitutes the initial flooding and transgressive modification of the sequence boundary that generated BD-2. F1 can be easily identified in core and on well logs where mudstone directly overlies BD-2. In landward positions, however, the facies above F1 are sandy and difficult to differentiate from the progradational elements lying above P1. In these positions, it is unreliable to differentiate between the sandy transgressive elements and the sandy progradational elements particularly using geophysical well logs. For this reason, the F1 surface is defined by the limit of the mudstone component. The sandstone/conglomerate body of Parasequence 1 is regarded to overlie P1 in its entirety, although the body actually consists of both a thin transgressive component that should lie below P1, and an overlying progradational component lying above P1 (Fig. 4).

Parasequence 1

Basal Discontinuity 2 (BD-2)

BD-2 is an amalgamated sequence boundary and flooding surface with a scarp-like geometry, that truncates the early transgressive shoreface deposits of Sequence 2. The surface shows evidence of erosion in all cored intervals. Although geophysical well log mapping of the surface shows it to be a broad, asymmetrical U-shaped trough. MacEachern et al. (1998) demonstrated that this shape was an artifact of stratigraphic pull-up imparted by the necessity of using an originally seaward-inclined stratigraphic surface as the datum horizon.

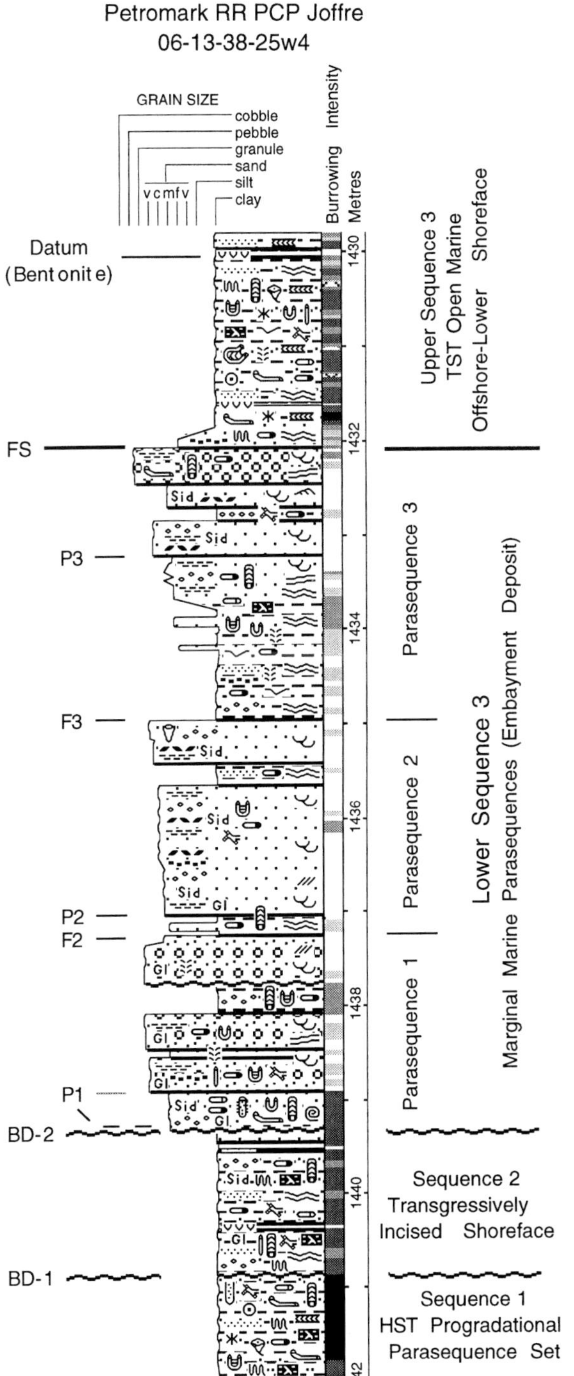

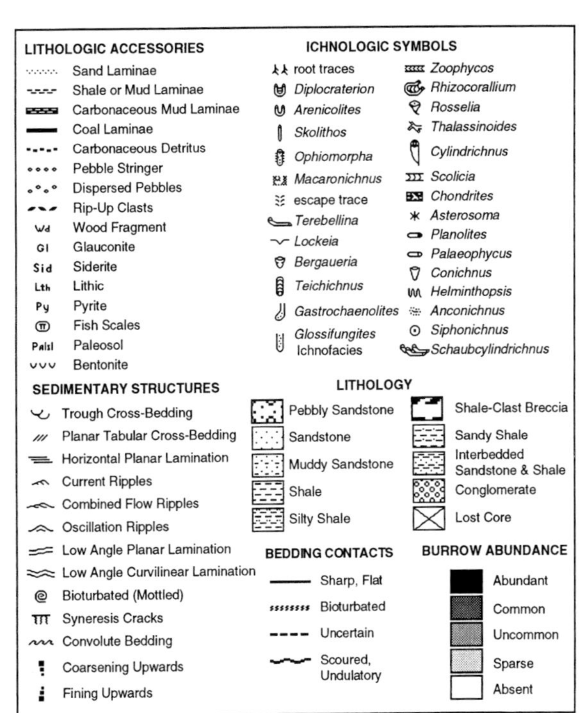

Fig. 4.—Representative core litholog for the Viking Formation of the Joffre area, showing the discontinuities. The dashed line below P1 represents the most practical selection for P1. The gray line represents the actual position of P1. Refer to the text for further discussion.

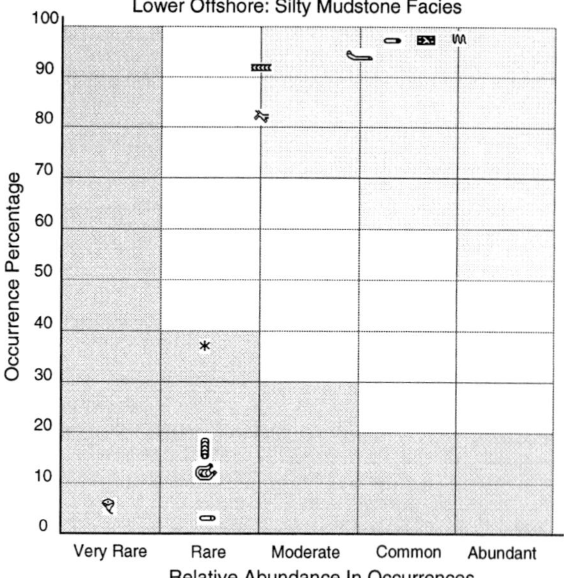

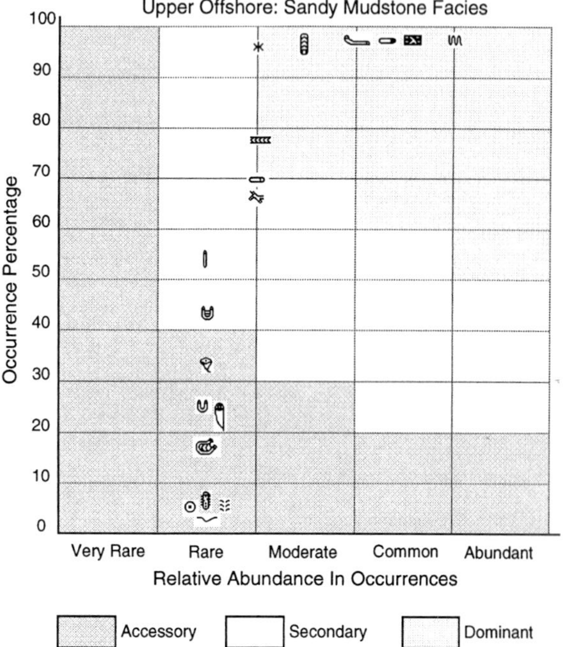

Fig. 5.—Cross-plots, showing ichnogenera occurrence percentage *vs.* abundance percentage for highstand marine offshore deposits of the regional Viking Formation (modified after MacEachern and Pemberton, 1994). Refer to Figure 4 for the legend of trace fossil symbols.

The BD-2 surface is locally mantled by a thin (1-5 cm thick) chert pebble lag and is commonly demarcated by a *Glossifungites* assemblage, dominated by firmground *Diplocraterion*, with local development of firmground *Thalassinoides* and *Skolithos*. The *Glossifungites* ichnofacies is a recurring, substrate-controlled assemblage of trace fossils that reflects the colonization of a firmground. The suite encompasses ichnogenera which are "pseudo-bored" into an underlying, semi-lithified substrate. Ichnogenera of the firmground assemblage are typically unlined, sharp-walled (and locally scratch marked), vertical to subvertical dwelling structures. The structures cross-cut the original resident softground trace fossil community, and are generally passively infilled with sediment overlying the discontinuity (Saunders and Pemberton, 1986; Savrda, 1991; MacEachern et al., 1992; Pemberton et al., 1992; Pemberton and MacEachern, 1995).

Facies Association Overlying Flooding Surface 1 (F1)

The transgressive limit of the fine-grained facies associated with F1 generally lies basinward of all other flooding surfaces (Fig. 6). In reality, however, the sandstone component of this initial transgression (mainly Facies A, described below) probably persists nearly to the base of the escarpment on BD-2 (Fig. 7). Fine-grained deposits associated with F1 mantle BD-2 in only six cored intervals and lie outboard of the Viking reservoir. In each of the six cases, the interval associated with F1 is quite thin, ranging from 0.1-0.6m, and averaging 0.4 m. Most cored intervals lie in the northwestern part of the field.

The facies has been designated Facies E (MacEachern et al., 1998; cf., Fig. 8), and is characterized by interbedded mudstone and sandstone that typically contain dispersed pebbles and granules of chert, glaucony, pyrite and carbonaceous detritus. Sandstone beds range from 1.0-5.0 cm in thickness and comprise between 5% and 15% of the facies. Individual sandstone beds tend to be well sorted, but may range in grain size from lower fine to lower medium. Sandstone beds are sharp based and are oscillation rippled, combined flow rippled, or contain wavy parallel laminations. Mudstone beds range from 1.0-20.0 cm in thickness, are typically silt and sand poor, and contain considerable carbonaceous detritus that imparts a dark color. Mudstone beds are locally siderite cemented or display displacive siderite nodule development.

Facies E is generally weakly burrowed with a sporadically distributed and low diversity trace fossil suite (Fig. 9). It is unlikely, however, that a complete trace fossil assemblage is known from this facies in light of the limited number of intervals and reduced thicknesses encountered. *Helminthopsis* occurs in only two of the six intervals and in very rare numbers. Ichnogenera with an occurrence of approximately 17% reflect only a single interval in which they were encountered. Most intervals only contain between 3 and 4 ichnogenera. Thus, although the assemblage contains a total of 13 ichnogenera, only *Planolites*, *Teichichnus*, *Diplocraterion*, and *Thalassinoides* can be considered as recurring elements of the suite. This constitutes a low diversity, low abundance assemblage, generated by facies-crossing (opportunistic) organisms. Such suites are typical of stressed environments.

Facies Association Overlying Progradational/ Depositional Surface 1 (P1)

The facies association overlying progradational/depositional surface 1 (P1) occurs in 49 cored intervals, and consists of 3 main facies. In most landward locations, P1 is regarded to directly overlie BD-2 (Fig. 4). The basal facies corresponds to Facies A of MacEachern et al. (1998), and comprises highly glauconitic, muddy and pebbly burrowed sandstone. These pass upward into trough cross-stratified, variably glauconitic and pebbly sandstones, corresponding to Facies B, C, and D of MacEachern et al. (1998). Locally, Facies B/C/D may be capped by a variably burrowed, interstratified conglomerate, pebbly sandstone and mudstone facies, herein referred to as Facies F. More commonly, however, Facies F occurs basinward (northeastward) of the trough cross-stratified coarse clastics, suggesting that there is a proximal-distal relationship between the two.

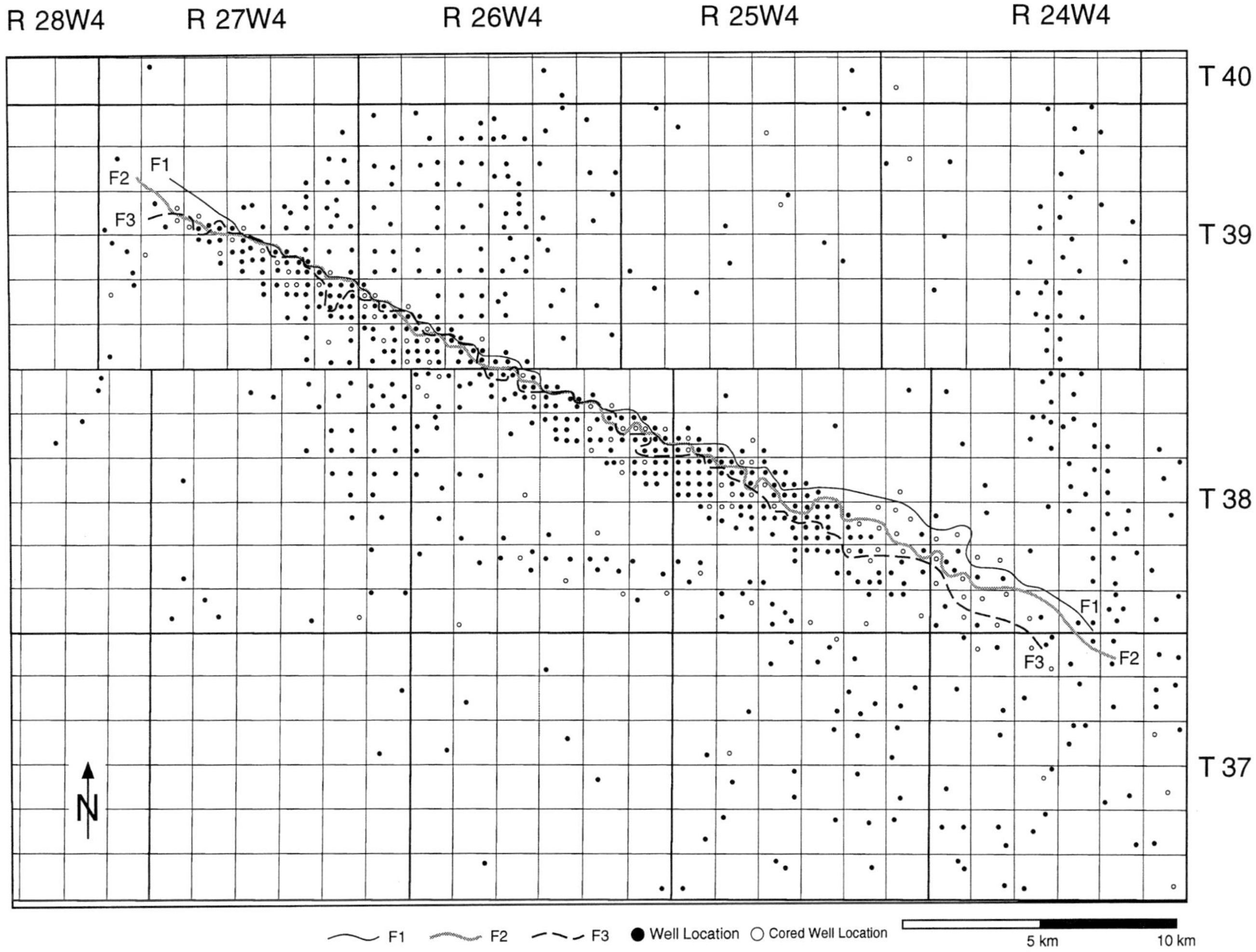

Fig. 6.—Landward transgressive limits of the marginal marine (bay) flooding surfaces F1, F2 and F3.

Facies A—In most locations, BD-2 is directly overlain by Facies A, although locally, a thin (0.5-10 cm) pebble lag, commonly glauconitic, mantles the discontinuity. Facies A is particularly well developed toward the southeast end of the field (Township 38, Ranges 24-25W4). These coarse clastics contain abundant, thin and locally siderite-cemented mudstone interbeds, as well as mud laminae and mudstone rip-up clasts (Fig. 10). Sand sizes range from lower medium to lower coarse, and typically contain very coarse sand and chert granule stringers. Basal units may be quite conglomeratic. Primary physical sedimentary structures are dominated by 3.0-5.0 cm thick, current ripple-laminated beds and 5.0-10.0 cm thick, small-scale trough cross-stratified beds. Low angle (<15°), planar stratification and rarer oscillation

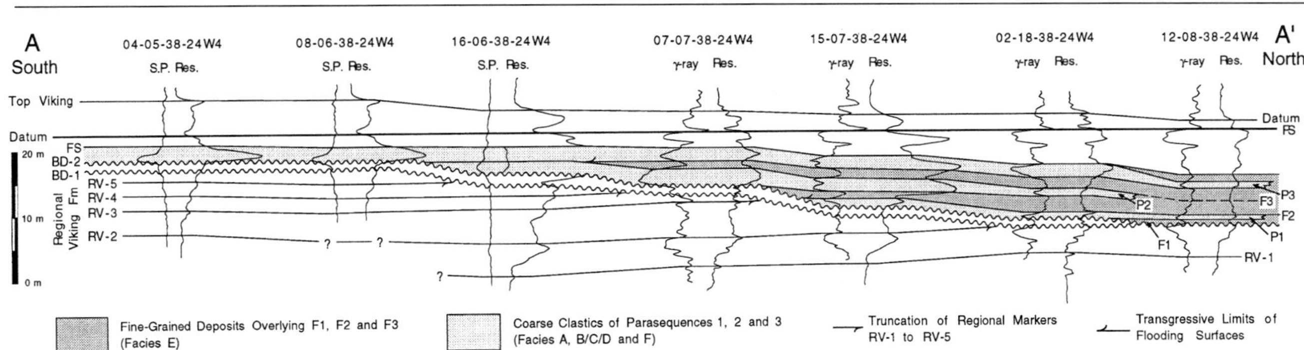

Fig. 7.—Cross-section A-A', showing the depositional onlap of Parasequences 1, 2 and 3 onto basal discontinuity BD-2, as well as their offlap to the northeast. The preserved transgressive limits of F1, F2 and F3 are also indicated. The line of section is displayed in the map of Figure 2.

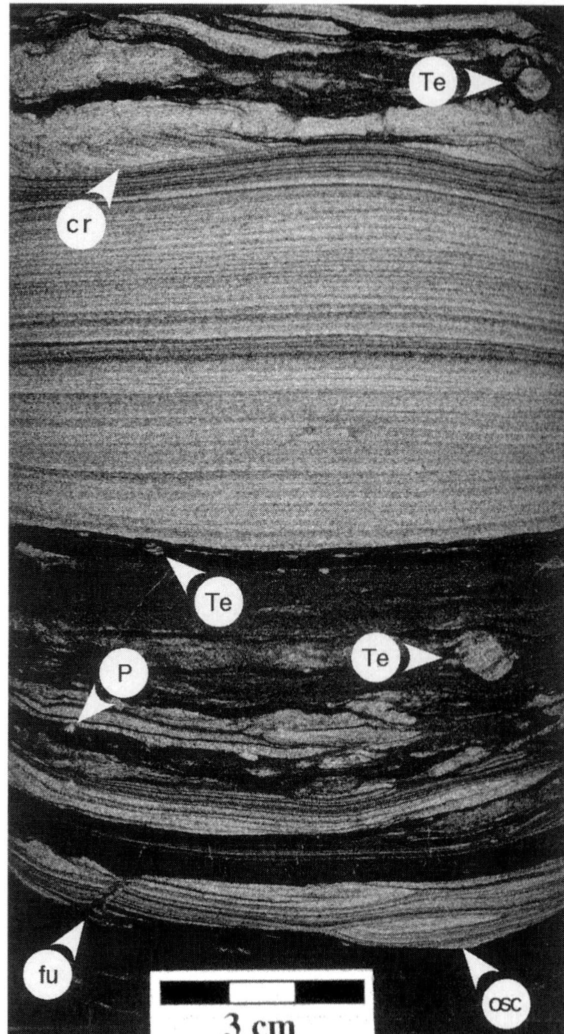

Fig. 8.—Photo of Facies E, consisting of interstratified mudstone and sandstone. Sandstone beds contain low angle wavy parallel lamination, oscillation ripple lamination (osc), and current ripple lamination (cr). Note the *Planolites* (P), *Teichichnus* (Te), and fugichnia (fu). Well 03-24-38-25W4; 1437.7m.

ripple lamination occur in some intercalated sandstone beds.

Facies A displays moderate to low degrees of burrowing, sporadically distributed and diminishing in intensity upward. Sandstone beds contain a trace fossil suite that is dominated by *Diplocraterion, Skolithos, Conichnus, Ophiomorpha, Palaeophycus* and *Rosselia*, with variable numbers of escape structures (fugichnia). The mudstone interbeds typically contain small numbers of *Planolites, Teichichnus*, and *Chondrites*. *Terebellina, Bergaueria, Siphonichnus, Asterosoma, Arenicolites* and *Helminthopsis* are very rare components of the assemblage. The overall trace fossil suite corresponds to a somewhat impoverished mixed *Skolithos-Cruziana* assemblage. The sandstone contains the *Skolithos* ichnofacies, while the interstratified mudstone possesses a proximal *Cruziana* suite. The bulk of all bioturbation within Parasequence 1 occurs within this facies, such that the trace fossil assemblage cross-plot illustrated in Figure 11 largely reflects the assemblage of Facies A. Although the assemblage appears quite diverse (18 ichnogenera) only 7 genera can be considered "characteristic" of the suite. The remaining forms are encountered in only a few intervals. As well, many of the ichnogenera are likely associated with the more marine units, and correspond to proximal facies related to initial transgression across BD-2, rather than to the progradational portion of the succession. Differentiating between these two genetically discrete sandstones is problematic, due to the "sand-on-sand" contact, the widespread cannibalization or intense reworking of the earlier transgressive sandstone by overlying progradational units, and the poorly recovered and/or mis-ordered character of many of the older cores.

Facies B, C and D—Facies A is typically overlain by moderately well- to well-sorted, unidirectional trough cross-stratified and lesser low-angle, planar stratified sandstone (Facies B), pebbly sandstone (Facies C), and rarer granule-rich conglomerate (Facies D). Contacts vary from gradational (rare) to sharp and erosive. Sand grain sizes range from lower medium to lower coarse, with variable concentrations of very coarse sand, granules, and small pebbles consisting mainly of quartz and chert. Glaucony is mainly restricted to Facies B, particularly near the southeastern portion of the field. Carbonaceous detritus locally marks stratification, and wood fragments, coalified *in situ*, are intercalated. Beds range from 5 cm to 25 cm in thickness, locally amalgamated into bedsets up to 0.3-0.8 m thick. The coarse clastics contain granule and pebble stringers as well as mudstone rip-up clasts and thin mudstone interbeds (Figs. 4 and 12). Mudstone beds are typically 5-15 cm thick, dark in colour and locally siderite cemented.

Burrowing, though present, is sporadically distributed, of low intensity, and generally of reduced diversity. Trace fossils are far more common within the sandstone and pebbly sandstone facies than they are in the conglomeratic facies.

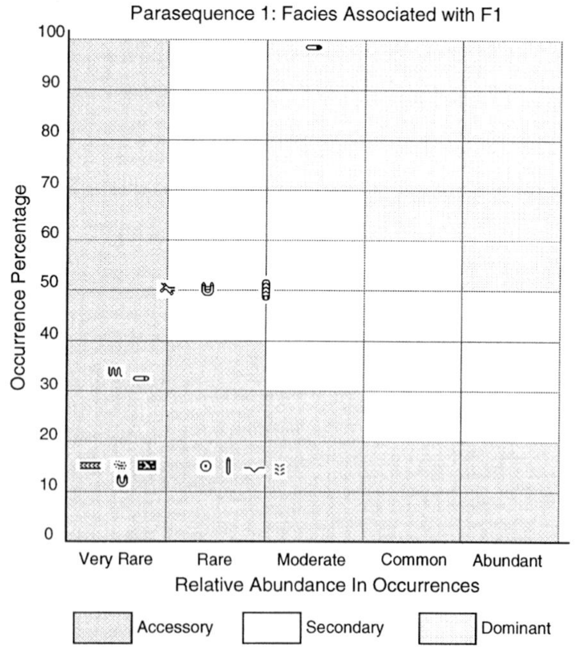

Fig. 9.—Cross-plot, showing ichnogenera occurrence percentage *vs.* abundance percentage for interstratified mudstone and sandstone of Facies E overlying F1. Refer to Figure 4 for the legend of trace fossil symbols.

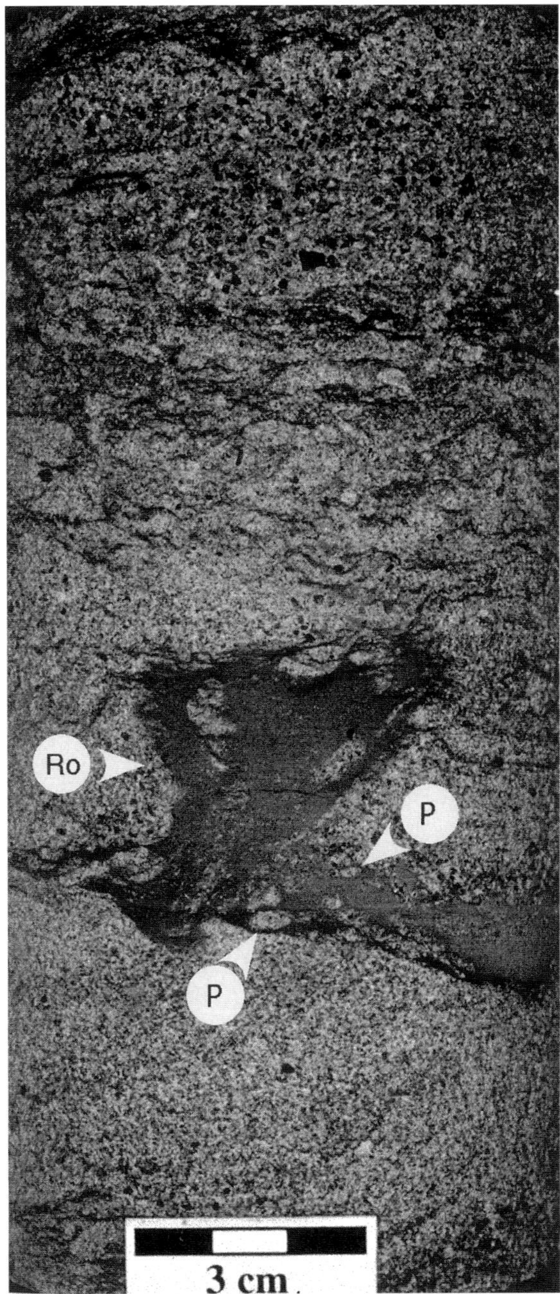

Fig. 10.—Photo of Facies A, consisting of glauconitic pebbly, muddy sandstone. Note the *Planolites* (P) and *Rosselia* (Ro). Well 14-05-38-24W4; 1372.3m.

Nonetheless, mudstone interbeds within conglomeratic units typically display evidence of biogenic reworking, attesting to their marginal marine origin. Ichnogenera are characterized by low numbers of *Diplocraterion*, *Skolithos*, *Palaeophycus*, and *Ophiomorpha* within the coarse-grained beds (cf., Fig. 13), with *Teichichnus*, *Planolites*, and *Terebellina* largely restricted to the mudstone interbeds. The remainder of the suite is exceedingly rare and consists of small numbers of *Arenicolites*, *Asterosoma*, *Thalassinoides*, and escape traces. Again, although the overall suite contains 10 ichnogenera, individual intervals contain four or fewer ichnogenera.

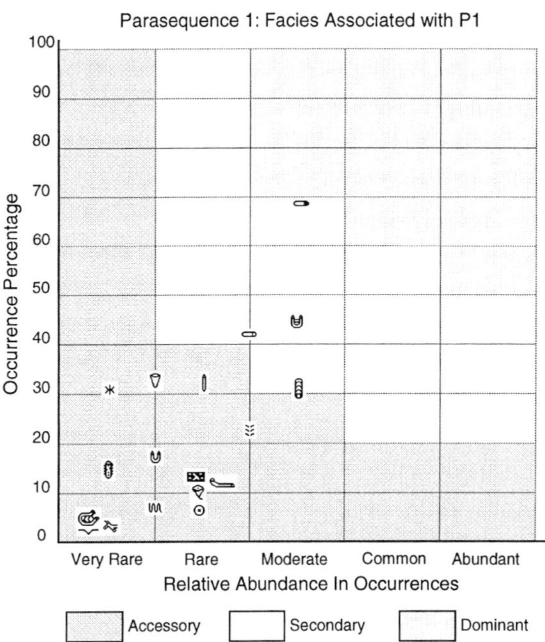

Fig. 11.—Cross-plot, showing ichnogenera occurrence percentage versus abundance percentage for facies overlying P1. Refer to Figure 4 for the legend of trace fossil symbols.

Facies F—Facies F overlies the trough cross-bedded facies in a few locations within Parasequence 1, but typically lies in a basinward position reflecting more distal depositional conditions. The facies consists of regularly interstratified granule conglomerate, sandstone and mudstone (cf., Fig. 14). In distal positions, Facies F grades upward out of the interstratified mudstone and sandstone of Facies E, and forms a depositional platform across which Facies B/C/D prograde. Facies F consists of beds 0.1-0.3 m thick. Glaucony is relatively uncommon, although carbonaceous detritus, mudstone rip-up clasts and thin mudstone interlaminae are locally abundant. Several mudstone beds are siderite cemented.

The sandstone and conglomerate are similar in character to beds in Facies B/C/D, but comprise beds less than 10 cm in thickness. Conglomerates are trough cross-stratified, but sandstone beds contain both current ripple lamination, trough cross-stratification, rare oscillation ripple lamination, combined flow ripple lamination and low angle, undulatory parallel lamination.

Facies F is characterized by rare to moderate bioturbation, and contains a trace fossil assemblage sporadically distributed and manifest by rare numbers of *Planolites*, *Teichichnus*, *Chondrites*, *Terebellina*, *Palaeophycus*, *Asterosoma*, *Diplocraterion*, *Skolithos*, *Arenicolites* and fugichnia. Although the suite encompasses a total of 9 ichnogenera, most intervals contain 4 or less. *Planolites*, *Teichichnus*, *Diplocraterion* and fugichnia comprise the recurring elements of the suite.

Geometry of Parasequence 1

The coarse clastic unit of Parasequence 1 displays a narrow, northwest to southeast trend (Fig. 15). The body preserves a depositional edge lying to the northeast (basinward), where it passes into interstratified mudstone and sandstone of Facies E. In the southwest (landward) direction, however, the surface varies from a depositional to

Fig. 12.—Photo of Facies B and C, consisting of trough cross-stratified sandstone and pebbly sandstone with mudstone interbed and mudstone rip-up clast. Well 06-14-38-25W4; 1422.1m.

an erosional edge. Preservation of the depositional edge is more common toward the southeastern end of the field, where Parasequence 1 onlaps relief on BD-2 (cross-section A-A'; Fig. 7). Although locally erosional throughout the entire study area, the landward edge is consistently erosional near the northwest end of the field where the parasequence has been truncated by coarse-grained deposits of Parasequence 2 (cross-section B-B'; Fig. 16). The coarse-grained deposits of Parasequence 1 are quite thin, ranging from approximately 10 cm to 2.1 m, and likely reflect their preservation as an erosional remnant below Parasequence 2.

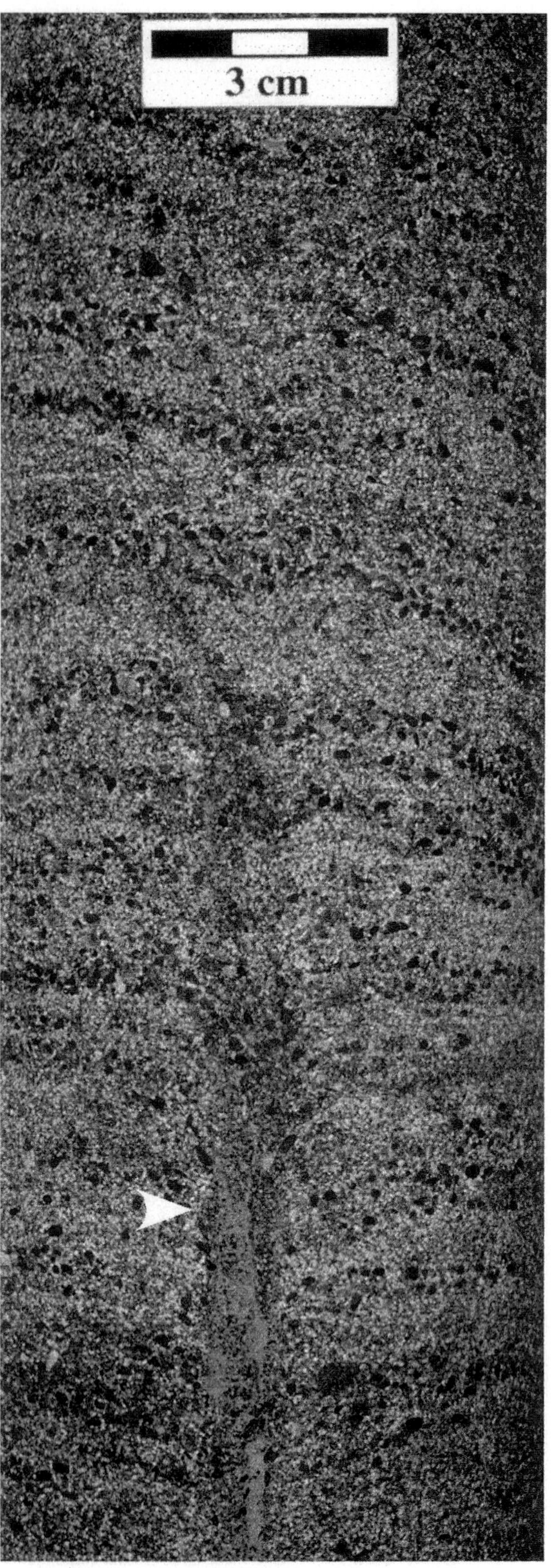

Fig. 13.—Photo of Facies B, consisting of trough cross-stratified sandstone with *Diplocraterion* (arrow). Well 11-07-39-26W4; 15481m.

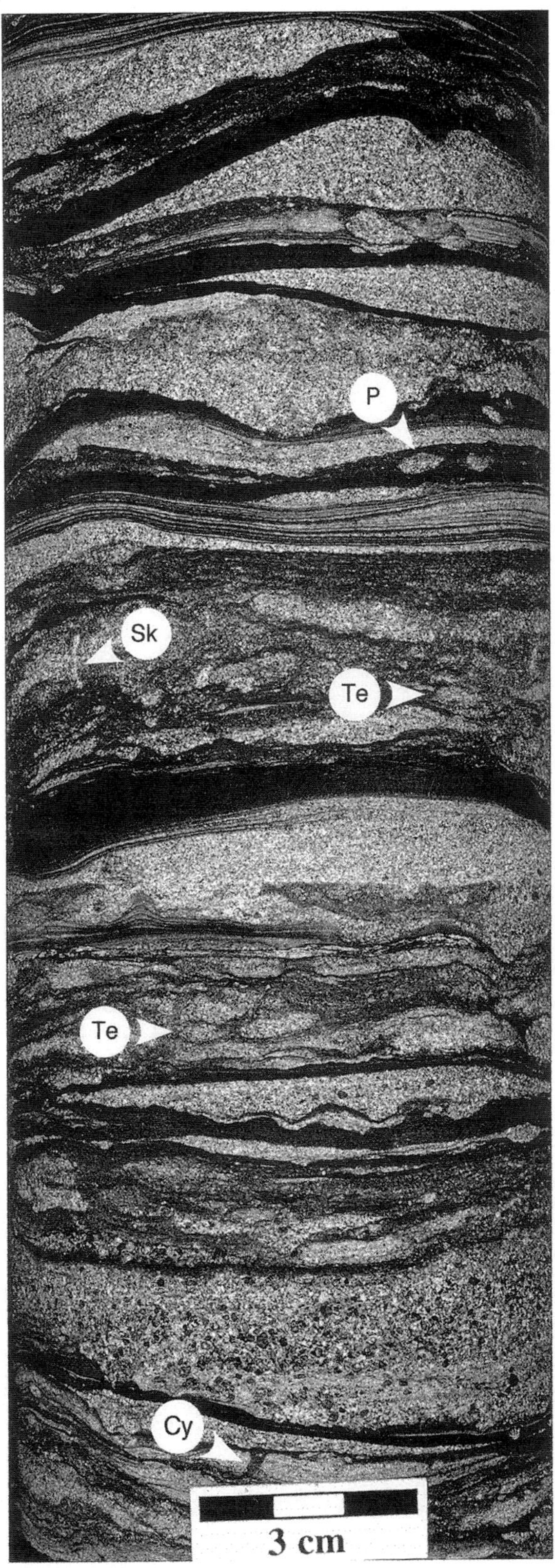

Fig. 14.—Photo of Facies F, consisting of interstratified, oscillation rippled, granule sandstone and mudstone. Note the sporadically distributed *Planolites* (P), *Teichichnus* (Te), *Skolithos* (Sk) and *Cylindrichnus* (Cy). Well 16-06-38-24W4; 1351.2m

Parasequence 2

Facies Overlying Flooding Surface 2 (F2)

The transgressive limit of the fine-grained facies associated with F2 generally lies landward of F1 and closely mimics the position of F3, except in the southeastern part of the Joffre Field (Fig. 6). Fine-grained facies associated with F2 drape Parasequence 1 in 23 cored intervals and correspond to Facies E. In most regards, the facies is identical to that associated with F1.

The facies is characterized by interbedded mudstone and sandstone (Fig. 8), typically with dispersed pebbles and granules of chert, glaucony, carbonaceous detritus and coalified wood fragments. Sandstone beds range from 2-10 cm in thickness and comprise between 5% and 20% of the facies. Individual sandstone beds tend to be well sorted, but may range in grain size from lower fine to lower medium. Sandstone beds are sharp based and display oscillation ripple, combined flow ripple, or wavy parallel lamination. Mudstone beds range from 1-10 cm in thickness, and are typically silt and sand poor. Mudstone beds are locally siderite cemented or display displacive siderite nodule development, and have variable pyrite contents.

Facies E in this interval is generally weakly burrowed with a sporadically distributed and low diversity trace fossil suite (Fig. 17). Although the facies contains a total of 14 ichnogenera, this serves to exaggerate the trace fossil diversity. Of the 23 intervals, only three contain very rare to rare numbers of *Helminthopsis* and *Palaeophycus*, and only single intervals contain *Anconichnus*, *Asterosoma*, *Thalassinoides*, *Terebellina*, *Arenicolites Diplocraterion*, *Skolithos*, *Ophiomorpha* and *Siphonichnus*. The suite is typified by *Planolites*, rarer *Teichichnus* and fugichnia. Only five ichnogenera occur in more than 10% of the intervals. Most intervals contain no more than 2-4 ichnogenera. This assemblage reflects a low diversity, low abundance suite generated by facies-crossing (opportunistic) organisms, typical of stressed depositional environments.

Facies Association Overlying Progradational/ Depositional Surface 2 (P2)

The facies association overlying P2 occurs in 55 cored intervals and is broadly similar to that overlying P1. The succession displays, however, significant variations in proximal, intermediate and distal positions. Burrowing diversity, as a whole, is lower than in the underlying parasequence (Fig. 18).

Facies A—In proximal and intermediate positions, Facies A directly overlies progradational/depositional surface P2. The facies is more common in the southeast portion of the field, in the vicinity of Township 38, Range 24W4 and Township 38, Range 25W4. The facies is broadly similar to those of Parasequence 1 (Fig. 10), but consists of beds 0.1-1.0 m thick, averaging 0.4 m thick. Locally, intervals are entirely cemented with siderite, or contain thin, siderite-cemented mudstone interbeds. Glaucony occurs in approximately 60% of the intervals.

Facies A units display moderate to low degrees of burrowing, consisting of a suite of 15 ichnogenera. The trace fossil suite is dominated by *Planolites*, *Palaeophycus* and *Diplocraterion*. Rare numbers of intervals contain uncommon *Terebellina*, *Teichichnus*, *Arenicolites*, *Asterosoma*, *Conichnus*, *Rosselia*, *Skolithos* and fugichnia. *Cylindrichnus*, *Thalassinoides*, *Ophiomorpha* and *Lockeia* occur in single intervals. Although some intervals contain up to six ichnogenera, most contain only three or four. The overall suite represents the stressed

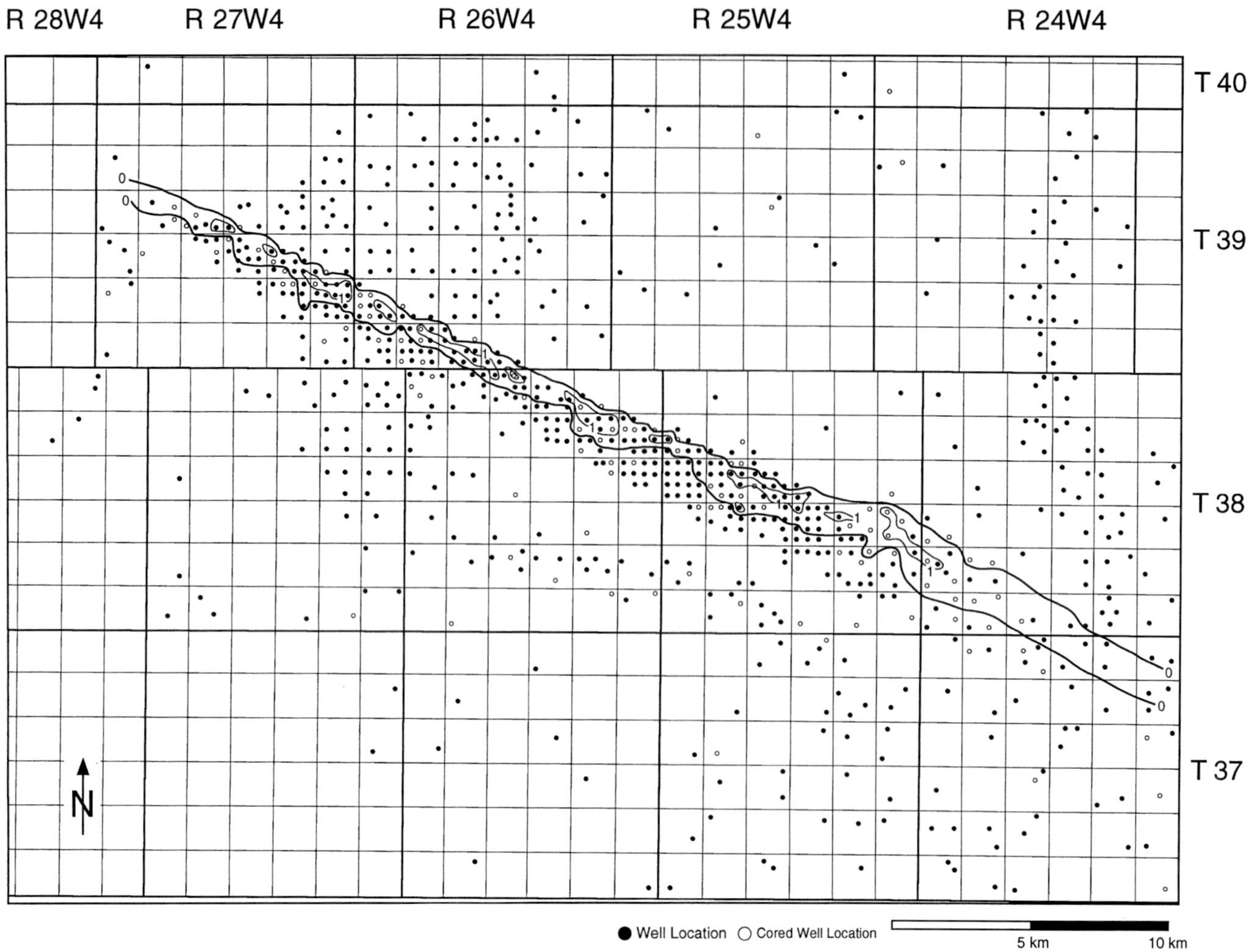

Fig. 15.—Isopach map of the coarse clastics of Parasequence 1. Contour interval is 1 m.

mixed *Skolithos-Cruziana* ichnofacies, characterized by facies crossing structures typical of opportunistic organisms.

Facies B, C and D—These facies are present in proximal, intermediate and distal positions, and comprise the dominant elements of Parasequence 2. The facies tend to be thinner in proximal positions, thickening in intermediate positions and thinning distally. In proximal positions, the facies range from 0.4-0.8 m, averaging 0.5 m. Intermediate intervals typically display intervals 0.6-1.8 m, averaging 0.7 m, while distal intervals are 0.2-0.5 m, averaging 0.3 m. Intervals are locally cut into Facies A units, or incised into mudstone of Facies E. In intermediate positions, the facies may be interstratified with Facies F. Facies B and C are generally more common in the southeast portion of the field, whereas Facies C and D dominate the northwestern end of the field.

In proximal positions, the facies are generally pebbly, with intercalated shale interbeds (locally siderite cemented), coalified wood fragments, rare glaucony, and minor carbonaceous detritus. Trough cross-beds tend to be small scale, with intercalated current ripple lamination. Bioturbation is rare to moderate in intensity but highly sporadic in distribution. A few intervals are completely unburrowed. The trace fossil assemblage is characterized by small numbers of *Diplocraterion, Skolithos, Conichnus, Palaeophycus, Ophiomorpha, Teichichnus*, and fugichnia. The bulk of the burrowing occurs in the southeastern portion of the field area, associated with Facies B. Most intervals contain only four ichnogenera.

In intermediate positions, the facies are characterized by larger scale trough cross-stratification, and contain dispersed granules and pebbles, pebble and granule stringers, and numerous mudstone rip-up clasts (Fig. 12). Burrowing intensity is low, sporadically distributed, and progressively weaker in a northwest direction. Trace fossils are more common in Facies B (Fig. 13) and become less abundant in Facies C and D, respectively. The suite consists of *Diplocraterion* (m-r), *Skolithos* (m-r), fugichnia (m-r), *Palaeophycus* (r), *Arenicolites* (r-vr), *Rosselia* (r), *Thalassinoides* (r), *Planolites* (r), *Teichichnus* (r), *Chondrites* (vr), *Terebellina* (vr), *Conichnus* (vr), *Ophiomorpha* (vr), *Lockeia* (vr), *Asterosoma* (vr). Eleven intervals, most lying northwest of Township 38, Range 25W4, do not contain trace fossils. Intervals rarely display more than four ichnogenera.

In distal positions, Facies B/C/D are relatively uncommon, occurring in only seven intervals. Like intermediate positions, they contain glaucony, sideritized mudstone interbeds and rip-up clasts, carbonaceous detritus, and chert granules or pebbles. Trace fossils are generally uncommon, with three of the intervals unburrowed. Trace fossils include *Diplocraterion* (r), *Planolites* (r), *Asterosoma*

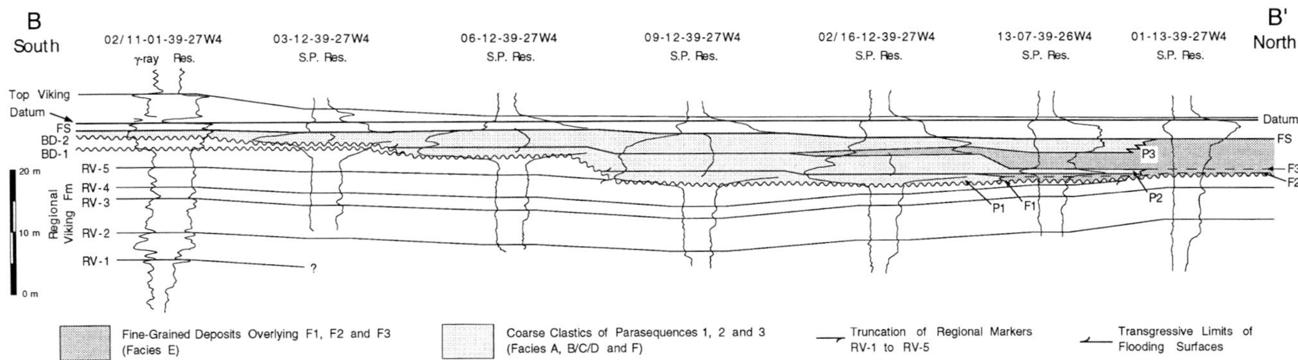

Fig. 16.—Cross-section B-B', showing the depositional onlap of Parasequences 1, 2 and 3 onto basal discontinuity BD-2, as well as their offlap to the northeast. The preserved transgressive limits of F1, F2 and F3 are also indicated. Note the greater degree of erosional amalgamation in this northwestern portion of the study area compared with the southeastern area shown in Figure 7. The line of section is located on the map of Figure 2.

(vr), *Skolithos* (vr), *Rhizocorallium* (vr), *Terebellina* (vr), and *Arenicolites* (vr). Intervals typically contain only two ichnogenera.

Facies F—Facies F (Fig. 14) occurs in only intermediate and distal positions. Units contain carbonaceous detritus, sideritized mudstone interbeds, glaucony, dispersed chert pebbles and granules, and coalified wood fragments. Mudstone rip-up clasts are exceedingly rare. Sandstone beds range from 0.2-1.3 m in thickness, averaging 0.5 m with intervening mudstone beds 0.5-5 cm thick. The facies group is rarely present northwest of Township 38, Range 25W4. Intervals consist of 30-40% sandstone beds with rare granule-rich beds, dominated by wavy parallel lamination and oscillation ripple lamination, with rarer combined flow ripple lamination and very rare current ripple lamination. Burrowing is typically rare to moderate in intensity and more uniformly distributed. The trace fossil assemblage is characterized by *Planolites* (a), *Teichichnus* (c), *Diplocraterion* (r), *Palaeophycus* (r), *Thalassinoides* (vr), *Chondrites* (vr), *Lockeia* (vr), *Skolithos* (vr), *Rosselia* (vr), *Arenicolites* (vr), fugichnia (vr), *Helminthopsis* (vr), *Conichnus* (vr), *Terebellina* (vr), *Siphonichnus* (vr), *Zoophycos* (vr) and *Anconichnus* (vr). Most intervals contain only 3-6 ichnogenera, although a single thick interval contained nine ichnogenera. *Zoophycos*,

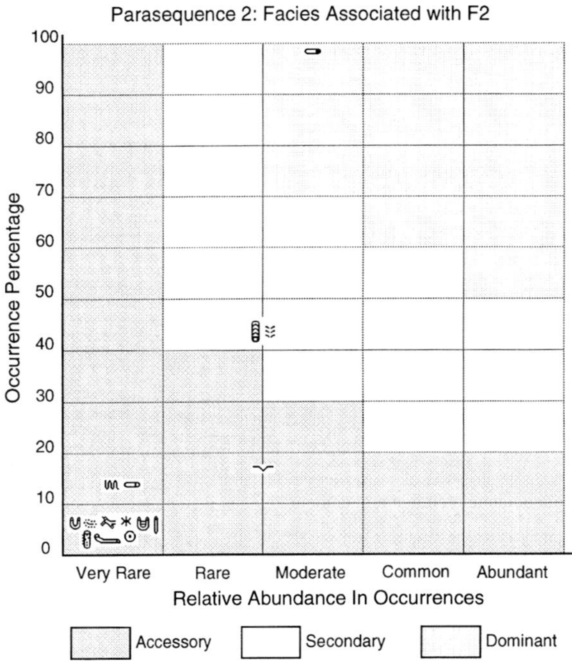

Fig. 17.—Cross-plot, showing ichnogenera occurrence percentage versus abundance percentage for interstratified mudstone and sandstone of Facies E overlying F2. Refer to Figure 4 for the legend of trace fossil symbols.

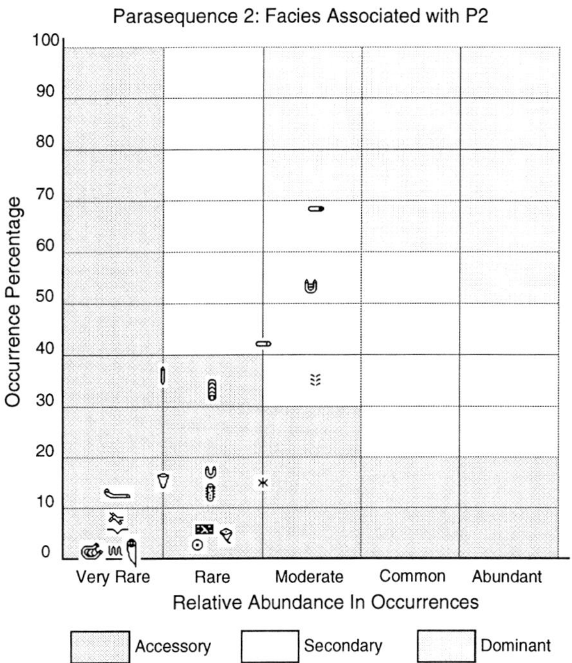

Fig. 18.—Cross-plot, showing ichnogenera occurrence percentage versus abundance percentage for facies overlying P2. Refer to Figure 4 for the legend of trace fossil symbols.

Helminthopsis, *Anconichnus*, *Rosselia* and *Conichnus* occur in only single intervals out of 24. The facies is more common toward the northwest end of the field, particularly in the vicinity of Township 39, Range 26W4.

Geometry of Parasequence 2

The coarse clastic unit of Parasequence 2 displays a northwest to southeast trend (Fig. 19), although not nearly so narrow a one as Parasequence 1. The interval broadens markedly toward the southeastern end of the field. Like Parasequence 1, the body preserves a depositional edge lying to the northeast (basinward), where it passes into the finer-grained deposits of Facies E. In contrast to the underlying parasequence, however, Parasequence 2 does not extend as far basinward, except in the southeastern part of the field. In the southwest (landward) direction, the zero edge is consistently erosional, associated with truncation by Parasequence 3. Parasequence 2 extends landward (southwest) of Parasequence 1, also onlapping the relief on BD-2 (Figs. 7 and 16). The coarse-grained deposits of Parasequence 2 vary markedly in thickness along the entire trend, ranging from approximately 20 cm to 3.9 m. These variable thicknesses are the result of differential erosion associated with the accumulation of Parasequence 3, which likely cannibalized much of its coarse clastic material from Parasequence 2. Hence, like the underlying parasequence, Parasequence 2 is also preserved largely as an erosional remnant.

Parasequence 3

Facies Overlying Flooding Surface 3 (F3)

The transgressive landward limit of the fine-grained facies associated with flooding surface F3 generally lies slightly seaward of those of F1 and F2, except in the southeastern part of the Joffre field, where it shifts markedly landward (Fig. 6). The facies associated with F3 overlie Parasequence 2 in 40 cored intervals and correspond to Facies E or more rarely, Facies F.

The interval is characterized by interbedded mudstone and sandstone (Figs. 4 and 8), typically with dispersed pebbles and granules of chert, glaucony, carbonaceous detritus and coalified wood fragments. Sandstone beds, ranging from 2-10 cm in thickness, comprise between 10% and 60% of the facies, although most units contain 10-30% sandstone. In landward positions, sandstone contents reach 30-60% and the facies corresponds to Facies F. Individual sandstone beds tend to be well sorted, and range in grain size from lower fine to lower medium. Sandstone beds are

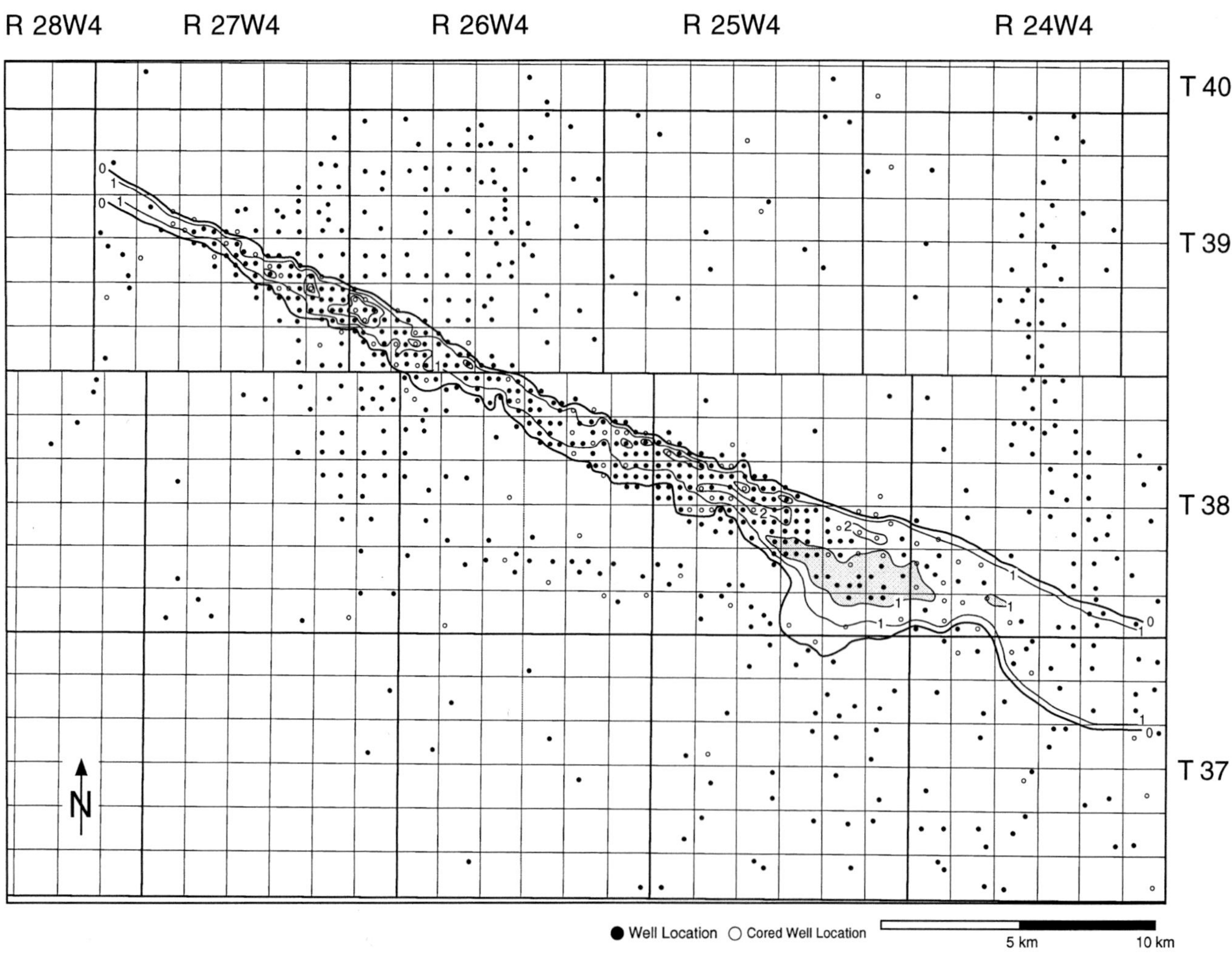

Fig. 19.—Isopach map of the coarse clastics of Parasequence 2. Contour interval is 1 m.

sharp based and show oscillation ripples, combined flow ripples, current ripples, and wavy parallel laminations. Convolute lamination, probably of dewatering derivation, occurs in two intervals. Mudstone beds range from 1-10 cm in thickness, are typically silt and sand poor, and locally are siderite cemented or display displacive siderite nodule development.

Burrowing intensity within these facies ranges from rare to common, though it is sporadically distributed. Trace fossil diversity (Fig. 20) is considerably greater than in similar facies associated with F1 and F2. The facies contains a total of 19 ichnogenera, but like that of facies associated with F2, this exaggerates the diversity of the suite. *Rosselia, Bergaueria, Rhizocorallium, Zoophycos,* and *Conichnus* occur in only single intervals, and only seven ichnogenera occur in 30% or more of the intervals. *Teichichnus, Planolites* and fugichnia constitute the only trace fossils that are ubiquitous.

On the other hand, most intervals display between two and 10 ichnogenera, though typically from four to seven. This, as well as the presence of *Helminthopsis, Chondrites* and *Diplocraterion* in more than 30% of the intervals suggests that these facies accumulated under a greater marine influence than those related to the underlying parasequences. The widespread presence of fugichnia within the succession attests to the episodic nature of sandstone deposition within the setting.

Facies Association Overlying Progradational/ Depositional Surface 3 (P3)

The facies association overlying P3 has the widest preserved distribution of lower Sequence 3, and occurs in 85 cored intervals. The succession overlying P3 also displays a higher degree of burrowing intensity and occurrence than those of the underlying parasequences (Fig. 21). The facies succession is broadly similar to that overlying P2, particularly in that the interval displays significant variations from proximal to distal positions. Unlike the facies associations overlying P1 and P2, this association typically lacks Facies A. The interval is wholly dominated by Facies B/C/D in proximal and intermediate positions, locally interfingering with and passing distally into Facies F toward the northeast. In proximal positions, the succession is characterized by Facies B/C/D locally intercalated with sand-dominated Facies F. In intermediate positions, the succession is characterized by relatively thick bedsets of Facies B/C/D with only very rare intervals of Facies F. In distal positions, the association is typified by sand-rich to sand-poor Facies F with minor intercalations of Facies B/C/D, reflecting the feather edges of coarse-grained deposition.

Facies B, C and D—Facies B/C/D units within Parasequence 3 are virtually identical to that of the underlying parasequences. The facies are encountered in 15 cored intervals within proximal positions. In these settings, the beds are thin, ranging from 0.2-1.7 m and averaging 0.5 m, and are interstratified with Facies F units of similar thickness. Within intermediate positions, these facies are encountered in 21 intervals and comprise bedsets 0.2-3.6 m in thickness, averaging 1.0 m. Facies F intervals are also intercalated, but considerably thinner than the trough cross-stratified facies. Trough cross-beds are of much larger scale than in proximal intervals. In addition, the bedsets are thinner toward the southeast end of the field, and thicken toward the northwest (averaging 1.5 m). The facies occurs in only a single core from a distal position, and is 0.3 m thick.

Bioturbation is generally of rare intensity, highly variable in distribution, and includes 13 ichnogenera. A significant number of intervals (5 of 15 intervals in proximal positions,

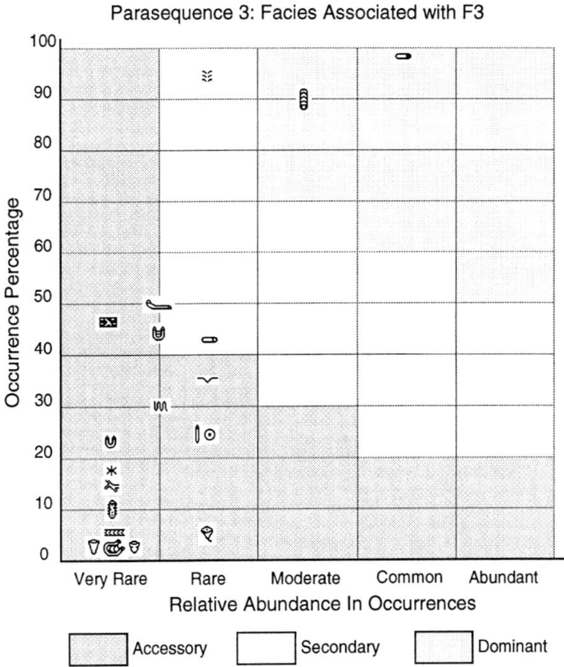

Fig. 20.—Cross-plot, showing ichnogenera occurrence percentage versus abundance percentage for interstratified mudstone and sandstone of Facies E overlying F3. Refer to Figure 4 for the legend of trace fossil symbols.

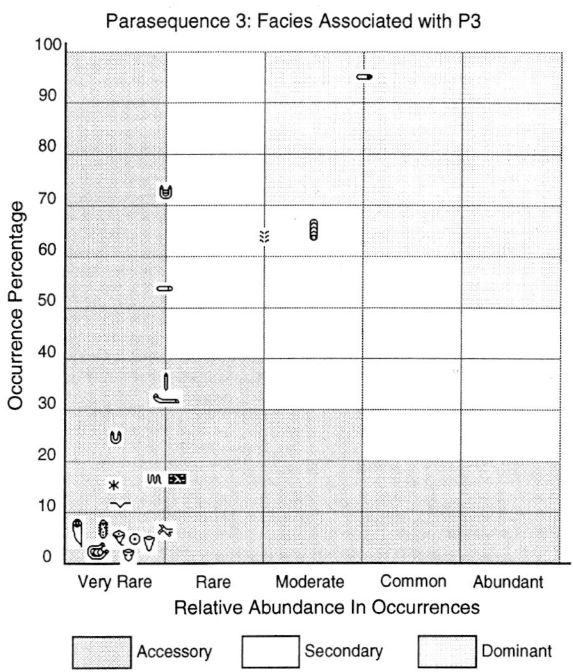

Fig. 21.—Cross-plot, showing ichnogenera occurrence percentage versus abundance percentage for facies overlying P3. Refer to Figure 4 for the legend of trace fossil symbols.

and 15 of 21 in intermediate positions) are entirely unburrowed. The trace fossil assemblage is characterized by *Diplocraterion* (m-c), *Planolites* (m-c), *Skolithos* (m), *Conichnus* (vr), *Palaeophycus* (vr-r), *Ophiomorpha* (vr), *Teichichnus* (vr), *Terebellina* (vr), *Arenicolites* (vr), *Rosselia* (vr), *Bergaueria* (vr), *Asterosoma* (vr), *Lockeia* (vr) and fugichnia (m-r). Proximal intervals typically contain 2-4 ichnogenera, and intermediate intervals contain 2-6 ichnogenera, though typically only three. Diversity of ichnogenera is actually quite low. In particular, most of the "very rare" trace fossils occur in only one or two intervals, greatly reducing the actual diversity of the assemblage. *Diplocraterion*, *Skolithos*, and *Planolites* can be considered the only recurring elements of the suite.

Facies F—Facies F occurs in proximal, intermediate and distal positions within Parasequence 3. The facies is virtually identical to equivalent facies in the underlying parasequences. In proximal positions, the facies was encountered in 17 intervals, ranging from 0.1-1.6 m in thickness, averaging 0.5 m. A total of 19 cored intervals contain Facies F in intermediate positions, ranging from 0.1-1.2 m, averaging 0.5 m. In distal positions, Facies F was encountered in 14 cores, and ranges from 0.2-2.1 m, averaging 0.9 m in thickness. Coarse-grained beds constitute between 10-60% of the units, though generally comprising 30-40% in most cases.

Burrowing intensity is typically rare to moderate in intensity, and becomes more pronounced in a basinward direction. Trace fossils are sporadically distributed throughout the intervals. The trace fossil assemblage in proximal, intermediate and distal settings is characterized by *Planolites* (m-a), *Teichichnus* (m-c), *Diplocraterion* (m-c; r in distal settings), *Palaeophycus* (m-c; r in distal settings), *Terebellina* (r-m), *Arenicolites* (vr-r), *Helminthopsis* (vr-r), *Chondrites* (vr-r), *Lockeia* (vr-r), *Skolithos* (vr-r), *Asterosoma* (vr), *Thalassinoides* (vr) and fugichnia (m-r). In intermediate and distal positions, the suite also includes *Siphonichnus* (vr), *Zoophycos* (vr), *Rhizocorallium* (vr), *Rosselia* (vr), *Ophiomorpha* (vr) and *Cylindrichnus* (vr). Most intervals in proximal positions contain between 3-9 ichnogenera, and typically 4-6. Intermediate intervals display 4-8 ichnogenera, though commonly six, while distal intervals range from 5-13 ichnogenera, and typically contain seven. The assemblage represents the most marine suite of all the successions within the lower portion of Sequence 3 (Fig. 4). Compared with the underlying highstand marine parasequences of the regional Viking Formation (Fig. 5), as well as the transgressive marine offshore deposits of Sequence 2 and the upper part of Sequence 3 (Fig. 22), however, the suite is impoverished and reflects environmental stress. Although containing an overall diversity of 17 ichnogenera, only *Planolites*, *Teichichnus*, *Diplocraterion*, *Palaeophycus*, *Terebellina*, and fugichnia can be considered recurring elements of the assemblage.

Geometry of Parasequence 3

The coarse clastic unit of Parasequence 3 displays a broad irregular apron in the study area (Fig. 23), although its depositional thicks lie along a northwest to southeast trend similar to the underlying parasequences. Likewise, the coarse clastic unit displays a depositional edge to the northeast (basinward), where it passes into Facies E interbedded mudstone and sandstone. Parasequence 3 extends further basinward than Parasequence 2, to approximately the limit of Parasequence 1. In the southwest (landward) direction, the zero edge is consistently erosional, associated with truncation by the overlying discontinuity FS, the regionally extensive wave ravinement surface that demarcates the return of open marine conditions to the study area. Parasequence 3 extends landward (southwest) of both underlying parasequences, and onlaps the remaining relief on BD-2 (Figs. 7 and 16). The coarse-grained deposits of Parasequence 3 are quite variable in thickness, ranging from approximately 10 cm to 4.1 m. The variable thicknesses are partly a function of: 1) differential erosion into Parasequence 2; 2) drape into lows developed on Parasequence 2; 3) variable erosional relief on BD-2; and 4) the magnitude of

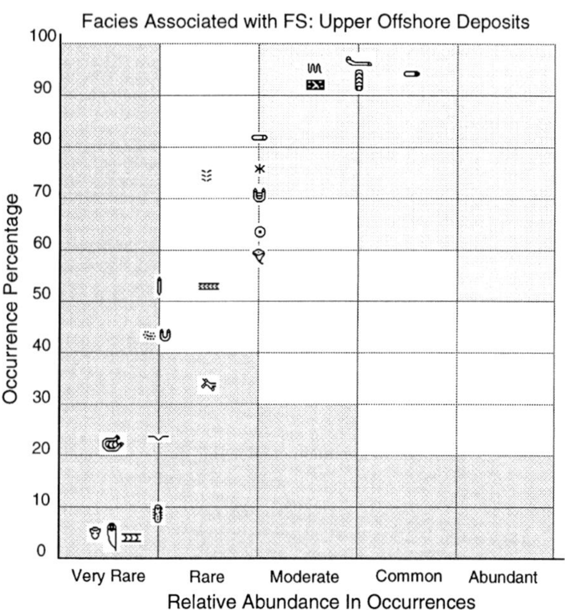

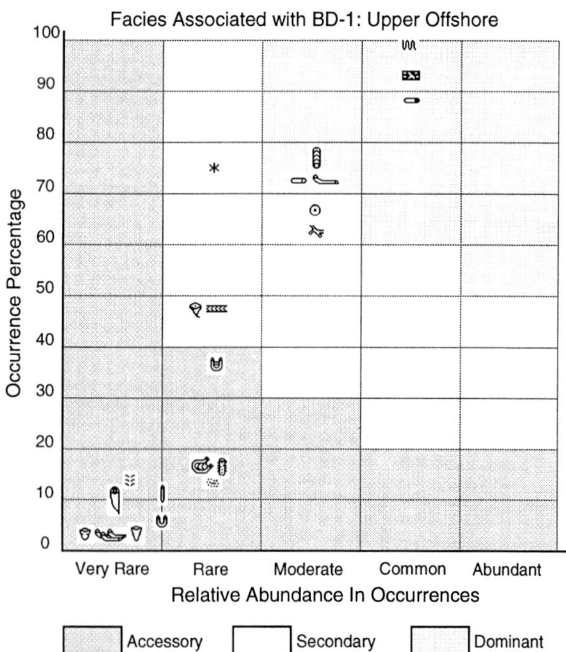

Fig. 22.—Cross-plots, showing ichnogenera occurrence percentage versus abundance percentage for transgressive offshore mudstone facies overlying BD-1 and FS. Refer to Figure 4 for the legend of trace fossil symbols.

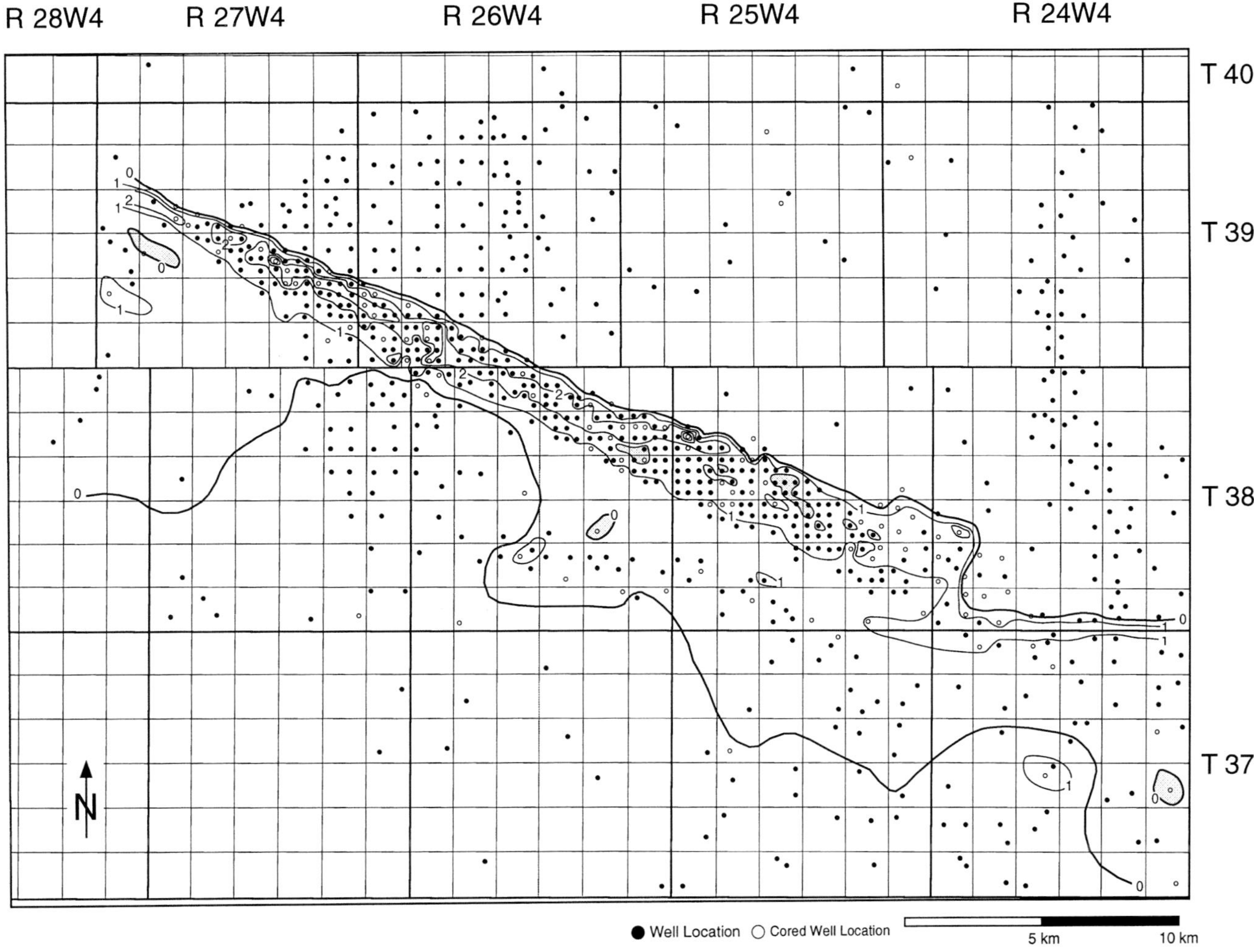

Fig. 23.—Isopach map of the coarse clastics of Parasequence 3. Contour interval is 1 m.

truncation by the overlying wave ravinement surfaces. Like the underlying parasequences, Parasequence 3 is preserved as an erosional remnant.

INTERPRETATION OF LOWER SEQUENCE 3

The interpretation of the facies within Sequence 3 has been addressed by MacEachern et al. (1998). They attributed the entire facies succession to deposition within a marginal marine bay-head delta/embayment complex, and provided compelling arguments dismissing both shoreface and incised valley interpretations as viable alternative depositional settings.

The basal discontinuity (BD-2) reflects a sequence boundary that was erosionally modified by, and amalgamated with, a wave-cut, bay ravinement surface. The glauconitic sandstone of Facies A is interpreted as a transgressive marine sand sheet, possibly consisting of original lowstand deposits reworked during transgressive modification of the sequence boundary during relative sea level rise that formed BD-2. Facies A contains the most diverse ichnological suite of the succession, presumably reflecting an initial period of largely marine conditions during ravinement.

The coarse clastics of Facies B/C/D are interpreted to reflect migrating dunes within channels. The restricted trace fossil suite demonstrates that channel deposition occurred in marginal marine conditions. The multiple scours, fining upward character, and uniform orientation of the trough cross-stratification, coupled with the presence of intercalated mudstone laminae, mudstone interbeds, and non-resistant mudstone rip-up clasts, are consistent with deposition in response to channelized flow. The mudstone interbeds and laminae attest to repeated fluctuations in flow strength. These deposits are interpreted to reflect marine-influenced distributary channels of bay-head delta systems, associated with the intertidal portions of the embayment. The distribution of Facies B, C, and D, which fringes the southwest edge of the embayment along the entire strike of the deposit, implies multiple point sources for clastic input. Shore-normal feeder systems do not appear to have been preserved in the area, and were likely removed during later wave ravinement associated with the flooding surface FS at the top of Parasequence 3 (Fig. 4).

Facies E is interpreted to reflect marginal marine, sandy (proximal) and muddy (distal) bay deposits, affected by wave processes, subordinate storm events and rare current processes. Upward increase in sandstone content within individual parasequences reflects shallowing of the bay during fill, and locally, may indicate close proximity to a bay-head delta, particularly where it grades upward or landward into Facies F intervals. The trace fossil assemblage is consistent with a salinity-stressed setting (Pemberton et al., 1992; MacEachern and Pemberton, 1994), characterized by pronounced fluctuations in salinity. The ichnological suite is intermediate in diversity and abundance between the more brackish-water, central basin deposits of the Viking Formation estuarine incised valley complexes (MacEachern and Pemberton, 1994), and the unstressed, fully marine highstand parasequences of the Regional Viking, the transgressively incised offshore/shoreface deposits of Sequence 2, and the overlying transgressive offshore deposits of upper Sequence 3. The palynological and foraminiferal paleoecology of these deposits strongly support this interpretation of an intermediate salinity condition (MacEachern et al., this volume).

The interstratified conglomerate, pebbly sandstone, sandstone, and mudstone of Facies F are interpreted to reflect progradation of the bay-head delta front into the embayment, and display primary structures that reflect a combination of oscillatory, storm event, combined flow and current depositional processes. In particular, the storm-induced stratification consists of thin (<15 cm thick) low angle, undulatory parallel lamination, reflecting either a distal depositional position, or a highly sheltered one. Given the interbedding of conglomeratic sandstone beds, presence of dispersed pebbles, and intercalation of current-generated structures, a sheltered (embayed) interpretation is favoured over a distal (basinal) one. The low diversity trace fossil suite, coupled with the close association of this facies with Facies E supports this environmentally stressed, inshore embayment interpretation. Facies F constitutes the depositional platform across which Facies B/C/D progrades. In many localities, the contact relationships between these facies are erosional, however, this is to be expected, given the channel interpretation afforded Facies B/C/D. The sharp erosional contact between these facies does not imply, therefore, a stratigraphic discontinuity of allocyclic origin. The interstratified character of these facies on a small scale strongly supports a genetic relationship, where autocyclic processes are responsible for their vertical and lateral juxtaposition.

Discussion of an Incised Valley Interpretation

Despite the marginal marine character of the succession, an estuarine incised valley complex is untenable, given the characteristics of the deposit. In the first place, the deposit is oriented parallel to the inferred shoreline trends during Viking time. Although some valleys may become re-oriented to parallel old shoreline trends during lowstand conditions (cf., Suter et al., 1987; Thomas and Anderson 1994; Sullivan et al., 1995), this has not been the case for any of the *known* incised valley complexes of Viking age in Alberta. All currently recognized Viking Formation valley complexes have orientations perpendicular to inferred paleoshoreline trends (Reinson et al., 1978; Boreen and Walker, 1991; Pattison, 1991; MacEachern and Pemberton, 1994; Pattison and Walker, 1998).

In the second place, fluvial deposits or fluvially-supplied deposits are entirely lacking in the vicinity of the stratigraphically lowest position of BD-2. Isopach thicks for the lower part of Sequence 3 correspond to predominantly muddy intervals within the succession, believed to consist of the brackish-water mudstone and thin sandstone of Facies E. If BD-2 had been cut by fluvial incision, it would require the valley to have operated as a zone of total coarse-sediment bypass, not only during lowstand conditions, but during early transgression and concomitant increasing accommodation space as well. Further, the succession cannot be accounted for using a terraced valley model of the type proposed by Blum (1992), because cross sections clearly demonstrate that the coarse-grained facies along the margins of the deposit interdigitate with brackish-water, fine-grained deposits to the northeast (Figs. 7 and 16), indicating a genetic relationship.

The remaining problem with an incised valley interpretation is that the three internal parasequences onlapping BD-2 possess orientations inconsistent with a shore-parallel incised valley. In valleys, parasequences are oriented with their strikes perpendicular to the valley trend, onlap the depositional surfaces in an up-valley direction, and downlap/offlap in a down-valley direction. In contrast, Parasequences 1, 2 and 3 strike *parallel* to the length of the deposit (NW-SE; Figs. 15, 19 and 23), while onlapping to the southwest and offlapping to the northeast (Figs. 7 and 16). This orientation is more characteristic of shoreline or intertidal parasequences.

Discussion of a Transgressive Shoreface Interpretation

Despite the shore parallel orientation, the succession is not consistent with the previously proposed transgressively incised shoreface interpretation of Downing and Walker (1988), or the high frequency forced regression shoreface model of Burton and Walker (this volume). In shoreface depositional models, trough cross-stratified coarse clastics are generally regarded to correspond to upper shoreface (nearshore) conditions subjected to high energy, wave-forced currents and longshore drift processes (cf., Clifton et al., 1971; Davidson-Arnott and Greenwood, 1976; Hunter et al., 1979). This seems particularly true for currents capable of transporting the gravels of Facies C and D. These gravels, however, are regularly and widely interstratified with marginal marine mudstone, and interbedded mudstone and sandstone at a variety of scales, ranging from millimetres to decimeters and demonstrate repeated fluctuations between traction transport and suspension deposition. In addition, the presence of mudstone rip-up clasts implies incision into, or erosion across the fine-grained deposits, atypical of upper shoreface settings. Such non-resistant rip-up clasts are unlikely to have survived transport or reworking for any significant period in the nearshore (surf zone) environment. The suggestion that these facies corresponds to lower or middle shoreface deposits is equivocal. We know of no modern setting or ancient shoreface succession that is dominated by current ripple lamination and trough cross-stratification.

Further, if Facies B/C/D reflect nearshore deposition, then one would expect these facies would fine and pass gradationally seaward into contemporaneous middle- and lower-shoreface burrowed to hummocky/swaley stratified sandstone. In contrast, these facies cut into or directly overlie the interstratified conglomerate, sandstone and mudstone of Facies F (Fig. 14), that displays a mixture of oscillatory structures, thin (and, we believe, sheltered rather than distal) storm-generated laminations, combined flow structures and current structures. The coarse clastics also pass basinward into, and become interstratified with the thinly interbedded oscillation rippled fine-grained sandstone and dark mudstone of Facies E (Fig. 8), which likewise reflect low energy (and, we believe, highly sheltered) settings (Figs. 7 and 16). Facies E and Facies F are unlikely to have been deposited in offshore or shelf conditions because:

1) the abrupt transition from coarse clastics to fine-grained deposits (e.g., Fig. 16) over distances of approximately

400 m constitutes depositional gradients that are too steep for a transition from the upper shoreface to the offshore;

2) Facies E contains a low diversity, stressed trace fossil suite consistent with reduced salinity settings. This suite is impoverished compared with the fully marine offshore deposits of the underlying transgressive shoreface succession and highstand regional Viking parasequences as well as the offshore deposits overlying FS (compare Figs. 5 and 22 with any of Figs. 9, 11, 17, 18, 20 or 21). Although various parameters may contribute to environmental stress, the nature of the assemblage generated is more easily accommodated by sporadic but overall reduced salinity conditions, rather than as a result of changes in oxygenation, food resources or substrate consistency; and

3) foraminiferal assemblages within the mudstone generally display a low diversity of forms, and a paucity of environmentally sensitive genera, consistent with the stressed conditions associated with reduced salinity settings (MacEachern et al., this volume).

Summary of Proposed Interpretation

An alternative model for lower Sequence 3 is presented in Figure 24, and better accounts for the problematic relationships already discussed. The BD-2 surface reflects transgressive ravinement that erosively modified the sequence boundary at the base of Sequence 3. This early stage of transgressive ravinement reworked available lowstand sediments landward to produce the basal, glauconitic, pebbly sandstone of Facies A, which onlap part of the relief on BD-2. This transgression introduced a broad, shallow brackish-water embayment in the Joffre area. Incremental cycles of progradation were punctuated by relative rises in sea level, and resulted in marginal marine flooding surfaces that separate three, coarse-grained, shallowing upward cycles comprising the stacked parasequences.

All three parasequences are broadly similar. They predominantly consist of sharp, erosionally based, trough cross-stratified sandstone, pebbly sandstone and conglomerate toward the southwestern margin of the deposit, which progressively overlie interstratified conglomerate, sandstone and mudstone in intermediate positions, and ultimately depositionally thin and/or become interstratified with interbedded mudstone and sandstone toward the northeast. Each parasequence is partially truncated by overlying deposits and each progressively onlaps relief developed on BD-2 in a landward (southwest) direction.

Each parasequence comprises the deposits of shore-normal and shore-parallel marginal marine channels and creeks, that fed coarse clastics to bay-head deltas fronting the elongate, shore-parallel brackish-water embayment (Fig. 24). The coarse clastics of Facies B/C/D accumulated along the southwestern (landward) margin of the embayment in the form of marginal marine distributaries that cut into and coalesced with Facies F bay-head delta front deposits, forming a broad, shore-parallel (NW-SE oriented), coarse-grained bay-head delta "apron". As the embayment filled, high sediment supply to the bay-head delta aprons permitted these systems to prograde northeastward into the bay, interfingering with the finer-grained brackish-water bay deposits of Facies E.

The marginal marine flooding surfaces (F1, F2 and F3) separating the parasequences are interpreted to reflect fluctuations in the rate of transgression, rather than as variations in sediment supply due to autocyclic channel switching/avulsion. Fluctuations in transgressive rate are favoured because of the along-strike persistence of all three flooding surfaces. Pulses of relative sea level rise allowed the brackish bay mudstone and interbedded sandstone and mudstone of Facies E to onlap the more proximal bay-head delta/channel complexes along the entire length of the embayment. The northeastward progradation of the coarse-grained deposits along the length of the embayment resulted in parasequences with shore-parallel strikes, offlapping/downlapping to the northeast and onlapping to the southwest. The fact that Parasequence 1 and Parasequence 3 have similar progradational limits while the intervening parasequence displays a landward progradational limit makes it impossible to define the succession as either a progradational or retrogradational parasequence set. Major transgression resulted in wave ravinement surface FS, and returned the area to open marine offshore conditions. This, coupled with the increased abundance and diversity of ichnogenera within Parasequence 3 supports increasing marine conditions, and indicates that the lower Sequence 3 succession is best placed within a transgressive systems tract.

The proposed model suggests the existence of a barrier complex lying northeastward of the zero edge of the Joffre embayment complex; a barrier whose deposits are conspicuously absent from the depositional record. Certainly there is indirect evidence of this barrier's existence. Facies E displays ichnological and sedimentological characteristics that indicate it was environmentally restricted and sheltered from open marine conditions, presumably by a barrier system. The preservation potential of a barrier complex is low, and resumed transgression which cut the ravinement surface FS across the top of the embayment deposit may have removed much, if not all, of the evidence of this depositional setting (cf., Rampino and Sanders, 1980; Nummedal and Swift, 1987; Walker, 1992). All that might remain of the barrier complex itself is the granule or pebble lag and thin sandstone deposit mantling FS. Alternatively, some of the upper portion of Parasequence 3, itself, may consist of reworked deposits derived from the backstepped barrier during initial transgression across the area. The paucity of core data available from immediately northeast of the Joffre Field further inhibits the identification of any barrier system remnants.

The major transgression, marked by FS, returned the study area to fully marine conditions and displaced the shoreline well to the south and southwest. The deposits overlying FS constitute the upper portion of Sequence 3.

Discussion of the Forced Regression Shoreface Model of Burton and Walker (this volume)

In light of some of the complexities within the succession at Joffre, Burton and Walker (this volume) have reassessed the original interpretation of Downing and Walker (1988). In their revised interpretation, two sharp-based conglomeratic shoreface successions are recognized, that reflect high frequency relative falls in sea level during an overall transgression. This model is broadly similar to that proposed by Davies and Walker (1993) for the Caroline/Garrington area. This revised model alleviates some of the difficulties in juxtaposing relatively thick mudstone-dominated units against conglomeratic deposits because the new model proposes that these facies are not contemporaneous and therefore do not reflect laterally adjacent environments. We have several concerns with the new interpretation, however.

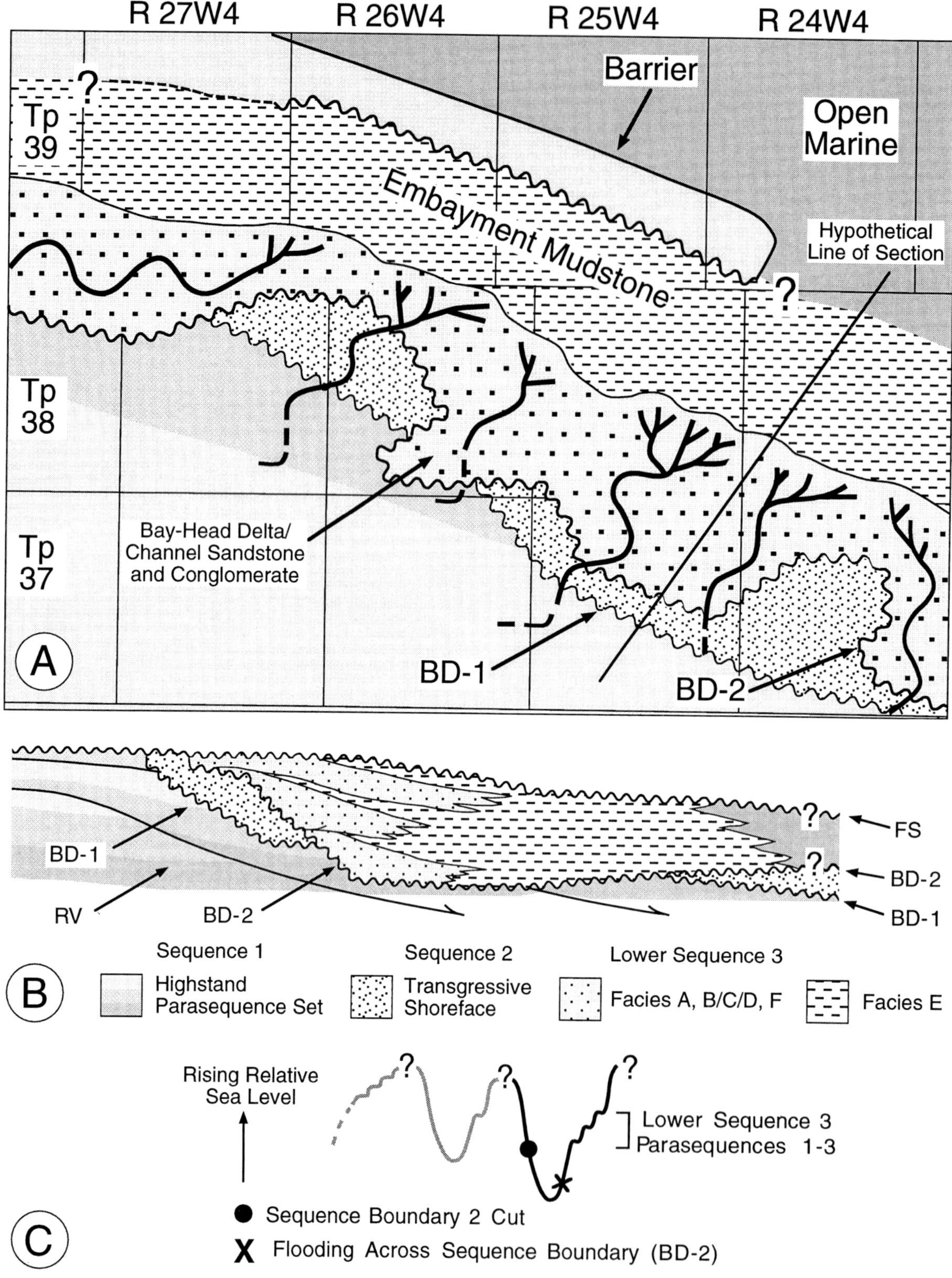

Fig. 24.—Depositional model for the lower portion of Sequence 3 (modified after MacEachern et al., 1998). A: Paleogeographic map of Parasequence 3 in lower Sequence 3. The hypothetical line of section is shown for the cross-section in B. The model shows the suggested position of a barrier complex (not preserved). B: A schematic cross-section showing incision of Sequence 1 and Sequence 2 by lower Sequence 3. C: Proposed relative sea level changes in response to Viking deposition in the study area. Relative sea level changes associated with lower Sequence 3 are shown in black.

1. The central position of their interpretation is that the conglomeratic sandstone represent lower shoreface deposition. We are aware of no shoreface complexes that consist exclusively of trough cross-stratified bedforms in the lower shoreface component of the system. Their contention that these are generated by longshore currents seems untenable, since longshore currents primarily operate in the nearshore zone (upper shoreface). Rip currents are oriented normal to the paleoshoreline, and as a mechanism, cannot explain the shore-parallel extension of cross-stratified sandstone. The suggestion that the facies might be formed by storm-forced currents leaves one wondering why the facies are exclusively current-generated and do not pass basinward into finer-grained hummocky stratified sandstone, particularly, since Facies E and F demonstrate that fine-grained sand does exist basinward of these coarser clastics. Further, the trough cross-stratified beds contain a variety of intervening mudstone at all scales, ranging from interlaminae to interbeds. Mudstone rip-up clasts are common and record erosional amalgamation of these current-generated bedforms and scouring of adjacent and intervening mudstone. We find these features unusual for lower shoreface deposits.

2. The criticism that the "juxtaposition of medium- to coarse-grained cross-bedded sandstone with mudstone horizons are unusual" (Burton and Walker, this volume) is not consistent with the available core and well log data. From the 91 cores that we have logged within lower Sequence 3, 86 of them (96%) consist of cross-bedded intervals containing mudstone interlaminae and/or beds 1-3 cm thick. Further, 47 cored intervals (52%) display the presence of mudstone beds 3 cm or thicker, and are most abundant in a basinward position. The common occurrence of mudstone rip-up clasts in sandstone lying landward of the step on BD-2 clearly demonstrates that such interbedding was more widespread but has been removed by erosional amalgamation of the sandbodies. We also disagree that there is any fundamental difference between Burton and Walker's "split" and the black mudstone of their Facies 4 (cf., their Fig. 11). Our analysis of the ichnological, palynological, and foraminiferal assemblages from the black mudstone shows that there is no significant difference between those they regard as the "split" and those lying basinward of it (MacEachern et al., 1998; this volume). The minor differences between the upper part of their "split" and black mudstone are attributable to variations along a proximal-distal trend.

3. Burton and Walker (this volume) contend that there are only two parasequences in the succession. This is based on their interpretation that the sandstone immediately below BD2 RT in the 10-13-38-25W4 well does not correlate with the upper sandstone in the 06-13-38-25W4 well (their Figs. 7 and 11). We have based the existence of the three parasequences on correlations using 18 cross-sections and comparison with more than 700 geophysical well logs across the length of the Joffre field. Further, the correlations employed in their Figures 7 and 11 are curious. The last three wells in Figure 7 and the last two cores in Figure 11 do not contain their underlying datum and therefore do not have a basis for their positioning. We also question the validity of employing an underlying datum, particularly one within the regional Viking that is separated from the interval in question by two sequence boundaries. The overlying datum we have employed is the widespread bentonite encased within offshore marine mudstone. It is highly unlikely that there was any paleotopographic relief on this datum. We believe that this inherently yields a better alignment of wells and favours superior correlations. Their correlation of Facies 2 and Facies 3 truncated by BD2-RT in Burton and Walker (this volume; their Fig. 11) is curious. The upward climb of these facies in a basinward direction, particularly where BD-2 becomes deeply incised, is unconventional and unnecessary, in our opinion. Our correlations in MacEachern et al. (1998; our Fig. 10) and in Figures 7 and 16 show three discrete prograding sandbodies that downlap in a basinward direction. The underlying datum employed by Burton and Walker (this volume) also appears to be the cause of the odd morphology on BD2 RT and BD3 depicted in their Figure 12. Their correlation illustrates ravinement/flooding surfaces that appear to rise and then fall (as it incises) in a progressively landward direction. In our estimation, this is an unlikely morphology for a transgressive surface.

4. Our paleontologic analysis of the intervening mudstone demonstrates salinity conditions intermediate between fully marine and estuarine (MacEachern et al., this volume). We feel that these are exactly the types of conditions that can be expected within a partially barred lagoonal embayment or a shallow bay. Although we find environmentally sensitive foraminifera present in some intervals, they occur in very low numbers and are consistent with stressed conditions compared to open marine facies. The paleoecology of the foraminiferal suite is most consistent with a salinity reduction, rather than substrate consistency or high sedimentation rates (MacEachern et al., this volume). We also noted that the palynological suites of the embayment facies are similar to the open marine, but attributed this to the ease to which palynomorphs are washed in from the offshore. Burton and Walker (this volume) mistakenly claim that we see individual parasequences become more marine upward, a feature inconsistent with the encroachment of the bay-head delta complexes. What we see, instead, is that each *successive* parasequence is more marine than the one underlying it, with the exception of the glauconitic transgressive sand sheet at the base. This progressive change heralds the onset of major transgression, marked by the development flooding surface FS (their BD2 RT).

5. Burton and Walker (this volume) reject the interpretation of an embayment complex because they see no preserved record of it. We argue that the preservation potential of the barrier complex, itself, is low (cf., Rampino and Sanders, 1980; Nummedal and Swift, 1987; Walker, 1992), particularly if the rate of sea level rise is slow. During subsequent ravinement, these barrier complexes are probably destroyed or preserved only as offshore to lower shoreface remnants resting on transgressive surfaces of erosion. The back-barrier mudstone and coarse clastics feeding into the embayment, however, have a higher preservation potential because they occupy a paleotopographic depositional low, and the ravinement surfaces rise topographically as the shoreline translates landward. We believe that the most likely indication of a barrier complex in the ancient record is the recognition of the lagoonal deposits themselves. The facies of lower Sequence 3 demonstrate an abundance of features that reflect brackish water sheltered conditions, consistent with a lagoonal origin.

Further, although we have postulated the existence of a barrier system as a means to generate a lagoonal/embayment complex, we do not believe that a barrier is essential. It is possible that the entire succession may reflect a shallow, open bay, markedly reduced in salinity due to freshwater dilution associated with coarse clastic input to the bay-head delta systems along the bay margin. A broadly similar model was employed by Mellere and Steel (1995) and Mellere (1996) to explain cross-stratified sandstone overlying shoreface parasequences from the Haystack Mountain Formation of Wyoming. An open bay system would facilitate some nearshore (longshore drift?) modification of the bay-head delta systems and straightening of the depositional edge within Lower Sequence 3 at Joffre.

SUMMARY

The Viking Formation of the Joffre area comprises parts of at least three discrete sequences. The Viking reservoir facies lie within the lower part of Sequence 3. During Sequence 3 time, there was a relative fall of sea level that permitted the excavation of a sequence boundary. This sequence boundary dissected the underlying marine shoreface deposits of Sequence 2 and locally incised through them and into the regional Viking parasequences of Sequence 1 (Figs. 7 and 16), reflecting a shift of the shoreline to the north and northeast of the study area. The sequence boundary was subsequently erosionally modified by wave ravinement during an ensuing transgression, forming a broad NW-SE trending asymmetric incision referred to as BD-2 (Fig. 24). Any overlying lowstand deposits were reworked to produce the transgressive lags and glauconitic transgressive sand sheets (Facies A) that mantle the surface. This transgression generated a broad, NW-SE trending embayment of the shoreline, possibly separated from the open marine seaway by a barrier complex lying in the northeastern portion of the study area. An alternative model is that the complex accumulated within a broad, shallow, open bay sheltered from the strong waves and storms developed in the interior seaway.

Pauses in the rate of flooding during transgression permitted periods of northeastward progradation of conglomeratic bay-head delta complexes into the brackish-water embayment. The bay-head deltas were supplied with sediment from small, shore-normal and shore-parallel marginal marine channels. The coarse-grained deposits coalesced to form broad, NW-SE trending aprons, consisting of distributary channel and bay-head delta front deposits along the southwestern margin of BD-2 (Facies B/C/D and Facies F, respectively), that interfinger with marginal marine, interstratified mudstone and sandstone (Facies E) in a basinward (northeastward) direction. Variations in sediment supply and autocyclic channel/delta abandonment may have caused much of the lithological heterogeneity within individual parasequences.

Allocyclically generated fluctuations in the rate of transgression resulted in the shifting of brackish-water bay deposits over the bay-head delta/channel complexes, and generated three discrete parasequences marked by marginal marine or bay flooding surfaces (F1, F2 and F3). Short-lived pulses of progradation (P1, P2 and P3) resulted in the deposition of bay-head delta front deposits over the bay mudstone, ultimately capped by distributary channel complexes. Each parasequence onlaps relief on BD-2 along its landward (southwestern) margin, and offlaps/downlaps to the northeast (Figs. 7 and 16). These parasequences constitute the Joffre embayment complex of lower Sequence 3. Although the parasequences do not stack into a retrogradational parasequence set, the upper most parasequence displays an increased marine influence, and the top of the succession is truncated by a regionally extensive transgressive ravinement surface (FS), suggesting that lower Sequence 3 belongs in a transgressive systems tract. The resumed transgression that cut the ravinement surface FS terminated brackish-water deposition and returned the study area to fully marine, offshore conditions.

Successions characterized by coarse clastics regularly interstratified with marine or marginal marine mudstone are commonly interpreted either as coarse-grained conglomeratic shoreface deposits (cf., Downing and Walker, 1988; Posamentier et al., 1992; Davies and Walker, 1993; Walker and Bergman, 1993; Posamentier and Chamberlain, 1993; Bergman, 1994; Bergman and Walker, 1995) or as estuarine incised valley complexes (cf., Reinson et al., 1988; Boreen and Walker, 1991; Pattison, 1991; Sullivan et al., 1995). The ichnological, sedimentological and sequence stratigraphic characteristics of the succession demonstrate that neither model is appropriate for lower Sequence 3 of the Viking Formation at Joffre, and point to an alternative model.

In modern settings, relative sea level rise has encouraged the development of highly embayed transgressive shorelines, commonly fronted by barrier systems. During subsequent wave ravinement, these barrier complexes are likely to be destroyed or preserved only as erosional remnants consisting of offshore to lower shoreface deposits that rest on marine flooding surfaces. The back-barrier mudstone and the coarse clastics feeding into the embayment, however, have a comparatively higher preservation potential during the transgression. This is because these deposits occupy a paleotopographic low, and the ravinement surface rises topographically during erosive shoreface retreat. Alternatively, a broad, shallow, open bay may be highly sheltered from open marine conditions even without a barrier complex, and could explain many of the characteristics of lower Sequence 3. In spite of the ubiquitous occurrence of brackish lagoonal and embayment environments in modern transgressive shoreline systems, current interpretations of ancient transgressive successions appear to ignore or fail to recognize the deposits of these environments.

ACKNOWLEDGMENTS

This paper derives from a post-doctoral project undertaken by the first author in collaboration with the other authors, as part of a Natural Sciences and Engineering Research Council (NSERC) Collaborative Research and Development (CRD) Grant 180563 awarded to S.G. Pemberton and an NSERC Operating Grant 184293 to J.A. MacEachern. The data was collected while the first author was engaged at PanCanadian Petroleum Ltd. The authors would like to thank PanCanadian Petroleum Ltd. for their financial and logistical support throughout the course of this study. We would particularly like to thank Jamie Burton and Roger Walker for sharing their views on this complex deposit, and for encouraging a congenial atmosphere in which to compare and contrast our varying interpretations. Like them, we have attempted to structure our paper in such a way as to allow readers to appreciate the differences between our two interpretations as well as the similarities. The project has benefited from discussions with Lee Krystinik, Ed Clifton, Ron Boyd, Bob Dalrymple, Dale Leckie, John Suter, Indraneel Raychaudhuri, and Jeff Peterson. Any shortcomings in the interpretation, however, are the authors' alone. Critical review by Karen Porter and Danny

Labelle greatly improved this paper and we gratefully acknowledge their efforts.

REFERENCES

BERGMAN, K.M., 1994, Shannon Sandstone in Hartzog Draw-Heldt Draw fields reinterpreted as detached lowstand shoreface deposits: Journal of Sedimentary Research, v. B64, p. 184-201.

BERGMAN, K.M. AND WALKER, R.G., 1995, High resolution sequence stratigraphic analysis of the Shannon sandstone in Wyoming, using a template for regional correlation, *in* Swift, D.J.P., Snedden, J.W. and Plint, A.G., eds., Tongues, ridges and wedges: highstand *versus* lowstand architecture in marine basins: Society of Economic Paleontologists and Mineralogists Research Conference, Powder River and Bighorn Basins, Wyoming, June 24-29, unpaginated.

BLOCH, J., SCHRÖDER-ADAMS, C., LECKIE, D.A., MCINTYRE, D.J., CRAIG, J. AND STANILAND, M., 1993, Revised stratigraphy of the lower Colorado Group (Albian to Turonian), Western Canada: Bulletin of Canadian Petroleum Geology, v. 41, p. 325-348.

BLUM, M.D., 1994, Genesis and architecture of incised valley fill sequences ... a late Quaternary example from the Colorado River, Gulf Coastal Plain, Texas, *in* Weimer, P. and Posamentier, H.W., eds., Siliciclastic Sequence Stratigraphy: Recent Developments and Applications: American Association of Petroleum Geologists Memoir 58, p. 259-283.

BOREEN, T. AND WALKER, R.G., 1991, Definition of allomembers and their facies assemblages in the Viking Formation, Willesden Green area, Alberta: Bulletin of Canadian Petroleum Geology, v. 39, p. 123-144.

CLIFTON, H.E., HUNTER, R.E. AND PHILLIPS, R.L., 1971, Depositional structures and processes in the non-bar high energy nearshore: Journal of Sedimentary Petrology, v. 41, p. 651-670.

COBBAN, W.A. AND KENNEDY, W.J., 1989, The ammonite *Metengonoceras* Hyatt, 1903, from the Mowry Shale (Cretaceous) of Montana and Wyoming: United States Geological Survey, Bulletin 1787-L, p. L1-L11.

DAVIDSON-ARNOTT, R.G.D. AND GREENWOOD, B., 1976, Facies relationships in a barred coast, Kouchibouguac Bay, New Brunswick, Canada, *in* Davis, Jr., R.A. and Ethington, R.L., eds., Beach and nearshore sedimentation: Society of Economic Paleontologists and Mineralogists, Special Publication 24, p. 149-168.

DAVIES, S.D. AND WALKER, R.G., 1993, Reservoir geometry influenced by high frequency forced regressions within an overall transgression; Caroline and Garrington fields, Viking Formation (Lower Cretaceous) Alberta: Bulletin of Canadian Petroleum Geology, v. 41, p. 407-421.

DOWNING, K.P, 1986, The depositional history of the Lower Cretaceous Viking Formation at Joffre, Alberta, Canada: Master's thesis, McMaster University, Hamilton, 137 p.

DOWNING, K.P. AND WALKER, R.G., 1988, Viking Formation, Joffre Field, Alberta: shoreface origin of long, narrow sand body encased in marine mudstones: American Association of Petroleum Geologists Bulletin, v. 72, p. 1212-1228.

ENERGY RESOURCES CONSERVATION BOARD (ERCB), 1994, Alberta's reserves of crude oil, oil sands, gas, natural gas liquids and sulphur: Energy Resources Conservation Board, Reserve Report Series, ERCB St 94-18, 32 edition, unpaginated.

GLAISTER, P., 1959, Lower Cretaceous of southern Alberta and adjoining areas: American Association of Petroleum Geologists Bulletin, v. 43, p. 590-640.

HUNTER, R.E., CLIFTON, H.E. AND PHILLIPS, R.L., 1979, Depositional processes, sedimentary structures, and predicted vertical sequences in barred nearshore systems, southern Oregon coast: Journal of Sedimentary Petrology, v. 49, p. 711-728.

LECKIE, D.A. AND REINSON, G.E., 1993, Effects of middle to late Albian sea level fluctuations in the Cretaceous Interior Seaway, western Canada, *in* Caldwell, W.G.E. and Kauffman, E.G., eds., Evolution of the Western Interior Basin: Geological Association of Canada, Special Paper 39, p. 151-175.

MACEACHERN, J.A., 1994, Integrated ichnological-sedimentological models: applications to the sequence stratigraphic and paleoenvironmental interpretation of the Viking and Peace River formations, west-central Alberta: Ph.D. thesis, University of Alberta, Edmonton, 618 p.

MACEACHERN, J.A. AND PEMBERTON, S.G., 1992, Ichnological aspects of Cretaceous shoreface successions and shoreface variability in the Western Interior Seaway of North America, *in* Pemberton, S.G., ed., Applications of ichnology to petroleum exploration, a core workshop: Society of Economic Paleontologists and Mineralogists, Core Workshop 17, p. 57-84.

MACEACHERN, J.A. AND PEMBERTON, S.G., 1994, Ichnological aspects of incised valley fill systems from the Viking Formation of the Western Canada Sedimentary Basin, Alberta, Canada, *in* Dalrymple, R.W., Boyd, R. and Zaitlin, B.A., eds., Incised valley systems: origin and sedimentary sequences: Society of Economic Paleontologists and Mineralogists, Special Publication 51, p. 129-157.

MACEACHERN, J.A., RAYCHAUDHURI, I. AND PEMBERTON, S.G., 1992, Stratigraphic applications of the *Glossifungites* ichnofacies: Delineating discontinuities in the rock record, *in* Pemberton, S.G., ed., Applications of ichnology to petroleum exploration, a core workshop: Society of Economic Paleontologists and Mineralogists Core Workshop 17, p. 169-198.

MACEACHERN, J.A., PEMBERTON, S.G. AND ZAITLIN, B.A., 1995, A late lowstand to early transgressive coarse-grained tongue from the Viking Formation of the Joffre Field, Alberta: embayment complex or shoreface wedge?, *in* Swift, D.J.P., Snedden, J.W. and Plint, A.G., eds., Tongues, ridges and wedges: highstand *versus* lowstand architecture in marine basins: Society of Economic Paleontologists and Mineralogists Research Conference, Powder River and Bighorn Basins, Wyoming, June 24-29, unpaginated.

MACEACHERN, J.A., ZAITLIN, B.A. AND PEMBERTON, S.G., 1995b, High resolution sequence stratigraphy of stacked early transgressive shoreface and early transgressive-related deposits of the Viking Formation, Joffre Field, Alberta, Canada, *in* Sandvik, K.O., ed., Predictive high resolution sequence stratigraphy: Norsk Petroleums-forening, Norwegian Petroleum Society, Stavanger Forum, Stavanger, Norway, unpaginated.

MACEACHERN, J.A., ZAITLIN, B.A. AND PEMBERTON, S.G., 1998, High-resolution sequence stratigraphy of early transgressive deposits, Viking Formation, Joffre Field, Alberta, Canada: American Association of Petroleum Geologists Bulletin, v. 82, p. 729-755.

MCGOOKEY, D.P., HAUN, J.D., HALE, L.A., GOODELL, H.G., MCCUBBIN, D.G., WEIMER, R.J. AND WULF, G.R., 1972, Cretaceous System, *in* Mallory, W.W., ed., Geological atlas of the Rocky Mountain region, U.S.A: Rocky Mountain Association of Geologists, Denver Colorado, p. 190-228.

MELLERE, D., 1996, Seminoe 3, a tidally influenced lowstand wedge and its relationships with subjacent highstand and overlying transgressive deposits, Haystack Mountains Formation, Cretaceous Western Interior, Wyoming (USA): Sedimentary Geology, v. 103, p. 249-272.

MELLERE, D. AND STEEL, R.J., 1995, Facies architecture and sequentiality of nearshore/shelf sandbodies (Haystack Mountains Formation–Wyoming, U.S.A.): Sedimentology, v. 42, p.

NORTH AMERICAN COMMISSION ON STRATIGRAPHIC NOMENCLATURE (NACSN), 1983, North American stratigraphic code: American Association of Petroleum Geologists Bulletin, v. 67, p. 841-875.

OBRADVICH, J.D., 1993, A Cretaceous time scale, *in* Caldwell, W.G.E. and Kauffman, E.G., eds., Evolution of the Western Interior Basin: Geological Association of Canada, Special Paper 39, p. 379-396.

PATTISON, S.A.J., 1991, Sedimentology and allostratigraphy of regional, valley-fill, shoreface and transgressive deposits of the Viking Formation (Lower Cretaceous), Central Alberta: Ph.D. thesis, McMaster University, Hamilton, 380 p.

PATTISON, S.A.J. AND WALKER, R.G., 1998, Multiphase transgressive filling of an incised valley and shoreface complex, Viking Formation, Sundance-Edson area, Alberta: Bulletin of Canadian Petroleum Geology, v. 46, p. 89-105.

PEMBERTON, S.G. AND MACEACHERN, J.A., 1995, The sequence stratigraphic significance of trace fossils: examples from the Cretaceous foreland basin of Alberta, Canada, *in* Van Wagoner, J.C. and Bertram, G., eds., Sequence stratigraphy of foreland basin deposits- outcrop and subsurface examples from the Cretaceous of North America: American Association of Petroleum Geologists Memoir 64, p. 429-475.

PEMBERTON, S.G., MACEACHERN, J.A. AND FREY, R.W., 1992, Trace fossil facies models: environmental and allostratigraphic significance, *in* Walker, R.G. and James, N., eds., Facies models: response to sea level change: Geological Association of Canada, St. John's, Newfoundland, p. 47-72.

POSAMENTIER, H.W. AND CHAMBERLAIN, C.J., 1993, Sequence stratigraphic analysis of Viking Formation lowstand beach deposits at Joarcam field, Alberta, Canada, *in* Posamentier, H.W., Summerhayes, C.P., Haq, B.U. and Allen, G.P., eds., Stratigraphy and facies associations in a sequence stratigraphic framework: International Association of Sedimentologists, Special Publication 18, p. 469-485.

POSAMENTIER, H.W., ALLEN, G.P., JAMES, D.P. AND TESSON, M., 1992, Forced regressions in a sequence stratigraphic framework: concepts, examples, and exploration significance: American Association of Petroleum Geologists Bulletin, v. 76, p. 1687-1709.

REINSON, G.E., CLARK, J.E. AND FOSCOLOS, A.E., 1988, Reservoir geology of Crystal Viking Field, Lower Cretaceous estuarine tidal channel-bay complex, south-central Alberta: American Association of Petroleum Geologists Bulletin, v. 72, p. 1270-1294.

SAUNDERS, T. AND PEMBERTON, S.G., 1986, Trace fossils and sedimentology of the Appaloosa Sandstone: Bearpaw-Horseshoe Canyon Formation transition, Dorothy, Alberta: Canadian Society of Petroleum Geologists Field Trip Guide Book, 117 p.

SAVRDA, C.E., 1991, Ichnology in sequence stratigraphic studies: An example from the Lower Paleocene of Alabama: *Palaios*, v. 6, p. 39-53.

STELCK, C.R. AND LECKIE, D.A., 1990, Biostratigraphy of the Albian Paddy Member (Lower Cretaceous Peace River Formation), Goodfare, Alberta: Canadian Journal of Earth Sciences, v. 27, p. 1159-1169.

SULLIVAN, M., VAN WAGONER, J., FOSTER, M., STUART, R., JENETTE, D., LOVELL, R. AND PEMBERTON, S.G., 1995, Lowstand architecture and sequence stratigraphic control on Shannon incised valley distribution, Hartzog Draw Field, Wyoming, *in* Swift, D.J.P., Snedden, J.W. and Plint, A.G., eds., Tongues, ridges and wedges: highstand *versus* lowstand architecture in marine basins: Society of Economic Paleontologists and Mineralogists Research Conference, Powder River and Bighorn Basins, Wyoming, June 24-29, unpaginated.

SUTER, J.R., BERRYHILL, H.L. AND PENLAND, S., 1987, Late Quaternary sea-level fluctuations and depositional sequences, southwest Louisiana continental shelf, *in* Nummedal, D., Pilkey, O.H. and Howard, J.D., eds., Sea-level fluctuation and coastal evolution: Society of Economic Paleontologists and Mineralogists, Special Publication 41, p. 199-219.

THOMAS, M.A. AND ANDERSON, J.B., 1994, Sea-level controls on the facies architecture of the Trinity/Sabine incised-valley system, Texas continental shelf, *in* Dalrymple, R.W., Boyd, R. and Zaitlin, B.A., eds., Incised valley systems: origin and sedimentary sequences: Society of Economic Paleontologists and Mineralogists, Special Publication 51, p. 63-82.

WALKER, R.G., 1997, High-resolution regional stratigraphy and sedimentology of the Viking Formation, central Alberta: Canadian Society of Petroleum Geologists Viking Advantage Course Notes, 105p.

WALKER, R.G. AND BERGMAN, K.M., 1993, Shannon Sandstone in Wyoming: a shelf-ridge complex reinterpreted as lowstand shoreface deposits: Journal of Sedimentary Petrology, v. 63, p. 839-851.

WALKER, R.G. AND WISEMAN, T.R., 1995, Lowstand shorefaces, transgressive incised shorefaces, and forced regressions: examples from the Viking Formation, Joarcam area, Alberta: Journal of Sedimentary Research, v. B65, p. 132-141.

WEIMER, R.J., 1984, Relation of unconformities, tectonics, and sea level changes, Cretaceous of the Western Interior, U.S.A., *in* Schlee, J.S., ed., Interregional unconformities and hydrocarbon accumulation: American Association of Petroleum Geologists Memoir 36, p. 7-35.

CRETACEOUS SUSSEX SANDSTONE IN HOUSE CREEK FIELD (WYOMING, USA): TRANSGRESSIVE INCISED SHOREFACE DEPOSITS

KATHERINE M. BERGMAN
Department of Geology, University of Regina, Regina, Saskatchewan, S4S 0A2, Canada

ABSTRACT: The Upper Cretaceous Sussex Sandstone at House Creek field (Powder River Basin, Wyoming) has previously been interpreted as a prograding succession of sandstones that were transported southeastward by currents along a relatively shallow marine shelf. These sandstones were deposited many tens of kilometers seaward of the shoreline and a few tens of kilometers from the inferred shelf-slope break. Sand was believed to have been reworked by shelf processes into an offshore bar complex. This interpretation poses difficult problems and fails to account for several of the features observed in the Sussex including 1) pebbly sandstone, 2) stratigraphic variability of the sandstone, 3) erosional base and top of the Sussex and 4) regionally extensive depositional discontinuities marked by juxtapositions of facies, eroded mudstone clasts, glauconite and chert pebbles. These features suggest that fluctuations in rates of relative sea level change were important during Sussex deposition.

In the House Creek area, the base of the Sussex is a shallow northwesterly dipping planar surface with respect to the lower silty marker and is characterized by the abrupt appearance of interbedded sandstone and bioturbated mudstone on top of black mudstone. Burrows of the *Glossifungites* ichnofacies commonly occur along this contact. The top of the Sussex is marked by an abrupt change from pebbly sandstone to conglomerate which passes upward through pebbly bioturbated sandstone and pervasively bioturbated pebbly mudstone into laminated black mudstone.

The presence of erosional discontinuities in the Sussex Sandstone is suggested by 1) a sharp basal erosion surface that truncates underlying parasequences in the Cody Shale, 2) burrows of the *Glossifungites* ichnofacies at the basal contact, 3) lateral continuity of the individual facies contained between the discontinuity surfaces, 4) abrupt vertical facies changes and the association of glauconite, chert pebbles and mudstone rip-up clasts with these changes, 5) stratigraphic variability of the sandstones and 6) erosional termination of sandstone deposition and progradation. These discontinuities are characterized by an erosional escarpment with steeper landward margins that flatten basinward. This geometry and the linearity of the overlying sandstone contained in these erosional escarpments suggest incised shoreface deposits. These asymmetrical escarpments and their overlying sandstone form a retrogradational stacking pattern suggesting shoreface incision during an overall transgression. The Sussex Sandstone in the House Creek area is reinterpreted here as six backstepping incised shoreface deposits formed during a transgression punctuated by periods of decreased rates of relative rise and/or relative stillstand.

INTRODUCTION

The Campanian Sussex Sandstone in the Powder River Basin, Wyoming consists of a series of en echelon sandbodies encased in marine mudstone (Fig. 1). Earlier interpretations of the Sussex Sandstone have suggested that it was deposited as a shelf-ridge complex 200 km from shoreline deposits of the correlative Eagle Sandstone (Berg, 1975; Hobson et al., 1982) in water depths of 30 to 60 m (Berg, 1975). The primary objective of this paper is to illustrate the characterisitics of trangressively incised shoreface deposits. The Sussex Sandstone in House Creek area (Fig. 2) was re-examined using high-resolution stratigraphic techniques as a means of identifying and correlating surfaces interpreted to have formed as a result of fluctuations in relative sea level. Other objectives are to illustrate the 1) morphology of the erosional discontinuities marked by anomalous juxtaposition of facies and 2) lateral continuity of facies preserved between the discontinuities.

There is no unique identifying characteristic in a single core or outcrop that defines whether the erosional base of a shoreface formed during a forced regression or a transgression. The interpretation of these erosional surfaces is based on the facies and facies associations, the geometry of the deposits, their relationship to key surfaces and stratal stacking patterns. In this interpretation the base of the Sussex steps basinward and drops stratigraphically truncating markers in the Cody Shale and is interpreted to have formed during a drop of relative sea level. The top of the individual sandbodies marking the base of the Sussex (older B sandstone) are truncated by the same younger erosion surface (T3). There is an apparent landward and upward shift of facies across this surface (Fig. 3) and therefore it is interpreted as a ravinement surface. A second regressive surface of erosion shifts the shoreline basinward and downward truncating the earlier Sussex deposits and aggradational parasequences in the Cody Shale. The correlation of the upper sandbodies (younger A

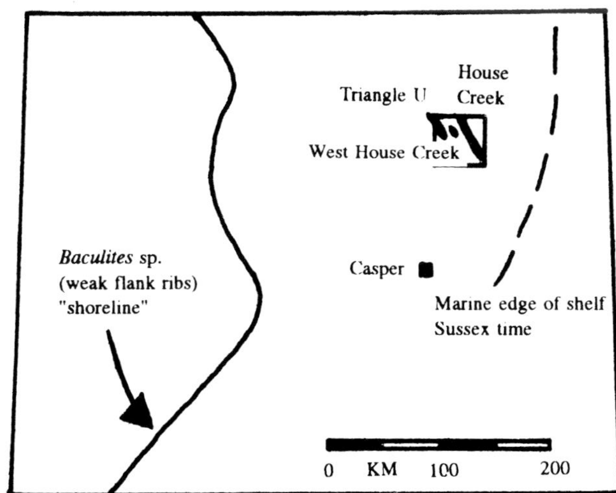

Fig. 1.—Map of Wyoming showing Sussex time-equivalent shoreline (Gill and Cobban 1973) and marine edge of shelf (Asquith 1970; Hobson et al. 1982). Rectangular box shows area examined in this study. Black areas are the major Sussex reservoirs Triangle U, West House Creek and House Creek. Both the eastern and western Wyoming State Borders trend north-south.

Isolated Shallow Marine Sand Bodies: Sequence Stratigraphic Analysis and Sedimentologic Interpretation.
SEPM Special Publication No. 64, Copyright © 1999
SEPM (Society for Sedimentary Geology), ISBN 1-56576-057-3, p. 297-319.

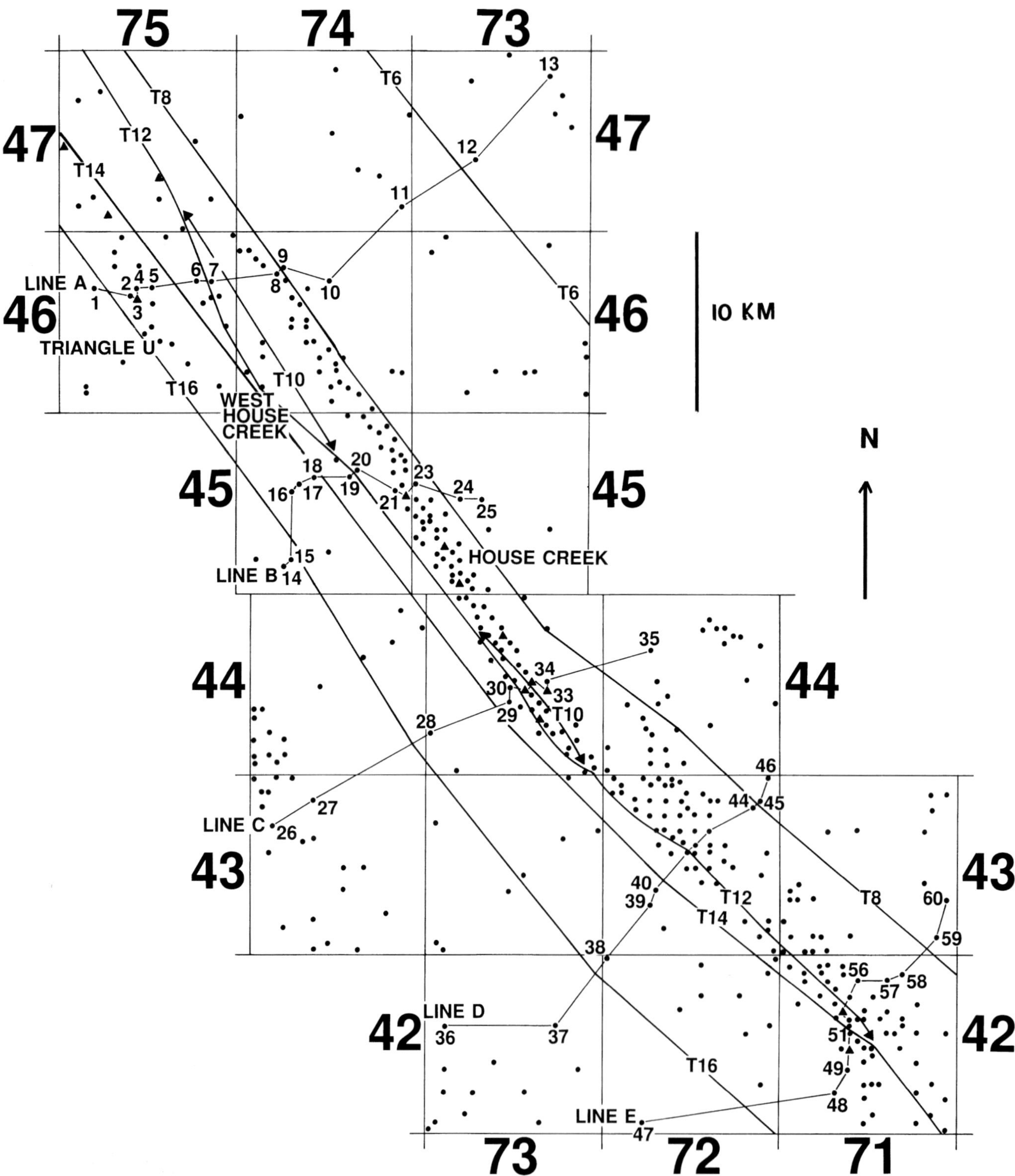

Fig. 2.—Detailed map of the study area (rectangle in Fig. 1). The location of the Sussex reservoirs and the positions of Lines A, B, C, D and E are shown. The lines labeled T indicate the positions of the incised shoreface profiles. Arrows at the end of the lines indicate T surfaces where one transgressive surface has been truncated by later incision. Closed circles represent the well log control used in the study; black triangles are the cored wells examined.

sandstone) shows a planar surface that truncates the top of the more seaward sandbody and forms the base of the next landward sandbody giving rise to a retrogradational stacking pattern and continuous erosion of any coastal plain and beach deposits that may have formed. Six backstepping sandbodies that record an overall upward fining succession were identified. Based on the stacking pattern and the overall fining upward succession these sandbodies were interpreted to have formed by shoreface incision during periods of stillstand in an overall transgression following a major lowstand of sea level.

STUDY AREA AND DATA BASE

Facies and facies successions were established by detailed correlation in House Creek field where there is core control (Townships 42N to 47N, Ranges 71W to 75W; Fig. 2). House Creek field is 45 km long and 1.6 km wide, except at the southern end. West House Creek and Triangle U fields are located approximately 4 km and 7 km to the west of House Creek field, and are 4 km and 11.5 km long, and 1 km and 1.6 km wide respectively (Figs. 1 and 2). The fields form an en echelon pattern trending N40°W (Fig. 1). The data base consists of 797 wells, of which 15 are cored (Fig. 2). Of these cored wells 11 intersect the lower contact of the Sussex Sandstone, 12 intersect the upper sandstone and shale contact and 9 intersect both contacts. All 15 cores were described and photographed. The facies associations are shown in core box photos (Fig. 3) and individual facies are illustrated in Figures 4 through 13.

The datum for the log and core sections is the Lower Silty Marker (LSM; Figs. 14, 15, 16, 17, 18, 19 and 20). Hobson et al. (1982) postulated water depths of 46 m for the deposition of LSM. The core cross section (Fig. 14) and a partial log section (Fig. 15) are used to illustrate the vertical relationships observed and lateral facies relationships interpreted in House Creek field. These relationships are used as a template to guide correlations in areas of limited core and log control.

The Ardmore Bentonite (UM4), Sussex bentonite (UM5) and other markers are shown on the cross sections. All of the markers are parallel to subparallel in the study area. Within the Sussex interval the numbers used on the cross sections are in inferred chronologic order with T indicating transgressive surfaces and P indicating progradation. Regional Sussex deposits were labeled as TST and HST, and refer to Transgressive and Highstand Systems Tract respectively. The strike parallel correlations shown on Figure 2 were made using all adjacent landward and seaward pairs of well logs along the trend.

STRATIGRAPHY AND PREVIOUS WORK

The Campanian Sussex Sandstone in the Powder River Basin is contained in the Cody Shale of the Montana Group and is assigned to the Baculites sp. (weak flank ribs) ammonite zone (Gill and Cobban, 1973). The Sussex is a 12 m thick sandstone interval encased in marine mudstone. There is no preserved evidence of emergence in the House Creek area. Anderman (1976) informally subdivided the Sussex into a younger A sandstone (main reservoir at Triangle U, Figs. 1 and 2), a middle shale and an older B sandstone (main reservoir at House Creek and West House Creek, Figs. 1 and 2). The stratigraphic top of the Susssex Sandstone is defined by a prominent bentonite marker informally named the Sussex bentonite (Kinnison, in Brenner, 1978). The Sussex bentonite (UM5 on the cross sections, Figs. 15 to 20) forms part of a 15 to 30 m thick bentonite rich mudstone termed the Ardmore Bentonite (Asquith, 1970, 1974; Table 1). The Sussex Sandstone is overlain by the *Bacculites obtusus* Range Zone (Gill and Cobban, 1973) and appears to be related in time to the transgression of the Claggett sea and subsequent reworking of the uppermost Eagle Sandstone and equivalents (Hobson et al., 1982). The Sussex Sandstone represents an early phase of deposition associated with this rapid transgression (Hobson et al., 1982). In this setting the Sussex appears to be consistent with the predictions of the third order sea level curve of Haq et al. (1988) suggesting that eustatic fluctuations were important during Sussex deposition.

The Sussex Sandstone is equivalent (Table 1) to part of the Eagle Sandstone of the Lower Mesaverde Group in the Bighorn Basin (Gill and Cobban, 1973). Based on paleontologic evidence, Gill and Cobban (1973) suggested that the shoreline moved eastward during Sussex time. However, the Sussex Sandstone at House Creek field lies more than 160 km east of the time equivalent shoreline in the Eagle Sandstone (Fig. 1), and was thought to be deposited in about 60 m of water (Asquith, 1970). The proposed paleogeography (Asquith, 1970) at the maximum extent of the Gammon (Table 1) shelf progradation (Fig. 1) shows the location of House Creek field about 48 km landward of the projected shelf edge (Asquith, 1970). In this reconstruction the Sussex was deposited on a nearly flat, horizontal midshelf (Asquith, 1970) and sandbody elongation is oblique (33° to 45°) to general trends of the defined shelf slope break (Fig. 1). Berg (1975) and Brenner (1978, 1980) suggested that the bedding and textural sequences indicated a southeasterly prograding succession that was deposited by shore-parallel currents. Brenner (1980) further suggested that this progradation occurred during a period of sea level stillstand. In these interpretations sediment was supplied from an undefined point source in the NW, possibly a deltaic system in central Montana. A further speculation would be that the Sussex Sandstone at House Creek field indicates a NE-SW transport route across the shelf along which sediment moved downslope toward the deep basin (Berg, 1975) and formed a fan shaped deposit of siltstone 60 to 90 m thick and covering 390 km^2 on the basin floor (Asquith, 1970). Alternatively, Hobson et al. (1982) suggested that rapid drowning of low-relief Sussex coastlines resulted in relict sand abandoned on the shelf by rapid shoreface retreat during transgression. These sediments were subsequently transported and reworked by storm and other marine processes into broad elongate offshore bar complexes. These bar complexes eventually became isolated and preserved in position because of deepening water.

Neither of these interpretations adequately explains many of the characteristics preserved in the Sussex Sandstone. Berg (1975) and Brenner (1980) suggested that the lithologies preserved in the Sussex resemble sediments present along a modern delta-front, but could not account for the lack of evidence of subaerial exposure and lack of lithologic continuity with known paleoshorelines and therefore placed the Susssex Sandstone in an offshore depositional setting. However, deposition of the Sussex as a marine shelf sandbody does not account for 1) thin and erratic occurrence of pebbly sandstone (Berg, 1975), 2) coarse grained chert granules present throughout the bioturbated siltstone that appears to have been a site of offshore marine deposition (Berg, 1975), 3) presence of large yellow claystone pebbles (Berg, 1975) or 4) presence of depostional discontinuities marked by a juxtaposition of facies, glauconite, mudstone rip-up clasts, chert pebbles and *Glossifungites* ichnofacies.

Fig. 3.—Core box photos of A) Well 22, Line B (northern half of House Creek) and B) Well 54, Line E (southern half of House Creek) showing the numbered transgressive surfaces to the left of each label and the facies contained in each progradational event. Stratigraphic top is to the upper left. BS, Base of Sussex; P, progradation number; T, transgressive surface of erosion of a given shoreface incision; MFS, maximum flooding surface. Core boxes are 1m. Well 22 is missing 5 feet of core.

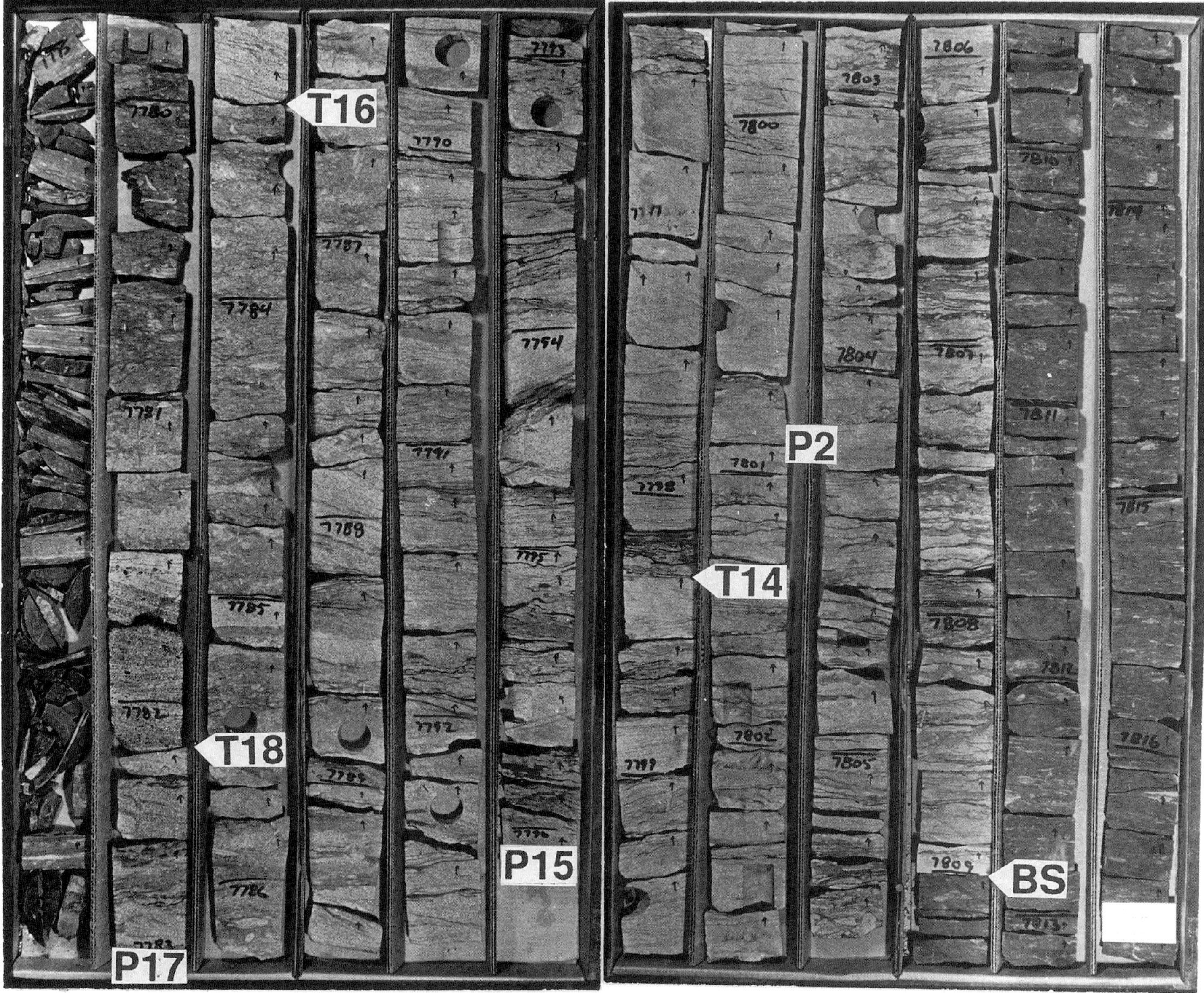

Fig. 3 (continued).

FACIES DESCRIPTIONS

Core box photos (Fig. 3A) for Well 22 (Line B; Fig. 17) and (Fig. 3B) Well 54 (Line E; Fig. 20) show the House Creek sandstone succession in the northern and southern part of the field respectively. The lithologs for Wells 31, 32 and 33 (Line C; Fig. 18) are shown in Figure 14. The facies are interpreted on the basis of their own characteristics, and their vertical and lateral relationship to each other. The interpretation of the facies will follow the description of the cross sections.

Facies 1: Black Mudstone

Facies 1 is below the stratigraphic base of the Sussex Sandstone (BS; Figs. 14 to 20). It is a black massive mudstone with silty lenses (Figs. 3, 4). The mudstone is intensely bioturbated, although *Helminthopsis* and *Terebellina* could be locally discerned (Fig. 4). The upper contact is sharp and is abruptly overlain by Facies 2. The contact between Facies 1 and 2 (labeled BS on the core box photos; Fig. 3) is commonly marked by *Thalassinoides* burrows filled with very fine to fine sandstone (G; Fig. 4B). In Core NE NW 21-42N-71W, 4 mm chert grains are found at this contact.

Facies 2: Interbedded Sandstone and Bioturbated Mudstone

This facies is composed of bioturbated mudstone (thickness averages 5-6 cm) separated by thin (average 1-3 cm thick; many are < 1cm thick), sharp-based beds of asymmetrical small scale cross laminated, very fine to fine sandstone beds (Fig. 5). Many of the sandstone beds have bioturbated tops. Burrow forms include *Teichichnus*, *Chondrites*, *Paleophycus*, *Planolites*, *Helminthopsis*, *Skolithos*, *Ophiomorpha* and *Thalassinoides*. Facies 2 has an average thickness of about 1 m and sharply overlies BS (lower part of P2; Figs. 3, 14). The upper contact is gradational with Facies 3.

Facies 3: Thin-Bedded Bioturbated Sandstone

Facies 3 is composed of thicker (average 3-5 cm thick), sharp based beds of asymmetrical small scale cross laminated very fine to fine sandstone separated by thin (< 1cm thick) shale partings (Figs. 3, 6). In Core NE NW 21-42N-71W, 1-2 mm chert grains are scattered throughout. Some sandstone beds contain mudstone rip-up clasts. In many thicker beds (> 8 cm) low angle inclined stratification is observed, although in general the stratification is poorly defined. Facies 3 is much less intensely bioturbated than Facies 2. Recognizable burrow forms include *Chondrites*, *Planolites*, *Paleophycus*, *Helminthopsis* and *Thalassinoides*. This facies has an average thickness of 4.5 m (upper part of P2; Fig. 3, 14) and the upper contact is sharp and is overlain by Facies 4 (T12 in Fig. 3A).

Facies 4: Medium Sandstone

This facies varies from <1 m to 2 m in thickness and is composed of small scale cross laminated, massive (Fig. 7A) to low angle inclined stratified (Fig. 7B) or rarely small scale (5-10 cm thick) trough cross bedded medium sandstone (occurs above P2; Figs. 3, 14). Rarely soft sediment deformation is observed. There is a general increase in bed thickness and a decrease in the number of mudstone partings upward. Mudstone and siderite rip-up clasts (up to 6 cm long) and coaly fragments are common. Core NE NW 21-42N-71W contained 1-2 mm scattered chert grains. Glauconite is disseminated throughout. Well defined, identifiable traces were rare although *Ophiomorpha* burrows occur in some cores. In general the well defined burrows tend to be associated with the muddier horizons and recognizable burrow forms include *Helminthopsis* and *Paleophycus*. The upper contact is commonly gradational with Facies 5.

Facies 5: Pebbly Cross-Bedded Sandstone

Facies 5 has an average thickness of 1 m and is composed of cross-bedded, pebbly, medium sandstone (Fig. 8). Mudstone and siderite rip-up clasts are abundant. Glauconite is disseminated throughout. Chert pebbles have an average long axis diameter of 4 mm (largest ~1cm). The upper contact is sharp and is characterized by some or all of the following features: a juxtaposition of facies associated with a grain size change, a marked increase in glauconite, coaly clasts, siderite clasts and mudstone rip-up clasts (Fig. 14; T16 in Figs. 3A and 10A; T18 in Figs. 3B and 9).

Facies 6: Conglomerate

This facies is composed of thin (up to 30 cm thick) chert pebble conglomerate (max pebble size ~ 1cm) with a medium sandstone matrix (Fig. 9). There are abundant mudstone and siderite rip-up clasts, and it is highly glauconitic. Cross-bedding (Fig. 9) and symmetrical small scale cross lamination are common in thicker deposits. Facies 6 is commonly overlain by Facies 7.

Facies 7: Pebbly Bioturbated Sandstone

Facies 7 has an average thickness of about 0.5 m and is composed of scattered up to 4 mm chert grains and mudstone rip-up clasts in a bioturbated, fine sandstone (above T16 in Figs. 3A, 10A; T14 in Fig. 14 or above Facies 6 in Figs. 3B and 14). At the base of the unit a few cm thick sandstone beds may be preserved otherwise the facies is thoroughly bioturbated. Recognizable burrows include *Thalassinoides*, *Helminthopsis*, *Teichichnus*, vertical mud filled burrows (*Rosselia*?) and *Skolithos*. The abundance and size of the pebbles and sandstone decreases upward. The upper contact is gradational with Facies 8.

Facies 7A: Bioturbated Sandstone

This facies is similar to Facies 7 except that there are no chert grains (Fig. 10B). Facies 7A is composed of a bioturbated very fine to fine sandstone and mudstone with rare thin (average 2 to 4 cm thick), sharp based sandstone beds. Burrows include *Chondrites*, *Helminthopsis*, *Skolithos*, *Teichichnus*, *Planolites*, *Schaubcylindrichnus*?, and *Paleophycus*. Glauconite is disseminated throughout.

Facies 8: Pervasively Bioturbated Pebbly Mudstone

This facies is 1 m thick and is composed of scattered 1 to 2 mm chert grains, isolated 0.5 cm chert pebbles and phosphatic fragments in a pervasively bioturbated mudstone (Figs. 3, 11, 14). Rare siltstone laminae are preserved. The facies is thoroughly bioturbated, although few recognizable burrows were identified. These include *Helminthopsis* and *Paleophycus*. In some wells this facies becomes sandier upward.

Facies 9: Dark Bioturbated Mudstone

Facies 9 is similar to Facies 1. It is composed of black mudstone with rare, thin (~1 cm) siltstone to very fine sandstone beds (Figs. 3, 12). The facies is thoroughly bioturbated. Burrow forms include *Helminthopsis*, *Terebellina* and *Planolites*. Shell fragments are scattered in the mudstone.

Facies 10: Laminated Black Mudstone

This mudstone contains laminae of very fine sandstone to siltstone (Figs. 3, 13). *Helminthopsis* are abundant (Fig. 13).

CONSTRUCTION OF A CORRELATION TEMPLATE

The sandstones in House Creek field could not be correlated continuously basinward on any of the cross sections. Three cores (Fig. 14) from Line C (Fig. 18) located on Figure 2 are used to illustrate the interpreted stratigraphic relationships based on the vertical facies associations observed in core. Four basic elements, discussed in detail below, describe the interpreted stratigraphic relationships 1) coarsening upward succession, 2) sharp based sandbodies, 3) apparent abrupt westward termination of the coarsening upward sandstone and 4) extension of surfaces seaward of the coarsening upward sandbody. These stratigraphic relationships (Fig. 14) serve as a template for correlation in areas of limited core control.

1) Coarsening Upward Sandbody

Sandbodies coarsen upward (e.g., P11, P13; Fig. 14) from medium sandstone containing low angle inclined stratification (Facies 4) to pebbly trough cross bedded sandstone (Facies 5). In some cores the medium sandstone is massive. The sandbodies have an average preserved thickness of about 2.5 m. The width of the sandbodies is more difficult to determine, but based on the log cross section (Line C Fig. 18) the sandbodies appear to be about 6 km wide (P13 sandbody in Well 31 can be traced northeastward to about Well 34). As the sandbodies are traced eastward (Line C; Fig. 18) they thin and pass laterally into mudstone. The coarsening upward nature of the sandbodies implies progradation.

2) Sharp Based Sandbodies

The bases of the sandbodies are sharp (T10 and T12; Fig. 14) and are commonly marked by abrupt changes in grain size (T12 in Fig. 3A) from fine to medium sandstone, mudstone and siderite rip-up clasts (T14; Fig. 3B), and/or glauconite. Isolated chert pebbles occur at these contacts. Based

Table 1.—Summary of the stratigraphy and K- Ar ages of the Upper Cretaceous Lower Montana Group and equivalent rocks in Wyoming (modified from Gill and Cobban 1973).

SUBSTAGE	ESTIMATED DATES (M.Y.A.)	AMMONITE ZONE	CENTRAL MONTANA		HOUSE CREEK POWDER RIVER BASIN WYOMING	EASTERN WYOMING
	79	*Baculites mclearni*	CLAGGETT SHALE		SHALE	SHARON SPRINGS MEMBER
			ARDMORE		ARDMORE	ARDMORE
		Baculites obtusus	BENTONITE		BENTONITE	BENTONITE
LOWER CAMPANIAN	80	*Baculites* sp. (weak flank ribs)		CODY SHALE (STEELE MEMBER)	SUSSEX SANDSTONE MEMBER	
			EAGLE SANDSTONE		SHALE	GAMMON-FERRUGINOUS MEMBER
		Baculites sp. (smooth)			SHANNON SANDSTONE MEMBER	
					SHALE	
	81	*Scaphites hippocrepis* (fine ribbed) III			FISHTOOTH SANDSTONE	
		Scaphites hippocrepis (coarse ribbed) II				
	82	*Scaphites hippocrepis* (coarse ribbed) I	TELEGRAPH CREEK FORMATION		SHALE	NIOBRARA FORMATION

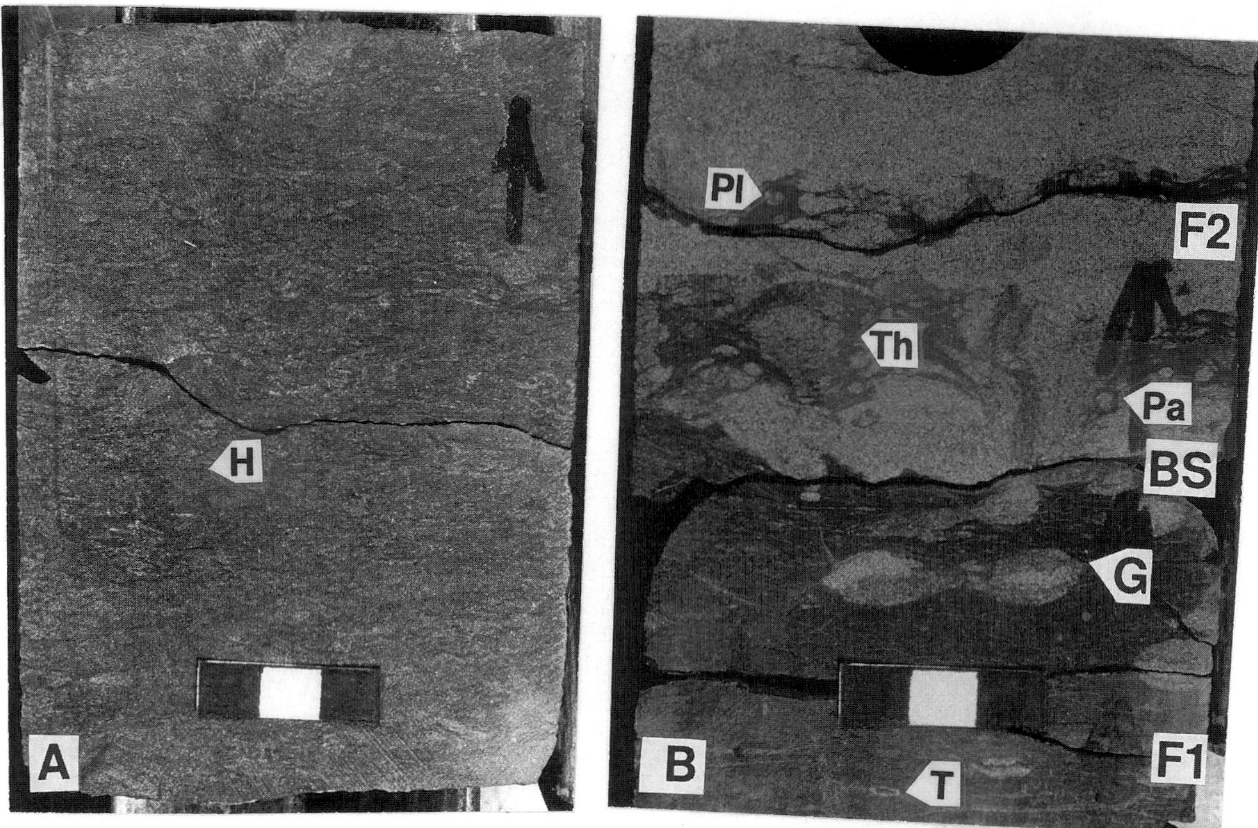

Fig. 4.—Detail core photo (SW NE 32-47-75) showing A) Black Mudstone (Facies 1), taken 0.25 m from the base of the core, and B) the Base of the Sussex (BS) in House Creek, taken 1.75 m from the base of the core. Scale bar is 3 cm long. Note the presence of the passively filled Thalassinoides burrow of the *Glossifungites* ichnofacies below the base of the contact. F1, Facies 1; F2, Facies 2; G, *Glossifungites*; T, *Terebellina*; H, *Helminthopsis*; Pa, *Paleophycus*; Pl, *Planolites*; Th, *Thalassinoides*; BS, Base of Sussex.

on the correlations as the base of the sandbody is traced basinward (T12; Fig. 14), it is commonly overlain by up to 30 cm of conglomerate (Facies 6) containing abundant mudstone and siderite rip-up clasts, and glauconite. The conglomerate (Facies 6) is found abruptly overlying the pebbly sandstone (Facies 5) or the medium sandstone (Facies 4) of the previous sandbody (T12 Wells 32 and 33, Fig. 14; T14 Well 22; Fig. 3A). The sharp base of the sandbodies is interpreted to have formed as the result of erosion at fairweather wave base during ravinement. The depth of erosion by fairweather wave base varies with the degree of consolidation of the underlying material (Anderson et al., 1991).

3) Westward Termination of the Sandbodies

The base of the sandbodies can be traced basinward along the line of section (Fig. 14) where their gradients flatten (T surfaces on Line C; Fig. 18). It is more difficult to trace the surface landward and the correlations depend on several arguments. The coarsening upward sandbody present in Well 32 (P11, Fig. 14) is absent in the adjacent Well 31 (Fig. 14). Thin-bedded bioturbated sandstone (Facies 3) is found at this horizon in core of Well 31 (Fig. 14). Bergman (1994) and Bergman and Walker (1987) interpreted a similar stratigraphic relationship in the Shannon Sandstone and Cardium formations respectively as areas of erosion and projected the erosion surface at the base of a prograding sandbody upward between these two wells giving rise to a one-sided scour morphology. These asymmetrical erosion surfaces can be traced alongstrike (Fig. 2) for several tens of kilometers at each stratigraphic horizon, where they form mappable, long linear escarpments.

4) Extension of Erosion Surfaces

These erosion surfaces can be correlated basinward beyond the immediate position of the incision. The position of the erosion surfaces may be difficult to pick solely on the basis of well log signatures, particularly in areas where it may be a sand-on-sand contact marked by a grain size change (T12 in Well 22; Fig. 3A) or a mud-on-mud contact (Well 35, Line C; Fig. 18) marked by coarse grains. Beyond the core cross section there may be log markers that are truncated or onlap these erosion surfaces. These are present on the cross sections (Figs. 15 to 20) but are not incorporated into this discussion.

APPLICATION OF THE TEMPLATE

Any interpretation of the Sussex Sandstone in this area must account for 1) juxtapositions of facies, commonly associated with marked grain size changes, 2) abrupt westward termination of the sandbodies and 3) sandbodies occurring at different stratigraphic horizons. The few published cross sections of the Sussex Sandstone (Berg, 1975; Hobson et al., 1982) imply that the sandbodies have been

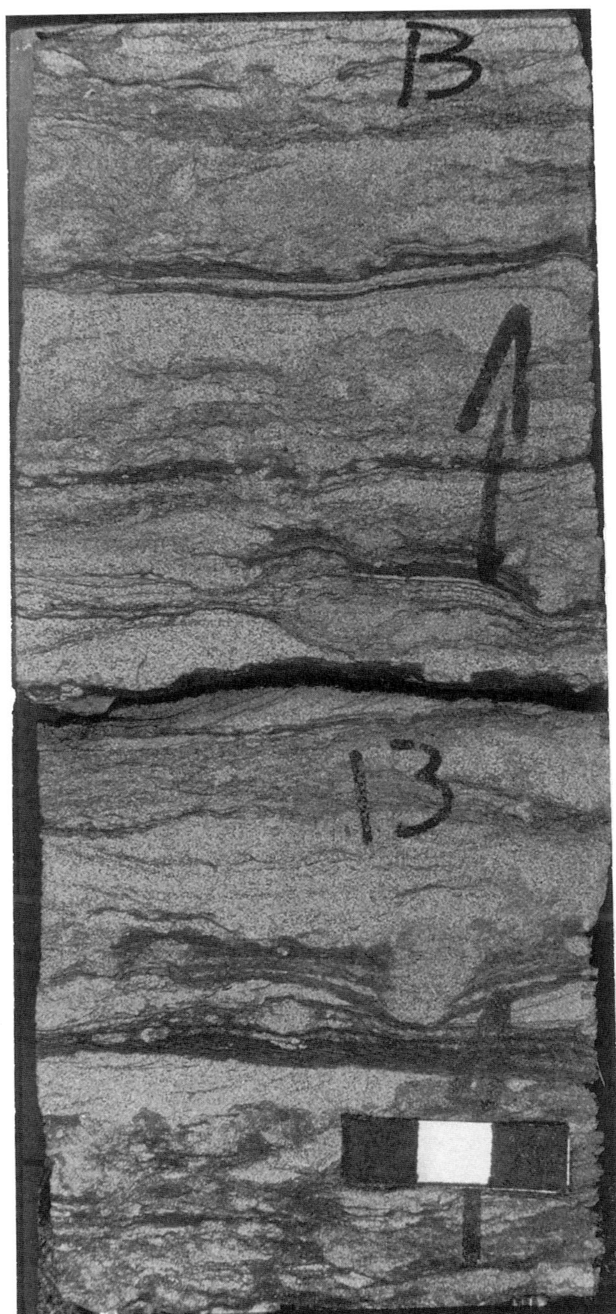

Fig. 5.—Detail core photo of Facies 2 (NW SE 32-45N-73W), taken 1.65 m from the base of the core. Scale bar is 3 cm long.

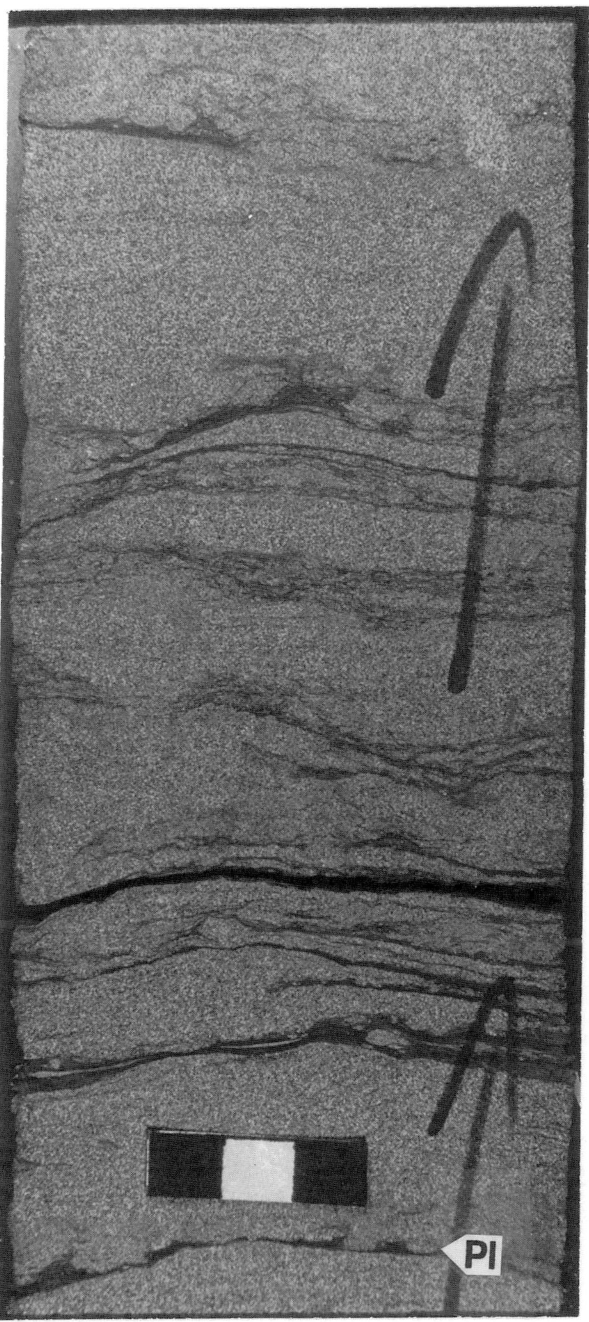

Fig. 6.—Detail core photo of Facies 3 (NW SE 32-45N-73W), taken 5.0 m from the base of the core. Scale bar is 3 cm long. Pl, *Planolites*.

truncated by the same trangressive surface of erosion (Fig. 15A). Although this type of correlation accounts for the abrupt truncation of the sandbodies it does not account for the 1) lateral relationship between sandstone and mudstone, 2) sharp base of the sandbodies and 3) stratigraphic variability of the sandbodies. To accommodate these relationships a modified correlation (Fig. 15B) is suggested using the stratigraphic relationships identified on the core cross section (Fig. 14) as a template. From this six separate sandbodies bounded by erosional discontinuities can be traced and mapped (Fig. 2). The base of the sandbodies on logs is marked by an abrupt decrease in the gamma ray and a corresponding increase in the resistivity signal (Fig. 15B). In core (Fig. 3A and 3B) this surface is marked by an abrupt shift in facies and an associated change in grain size. The surfaces forming the base of these coarsening upward successions and their associated sandstone deposits can be correlated eastward along the cross sections (Figs. 16 to 20) unless removed by subsequent erosion. As the sandbodies are correlated eastward they thin and grade laterally into

Fig. 7.—Detail core photo of Facies 4. A) Massive sandstone (Well 22, Line B) taken 8.84 m from the base of the core. B) Low angle inclined stratified sandstone (Well 50, Line E), possibly cross bedding taken 8.75 m from the base of the core. Scale bar is 3 cm long.

bioturbated sandstone and mudstone (e.g., from Wells 5/6 and 8/9, Line A, Fig. 16; 28/29, Line C, Fig. 18; 38/39 Line D, Fig. 19; 55/56, Line E, Fig. 20).

The westward terminations of the sandbodies and their bounding surfaces are more difficult to trace. The sandstone is laterally adjacent to mudstone and the erosion surface cannot be traced into the laterally adjacent well (Figs. 14 and 15B). This relationship is illustrated on the cross sections (e.g., between Wells 1/2, 11/12, Line A, Fig. 16; 15/16, Line B; Fig. 17; 28/29, 31/32, Line C; Fig. 18; 37/38, Line D, Fig. 19; 54/55, Line E; Fig. 20). The surfaces can be traced along the length of the House Creek area (Fig. 2) except where truncated by subsequent erosional events suggesting that these surfaces are regionally extensive.

INTERPRETATION OF BOUNDING SURFACES AND FACIES

The correlations described above are based on the recognition and correlation of bounding discontinuities. The morphology of the erosion surfaces and the facies relationships contained between these surfaces provide the framework for reinterpreting the Sussex Sandstone at House Creek as a series of backstepping transgressive incised shoreface deposits. As identified in the Cardium Conglomerates at Carrot Creek (Bergman and Walker, 1987) and Shannon Sandstone at Hartzog Draw (Bergman, 1994) three features suggest shoreface incision rather than general erosion on the shelf or fluvial incision: 1) long linear erosion surfaces with an asymmetric morphology; 2) progressive westward shift and stratigraphic rise of each successive sandbody and 3) facies relationships across the discontinuities.

The Sussex Sandstone rests abruptly on the black mudstone (Facies 1) of the Cody Shale (Figs. 3 and 4B). The black mudstone (Facies 1) is interpreted to have been deposited on the mid- to outer-shelf. This interpretation is consistent with previous studies (Berg, 1975; Brenner, 1978, 1980; Hobon et al., 1982). Several aggradational parasequences can be identified on the log sections in the underlying Cody Shale (Figs. 16 to 20).

BS appears as a shallow northwesterly dipping planar surface with respect to the datum (LSM). BS removes marker beds in the Cody Shale between Wells 2 and 3, and Wells 11 and 12 (Line A; Fig. 16); Wells 15 and 16 (Line B; Fig 17); Wells 27 and 28 (Line C; Fig. 18). From the cross sections it appears

Fig. 8.—Detail core photo of Facies 5 (Well 31, Line C), taken 7.70 m from the base of the core. Scale bar is 3 cm long.

Fig. 9.—Detail core photo of Facies 6 (Well 54, Line E), taken 10.25 m from the base of the core showing wave ripples in the conglomerate. T18 is the surface at the base of the conglomerate shown in (Fig. 3B). Scale bar is 3 cm long. F5, Facies 5; F6, Facies 6.

that there is an increased rate of subsidence toward the southeastern part of the House Creek area because BS does not remove these marker beds on the southern cross sections (Line D; Fig. 19 and Line E; Fig. 20). BS is directly overlain by interbedded sandstone and bioturbated mudstone (Facies 2). The trace fossil assemblage in Facies 2 is similar to the Cruziana ichnofacies (MacEachern and Pemberton, 1992; Howard and Frey, 1984) suggesting offshore marine deposits. The sharp based, fine sandstone beds containing current ripples are interpreted as storm beds deposited in this offshore environment. Facies 2 is interpreted to be deposited in the upper offshore below fairweather wave base and above storm wave base.

The contact (BS) between Facies 1 and Facies 2 is marked by *Thalassinoides* burrows (Fig. 4B) which appear to be passively filled by the fine sandstone of Facies 2 and are interpreted to be part of the *Glossifungites* ichnofacies as described by MacEachern and Pemberton (1992). Chert granules (4 mm)

Fig. 10.—Detail core photo of Facies 7. A) Pebbly Bioturbated Sandstone (Well 22, Line B) abruptly overlying Facies 5, taken 11.25 m from the base of the core. T16 marks the contact between Facies 5 and Facies 7 (Fig. 3A) and is characterized by an abrupt juxtaposition of facies and grain size change. B) Bioturbated Sandstone (Facies 7A; Well 50, Line E) taken 6.25 m from the base of the core. Scale bar is 3 cm long. F7, Facies 7; F5, Facies 5.

are commonly found at this contact. The juxtaposition of facies (Fig. 3), the apparent truncation of log markers in the Cody Shale (Figs. 16, 17 and 18), and 4 mm chert granules all suggest the erosive nature of this contact. The *Glossifungites* ichnofacies (Fig. 4B) at this surface is consistent with the erosional nature of BS (MacEachern and Pemberton, 1992).

Based on the cross sections (Figs. 16 to 20) BS appears to be an amalgamated erosion surface. There is an asymmetric scour (e.g., Wells 27 and 28, Line C; Fig. 18) into marine mudstone of the Cody Shale. The surface forming the base of the Sussex (BS) steps basinward and drops stratigraphically truncating underlying parasequences in the Cody Shale (Figs. 16 to 20) and therefore is interpreted as a regressive surface of erosion. The asymmetric incision is interpreted to have formed by shoreface erosion during relative stillstand in a fall of relative sea level (forced regression).

BS is overlain by the deposits of P2 (Figs. 3 and 14). P2 is a coarsening and sandier upward succession (Well 31, Fig. 14) from the upper offshore deposits of Facies 2 to thin-bedded bioturbated sandstone (Facies 3). There is a general decrease in the thickness of mudstone partings and intensity of bioturbation, and a corresponding increase in the thickness of sharp based fine sandstone beds suggesting a closer proximity to source. The low angle inclined stratification is interpreted as HCS. The thin-bedded bioturbated sandstone (Facies 3) is interpreted to have been deposited in the transition zone between the shoreface and offshore. The trace fossil assemblage is consistent with this interpretation. P2 is recog-

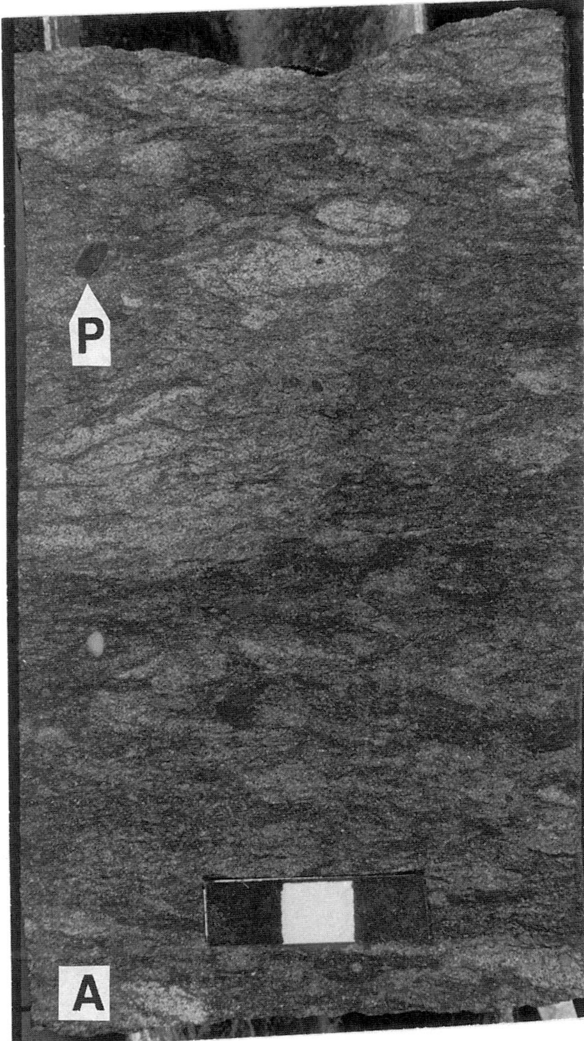

Fig. 11.—Detail core Photo of Facies 8. A) Pervasively Bioturbated Pebbly Mudstone (Well 22, Line B) with 0.5 cm long chert pebbles, taken 10.40 m above the base of the core. B) Pervasively Bioturbated Pebbly Mudstone (NW NW 19-47N-75W) with 1 to 2 mm chert grains, taken 13.5 m above the base of the core. Scale bar is 3 cm long. P, Pebbles; g, mm size cherts grains.

nized throughout the study area except where removed by subsequent erosion. P2 has an average thickness of 5 m. Based on the asymmetrical basal erosion surface and the overlying coarsening upward succession from upper offshore bioturbated deposits to transitional shoreface deposits P2 was interpreted as the distal part of a prograding shoreface formed during a forced regression. Sediment supplied to this incised shoreface resulted in progradation at least as far as the eastern margin of House Creek Field.

T3 truncates the amalgamated BS shoreface incision and prograding sandbody (P2). T3 is overlain (Well 50 Line E, Fig. 20) by interbedded sandstone and bioturbated mudstone (Facies 2) that was interpreted as upper offshore. In core T3 is marked (Well 50, Line E; Fig. 20) by a juxtaposition of facies from distal shoreface deposits below (P2) to offshore marine deposits above suggesting a landward shift in facies. T3 is overlain by at least one regionally extensive onlapping marker and is interpreted here as a transgressive surface of erosion formed by erosion at fairweather wave base as the shoreline moved westward out of the House Creek area. Based on the gamma ray log signature T3 is overlain by mudstone rich deposits in the study area (one core through this interval) and in this setting the mudstone is interpreted to have formed during transgression and by definition is part of the transgressive systems tract (TST4). The transgressive mudstone gradually becomes sandier upward on the gamma ray log signature. There is one core through this interval (Well 50 Line E; Fig. 20). In core this interval is composed of bioturbated sandstone (Facies 7A). The trace fossil assemblage of Facies 7A is consistent with the *Cruziana* ichnofacies and is suggestive of offshore marine conditions (MacEachern and Pemberton, 1992; Howard and Frey, 1984). The thin sharp based fine sandstone beds are interpreted as storm beds. Facies 7A was interpreted as upper offshore deposits of a shoreface that lay to the west of the House Creek study area. The overall sandier upward succession above transgressive mudstone led to the interpretation of this interval as highstand systems tract (HST5). The surface (Line A; Fig. 16) separating the transgressive (TST4) and highstand (HST5) systems tract by definition is the Maximum Flooding Surface.

Fig. 12.—Detail core photo of Facies 9 (NE NW 34-47N-76W), taken 8.5 m from the base of the core. Scale bar is 3 cm long. H, *Helminthopsis*

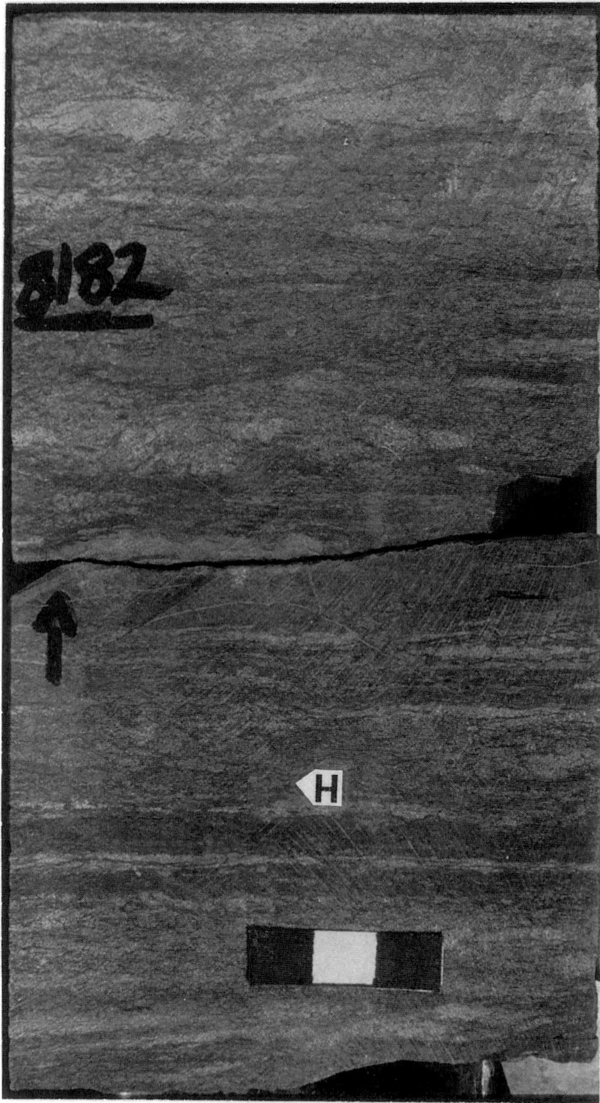

Fig. 13.—Detail core photo of Facies 10 (SE SW 27- 47N-75W), taken 6.75 m from the base of the core. Note the presence of abundant *Helminthopsis* in the mudstone. Scale bar is 3 cm long. H, *Helminthopsis*

Highstand Systems Tract deposition was truncated by a major drop of relative sea level (Forced Regression) that shifted the shoreline position east of House Creek Field. Across this surface there is a basinward step and downward shift of facies. This drop truncated the earlier P2 deposits, BS and aggradational parasequences in the Cody Shale east of House Creek Field. The eastward progradation of these regional Sussex deposits is truncated on each cross section (Figs. 16 to 20). This regressive surface of erosion is interpreted to be modified by subsequent transgressive ravinement that eroded and reworked the underlying deposits as the facies step landward and rise stratigraphically. This overall transgression is interpreted to have been interrupted by a number of relative stillstands giving rise to a series of backstepping erosion surfaces (labeled T6 to T16) each of which is overlain by a progradational sandbody (labeled P7 to P17) respectively. During these periods of stillstand erosion at fairweather wave base formed the regionally extensive one-sided erosional surfaces labeled T6 through T16. All of these asymmetric erosion surfaces can be correlated along strike (Fig. 2) except where removed by subsequent erosion, and are interpreted as incised shoreface profiles. The bases of these sandbodies are commonly marked by an abrupt grain size change from fine sandstone below to medium sandstone above (Fig. 14). Mudstone rip-up clasts (Figs. 3 and 14) are also common at the bases of the sandbodies.

Sediment was supplied to these incised shorefaces most probably from transgressive reworking of underlying material and resulted in progradation (Fig. 14). This is supported by the increased glauconite content observed in Facies 5 and 6. Close to the incision the T surfaces are overlain by massive to low angle inclined medium sandstone (Facies 4) which is gradationally overlain by pebbly cross bedded sandstone (Facies 5). There is very little visible bioturbation associated with these sandstone deposits. However, the massive aspect (Fig. 7A) of the medium sandstone (Facies 4) is interpreted to be the result of the burrowing activities of meiofauna (Cullen, 1973; Saunders et al., 1994). The low angle inclined stratification observed in Facies 4 (Fig. 7B) may represent wave ripples in medium to coarse sediment. Based on the stratigraphic position, the inferred trace fossil assemblage and the vertical facies relationships Facies 4 was interpreted as the distal part

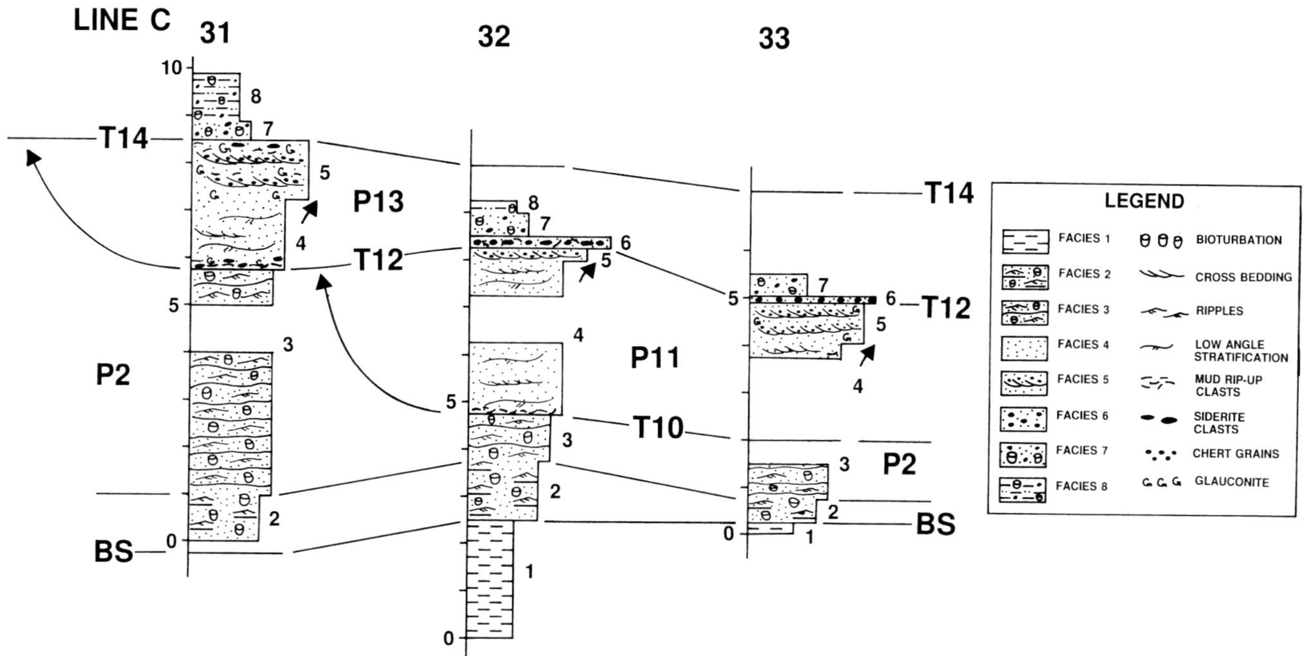

Fig. 14.—Three cores from Line C located on Figure 2. The well log correlation is shown on the western end of Line C (Fig. 18). Wells 32 and 33 contain the shoreface deposits of P11, and Well 31 contains the shoreface deposits of P13. The facies are labeled as 1-8 on the right side of the litholog and are described in the text. The transgressive surfaces and progradational events are labeled as referred to on Line C (Fig. 18) and in the text. The core lithologs are positioned with respect to the LSM datum. In Well 31 the base of the Sussex is not cored. Its position as determined from the well logs is shown by a horizontal bar. Closed arrows represent truncated horizons. Vertical scale is in meters.

of a shoreface deposit. The coarse, cross bedded aspect of Facies 5 gradationally overlying Facies 4 suggests more proximal deposition. The cross bedding may reflect transport in the longshore drift system. Paleoflow directions from sandstone deposits in the Sussex outcrop at Salt Creek Anticline (Fig. 1) trend south-southeast. The sandstone facies thin basinward and pass laterally into mudstone. Based on the asymmetric morphology of the erosion surface and the progradational nature of the sandbody, the medium sandstone (Facies 4) and pebbly cross bedded sandstone (Facies 5) are interpreted as shoreface deposits that prograded during a reduced rate of sea level rise or stillstand over their offshore deposits (Fig. 14). The absence of HCS from the medium (Facies 4) and pebbly cross bedded sandstone (Facies 5) may be a function of grain size control. HCS is commonly found in very fine to fine sand while medium to coarse sand is typically cross bedded (Walker and Plint, 1992). Similar cross bedded deposits have been interpreted as shoreface deposits in the Viking Formation (Davies and Walker 1993; Walker 1995) and the Shannon Sandstone (Walker and Bergman 1994; Bergman, 1995).

Basinward the T surface is overlain by conglomerate (Facies 6), pebbly bioturbated sandstone (Facies 7) and pervasively bioturbated pebbly mudstone (Facies 8). The conglomerate (Facies 6) is interpreted to be winnowed from the underlying coarse and pebbly sandy shoreface deposits of the previous stillstand during subsequent transgressive erosion. Longshore currents formed cross bedding in the lag. The conglomerate (Facies 6) is interpreted as a transgressive lag. Similar lag deposits were described by Walker (1995) overlying the VE4 ravinement surface in the Viking Formation. The pebbly bioturbated sandstone (Facies 7) and pervasively bioturbated pebbly mudstone (Facies 8) represent deposition during resumed transgression (Fig. 14). The overall fining and muddier upward succession resulted from continued retreat of the shoreline. Trace fossil assemblages contained in both the pebbly bioturbated sandstone (Facies 7) and pervasively bioturbated pebbly mudstone (Facies 8) are suggestive of upper offshore to lower offshore marine conditions (MacEachern and Pemberton, 1992; Howard and Frey, 1984). During storms pebbles could be transported from the retreating shoreface and deposited in the accumulating offshore deposits. Similar relationships were observed in the Cardium conglomerates at Carrot Creek (Bergman and Walker, 1987, 1988). The surface separating the pebbly deposits from the next prograding sandstone is interpreted as a local maximum flooding surface (MFS on Fig. 3A) and marks the onset of progradation associated with the next incised shoreface. In some areas dark bioturbated mudstone (Facies 9) may be preserved at this transition. A similar relationship has been observed in the Viking Formation at Joarcam (Walker and Wiseman, 1995) where the incised shoreface progrades over its own offshore deposits. Based on the stratigraphic context and the overall similarity to Facies 1 the dark bioturbated mudstone (Facies 9) is interpreted to represent offshore marine deposition during maximum transgression. The sediments labeled P between each transgressive event contain the deposits associated with the resumed transgression and the subsequent progradation (Fig. 14).

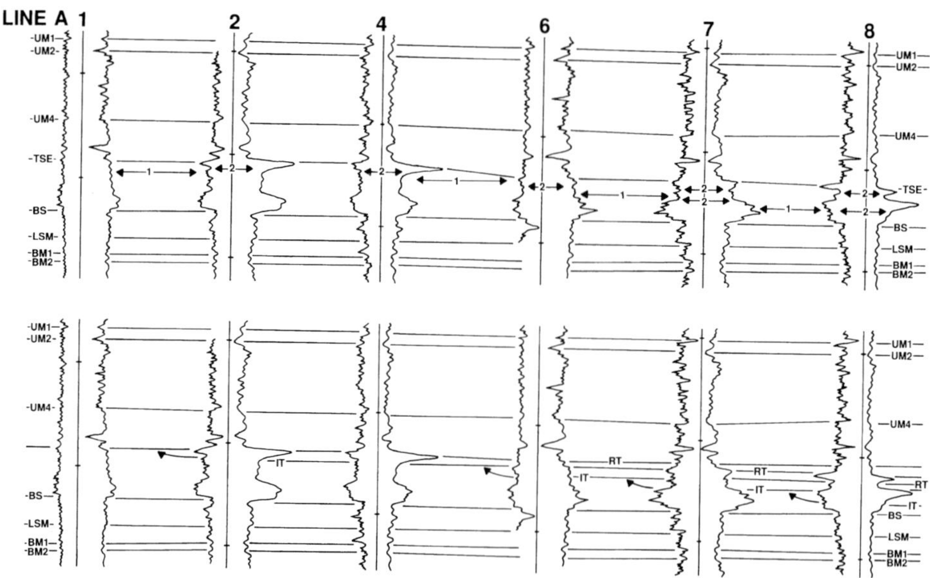

Fig. 15.—Six wells taken from Line A to illustrate the differences in correlation methods. The details of the figure are discussed in the text. For land locations see Appendix 1. Ticks on the central axis are 100 feet (because the original logs are in feet; 1 ft ~ 0.3 m) and there is no implied horizontal scale. Correlation lines are labeled as referenced in the text. Both sections are hung on the lower silty marker (LSM). BM, Base Marker; BS Base of Sussex; UM, Upper Marker; 1, Sandstone laterally adjacent to mudstone; 2, sharp based sandbody; TSE, transgressive of erosion; IT, initial transgression; RT, resumed transgression

There are 6 separate transgressive incised shoreface profiles which can be correlated along strike (Fig. 2), except where removed by erosion associated with shoreface incision during the next pause in the overall transgression. This gives rise to the overall fining upward succession and the backstepping pattern recorded in the House Creek area. This correlation imposes pauses in the overall transgression. During periods of stillstand in the overall transgression (initial transgression, IT) incision of a new asymmetric erosion surface followed by progradation will occur. This will continue until sea level begins to rise again (resumed transgression, RT) and the coarsening upward succession is truncated. During the next pause in sea level the process is repeated giving rise to a series of backstepping shoreface sandbodies. The shoreface sandbodies are bounded by concave upward asymmetrical erosion surfaces at their bases and are truncated by planar erosion surfaces that when traced landward form the concave upward asymmetrical erosion surface at the landward margin of the next sandbody. Based on the facies relationships, the overall fining upward succession and the backstepping pattern of the sandbodies (Fig. 14), these erosion surfaces are interpreted to have formed during periods of stillstand in an overall transgression. Similar relationships have been described in the Albian Viking Formation (Walker and Wiseman, 1995; Davies and Walker, 1993), Turonian Cardium Formation (Bergman and Walker, 1988), and from the Holocene sand banks on the east Texas inner continental shelf (Anderson et al., 1991). The Holocene shelf sand banks represent overstepped coastal lithosomes which have been reworked during transgression and the backstepping parasequences described from these banks have been attributed to episodic or step-like rises of sea level (Anderson et al., 1991).

This punctuated transgressive pattern is terminated by T18 which marks the end of Sussex sandstone deposition in the House Creek area. The erosional nature of T18 is suggested by the presence of mudstone rip-up clasts, conglomerate and the abrupt juxtaposition of facies (Well 54, Fig. 3B) at this horizon. Little topographic relief is associated with this erosion surface. T18 is overlain by a conglomeratic transgressive lag (Facies 6) which passes vertically upward into pebbly bioturbated sandstone (Facies 7) and pervasively bioturbated pebbly mudstone (Facies 8). These are subsequently blanketed by laminated black mudstone (Facies 10). The muddy nature and lack of sedimentary structures and bioturbation other than *Helminthopsis* suggests deposition on the mid- to outer-shelf, with emplacement of the sharp based very fine sandstone and siltstone laminae by storms. Based on the morphology of the surface and the overlying facies, T18 is interpreted as a transgressive surface of erosion, characterized by the presence of at least three regionally extensive onlapping markers.

CONCLUSIONS

Previous lithostratigraphic correlations of the Sussex sandstone in the House Creek area have identified two distinct sandstone horizons. In this interpretation the Sussex can be subdivied into two discrete packages separated by a regressive surface of erosion. The lower part preserves the initial lowstand, TST and distal HST deposits of the Sussex (Fig. 21A). A major drop in relative sea level (Fig. 21A) shifts the shoreline position downward (erodes into Cody Shale) and basinward of House Creek (Fig. 21B). This regressive surface of erosion is modified during subsequent ravinement as the shoreline retreats landward. Recognition and correlation of regionally extensive erosional discontinuities allowed the definition of 6 distinct progradational sandbodies in the upper sandstone (Fig. 21B). These sandbodies are contained between asymmetrical erosion surfaces and show a

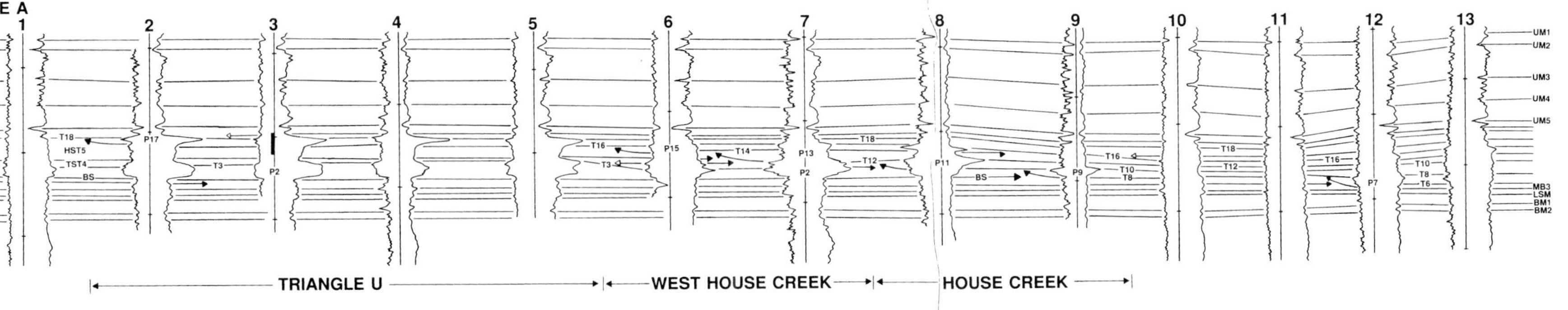

Fig. 16.—Line A located in Figure 1. This section links the northern part of House Creek Field with both West House Creek and Triangle U Fields to the west (Fig. 2). There is limited core control. Northern House Creek Field is characterized by the presence of 6 progradational sandbodies (labeled P7-P17) whose western margins (labeled T6-T16) are laterally adjacent to mudstone as seen between Wells 1 and 2 (P17), 5 and 6 (P15), 6 and 7 (P13), 7 and 8 (P11), 8 and 9 (P9), and 11 and 12 (P7). T6 was recognized only on Line A presumably because the other lines do not extend far enough to the east (Fig. 1). Well logs are gamma ray and resistivity. Wells are numbered sequentially. For land locations see Appendix 1. Ticks on the central axis are 100 feet (because the original logs are in feet; 1 ft ~ 0.3 m) and there is no implied horizontal scale. The black bar between the well logs represents the cored interval. Correlation lines are labeled as referenced in the text. Open arrows represent onlapping markers and closed arrows represent truncated horizons. The section is hung on the lower silty marker (LSM). BM, Base Marker; MB, Marker Bed; BS Base of Sussex; UM, Upper Marker; P, progradational event; T, transgressive surface; TST, transgressive systems tract; HST, highstand systems tract.

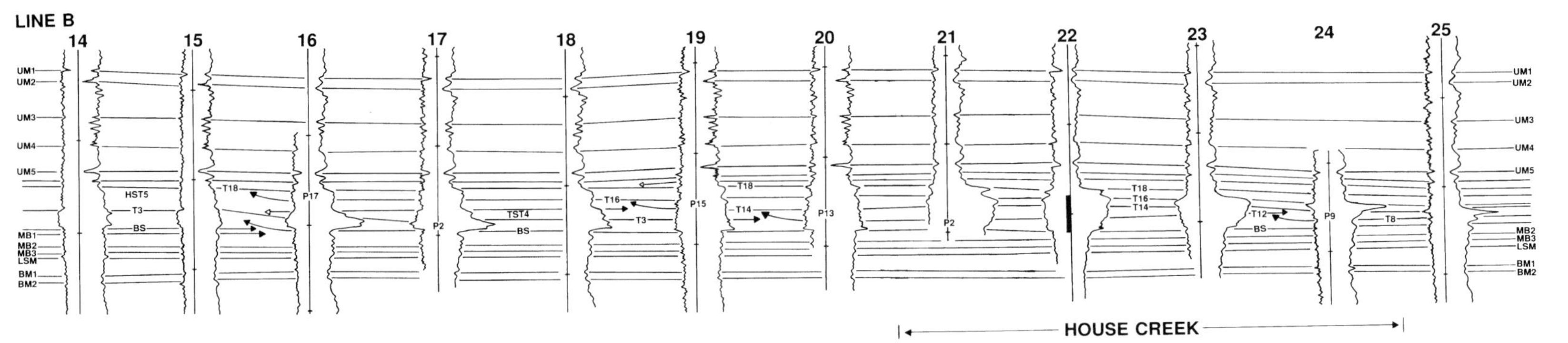

Fig. 17.— Line B located in Figure 1. This section is located immediately south of Triangle U and West House Creek fields (Fig. 2). The same pattern of log signatures described on Line A for the House Creek Field is observed on this section. Four progradational sandbodies were identified whose western margins are laterally adjacent to mudstone as seen between Wells 15 and 16 (P17), 18 and 19 (P15), 19 and 20 (P13), and 23 and 24 (P9). The base of these sandbodies is marked by an abrupt shift in the log signatures. In core (Well 22; Fig. 3A) these surfaces are characterized by an abrupt change in facies and grain size from fine to medium sandstone. Mudstone rip-up clasts commonly occur at the base of these sandbodies. T14 is marked by an abrupt change from pebbly sandstone (Facies 5) to pervasively bioturbated pebbly mudstone (Fig. 3A). T6 was not recognized on this cross section presumably because the line does not extend far enough to the east. T10 appears to have been removed by erosion associated with T12 (Fig 2). All conventions are explained in the caption of Figure 16.

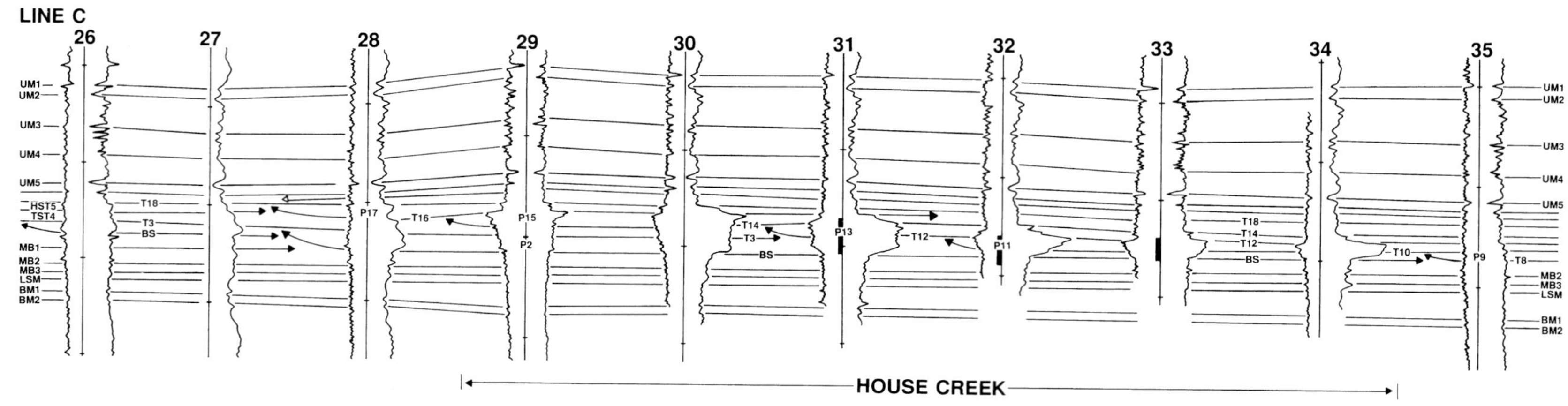

Fig. 18.—Line C located in Figure 1. This section is located across the central portion of House Creek (Fig. 2). There is core control in House Creek Field along this line. The core section (Fig. 14) was discussed in detail above. The same pattern of log signatures described above for Lines A and B (Figs. 16 and 17) for Northern House Creek Field are observed in the Central portion of the field. On this cross section 5 progradational sandbodies were identified whose western margins are laterally adjacent to mudstone as seen between Wells 27 and 28 (P17), 28 and 29 (P15), 30 and 31 (P13), 31 and 32 (P11), and 34 and 35 (P9). The lack of sandstone in Well 35 is presumably due to the basinward position of the well. Wells closer to the inferred position of the incision alongstrike (Fig. 2) preserve sandstone at this stratigraphic horizon. The bases of the sandbodies are marked by abrupt shifts in the log signatures. All conventions are explained in the caption of Figure 16.

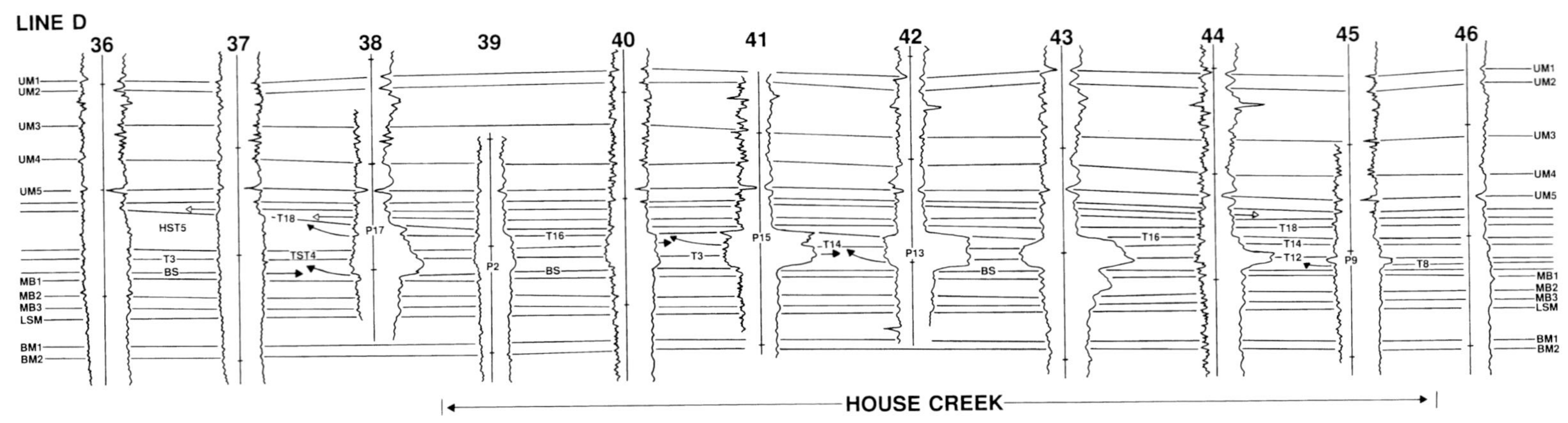

Fig. 19.—Line D located in Figure 1. This section is located across the central portion of House Creek (Fig. 2). There is limited core control in this area. The marker beds (MB1, MB2 and MB3) below BS can be traced across the line of section. They are not truncated as in the previous sections. The same pattern of log signatures described in Line A is observed on this section. On this line, four sandbodies were identified whose western margins are laterally adjacent to mudstone as seen between Wells 37 and 38 (P17), 40 and 41 (P15), 41 and 42 (P13), and 44 and 45 (P9). T10 appears to have been removed by erosion associated with T12 (Fig. 2). All conventions are explained in the caption of Figure 16.

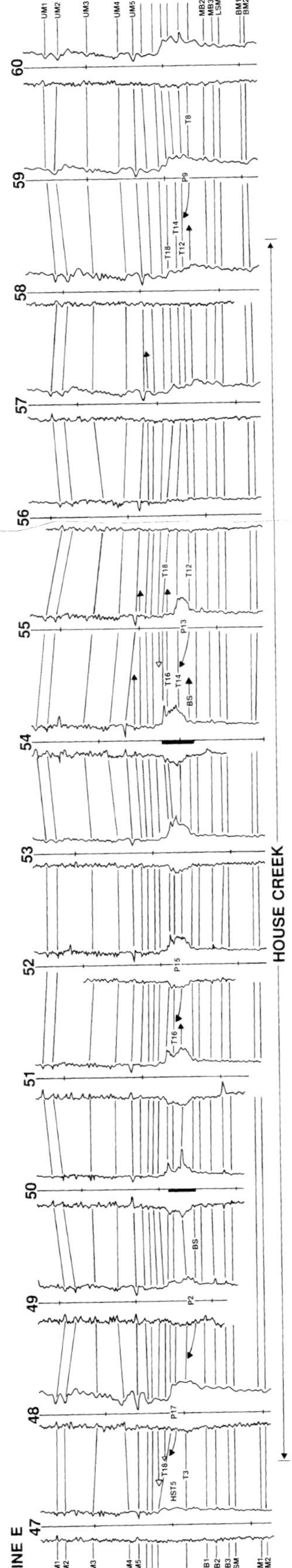

Fig. 20.—Line E located in Figure 1. Line E is located across the southern portion of House Creek (Fig. 2). There is limited core control. The marker beds below the base of the Sussex can be traced along the line of section. The UM5 marker bed appears to be an erosion surface removing several onlapping markers between Wells 54 and 55, Wells 55 and 56, and Wells 57 and 58. These onlapping markers below UM5 were not found on the other lines and were presumably removed by erosion at this horizon. This erosion surface is blanketed by the Sussex Bentonite (Brenner 1978; Hobson et al. 1982). The same pattern of log signatures described in the previous sections for the House Creek Field is observed on this section. On Line E 4 sandbodies were identified whose western margins are laterally adjacent to mudstone as seen between Wells 47 and 48 (P17), 51 and 52 (P15), 54 and 55 (P13), and 58 and 59 (P9). The base of the sandbodies are characterized by an abrupt shift in the log signatures. In core (Well 54; Fig. 3B) these shifts are characterized by an abrupt change in facies and grain size (T14, T16 and T18). T10 appears to have been removed by erosion associated with T12 (Fig. 2). Immediately south of this line T12 appears to have been removed by erosion associated with T14 (Fig. 2) All conventions are explained in the caption of Figure 16.

retrogradational stacking pattern giving rise to an overall fining upward succession. The scour surfaces (T6 to T16) are interpreted to have formed as a result of wave erosion during erosional shoreface retreat during periods of relative stillstand in an overall transgression. It is explicit in this interpretation that the sandbodies contained between these scours are not contemporaneous.

Defining the sandbodies on the basis of regionally extensive discontinuities alters not only the inferred sandbody geometry but also the number of sandbodies identified in the House Creek study area. This interpretation of the Sussex Sandstone at House Creek as a series of backstepping transgressive shoreface deposits overcomes the problems identified in the shelf ridge interpretation (Berg, 1975) and is consistent with both the physical and biological structures contained in the Sussex Sandstone and the overall stratigraphic position. The complex dissection of the reservoir sandstone described in this interpretation particularly in the House Creek field may have implications on fluid migration and reservoir compartmentalization. Furthermore high-resolution stratigraphic analyses provides a means of defining the stratigraphic relationships not only between the oil fields, but also between the outcrop and the subsurface both locally and regionally. From these types of correlations the history of relative sea level during Sussex time can be determined.

ACKNOWLEDGMENTS

This work was supported by the Natural Sciences and Engineering Research Council of Canada Research Grant to the author. I thank the USGS Federal Core Storage Facility in Denver for access to the Sussex cores at no charge. This manuscript has been greatly improved by comments from Roger Walker and reviewers Dag Nummedal, Dale Leckie, Morgan Sullivan, Bill Arnott, Janok Bhattacharya and John Carey.

REFERENCES

Anderman, G.G., 1976, Sussex sandstone production, Triangle U field, Campbell County, Wyoming: Wyoming Geological Association, 28th Annual Field Conference Guidebook, p. 107-13.

Anderson, J.B., Siringan, F.P., Smyth, W.C. and Thomas, M.A., 1991, Episodic nature of Holocene sea-level rise and the evolution of Galveston Bay: Gulf Coast Section, Society of Economic Paleontologists and Mineralogists Foundation 12th Annual Research Conference, Program and Abstracts, p. 8-14.

Asquith, D.O., 1970, Depositional topography and major marine environments, Late Cretaceous, Wyoming: American Association of Petroleum Geologists Bulletin, v. 54, p. 1184-1226.

Asquith, D.O., 1974, Sedimentary models, cycles, and deltas, Upper Cretaceous, Wyoming: American Association of Petroleum Geologists Bulletin, v. 58, p. 2274-2283.

Berg, R.R., 1975, Depositional environment of Upper Cretaceous Sussex Sandstone, House Creek Field, Wyoming: American Association of Petroleum Geologists Bulletin, v. 59, p. 2099-2110.

Bergman, K.M., 1994, Shannon Sandstone in Hartzog Draw-Heldt Draw fields (Cretaceous, Wyoming, USA) reinterpreted as lowstand shoreface deposits: Journal of Sedimentary Research, v. B64, p. 184-201.

Bergman, K.M. and Walker, R.G., 1995, High-resolution sequence stratigraphic analysis of the Shannon Sandstone in Wyoming, using a template for regional correlation: Journal of Sedimentary Research, v. B65, p. 255-264.

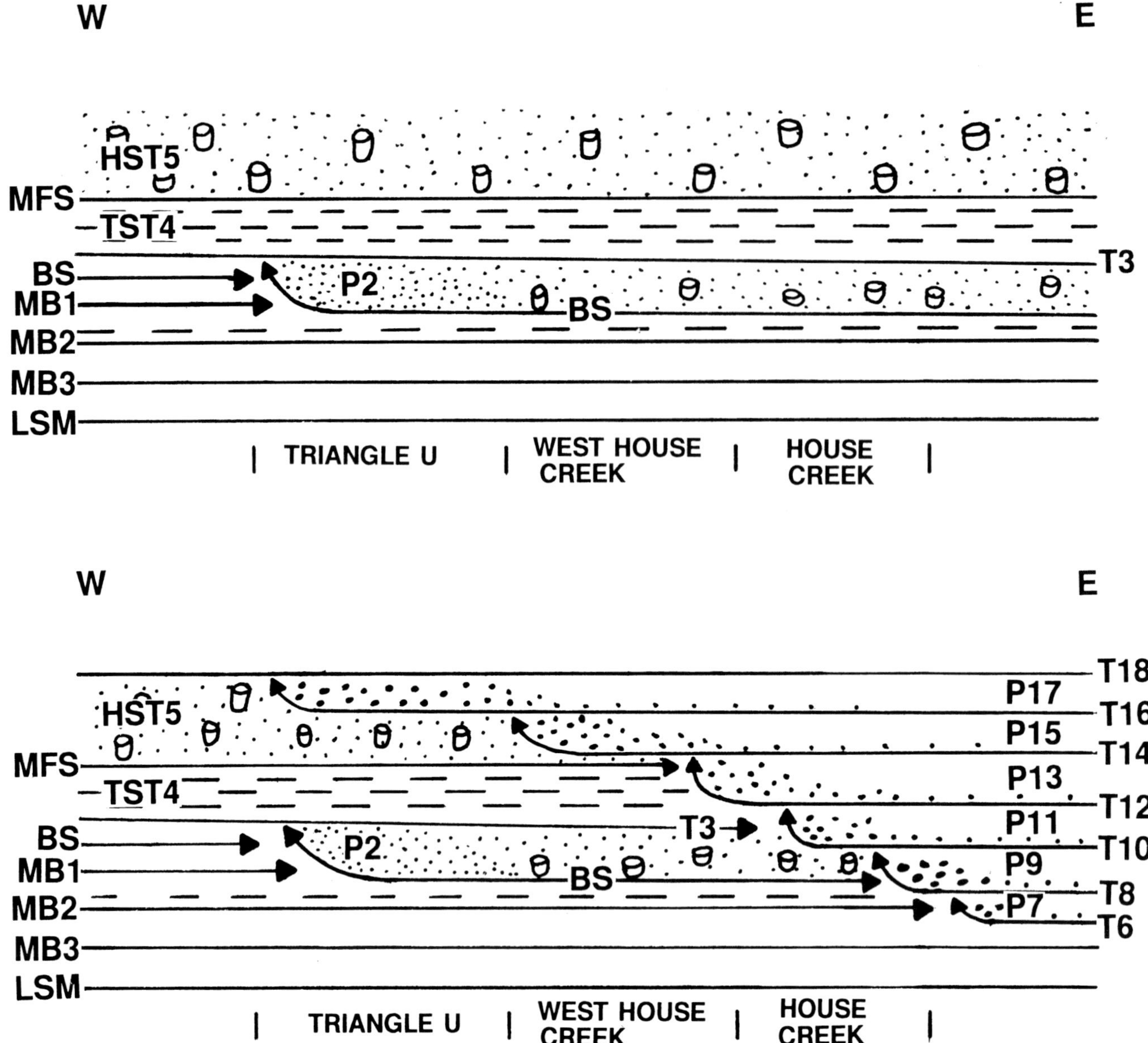

Bergman, K.M. and Walker, R.G., 1988, Formation of Cardium erosion surface E5, and associated deposition of conglomerate: Carrot Creek field, Cretaceous Western Interior Seaway, Alberta, *in* James, D.P. and Leckie, D.A., eds., Sequences, Stratigraphy, Sedimentology: Surface and Subsurface: Calgary, Canadian Society of Petroleum Geologists Memoir 15, p. 15-24.

Bergman, K.M. and Walker, R.G., 1987, The importance of sea level fluctuations in the formation of linear conglomerate bodies: Carrot Creek Member, Cretaceous Western Interior Seaway, Alberta, Canada: Journal of Sedimentary Petrology, v. 57, p. 651-665.

Brenner, R.L., 1978, Sussex sandstone of Wyoming—Example of Cretaceous offshore sedimentation: American Association of Petroleum Geologists Bulletin, v. 62, p. 181-200.

Brenner, R.L., 1980, Construction of process-response models for ancient epicontinental seaway depositional systems using partial analogs: American Association of Petroleum Geologists Bulletin, v. 64, p. 1223-1244.

Cullen, D.J., 1973, Bioturbation of Superficial Marine Sediments by Interstitial Meiobenthos: Nature, v. 242, p. 323-324.

Davies, S.D. and Walker, R.G., 1993, Reservoir geometry influenced by high-frequency forced regressions within an overall transgression: Caroline and Garrington fields, Viking Formation (Lower Cretaceous), Alberta: Bulletin of Canadian Petroleum Geology, v. 41, p. 407-421.

Gill, K.R. and Cobban, W.A., 1973, Stratigraphy and geologic history of the Montana Group and equivalent rocks, Montana, Wyoming and North Dakota: Denver, United States Geological Survey Professional Paper 776, 37p.

←

Fig. 21.—Summary cartoon showing the depositional history of the Sussex sandstone in the the House Creek Area. Panel A shows the development of P2 during a relative sea level fall and its subsequent truncation by transgression (T3). Above T3 the deposits of the TST and HST form giving rise to regional Sussex. These regional Sussex deposits are truncated by a regressive surface of erosion that shifts the facies downward and basinward. This surface is modified by a series of backstepping incised shoreface deposits formed during periods of stillstand in an overall transgression (Panel B).

Aggradation: Aggradation of the basin floor gave rise to the lower markers BM1 and BM2, the silty marker LSM and the parasequences capped by flooding surfaces labeled MB1, MB2 and MB3.

Relative Sea Level Fall: During falling stage of relative sea level, BS eroded into the aggradational parasequences of the underlying Cody Shales and a new shoreface profile was established immediately west of House Creek. This shoreface prograded eastward (P2) into the House Creek area.

Transgression: P2 deposition was truncated by T3 during a rising stage of relative sea level, shifting the shoreline basinward west of the House Creek study area. During this time several aggrading parasequences accumulated in the study area during transgression and subsequent highstand.

Relative Sea Level Fall: Renewed sealevel fall shifted the shoreline eastward of the House Creek study area and downward truncating the underlying Sussex deposits and parasequences in the Cody Shale. This erosion surface was modified by erosion during the following transgression.

Punctuated Transgression: Subsequent transgression resulted in a series of backstepping incised shoreface deposits giving rise to the producing sediments at House Creek. Each T surface incises a shoreface profile and is followed by progradation. T6 incises and is overlain by P7. T8 truncates P7 and incises a shoreface profile to the west of P7 giving rise to P9 (Fig. 2). T10 truncates P9 and incises a shoreface profile west of P9 (Fig. 2) giving rise to P11. T12 truncates P11 giving rise to P13 west of P11 (Fig. 2). T14 truncates P13 giving rise to P15. T14 is not continuous along the length of House Creek Field. It is removed by erosion associated with T16 (Fig. 2). P15 is truncated by T16 which incises a profile west of P15 (Fig. 2) giving rise to P17. T18 is a regional transgressive surface of erosion which truncates P17 and marks the end of sandstone deposition during Sussex time in the House Creek area. T18 is characterized by at least three regionally extensive onlapping markers.

HAQ, B.U., HARDENBOL, J. AND VAIL, P.R., 1988, Mesozoic and Cenozoic chronostratigraphy and Eustatic cycles, *in* Wilgus, C.K., Hastings, B.S., Kendall, C.G. St.C, Posamentier, H.W., Ross, C.A. and Van Wagoner, J.C., eds., Sea-Level Changes: An Intergrated Approach: Tulsa, Soceity of Economic Paleontologists and Mineralogists Special Publication 42, p. 71-108.

HOBSON, J.P. JR., FOWLER, M.L. AND BEAUMONT, E.A., 1982, Depositional and statistical exploration models, Upper Cretaceous offshore sandstone complex, Sussex Member, House Creek Field, Wyoming: American Association of Petroleum Geologists Bulletin, v. 66, p. 689-707.

HOWARD, J.D. AND FREY, R.W., 1984, Characteristic trace fossils in neashore to offshore sequences, Upper Cretaceous of east-central Utah: Canadian Journal of Earth Sciences, v. 21, p. 200-219.

MACEACHERN, J.A. AND PEMBERTON, S.G., 1992, Ichnological aspects of Cretaceous shoreface successions and shoreface variability in the Western Interior Seaway of North America, *in* Pemberton, S.G., ed., Applications of Ichnology to Petroleum Exploration: A Core Workshop: Tulsa, Society of Economic Paleontologists and Mineralogists Core Workshop 17, p. 57-84.

PATTISON S.A.J. AND WALKER, R.G., 1992, Deposition and interpretation of long, narrow sand bodies underlain by a basinwide erosion surface: Cardium Formation, Cretaceous Western Interior Seaway, Alberta, Canada: Journal of Sedimentary Petrology, v. 62, p. 292-309.

POSAMENTIER, H.W., AKKEBM G.P., JAMES, D.P. AND TESSON, M., 1992, Forced regressions in a sequence stratigraphic framework: concepts, examples and exploration significance: American Association of Petroleum Geologists Bulletin, v. 76, p. 1687-1709.

WALKER, R.G., 1995, Sedimentary and Tectonic Origin of a Transgressive Surface of Erosion: Viking Formation, Alberta, Canada: Journal of Sedimentary Research v. B65, p. 209-221.

WALKER, R.G. AND EYLES, C.H., 1991, Topography and significance of a basinwide sequnce-bounding erosion surface in the Cretaceous Cardium Formation, Alberta, Canada: Journal of Sedimentary Petrology, v. 61, p. 473-496.

WALKER, R.G. AND PLINT, A.G., 1992, Wave- and storm-dominated shallow marine systems, *in* Walker, R.G. and James, N.P., eds., Facies Model: Response to Sea Level Change, Ottawa, Geological Association of Canada: p. 219-238.

WALKER, R.G. AND WISEMAN, T., 1995, Lowstand shorefaces, transgressive incised shorefaces, and forced regressions: examples from the Viking Formation, Joarcam area, Alberta: Journal of Sedimentary Research, v. B65, p. 132-141.

APPENDIX 1.—List of numbered well locations used in the cross sections

Line A:

1 PCP
#1-14-8 HDU
SW SW 8-46N-75W

2 CONOCO
#1 Conoco Champlin
NE NW 16-46N-75W

3 CONOCO
#1-16 Conoco State
NW NE 16-46N-75W

4 Inexco
#3-9 Saunders
SW SE 9-46N-75W

5 Davis Oil
#1 Schnier Federal
SW SW 10-46N-75W

6 Edwin Cox
#1-11 Nova Federal
NW SE 11-46N-75W

7 Diamond Shamrock
#13-12 Philrock
NW SW 12-46N-75W

8 Inexco
#2-8 McBeth
SE NW 8-46N-74W

9 Inexco
#1 Federal Wagstaff
NW NE 8-46N-74W

10 Davis Oil
#1 Badger
NW SW 10-46N-74W

11 Smokey
#31-36 Fry
NW NE 36-47N-74W

12 Dekalb
#11-21 Davis
NW SW 21-47N-73W

13 Depco
#34-2 Heiland
SW SE 2-47N-73W

Line B:

14 Louisiana Land and Exploration
31-32 Louisiana #1
NW NE 32-45N-74W

15 Union Texas
#29-1 Bounty
SE SE 29-45N-74W

16 Davis Oil
#1-13 Colonies
NE SE 17-45N-74W

17 Davis Oil
#1 Virginia State
SW NW 16-45N-74W

18 Davis Oil
#1 Zicar State
NW NE 16-45N-74W

19 Shell
#41-15 Schlautmann
NE NE 15-45N-74W

20 Louisiana Land and Exploration
#14-11 Schlautmann
SW SW 11-45N-74W

21 Walter Duncan
#1-13 Cullens
NE SW 13-45N-74W

22 Southland Royalty
#2 House Creek
E SE 13-45N-74W

23 Inexco
#2-18 Echo Whipple
SW NW 18-45N-73W

24 True Oil
#34-17 Echo Federal
SW SE 17-45N-73W

25 Butte Resources
#1-16 Inexco State
SE SW 16-45N-73W

Line C:

26 Houston
#43-X-7 Doll
C SE 7-43N-74W

27 Diamond Shamrock
#14-4 Hart
SW SW 4-43N-74W

28 Davis Oil
#1A Marquis
NW SW 30-44N-73W

29 Kerr McGee
#74-1 House Creek Unit
NE SE 21-44N-73W

30 Woods Petroleum
#1A Empire Federal
NE NE 21-44N-73W

31 Woods Petroleum
#1 Mandel Federal
NE NW 22-44N-73W

32 Woods Petroleum
#1 Hurd Federal
SW SE 15-44N-73W

33 Universal Resources Corporation
#1-73 Federal
NW NW 23-44N-73W

34 Universal Resources Corporation
#1-14 Federal
SW SW 14-44N-73W

35 Tenneco
#1-8 Bader & USA
SW SE 8-44N-72W

Line D:

36 Champlin
#3-32-18 Spring Creek
SW NE 18-42N-73W

37 Champlin
#22-14 Kamon
SE NW 14-42N-73W

38 Yates
#1 Reno Flat
NW NW 6-42N-72W

39 Woods Petroleum
#29-1 Cosner
SW NE 29-43N-72W

40 Crawley
#20-1 Cosner
SE SE 20-43N-73W

41 Inexco
#2-16 State
NE SE 16-43N-72W

42 Conoco
#1 Cosner Federal
SW NW 15-43N-72W

43 Woods Petroleum
#10-2A Cosner Federal
SW SE 10-43N-72W

APPENDIX 1 (continued).

44 Davis Oil
#1 Stuart Ranch
NW NW 12-43N-72W
45 Duncan
#1-1 Stuart Brothers
SE SW 1-43N-72W
46 H & M Oil
#1 Conoco Stuart
NW NE 1-43N-72W

Line E:
47 DSEC
#23-32 Rattle Snake
NE SW 32-42N-72W
48 Ladd & Luckow
#1-29 Matheson
NE SE 29-42N-71W
49 Ladd & Lukowicz
#2-21 Federal
SE SW 21-42N-71W
50 Amoco
#1 Campbell
NE NW 21-42N-71W
51 Meridian Oil
#23-16 State
NE SW 16-42N-71W
52 Burlington North
#22-16 State
SE NW 16-42N-71W
53 Milestone
#21-16 State
NE NW 16-42N-71W
54 Milestone
#14-9 Federal
SW SW 9-42N-71W
55 Burlington North
#22-9 Federal
SE NW 9-42N-71W
56 Burlington North
#34-4 Federal
SW SE 4-42N-71W
57 Conoco
#3-1 Birdsall
SW SE #-42N-71W
58 Industrial
#1 Federal Porcupine
NW SW 2-42N-71W
59 Woods Petroleum
#1 Bradshaw State
SE NW 36-43N-71W
60 Texaco
#2 Gov't Putnam
NW NE 25-43N-71W

THE SHANNON SANDSTONE AND ISOLATED LINEAR SAND BODIES: INTERPRETATIONS AND REALIZATIONS

JOHN R. SUTER AND H. EDWARD CLIFTON
Conoco Inc., P.O. Box 2197, Houston, TX 77252-2197

ABSTRACT: Isolated linear sand bodies in the Cretaceous Western Interior Seaway have been remarkably resistant to a consensus interpretation. The best known of these deposits is the Shannon Sandstone, the primary focus of this volume and the SEPM Research Conference from which it derived. Our purpose in this summary paper is to attempt an objective evaluation of the various interpretations for the Shannon, given our own backgrounds, experience, and resultant biases. We are not charged with coming up with our own model for this enigmatic deposit and its isolated linear sand body siblings. We compare and find substantial differences in the investigators' perceptions of the Shannon in outcrop in the Salt Creek anticline area in northeastern Wyoming and in the subsurface in the relatively nearby Hartzog Draw Field. We examine their interpretations in light of modern analogues and paleogeographic constraints and conclude that no interpretation proposed to date adequately explains all aspects of the Shannon. We suspect that the Shannon Sandstone may be more complex than previously thought and that a single interpretation may not be sufficient to explain the whole system. We also came away with renewed respect for the principle of multiple working hypotheses.

INTRODUCTION

The origin of isolated marine sand bodies, exemplified by the Campanian Shannon Sandstone of the Powder River Basin in the Western Interior of the United States, is highly controversial. As pointed out by the editors of this volume in the lead paper, deposits of this general class have considerable economic importance, but the problem has taken on a greater significance to sedimentary geology. As the Cretaceous is widely believed to have been a period of high average sea level, and these sand bodies are encased in marine shales, they were initially interpreted as shelf sands, becoming the archetypal "offshore bars." Later, with the advent of sequence stratigraphy and its emphasis on relative sea level change and incised valleys, "offshore bar" became a term of scorn, and shelf sands fell into deep disfavor, although some adherents kept the faith.

A series of mutually incompatible interpretations have been put forward to explain the occurrence of isolated marine sand bodies, in many cases using the same data for the same unit.

1. Storm-dominated shelf sand bodies (Spearing, 1976; Tillman and Martinsen, 1984; Swift and Parsons, this volume)
2. Tide-dominated delta within an incised valley (Sullivan et al., 1995, 1997; the "Exxon model")
3. Transgressed forced regressive and lowstand shorelines (Bergman, 1994; Bergman and Walker, 1995, and this volume)
4. Tidal sand ridges within a mixed wave-tidal regime open embayment (Tillman, this volume) and/or estuary-mouth shoals (Elliot, 1995).
5. Some combination of any or all of the above.

These interpretations vary in their reliance on outcrop and subsurface (well log motifs and correlations, and core description and interpretation) data and modern analogs. Each requires a particular set of paleogeographies, sea-level conditions, and dominant processes. Each also has implications for other types of deposits that should exist in the basin.

In 1995, a SEPM Research Conference, "Tongues, Ridges, and Wedges," was convened to consider the evidence for these multiple interpretations and move toward a resolution of the question of isolated marine sand bodies. The success of that effort (or lack thereof) can be judged by the multiple interpretations still extant in this volume and elsewhere. The major model proponents left with their interpretations intact, largely unconvinced by other concepts. What does this say about the state of the art of sedimentary geology at the close of the 20th century?

Sometimes in the middle of heated debates over interpretations, we lose sight of the real purpose of the exercise. Is the mode of formation of a set of rocks in Wyoming worth the expenditure of so much physical, mental, and emotional energy? At present, it seems there are several justifications for the debate swirling around the Shannon and its "isolated" marine sand body siblings:

1. The immediate economic significance of similar deposits in the Western Interior.
2. General exploration models. Depending on how we interpret the Shannon, it has implications for exploration in frontier foreland basin settings, unless we believe that deposits of this type are unique to the Cretaceous Western Interior.
3. Usefulness as analogs for field development and reservoir modeling. Tillman and Martinsen (1984) stated that application of their model was highly successful in the original drilling of the Hartzog Draw Field. Sullivan et al. (1997) claimed a 40% increase in the success of infill drilling using their considerably different interpretation of the same field. This success is especially interesting in light of the differences between the two models.
4. The importance to sequence stratigraphic models.
5. The intellectual exercise! Deciphering the Earth's history is important in its own right, and we all want to improve our understanding of different deposits, so as to be better prepared for the next interpretation challenge.

Currently, the controversy seems more focused on intellectual exercise and importance to sequence-keyed models, although the greatest economic significance presumably lies in models for exploration, field development, and reservoir modeling. Our purpose in this summary paper is to attempt an objective evaluation of the various interpretations, given our own backgrounds, experience, and resultant biases. One of us (JRS) has worked on the Shannon sandstone in the outcrop and subsurface and participated in the SEPM Re-

Isolated Shallow Marine Sand Bodies: Sequence Stratigraphic Analysis and Sedimentologic Interpretation.
SEPM Special Publication 64, Copyright © 1999
SEPM (Society for Sedimentary Geology), ISBN 1-56576-057-3, p. 321-356.

search Conference. The other (HEC) has never seen the outcrop or subsurface data, except as presented in the literature. In our analysis of the controversy, we have tried to identify differences in perception and interpretation. We have considered the evidence and interpretations drawn from the outcrop, subsurface, and modern analogs as systematically and impartially as possible, including the facies that have been recognized, the geometry of the sand bodies, bounding surfaces, internal architectures, directional structures, inferred processes, and mineralogy. We are not charged with coming up with our own model for the Shannon Sandstone and its siblings. Readers will note, however, that we have not been able to resist the occasional suggestion.

PERCEPTIONS OF THE SHANNON

The extent of disagreement about the origin of the Shannon linear sand bodies implies that the various investigators either: 1) see different features in the rock; 2) share common observations, but interpret them differently; or 3) engage in some combination of the two. Commonly, perceptual and interpretive differences are inseparable. Laminae that one observer sees as hummocky cross-stratification, for example, may be seen as something quite different by another. In our analysis of the controversy, we will try to identify differences in perception and interpretation. We will consider the facies that have been recognized, the geometry of the sand bodies, internal architectures, directional structures, biofacies, and mineralogy.

Lithofacies Descriptions

The most obvious perceptual differences are likely to be found in the different descriptions of the lithofacies that compose the Shannon (Spearing, 1976; Tillman, this volume; Tillman and Martinsen, 1984, 1987; Walker and Bergman, 1993; Bergman, 1994; Gaynor and Swift, 1988; Swift and Parsons, this volume; and Sullivan et al., 1995, 1997). All generally agree on the overall character of the facies, which fall into one of four categories: 1) cross-bedded sandstone; 2) facies with mixed lithologies; 3) bioturbated sandstone/siltstone; and 4) siltstone/shale (Tables 1-4). Within these facies sets, however, the descriptions can differ substantially. For example, in the cross-bedded facies category (Table 1), Tillman (this volume) specifies that mud drapes are abundant in his "high energy ridge margin facies" and Gaynor and Swift find them to be common in their trough cross-bedded Subfacies A. In contrast, Walker and Bergman (1993) either do not mention mud drapes or specifically say they find "no convincing example of mud drapes on the foresets" (Unit 10, p. 848) or "no unambiguous single or double mud drapes on foresets" (Unit 5, p. 844).

In the "facies with mixed lithologies" category (Table 2), Walker and Bergman (1993) find hummocky cross-stratification (HCS) to be abundant, common, or at least present in three of the four units that fall into this category. In contrast, Gaynor and Swift (1988, p. 873) specify that "Hummocky cross-strata sets..........have not, as yet, been documented in the study area" and Tillman (this volume) states that "less that 5% of the Shannon sandstone is interpreted..... to be wave dominated" and that by inference seems to be in the form of symmetric ripples. Sullivan et al. (1997) note a general lack of wave-generated stratification indicative of shoreface deposition. Neither Spearing (1976) nor Sullivan et al. (1997) mentions HCS, but Elliott (1995) notes that HCS is occasionally preserved in the Shannon.

The abundance of wave ripples is also perceived differently in the facies with mixed lithologies. Walker and Bergman (1993) find wave ripples to be abundant, common, or at least present in all of their units with mixed lithologies. In contrast, Gaynor and Swift (1988) note that current ripples predominate over oscillation ripples, a view shared by Tillman (this volume). Spearing (1976) states that wave ripples predominate in his "thin-bedded sandstone facies."

In the "predominantly bioturbated sandstone" category (Table 3), Bergman (1994) recognizes a coarse bioturbated sandstone with common glauconite, at the top of the Shannon section in Hartzog Draw, that is not noted by the other investigators. In the "fine-grained (shaley) facies" category (Table 4), the main differences relate to the degree of bioturbation, which may relate to specific facies defined within this category.

Can the Shannon controversy derive in part from the lack of a common database among the facies descriptions? Table 5 summarizes the data used by the various investigators. One study focused almost entirely on Shannon outcrops, mostly in the Salt Creek area, whereas others were concerned primarily with the subsurface, notably in Hartzog Draw (Fig. 1). A few studies integrated observations from both outcrop and subsurface. The number of measured sections and cores examined should be sufficient to ensure a reasonably common database within either the outcrop or subsurface. The possibility exists, however, that differences between outcrop and subsurface have fueled the controversy.

Fortunately, an excellent set of data provides a basis for evaluating this possibility. Tillman and Martinsen (1984) published a detailed and partly quantified description of the Shannon cropping out in the Salt Creek area and in 1987 followed with a parallel review of the Shannon in Hartzog Draw. Table 6, which compares their surface and subsurface observations, shows some fairly significant differences between the two. Some of these may reflect the natural variability between outcrop observations (which typically provide a laterally extensive view of the rock, but with potentially cryptic lithologic details) and core (a limited lateral view of the rock, but with generally well-defined lithologic detail). It also may reflect differences imposed by weathering at the outcrop, which might obscure some of the finer facies. Nonetheless, it appears that most of the facies in Hartzog Draw are less sandy and generally finer than their counterparts in the outcrops. The subsurface facies also have less high-angle cross-bedding and planar lamination. The degree of bioturbation is about the same or, in two facies ("inter-ridge shale" and "planar-laminated sandstone"), distinctly greater in the subsurface. Only in the underlying Cody shale does the bioturbation significantly decrease in the subsurface. Three sandstone facies identified in outcrop ("inter-ridge sandstone," "bioturbated sandstone," and "shelf sandstone") were not recognized in the subsurface. The core also consistently has less glauconite than was seen in outcrop.

It is difficult to attribute the differences shown in Table 6 solely to the observational character of outcropping rock versus subsurface core. The comparison suggests that the subsurface Shannon in Hartzog Draw reflects somewhat lower energy conditions than do those in the outcrops around the Salt Creek anticline. If so, the two conceivably represent genetically unrelated sand bodies.

Geometry

The geometry of the Shannon in Hartzog Draw is well defined and all investigators agree that the sand forms a

Table 1.—Comparison of lithologic descriptions of predominantly cross-bedded sand facies in the Shannon Sandstone, Central Wyoming. Significant differences shown in boldface.

texture	cross-bedding			ripples			Parallel lamination	HCS	mud clasts	mud drapes	burrows	glauconite
	trough	tabular	N/S	wave	current	N/S						
clay, sls	A T_{HE} GS_A GS_C	A GS_B	A T_{CR} S_7 S_8 WB_{5a} WB_{10} B_6	A	A	A	A	A	A T_{HE} GS_C WB_{5a} WB_{10} B_4	A **T_{HE}**	A WB_{5a}	A T_{HE} GS_C WB_{5a} B_4
vf	C	C	C	C WB_{5a}	C T_{HE} GS_A GS_B	C	C	C	C GS_A S_8	C **GS_A**	C	C T_{CR}
f, m (T_{CR}, T_{HE}, GS_B, GS_A, GS_C, S_7, S_8, WB_{10}, WB_{5a}, B_6, B_4)	R	R	R	R	R	R	R GS_B	R	R T_{CR}	R GS_B B_6	R T_{CR} T_{HE} S_7 S_8 B_4 B_6	R
c	N/S B_4	N/S	N/S	N/S	N/S WB_{10}	N/S B_4	N/S WB_{10}	N/S WB_{10}	N/S B_6	N/S GS_C S_7	N/S GS_A GS_B GS_C WB_{10}	N/S
vc	N	N	N	N	N	N	N	N GS_A GS_B	N	N **WB_{5a} WB_{10}**	N	N
N/S	N/M	N/M	N/M	N/M	N/M T_{CR}	N/M S_8 S_7 GS_C B_6	N/M T_{CR} T_{HE} GS_A GS_C S_7 S_8 WB_{5a} B_4 B_6	N/M T_{CR} T_{HE} GS_C S_7 S_8 WB_{5a} B_4 B_6	N/M GS_B S_7	N/M T_{CR} S_8 WB_4 B_4	N/M B_4	N/M GS_A GS_B S_7 S_7 WB_{10} B_6

Predominantly Crossbedded Sand Facies

T_{CR} Tillman (this volume) Central ridge ss
T_{HE} Tillman (this volume) High energy ridge margin ss

GS_A Gaynor and Swift (1988) Crossbedded ss A
GS_B Gaynor and Swift (1988) Crossbedded ss B
GS_C Gaynor and Swift (1988) Crossbedded ss C

S_8 Sullivan et al. (1997) 8
S_7 Sullivan et al. (1997) 7

WB_4 Walker and Bergman (1993) Unit 4
WB_{5a} Walker and Bergman (1993) Unit 5 outcrop
WB_{10} Walker and Bergman (1993) Unit 10

B_4 Bergman (1994) glauconitic ss
B_6 Bergman (1994) cross-bedded ss

Abundance of feature:
A = Abundant
C = Common
R = rare
N/S = Present, type or abundance not specified
N = noted as not present
N/M = Not mentioned

Degree of bioturbation:
A = thoroughly bioturbated
C = some physical structure visible
R = minor burrowing
N/S = Present, type or abundance not specified
N = none noted
N/M = Not mentioned

Table 2.—Comparison of lithologic descriptions of facies of mixed lithologies in the Shannon Sandstone, Central Wyoming. Significant differences shown in boldface.

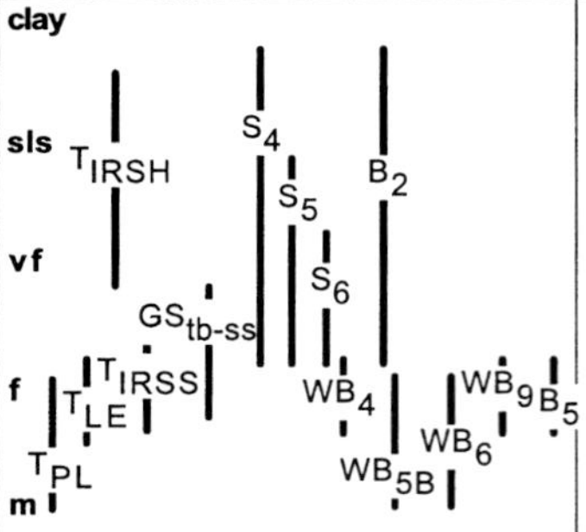

texture	cross-bedding			ripples			Parallel lamination	HCS	mud clasts	mud drapes	burrows	glauconite
	trough	tabular	N/S	wave	current	N/S						
clay sls	A	A	A	A $\mathbf{WB_9}$	A T_{IRSS}	A	A T_{PL} S_4 S_6 S_5	A $\mathbf{WB_4}$	A	A	A S_4	A WB_6
vf	C T_{LE}	C	C WB_6	C $\mathbf{WB_4}$ $\mathbf{WB_6}$	C WB_9	C T_{LE} WB_6	C WB_6 WB_9	C $\mathbf{WB_9}$	C T_{LE}	C GS_{tb-ss} S_4 S_6 B_2 B_5	C T_{IRSS} T_{IRSH} WB_6 WB_{5B} S_5 WB_9 B_2 B_5 S_6	C T_{LE} B_2 B_5
f m	R	R	R	R $\mathbf{T_{IRSS}}$ $\mathbf{T_{IRSH}}$	R	R T_{PL}	R GS_{tb-ss}	R	R	R	R T_{PL} T_{LE} WB_4	R T_{IRSS} T_{IRSH}
c	N/S WB_6	N/S	N/S B_5	N/S GS_{tb-ss} WB_{5B}	N/S GS_{tb-ss}	N/S B_2 B_5	N/S	N/S WB_6	N/S	N/S	N/S GS_{tb-ss}	N/S
v c	N	N	N WB_{5B}	N	N	N	N	N $\mathbf{GS_{tb-ss}}$	N	N	N	N
N/M	N/M	N/M	N/M T_{IRSS} T_{IRSH} T_{PL} GS_{tb-ss} S_4 S_5 S_6 WB_4 B2	N/M	N/M	N/M S_4 S_5 S_6	N/M T_{IRSS} T_{IRSH} T_{LE} WB_4 WB_{5B} B_2 B_5	N/M T_{IRSS} T_{IRSH} T_{PL} T_{LE} S_4 S_5 S_6 WB_{5B} B_2 B_5	N/M T_{IRSS} T_{IRSH} T_{PL} GS_{tb-ss} S_4 S_5 S_6 WB_4 WB_{5B} WB_6 WB_9 B_2 B_5	N/M T_{IRSS} T_{IRSH} T_{PL} T_{LE} S_5 WB_4 WB_{5B} WB_6 WB_9	N/M	N/M T_{PL} GS_{tb-ss} S_4 S_5 S_6 WB_4 WB_{5B} WB_6 WB_9

Facies with mixed lithologies

T_{PL} Tillman (this volume) Planar laminated ss
T_{LE} Tillman (this volume) Low energy ridge margin ss
T_{IRSS} Tillman (this volume) Inter-ridge ss
T_{IRSH} Tillman (this volume) Inter-ridge shaley facies

S_4 Sullivan et al. (1997) 4
S_5 Sullivan et al. (1997) 5
S_6 Sullivan et al. (1997) 6

WB_4 Walker and Bergman (1993) Unit 4
WB_{5B} Walker and Bergman (1993) Unit 5B (core)
WB_6 Walker and Bergman (1993) Unit 6
WB_9 Walker and Bergman (1993) Unit 9

GS_{tb-ss} Gaynor and Swift (1988) thin-bedded ss

B_2 Bergman (1994) thin-bedded and bioturbated ss
B_5 Bergman (1994) thin-bedded ss

Abundance of feature:
A = Abundant
C = Common
R = rare
N/S = Present, type or abundance not specified
N = noted as not present
N/M = Not mentioned

Degree of bioturbation:
A = thoroughly bioturbated
C = some physical structure visible
R = minor burrowing
N/S = Present, type or abundance not specified
N = none noted
N/M = Not mentioned

Table 3.—Comparison of lithologic descriptions of predominantly bioturbated sandstone facies in the Shannon Sandstone, Central Wyoming. Significant differences shown in boldface.

texture	cross-bedding			ripples			Parallel lamination	HCS	mud clasts	mud drapes	burrows	glauconite
	trough	tabular	N/S	wave	current	N/S						
clay	A	A	A	A	A	A	A	A	A	A	A T_{b-ss} T_{b-sls} GS_{b-ss} GS_D S_3 WB_2 WB_3 WB_7 WB_8 B_3 B_7	A
sls												
vf	C	C	C	C	C	C	C	C	C	C		C $\mathbf{B_7}$
f	R	R	R GS_D	R WB_2	R	R T_{b-sls} T_{b-ss}	R T_{b-sls} S_3 WB_2	R WB_2	R	R GS_{b-ss}	R	R $\mathbf{T_{b-ss}}$ $\mathbf{T_{b-sls}}$
m												
c	N/S	N/S	N/S	N/S	N/S	N/S	N/S	N/S	N/S	N/S	N/S	N/S
vc	N	N	N	N	N	N	N	N	N	N	N	N
N/M B_3	N/M	N/M	N/M T_{b-ss} T_{b-sls} GS_{b-ss} GS_D S_3 WB_2 WB_3 WB_7 WB_8 B_3 B_7	N/M	N/M	N/M GS_{b-ss} GS_D WB_3 WB_7 WB_8 S_3 B_3 B_7	N/M T_{b-ss} GS_{b-ss} GS_D WB_3 WB_7 WB_8 B_3 B_7	N/M T_{b-ss} T_{b-sls} GS_{b-ss} GS_D S_3 WB_3 WB_7 WB_8 B_3 B_7	N/M T_{b-ss} T_{b-sls} GS_{b-ss} GS_D S_3 WB_2 WB_3 WB_7 WB_8 B_3 B_7	N/M T_{b-ss} T_{b-sls} GS_D S_3 WB_2 WB_3 WB_7 WB_8 B_3 B_7	N/M	N/M GS_{b-ss} GS_D WB_2 WB_3 WB_7 WB_8 S_3 B_3

Predominantly Bioturbated Sandstone facies

T_{b-ss} Tillman (this volume) bioturbated ss
T_{b-sls} Tillman (this volume) bioturbated sls

GS_D Gaynor and Swift (1988) X-bdd ss D
GS_{b-ss} Gaynor and Swift (1988) bioturbated silty ss

S_3 Sullivan et al. (1997) 3

WB_2 Walker and Bergman (1993) Unit 2
WB_3 Walker and Bergman (1993) Unit 3
WB_7 Walker and Bergman (1993) Unit 7
WB_8 Walker and Bergman (1993) Unit 8

B_3 Bergman (1994) bioturbated ss
B_7 Bergman (1994) coarse bioturbated ss

Abundance of feature:
A = Abundant
C = Common
R = rare
N/S = Present, type or abundance not specified
N = noted as not present
N/M = Not mentioned

Degree of bioturbation:
A = thoroughly bioturbated
C = some physical structure visible
R = minor burrowing
N/S = Present, type or abundance not specified
N = none noted
N/M = Not mentioned

Table 4.—Comparison of lithologic descriptions of predominantly shaley facies in the Shannon Sandstone, Central Wyoming. Significant differences shown in boldface.

texture	cross-bedding			ripples			Parallel lamination	HCS	mud clasts	mud drapes	burrows	glauconite
	trough	tabular	N/S	wave	current	N/S						
clay T_{SH} $GS_{sb/l-sls}$ S_1 S_2 WB_1 B_1 B_8; **sls**	**A**	**A**	**A**	**A**	**A**	**A**	**A**	**A**	**A**	**A**	**A** **$GS_{sb/l-sls}$ WB_1 B_1**	**A**
vf	**C**	**C**	**C**	**C**	**C**	**C**	**C** T_{SH} S_2 B_8	**C**	**C**	**C**	**C** **T_{SH} S_2**	**C**
f; **m**	**R**	**R**	**R**	**R**	**R** T_{SH}	**R** $GS_{sb/l-sls}$	**R**	**R**	**R**	**R**	**R** **S_1 B_8**	**R** T_{SH}*
c	**N/S**	**N/S**	**N/S**	**N/S**	**N/S**	**N/S**	**N/S**	**N/S**	**N/S**	**N/S**	**N/S**	**N/S**
vc	**N**	**N**	**N**	**N**	**N**	**N**	**N** WB_1	**N** WB_1	**N**	**N**	**N**	**N**
N/M	**N/M**	**N/M**	**N/M** T_{SH} $GS_{sb/l-sls}$ S_1 S_2 WB_1 B_1 B_8	**N/M**	**N/M**	**N/M** S_1 S_2 WB_1 B_1 B_8	**N/M** $GS_{sb/l-sls}$ S_1 B_1	**N/M** T_{SH} $GS_{sb/l-sls}$ S_1 S_2 B_1 B_8	**N/M** T_{SH} $GS_{sb/l-sls}$ S_1 S_2 WB_1 B_1 B_8	**N/M** T_{SH} $GS_{sb/l-sls}$ S_1 S_2 WB_1 B_1 B_8	**N/M**	**N/M** $GS_{sb/l-sls}$ S_1 S_2 WB_1 B_1 B_8

Fine-grained (shaley) facies

T_{SH} Tillman (this volume) Silty sh
$GS_{sb/l-sls}$ Gaynor and Swift (1988) bioturbated/lenticular-bedded sls

S_1 Sullivan et al. (1997) 1
S_2 Sullivan et al. (1997) 2

WB_1 Walker and Bergman (1993) Unit 1
B_1 Bergman (1994) black mudstone
B_8 Bergman (1994) laminated mudstone

Abundance of feature:
A = Abundant
C = Common
R = rare
N/S = Present, type or abundance not specified
N = noted as not present
N/M = Not mentioned

Degree of bioturbation:
A = thoroughly bioturbated
C = some physical structure visible
R = minor burrowing
N/S = Present, type or abundance not specified
N = none noted
N/M = Not mentioned

* Reported in the same facies by Tillman and Martinsen (1987)

Table 5.—Comparison of data bases used to describe facies in the Shannon Sandstone, Central Wyoming.

Location of observations

outcrop	outcrop and core	core
Tillman and Martinsen (1984) [1]	Gaynor and Swift (1988) [2]	Tillman and Martinsen (1987) [4]
	Walker and Bergman (1993) [3]	Bergman (1994) [5]
		Sullivan et al. (1997) [6]

1. 14 measured sections in the Salt Creek area
2. 7 measured sections in the Salt Creek area; 6 cores from Hartzog Draw
3. 14 measured sections on the Salt Creek anticline; 18 cores from Teapot Dome
4. 5 cores from Hartzog Draw
5. 55 cores from Hartzog Draw-Heldt Draw fields
6. 46 cores from Hartzog Draw

northwest-southeast trending body about 35 km long and 2-4 km wide. The northeastern side of the sand body is slightly steeper than the southwestern side (slope gradients of about 1:100 and 1:180, respectively). The investigators disagree, however, as to whether this geometry reflects an original depositional morphology or an erosional remnant.

In the outcrop area of the Salt Creek anticline, geometric trends are more difficult to define. Tillman and Martinsen (1984, Figs. 38 and 39) combine surface and subsurface data to delineate a pronounced north-south trend for the lower Shannon, and a more diffuse north-south trend for the upper succession. Gaynor and Swift (1988, Figs. 5 and 6) infer that the Shannon here comprises multiple sandstone bodies, 5-15 m thick and several kilometers across. Although the long-axis orientation of these sand bodies is not specified, most are shown to be asymmetric with steeper northern sides. Walker and Bergman (1993, Figs. 4, 5, and 24) show their lower coarsening-up succession thinning toward the west in the southern Salt Creek anticline area, apparently as the result of onlap onto a westerly-rising basal surface. Their upper cross-bedded sandstone is thicker in two of their measured sections (6 and 12), but no geometric trends are implied.

Internal Architecture

There is relatively little disagreement regarding the bounding or external surfaces of the Shannon. The basal contact at the boundary with the Cody Shale is generally thought to be an unconformity (Swift and Parsons, this volume; Sullivan et al., 1997; Bergman and Walker, this volume), although Tillman (this volume) describes that contact as "conformable and slightly transitional." The upper Shannon contact is accepted by all as a marine flooding surface (on the sides of the Hartzog Draw sand body. Sullivan et al. (1997, Fig. 5) place the flooding surface within the fine-grained interval above the sandstone).

Perceptual differences, however, of the distribution and significance of internal erosional surfaces are sharp and underlie much of the contention over the Shannon. Figure 2 summarizes the differences in perception of the internal character of the Shannon at Hartzog Draw. Gaynor and Swift (1988) and Swift and Parsons (this volume) identify no through-going erosional breaks within either outcrop or subsurface (Fig. 2A), nor (Fig. 2B) do Tillman and Martinsen (1984, 1987) or Tillman (this volume). Bergman and Walker (this volume), in contrast, identify six erosional surfaces in the Heldt Draw-Hartzog Draw area (Fig. 2C), three of which were produced by relative sea-level fall and three transgressive surfaces of erosion. Sullivan et al. (1997) carry three sequence boundaries through Hartzog Draw: (1) "Copenhagen Blue," of limited relief at the base of the Shannon, (2) "Crimson Red," also of limited relief just below the upper coarsening-up succession, and (3) "Canary Yellow," of substantial relief truncating that succession and defining the ridge-like shape of the Shannon (Fig. 2D). Comparison of the surfaces suggests that the "Copenhagen Blue" surface of Sullivan et al. (1997) coincides with Bergman and Walker's (this volume) BD-1, atop the Cody Shale and also with Gaynor and Swift's possible erosion surface at the base of the Shannon. The "Canary Yellow" surface seems to equate in part to Bergman and Walker's (this volume) transgressive surfaces of erosion BD-6 (T-17) and, possibly, BD-4 (T-15). Bergman and Walker's BD-5 surface might locally correspond with the "Crimson Red" surface. Otherwise, there seems to be little coincidence of inferred surfaces.

Criteria for recognition of erosional surfaces include sharp contacts with abrupt facies change and local lags of pebbles or other clasts. The basal Shannon contact seems to qualify as such a surface. Where present in core photographs (Tillman and Martinsen, 1987, Fig. A11; Bergman, 1994, Figs. 3 and 4), the contact marks a very abrupt change from silty shale to burrowed, flaser-bedded sandstone/siltstone. Walker and Bergman (1993) report phosphatic pebbles at the contact. Tillman and Martinsen (1987, Fig. 19) show a contact at the southeastern end of Hartzog Draw between their "Disseminated Silty Shale Subfacies" (which everywhere lies below the ridge complex) and their "Shelf Silty Shale Subfacies" (which either lies above or lateral to the ridge complex). The contact appears to be sharp, but the two subfacies are so lithologically similar that the distinction is not completely

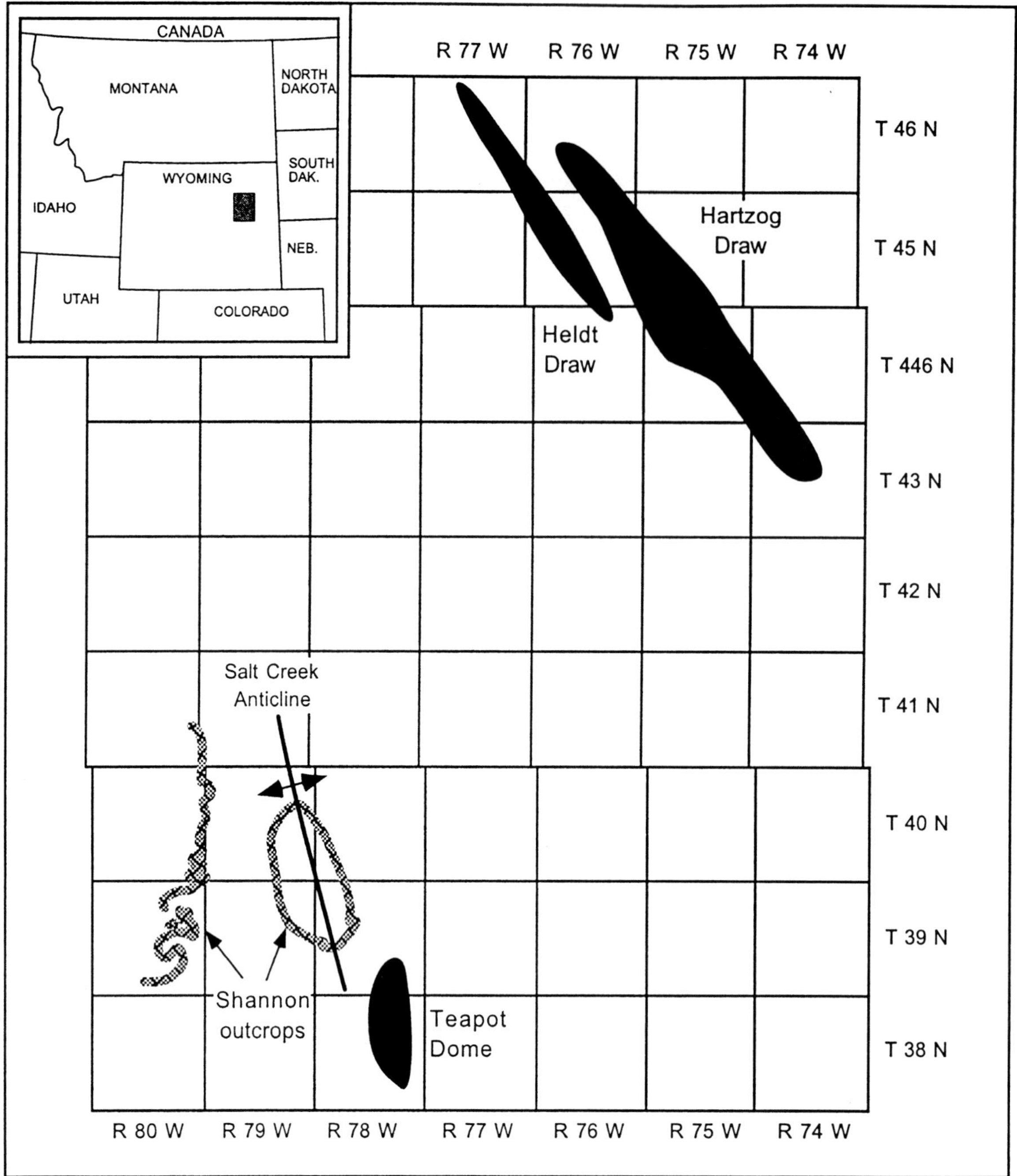

Fig. 1.—Location of Hartzog Draw, Heldt Draw, and Teapot Dome Fields and Shannon outcrops discussed in this report. Compiled from illustrations in Tillman and Martinsen (1987), Gaynor and Swift (1988), and Walker and Bergman (1993).

clear. Possibly this contact represents a merger of the "Copenhagen Blue" sequence boundary and the "Canary Yellow" flooding surface of Sullivan et al. (1997) and/or a merger of Bergman and Walker's (this volume) T-17 and BD-1 surfaces.

The interpretations of both Sullivan et al. (1997) and Bergman and Walker (this volume) are predicated on the existence of extensive erosional surfaces produced by changes in relative sea level. A critical issue is whether sharp contacts at the base of the cross-bedded sandstone facies reflect base-level change or are local features cut by the currents that generate the cross-bedding. Core photographs in Tillman and Martinsen (1987, Figs. A3, A7, and A11) show the nature of contacts at the base of their cross-bedded facies at the stratigraphic position that we would infer for the "Crimson Red" sequence boundary. Core from one well (Bud Christensen #2, Fig. A-7) contains a convincing erosional surface at the base of the cross-bedded sandstone in its upper part (at 9182.9'), which may be the "Crimson Red" contact. In the other two wells (Cities Service Federal AE-1 and Cities Service Federal AS-1), the position of the "Crimson Red" sequence boundary seems well-defined in the accompanying

Table 6.—Comparison of features in the Shannon Sandstone outcrops and cores from Hartzog Draw field as reported by Tillman and Martinsen (1984, 1987). Facies terminology of Tillman (this volume).

Facies	% Sand	% Silt	% Clay	Average sand grain size (microns)	Cross-bedding >20°	Cross-bedding 10-20°	Cross-bedding <10°	Horizontal laminae (sand)	Ripples	Ripples on troughs	Mud clasts	Mud laminae	Bio-turbation	Glauconite
Central Ridge ss - outcrop	96		4	245	14	70	5	7	7		2	2	10	12
Central Ridge ss - core	92	1	5	215	13	33	19	0	10	10	3	2	4	8
Planar laminated ss - outcrop	100			205			tr	98	3				0	
Planar laminated ss- core	84	5	11	150	0	0	4	25	30		tr		20	4
High-energy ridge margin ss- outcrop	94		6	220	40	56	23		31	20	9	9	5	20
High-energy ridge margin ss- core	75	3	22	225	4	27	16	1	11	15	6	1	4	16
Low-energy ridge margin ss- outcrop	97		3	215	17	54			34		2	2	10	20
Low-energy ridge margin ss- core	64	6	30	165	3	15	25	0	44	10	6	3	13	15
Inter-ridge ss - outcrop	80	12	8	190		5	10	9	70			8	25	7
Not recognized in core														
Inter-ridge facies (shaley) - outcrop	?*	?*	?*	175					95				5	10
Inter-ridge facies (shaley) - core		13	31	135	0	0	2	2	61	0		31	34	3
Bioturbated ss - outcrop	58	32	10	150			7	5	11		?	?	90	12
Not recognized in core														
Bioturbated siltstone - outcrop	5	75	20	105				9	10		?	?	87	4
Bioturbated siltstone - core	22	30	39	130			2	1	16	0		?	79	3
Shelf ss -outcrop	?	?	?	163								5	10	
Not recognized in core														
Shelf silty sh - outcrop**		75	25	113									60	
Shelf silty sh_1 - core**	1	22	77	85					23			30	46	1
Shelf silty sh_2 - core***	6	27	67	100					68	0		19	14	tr

* Only subsurface data provided

** Cody Shale below the ridge complex

*** Silty shale within the ridge complex

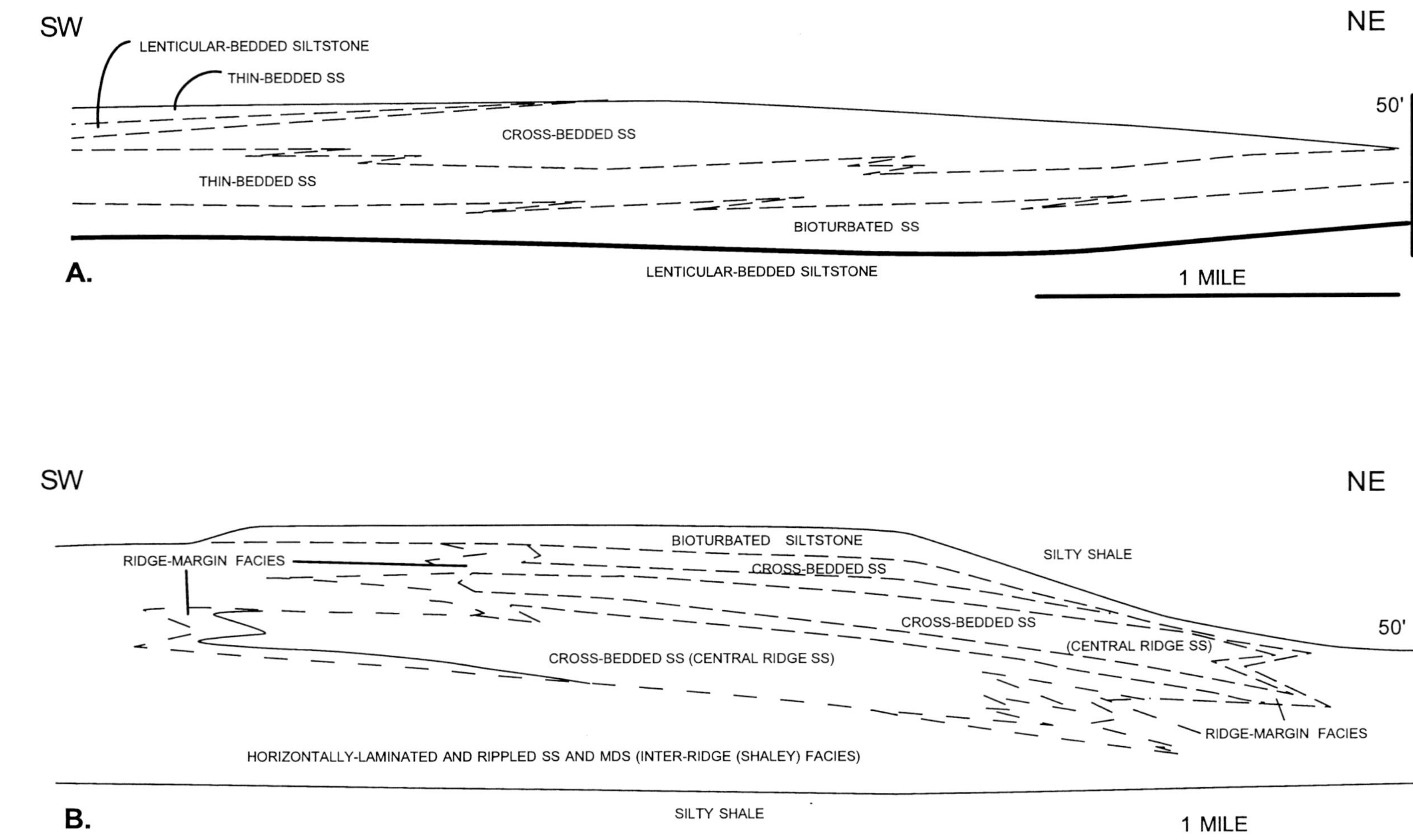

Fig. 2.—Comparison of inferred internal architecture of the Shannon Sandstone in Hartzog Draw, Wyoming, in northeast-southwest section showing contacts and lithofacies. A) Gaynor and Swift, (1988, Fig. 8); B) Tillman (this volume, Fig. 28); C) Bergman and Walker (this volume) and Bergman (1994); D) Sullivan et al. (1997, Figs. 8, 9 were adjusted to conform to a northeast-southwest section). Dashed lines: gradational contacts and Light solid lines: transgressive contacts (may be erosional). Heavy solid lines: regressive erosional surfaces (sequence boundaries). Differences in thickness may reflect sections through different parts of Hartzog Draw.

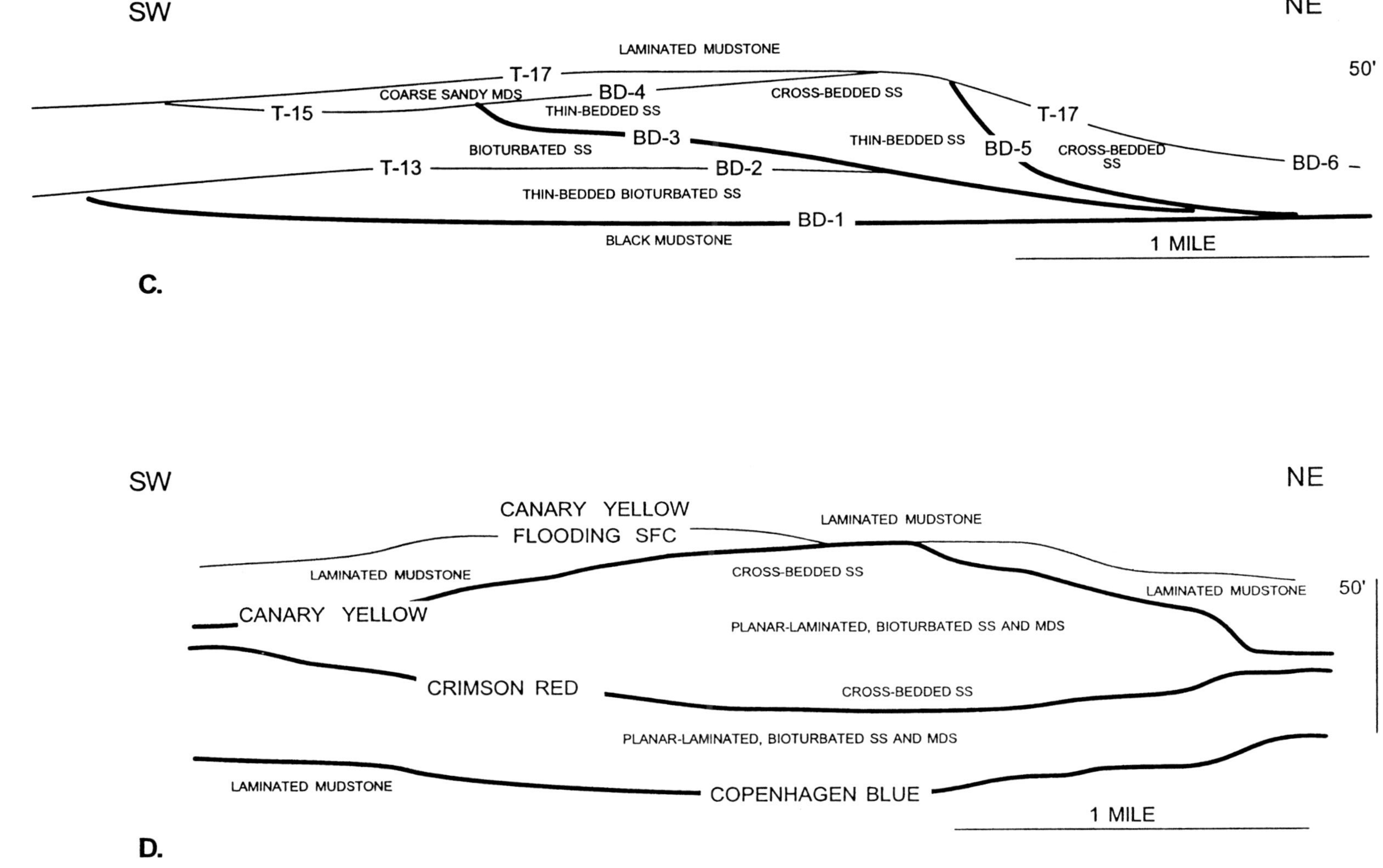

Fig. 2.—Continued.

electric logs (at 9193.5' in Fig. A4 and 9475.7' in Fig. A12, respectively, in Tillman and Martinsen, 1987). In both cases, however, the corresponding core shows an apparent upward transition through an alternation of facies; in neither case is there a convincing erosional surface.

We consider the "Crimson Red" sequence boundary to be particularly suspect. In both longitudinal and transverse sections across the field (Sullivan et al., 1997, Figs. 7 and 8), it shows no evidence of incision relative to the underlying "Copenhagen Blue" boundary. Furthermore, placing the boundary below the lower "proximal tidal bar facies," creates an inconsistency whereby this facies is relegated to the basal part of their inferred southeasterly progradational clinoforms, depositionally beneath their "distal tidal bar facies" (Sullivan et al., 1997, Fig. 8).

There is moderate agreement regarding the distribution of lithofacies in the Shannon. All investigators agree on a general stacking pattern of facies, akin to the succession (bioturbated sandstone\thin-bedded sandstone\cross-bedded sandstone) described initially by Spearing (1976). In both surface and subsurface, two superposed stacks are perceived. Considerable difference exists, however, regarding the lateral distribution of facies in the individual sand bodies and the nature of the bounding surfaces that separate them.

In outcrop, Tillman and Martinsen (1984, Figs. 35, 36, and 37) show pronounced lateral facies changes in both east-west and north-south outcrop cross-sections in the Salt Creek area. In both sets of sections, their cross-bedded "central bar facies" are replaced laterally by "bar margin" or "bioturbated shelf siltstone" or "bioturbated shelf sandstone facies." No directionally systematic trends are evident. Gaynor and Swift (1988, Figs. 5 and 6) infer that within the same area, the multiple sand bodies in the Shannon show a systematic facies trend whereby cross-bedded sandstone is directly overlain by bioturbated sandstone on the northern, steeper sides, but passes laterally into thin-bedded sandstone and thence into bioturbated sandstone on the southern margins. Walker and Bergman (1993, Figs. 4, 5, and 24) correlate the sandstone in their measured sections in the same area more continuously and show no lateral facies change.

The subsurface interrelation of facies in Hartzog Draw is a source of much disagreement (Fig. 2). All investigators show a cross-bedded sandstone facies stratigraphically high in the central part of the field, somewhat offset toward its northeastern side. Gaynor and Swift (1988, Fig. 8) infer that this cross-bedded sandstone facies lenses out laterally to the southwest into their "thin-bedded sandstone" and "lenticular-bedded siltstone" facies (Fig. 2A). Tillman and Martinsen (1987) also show gradational facies changes, but in a more complex arrangement (Fig. 2B). Sullivan et al. (1997), in contrast, map the cross-bedded facies as a remnant, truncated on both sides by a single surface of erosion (their "Canary Yellow" sequence boundary) (Fig. 2D). They infer, however, that it is the upper part of a downlapping succession that progrades to the southeast along the axis of the sand body (Sullivan et al., 1997, Fig. 10). Bergman (1994, Fig. 13) and Bergman and Walker (this volume) perceive the cross-bedded facies to lie between regressive and transgressive erosional surfaces (Fig. 2C). In their template, Bergman and Walker (this volume, Fig. 3) also describe a southwesterly (landward) lateral transition of cross-bedded sandstone into thin-bedded sandstone above an incision caused by a fall of relative sea level. Presumably this transition reflects a basinward downstepping of facies associated with the forced regression followed by erosion of the topographically higher portion of the cross-bedded sandstone during subsequent transgression.

Much of the interpretation of the internal architecture is based on subsurface correlation. The Shannon seems to lack unique attributes that allow for unequivocal correlations, and as a result, all correlations are in essence realizations, that is, possible interpretations that are consistent with the evidence, but ultimately unprovable. Bergman and Walker (1995, and this volume) acknowledge the use of a template, based on a model for shoreline response under conditions of rising and falling relative sea level, to guide their correlations. Although the other investigators of the Shannon do not specify a template, all have tacitly employed one in their own correlations. Each of the different Shannon interpretations thereby relies, to one degree or another, on the resulting realizations. The greater that reliance, particularly as opposed to physical evidence from outcrop or core, the weaker the argument and the more likely the potential for circular reasoning. Readers are free to draw their own conclusions.

Directional Structures and Paleogeography

As summarized by Swift and Parsons (this volume), much of the cross-bedding in outcrop dips toward the south-southwest (about 190E), although Walker and Bergman (1993) found mean flow directions of 165E and 180E in their lower and upper sand units, respectively, in the Salt Creek area. Subsurface cross-bedding directions derived from oriented core show a similar orientation as that in outcrop in one core (Tillman, this volume, Fig. 16) but are directed toward the east-southeast (131E, 166E, 169E) in three others (Hearn et al., 1984). The subsurface measurements shown by Tillman appear to be more widely scattered than those in outcrop (perhaps as a result of the difficult measurement technique). None of the current workers comment on Spearing's observation of a small number of opposing cross-beds in the Shannon (which are also evident in Tillman's (this volume) subsurface data). Tillman (this volume) does note that, "Flow directions that have other than a southerly component are rare in the Shannon and commonly occur only at the top of a bed sequence."

Recent interpretations generally ignore the implications of wave ripple crest orientations. Spearing's (1976) measurement of 171 symmetric ripple crests indicate a strong northwest-southeast trend (summarized as 128E by Swift and Parsons, this volume). Walker and Bergman (1993) measured the strike of 14 ripple crests that averages a trend toward 113-293E and Tillman (this volume) shows five wave ripples with an average crest trend of 135E. The implications of the directional structures relative to the different interpretations are discussed under the section "Paleogeography".

Biofacies

Actual fossils are rare to nonexistent in the Shannon. Walker and Bergman (1993) cite Tillman and Martinsen's (1984) statement that "samples collected in the shaley portions interbedded with the Shannon were barren" and conclude that foraminiferal control of depositional depths within the Shannon is lacking. Tillman and Martinsen (1987) describe shelf-depth foraminfera "from shales immediately above and below the shelf-ridge facies." Tillman (this volume), however, asserts that "foraminiferal data collected, primarily from shales immediately above, below, and between the vertically stacked Shannon sand ridges" indicate mid- to outer-shelf species. Such forams in the shaley sediment within the sand ridge complex would be highly significant, but Tillman (this volume) does not elaborate further.

Trace fossils figure prominently in the debate over the Shannon. Tillman and Martinsen (1984) postulate that "the wide diversity in size, orientation, and type of fill material [of the traces] suggests a relatively hospitable environment," and Swift and Parsons (this volume) describe the trace listed by Tillman and Martinsen (1984) as "shelf trace fossils." In contrast, Sullivan et al. (1997) argue that the assemblages are restricted, suggesting environmental stresses common to "tidal (brackish water) settings."

We tend to agree with Swift and Parsons (this volume) in their analysis of the assemblages; the recognition of restricted trace fossil assemblages, although potentially of great importance, has probably not yet been tested sufficiently to be unconditionally accepted at present as evidence for a restricted embayment or tidal setting. Interestingly, Tillman and Martinsen (1984) describe the microfossil assemblage as impoverished, which they suggest may be due to sub-normal salinity or other factors. We also note that Sullivan et al. (1997) equate tidal and brackish settings. Although the two aspects commonly are associated, brackish environments may experience little or no tides, many tidal settings contain water of normal oceanic salinity, and many otherwise marine settings are subjected to regular or intermittent fresh water stress.

Another question involves the trace *Macaronichnus*. Walker and Bergman (1993) argue that this trace indicates a foreshore or upper shoreface setting and is absent in regressive successions "that do not build right up to the upper shoreface or foreshore." In contrast, Swift and Parsons (this volume) contend that *Macaronichnus* is not specific to a shoreline environment; instead "the responsible organism was viable on a mobile sand bottom at any shelf depth." They postulate that the absence of *Macaronichnus* in Cretaceous Western Interior Basin shelf sands reflects the general lack of preservation of these sands.

Commonly overlooked, in the application of *Macaronichnus* to environmental interpretations, is the fact that the trace was first found in modern sand in a tidal setting (Clifton and Thompson, 1978). It is not specific to a shoreface, although it commonly occurs there. Where shelf sands are well-preserved, as on the coast of California, the trace is not present, even though it is common in shoreface and foreshore deposits of the same succession (Clifton, 1988). We reject Swift and Parson's (this volume) suggestion that *Macaronichnus* is suspect as a surf zone indicator because it is a horizontal rather than vertical trace. Probably the trace owes any environmental significance to the fact that it is a horizontal trace, made today by errant polychaetes that dwell several tens of centimeters below the sediment-water interface. Such an existence requires specific conditions of texture and energy whereby oxygen is continuously supplied to the substrate. These conditions are likely to be met only in high-energy, shallow-water settings such as a shoreface, foreshore, or an area with strong tidal flow.

Although bearing only obliquely on the controversy, the occurrence of the trace fossil *Ophiomorpha* provides a good example of how differently these rocks are perceived. This trace is one of the most distinctive, easily recognizable traces in the rocks. Some interpreters have considered it as typical of the shoreface (Tillman, this volume), although it has been identified in other settings, including submarine canyon floors well below the influence of waves (Hill, 1981).

Walker and Bergman (1993) include *Ophiomorpha* among the most prominent traces in the cross-bedded sandstone facies and further include it among the most common traces in their bedded sand facies. In direct contrast, Tillman (this volume) states that *Ophiomorpha* is almost completely absent. Sullivan et al. (1997) note the presence of *Ophiomorpha* in their cross-bedded "proximal tidal bar" facies but state that these and related vertical dwelling structures are less common than the horizontal deposit-feeding traces. Gaynor and Swift (1988) do not include *Ophiomorpha* in their list of trace fossils in the Shannon, and Bergman (1994) does not cite *Ophiomorpha* among the recognizable traces in core from the Hartzog Draw-Heldt Draw Fields.

Conceivably, some of these differences (i.e., Walker and Bergman, 1993, and Sullivan et al., 1997) simply reflect a variation between outcrop and subsurface facies. The diametrically opposed conclusions by Tillman (this volume) and Walker and Bergman (1993), however, seem to be a striking difference in perception.

Glauconite and Siderite

Glauconite and, to a lesser degree, siderite (in the form of clasts and layers) are significant components of the Shannon Sandstone. They also potentially bear significantly on its interpretation. According to Swift and Parsons (this volume), "The glauconite content of the Shannon is astonishing and constitutes a major constraint for any actualistic depositional model."

Locally comprising as much as 75% of the rock, glauconite is present in nearly all of the Shannon facies (Tillman and Martinsen, 1987, Table A2). Table 7 summarizes the mean glauconite concentration among the different facies described by Tillman (this volume). Because glauconite is most likely to occur in the sand fraction, Table 7 also shows the normalized concentrations of the mineral with respect to the sand content. Although the absolute amounts of glauconite in the subsurface samples are less than those in outcrop, the difference probably reflects variations in sand content. The normalized glauconite values show far less difference between surface and subsurface samples.

The concentrations of glauconite, both normalized and absolute, differ with facies. The mineral is most abundant in the "High-Energy and Low-energy Ridge Margin" facies that contains between two and six times as much glauconite as the other sandstone/siltstone facies. The "Central Ridge" and "Planar-Laminated" facies contain significantly less, as do the "Inter-ridge Sandstone" and "Inter-ridge (shaley) facies," even after normalization. The bioturbated facies (both sandstone and siltstone), in contrast, contain fairly high amounts of glauconite after normalization. Small quantities of glauconite are reported in the shelf silty shale that lies above, below, or lateral to the Hartzog Draw sand body (Tillman and Martinsen, 1987).

All of the investigators, except Sullivan et al. (1995, 1997), note the presence of glauconite in the sandstone, but their interpretations of its occurrence differ widely. Swift and Parsons (this volume) view the glauconite as originating in a shelf setting concurrently with the development of the sand ridge. Walker and Bergman (1993) and Bergman (1994), in contrast, infer that the glauconite originated at shelf depths, but was reworked and incorporated into the sands following a forced regression. Tillman (this volume) considers the glauconite accumulation to be considerably greater than that on most shoreface deposits.

Stonecipher (this volume) notes the possibility of two "glauconitic" facies: a "glaucony" facies composed of true glauconite minerals and a "verdine" facies, which includes other green pelleted minerals (chamosite, an iron-rich chlorite, and berthierine, an iron-rich kaolinite). Following Odin

Table 7.—Textural character and mean and normalized concentration of glauconite in the different facies of Tillman (this volume).

Facies	% Sand	% Silt	% Clay	Average sand grain size (microns)	Glauconite %	Normalized * glauconite (%)
Central Ridge - outcrop	96		4	245	12	12.5
Central Ridge - core	92	1	5	215	8	8.7
Planar laminated - outcrop	100			205		
Planar laminated - core	84	5	11	150	4	4.8
High-energy ridge margin - outcrop	94		6	220	20	21.3
High-energy ridge margin - core	75	3	22	225	16	21.3
Low-energy ridge margin - outcrop	97		3	215	20	20.6
Low-energy ridge margin - core	64	6	30	165	15	23.4
Inter-ridge ss - outcrop	80	12	8	190	7	8.8
Not recognized in core						
Inter-ridge shaley - outcrop	?**	?**	?**	175	10	?*
Inter-ridge shaley- core	58	13	31	135	3	5.4
Bioturbated shelf ss - outcrop	58	32	10	150	12	20.7
Not recognized in core						
Bioturbated shelf sls - outcrop	5	75	20	105	4	80.0
Bioturbated shelf sls - core	22	30	39	130	3	13.6
Shelf silty sh - outcrop***		75	25	113		
Shelf silty sh_1 - core***	1	22	77	85	1	
Shelf silty sh_2 - core****	6	27	67	100	tr	

* Normalized with respect to percent sand (% glauconite / % sand)

** Only subsurface data provided

*** Cody Shale below the ridge complex

**** Silty shale above or lateral to the ridge complex

and Matter (1981), she attributes the typical formation of the glaucony facies to shelf depths of 60-550 m or the tops of still-deeper isolated mid-ocean highs. The verdine facies, in contrast, is found on the inner shelf and is "particularly abundant in estuarine facies or immediately offshore from fluvial deltas." In a petrographic study of the Shannon, Hansley and Whitney (1990) describe neither chamosite nor berthierine, suggesting that the glauconite in the rock is not part of the coastal verdine facies.

Stonecipher (this volume) recapitulates Odin and Matter's (1981) theory, which holds that the process of glauconitization occurs at the boundary of oxidizing seawater and reducing interstitial waters. An abundance of glauconite implies a combination of continual reworking of the sediment and slow rates of clastic input in the area of generation. The mineral, however, can be redeposited as a detrital component into subsequent sediment. Citing Amorosi (1995), Stonecipher (this volume) notes that glaucony can exist in almost all shallow-marine environments due to various combinations of authigenesis or reworking.

The distribution of the glauconite within the Shannon may be significant. Swift and Parsons (this volume) propose that glauconite is concentrated in cross-bedded sandstone on the up-current side of the ridge, where the rate of reworking is likely to be highest. In their model, the concentration of glauconite decreases in a down-current direction across the ridge where the reworking rate decreases. The distribution they infer for the mineral (Fig.3A) can be compared to the facies (and thereby glauconite) distribution (Fig. 3B, C) recorded in Hartzog Draw by Tillman and Martinsen (1987), who indicate that percentage glauconite is highest (mean 16%) in their "High-Energy Ridge Margin" facies. This facies lies on the northeastern side of the sand body in Hartzog Draw, as Swift and Parsons predict. Nearly equal values of glauconite (mean 15%) are present, however, in a "Low-Energy Ridge Margin" facies on its opposite side, and separating these two facies is the cross-bedded "Central Ridge" facies, which has a mean of only 8% glauconite. Although Swift and Parson's (this volume) model is broadly consistent with the glauconite distribution in the Shannon, it differs strikingly in detail. It also begs the question as to why there should be a relatively thick cross-bedded sandstone preserved on the up-current, and therefore erosional, side of the ridge.

If the glauconite in the Shannon is detrital, as suggested by Walker and Bergman (1993), a significant volume of shelf facies must be eroded to generate the observed amount of glauconite. Using the data of Tillman and Martinsen (1987), and assuming a consistent distribution of glauconite among their different facies throughout the sand body, the Hartzog Draw sandstone contains approximately 0.03 km^3 of glauconite. Assuming an average value of 1% glauconite in the underlying shelf siltstone and no significant loss due to attrition or to removal by transport, approximately 1 m of substrate would have to be eroded from an area about 30-40 times larger than that underlying the ridge (about 3,000 km^2) (or 0.5 m from an area of about 6,000 km^2, etc.). Such erosion is not inconceivable during forced regression, valley incision, or transgressive ravinement. Erosion within incised valleys, as suggested by Sullivan et al. (1997), might require even more erosion, in that much of the eroded glauconite would be carried basinward.

In conclusion, based on Stonecipher's review (this volume), there seems to be little basis for opting for either autochthonous or allochthonous origins for the Shannon glauconite. The data do suggest the following:

1. The environment produced glauconite before and after formation of the sand bodies.
2. Glauconite was available for reworking into the sandstone, and volumetric analysis implies that reworking could account for the observed glauconite in Hartzog Draw.
3. There is no unequivocal evidence either for or against the generation of authigenic glauconite during deposition of the sandy facies.
4. Smaller concentrations of glauconite in central ridge facies suggest possible mineral loss due to abrasion (reworking) under the highest energy conditions.

The presence of siderite as clasts and layers also enters the arguments over the origin of the Shannon. Walker and Bergman (1993) and Bergman (1994) cite Hansley and Whitney's (1990) conclusion that the association of siderite, which they feel is most likely to form in "swampy, coal-bearing lagoonal or deltaic-estuarine settings" and glauconite, formed on a mid- to outer-shelf, support the concept of substantial sea-level change. Sullivan et al. (1995) associate the sideritized mudstone clasts with soil development in a fresh-water environment.

Swift and Parsons (this volume) interpret the siderite as fully marine, formed during early diagenesis. Hansley and Whitney (1990) acknowledge that siderite can form in this way in the process of methanogenesis. They note that the $\delta^{13}C$ values found in the Shannon siderite pebbles and beds lie in the lower part of the range of methanogenic siderite. The combination of $\delta^{13}C$ and $\delta^{18}O$ values presented by Hansley and Whitney (1990) fall into a field dominated by, but not restricted to, a marine setting (Stonecipher, this volume, Fig. 10). We conclude that the siderite in the Shannon is not incontrovertible evidence of a brackish water setting.

INFERRED PROCESSES

Despite the controversy, a measure of agreement exists regarding the processes involved in the formation of the Shannon. No one has, as yet, invoked fluvial, eolian, turbidite, or glacial processes for the Shannon in this area. The consensus ends there, however. The crux of disagreement lies in contention as to whether tidal influences or waves and storm processes dominated Shannon deposition.

Tidal Influence

How much did the tidal currents influence Shannon deposition? That question is central to the controversy and is the source of considerable disagreement. Sullivan et al. (1997) cite sigmoidal cross beds, unidirectional and bidirectional reactivation surfaces, mudstone drapes, and mudstone couplets as being characteristic of tidal settings and infer that the ridges are complexes of stacked tidal bars. Tillman (this volume) states that, based on the 1995 field conference, "This author was convinced that tides as well as storms were probably important in deposition of the Shannon sand ridges," but cites no specific evidence. Swift and Parsons (this volume) believe that, "while tidal forcing may have contributed to the hydraulic regime, it does not appear to have played a dominant role." They assert this largely because they find diagnostic evidence of tides such as double clay drapes and tidal bundling to be scarce and that other features, such as mud drapes and sigmoidal cross-beds, can be produced by other processes and are not reliable criteria. Bergman and Walker (this volume) do not mention the possibility of tidal influence, but Walker and Bergman (1993) noted that they found no unambiguous evidence of tidal processes.

We feel that there are few, if any, unequivocal criteria for identifying tidal influences. Bundles of laminae or cross-strata can be produced by other processes (see for example, Rubin, 1987, Fig. 46). An individual double mud drape can be produced by happenstance in the absence of tides, and, commonly, apparently good couplets extend laterally into a single layer, or bifurcate into several laminae. Other criteria, such as rhythmic stratification, mud drapes, and sigmoidal cross-beds are even less reliable. Nonetheless, when these features are found in association, and where double mud drapes occur repeatedly within a section, the evidence for tidal flow can be overwhelming. We find double mudstone drapes (including muddy layers with a discontinuous internal sand lamina) to be remarkably common in the published photographs of the Shannon. Rhythmically repeated mudstone layers are also very common. Table 8 summarizes the evidence for tidal influence that we find in the figures of core and outcrop of the Shannon. Equally good evidence for tides (an association of double mudstone drapes, rhythmically repeated mudstone layers, and rhythmic sandstone lamination) exists in the core photographs of Tye et al. (1986). In short, we find the evidence for tidal influence on Shannon deposition to be pervasive and compelling.

The best tidal evidence lies in the thin-bedded and finer facies. The role of tides in the cross-bedded facies is less obvious in the published photographs. Elliott (1995) presents evidence for tidal influence in the cross-bedding in the Salt Creek area, and Sullivan et al. (1997) mention sigmoidal cross-beds in their summation of tidal evidence. The directional aspects of the cross-bedding must be explained, however. Although opposing cross-bedding exists in the Shannon (Spearing, 1976; Tillman and Martinsen, 1984, 1987; Elliott, 1995), cross-strata measurements consistently show a strong predominance of southerly dipping foresets. Such a nearly unidirectional pattern can be explained using tidal currents alone, but requires a strong asymmetry of the tidal currents (a concept supported by the prevalence of double mud drapes separated by a thin, discontinuous sand parting). One possibility is time or velocity asymmetry imposed by the shape of the basin, whereby either the ebb or flood current

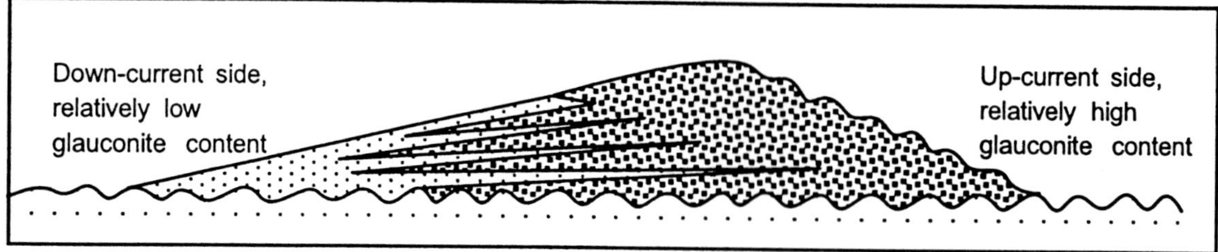

A. Predicted distribution of glauconite according to the model proposed by Swift and Parsons (this volume).

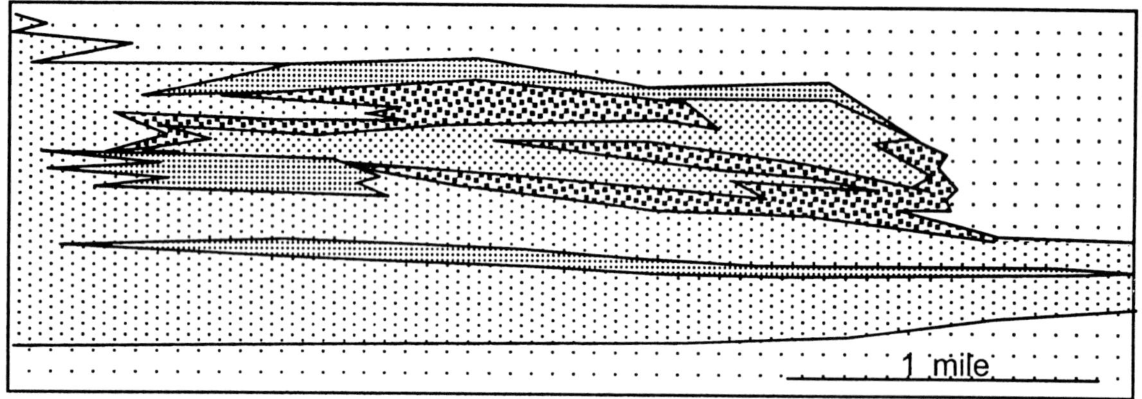

B. Inferred distribution of glauconite according to the facies distribution of Tillman and Martinsen (1987) in northern Hartzog Draw.

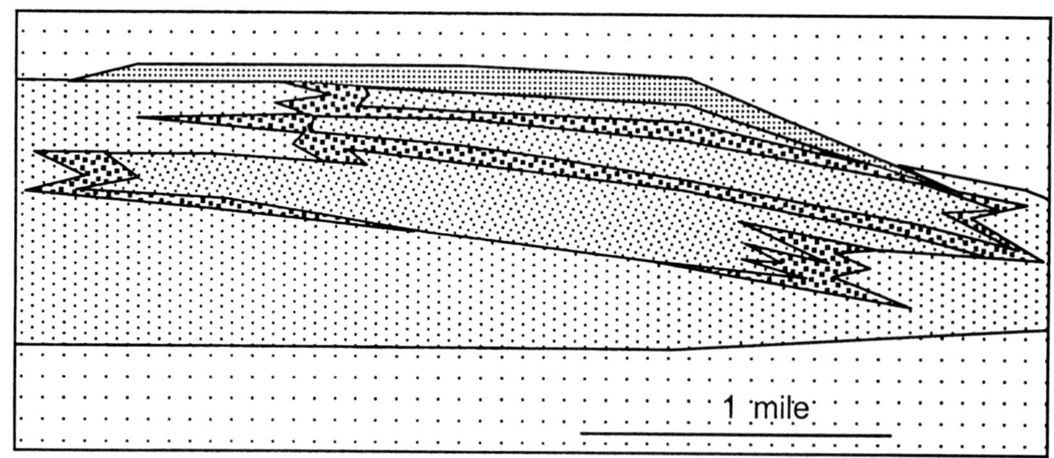

C. Inferred distribution of glauconite according to the facies distribution of Tillman and Martinsen (1987) in north-central Hartzog Draw.

20-25% (Ridge-Margin Facies)

13-20% (Bioturbated Facies)

9-13% (Central-Ridge Sandstone)

5-9% (Inter-Ridge Shaley Facies

trace-1% (Silty Shale)

Fig. 3.—Comparison with predicted and inferred distributions of glauconite in Hartzog Draw Field.

Table 8.—Evidence for tidal influence in published photographs of the Shannon Sandstone, Central Wyoming.

Evidence	Tillman and Martinsen (1984)	Tillman and Martinsen (1987)	Walker and Bergman (1993)	Bergman (1994)	Sullivan et al. (1997)
Double mudstone drapes					
well-developed		Figs. A3; A7, A11		Fig. 3	Fig. 3
probable		Fig. 14b		Fig. 8	
Rhythmites					
parallel laminae	Fig. 26c	Figs. 20a,b,c, 15a, A3, A7		Fig. 3	
mudstone drapes	Fig. 26b	Figs. 20d, A3, A7, A11	Fig. 15		Fig. 3
foresets	Fig. 16c				Fig. 3
Bundles	Fig. 23a,b	Figs. 14c, A7		Fig. 5	
Opposing bedforms					
crossbeds	Fig.15b				
ripples				?Fig. 4	

strongly predominate in this area as described by Belderson (1986) on the northwest European shelf. Another is the influence of Coriolis force, which might be expected to augment south-flowing tidal currents on the western side of the seaway. Additionally, a south-flowing geostrophic current, driven by wind or water-density contrasts within the seaway, could reinforce a southerly tidal flow (see Slingerland and Keen, this volume). Dalrymple et al. (1992a) document such an influence by the relatively weak Labrador Current on the orientation of wave-generated structures on the western Grand Banks of Newfoundland. They note that, "Relatively weak oceanic currents, when superimposed on storm (or tidal) currents have an influence that outweighs the strength of the oceanic current by itself" and suggest that a similar current could be responsible for the unidirectionality of cross-bedding in the Shannon Sandstone. Belderson (1986) notes similar effects around the margins of the North Sea.

Conceivably, the uni-directional currents could be unrelated to tides, driven by storm forcing as suggested by Swift and Parsons (this volume), although the observed evidence for tides in the cross-bedded facies argues for at least some tidal influence.

Wave and Storm Effects

Most of the authors accept the effects of storm waves on the Shannon, but, as noted in the section on lithofacies, they view abundance of these features quite differently. Perhaps as important as their relative abundance is the stratigraphic position of storm-influenced facies relative to other facies. Shoreface successions commonly have hummocky cross-stratified sandstone stratigraphically beneath cross-bedded upper shoreface deposits. Elliott (1995) suggests a similar relationship where tidally cross-bedded sandstone on regressive ebb-tidal deltas prograde over storm-dominated sediment deposited in somewhat deeper water. On the other hand, shelf or embayment sandstone ridges are likely to have the storm-dominated facies atop the cross-bedded sandstone, as a consequence of subsequent reworking of the ridge tops by storms. A complete estuary mouth succession in a mixed energy setting also should show underlying cross-bedded tidal inlet or channel facies overlain by the mixed storm and tidal deposits of a shoal.

Both types of successions are reported in the Shannon. Walker and Bergman (1993) identify hummocky cross-stratification in two of their units immediately below a cross-bedded facies (in Unit 4 which is below Unit 5 and in Unit 9 below Unit 10). They also, however, note its presence in Unit 6, which is composed of glauconitic sandstone and mudstone and forms a transitional facies between underlying cross-bedded sandstone and overlying bioturbated sandstone. Elliott (1995) also recognizes a sharp-based, storm-dominated sheet sandstone, containing "occasionally preserved HCS," that underlies a cross-bedded facies. He, however, also identifies storm features associated with tidal features within the overlying cross-bedded sandstone. Tillman and Martinsen (1984, 1987) do not specify storm features, but their "planar laminated facies," dominated by subhorizontal plane-parallel laminated sandstone, might be attributed to storm waves. Where shown in outcrop (Tillman and Martinsen, 1984, Fig. 8), this facies caps cross-bedded sandstone and underlies bioturbated shelf siltstone. In the subsurface, the one illustrated example of the planar laminated facies is also at the top of a cross-bedded succession (Tillman and Martinsen, 1987, Fig. A3). There appears to be little consistency to the stratigraphic position of storm-influenced sediment in the Shannon.

Both Walker and Bergman (1993) and Spearing (1976) note a prevalence of wave ripples in the outcropping Shannon. Because the stratigraphic level within shallowing-up successions depends on water depth, the position of symmetric ripples within the succession can give a rough indication of the size of formative waves. Occurrence low in the succession implies wave reworking of the sea bed by fairly sizeable waves. Spearing (1976, Fig. 8) implies the presence of wave ripples 15 m below the top of one of the coarsening-up successions in the Shannon. Assuming that this represents a minimum water depth, such ripples are likely to be formed only under storm conditions.

None of the investigators identified gutter casts or tool marks in the Shannon. We have found such features to be common in shoreface deposits of Cretaceous progradational successions elsewhere in the western Interior Seaway (e.g., the Blackhawk Formation in the Book Cliffs, and the Point Lookout Formation in southwestern Colorado). Walker and Plint (1992) identify gutter casts as characteristic features at the base of shoreface deposits that accumulate during a forced regression. Their presence in the Shannon would provide useful evidence of the influence of storm waves on a shoreline system (Duke, 1990). Given the abundance of appropriate substrates (mudstone layers) in the Shannon, the absence of these features, if real, casts doubt on the influence of shoaling storm waves during deposition.

PALEOGEOGRAPHY

Fundamental to paleogeographic issues are 1) the origin of the sand body linearity, 2) lack of a well-established regional paleogeographic framework, and 3) the relationship of the observed directional structures to the individual interpretations.

Origin of Sand Body Linearity

As noted in our preceding discussions of the geometry and internal architecture of the Shannon, disagreement exists as to whether the linear sand bodies of the Shannon system reflect the morphology of original sand ridges (Tillman, this volume; Swift and Parsons, this volume), erosional remnants of a more laterally extensive sand system (Sullivan et al., 1997), or an accumulation of sand with the general trend of the present bodies, modified by subsequent erosion (Bergman and Walker, this volume). In the first case, the orientation of the sand bodies results from the combination of processes during deposition. In the second case, the orientation reflects a subsequent pattern of incision. In the third case, the orientation parallels ancient shoreline trends. Each of these interpretations carries profound differences in the interrelationship of paleogeography and process. In the following discussions, we examine these differences as they relate to the overall paleogeographic framework and the observed directional features in the Shannon.

Regional Paleogeography

The absence of an accepted paleogeographic framework has undoubtedly fueled the Shannon controversy. There is general agreement that the presumed highstand shorelines lay to the west, but there is no agreement on the orientation or general configuration of any Shannon shorelines themselves. Did they trend north-south, east-west, or some other direction? Were they straight or did they form a bight? Did streams dissect them? Was the sand derived from the foreland and thrust belt to the west, from deltaic systems in what is now Montana, from the basin (if a tidal shoal), or from shoal areas to the east? It has long been assumed that the marine shales that encase the Shannon sandstone prograded eastward (e.g., Swift and Parsons, this volume; Bergman and Walker, this volume). However, Olson et al. (1998) found that while strata within the Cody Shale in the Powder River Basin thinned eastward, they did not display eastward-prograding clinoform geometries. Southerly progradational architecture was observed. The ambiguity raised by these questions allows for the current wide range of explanations for the Shannon; they also raise questions regarding any one of them.

None of the investigators of the Shannon consider the possibility of reciprocal stratigraphy in foreland basins as suggested by Catuneanu et al. (1997). According to their model, episodes of thrusting generate subsidence adjacent to the orogenic belt and coincident uplift on a peripheral bulge further into the basin. Subsequent unloading in the orogenic belt produces concurrent regression in the proximal area and subsidence of the peripheral bulge. As a result, episodes of transgression and marine shale deposition in areas proximal to the orogen have corresponding episodes of sand accumulation on the peripheral bulge; episodes of regression and shoreline progradation in the proximal areas are coincident with drowning and shale deposition where the bulge existed.

If this model is valid, it offers the possibility that deposition of the Shannon sands is unrelated to any presumably corresponding highstand shorelines to the west. Instead, the Shannon sandstone could have accumulated on a broad northwest-trending shoal. Sand carried along the crest of the shoal by south-flowing geostrophic currents (Slingerland and Keen, this volume) could be reworked into ridges by storm or tidal currents. Shorelines might be intermittently possible. Internal erosional surfaces of the kind postulated by Sullivan et al. (1997) could develop during episodes of uplift. In short, many of the features observed in the Shannon could be produced in complex association in such a setting. No fairway, however, has been identified that corresponds to a northwest-trending shoal, and the abundant mud in the Shannon could be difficult to explain by this mechanism.

Implications of Shannon Directional Structures

The mutual obliquity of the orientations of the long axis of the sand bodies, the cross-bedding, and the wave ripple crests (Fig. 4) impose constraints on various interpretations. These constraints are discussed as they pertain to the different modern analogs.

MODERN ANALOGS

The application of modern analogs to the isolated marine sandbody question in general and the Shannon Sandstone question in particular is somewhat spotty. The advocates of the shelf sand ridge theory (e.g., Tillman and Martinsen (1984), Tye et al. (1986), Swift and Parsons (this volume) have made extensive reference to Quaternary sand ridges on the eastern United States continental shelf as analogs. Adherents of incised valleys and lowstand shorelines/forced regressive models have not utilized modern analogs to any great extent, if indeed they reference them at all.

The usefulness of Quaternary analogs basically boils down to two factors. The modern analog must be well understood and documented, and the deposit being "analogized" should be a close match in terms of lithofacies, tectonic setting, paleogeography, geometry, and inferred depositional processes. A given Quaternary deposit may not be an exact match for a particular ancient example in all of these categories, but may be a suitable analog in one or more of them (cf., Walker and James, 1992). Presumably, we should expect that for a given ancient deposit to be interpreted as, for example, a tidal sand ridge, it should match the lithofacies, geometric, and directional characteristics of modern, observable tidal sand ridges.

There are fundamentally two approaches to the application of modern analogs in sedimentary geology. The first consists of simple comparison of a particular unit to a modern deposit. This approach involves the collection of facts and observations, which are compared with the experience base of the interpreter. If enough characteristics match, we assume that similar processes and depositional environments produced the analogous response. The second approach is more conceptual: reasoning as to what should be present within a given deposit if certain dynamic processes were operating during its deposition. If enough characteristics match the hypothetical results of the thought experiment, the favored process and environments are invoked as analogs. This is particularly useful if aspects of the modern analogs are insufficiently known. Each of these approaches appears to introduce an element of bias into the process of interpretation. Inevitably, a certain amount of "cherry picking" of characteristics and correlations takes place. However, we can still test the validity of a particular interpretation by comparing it to Quaternary examples. In the following section, we have tried to evaluate the proposed models for the Shannon Sandstone in light of what is known about analogous Quaternary deposits.

Storm-dominated Shelf Sand Bodies

The principal application of modern analogs to the "isolated marine" sand body question has been those drawn from storm-dominated shelf deposits. This in itself reflects a certain bias, as the early interpretations of the Shannon as a continental shelf deposit were the impetus for some of the original studies of Quaternary shelf sands. Linear sand bodies are nearly ubiquitous on Quaternary continental shelves, and have been the subject of considerable research and publication. The most recent and comprehensive review of these deposits is by Snedden and Dalrymple (this volume). These features occur anywhere from just beyond the shoreface to more than 100 km from the "coeval" highstand shoreline. Shoreface attached ridges are generally oblique to the coeval shoreline, while offshore ridges tend to be subparallel to the coast, thus providing geometric and apparent paleogeographic analogs to the Shannon Sandstone and its siblings. Modern continental shelves have experienced repeated sea-level fluctuations during Quaternary glaciation. With an ambulatory shoreline, there is no need to speculate about the mechanism for delivery of sand to an otherwise fine-grained shelf. This solves one of the great mysteries of the isolated sand body problem and thus has considerable appeal.

However, despite decades of research, there is a great deal we do not know about these deposits. For example, the shelf shoals of the Atlantic and Gulf of Mexico continental shelves

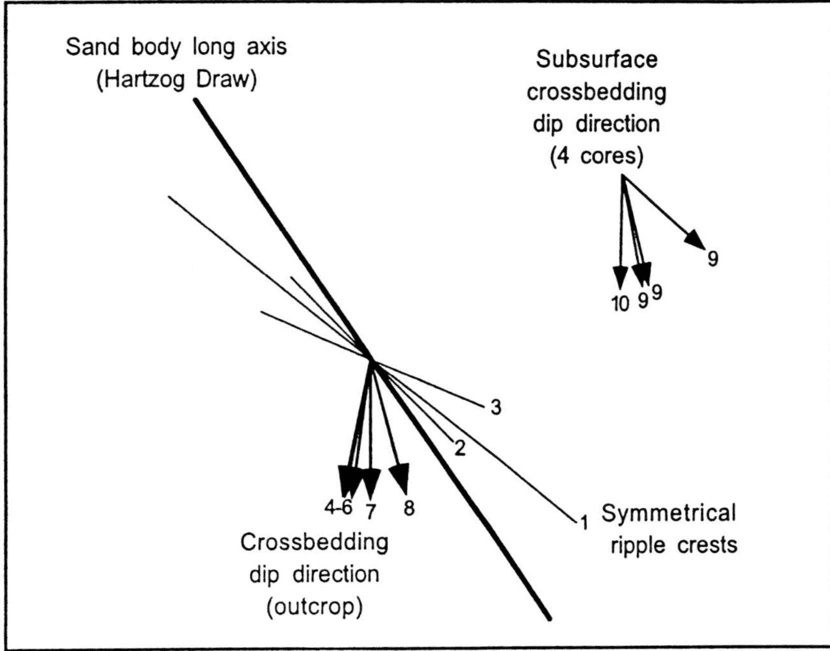

Fig. 4.—Observed directional relationships for sand body long axis (Hartzog Draw), orientation of symmetric ripple crests (1-Spearing, 1976; 2-Tillman and Martinsen, 1984; 3-Walker and Bergman, 1993) and predominant dip direction of crossbeds in outcrop (4-Spearing, 1976; 5-Gaynor and Swift, 1988; 6-Tillman and Martinsen, 1984; 7- and 8-Walker and Bergman, 1993), and subsurface (9-Hearn et al., 1984; 10-Tillman and Martinsen, 1987).

of the United States have been consistently invoked as analogs for the Shannon and other units of this type [e.g., Tillman and Martinsen (1984), Gaynor and Swift (1988), Swift and Parsons (this volume)]. These modern shoals and ridges are subparallel to the shoreline, and can be reliably considered to be currently subaqueous and on the continental shelf, but the actual mode of their formation, constituent lithofacies, nature of internal surfaces, and ultimate fate are still matters of debate and ongoing research. The general consensus today is that there are various types of continental shelf sand bodies, formed by a variety of mechanisms largely related to transgression (Snedden and Dalrymple, this volume). Many of these deposits are still in active process environments and are being transformed by current shelf processes.

According to Stubblefield et al. (1984), Atlantic margin sand ridges vary from 4 m to 6 m in thickness, and average 3 km in width. Gulf coast sand bodies are of similar size and orientation. Mississippi Delta shoals are from 4 m to 6 m thick, 8-10 km wide, and up to 50 km in length (Penland et al., 1989). Shoals along the Mississippi-Alabama, western Louisiana and eastern Texas shelves (Curray, 1960; Nelson and Bray, 1970; Frazier, 1974; Suter, 1986; McBride et al., this volume; Rodriguez et al., this volume) fall within this range. Individual Shannon "ridges" are substantially thicker, reaching close to 100 ft (~ 30 m) (Tillman and Martinsen, 1984). While this may reflect variations in sand abundance, accommodation space, and/or energy regimes, it is still apparent that there are some significant differences between the geometries of these modern storm-dominated shelf ridges and the Shannon deposits.

Vibracores through many of the Quaternary features (e.g., Penland et al., 1986; Swift et al., 1986; Rine et al., 1986, 1991; Hoogendoorn, 1989; Snedden et al., 1994; Dalrymple and Hoogendoorn, 1997; Snedden et al., this volume) show the ridge lithofacies dominated by sand. Where not wholly deformed by the coring process, these sands are either dominated by low-angle laminae, cross-bedding, or bioturbation. Subordinate amounts of mud and silt are present, (e.g., Snedden et al., 1994, their Fig. 14) but as individual beds within distinct units, not distributed throughout the sand bodies (cf., Rine et al., 1991). These may reflect pre-existing deposits upon which the shoals nucleated, (cf., Stubblefield et al., 1984, Penland et al., 1988; Rodriguez et al., this volume; Snedden and Dalrymple, this volume). Within the vibracore examples cited, significant amounts of sigmoidal cross-stratification, mud drapes, double mud drapes, elongate rip-up clasts, current ripples, or rhythmites that characterize the Shannon lithofacies as shown in core photographs are notably absent. The ichnofacies of Quaternary shelf sands have not been studied to any great extent, but shallow-marine trace fossils do occur in the vibracores, for example, the large *Ophiomorpha* burrows in Ship Shoal (Penland et al., 1986, their Fig. 12). As noted previously, a brackish trace fossil suite does not necessarily imply an estuarine or tidal setting. A southerly directed coastal current (cf., Slingerland et al., this volume) might create freshwater stress in an otherwise marine environment. Thus, a deposit like Ship Shoal, ostensibly in a wholly marine setting, might be expected to show a brackish trace fossil suite because it is subjected to considerable freshwater stress from the Mississippi River. This represents an area of possible future research, and tends to support the contention that application of ichnology to facies interpretation remains an evolving field (cf. Swift and Parsons, this volume).

The lithofacies seen in vibracores from Quaternary shelf sands are distinctly different from those of the Shannon Sandstone, either in outcrop or subsurface data. If our criteria for application of modern analogs are valid, and supposing our relatively limited sample of modern shoal sands is representative, we can rule out these sands in their current form as direct lithologic analogs for the Shannon. This does not preclude formation of the Shannon Sandstone by submergence and reworking of pre-existing shoreline features on a continental shelf, but does imply a final process environment different from than that of the storm-dominated sand bodies detailed above.

The basal contacts of modern ridges are variable. Snedden and Dalrymple (this volume) state that all modern ridges overlie a ravinement surface. Parts of the sand bodies lie on a wave ravinement surface (the "flooding surface-sequence boundary"), or the diastem formed by an associated erosional trough, while others are gradational. Penland et al. (1988) proposed that the Mississippi delta shoals undergo variable amounts of reworking and landward migration during transgression, and preserve portions of their origin as barrier shoreline deposits. Thus, the shelf sand facies per se will have an erosional lower boundary, while the basal contact of the ridge itself may be gradational. Most upper contacts are erosional, while some are beginning to be mantled by muds.

The directional aspects of the Shannon are compatible with the postulated storm-generated shelf sand ridge interpretation (Fig. 5). Southerly directed, wind-driven geostrophic currents are consistent with wind-driven waves from the northeast, producing the observed wave ripples, and the oblique orientation of the cross-beds relative to the sand-body elongation is consistent with the postulated formative processes of sand ridges (Huthnance, 1982).

However, the directional aspects of Quaternary shelf sands are quite variable. Bathymetric map analyses, seismic profiling, and side-scan sonar imagery have shown that the shoals are migratory. Side-scan sonar images show that bedforms migrate obliquely along the Atlantic ridges (e.g., Swift et al., 1986). Ship Shoal has moved landward more than 1.5 km in the last 100 years (Penland et al., 1988), and seismic profiles show low angle landward-dipping surfaces within the shoal. Seismic reflection profiles have shown seaward-dipping reflections, landward-dipping reflections, as well as channeling and other features within various units (e.g., Rodriguez et al., this volume). Shoreface-attached ridges are known to migrate along shore (McBride and Moslow, 1991; Dalrymple and Hoogendoorn, 1997; Snedden et al., this volume), and thus should produce accretion sets that dip subparallel to the coast, but at a high angle to the ridge body itself.

It is unlikely, however, that the linear sand bodies of the Shannon represent shoreface-attached ridges derived from ebb-tidal deltas (cf., Snedden et al., this volume). The direction of wave approach indicated by the orientation of the wave ripple crests implies that any net longshore drift and inlet migration would be to the south, which, combined with a westerly transgression, would generate northeast-southwest trending ridges (Fig. 6). To reach the present northwest-southeast trend, the sand bodies would require a subsequent rotation of nearly 90°E.

Incised Valleys

The leading proponents for incised valley interpretations for the "isolated marine" sand bodies have come from Exxon Production Research (Van Wagoner et al., 1990; Jennette and Jones, 1995; Sullivan et al., 1995, 1997). These interpretations

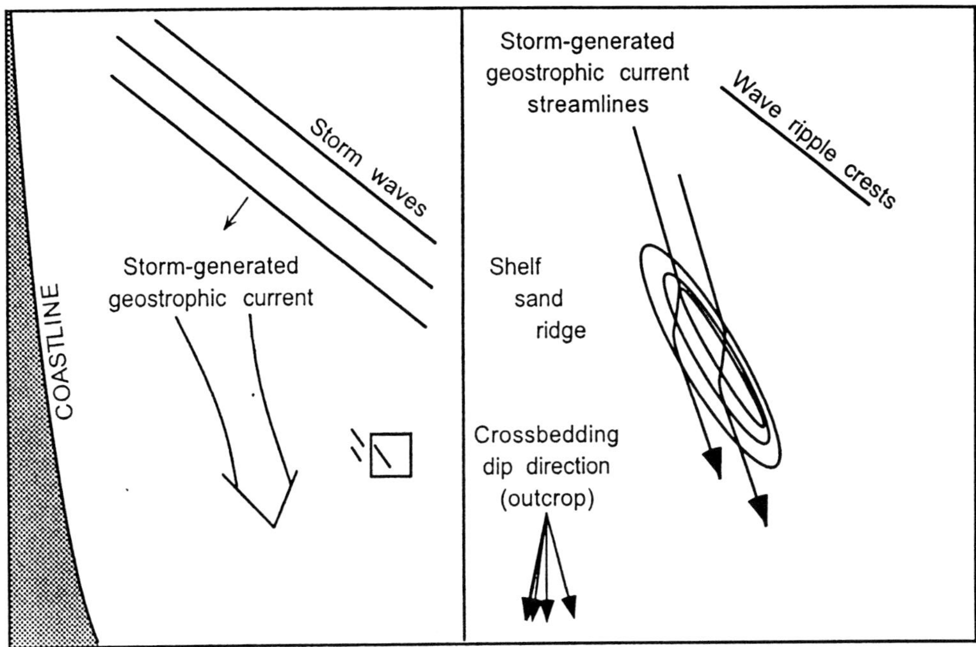

Fig. 5.—Inferred paleogeographic setting of the Shannon sandstone if the sand bodies represent shelf sand ridges as postulated by Swift and Parsons (this volume). Flow streamlines across ridge as postulated by Gaynor and Swift (1988, Fig. 11)

have been largely based on outcrop geology and subsurface well-log correlations, typically without direct reference to modern analogs. Well-studied incised valleys on Quaternary continental shelves are largely found along passive margins (e.g., Suter et al., 1987; Thomas and Anderson, 1994; Foyle and Oertel, 1992). These may not represent good paleogeographic analogs to Campanian incised valleys in a foreland basin setting, but if similar processes are involved in their incision and filling (i.e., sea level fall and rise, fluvial erosion, and marine transgression) we can reasonably expect similar responses.

There is still considerable debate about the origin and evolution of Quaternary incised valleys, both onshore and on the continental shelf. Following is a brief summary of the current consensus of the characteristics of these deposits:

1. They initiate by fluvial erosion under periods of falling base level
2. They are modified by fluvial, tidal, and marine processes under rising base level conditions;
3. Fill within the systems is complex, and is controlled by base level fluctuations, sediment supply, fluvial discharge and caliber, gradient, tidal range, to name a few processes. Documented valley fill types include marine, estuarine (wave- and tide-dominated), fluvial, and deltaic facies.

Dalrymple et al. (1992b) and Zaitlin et al. (1994) have constructed facies models for estuaries formed in late Quaternary incised valleys. These models incorporate data from modern highstand estuaries, as well as their submerged counterparts on the continental shelves in various parts of the world. Thomas and Anderson (1994) provide an excellent example of the distribution of lithofacies in a transgressed incised valley system on the continental shelf.

Does the large-scale lithofacies architecture of the Shannon Sandstone match the organization suggested for incised valley fills by the estuarine facies models? Insufficient attention has been paid to the distribution of fine-grained facies within the Shannon system to make definitive comparisons with postulated modern counterparts. The fine-grained facies within the Shannon are distinct from the underlying Cody Shale, but have typically only been considered where they are intimately associated with the "ridges." Sullivan et al. (1997) describe the facies overlying the Copenhagen Blue sequence boundary at Hartzog Draw as "estuarine" on the basis of lithofacies and ichnology. They do not specify which type of estuarine facies they identified, or provide maps to evaluate the paleogeographic distribution. Generally speaking, the fine-grained facies of the Shannon Sandstone do not seem to reflect an obvious wave-dominated estuarine organization. There is even less evidence for fine-grained tidal estuarine facies. No one has identified tidal flat, salt marsh, or tidal-fluvial deposits within the Shannon sandstone, and no obvious examples of these are apparent in the published core photographs or outcrop examples.

Elliot (1995) interpreted the upper parasequence at Castle Rock in the Salt Creek Anticline area as an "estuary mouth shoal," containing both wave-formed structures and cross-bedding resulting from southerly directed ebb tidal currents. Such an estuary would necessarily have been oriented approximately north-south, opening southward. Estuary mouth shoals are termed "sand plugs" in the Zaitlin et al. and Dalrymple et al. models. The forms vary, but for a mixed energy estuary as implied above, the "estuary mouth shoal" would likely take the form of an ebb tidal delta, part of a larger shoreline complex. Upon transgression, such a deposit might form a dip-oriented sand body confined to the trend of the ancestral estuary or incised valley. Such deposits have been interpreted on the Atlantic continental shelf [the shoal retreat massif of Swift (1973)]. However, there are several difficulties with this scenario. The shoreline system would necessarily have been oriented approximately east-west. No Shannon investigator has reported east-west oriented shoreline facies. To date, no tidal inlet or channel facies have been identified

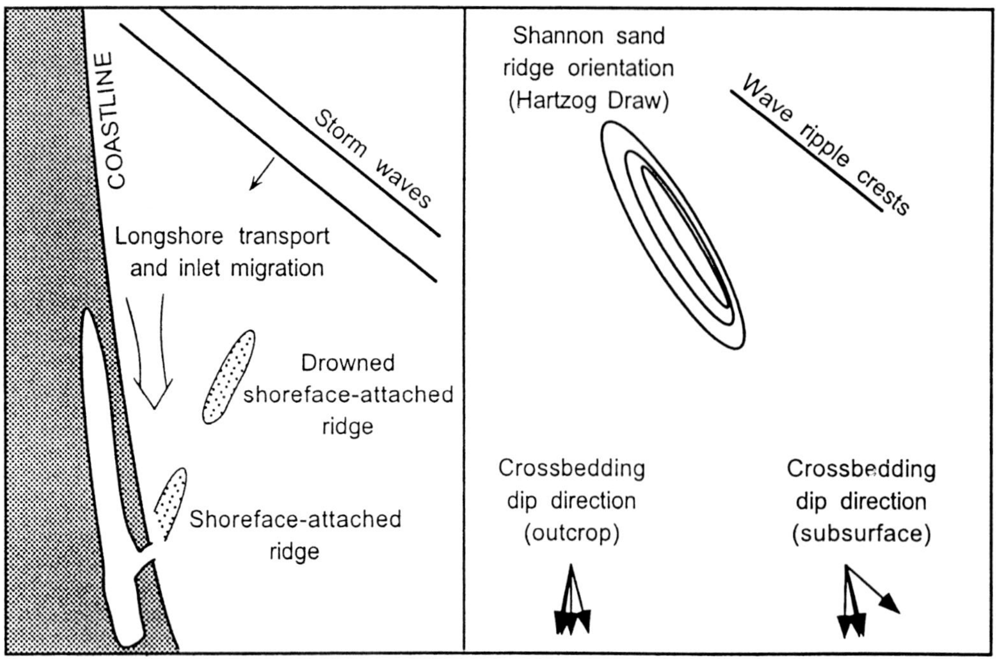

Fig. 6.—Development of shoreface attached ridges as a result of inlet migration in the paleogeographic setting inferred for the Shannon sandstone. The orientation of the shoreface-attached ridges does not correspond to the orientation of the Shannon sand ridges.

within the Shannon. These are very significant in modern mixed energy estuaries, and would have the highest preservation potential in a system such as envisioned above. Wave-ripple trends within the Shannon (Spearing, 1976; Walker and Bergman, 1993) imply wave approach from the northeast. For the estuary mouth shoal model above, such waves would be approaching from the estuarine or onshore direction, and thus are unlikely to generate significant energy. Presumably, any waves that did approach from the seaward side would set up longshore currents, oriented at high angles to the tidal currents. This is at odds with dominantly southerly directional features in the Shannon. Although the evidence is mostly negative, we conclude the distribution of candidate sand plug facies does not match that expected for a north-south oriented mixed energy estuary.

Bay head deltas, a prominent component of modern mixed energy estuaries (Dalrymple et al., 1992b; Zaitlin et al., 1994), have not been explicitly mentioned by any of the Shannon workers, although the concept is implicit in the interpretation of Sullivan et al. (1997). Bayhead delta progradation from a northerly fluvial source would produce a deposit with largely south-oriented paleocurrents, matching the paleocurrents of the Shannon. Such a deposit might well be tide-dominated, producing a series of tidal ridges such as seen in the mixed-energy Gironde estuary (Allen, 1991). However, tidal sand ridges in the Gironde comprise coarsening-upward successions 6 m -7 m thick with complex sediment transport patterns. Cross-beds are sigmoidally shaped, but clay drapes are rarely preserved (Allen, 1991). The Gironde also contains extensive tidal flats and salt marshes, central basin facies, and a well-developed sand plug, all of which are apparently lacking in the Shannon.

Sullivan et al. (1997) interpret the Shannon as a "tide-dominated delta in which thickness trends result from differential erosion in a southeast-trending incised valley." The delta is inferred (Sullivan et al., 1995, 1997) to have prograded to the southeast, parallel to the Hartzog Draw field. The direction of progradation (Fig. 7B) is reasonably close to the direction of cross-bedding dips in the subsurface as measured by Hearn et al. (1984), although it is not clear if the measured cross-beds were in the prograding unit ("Crimson Red"). One problem with the direction of progradation implied by Sullivan et al. (1995, 1997) is that it parallels the long axes of the inferred sand bars. Studies of modern systems show, to the contrary, that tidal sand bars typically accrete laterally rather than longitudinally (Dalrymple, 1992). Such accretion should produce clinoforms highly oblique, rather than parallel, to the tidal flow.

The orientation of the Hartzog Draw field is attributed to valley incision associated with the "Canary Yellow" sequence boundary (Sullivan et al., 1997). These authors do not speculate as to whether the outcropping Shannon is also part of the "Canary Yellow" sequence, but the cross-bedding in the outcrop is distinctly oblique to the inferred erosional trends implied by Hartzog Draw (Fig. 7C). The orientation and distribution of wave ripples in outcrop also seems inconsistent with the "narrow southeast-trending incised valleys" postulated for the "Canary Yellow" interval (Sullivan et al., 1997), in that the generating waves would be required to approach across the short axis of the valley (Fig. 7C). Such waves seemingly have insufficient fetch to generate ripples at depths of at least 15 m (Spearing, 1976, Fig. 8). Either the valleys were much broader (Fig. 7D), or the outcropping Shannon is unrelated to the sequences defined in the subsurface.

The overall paleogeographic setting implied by Sullivan et al. (1997) warrants review. The corresponding highstand shorelines for the Shannon Sandstone are oriented roughly north-south (Fig. 1). Fluvial systems feeding such shorelines would generally have an east-west orientation, even allowing for structural complications. Incision of valleys by these rivers

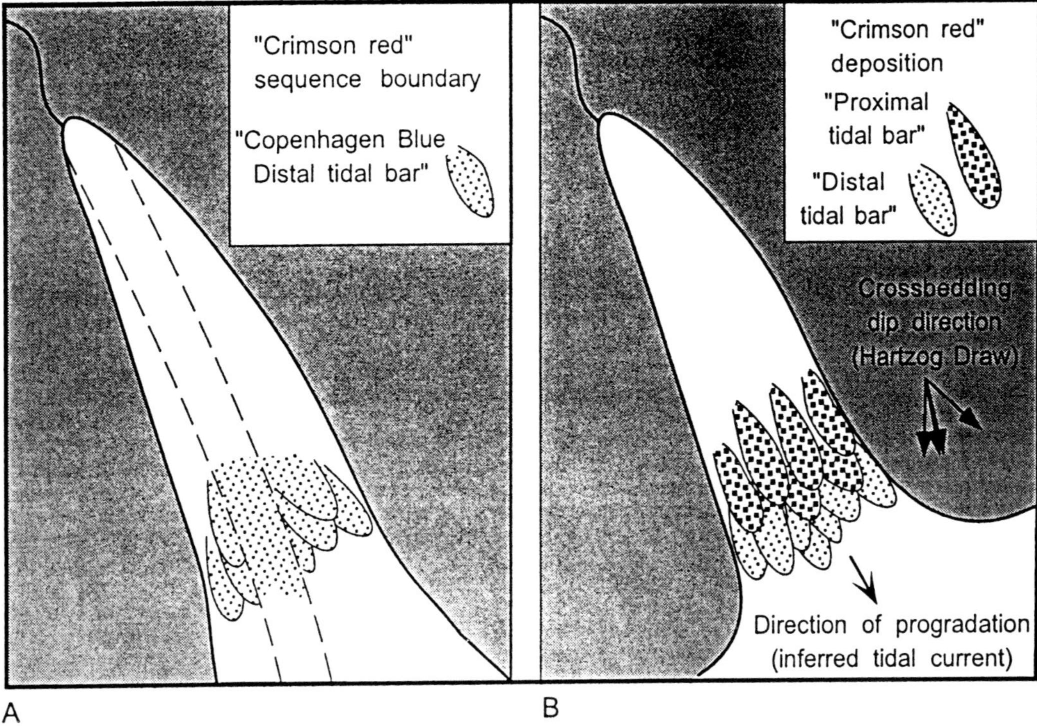

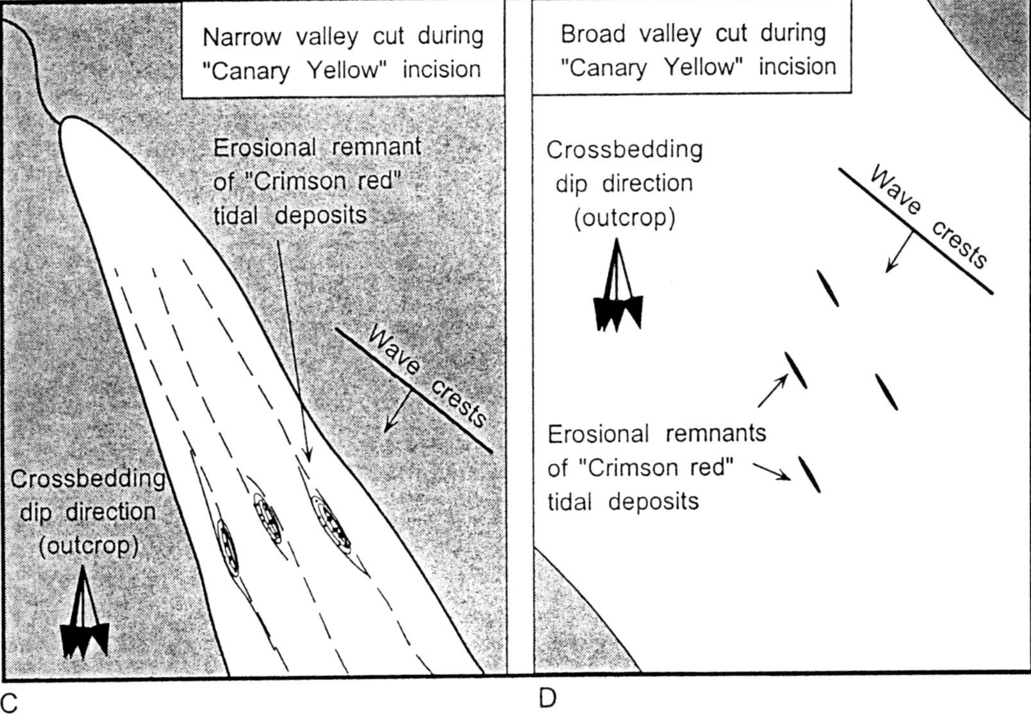

Fig. 7.—Inferred paleogeographic setting of the Shannon sandstone if the sand bodies represent erosional remnants of tidal bar complexes within an incised-valley system as postulated by Sullivan et al. (1997). A) "Copenhagen Blue" distal tidal bars are dissected by "Crimson Red" sequence boundary. B) Progradation of "Crimson Red" tidal bars. Subsurface cross-bedding dips slightly obliquely to inferred direction of progradation. C) Narrow valley incision ("Canary Yellow"). Observed wave ripples in outcrop require waves that cross the short axis of the valley and outcrop cross-bedding is oblique to the inferred direction of valley incision. D) Broad valley incision ("Canary Yellow"). Problem imposed by wave ripples is largely obviated, but cross-bedding in outcrop remains oblique to the orientation of the inferred erosional remnants.

during falling base level would necessarily initially follow the original orientation of the ancestral fluvial systems, before following the topographic gradient on the exposed shelf. Presumably such incision would leave a stratigraphic record, in the form of an incised valley network connecting the highstand stream courses to their lowstand counterparts. To date, none of the Shannon workers have found any fluvial deposits in the system, either in the outcrop or subsurface. No east-west oriented shoreline deposits of Campanian age have been interpreted in the area of the Shannon Sandstone. Also missing are shoreline features that would reflect rotation of the shoreline from east-west during the lowstand to north-south during the highstand. R. Fitzsimmons (pers. comm., 1999) has interpreted incised valley deposits of presumably similar ages between the Shannon outcrops and the Wyoming thrust belt. These deposits are also oriented north-south, and are incising into shoreline deposits that prograded east-west. At least one of the following criteria must be met to satisfy these paleogeographic constraints:

1. All evidence of connecting tributary systems and/or falling stage and transgressive system tract shorelines must be completely removed by either ravinement or by subsequent subaerial erosion.
2. The evidence for fluvial incision and shoreline rotation has simply been missed in the previous detailed studies.
3. The Shannon outcrop and subsurface deposits are far enough from the "coeval"shorelines to the west to be out of the zone of transverse drainages and re-oriented shorelines.
4. The outcrop and subsurface database is insufficient for the task.
5. The presumed paleogeography is wrong.

Fluvial drainages in modern thrust belts and foreland basins shed some light on this issue. As illustrated in Figure 8, fluvial systems in modern forelands follow a variety of orientations relative to their thrust belts. The incised valley orientation for the Shannon Sandstone interpreted by Sullivan et al. (1995, 1997) is thus plausible, without even requiring changes in drainage orientation from highstand to lowstand. The Fly River of Papua, New Guinea (Fig. 8C) provides an interesting paleogeographic analog. The highstand Fly River forms a tide-dominated delta in the Gulf of Papua in an axial orientation to the mountain front. Figure 9 shows the map of Papua, New Guinea inverted and reoriented, superimposed on the present day geography of the Western Interior. The orientations and locations of the thrust front, shorelines, fluvial systems, and tidally dominated deltaic facies are similar to those required for the incised valley interpretation of the Shannon sandstone. Thus, the Shannon sandstone might originate by deposition during lowered sea level in a valley incised from an original, axial highstand position. This analog solves the paleogeographic conundrum of the missing incised valley network and the need for reorienting the shoreline at various sea-level positions.

One common aspect of the Quaternary incised valleys that does match the Exxon Hartzog Draw interpretation is the multiple sequence architecture. Mapping from seismic reflection profiles of Quaternary incised valleys (e.g., Suter, 1986; Suter et al., 1987; Thomas and Anderson, 1994; Foyle and Oertel, 1992), as well as onshore drilling (Blum, 1994) clearly shows the re-occupation of drainages and re-incision during multiple downcutting events (Fig. 10). These events result from a number of causes, including base-level falls (sequence boundaries), climatic fluctuations, tidal erosion on transgression (ravinement), as well as autocyclic fluvial and estuarine geomorphologic processes.

Lowstand Shorelines

Living as we do at a glacio-eustatic highstand, it should be self-evident that there are few direct Holocene analogs for "lowstand shorelines." This may help to explain the virtual absence of such in the interpretations of the Shannon that invoke lowstand shorelines and forced regressions. Such information as is available for Quaternary examples comes from high resolution seismic profiling of submerged deposits along the continental margin, supplemented by limited drilling data. Analogs here are largely geometric and paleogeographic. Because of the greater water depth, these data are still more limited than those for Quaternary shelf sands. What core information has been collected shows that lithofacies and sedimentary structures match those expected for the depositional environment (e.g., Sydow and Roberts, 1994).

The literature is replete with studies of Quaternary wave-dominated shoreline/shoreface systems (e.g., Bernard et al., 1959, 1970; Curray et al., 1967; Clifton et al., 1971; Short, 1984). The thicknesses of such deposits are quite variable, depending upon wave regime, but it is easily possible to have a wave-dominated shoreline succession as thick as an individual Shannon sand body. The general lithofacies, sedimentary structures, and ichnology of Quaternary wave-dominated shoreline/shoreface systems are well known: an overall coarsening upward succession of wave formed sedimentary structures with variable amounts of bioturbation. Notably absent are significant amounts of mud drapes, double mud drapes, elongate rip-up clasts, current ripples, tidal bundles, or rhythmites. We are unaware of any Holocene wave-dominated shoreline/shoreface system that produces the types of sedimentary structures shown in the Hartzog Draw cores.

Bergman and Walker (Fig. 2, this volume) define their shoreline trends parallel to the long axis of the linear sand bodies. In this case, the wave ripples would imply waves approaching the shore obliquely from the northeast (Fig. 11). The orientation of the cross-bedding is difficult to explain in this context, however. Modern shorefaces are complex, dynamic environments. Currents in such environments are hardly monotonous or unidirectional. Dominant current directions in these deposits are oriented longshore, often showing two directions of transport caused by seasonal climatic changes, or variations in wave refraction patterns around shoreline features. In areas of tidal influence, bedforms are oriented normal or oblique to the shoreline orientation. Rip currents produce offshore orientations and have a fairly high preservation potential. We know, however, of no shoreface mechanism for generating cross-strata that is directed so consistently obliquely landward.

The Quaternary examples from which the general shoreline/shoreface facies model has been established are all transgressive to highstand systems tract deposits. These should not be expected to show the development of a forced regressive surface or candidate sequence boundary, either within the deposit or at its base. Geometric analogs for forced regression may be drawn from high-resolution seismic data of Pleistocene deposits on the outer continental shelf (e.g., Posamentier et al., 1992; Berne et al., 1998; Rabineau et al., 1998). Shoreline systems from areas of glacial rebound, such as FennoScandia (e.g., Martinsen et al., 1995) and Canada (e.g., Hart and Long, 1996) provide some insight into processes.

A partial Quaternary lithofacies and process analog to a Shannon forced regressive shoreline may be found along

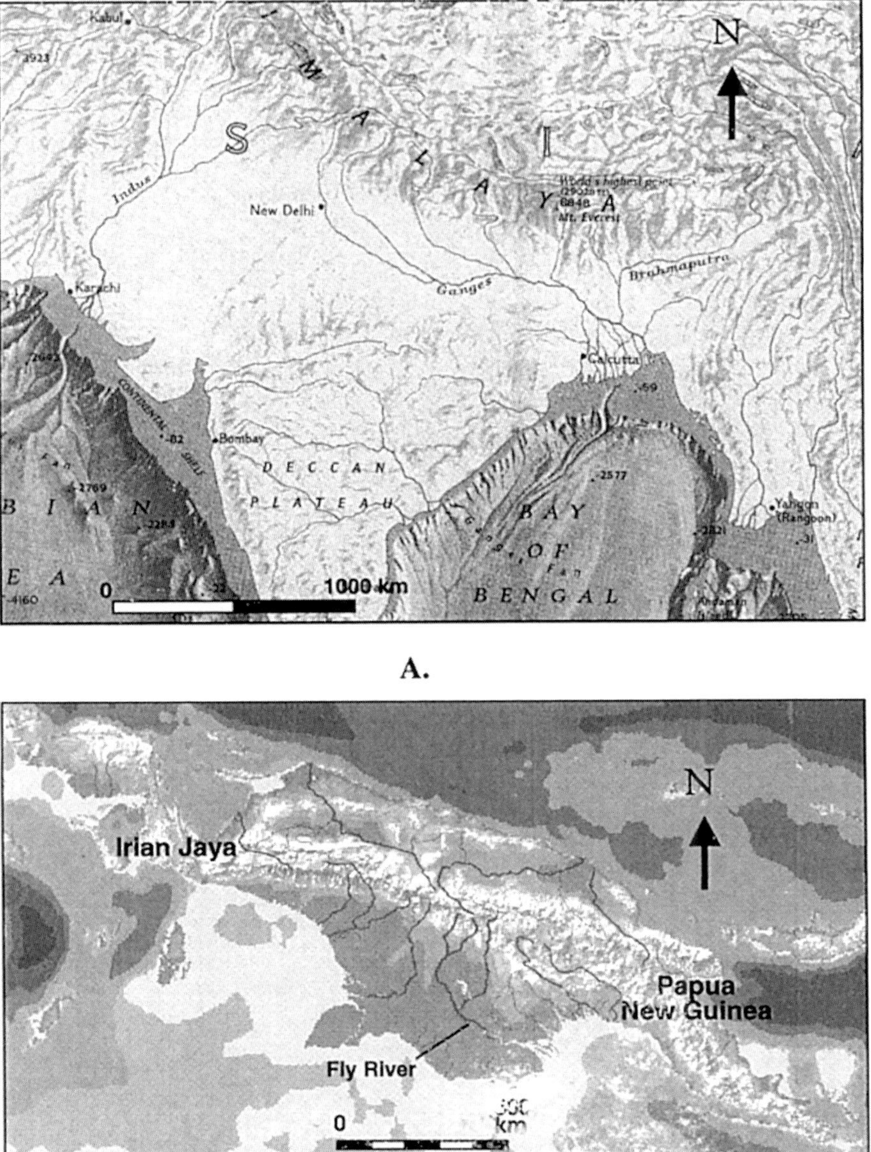

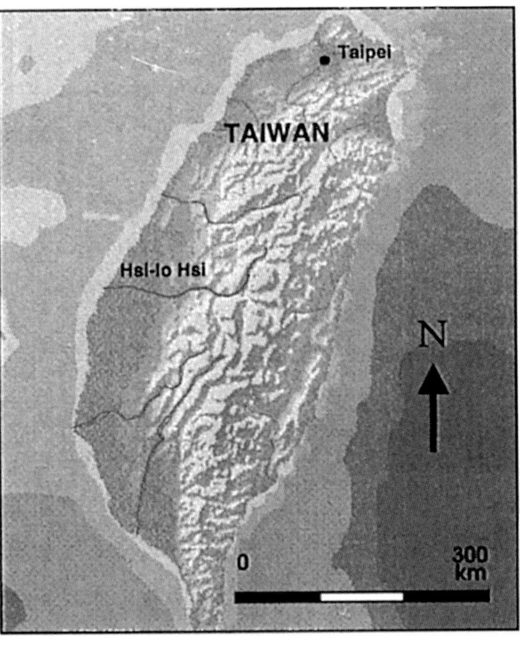

Fig. 8.—Fluvial systems in modern thrust and foreland settings show a variety of orientations relative to their mountain fronts. A) The Indus and Ganges-Brahmaputra rivers. B) Hsi-lo- Hsi in Taiwan, and C) the Fly River in Papua, New Guinea.

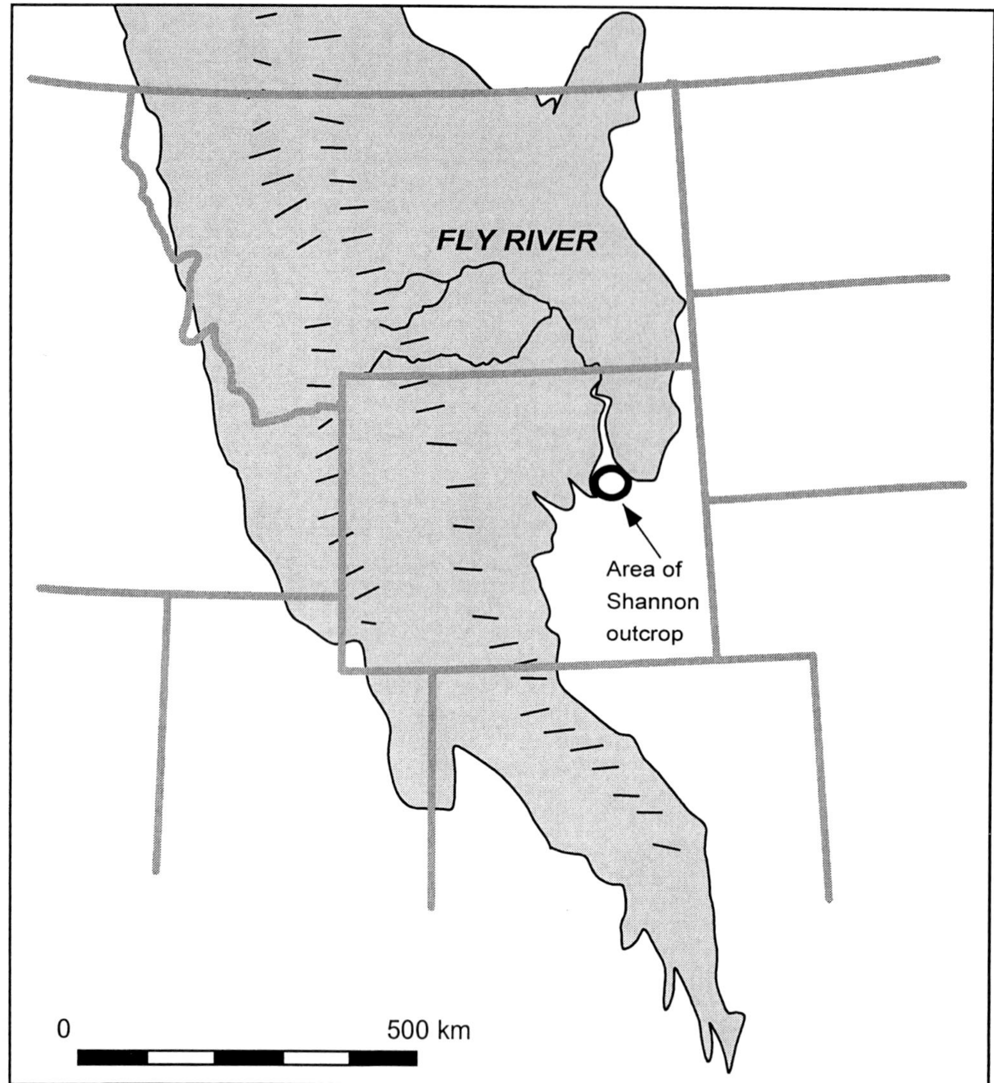

Fig. 9.—Superimposition of the Holocene Fly River of Papua, New Guinea, onto the presumed paleogeography of the thrust belt and foreland basin of the Cretaceous Western Interior. From its headwaters in the thrust belt, the Fly flows at a high angle across its coastal plain, before turning sharply to form an axial tide-dominated delta. A similar type of behavior would be required of streams draining the presumably corresponding Shannon highstand shorelines if any were to form north-south oriented incised valleys.

the eastern coast of Brazil (e.g., Dominguez et al., 1987, 1991, and 1992). Here falling relative sea level since the maximum Holocene transgression has resulted in the progradation of a series of beach ridge plains. Unfortunately, the published literature on these deposits is largely geomorphologic. Dominguez and Wanless (1991) present a schematic vertical succession for these deposits based on limited vibracores and river exposures (Dominguez and Wanless, 1991, their Fig. 8). This highly schematic succession shows an erosional surface between upper shoreface and lower deposits reminiscent of that advocated as a forced regression surface in the lower parasequence of the Shannon outcrop at Castle Rock (Bergman and Walker, this volume). However, the lithofacies and sedimentary structures of the Brazilian example differ markedly from those shown in the Hartzog Draw and other core studies of the subsurface Shannon.

Aspects of the stratigraphic architecture of Quaternary shorelines lend some support to the compartmentalized architecture of the Shannon envisioned by Bergman and Walker (this volume). Numerous examples, including the Texas and Brazilian coasts cited above (Fig. 12), the east coast of the United States, as well as the eastern coast of Australia (Roy et al., 1994; Roy and Boyd, 1996) display sets of highstand shorelines deposited at different sea level stages. These composite shoreline deposits reflect a history of glacio-eustatic fluctuations, and transition from regressive to transgressive regimes, similar to that advocated by Bergman and Walker (this volume). However, the associated deposits, such as incised valleys and continental shelf sands, show a more complicated stratigraphic picture than that envisioned for an amalgamated Shannon shoreline. Berne et al. (1998) and Rabineau et al. (1998) studied transgressed, presumably lowstand shorelines on the outer continental shelf in the Gulf

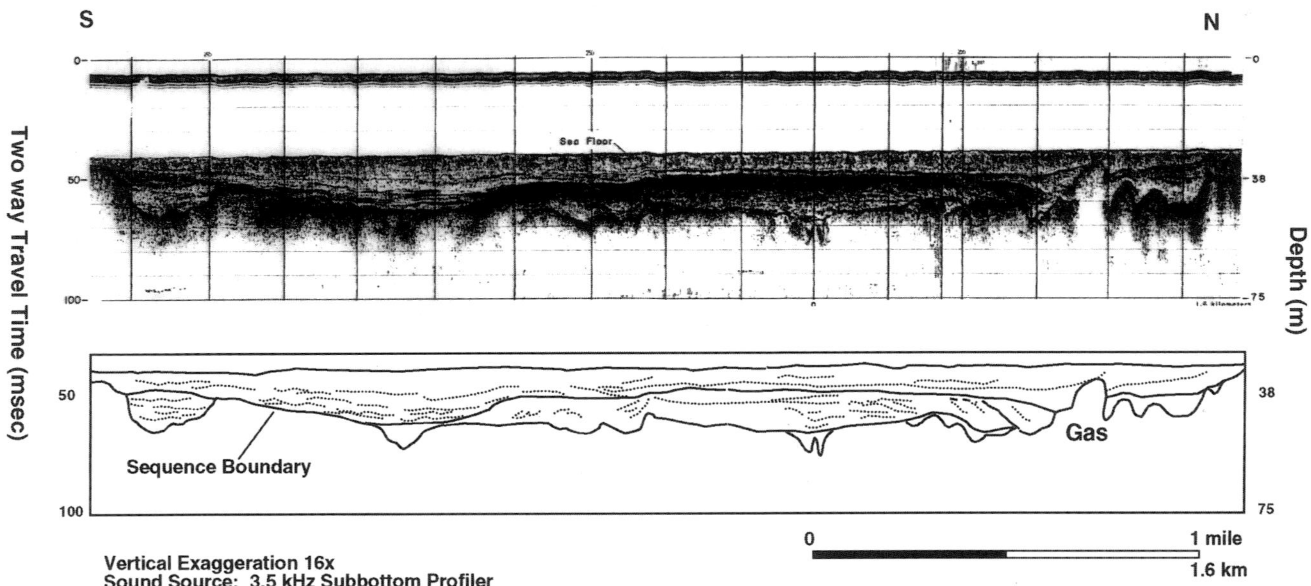

Fig. 10.—High resolution seismic profile and interpretive line drawing of a late Quaternary incised valley on the Louisiana continental shelf, Gulf of Mexico, USA. This profile illustrates the multi-story architecture common to Quaternary incised valleys on the continental shelf (after Suter, 1986).

of Lion, portions of which they suggested as possible analogs for sharp-based shoreface deposits such as the Shannon. Seismic lines through these features show a complex stratigraphic architecture, including lowstand shorefaces, delta lobes, sand waves, incised valleys, and submarine canyons, of various scales, orientations, and interpreted origins.

Again we note the importance of the type and distribution of fine-grained facies. A seeming requirement of fluctuating shorelines and rising base level would be the formation of back barrier basins with consequent tidal exchange, resulting in tidal inlets, tidal deltas, and associated central basin and bayhead delta facies. Even in microtidal settings, tidal inlets are frequently eroded well beneath the depth of shoreface erosion, and in fact have the highest preservation potential in barrier island systems. These would inevitably produce paleocurrents oriented at a high angle to the shoreline trend.

Tidal Sand Ridges

Sullivan et al. (1995, 1997) have interpreted the Shannon as comprising tidal bars deposited in a tide-dominated deltas within multiple incised valleys. Modern examples of such deposits form as bayhead deltas at the upper end of a mixed-energy estuary (e.g., the Gironde). Other modern tide-dominated deltas form in open embayments (e.g., the Fly River in Papua, New Guinea or the Colorado delta in the Sea of Cortez). These might represent variants of the "open bay" concept proposed by Tillman (this volume). Formation of the linear sand bodies in a large open bay, where storm waves from the northeast could prevail, would be consistent with the directional data in the Shannon Sandstone.

Most descriptions of modern tide-dominated deltas are largely geomorphologic. Some authors even dispute the existence of these features, believing them all to be tidally dominated estuarine deposits (Walker and James, 1992). Irrespective of this debate, the major framework facies in these environments are dip-oriented, elongate sand ridges. These ridges grade basinward into the delta front and prodelta, which are typically finer-grained. The delta plain comprises various types of inter-tidal and supratidal flats and marshes. In general, progradation of such environments should produce a coarsening-upward succession, ultimately culminating in the extension of distributary channels into the sand ridge field. Sharp-based bars may occur in distributary channels, as is the case for the tide-influenced Mahakam delta (Allen et al., 1979). Paleocurrents in this setting are likely to show both ebb and flood orientations, although one direction may predominate.

There are relatively limited lithofacies data available, although bathymetric mapping, seismic profiling, surface sampling, and some core data exist for various examples. The vertical facies successions that have been put forward so far are hypothetical (e.g., Dalrymple, 1992), and are based on composite stratigraphic columns (Coleman and Wright, 1975) and cores from related environments (Meckel, 1975; Allen et al., 1979). Meckel's (1975) work on the Colorado delta in the Sea of Cortez shows core photographs and lithofacies descriptions of tidal bars, which are actually taken from buried, presumably analogous deposits near the more wave-dominated portion of the embayment. Since they are not taken directly from the active tidal ridges, these are already one interpretive step from a modern analog. Some vibracore information is available for the muddy Fly Delta (Fig. 8C). Harris et al. (1996) show photos of a few vibracore examples from various facies. Dalrymple (1997) reported current-parallel, elongate bars comprising rhythmites of fine, rippled sands and massive muds, as well as elongate distributary mouth bars of clean sand characterized by HCS, cross-bedding, and an overall fining-upward succession.

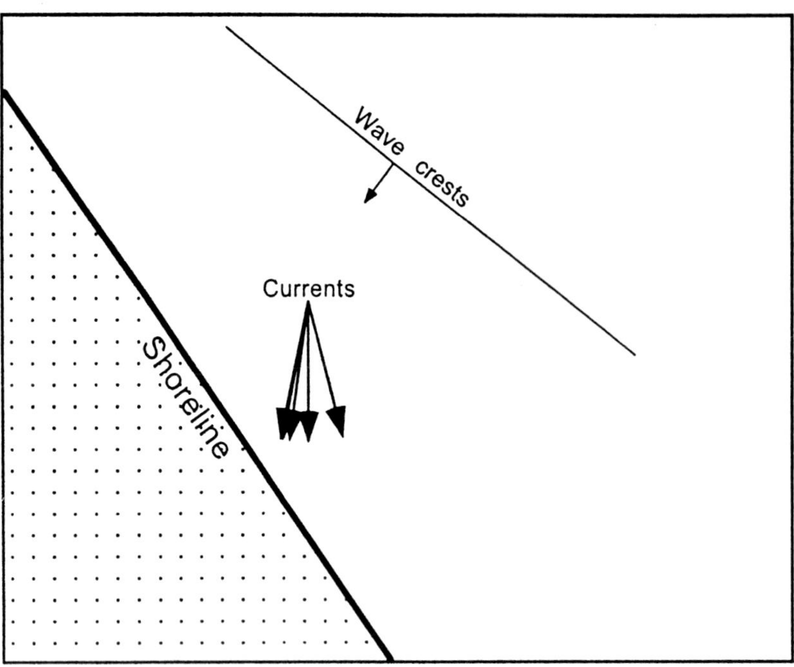

Fig. 11.—Inferred geometric relationships for the forced regressive shoreline model (Bergman and Walker, this volume). Obliquely shoreward currents are inferred from cross-bedding dip direction in outcrop.

A variant on the open-bay or tide-dominated delta and shelf sand ridge models would be formation of the Shannon as tidal ridges on a continental shelf. Snedden and Dalrymple (this volume) made no fundamental distinction between storm-built and tidally formed shelf ridges. However, Quaternary examples have different characteristics. Shelf sands in storm-dominated settings are oriented subparallel to the overall shoreline trend. Tidal sand ridges are oriented subparallel to the prevailing tidal currents, which may have a variety of orientations to the shoreline depending on the configuration of the basin. Tidal sand ridges occur in a number of places in the world, including the North Sea (Stride et al., 1982; Belderson et al., 1982), Georges Bank (Twichell, 1983), and the East China Sea (Yang and Sun, 1988). Their corresponding shorelines comprise a variety of types, including wave-dominated, mixed-energy, and macrotidal. Additionally, tidal sand ridges can be considerably larger than their storm-dominated counterparts, reaching thicknesses of 40 m, lengths of 100 km, and widths of as much as 10 km. These thicknesses are a closer match to those of individual Shannon sand bodies than those of Quaternary storm-dominated shelf sands.

Yang (1989) and Yang and Sun (1988) interpret the East China Sea ridges to represent the retreat path of the Changjiang River Delta during the Holocene transgression (Fig. 13A). Individual ridges are from 10 km to 60 km long, 2-5 km wide, and up to 20 m in height. High-resolution seismic lines document a complex internal stratigraphy and multi-stage history (Fig. 13B). Packages of clinoforms dip obliquely to the long axis of the ridges, and parallel to their steeper sides at angles of about 2°. Individual packages are about 1 km in width and about 10 m thick. Note that these are similar in configuration and orientation to the sand body trend as the correlation units of Bergman and Walker (this volume). The direction of migration is different relative to the sand body than that in the model of Sullivan et al. (1997). No paleocurrent data are available from the ridges, but they are oriented subparallel to the tidal currents in the East China Sea (Yang and Sun, 1988).

Fundamentally little is known of the internal lithofacies of Quaternary tidal sand ridges. Grain-size sampling and analyses, side-scan sonar images, and seismic profiling have shown some of the surficial bedforms and internal structures of Quaternary tidal sand ridges. Hypothetical facies models have been proposed (Stride et al., 1982), but data are still lacking to determine their applicability to the ancient record. Davis and Balson (1992) drilled a 35 m thick, moribund tidal sand ridge in the East Bank complex of the North Sea. Recovery was only about 16% of the total thickness, distributed throughout the ridge. The overall deposit comprised two coarsening upward successions of well-sorted fine sand, with bioturbation limited to the top meter of the upper unit. Limited information was obtained regarding sedimentary structures. Berné et al. (1994) found the Middelkerke bank of the North Sea to be a complex assemblage of depositional units, consisting of erosional remnants, and the combination of tidal and storm deposits. Similarly, the Banc de Kaiser-I-Hind sand ridge in the Celtic Sea appears to comprise lowstand estuarine-deltaic deposits, reworked by tidal and/or wave processes during and after transgression, and may even represent an amalgamation of different sequences (Berné et al., 1998). Many of the expected characteristics of subtidal facies, such as might be present on a tidal sand ridge, are based on the studies of the Oosterschelde sand pits in the North Sea (e.g., Nio and Yang, 1991). The large scale sigmoidal cross beds there are actually interpreted as tidal channel deposits, not ridge or shoal facies.

The Efficacy of Modern Analogs

Collectively, geologists have varying degrees of confidence in the applications of facies models that have been built

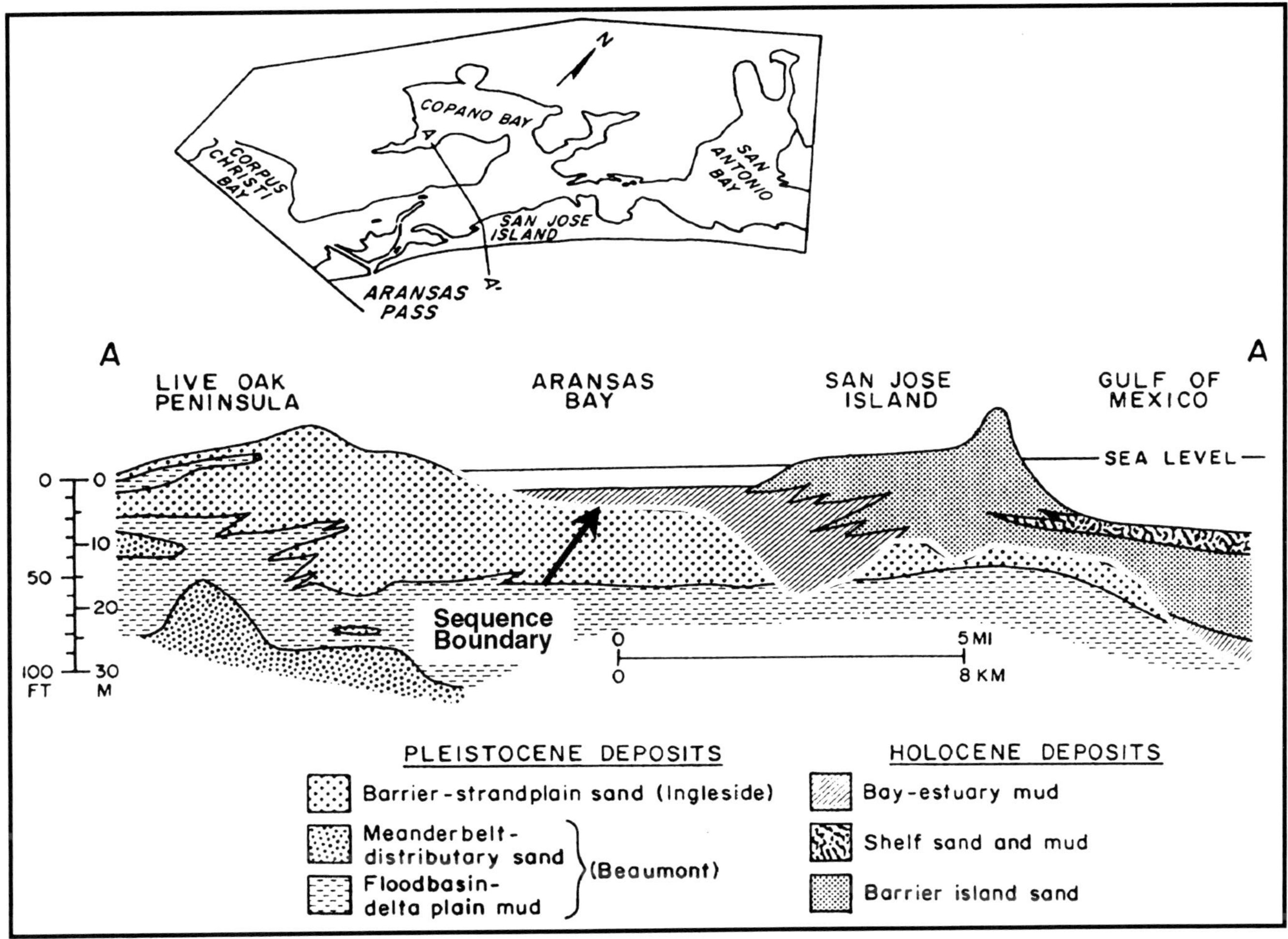

Fig. 12.—Stratigraphic cross-section on the Texas coast, Gulf of Mexico, USA, illustrating the multiple sequence architecture of wave-dominated highstand shorelines in the Quaternary (after Morton, 1994). The sequence boundary separating the Pleistocene and Holocene shorelines is a composite surface that combines multiple sea-level falls and wave and bay ravinement diastems. The Pleistocene shoreline may itself be a composite of multiple sequences. Note the preservation of associated estuarine and coastal plain environments.

from modern environments. Particularly successful examples would be the fining-upward point bar succession (Allen, 1965; Bernard et al., 1959, 1970), and the coarsening-upward shoreface model (e.g., Clifton et al., 1971). Deltaic facies models have a long history, although the greater complexity and variability of deltaic systems makes the modeling approach less straightforward. Estuarine facies models are a recent addition to this roster. Typically we can say that we have a reasonably good handle on processes, geometries, and lithofacies of the more accessible modern depositional environments. Deposits that are more difficult to study, because of water depth, energy levels, remote location or harshness of climate (deep sea fans and shelf margin deposits, for example) are less well known.

It is important to draw a distinction between specific, still-active Holocene and the more general Quaternary analogs. There are a number of advantages to the use of Holocene depositional analogs. The primary interpretation is usually safe. We can directly study the lithofacies and geomorphology, observe the processes, and document the response. Chronostratigraphic control is generally excellent. Finally, tectonic setting, regional trends, and associated facies relationships are unequivocal. Extending this control to the remainder of the Quaternary is more problematic, and our analogs become less direct, as there is usually, though not always, an interpretive step in the utilization of a Pleistocene deposit. The Quaternary provides us with different types of analogs, chiefly relating to relative sea-level changes, paleogeography, geometry and, to a lesser extent, lithofacies. We know that rapid, high-magnitude sea-level fluctuations and climatic changes have occurred. Regional relationships and tectonic settings are well constrained, but chronostratigraphy is less reliable than for the Holocene. Lithofacies information is more limited, and processes are usually inferred from the response rather than directly measurable. Pleistocene examples go part way toward indicating the likelihood and style of preservation.

The shortcomings of both Holocene and other Quaternary models include the fact that the Holocene is a glacio-eustatic highstand, a pause in the cycle of rapid, high-magnitude glacio-eustatic fluctuations that characterize the Quaternary and late Tertiary. Floral and faunal evolution have produced different conditions in the modern world than existed previously-this is particularly important for pre-Tertiary deposits. There are certainly deposits in the ancient record for which there may be no modern analog. Finally, we must be aware of the biggest pitfall in the use of modern analogs: preservation potential.

The time involved in the accumulation of a deposit is commonly overlooked in the application of modern analogs. Most Holocene environments represent a depositional interval of no more than a few hundred (e.g., Mississippi delta lobes) or a few thousand years. Preservation in their current form is often problematic. Surprising changes can occur as the result of transgression. For example, one of the largest sand bodies in the abandoned Mississippi River delta plain, a microtidal, presumably deltaic area, is a tidal inlet fill (Cat Island Pass; Suter and Penland, 1987). This points out that many deposits, particularly those in the critical zone of the coastal plain/continental shelf interface, commonly pass into the stratigraphic record in a modified form. Pleistocene accumulations may encompass tens or even hundreds of thousands of years. The advent of high-resolution seismic profiling and offshore vibracoring has improved access to early Holocene and Pleistocene deposits on the continental shelves. Together with the advent of a range of new dating techniques (e.g., amino acid racemization and thermoluminescence) that bridge the gap between ^{14}C and longer time-span radiometric methods (e.g., U-Pb), these have greatly improved understanding of the effects of sea-level fluctuations and the preservation potential of shoreline systems.

To cite another example, most Quaternary shelf sands, whatever their origin, are not in their final process environments. Simple transgression to below the depth of wave or tidal ravinement does not constitute preservation. Only burial can turn off the process clock. The Mississippi delta shoals (Penland et al., 1986, 1988, 1989; Pope et al., 1990) present a partial analog in this case: Trinity Shoal is currently being buried by the sediments of the newly active Atchafalaya delta, whereas its counterparts further east (e.g., Ship Shoal) continue to undergo marine reworking. Upon sea-level fall, it is much more likely that Trinity Shoal will be preserved in something like its current form than Ship Shoal. By contrast, in lower sediment supply areas, such as the Atlantic margin or the Mississippi-Alabama shelf (McBride et al., this volume), shelf sands are unlikely to be buried before there is a significant fall in base level, whereupon they will be exposed to shoreface dynamics and subsequent subaerial erosion.

Perhaps the best usage of modern analogs is as a reality check for a given interpretation. Modern analogs provide guidelines and can enhance or decrease confidence levels in a particular interpretation. Departure from a direct modern analog, be it depositional or paleogeographic, decreases the level of confidence in an interpretation of any deposit. However, the lack of a direct modern analog does not invalidate any given interpretation. Ancient depositional systems such as the Shannon typically reflect broader time intervals (one to several millions of years) than Quaternary deposits. Over longer time spans, the depositional regime may be repeatedly and substantially reorganized. Accumulations at the scale of modern and Pleistocene facies are likely to be totally or partly lost, compressed, or stacked in multiple ways with other facies. Any attempt to characterize ancient deposits solely in terms of observed successions in a Quaternary world may be ultimately misleading.

CONCLUSIONS

1. We were not charged in this paper to decide between the competing interpretations or suggest our own. Indeed, we cannot, as the result of our analysis, unequivocally disprove any of the interpretations proposed for the Shannon, and we suspect that a collective inability to do so convincingly is the fuel on which this controversy runs.

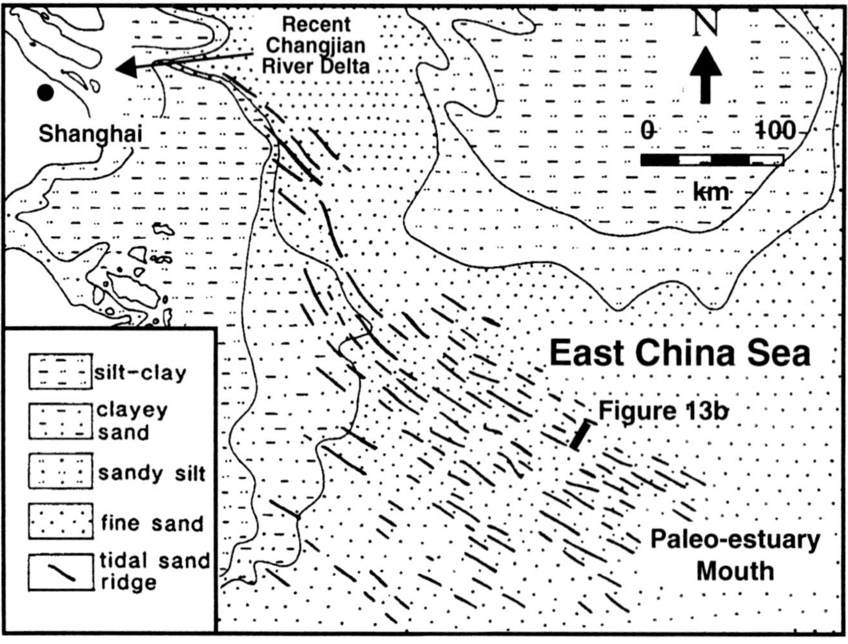

Fig. 13.—A) Location and lithology of East China Sea tidal sand ridges (after Yang, 1989). B) High-resolution seismic profile and line drawing of an East China Sea sand ridge. Location shown in Figure 13A. Ridge migration is parallel to its steeper side, oblique to its long axis. Numbers relate to seismic facies interpreted by Yang (1989). Compare this stratigraphic architecture with those of the various interpretations in Figure 2 (data from Yang, 1989).

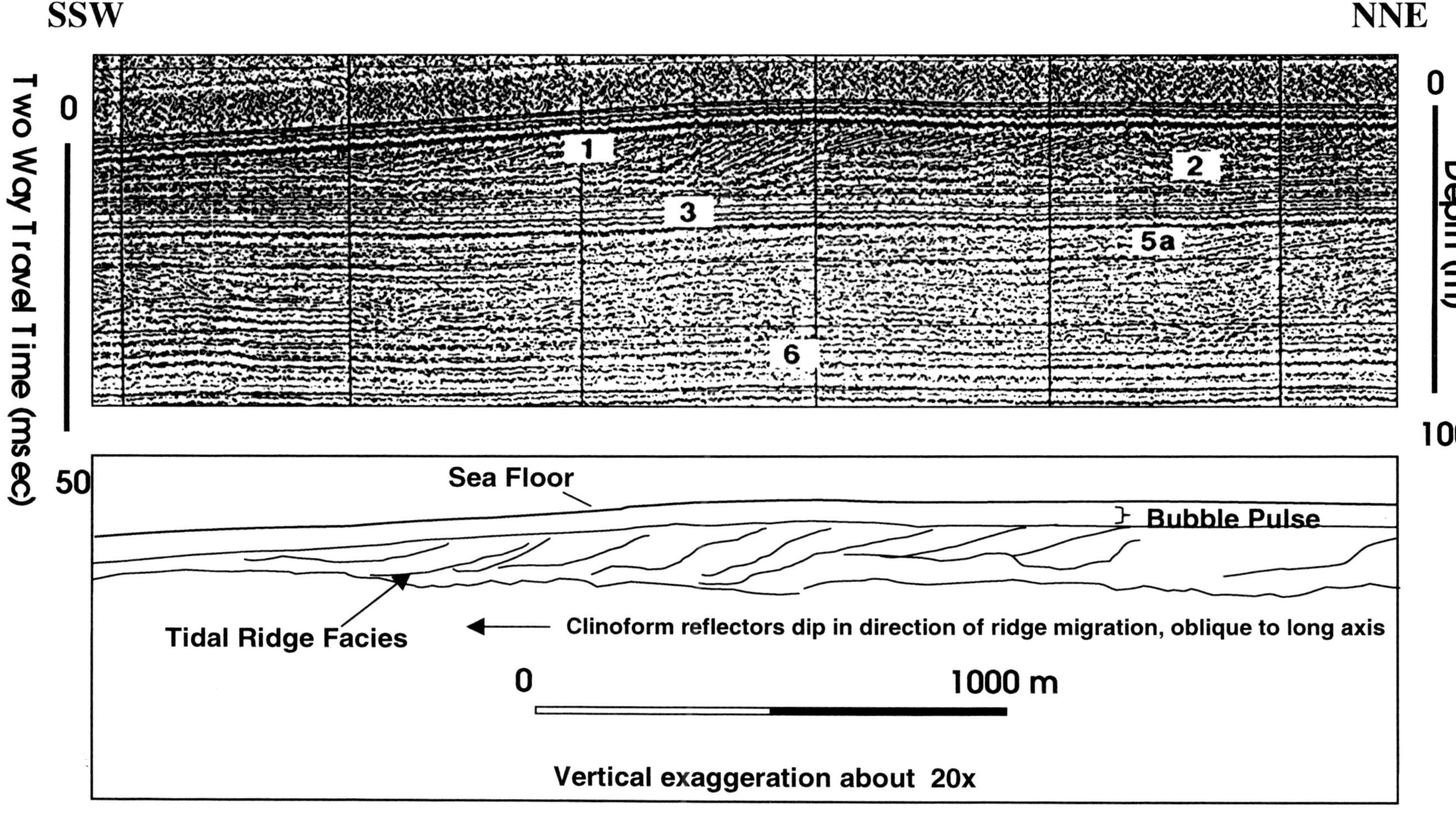

Fig. 13.—Continued.

We do feel, based on the presentations of the different investigators, that although each of the current interpretations accounts for specific parts of the unit, none adequately explains all aspects of the Shannon Sandstone.

- Although the Shannon lithofacies resemble those in the conceptual model for a storm-dominated shelf sand ridge postulated by Swift and Parsons (this volume), they are dissimilar from those in any such sand ridge that has actually been cored. Specifically, the abundance of mud, mud drapes, and tidal features has yet to be found in the ridge facies of a storm-dominated shelf sand ridge. Although the directional relationships tend to rule out a shoreface-attached ridge, they are compatible with a ridge that is either initiated on the shelf or fully reorganized by storm processes.
- The incised-valley fill model provides an in-vogue explanation for the isolated, linear aspect of the Shannon sand bodies. We feel the interpretation suffers from a lack of supporting evidence in the form of fluvial, intertidal, or other facies typically associated with known incised-valley fill or the systems that must have connected the valleys with corresponding high-stand shorelines. Particularly difficult to explain is the presence of "offshore laminated mud" above the Canary Yellow Sequence Boundary, which is inferred to be the floor of a fluvially cut valley. Directional problems include the inferred direction of progradation of the postulated tidal sand bars and orientations of wave ripples.
- The lowstand shoreface interpretation encounters difficulty with both facies and directional relations. The abundant mud drapes, mud clasts, and tidal rhythmites and couplets displayed in core photographs of the Shannon in Hartzog Draw, (Tye et al., 1986; Tillman and Martinsen, 1987; Sullivan et al., 1997; and Bergman, 1994) would, in our experience, be anomalous on a shoreface. The pronounced unidirectionality of the cross-bedding and its orientation relative to the inferred shoreline trend are also very difficult to explain in a shoreface setting.
- The suggestion that the Shannon represents estuarine sand ridges influenced by both waves and tides in an "open embayment" satisfies the lithofacies comparison, and much of the directional data. It does not account *a priori* for the unidirectional character of the cross-bedding and, without a more specific paleogeographic context, offers a fairly sweeping generalization.
- A prograding estuary mouth shoal would generate many of the lithofacies observed in the Shannon. The unidirectionality of cross-bedding dips can be attributed to a dominance of ebb tidal currents (Elliott, 1995). The paleogeographic context, which implies an embayment that extends toward the north-northeast, is difficult to reconcile with the orientation of the wave ripples. Perhaps the greatest difficulty lies in the absence of tidal channel deposits, which should have the highest potential for preservation in an estuary-mouth shoal system; sharp-based, fining-up successions of tidal sand deposits have not been observed in the Shannon. The estuary-mouth shoal interpretation was proposed for the outcropping Shannon in the Salt Creek Anticline area (Elliott, 1995); extrapolating the same depositional system to Hartzog Draw would encompass an area on the order of a thousand square miles (about three times the size of the inlet area of Chesapeake Bay). The linearity and orientation of the Hartzog Draw sand body is also difficult to explain in the context of an estuary mouth shoal.

2. We note that the participants in the Shannon controversy tend to interpret a large, complex system as forming under a single set of conditions. Much of the discussion in this volume is directed toward defending one or the other of these positions. Only Walker and Bergman (1993) subdivide the Shannon outcrop into stratigraphic units and couch their observations in terms of this subdivision. We have commented on the possibility that the Shannon outcrop and its subsurface counterpart in Hartzog Draw are dissimilar enough to warrant consideration as two separate depositional systems. Sullivan et al. (1997) provide a possible framework for such a difference, but do not pursue it.
   Given the size and complexity of the Shannon system in the Salt Creek anticline/Hartzog Draw area, we feel that the possibility of multiple origins should not be ignored. If so, it is equally possible that any of the interpretations are valid for any specific part of the unit and that all might be valid for some part of it. Wave-generated features seem far more common in the outcropping Shannon than in the subsurface at Hartzog Draw. Nearly all investigators agree that the Shannon rests on an extensive erosional surface cut into shelf mudstone of the Cody shale. This surface implies downcutting by wave or other currents associated with a fall of relative sea level, which would set the stage for shoreface deposition or the filling of incised valleys. Even if the bulk of the Shannon were composed of shelf sand ridge deposits, an association with shallower water deposits does not seem unreasonable.

3. If the Shannon is of compound origin, it may serve as a representative model for all isolated linear sand bodies only in its complexity. The Shannon siblings, in this case, will each require their own analysis and a separate interpretation.

4. The various investigators of the Shannon have amassed a remarkable amount of data. Nonetheless, a few areas remain where additional careful observation might help to resolve some of the outstanding issues.

- A fundamental issue is whether the linear geometry of the subsurface sand bodies reflects ridge morphology or an erosional remnant. The contact at the top of the sand may provide a key to this question. Ridges typically are preserved through drowning and encasement in shelf mud. This process should be sufficiently gradual as to create a transitional upper contact. Erosional surfaces might exist within the transition, owing to storms, but should be part of a general gradation into shelf mudstone. Examination of published core photographs that include the top of the Shannon (Tye et al., 1986, Fig. 9; Tillman and Martinsen, 1987, Fig. 3A; Bergman, 1994, Fig. 3; Sullivan et al., 1997, Fig. 3) suggest that the upper contact may be more sharp and erosional than gradational, but a systematic study of the contact throughout the field (and in outcrop) is needed to resolve this issue.
- A detailed study of the fine-grained facies associated with the Shannon could prove useful. Paleogeographic maps of the finer-grained facies are critical to evaluating the environmental interpretations. In particular, careful paleontologic analysis of the shalier facies within the "ridge complex" might better define the environment in which the sands accumulated. Study of the "offshore mudstone" within the "Canary Yellow" sequence of Sullivan et al. (1997) in Hartzog Draw might resolve whether this unit

accumulated in an incised valley or in a shelf setting. Examination of the shale at the top of the Shannon could better define the circumstances of preservation of the sand bodies.

- Systematic analysis of the occurrence and orientation of wave ripples in the outcropping Shannon would document the paleogeographic implications of these structures and provide a basis for a comparison with their abundance in the Hartzog Draw cores. The issue of the presence, absence, or abundance of hummocky cross-bedding in outcrop and whether it occurs only in outcrop warrants resolution. A systematic comparison of the abundance of wave-generated structures in outcrop and core might resolve the question as to whether the outcropping and subsurface Shannon represent two different depositional environments.
- A regional analysis might better define the paleogeographic setting. Two interpretations (incised-valley fill and estuary mouth shoal) require an overall shoreline more or less normal to the trend of the subsurface sand bodies; two others (shelf sand ridge and lowstand shoreface) invoke a shoreline trend parallel to the fields. Resolution of this issue would narrow the range of possible interpretations. Regional 2-D seismic data would be very helpful in such an exercise. 3-D data from Hartzog Draw or other fields might resolve some of the questions regarding orientation and directional features of sand bodies.

5. Substantial differences exist in the way the different investigators perceive the Shannon. We do not attempt to judge anyone's observational skills, but we do note that some of the perceptions are mutually exclusive. It is apparent to us that all of the papers dealing with the Shannon, including this one, are colored by the prior experiences of the investigators. We can only encourage the interested reader to critically analyze all of the evidence and resulting arguments, including those presented in this paper.

FINAL REFLECTIONS ON THE SHANNON CONTROVERSY

Perhaps more important than the validity of any interpretation of the Shannon are the implications of the controversy relative to the interpretation of depositional environments in general. We who decipher ancient depositional systems prefer to think that we apply rigorous scientific methods to our efforts. Yet we must acknowledge that woefully few incontrovertible diagnostic tools lie at our disposal. Typically, at best we can strive only to develop the least complicated hypothesis that is supported by the preponderance of evidence and is consistent internally and with proposed modern analogs. Each of the investigators of the Shannon rather clearly feels that his or her interpretation meets these criteria. Their disagreement demonstrates to us the degree that modern-day environmental interpretation is not only science, but also art.

There is also a lesson to be learned on the objectivity of observation. It is surely no accident that the features in the Shannon that are seen, not seen, or not discussed by each set of investigators reflect their interpretive stance. This observation is in no way intended as an indictment of their capabilities. Although all of us strive for objectivity in our observations, what we see in a rock is quite naturally governed by our prior experience. The lack of agreement by a group of seasoned scientists on the lithologic details of the Shannon demonstrates the range of viewpoints that can be possible owing to diverse backgrounds. It also casts a sobering reflection on the ability of the sedimentological community to shed unintentional bias in our observations. It also reminds us that the time-honored principle of carrying multiple working hypotheses still retains its currency.

ACKNOWLEDGMENTS

The order of authorship for this paper was decided by the flip of a coin. The senior author acknowledges losing the toss. We thank Conoco, Inc., for permission to publish this paper. In addition, we gratefully acknowledge the seemingly limitless patience of the editors of the volume. Writing this paper was an arduous task; painful, but we hope ultimately worthwhile. The manuscript was greatly improved by reviews by Bob Dalrymple, John Snedden, Katherine Bergman, and Roy Fitzsimmons. We also benefited from discussions with George Pemberton, Brian Zaitlin, Gus Gustafson, and Ron Boyd. They did not agree with everything we said, and we did not agree with everything they said, which seems appropriate.

We also greatly appreciate the efforts of all the Shannon interpreters, each of whom has marshaled a wealth of observation and thoughtful, creative analysis to their separate arguments. Although we do not necessarily agree with any of them in all regards, we respect their attempts to resolve a long-standing, troublesome sedimentologic/stratigraphic problem.

REFERENCES

ALLEN, J.L.R., 1965, A review of the origin and characteristics of recent alluvial sediment: Sedimentology, v. 5, p. 89-191.

ALLEN, G.P., LAURIER, D., AND THOUVENIN, J., 1979, Etude sedimentologique du delta de la Mahakam: Compagnie Francaise des Petroles, Notes et Memoires 15, 156 p.

ALLEN, G.P., 1991, Sedimentary processes and facies in the Gironde estuary: A recent model for macrotidal estuarine systems, *in* Smith, D.G., Reinson, G.E., Zaitlin, B.A., and Rahmani, R.A., eds., Clastic Tidal Sedimentology: Calgary, Canadian Society of Petroleum Geologists Memoir 16, p. 29-39.

AMOROSI, A., 1995, Glaucony and sequence stratigraphy: a conceptual framework of distribution in siliciclastic sequences: Journal of Sedimentary Research, v. B65, p. 419-425.

BELDERSON, R.H., 1986, Offshore tidal and non-tidal sand ridges and sheets: Differences in morphology and hydrodynamic setting, *in* Knight, R.J., and McLean, J.R., eds., Shelf Sands and Sandstones: Calgary, Canadian Society of Petroleum Geologists Memoir 11, p. 293-301.

BELDERSON, R.H., JOHNSON, M.A., AND KENYON, N.H., 1982, Bedforms, *in* Stride, A.H., ed., Offshore Tidal Sands: Processes and Deposits: New York, Chapman and Hall, p. 27-57.

BERGMAN, K.M., 1994, Shannon Sandstone in Hartzog Draw-Heldt Fields (Cretaceous) Wyoming, USA) reinterpreted as lowstand shoreface deposits: Journal of Sedimentary Research, v. 64, p. 184-210.

BERGMAN, K.M., AND WALKER, R.G., 1995, High resolution sequence stratigraphic analysis of the Shannon Sandstone in Wyoming, using a template for regional correlation: Journal of Sedimentary Research, v. B65, p. 265-264.

BERNARD, H.A., MAJOR, C.F., JR., AND PARROTT, B.S., 1959, The Galveston barrier island and environs: A model for predicting reservoir occurrence and trend: Transactions Gulf Coast Association of Geological Societies, v. 9, p. 221-224.

BERNARD, H.A., MAJOR, C.F., JR., PARROTT, B.S., AND LEBLANC, R.J., 1970, Recent Sediments of Southeast Texas, A field guide to the Brazos Alluvial and Deltaic Plains and the Galveston Barrier island complex: Austin, Texas Bureau of Economic Geology, Guidebook 11.

BERNÉ, S., TRENTESAUX, A., STOLK, A., MISSIAEN, T., AND DE BATIST, M., 1994, Architecture and long term evolution of a tidal sandbank: The Middelkerke Bank (southern North Sea): Marine Geology, v. 121, p. 57-72.

BERNÉ, S., LERICOLAS, G., MARSSET, T., BOURILLET, J-F., and DE BATIST, M., 1998, Erosional offshore sand ridges and lowstand shorefaces:examples from tide- and wave-dominated environments of France: Journal of Sedimentary Research, v. 68, p. 540-555.

BLUM, M.L., 1994, Genesis and architecture of incised valley fill sequences: A Late Quaternary example from the Colorado River, Gulf coastal plain of Texas, *in* Weimer, P., and Posamentier, H.W., eds., Siliciclastic Sequence Stratigraphy: Recent Developments and Applications: Tulsa, American Association of Petroleum Geologists Memoir 58, p. 259-284.

CATUNEANU, O., SWEET, A.R., AND MIALL, A.D., 1997, Reciprocal architecture of Bearpaw and Post-Bearpaw T-R sequences, uppermost Cretaceous, Western Canada Sedimentary Basin: Bulletin of Canadian Petroleum Geology, v. 45, 75-94.

CLIFTON, H.E., 1988, Sedimentologic approaches to paleobathymetry: Palaios, v. 3, p. 507-522.

CLIFTON, H.E., HUNTER, R.E., AND PHILLIPS, R.L., 1971, Depositional structures and processes in the non-barred, high-energy nearshore: Journal of Sedimentary Petrology, v. 41, p. 165-184.

CLIFTON, H.E., AND THOMPSON, J.K., 1978, *Macaronichnus segregatis*: A feeding structure of shallow marine polychaetes: Journal of Sedimentary Petrology, v. 48, p. 1293-1302.

COLEMAN, J.M., AND WRIGHT, L.D., 1975, Modern river deltas: Variability of processes and sand bodies, *in* Brousssard, M.L., ed., Deltas: Models for Exploration: Houston Geological Society, p. 99-149.

CURRAY, J.R., 1960, Sediments and history of Holocene transgression, continental shelf, northwest Gulf of Mexico, *in* Shepard, F.P., Phleger, F.B., and van Andel, T.H., eds., Recent Sediments, Northwest Gulf of Mexico: Tulsa, American Association of Petroleum Geologists, p. 221-266.

CURRAY, J.R., EMMEL, F.J., AND CRAMPTON, P.J.S., 1967, Holocene history of a strand plain, lagoonal coast, Nayarit, Mexico, *in* Castaneres, A.A., and Phleger, F.B., eds., Lagunas Costeras, un Simposio: Universidad Nacional Autonoma, Mexico, p. 63-100.

DALRYMPLE, R.W., 1992, Tidal depositional systems, *in* Walker, R.G., and James, N.P., eds., Facies Models: Response to Sea Level Change: Ottawa, Geological Association of Canada, p. 195-218.

DALRYMPLE, R.W., 1997, Sand body types in the Fly River Delta: Tulsa, American Association of Petroleum Geologists-Society for Sedimentary Geology (SEPM) Annual Convention Abstracts, p. A25.

DALRYMPLE, R.W., LEGRESLEY, E.M., FADER, G.B.J., AND PETRIE, B.D., 1992a, The western Grand Banks of Newfoundland: Transgressive Holocene sedimentation under combined influence of waves and currents: Marine Geology, v. 105, p. 85-118.

DALRYMPLE, R.W., ZAITLIN, B.A., AND BOYD, R., 1992b, Estuarine facies models: Conceptual basis and stratigraphic implications: Journal of Sedimentary Petrology, v. 62, p. 1130-1146.

DALRYMPLE, R.W., AND HOOGENDOORN, E.L., 1997, Erosion and deposition on migrating shoreface-attached ridges, Sable Island, Eastern Canada: Geoscience Canada, v. 24, p. 25-36.

DAVIS, R.A., AND BALSON, P.S., 1992, Stratigraphy of a North Sea tidal sand ridge: Journal of Sedimentary Petrology, v. 62, p. 116-121.

DOMINGUEZ, J.M.L., AND WANLESS, H.R., 1991, Facies architecture of a falling sea-level strandplain, Doce River coast, Brazil, *in* Swift, D.J.P., Oertel, G.F., Tillman, R.W., and Thorne, J.A., eds., Shelf Sand and Sandstone Bodies: Geometry, Facies, and Sequence Stratigraphy: Oxford, International Association of Sedimentologists Special Publication 14, p. 259-282.

DOMINGUEZ, J.M.L., MARTIN, L., AND BITTENCOURT, A.C.S.P., 1987, Sea-level history and Quaternary evolution of river mouth-associated beach-ridge plains along the east-southeast Brazilian coast, *in* Nummedal, D., Pilkey, O.H., and Howard, J.D., eds., Sea-level Fluctuation and Coastal Evolution: Tulsa, Society for Sedimentary Geology (SEPM) Special Publication 41, p. 115-127.

DOMINGUEZ, J.M.L., BITTENCOURT, A.C.S.P., AND MARTIN, L., 1992, Controls on Quaternary coastal evolution of the east-northeastern coast of Brazil: Roles of sea-level history, trade winds and climate, *in* Donoghue, J.F., Davis, R.A., Fletcher, C.H., and Suter, J.R., eds., Quaternary Coastal Evolution: Sedimentary Geology, v. 3 & 4, p. 213-232.

DUKE, W.L., 1990, Geostrophic circulation or shallow marine turbidity currents? The dilemma of paleoflow patterns in storm-influenced prograding shoreline systems: Journal of Sedimentary Petrology, v. 60, p. 870-883.

ELLIOTT, T., 1995, Physical processes, depositional settings and stratigraphic context of the Shannon Sandstone, Wyoming, U.S.A., *in* Swift, D.J.P., Snedden, J., and Plint, A.G. (conveners), Tongues, Ridges and Wedges: Highstand versus lowstand in Marine Basins: Tulsa, Society for Sedimentary Geology (SEPM) Research Conference Program, Powder River and Bighorn Basins, June 24-29, 1995.

FOYLE, A.M., AND OERTEL, G.F., 1992, Seismic stratigraphy and coastal drainage patterns in the Quaternary section of the southern Delmarva Peninsula, Virginia, USA, *in* Donoghue, J.F., Davis, R.A., Fletcher, C.H., and Suter, J.R., eds., Quaternary Coastal Evolution: Sedimentary Geology, v. 3 & 4, p. 261-278.

FRAZIER, D.E., 1974, Depositional Episodes: Their Relationship to the Quaternary Stratigraphic Framework in the Northwestern Portion of the Gulf Basin: Austin, Texas Bureau of Economic Geology Geological Circular 74-1, 28 p.

GAYNOR, G.C., AND SWIFT, D.J.P., 1988, Shannon Sandstone depositional model: Sand-ridge formation on the Campanian Western Interior Shelf: Journal of Sedimentary Petrology, v. 58, p. 868-880.

HANSLEY, P.L., AND WHITNEY, C.G., 1990, Petrology, diagenesis, and sedimentology of oil reservoirs in Upper Cretaceous Shannon Sandstone Beds, Powder River Basin, Wyoming: Denver, U.S. Geological Survey Bulletin 1917-C, 33 p.

HART, B.S., AND LONG, B.F., 1996, Forced regressive and lowstand deltas: Holocene Canadian examples: Journal of Sedimentary Research, v. 66, p. 820-829.

HARRIS, P.T., PATTIARATCHI, C.B., KEENE, J.B., DALRYMPLE, R.W., GARDNER, J.V., BAKER, E.K., COLE, A.R., MITCHELL, D., GIBBS, P., AND SCHROEDER, W.W., 1996, Late Quaternary deltaic and carbonate sedimentation in the gulf of Papua foreland basin: Response to sea level change: Journal of Sedimentary Research, v. 66, p. 801-819.

HEARN, C.L., EWBANKS, W.J., JR., TYE, R.S., AND RANGANATHAN, V., 1984, Geological factors influencing reservoir performance of the Hartzog Draw Field, Wyoming: Journal of Petroleum Technology, v. 36, p. 1335-1344.

HILL, G.W., 1981, Ichnocoenoses of a Paleocene submarine-canyon floor, Point Lobos, California, *in* Frizzell, V., ed., Upper Cretaceous and Paleocene Turbidites, Central California Coast: Tulsa, Society of Economic Paleontologists Mineralogists Pacific Section Field Trip Guidebook 6, San Francisco Annual Meeting, p. 79-92.

HOOGENDOORN, E.L., 1989, Sedimentology and dynamics of shoreface-attached ridges, Sable Island Bank, Nova Scotia: Unpublished Ph.D. Dissertation, Queen's University, Kingston, Ontario, Canada, 483 p.

HUTHNANCE, J.M., 1982, On one mechanism forming linear sand banks: Estuarine Coastal Marine Science, v. 14, p. 79-99.

JENNETTE, D.C., AND JONES, C.R., 1995, Sequence Stratigraphy of the Upper Cretaceous Tocito Sandstone: A Model for Tidally Influenced Incised Valleys, San Juan Basin, New Mexico, *in* Van Wagoner, J.C., and Bertram, G.T., eds., Sequence Stratigraphy of Foreland Basin Deposits: Outcrop and Subsurface Examples from the Cretaceous of North America: Tulsa, American Association of Petroleum Geologists Memoir 64, p. 311-347.

MARTINSEN, O., NOTTVEDT, A., AND HELLAND-HANSEN, W., 1995, Forced regressions: Processes and products with examples from the Holocene Sandfjorden Basin, Norwegian Arctic, *in* Swift, D.J.P., Snedden, J., and Plint, A.G. (conveners), Tongues, Ridges and Wedges: Highstand versus lowstand in Marine Basins: Tulsa, Society for Sedimentary Geology (SEPM) Research Conference Program, Powder River and Bighorn Basins, June 24-29, 1995.

MCBRIDE, R.A., AND MOSLOW, T.F., 1991, Origin, evolution, and distribution of shoreface sand ridges, Atlantic inner shelf, USA: Marine Geology, v. 97, p. 57-85.

MECKEL, L.D., 1975, Holocene sand bodies in the Colorado Delta area, Northern Gulf of California, *in* Broussard, M.L., ed., Deltas, Models for Exploration: Houston Geological Society, p. 239-266.

MORTON, R.A., 1994, Texas barriers, *in* Davis, R.A., ed., Geology of Holocene Barrier Island Systems: New York, Springer Verlag, p. 75-114.

NELSON, H.F., AND BRAY, E.B., 1970, Stratigraphy and history of the Holocene sediments in the Sabine-High Island area, Gulf of Mexico, *in* Morgan, J.P., ed., Deltaic Sedimentation, Modern and Ancient: Tulsa, Society of Economic Paleontologists and Mineralogists Special Publication 15, p. 48-77.

NIO, S.D., AND YANG, C.S., 1991, Diagnostic attributes of clastic tidal deposits: A review, *in* Smith, D.G., Reinson, G.E., Zaitlin, B.A., and Rahmani, R.A., eds., Clastic Tidal Sedimentology: Calgary, Canadian Society of Petroleum Geologists Memoir 16, p. 3-27.

ODIN, G.S., AND MATTER, A., 1981, De glauconarium origine: Sedimentology, v. 28, p. 611-641.

OLSON, M., CHISM, L., AND MARTINSEN, R., 1998, Tectonically influenced architecture of mud dominated clastic successions in the Cody Shale, Powder River Basin, Wyoming: Tulsa, American Association of Petroleum Geologists Expanded Abstracts.

PENLAND, S., BOYD, R.L., AND SUTER, J.R., 1988, Transgressive depositional systems of the Mississippi delta plain: A model for barrier shoreline and shelf sand development: Journal of Sedimentary Petrology, v. 6, p. 932-949.

PENLAND, S., SUTER, J.R., AND MOSLOW, T.F., 1986, Inner shelf shoal sedimentary facies and sequences: Ship Shoal, northern Gulf of Mexico, *in* Moslow, T.F., and Rhodes, E.G., eds., Modern and Ancient Shelf Clastics: A Core Workshop: Tulsa, Society of Economic Paleontologists and Mineralogists Core Workshop 9, p. 73-122.

PENLAND, S., SUTER, J.R., MCBRIDE, R.A., WILLIAM, S.J., KINDINGER, J.L., AND BOYD, R., 1989, Holocene sand shoals offshore of the Mississippi River Delta Plain: Transactions Gulf Coast Association of Geological Societies, v. XXXIX, p. 471-480.

POPE, D.L., PENLAND, S., SUTER, J.R., AND MCBRIDE, R.A., 1990, Holocene geologic framework of the Trinity Shoal region, Louisiana continental shelf: Tulsa, Society for Sedimentary Geology (SEPM)Gulf Coast Section, v. XII, p. 191-201.

POSAMENTIER, H.W.P., ALLEN, G.P., JAMES, D.J.P., AND TESSON, M., 1992, Forced regressions in a sequence stratigraphic framework: Concepts, examples, and exploration significance: American Association of Petroleum Geologists Bulletin, v. 76, p. 1687-1709.

RABINEAU, M., BERNÉ, S., LEDREZEN, E., LERICOLAIS, G., MARSSET, T., AND ROTUNNO, M., 1998, 3D architecture of lowstand and transgressive Quaternary sand bodies of the Gulf of Lion, France: Marine and Petroleum Geology, v. 15, p. 439-452.

RINE, J.M., TILLMAN, R.W., STUBBLEFIELD, W.L., AND SWIFT, D.J.P., 1986, Lithostratigraphy of Holocene Sand Ridges from the nearshore and middle continental shelf of New Jersey, *in* Moslow, T.F., and Rhodes, E.G. eds., Modern and Ancient Shelf Clastics: A core Workshop: Tulsa, Society of Economic Paleontologists and Mineralogists Core Workshop 9, p. 1-72.

RINE, J.M., TILLMAN, R.W., CULVER, S.J., AND SWIFT, D.J.P., 1991, Generation of late Holocene sand ridges on the middle continental shelf of New Jersey, USA-evidence for formation in mid-shelf setting based on comparisons with a nearshore ridge, *in* Swift, D.J.P., Oertel, G.F., Tillman, R.W., and Thorne, J.A., eds., Shelf Sand and Sandstone Bodies: Geometry, Facies, and Sequence Stratigraphy: Oxford, International Association of Sedimentologists Special Publication 14, p. 395-426.

ROY, P.S., AND BOYD, R.L., 1996, Quaternary geology of Southeast Australia: A tectonically stable, wave-dominated, sediment-deficient margin: International Geological Correlation Program Project #367 Field Guidebook, 174 p.

ROY, P.S., CROWELL, P.J., FERLAND, M.A., AND THOM, B.G., 1994, Wave-dominated coasts, *in* Carter, R.W.G., and Woodroffe, C.D., eds., Coastal Evolution: Late Quaternary Shoreline Morphodynamics: Cambridge, Cambridge University Press, p. 121-186.

RUBIN, D.M., 1987, Cross-bedding, bedforms and paleocurrents: Tulsa, Society for Sedimentary Geology (SEPM) Concepts in Sedimentology and Paleontology, v. 1, 187 p.

SHORT, A.D., 1984, Beach and nearshore facies, SE Australia: Marine Geology, v. 60, p. 261-282.

SNEDDEN, J.W., TILLMAN, R.W., KREISA, R.D., SCHWELLER, W.J., CULVER, S.J., AND WINN, R.J., 1994, Genesis and stratigraphy of a shoreface-attached sand ridge, Peahala Ridge, New Jersey: Journal of Sedimentary Research, v. B64, p. 560-581.

SPEARING, D.R., 1976, Upper Cretaceous Shannon Sandstone, an offshore shallow-marine sandbody: Casper, Wyoming Geological Association Guidebook, 28th Annual Field Conference, p. 65-72.

STRIDE, A.H., BELDERSON, R.H., KENYON, N.H., AND JOHNSON, M.A., 1982, Offshore tidal deposits: Sand sheet and sand bank facies, *in* Stride, A.H., ed., Offshore Tidal Sands: Processes and Deposits: New York, Chapman and Hall, p. 95-125.

STUBBLEFIELD, W.L., MCGRAIL, D.W., AND KERSEY, D.G., 1984, Recognition of transgressive and post-transgressive sand ridges on the New Jersey continental shelf, *in* Tillman, R.W., and Siemers, C.T., eds., Siliciclastic Shelf Sediments: Tulsa, Society of Economic Paleontologists and Mineralogists Special Publication 34, p. 1-24.

SULLIVAN, M.D., VAN WAGONER, J.C., FOSTER, M.E., STUART, R.M., JENNETTE, D.C., LOVELL, R.W., AND PEMBERTON, S.G., 1995, Lowstand architecture and sequence stratigraphic control on Shannon incised valley distribution, Hartzog Draw Field, Wyoming, *in* Fitzsimmons, R.J., Parsons, B.S., and Swift, D.J.P., eds., Tongues, Ridges and Wedges: Highstand Versus Lowstand Architecture in Shallow Marine Basins: Tulsa, Society for Sedimentary Geology (SEPM) Research Conference Field Guide, Appendix A.

SULLIVAN, M.D., VAN WAGONER, J.C., JENNETTE, D.C., LOVELL, R.W., FOSTER, M.E., AND STUART, R.M., 1997, Sequence stratigraphic and tectonic controls on Shannon incised valley distribution, Hartzog Draw Field, Wyoming, in Fitzsimmons, R.J., Parsons, B.S., and Swift, D.J.P., eds., Tongues, Ridges and Wedges: Highstand Versus Lowstand Architecture in Shallow Marine Basins: Tulsa, Society for Sedimentary Geology (SEPM) Research Conference Field Guide, Appendix A.

SUTER, J.R., 1986, Ancient fluvial systems and Holocene deposits, southwestern Louisiana continental shelf, *in* Berryhill, H.L., ed., Late Quaternary Facies and structure, northern Gulf of Mexico:

Tulsa, American Association of Petroleum Geologists Studies in Geology 23, p. 81-129.

Suter, J.R., Berryhill, H.L., and Penland, S., 1987, Late Quaternary sea-level fluctuations and depositional sequences, southwest Louisiana continental shelf, *in* Nummedal, D., Pilkey, O.H., and Howard, J.D., eds., Sea-level Fluctuation and Coastal Evolution: Tulsa, Society of Economic Paleontologists and Mineralogists Special Publication 41, p. 199-222.

Suter, J.R., and Penland, S., 1987, Evolution of Cat Island Pass, Louisiana, *in* Coastal Sediments '87: Keston, Virginia, American Society of Civil Engineers Waterways Division, p. 2078-2093.

Swift, D.J.P., 1973, Delaware Shelf Valley: Estuary retreat path, not drowned river valley: Geological Society of America Bulletin, v. 84, p. 2743-2748.

Swift, D.J.P., Thorne, J.A., and Oertel, G.F., 1986, Fluid processes and sea-floor response on a modern storm-dominated shelf: Middle Atlantic shelf of North America. Part II: Response of the shelf floor, *in* Knight, R.J., and McLean, J.R., eds., Shelf Sands and Sandstones: Calgary, Canadian Society of Petroleum Geologists Memoir 11, p. 191-211.

Sydow, J.A., and Roberts, H.H., 1994, Stratigraphic framework of a Late Pleistocene shelf edge delta, northeast Gulf of Mexico: American Association of Petroleum Geologists Bulletin, v. 78, p. 1276-1312.

Thomas, M.A., and Anderson, J., 1994, Sea-level controls on the facies architecture of the Trinity/Sabine incised valley system, *in* Dalrymple, R.W., Boyd, R., and Zaitlin, B.A., eds., Incised Valley Systems: Origin and Sedimentary Sequences: Tulsa, Society for Sedimentary Geology (SEPM) Special Publication 51, p. 63-82.

Tillman, R.W., and Martinsen, R.S., 1984, The Shannon shelf-ridge sandstone complex, Salt Creek Anticline area, Powder River Basin, Wyoming, *in* Tillman, R.W., and Siemers, C.T., eds., Siliciclastic Shelf Sediments: Tulsa, Society of Economic Paleontologists and Mineralogists Special Publication 34, p. 85-142.

Tillman, R.W., and Martinsen, R.S., 1987, Sedimentologic characteristics and production model of Hartzog Draw Field, Wyoming, a Shannon shelf-ridge sandstone, *in* Tillman, R.W., and Weber, K.J., eds., Reservoir Sedimentology: Tulsa, Society of Economic Paleontologists and Mineralogists Special Publication 40, p. 15-112.

Twichell, D.C., 1983, Bedform distribution and inferred sand transport on Georges Bank, United States Atlantic Continental Shelf: Sedimentology, v. 30, p. 695-710.

Tye, R.S., Ranganathan, V., and Ewbanks, W.J., Jr., 1986, Facies analysis and reservoir zonation of a Cretaceous shelf sand ridge: Hartzog Draw Field, Wyoming, *in* Moslow, T.F., and Rhodes, E.G., eds., Modern and Ancient Shelf Clastics: A Core Workshop: Tulsa, Society of Economic Paleontologists and Mineralogists Core Workshop 9, p. 169-216.

Van Wagoner, J.C., Mitchum, R.M., Campion, K.M., and Rahmanian, V.D., 1990, Siliciclastic sequence stratigraphy in well logs, cores and outcrops: Concepts for high-resolution correlation of time and facies: Tulsa, American Association of Petroleum Geologists Methods in Exploration Series, v. 7.

Walker, R.G., and James, N.P., 1992, Facies Models: Response to Sea Level Change: Ottawa, Geological Association of Canada, 409 p.

Walker, R.G., and Bergman, K.M., 1993, Shannon sandstone in Wyoming; a shelf-ridge complex reinterpreted as lowstand shoreface deposits: Journal of Sedimentary Petrology, v. 63, p. 839-951.

Walker, R.G., and Plint A.G, 1992, Wave- and storm-dominated shallow marine systems, *in* Walker, R.G., and James, N.P., eds., Facies Models: Response to Sea Level Change: Ottawa, Geological Association of Canada, p. 219-238.

Yang, C.S., 1989, Active, moribund, and buried tidal sand ridges in the East China Sea and the southern Yellow Sea: Marine Geology, v. 88, p. 97-116.

Yang, C.S., and Sun, J.S., 1988, Tidal sand ridges on the East China Shelf, *in* de Boer, P.L., van Gelder, A., and Nio, S.D., eds., Tide-influenced Sedimentary Environments and Facies: Dordrecht, D. Reidel Publishing Company, p. 23-38.

Zaitlin, B.A., Dalrymple, R.W., and Boyd, R., 1994, The Stratigraphic organization of incised-valley systems associated with relative sea-level changes, *in* Dalrymple, R.W., Zaitlin, B.A., and Boyd, R., eds., Incised Valley Systems: Origin and Sedimentary Sequences: Tulsa, Society for Sedimentary Geology (SEPM) Special Publication 51, p. 45-60.

INDEX

A

B

C

D

E

F

G

H

I

J

K

L

M

N

O

P

Q

R

S

T

U

V

W

Y

Z